AA003762

7th EuroGeo Conference

IOP Conference Series: Materials Science and Engineering
Volume 1260

Warsaw, Poland
3 - 7 September 2022

ISBN: 978-1-7138-6156-0
ISSN: 1757-8981

Printed from e-media with permission by:

Curran Associates, Inc.
57 Morehouse Lane
Red Hook, NY 12571

Some format issues inherent in the e-media version may also appear in this print version.

Institute of Physics
Dirac House, Temple Back
Bristol BS1 6BE UK

Phone: 44 1 17 929 7481
Fax: 44 1 17 920 0979

techtracking@iop.org

Additional copies of this publication are available from:

Curran Associates, Inc.
57 Morehouse Lane
Red Hook, NY 12571 USA
Phone: 845-758-0400
Fax: 845-758-2633
Email: curran@proceedings.com
Web: www.proceedings.com

TABLE OF CONTENTS

IOP Conf. Series: Materials Science and Engineering 1260 (2022) 011001 doi:10.1088/1757-899X/1260/1/011001

7th European Conference on Geosynthetics EUROGEO-7

Jacek Kawalec [1]

[1] Tensar International sro, Cesky Tesin, 737 01, Czech Republic; Silesian University of Technology, Gliwice, Poland

Email: kawalec@tensar.pl

Abstract. Within the framework of the presented material as result of Scientific Committee decision paper accepted for publication and presentation at the 7th European Conference on Geosynthetics EuroGeo-7.

We are pleased to present a collection of scientific papers accepted for publication and presentation at 7th European Conference on Geosynthetics (EUROGEO-7) organised on 3-7th September 2022 in Warsaw, Poland. Initially conference was planned for September 2020 but due to COVID-19 situation event was postponed for two years.

Below topics of the conference are listed:

- Agricultural Applications
- Coastal Protection
- Direct and Life-Cycle Cost Savings
- Drainage and Filtration
- Durability
- Embankments on Soft Soils
- Environmental Benefits
- Erosion Control
- Geosynthetics as Formwork
- Hydraulic Applications
- Innovations and New Developments
- Landfills
- Lightweight Construction
- Mining
- Monitoring
- Pavements
- Physical and Numerical Models
- Polymeric and Clay Geosynthetic Barriers
- Properties and Testing
- Quality Control and Quality Assurance
- Reinforced Walls and Slopes
- Roads, Railroads and Other Transportation Applications
- Seismic Applications
- Sustainability
- Stormwater Management

- Unpaved Roads
- Wastewater and Fresh Water Storage

Representatives of scientific and research as well as leaders from industry from all continents were invited for revision and verification of submitted papers. The group of reviewers who created Scientific Committee of the Conference is listed below:

- Dr Dimiter Alexiew, Germany
- Dr Talal Awwad, Syria
- Dr Radovan Baslik, Slovakia
- Prof. Richard Bathurst, Canada
- Prof. Abdelmalek Bouazza, Australia
- Prof. Heinz Brandl, Austria
- Prof. Gerhard Brau, Germany
- Prof. Joanna Bzówka, Poland
- Dr Danielle Cazzuffi, Italy
- Dr Marek Chalecki, Poland
- Prof. Jacques Cote, Canada
- Dr Jolanta Dąbrowska, Poland
- Prof. Eugeniusz Dembicki, Poland
- Dr Oliver Detert, Germany
- Prof. Neil Dixon, UK
- Dr Katarzyna Dołżyk-Szypcio, Poland
- Dr Angelika Duszyńska, Poland
- Prof. Valentin Feodorov, Romania
- Prof. Szabolcs Fisher, Hungary
- Prof. Kazimierz Garbulewski, Poland
- Prof. JP Giroud, France/USA
- Prof. Antonio Gomes Correia, Portugal
- Dr Jarosław Górszczyk, Poland
- Dr Marcin Grygierek, Poland
- Dr Joanna Grzybowska-Pietras, Poland
- Prof. Erol Guler, Turkey
- Prof. Chiwan Wayne Hsieh, Taiwan
- Dr Leos Hornicek, Czech Republic
- Prof. Jie Han, USA
- Dr Grzegorz Horodecki, Poland
- Prof. Colin J.F.P. Jones, UK
- Prof. Takeshi Katsumi, Japan
- Dr Jacek Kawalec, Poland
- Prof. Krystyna Kazimierowicz-Frankowska, Poland
- Dr Agnieszka Kiersnowska, Poland
- Prof. Eugeniusz Koda, Poland
- Prof. Robert Koerner, USA
- Dr Elżbieta Kokocińska-Pakiet, Poland
- Dr Andrew Lees, Cyprus
- Prof. Krzysztof Lejcus, Poland
- Dr Stanislav Lenart, Slovenia

- Prof. Maria de Lurdes Costa Lopez, Portugal
- Dr Marian Łupieżowiec, Poland
- Dr Konrad Malicki, Poland
- Prof. Nicola Moraci, Italy
- Prof. Mensur Mulabdic, Croatia
- Prof. Jose Manuel Neves, Portugal
- Prof. Giang Nguyen, Slovakia
- Dr Nejan Huvaj, Turkey
- Prof. Ennio Palmeira, Brasil
- Dr Andrei Petriaev, Russia
- Dr Zikmund Rakowski, Czech Republic
- Prof. Leszek Rafalski, Poland
- Prof. K. Rajagopal, India
- Eng. Pietro Rimoldi, Italy
- Prof. Kerry Rowe, Canada
- Prof. Aleksandr Rudkovsky
- Dr Jarosław Rybak, Poland
- Prof. Wojciech Sas, Poland
- Prof. Eun Chul Shin, South Korea
- Dr Witold Sterpejkowicz-Wersocki, Poland
- Prof. Zenon Szypcio, Poland
- Prof. Fumio Tatsuoka, Japan
- Dr Natalie Touze – Foltz, France
- Dr Roman Trach, Poland
- Prof. Krzysztof Trojnar, Poland
- Prof. Erol Tutumluer, USA
- Dr Andrzej Urbanik, Poland
- Dr Suzanne van Eekelen, Holland
- Prof. Ivan Vanicek, Czech Republic
- Prof. Magdalena Daria Vaverkova, Poland
- Dr Lars Vollmert, Germany
- Prof. Chao Xu, China
- Dr Katarzyna Zamara, New Zeeland/UK
- Dr Sabid Zekan, Bosnia&Hercegovina
- Prof. Katarzyna Zabielska-Adamska, Poland
- Prof. Askar Zh. Zhussupbekov, Kazakhstan
- Prof. Martin Ziegler, Germany
- Dr Zofia Zięba, Poland
- Prof. Jorge G. Zornberg, USA

Acknowledgments

The organizing committee of the conference expresses its gratitude to the staff of IOP Publishing, who helps to place the proceedings of scientific conferences in international journals and provide support for the publication of materials. In addition, the organizing committee would like to thank the conference participants; reviewers and everyone who helped organize this conference and form the presented volume for publication in the IOP Conference Series: Materials Science and Engineering.

Dr Jacek Kawalec

EUROGEO-7 Conference Chair

IOP Conf. Series: Materials Science and Engineering 1260 (2022) 011002 doi:10.1088/1757-899X/1260/1/011002

Peer Review Statement

All papers published in this volume have been reviewed through processes administered by the Editors. Reviews were conducted by expert referees to the professional and scientific standards expected of a proceedings journal published by IOP Publishing Publishing.

- **Type of peer review:** Single Anonymous
- **Conference submission management system:** By "syskonf.pl" (https://syskonf.pl)
- **Number of submissions received:** 118
- **Number of submissions sent for review:** 106
- **Number of submissions accepted:** 55
- **Acceptance Rate (Submissions Accepted / Submissions Received × 100):** 46.6
- **Average number of reviews per paper:** 2
- **Total number of reviewers involved:** 65
- **Contact person for queries:**
  Name: Tomasz Wrobel
  Email: t.wrobel@mazurkas.com.pl
  Affiliation: Mazurkas Travel

IOP Conf. Series: Materials Science and Engineering 1260 (2022) 012001 doi:10.1088/1757-899X/1260/1/012001

Short-term design of geosynthetic-reinforced embankments on soft foundation soils

D Romeo[1], P Pezzano[2] and P Rimoldi[3]

[1]Consultant, Bologna, Italy
[2]Officine Maccaferri Spa, Bologna, Italy
[3]Consultant, Milano, Italy

Email: demetrio.romeo.dr@gmail.com

Abstract. The stability of an embankment on soft foundation soil is most critical during the construction period, because of the relatively low permeability of the soft foundation. Since the stability increases as the foundation consolidates, related design is controlled by short-term constraints. Geosynthetics for basal reinforcement enhance the stability of an embankment over soft soil by preventing lateral sliding of the fill, extrusion of the foundation and rotational failure. The stabilizing force developed by the reinforcement is generated by shear stresses mobilized along both the interface surfaces of the reinforcement, with the foundation soil and the embankment fill respectively. In fact, the reinforcement has the double function of carrying the outward shear stresses transmitted by the embankment fill and of providing inward shear stresses to restrain the foundation soil from lateral squeezing-out. For each of the limit states mentioned above, the associated reinforcement strength and bond length should be checked to ensure that the limit state tensile load can be generated in the reinforcement. Starting from considerations proposed by the British Standard 8006:2010, this paper focuses on lateral sliding and foundation extrusion limit states by presenting a design method to evaluate the reinforcement strength and bond length required to provide the maximum improvement for an embankment on soft soil.

1. Introduction

In a reinforced embankment on soft soil the reinforcement is used to improve the stability of the embankment – foundation soil system. The stability is most critical at (or immediately following) the end of the construction stage, when the new loading has been applied rapidly with respect to the drainage time in the foundation. The clay foundation soil can only drain, consolidate and improve in strength with time.

To improve stability, reinforcement can be placed across the base of the embankment close to the foundation surface. If the reinforcement were absent, the factor of safety at the end of construction may fall to below unity (i.e. the desired cross-section could not be built without the reinforcement), before increasing as the strength in the foundation improves with consolidation. Thus, at short term, the reinforcement is needed to improve the stability during construction and subsequent consolidation.

2. Improvement of embankment foundation bearing conditions

The stability of an embankment on soft soil is influenced significantly by the foundation bearing capacity. An unreinforced embankment exerts not only vertical loading from self-weight, but also the lateral thrust in the fill produces a horizontal outward shear stress on the foundation that reduces the bearing capacity.

The primary role for reinforcement, therefore, is to hold the lateral thrust in equilibrium with a tensile force. If this can be achieved with no relative outward displacement of the reinforced embankment with respect to the surface of the underlying soft soil, the foundation will be subjected only to the vertical loading from the embankment. This primary mechanism of reinforcement improves the stability for all foundation cases.

A second role for the reinforcement is to resist lateral displacement in the foundation soil. Within the limits imposed by the shear contact with the foundation surface, and by the reinforcement extensibility, a reinforced embankment can provide restraint at the foundation surface in a manner similar to a rough footing. This improves the bearing capacity when the strength in the foundation increases with depth, or where the soft clay is of limited depth, above a hard stratum.

3. Lateral sliding

Lateral sliding is a limit state which involves only the equilibrium entirely within the embankment fill: the preferential direct sliding along the upper surface of the reinforcement layer has to be examined.

Sliding would result if the available resistance at the reinforcement interface were insufficient to support the lateral thrust from the fill.

According to BS8006:2010 [1], the tensile force per metre run in the reinforcement,T_{ds}, can be determined as:

$$T_{ds} = 0.5 K_h H (f_{fs} \gamma_1 H + 2 f_q w_s) \tag{1}$$

where: K_h is the horizontal earth pressure coefficient; H is the height of the embankment; γ_1 is the unit weight of the embankment fill; w_s is the surcharge intensity on top of the embankment; f_{fs} is the partial factor for soil unit weight; f_q is the partial factor for external applied loads.

While generating the tensile load T_{ds} in the reinforcement, the embankment fill should not slide outwards over the reinforcement. Hence, according to BS8006:2010 [1], to prevent the horizontal sliding the minimum reinforcement bond length, L_e, should be:

$$L_e = \frac{0.5 K_h H (f_{fs} \gamma_1 H + 2 f_q w_s) f_s f_n}{\gamma_1 h \frac{a' \tan \varphi'_{cv}}{f_{ms}}} \tag{2}$$

where: f_s is the partial factor for reinforcement sliding resistance; f_n is the partial factor governing the economic ramifications of failure; h is the average height of embankment fill above the reinforcement length L_e; a' is the interaction coefficient relating the embankment fill/reinforcement bond angle to $\tan \varphi'_{cv}$; φ'_{cv} is the angle of friction at large strains of the embankment fill under effective stress conditions; and f_{ms} is the partial material factor applied to $\tan \varphi'$.

Note that a' is otherwise called the direct shear factor f_{ds}, and it can be obtained from direct shear tests, according to EN ISO 12957-1 [3], or derived from published default values for the specific reinforcement and fill (see [4], among others).

In Eqs. (1) and (2) BS8006:2010 [1] assumes that $K_h = K_a$, but in this way the side slope of the embankment is considered to be vertical; hence T_{ds} results to be the same tensile strength that would be required for a vertical reinforced soil wall, with the difference that in a reinforced soil wall the total tensile force is distributed among several reinforcement layers, usually equally spaced along the height, while for basal reinforcement such total tensile force would be assigned to the only reinforcement layer at the base of the embankment; while such assumption is surely on the safe side, it can be easily argued that it is even too much on the safe side, resulting in a tensile strength of the basal reinforcement which can be 5÷10 times more than strictly required.

Rationally, the horizontal earth pressure coefficient K_h should be evaluated by considering the actual side slope of the embankment. Based on the scheme in Figure 1 (from [2]), design charts (from [2]) for steep reinforced embankments have been obtained to evaluate the horizontal earth pressure coefficient K_h (see Figure 2 left) and the reinforcement length to height ratio L_R/H (see Figure 2 right), with respect to the direct sliding mechanism, as function of the friction angle of the fill in drained conditions, φ', and of the slope angle β; charts are available for pore pressure ratio r_u values of 0, 0.25, and 0.50. Figure 2 (modified from [2]) reports the chart valid for pore pressure ratio $r_u = 0$ and cohesionless fill.

Note that, to be consistent with equations (1) and (2), the charts in Figure 2 shall be used with the friction angle of the fill equal to φ'_{cv}.

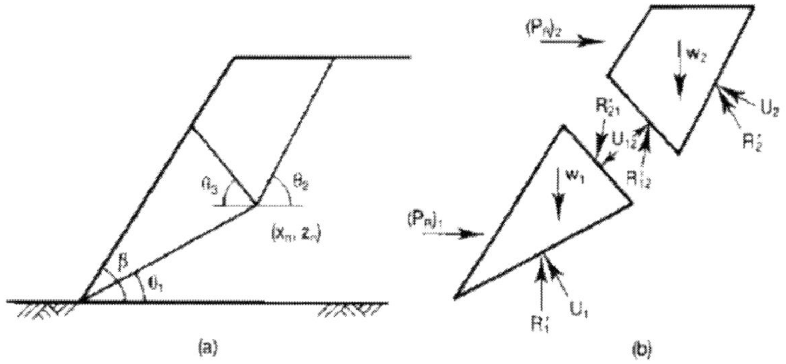

Figure 1. Scheme of steep reinforced embankment (from [2]).

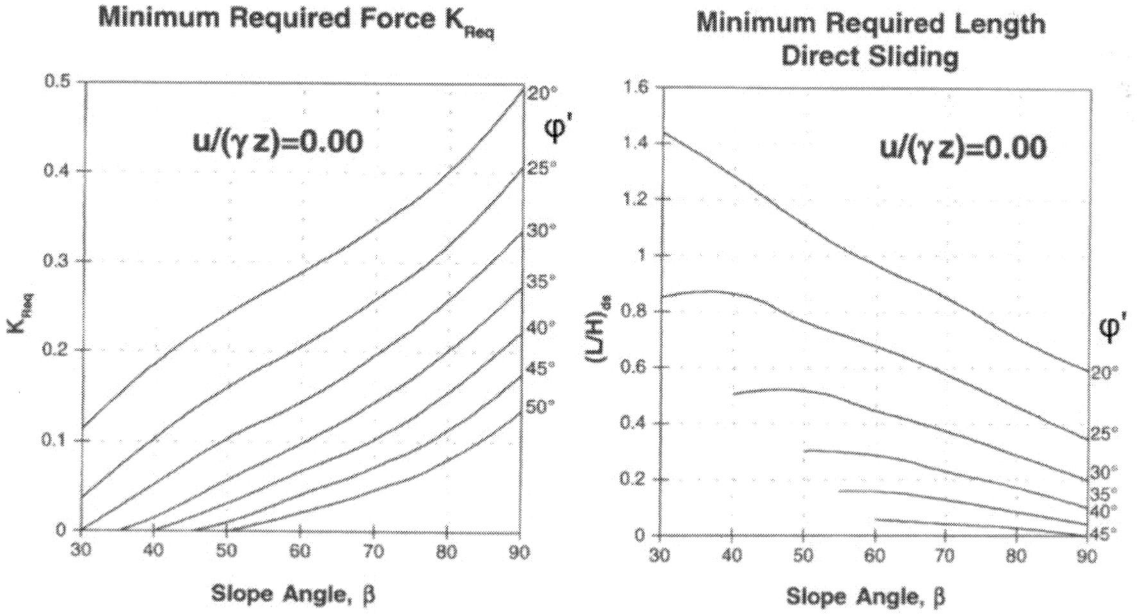

Figure 2. Design charts for steep reinforced embankments, valid for pore pressure ratio $r_u = 0$ and cohesionless fill (modified from [2]).

Hence the required reinforcement strength shall be obtained from Eq. (1) by setting K_h as:

$$K_h = K_{Req} \tag{3}$$

where K_{Req} is obtained from the chart in Figure 2 left.

Using the reinforcement length to height ratio $(L/H)_{ds}$ from the chart in Figure 2 right, the reinforcement length required to resist the sliding mechanism is:

$$L_e = H \cdot (L/H)_{ds} \cdot f_s f_n \tag{4}$$

Such length obviously starts from the toe of the side slope: since the basal reinforcement extends to the whole embankment base, the available reinforcement length is usually well in excess of L_e; hence the required length to resist direct sliding usually is not a limit state.

Another way for evaluating K_h is to use the well-known Coulomb's formula for the coefficient of active thrust with inclined wall face; with reference to Figure 3, in case of horizontal top ($\alpha = 0$) the formula can be written as:

$$K_h = \frac{cos^2(\varphi' - \lambda)}{cos^2\lambda \cdot \cos(\lambda + \delta) \cdot \left[1 + \sqrt{\frac{\sin(\delta+\varphi')\cdot\sin\varphi'}{\cos(\lambda+\delta)\cdot\cos\lambda}}\right]^2} \cdot \cos(\lambda + \delta) \tag{5}$$

where: λ is the angle of the slope to the vertical, δ is the friction angle at the interface between wall and soil, and α is the inclination of soil at top of embankment. For the embankment the slope angle λ in Eq. (5) is negative and is equal to (β− 90°), while the friction angle δ at the interface between wall (which in this case is the slope of an embankment made of fill) and fill can be assumed equal to φ'.

Note that, to be consistent with equations (1) and (2), Eq. (5) shall be used with the friction angle of the fill equal to φ'_{cv}.

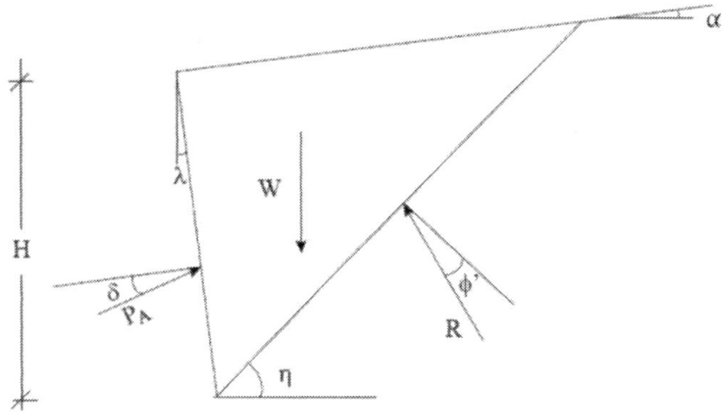

Figure 3. Scheme and symbols for the Coulomb's formula for active thrust.

4. Foundation extrusion

Vertical embankment loading causes an increase in the vertical stress in the foundation soil and a corresponding increase in the horizontal stress. Therefore, a lateral thrust develops in the foundation soil beneath the embankment crest which can eventually cause the foundation soil beneath the embankment side slope to displace laterally. If we assume that extrusion starts from the slope crest (where the vertical load decreases), to prevent this extrusion the side slope length of the embankment, L_s, should be great enough to develop adequate lateral confinement and prevent mobilization of this lateral thrust. The simplified scheme in Figure 4 (modified from [1]) illustrates this behaviour.

The forces acting on the soil block beneath the embankment side slope are: the horizontal thrust developed by the weight of the embankment, $\gamma_1 H$, and the surcharge, w_s; the passive resistance developed at the toe of the embankment; the shear resistance of the foundation soil at the underside of the reinforcement, $\alpha_{bc} c_u$; the shear resistance of the foundation soil at depth z_c. For ultimate limit states, partial factors are applied to loads, materials and resistances, as per indication of the British Standard BS 8006:2010, Section 8 [1].

For the foundation extrusion, analytical solutions are useful for preliminary design investigations, and for making checks on designs, as an adjunct to limit equilibrium methods. Their benefit is that they can be applied quickly and without long computation. Analytical solutions have been derived for foundations on cohesive soft soil of uniform cohesion and limited depth, and for soils with cohesion increasing with depth ([5], [6]).

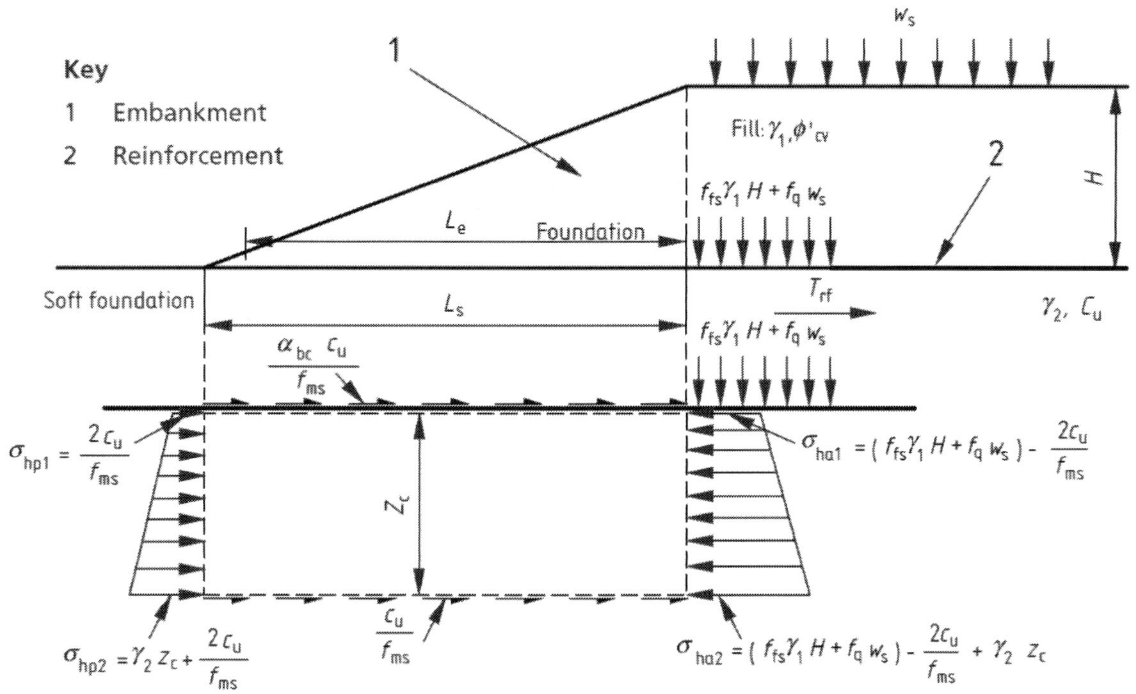

Figure 4. Scheme of foundation extrusion stability on cohesive soft soil (modified from [1]).

4.1. Foundation soil of uniform strength and limited depth

With reference to Figure 4, the following equation expresses the horizontal equilibrium balance for a soft foundation soil with uniform undrained shear strength, c_u, and limited depth, D (with respect to the embankment width), underlain by a rigid stratum:

$$\int_0^D \left(f_{fs}\gamma_1 H + f_q w_s - \frac{2c_u}{f_{ms}}\right) dz = \int_0^D \frac{2c_u}{f_{ms}} dz + \frac{\alpha_{bc} c_u}{f_{ms}} \cdot L_{s,min} + \frac{c_u}{f_{ms}} \cdot L_{s,min} \quad (6)$$

where: α_{bc} is the interface coefficient related the soil/reinforcement adherence to c_u, and $L_{s,min}$ is the minimum distance from the embankment toe to the edge of the embankment crest.

Integration of Eq. (6), with the boundary conditions $z = 0$ and $z = D$, provides the minimum side slope length $L_{s,min}$ required to support the weight of the embankment, $\gamma_1 H$, and the surcharge, w_s; adding the partial factor f_n governing the economic ramifications of failure, $L_{s,min}$ is given by:

$$L_{s,min} = \frac{\left(f_{fs}\gamma_1 H + f_q w_s - \frac{4c_u}{f_{ms}}\right) D}{\frac{c_u(1+\alpha_{bc})}{f_{ms}}} f_n \tag{7}$$

For obvious physical and geometrical reasons, if Eq. (7) provides a minimum side slope length $L_{s,min} \leq 0$, the foundation extrusion cannot occur. This condition can be given as follows:

$$f_{fs}\gamma_1 H + f_q w_s \leq \frac{4c_u}{f_{ms}} \tag{8}$$

Eq. (8) represents the cut-off value for the allowable bearing pressure for the soft foundation soil of uniform strength c_u.

If the inequality (8) is satisfied, the short-term stability of the embankment can be considered satisfied and basal reinforcement is not needed for the foundation extrusion limit state.

Actually, the general solution for the short-term bearing capacity of a strip foundation on uniform clay of strength c_u is:

$$f_{fs}\gamma_1 H + f_q w_s \leq \frac{N_c c_u}{f_{ms}} \tag{9}$$

where: $N_c = 2 + \pi = 5.14$ is the undrained bearing capacity factor for a strip foundation on clay soils.

Hence Eq. (9) could be used in place of Eq. (8), even if it is less conservative.

It is important to note that Eq. (8) can be used also to determine when the basal reinforcement will be no more required: since c_u increases while consolidation progresses, the limit time t_{lim} for the need of basal reinforcement can be obtained from the consolidation theory:

$$t_{lim} = t_c \tag{10}$$

where t_c is the consolidation time at which:

$$c_u(t_c) = \frac{f_{ms}}{4}(f_{fs}\gamma_1 H + f_q w_s) \tag{11}$$

From Figure 4, the tensile load per metre run generated in the basal reinforcement, T_{rf}, due to outward foundation shear stress is:

$$T_{rf} = \frac{\alpha_{bc} c_u}{f_{ms}} \cdot L_{s,min} \tag{12}$$

where: $L_{s,min}$ is the length of the side slope/reinforcement required to prevent foundation extrusion, c_u is the undrained shear strength of the foundation soil at the underside of the reinforcement, and α_{bc} is the interaction coefficient relating the soil/reinforcement adherence to c_u.

Note that α_{bc} could be obtained only from large scale unconsolidated undrained triaxial tests on both reinforced and unreinforced soil specimens, which are not easily available; moreover, no default values for α_{bc} are included in [1]. Hence, in absence of direct measurements or published default values of α_{bc}, it is suggested to assume $\alpha_{bc} = 0.50$.

Adopting Eq. (7) for $L_{s,min}$, the tensile load T_{rf} generated in the basal reinforcement due to foundation extrusion becomes:

$$T_{rf} = \frac{\alpha_{bc} c_u}{f_{ms}} \cdot \frac{\left(f_{fs}\gamma_1 H + f_q w_s - \frac{4c_u}{f_{ms}}\right) D}{\frac{c_u(1+\alpha_{bc})}{f_{ms}}} f_n \tag{13}$$

The force exerted by the foundation soil on the reinforcement depends on the undrained shear strength c_u of the soft foundation soil. In particular, the tensile load to be supported by the reinforcement decreases as the undrained shear strength grows and increases as the undrained shear strength decreases. This is what Eq. (13) say and what is expected from a physical point of view.

Note that this solution is different from what proposed in the BS8006:2010 [1], where an equation of the tensile load generated in the basal reinforcement, directly proportional to the undrained shear strength c_u, is proposed

$$T_{rf} = \frac{\alpha_{bc} c_u}{f_{ms}} \cdot L_e \tag{14}$$

where: L_e is the length of the reinforcement required to prevent the lateral sliding of the embankment fill.

The criticism of the authors to the formula proposed in the BS8006:2010 [1] is for the use of a reinforcement length, L_e, that is not the one calculated for the considered failure mechanism. In fact, the length L_s cannot be directly proportional to the undrained shear strength and, in general, to the parameters of the soft foundation soil. Therefore, the authors do not agree in using Eq. (14) when the foundation extrusion mechanism is investigated. The formula proposed in the BS8006:2010 [1] denies what is physically expected, that is the need of high strength reinforcement for foundation soil with low undrained strength (and vice versa), and contradicts the horizontal equilibrium balance itself that is set to calculate the minimum side slope length $L_{s,min}$, as indicated in Eq. (7). Hence both the length L_s and the tensile load T_{rf} shall be inversely proportional to the undrained shear strength, according to Eq. (7) and Eq. (13).

Note that if the minimum side slope length $L_{s,min}$, from Eq. (7), is equal to zero, also the tensile load T_{rf} generated in the basal reinforcement, from Eq. (13), is equal to zero.

4.2. Foundation soil with strength increasing with depth

The solution described above can be adapted to the problem of soft soil with strength increasing linearly with depth, that is with $c_u(z) = c_{uo} + \rho z$. The overall equilibrium between the unloaded toe and the load to be supported at a distance L_s from the toe is examined, with reference to Figure 4.

In this case the depth of the equilibrium block has to be determined. The same assumption about the horizontal and shear stresses acting on the block are made. As the strength increases with depth these stresses also vary with depth.

For a block of depth z the horizontal equilibrium balance is:

$$\int_0^z \left[f_{fs}\gamma_1 H + f_q w_s - \frac{2(c_{uo} + \rho z)}{f_{ms}} \right] dz = \int_0^z \frac{2(c_{uo} + \rho z)}{f_{ms}} dz + \frac{\alpha_{bc} c_{uo}}{f_{ms}} \cdot L_s + \frac{c_{uo} + \rho z}{f_{ms}} \cdot L_s \tag{15}$$

Integration of Eq. (15) provides the side slope length L_s required to support the external loads. In this case, the side slope length L_s depends on the depth z as follows:

$$L_s(z) = \frac{(f_{fs}\gamma_1 H + f_q w_s)z - \frac{4}{f_{ms}}\left(c_{uo} z + \rho \frac{z^2}{2}\right)}{\frac{c_{uo}(1+\alpha_{bc}) + \rho z}{f_{ms}}} = \frac{N(z)}{D(z)} \tag{16}$$

The critical value of the minimum side slope lengths can be obtained by differentiating Eq. (16) with respect to z and setting $L_s'(z) = \mathrm{d}L_s(z)/\mathrm{d}z = 0$

$$L_s^{'}(z) = \frac{N^{'}(z) \cdot D(z) - N(z) \cdot D^{'}(z)}{D^2(z)} = 0 \Leftrightarrow N^{'}(z) \cdot D(z) = N(z) \cdot D^{'}(z) \quad (17)$$

For reasons of length, all calculations are not reported here but it is easy to demonstrate that the equation below can be obtained:

$$\frac{2\rho^2 z^2}{f_{ms}^2} + \frac{4c_{uo}(1+\alpha_{bc})\rho z}{f_{ms}^2} - (f_{fs}\gamma_1 H + f_q w_s)\frac{c_{uo}(1+\alpha_{bc})}{f_{ms}} + \frac{4c_{uo}^2(1+\alpha_{bc})}{f_{ms}^2} = 0 \quad (18)$$

The resulting equation is arranged in the form of the quadratic equation $az^2 + bz + c = 0$, were the coefficients a, b, and c can be easily deducted from Eq. (18):

$$a = \frac{2\rho^2}{f_{ms}^2};\ b = \frac{4c_{uo}(1+\alpha_{bc})\rho}{f_{ms}^2}; \quad c = -(f_{fs}\gamma_1 H + f_q w_s)\frac{c_{uo}(1+\alpha_{bc})}{f_{ms}} + \frac{4c_{uo}^2(1+\alpha_{bc})}{f_{ms}^2} \quad (19)$$

Note that it is $a > 0$ for $\forall\rho > 0$, and $b > 0$ for $\forall\rho, c_{uo} > 0$ and $\forall\alpha_{bc} > -1$.
The resulting critical depth z_c is obtained from the solution of the quadratic equation

$$z_c = \frac{-b + \sqrt{\Delta}}{2a} \quad (20)$$

where: $\Delta = b^2 - 4ac$.

A critical depth $z_c > 0$is obtained if $-b + \sqrt{\Delta} > 0 \Leftrightarrow \Delta = b^2 - 4ac > b^2 \Leftrightarrow c < 0$.
The condition on the coefficient c can be given as follows:

$$f_{fs}\gamma_1 H + f_q w_s > \frac{4c_{uo}}{f_{ms}} \quad (21)$$

and it represents the cut-off value for the bearing capacity of the soft foundation soil. This is entirely similar to what found for the case of foundation with uniform strength and limited depth:reinforcement is required, for the foundation extrusion limit state, only if the inequality (21) is satisfied.

As mentioned in the previous chapter, the general solution for the bearing capacity of a strip foundation on clay of strength c_{uo}could be used as cut-off value, in place of Eq. (21), even if it is less conservative:

$$f_{fs}\gamma_1 H + f_q w_s > \frac{(2+\pi)c_{uo}}{f_{ms}} \quad (22)$$

Hence, adding the partial factor f_n governing the economic ramifications of failure, the minimum side slope length is:

$$L_{s,min} = L_s(z_c) = \frac{(f_{fs}\gamma_1 H + f_q w_s)z_c - \frac{4}{f_{ms}}\left(c_{uo}z_c + \rho\frac{z_c^2}{2}\right)}{\frac{c_{uo}(1+\alpha_{bc})+\rho z_c}{f_{ms}}} f_n \quad (23)$$

where z_c shall be calculated with Eqs. (19) and (20).

It is important to note that Eq. (21) can be used also to determine when the basal reinforcement will be no more required: since c_{uo} increases while consolidation progresses, the limit time t_{lim} for the need of basal reinforcement can be obtained from Eqs. (10 and (11), using $c_{uo}(t_c)$ in place of $c_u(t_c)$ in Eq. (11).

From Figure 4, the tensile load per metre run generated in the basal reinforcement T_{rf} due to outward foundation shear stress is:

$$T_{rf} = \frac{\alpha_{bc} c_{uo}}{f_{ms}} \cdot L_{s,min} = \frac{\alpha_{bc} c_{uo}}{f_{ms}} \cdot L_s(z_c) \tag{24}$$

where: $L_{s,min}$ is the length of the side slope required to prevent foundation extrusion, c_{uo} is the undrained shear strength of the foundation soil at the underside of the reinforcement and α_{bc} is the interaction coefficient relating the soil/reinforcement adherence to c_{uo}.

As in the previous case, in absence of direct measurements or published default values of α_{bc}, it is suggested to assume $\alpha_{bc} = 0.50$.

Adopting Eq. (23) for $L_{s,min}$, the tensile load in the basal reinforcement T_{rf} becomes:

$$T_{rf} = \frac{\alpha_{bc} c_{uo}}{f_{ms}} \cdot \frac{(f_{fs}\gamma_1 H + f_q w_s) z_c - \frac{4}{f_{ms}}\left(c_{uo} z_c + \rho \frac{z_c^2}{2}\right)}{\frac{c_{uo}(1+\alpha_{bc}) + \rho z_c}{f_{ms}}} f_n = \frac{\alpha_{bc} c_{uo}}{f_{ms}} \cdot L_s(z_c) \tag{25}$$

The force exerted by the foundation soil on the reinforcement depends on both the undrained shear strength parameters of the soft foundation soil, $c_{uo}$and ρ. In this case, it is more difficult to see how T_{rf} varies with the varying of c_{uo}.To solve this issue, Eq. (25) has to be differentiated with respect to the undrained shear strength c_{uo}. Then, the study of the sign is needed. In particular:

$$T'_{rf} = T'_{rf}(c_{uo}) = \frac{\alpha_{bc}}{f_{ms}} \cdot [x(c_{uo}) + c_{uo} \cdot x'(c_{uo})] \tag{26}$$

For the reinforced case $\alpha_{bc} > 0$, then it is $T'_{rf}(c_{uo}) > 0 \Leftrightarrow x(c_{uo}) + c_{uo} \cdot x'(c_{uo}) > 0$.

For reasons of length, all calculations are not reported here but it can be proven that the inequality above is equivalent to:

$$\frac{4z_c(1+\alpha_{bc})}{f_{ms}^2} \cdot c_{uo}^2 + \frac{8\rho z_c^2(1+\alpha_{bc})}{f_{ms}^2} \cdot c_{uo} + \frac{2\rho^2 z_c^3}{f_{ms}^2} - (f_{fs}\gamma_1 H + f_q w_s)\frac{\rho z_c^2}{f_{ms}} < 0 \tag{27}$$

The resulting equation is arranged in the form of the quadratic equation $d c_{uo}^2 + e c_{uo} + f = 0$. Note that it is $d, e > 0$ for $\forall \alpha_{bc} > -1$. The resulting ranges of c_{uo} are obtained from the solution of the quadratic inequality. Considering that it shall be $c_{uo} > 0$, the ranges are:

$$T'_{rf}(c_{uo}) > 0 \Leftrightarrow 0 < c_{uo} < \frac{-e + \sqrt{e^2 - 4df}}{2a} \tag{28}$$

And, on the other side, it is

$$T'_{rf}(c_{uo}) < 0 \Leftrightarrow c_{uo} > \frac{-e + \sqrt{e^2 - 4df}}{2d} \tag{29}$$

Hence the tensile load generated in the basal reinforcement increases as the undrained shear strength c_{uo}grows within the range of Eq. (28). The reason is that the tensile load is directly proportional to the mobilised shear stresses at the reinforcement/foundation soil interface. But the mobilised shear stresses are first unknown and, for simplicity, expressed as percentage of the undrained shear strength $\alpha_{bc}c_{uo}$. The soil cannot support a higher shear stress than the undrained shear strength and this is also the case at the reinforcement/soil interface. In this range, the incremental effect given by $\alpha_{bc}c_{uo}$ is more dominant than the reducing effect given by the critical length $L_{s,min} = L_s(z_c)$.

On the contrary, the tensile load decreases with the increasing of the undrained shear strength $c_{uo}$when the latter is higher than the value in Eq. (29). This is what expected from a physical point of view. The reason is that the tensile load also depends on the length $L_{s,min} = L_s(z_c)$ whose reducing effect is, in this case, more dominant than the incremental effect given by $\alpha_{bc}c_{uo}$.

Finally, when $\rho = 0$, Eq. (29) gives $T'_{rf}(c_{uo}) < 0$ for $\forall c_{uo} > 0$, that is the case of foundation soil with uniform undrained strength.

Therefore, in the case of soft soil with strength increasing linearly with depth, the required tensile strength of basal reinforcement can increase or decrease, in a non-linear way, as function of the values of the undrained shear strength parameters of the soft foundation soil, c_{uo} and ρ.

Again, Eq. (25) is different from what proposed in the BS8006:2010 [1], whose equation of the tensile load generated in the basal reinforcement is directly proportional to the undrained shear strength c_u, whatever is the value of c_u.

5. Design of the basal reinforced embankment at short-term

As shown in previous sections, there are two limit states affecting the short-term stability of geosynthetic reinforced embankments on soft soil: direct sliding of the fill above the basal reinforcement, and foundation extrusion of the soft foundation soil below the basal reinforcement.

The design of geosynthetic reinforced embankments on soft soil, at short-term, requires the definition of the minimum side slope length and of the tensile strength of reinforcement.

As above said, the side slope length does not depend on the direct sliding mechanism, hence is shall be assumed equal to $L_{s,min}$, that is the length of the side slope required to prevent foundation extrusion.

The design tensile strength of reinforcement, T_D, is the sum of the tensile strengths required for the direct sliding and foundation extrusion mechanisms:

$$T_D = T_{ds} + T_{rf} \tag{30}$$

The design tensile strength of reinforcement, T_D, is obviously less than the ultimate strength of the geosynthetic reinforcement. In fact, to get the required ultimate strength, T_{ult}, the design strength T_D should be multiplied by such factors as installation damage, creep and long-term degradation.

Hence, according to ISO/TR 20432 [7], the minimum required T_{ult} is:

$$T_{ult} = T_D \cdot (RF_{ID} \cdot RF_{CR} \cdot RF_W \cdot RF_{CH}) \tag{31}$$

where: T_D = design tensile strength of the reinforcement (kN/m); T_{ult} = ultimate tensile strength of the reinforcement (kN/m); RF_{ID} = reduction factor for installation damage; RF_{CR} = reduction factor for creep; RF_W = reduction factor for long term weathering exposure; RF_{CH} = reduction factor for long term chemical and biological degradation.

Note that: it is always $RF_W = 1$ if the reinforcement is covered with soil within 1 day exposure time; RF_{ID} depends on the type of reinforcement and the type of fill; while RF_{ID} and RF_{CH} depend also on the elapsed time after the end of construction: hence their values should be set considering the required design life of reinforcement, which may be taken equal to the limit time t_{lim} from Eqs. (10) and (11).

The Reduction Factors should be sourced from the technical data sheet of the specific reinforcement under consideration or derived from published default values for the specific type of reinforcement (see [4], among others).

6. Conclusions

The stability of an embankment on soft foundation soil is most critical during the construction period, because of the relatively low permeability of the soft foundation. Since the stability increases as the foundation consolidates, related design is controlled by short-term constraints. The reinforcement has the double function of carrying the outward shear stresses transmitted by the embankment fill and of providing inward shear stresses to restrain the foundation soil from lateral squeezing-out. For each of the limit state mentioned above, the associated reinforcement strength and bond length should be checked.

Starting from considerations proposed by the British Standard BS8006:2010 [1], this paper focuses on lateral sliding and foundation extrusion limit states for an embankment on soft soil, by presenting a design method to evaluate the required reinforcement force and bond length.

The limit time at which the basal reinforcement will be no more required has been determined.

The theoretical and analytical solution herein presented affords different formulas than in [1].

To the authors' opinion the new solution is consistent with what would be physically expected, and therefore it represents an important improvement for the design of basal reinforcement of embankments on soft soil.

References

[1] BS8006:2010. Code of practice for strengthened/reinforced soils and other fills. *British Standard Institution*. London, UK
[2] Jewell R A, Paine N, Woods R I 1985 Design method for steep reinforced embankments. Polymer grid reinforcement. *Thomas Telford Ltd*. London. UK
[3] EN ISO 12957-1 Geosynthetics - Determination of friction characteristics - Part 1: Direct shear test. *ISO*, Geneve, Switzerland
[4] Koerner R M 2012 Designing with Geosynthetics. 6th Ed., *Xlibris LLC*, Bloomington, IN, USA
[5] Jewell R A1988 The Mechanics of Reinforced Embankments on Soft Soils. *Geotextiles and Geomembranes* **7** pp 237-73
[6] Jewell R A 1996 Soil reinforcement with geotextiles. *CIRIA*, Special Publication **123** London, UK
[7] ISO/TR 20432:2007. Guidelines for the determination of the long-term strength of geosynthetics for soil reinforcement. *ISO*, Geneve, Switzerland

Reduction of carbon footprint of roads construction thanks to the Pavement Optimisation with hexagonal geogrids

P Mazurowski[1]

[1] Tensar Polska, Azymutalna 9, 80-298 Gdańsk, Poland

Email: pmazurowski@tensar.pl

Abstract. There is a growing demand on reduction of greenhouse gases emission from all branches of industry. Roads construction is responsible for a substantial part of building industry carbon footprint. One of the ways to reduce its carbon footprint is the use of Pavement Optimisation (PO) with hexagonal geogrids technology. This technology utilizes the phenomenon of aggregate stabilisation by interlock within apertures of geogrid. When aggregate is placed and compacted on a layer of geogrid, a geogrid/aggregate composite – called Mechanically Stabilised Layer (MSL) - is created. Stiff geogrid ribs resist particle movement reducing layer deformation, which increases its stiffness. Increased stiffness of aggregate can be utilized in design, and has a substantial influence on pavement life. This provides the possibility to reduce the thickness of a pavement, including thickness of asphalt. Thinner pavement will require less materials, workforce, time and energy. It will be cheaper, and its construction will result in lesser emission of CO2. This paper presents the overview of Pavement Optimisation technology and results of full scale accelerated pavement tests confirming the influence of hexagonal geogrids. Case studies of projects with optimised pavements and calculations of carbon footprint savings achieved thanks to the use of PO are presented.

1. Introduction

It seems that today there is a common understanding that greenhouse gases (GHG) emission resulting from different activities of humanity is an important factor influencing harmful changes to the Earth climate, i.e. global warming. Construction industry is responsible for significant part of global energy consumption and GHG emission. And a significant part of construction industry is transportation engineering, especially in rapidly developing regions like Central and Eastern Europe, Middle East, Asia and Africa.

Road construction industry puts a great effort to develop new technologies which could lead to reduction of energy consumption and GHG emission, while assuring achieving required quality and longevity of pavements. Good examples are use of WMA (Warm Mix Asphalt), which requires lower temperatures of production and installation, or use of RAP (Recycled Asphalt Pavement) as an element of new asphalt mixes compositions [1].

Similar effect can be achieved with different approach. It is to reduce energy consumption and GHG emission by reducing the thickness of pavements. There are different ways to achieve this, like development of new polymer modified bitumen, or asphalt mixes types (i.e. high modulus asphalt), and also – the approach which is covered in this paper – the use of geosynthetics in pavement structures.

2. Pavement Optimisation

One geogrid manufacturer developed a technology called "Pavement Optimisation" (PO), which utilizes the benefits of stabilisation of unbound aggregate layers with stiff hexagonal geogrids to optimize the pavement structure. Aggregate layer stabilised with hexagonal geogrid (called Mechanically Stabilised Layer – MSL) has higher parameters – stiffness and compaction – and durability compared to the same layer without geogrid. Increased parameters of one layer influence the design life of the whole pavement, which leads to the possibility of either the reduction of pavement thickness (including asphalt layers), substantial increase of pavement life or combination of both. PO technology results in reduction of pavement construction and maintenance cost as well as carbon emission. The basic idea of Pavement Optimisation is illustrated on Picture 1.

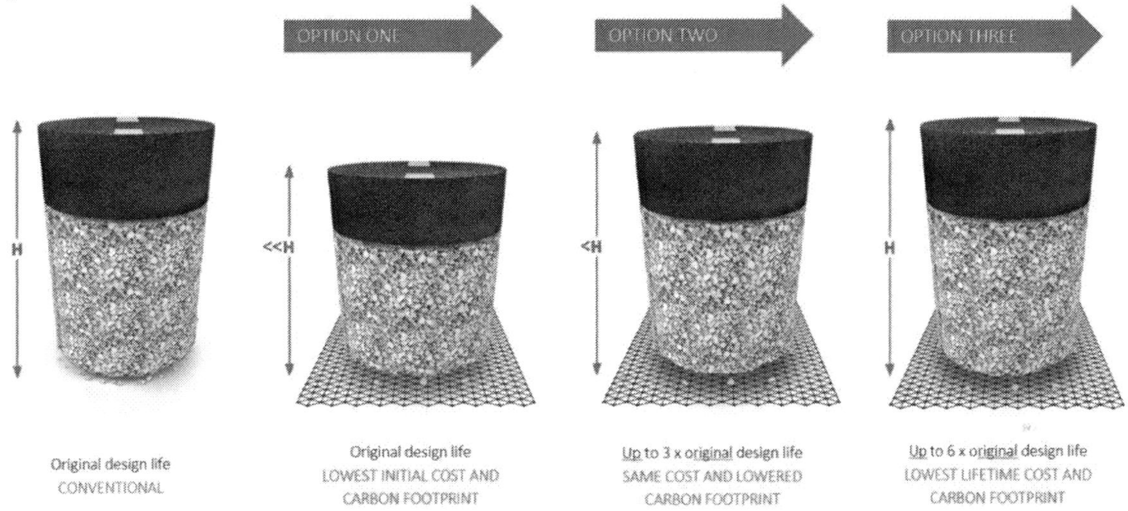

Figure 1. Options available with Pavement Optimisation

For a given original, conventional pavement structure, there are three options for PO with hexagonal geogrids. First one is to reduce the pavement thickness in maximum possible way, while keeping the required pavement life. Resulting pavement construction will have the lowest possible construction cost and carbon emission. Second option is to reduce the pavement thickness only slightly, usually to cover the additional cost of geogrid, while increasing the life of the pavement. In this option the cost of the construction of the pavement remains the same as the conventional one, however the reduction of lifetime costs and carbon emission due to increased life and reduction of thickness is obtained. And third option is to maintain the original thickness of the pavement, while adding the geogrid to the aggregate base. In this option both costs and carbon emission are slightly increased at the time of construction, but thanks to substantial increase of life (up to 6 times in some cases) of the pavement, both are substantially reduced during operation of the pavement, as maintenance costs and embodied carbon emission are reduced.

3. The concept of mechanical stabilisation with hexagonal geogrid

When unbound aggregate is placed and compacted on a layer of stiff geogrid, aggregate particles interlock within the geogrid apertures and are confined by its stiff ribs. A geogrid/aggregate composite – MSL - is created, in which lateral restraint provided by the geogrid increases the stiffness of the layer compared to the same layer without geogrid. An MSL can be equivalent – in terms of bearing capacity of the layer and its influence on pavement life – to 1,5 to 2 times thicker layer of the same aggregate without geogrid.

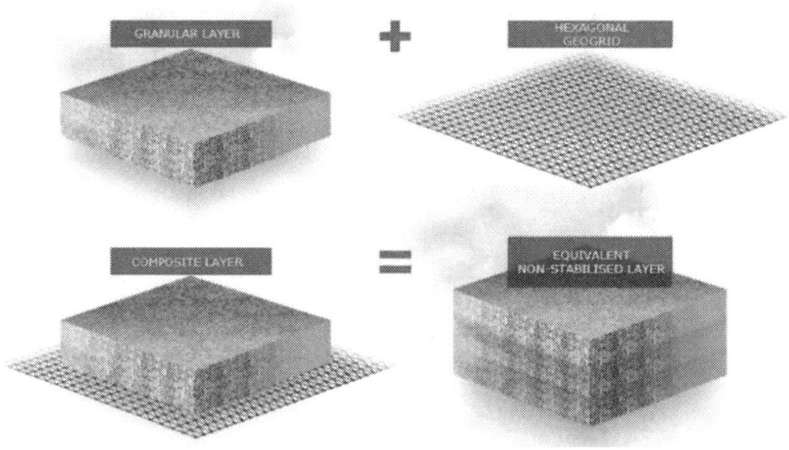

Figure 2. Concept of Mechanically Stabilised Layer

"Stabilisation" is one of the functions of geogrids, recognized by ISO [2] as a different function than Reinforcement, which until now was traditionally linked to the use of all geogrids.

4. Full scale trafficking tests

To confirm the influence of hexagonal geogrid on pavement life, and to get data necessary to modify pavement design methods, series of three full scale accelerated pavement tests was performed at the U.S. Army Engineer Research and Development Center (ERDC).

A Heavy Vehicle Simulator (HVS) device (Figure 3) was used to apply load on 8 test sections in total - 4 with hexagonal geogrid stabilised aggregate base, and 4 control sections without geogrid. The subgrade of test sections was prepared to achieve bearing capacity of either 3% or 6% CBR. The surfacing course consisted of 5, 7.5 or 10 cm layer of hot mix asphalt (HMA) on six sections, and double bituminous surface treatment (DBST) on two sections. The base consisted of 15 or 20 cm layer of crushed aggregate. Build up of sections with HMA surfacing is presented on Figure 4.

Figure 3. Heavy Vehicle Simulator device used in ERDC tests

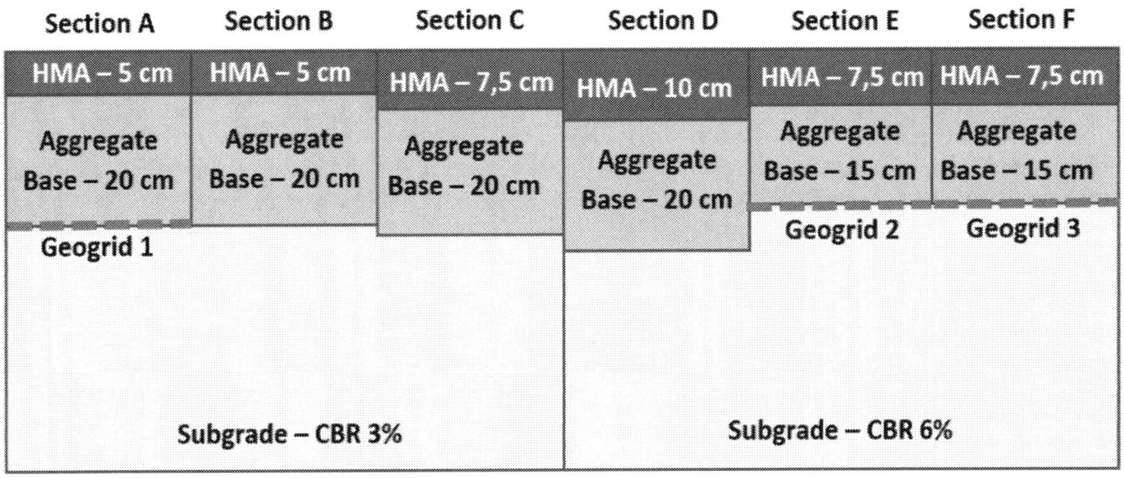

Figure 4. Build up of tests sections with HMA surfacing in ERDC tests

The number of Equivalent Standard Axle Loads (ESALs) applied during the three phases of the test was 100 000 ESALs for pavements on CBR 3% subgrade and 800 000 ESALs for pavements on CBR 6% subgrade. Results of trafficking for sections with HMA surfacing are presented in Table 1. Number of ESALs to reach 12,5 mm (0,5") rut depth represents the life of the pavement – this is the critical rut depth marking the failure of the pavement according to Asphalt Institute design method [3].

Table 1. A summary of trafficking results for test sections with HMA surfacing

Section	Subgrade CBR [%]	Subbase [cm]	Asphalt [cm]	No. of ESALs to 12.5 mm rut APT
B – Control	3	20	5	5400
C – Control	3	20	7,5	12640
D – Control	6	20	10	500000
A – Stabilised	3	20	5	100000
E – Stabilised	6	15	7,5	2000000[a]
F – Stabilised	6	15	7,5	2200000[a]

[a] Extrapolated, as rut depth in sections E and F only reached 6-7 mm rut depth after 800 000 ESALs, when the test was concluded

All sections with base stabilised with hexagonal geogrid heavily outperformed their control counterparts. Number of ESALs on stabilised sections was 4 to almost 20 times higher than on control sections, even in cases when pavement layers were thinner in stabilised sections.

Detailed description of test procedure, materials, loads applied and results is presented in the reports: Jersey et al. [4], Norwood et al. [5], Robinson et al. [6].

Based upon back calculations of results of ERDC tests, and supporting results from many other smaller scale tests, manufacturer of hexagonal geogrids developed a modifications of two pavement design methods – empirical AASHTO'93 method and mechanistic-empirical method. Details of modification of mechanistic-empirical method were presented by Mazurowski et al [7]. Both modified design procedures were independently verified [8,9].

Pavement optimisation with hexagonal geogrids offers two novelties in geogrids application. First is that in this approach full pavement structure is analyzed, from the subgrade to the wearing course, and the influence of geogrid is taken into account in the design of all layers, not only soil improvement layers as it was usually done earlier. Second is that with PO technology geogrids can be used both on weak and firm soils – good confirmation of the late are ERDC tests described above for 6% CBR subgrade. This is relatively good subgrade, and traditionally one wouldn't consider using geogrids for subgrade of such high bearing capacity.

5. Case studies

5.1. Voivodeship Road 507 Braniewo – Pieniężno (Poland)

Good example of substantial savings in carbon emission thanks to the use of PO is reconstruction of 29 km long section of Voivodeship Road no 507 Braniewo – Pieniężno in Poland. Project was divided into two parts: Braniewo – Wola Lipowska (~15 km) and Wola Lipowska - Pieniężno (~14 km). First part has been completed in 2019, second part is under construction at the moment.

Existing pavement was heavily damaged, and the Clients decision was to completely demolish it and construct a new one. A new pavement was designed for the whole section. To reduce costs and utilize recycled materials, a pavement type with base made of Recycled Asphalt with bitumen emulsion and cement mix (RAP) was chosen.

About 8 km of 15 km of the first section had poor ground conditions, with high plasticity clay of estimated bearing capacity of E2 = 15 MPa directly under the pavement. Pavement on this section have been optimised with the use of hexagonal geogrids.

Typical pavement for prognosed medium traffic category KR3 and stiff clay subgrade designed according to Polish Catalogue of Typical Flexible and Semi-Rigid Pavement Structures [10] consists of 12 cm of asphalt, 20 cm of RAP base, 24 cm of CBR60% aggregate subbase and 40 cm of CBR20% aggregate capping layer. As the catalogue does not cover the cases of subgrade consisting of high plasticity cohesive soils, an additional capping layer has to be added on the subgrade, in this case 15 cm of in-situ chemically stabilised layer. The full build up of the typical pavement for this case is presented on Figure 5, left side.

Optimisation with hexagonal geogrid allowed for substantial reduction of thickness of layers. Asphalt layers thickness was reduced by 1 cm, RAP base by 2 cm, and CBR20% aggregate capping layer by 20 cm. It was also possible to remove whole 15 cm in-situ chemically stabilised layer.

It should be noted that so big reduction of thickness was a result of combination of two factors: optimisation with geogrid and a fact that typical pavement from catalogue is slightly overdesigned for the volume of traffic prognosed for this section of a road.

Table 2 below summarizes carbon emission resulting from construction of 1 m2 of both types of structures – typical and optimised. Embodied CO2 emission (eCO2) – an equivalent measure of all GHG emissions related to production, transport and installation of material - values for each type of material were adopted from publications by Chappat and Bilal [11] and Raja et al [12]; some of the values had to be estimated as not all materials are discussed in these publications.

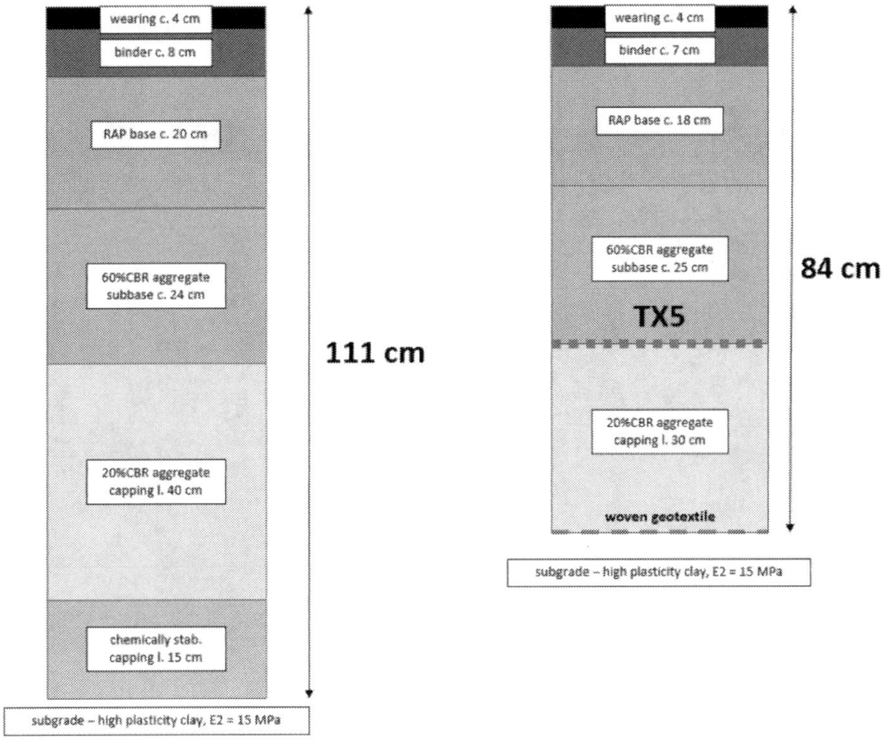

Figure 5. Typical and optimised pavements designed for road no. 507

Table 2. A summary of carbon emissions for road no. 507 pavements

	embodied CO2 emission		typical pavement		optimised pavement	
layer	eCO2 [kg/t]	eCO2 [kg/m3]	thickness [m]	eCO2 [kg/m2]	thickness [m]	eCO2 [kg/m2]
asphalt concrete	54	135	0.12	16.2	0.11	14.85
RAP base	30	66	0.2	13.2	0.18	11.88
granular base	15	27	0.24	6.48	0.25	6.75
granular capping layer	15	27	0.4	10.8	0.3	8.1
chemically stabilised capping	14	28	0.15	4.2	0	0
geogrid						0.6546
geotextile						0.235
total eCO2 – kg per m2 of pavement			50.88		42.47	

In case of this project Pavement Optimisation reduced the carbon emission by about 8,4 kg (or 17%) of eCO2 per every square meter of the pavement. For 8 km long section, on which this optimised pavement structure was used, it results in more than 470 T of eCO2 less introduced to the atmosphere compared to typical catalogue structure.

Figure 5. Installation of hexagonal geogrid on road no. 507

5.2. Local Road D148030G Czarna Dąbrówka (Poland)

Another example of environmental benefits due to PO is a local road D148030G near Czarna Dąbrówka village in Poland, constructed in 2018 (Figure 8). It was a relatively small project – 1 km long section of 6 m wide local road – but it is a good example of optimisation of a pavement constructed on a firm subgrade. It was a Design&Build type of project, which confirms that this technology offers economic benefits, as in these type of projects Contractors are usually looking for cheapest – yet still meeting all requirements – solutions.

Subgrade on this sections consisted of fine and medium sands of estimated bearing capacity of E2 = 80 MPa, so no capping layer or subbase war required. The original pavement structure, adopted from Catalogue [10] for light traffic category KR2, consisted of 12 cm of asphalt layers and 20 cm of crushed aggregate base. Optimised section, with aggregate base stabilised with hexagonal geogrid, consisted of 10 cm of asphalt and 15 cm of aggregate (Figure 7).

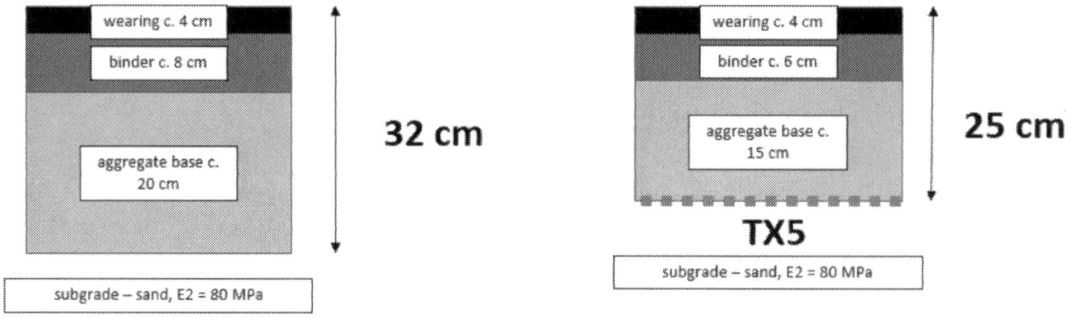

Figure 7. Typical and optimised pavements designed for road no. D148030G

Table 3. A summary of carbon emissions for road no. D148030G pavements

	embodied CO2 emission		typical pavement		optimised pavement	
layer	eCO2 [kg/t]	eCO2 [kg/m3]	thickness [m]	eCO2 [kg/m2]	thickness [m]	eCO2 [kg/m2]
asphalt concrete	54	135	0.12	16.2	0.1	13.5
granular base	15	27	0.2	5.4	0.15	4.05
geogrid						0.6546
total eCO2 – kg per m2 of pavement				21.6		18.2

A summary of eCO2 emission for the typical and optimised sections is presented in Table 3. PO allowed for 3,4 kg (or 16%) of eCO2 emission reduction compared to typical solution. It does not seem to be a big amount, but multiplied by area of the road it still means reduction of more than 20 T of eCO2 emission for such relatively small project.

Figure 8. Installation of hexagonal geogrid on road no. D148030G

6. Spectra Value Calculator

A manufacturer of hexagonal geogrid developed a relatively simple, yet very useful, software called Spectra Value Calculator. This software allows for calculation of whole set of data representing benefits of Pavement Optimisation with hexagonal geogrids. When details like pavement structure – both typical and optimised with geogrid, materials and activities costs and transportation distances are entered, the software calculates: reduction of construction costs, reduction of construction program, life cycle cost benefits and reduction of carbon footprint of the project. For the purpose of calculating the carbon footprint of structures the software uses methodology provided in the BSI reference [13], and it takes into account all activities required to produce, transport and install all materials needed to construct a pavement. An example of output from Spectra Value Calculator is presented on Figure 9.

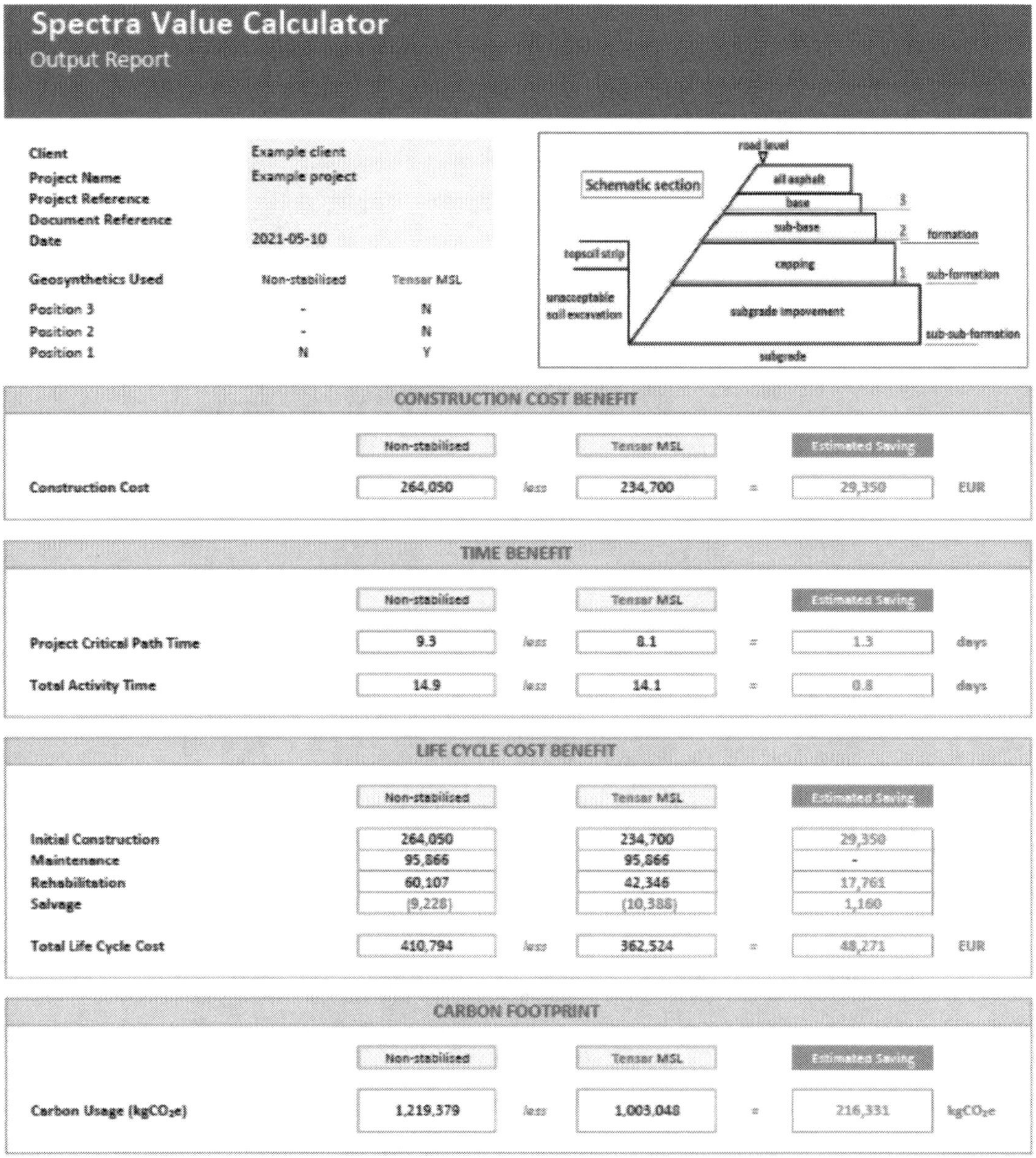

Figure 9. An example of output from Spectra Value Calculator

7. Conclusions

Developments of technologies in highway construction should not only be focused on obtaining safe, durable, and economical solutions for pavements but should also lead to solutions that minimize the negative impact of projects on the environment. Use of the geosynthetics in road construction is one of the way to reduce GHG emission compared to traditional solutions.

This paper focuses on the use of one technology involving the use of hexagonal geogrids to stabilise aggregate layers in pavements. Pavement Optimisation with hexagonal geogrids technology can help save a substantial amount of CO2 emission to the atmosphere, while providing pavements of the same or even higher life than traditional ones.

References

[1] Al Adday F 2019 Study of Reducing the Environmental Impact of CO2 Emissions of Flexible Pavement Materials: A Critical Review *International Journal of Development Research*

[2] EN ISO 10318-1:2015+A1:2018 Geosynthetics. Terms and definitions

[3] The Asphalt Institute 1982 Research and Development of the Asphalt Institute's Thickness Design Manual (MS-1), USA

[4] Jersey S R, Tingle J S, Norwood G J, Kwon J and Wayne M 2012 Full-Scale Evaluation of Geogrid-Reinforced Thin Flexible Pavements. *Transportation Research Record, Journal of the Transportation Research Board*, No. 2310, USA

[5] Norwood G J and Tingle J S 2014 Performance of Geogrid-Stabilized Flexible Pavements. Final Report. EDRC/GSL TR-14-28. U.S. Army ERDC, USA

[6] Robinson W J, Tingle J S and Norwood G J 2017 Full-Scale Accelerated Testing of Multi-Axial Geogrid Stabilized Flexible Pavements. Draft Final Report. EDRC/GSL TR-17-X. U.S. Army ERDC, USA

[7] Mazurowski P, Buckley J, Kawalec J 2019 Modification of mechanistic-empirical pavement design method to incorporate the influence of hexagonal stabilisation geogrids, 18th AAPA International Flexible Pavements Conference, Sydney, Australia

[8] Applied Research Associates 2015 Independent Review of Spectra M-E Pavement Design Software, Champaign, USA

[9] Applied Research Associates, Ryan R. Berg & Associates, Inc. 2013 Independent Review and Validation of Tensar's Modified 1993 AASHTO Pavement Design Procedure and Verification of Spectrapave4-Pro Software, USA

[10] General Directorship of National Roads and Highways 2014 Catalogue of Typical Flexible and Semi-Rigid Pavement Structures, Warsaw, Poland

[11] Chappat M and Bilal J 2003 The Environmental Road of the Future – Energy Consumption & Greenhouse Gas Emission, Colas, France

[12] Raja J, Dixon N, Fowmes G, Frost M and Assinder P 2015 Obtaining Reliable Embodied Carbon Values for Geosynthetics, *Geosynthetics International 22(5):1-9*

[13] Publicly Available Specification PAS 2050:2011 Specification for the assessment of the life cycle greenhouse gas emissions of goods and services http://shop.bsigroup.com/upload/shop/download/pas/pas2050.pdf

IOP Conf. Series: Materials Science and Engineering 1260 (2022) 012003 doi:10.1088/1757-899X/1260/1/012003

Stress-dilatancy relationship of sand-rubber mixtures

Z Szypcio[1], K Dołżyk-Szypcio[1], A Nurgaliyev[2]

[1]Faculty of Civil Engineering and Environmental Sciences,Bialystok University of Technology, Poland
[2]Kazakh Leading Academy of Architecture and Civil Engineering, Kazakh Republic

E-mail:z.szypcio@pb.edu.pl

Abstract. The results of drained triaxial compression tests and DEM simulations of sand-rubber mixtures presented in the literature were analysed using of frictional state concept. The stress-plastic dilatancy relationship at different shear stages can be approximated by straight lines determined by critical state angle and two parameters of the frictional state concept. The points representing failure state lie on the frictional state line as for pure sand without rubber. The frictional state concept can be used to simply describe the stress-dilatancy relationship of various sand-rubber mixtures.

1. Introduction

The amount scrap tyres is growing rapidly due to the continuous increase in number of cars, vans, trucks, agricultural and constructions machines. The ban of scrap tyres disposal in landfills in many countries with scrap tyres management programs have increased the scrap tyres recovery rate. At present, the recovery rate of scrap tyres has increased to 80%-95% in USA, Japan, and EU [1]. The main recovery directions are contribution in special furnaces, pyrolysis and material recycling. The use of rubber in geotechnical projects is especially beneficial by [2, 3]:

- reducing settlement of highway embankment, constructed on weak subsoil due to low weight of rubber or soil-rubber mixtures,
- reducing the lateral earth pressure on retaining walls,
- slope stabilization,
- erosion prevention,
- providing filter layers,
- thermal insulation,
- seismic insulation of foundations.

Rubber granulates are used in road pavement constructions [4, 5, 6].

In 2010 year only 7.4% of scrap tyres in UE and 7.8% in USA were used in civil engineering application [7]. This may be because of lack of environmentally friendly recycling of scrap tyres for reuse in civil engineering, and difficulties of laboratory tests and correct interpretation of tests results. Many laboratory tests results have shown that the inclusion of rubber granulates into the sand increases the effective angle of shearing resistance at the critical state and reduces dilatancy. The shear stiffness of sand-rubber mixtures depends on rubber size and content [8,9]. Shreds and larger chips often increase peak shear strength but smaller chips, crumbs have no effect on strength or negative effect [10, 11]. The rubber fibres inclusion into clays gives benefits at low stress levels and no benefits at high stress levels [10, 12].

The very large amount of research work has been undertaken to describe the behaviour of soil-rubber mixtures within a critical and steady state framework [10, 13]. The generally comparison between the various data sets are generally not possible.

Discrete element method (DEM) simulations allow the tracking of the distribution and magnitude of stress and strains at all test stages which is not possible in conventional laboratory experiments. The contribution each type of contact i.e. sand-sand, sand-rubber, rubber-rubber on the behaviour of sand-rubber mixtures can be described by DEM simulations [14].

The linear stress-dilatancy relationships for various granular materials at different stages of shearing can be described by use of Frictional State concept [15, 16, 17].

The paper presents the stress-dilatancy relationships for conventional triaxial compression tests of sand -rubber mixtures conducted by Rouhanifar end Ibraim [3] and DEM simulations [14]. It is shown that stress-dilatancy relationships at different stages of shearing may be approximated by straight lines defined by critical frictional state angle and two parameters of Frictional State concept [15]. The results may be helpful for analysis of tests results and modelling of sand-rubber mixtures.

2. Stress-plastic dilatancy relationship for drained triaxial compression

General stress-plastic dilatancy relationship for drained triaxial compression has the form [15, 16]

$$\eta = Q - AD^p \tag{1}$$

where

$$\eta = q/p' \tag{2.a}$$

$$Q = M_c^o - \alpha A_c^o \tag{2.b}$$

$$A = \beta A_c^o \tag{2.c}$$

$$D^P = \delta\varepsilon_v^p/\delta\varepsilon_q^p \tag{2.d}$$

$$q = \sigma_1' - \sigma_3' \tag{2.e}$$

$$p' = (\sigma_1' + 2\sigma_3')/3 \tag{2.f}$$

$$M_c^o = (6sin\varphi^o)/(3 - sin\varphi^o) \tag{2.g}$$

$$A_c^o = 1 - M_c^o/3 \tag{2.h}$$

$$\delta\varepsilon_v^p = \delta\varepsilon_1^p + 2\delta\varepsilon_3^p \tag{2.i}$$

$$\delta\varepsilon_q^p = 2(\delta\varepsilon_1^p - \delta\varepsilon_3^p)/3 \tag{2.j}$$

φ^o - critical frictional state angle, α and β frictional state concept parameters. For sands $\varphi^o \approx \varphi'_{cv}$ [16].

The plastic parts of volumetric and shear strain increments are calculated from equations

$$\delta\varepsilon_v^p = \delta\varepsilon_v - \delta\varepsilon_v^e \tag{3.a}$$

$$\delta\varepsilon_q^p = \delta\varepsilon_q - \delta\varepsilon_q^e \tag{3.b}$$

where

$$\delta\varepsilon_v^e = \delta p'/K \tag{4.a}$$

$$\delta\varepsilon_q^e = \delta q/3G \tag{4.b}$$

K, G - bulk and shear modulus of material.

The characteristic stress-plastic dilatancy relationships can be approximated by straight lines for different stages of shearing. The straight line for each stages are defined by critical frictional state angle (φ°), α and β parameters (Fig.1).

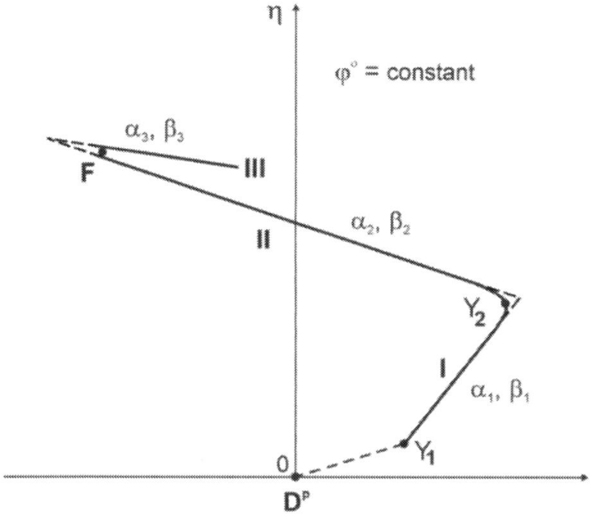

Figure1. Scheme of characteristic stress-plastic dilatancy relationship

The point Y_1 (Fig.1) represents the onset of plastic yielding - stage I of shearing. The plastic parts of volumetric and shear strains increments have suddenly appeared. At this stage of shearing, the observed stress-plastic dilatancy relationship is generally curvilinear but may be approximated by the straight line defined by φ°, α_1 and β_1. At the initial stage of shearing (0 - Y_1), the elastic behaviour is observed. The point Y_2 (Fig.1) represents the onset of plastic yielding stage II for which contribution of plastic parts of strain increments is small in the global increment. In this stage of shearing plastic dilatancy almost equal dilatancy ($D^p = D$, $D = \delta\varepsilon_{\upsilon} / \delta\varepsilon_q$). Point F represents a failure state for dilative behaviour of the material. At this stage of shearing, the stress-plastic dilatancy relationship is generally linear and defined by φ°, α_2 and β_2. For ideal granular material point F, lies on frictional state line (FSL) defined by φ°, $\alpha=0$ and $\beta=1$ [15]. Post-failure stage of shearing is generally represented by straight-line defined by φ°, α_3 and β_3. Ideal granular material exhibits purely frictional behaviour ($\alpha_3=0$, $\beta_3=1$). For many experiments $\alpha_3 \neq 0$ and $\beta_3 \neq 1$ [e.g.17] due to nonhomogeneous deformations of the sample at this stage of shearing, grain crushing effect and imperfections of the experiment.

3. Stress-plastic dilatancy of sand -rubber mixtures

3.1. Drained triaxial compression tests

The mixture of silica Leighton Buzzard fraction A sand with rubber chips was tested by Rouhanifar end Ibraim [3]. The rubber chips were sieving and sorting to reach equivalent granular rubber type (R) material with very similar particle size distribution to Leighton Buzzard sand (LBS). The rubber fraction in the mixture (FR) was defined as [3].

$$FR = \frac{V_{Rubber\ Particles}}{V_{Rubber\ Particles} + V_{Sand\ Particles}} \tag{5}$$

Cylindrical samples with 70mm in diameter and 70mm height with different values of rubber fraction FR values of 0%, 5%, 10%, 20%, 30% and 100% were made with moist tamping method. Samples were sheared under different confining pressures of 50kPa, 100kPa and 200kPa. The initial void ratio, e_0 =0.64 has been kept identical for all samples.

In this paper, only four experimental tests results were analysed. The tests with confining pressure σ'_3 =100kPa and rubber fracture content FR = 0%, 10%, 30%, 100%. In the original paper [3] the tests are named LB100R0c, LB100R10c, LB100R30c, and LB100R100c, respectively.

3.2. Methodology and results.

The tests results graphs q-ε_a and ε_υ-ε_a (Fig.2) were sectionally approximated by high degree polynomials. The values of tangent shear modulus

$$G_t = \delta q/3\delta\varepsilon_q \tag{6}$$

were calculated (Fig.3).

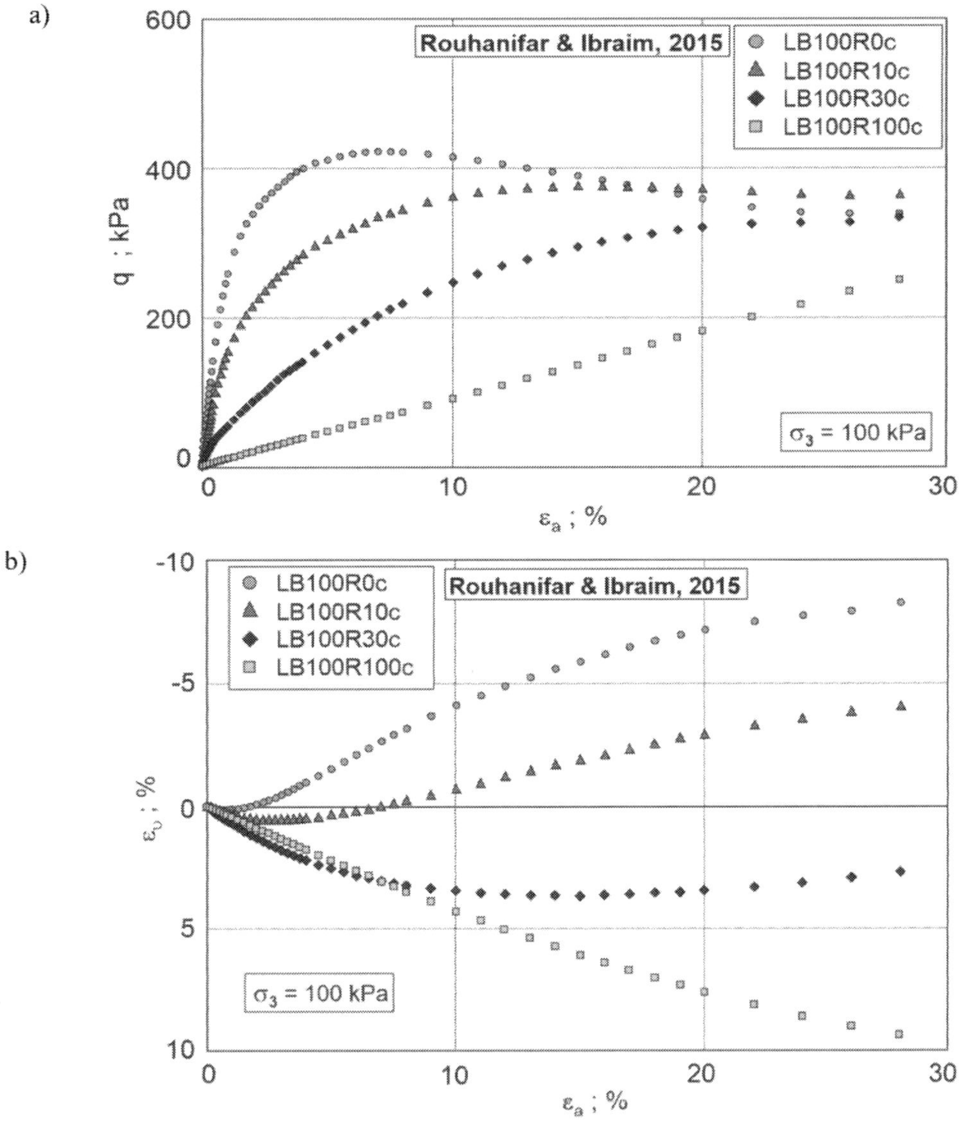

Figure 2. Tests results: a) changes of shear stress with axial strain, b) changes of volume with axial strain

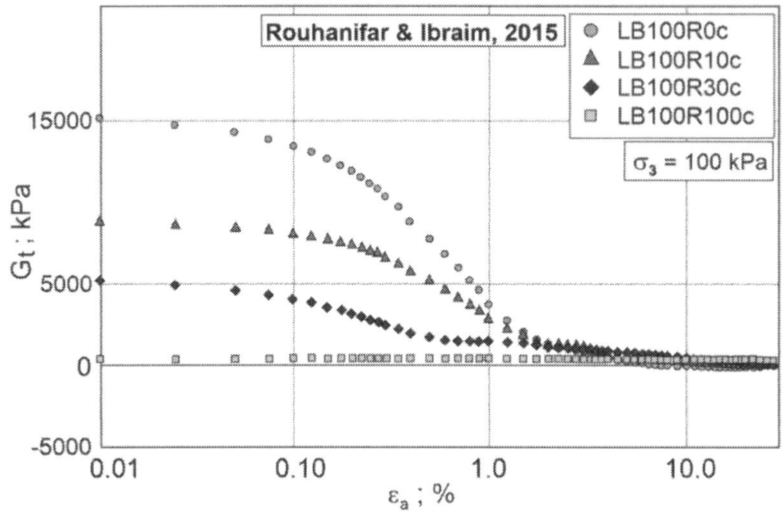

Figure 3. Changes of tangent shear modulus with axial strain

For all analysed tests values of tangent shear modulus are almost constant for $\varepsilon_a < 0.1\%$ ($\varepsilon_3 < 0.1\%$), steeply drop for $0.1\% < \varepsilon_a < 1\%$ (2%) and reach very small values for higher axial strains. The value of shear modulus G, taken for calculation, is assumed equal tangent shear modulus at the initial stage ($G=G_t$ for $\varepsilon_a < 1\%$). As ordinary, it is also assumed that shear modulus is a function of current void ratio e and mean normal stress (p')

$$G = G_O \frac{(2.97-e)}{(1+e)^2}\left(\frac{p}{p_a}\right)^{\xi} \tag{7}$$

For calculation elastic parts of strain increments, the constant Poisson's ratio for each sand-rubber mixtures is assumed. The adopted values Go, ξ, ν taken for calculations are shown in Table 1. The η-D^p relationships obtained from the calculation are shown in Fig.4.

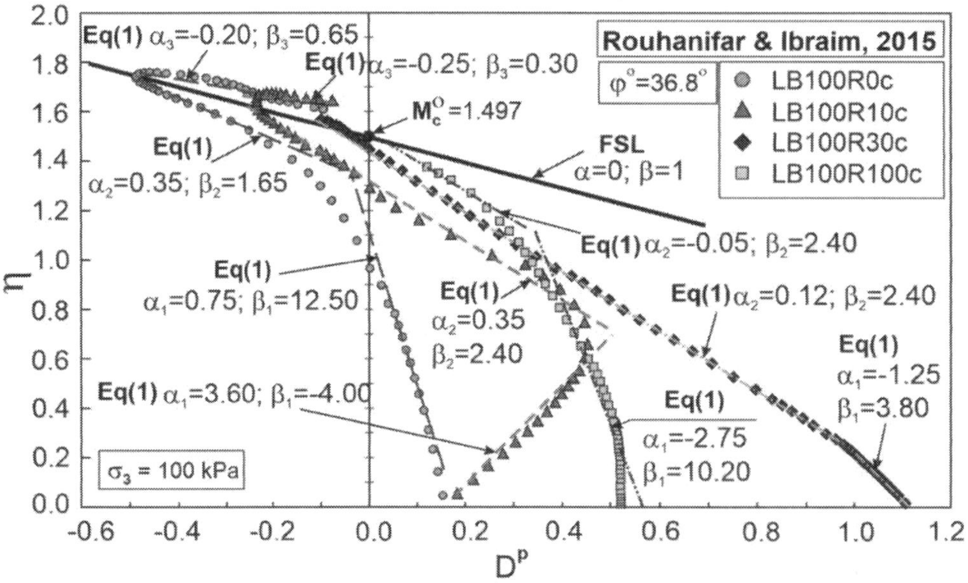

Figure 4. Stress ratio-plastic dilatancy relationship

The linear approximations of stress-plastic dilatancy relationships for different stages of shearing are also shown in Fig.4. We can see that for three tests, sand-rubber mixtures with rubber fraction FR≤30% the points F are approximately lying on the frictional state line (FSL) for $\varphi°=36.8°$, $\alpha=0$ and $\beta=1$. Sand-rubber mixture with FR=100% during shear shows contractive behaviour and the experiment was interrupted before reaching critical frictional state.The values of α and β are shown in Table 1.

Table 1. Values of elastic and frictional state parameters

Test	Go	ξ	v	$\varphi°$	α_1	β_1	α_2	β_2	α_3	β_3
	(kPa)	(-)	(-)	(°)	(-)	(-)	(-)	(-)	(-)	(-)
LB100R0c	120	0.5	0.30	36.8	0.75	12.50	0.35	1.65	-0.20	0.65
LB100R10c	110	0.5	0.25	36.8	3.60	-4.00	0.35	2.40	-0.25	0.30
LB100R30c	70	0.5	0.20	36.8	-1.25	3.80	0.12	2.40	-	-
LB100R100c	50	0.5	0.20	36.8	-2.75	10.20	-0.05	2.40	-	-

The $\varphi°=36.8°$ is higher than for Leighton Buzzard sand critical state angle shown in the literature [18]. It may be a sample shape effect. The shear strength of sand-rubber mixtures drops with grows of a rubber fraction. At stage I of shearing the rubber fraction strongly influence stress-plastic dilatancy relationship. At stage II the straight lines approximated stress ratio-plastic dilatancy relationships are almost parallel ($\beta_2=2.40$) and rubber fraction only displaces that lines in η-D^p plane. Post-failure behaviour should be fully frictional. The non-frictional behaviour at this stage is affected by non-homogeneous deformation of the sample. The presented methodology of analysis can be used only for homogeneous deformation of the sample.

4. DEM simulation

The drained triaxial tests of sand-rubber mixtures were simulated by the DEM method by Lopera Perez et al. [14]. Drained triaxial shear of sand-rubber mixtures with different rubber size ratio 1, 2.5, and 5 and content of 0%, 10%, 20%, and 30% by weight were DEM simulated up to large axial strains to reach a critical state. DEM simulation shows that the slope of critical state line in q-p' plane is affected by rubber size and content. The critical state angle of sand-rubber mixtures ranges between 17.6° to 19.8°[14]. The sand-rubber mixtures strength increases with rubber content for the higher size of rubber granules. The stress ratio-strain increment ratio (η- $\delta\varepsilon_v/\delta\varepsilon_1$) obtained for different sand-rubber mixtures are shown in Fig.5.

For clarity, only FSL are shown. The calculation was made with $\varphi°=17.6°$. Neglecting elastic part of strains increments the FSL in-plane η- $\delta\varepsilon_v/\delta\varepsilon_1$ has the form

$$\eta = M_c^o - A_c^o \frac{3(\delta\varepsilon_v/\delta\varepsilon_1)}{3-(\delta\varepsilon_v/\delta\varepsilon_1)} \tag{8}$$

where $M°_c= 0.672$, $A°_c =0.776$. The FSL in this plane is a convex curve.

It is unexpected that stress ratio - strain increment ratios at post-failure stages obtained from DEM simulations give almost fully frictional behaviour independently on rubber size and content (Fig.5). For clean sand $\varphi°=17.6°$ obtained from DEM simulations is much smaller than shown in literature [18]. This fact cannot be explained on the basis of data sets shown in the paper [14].

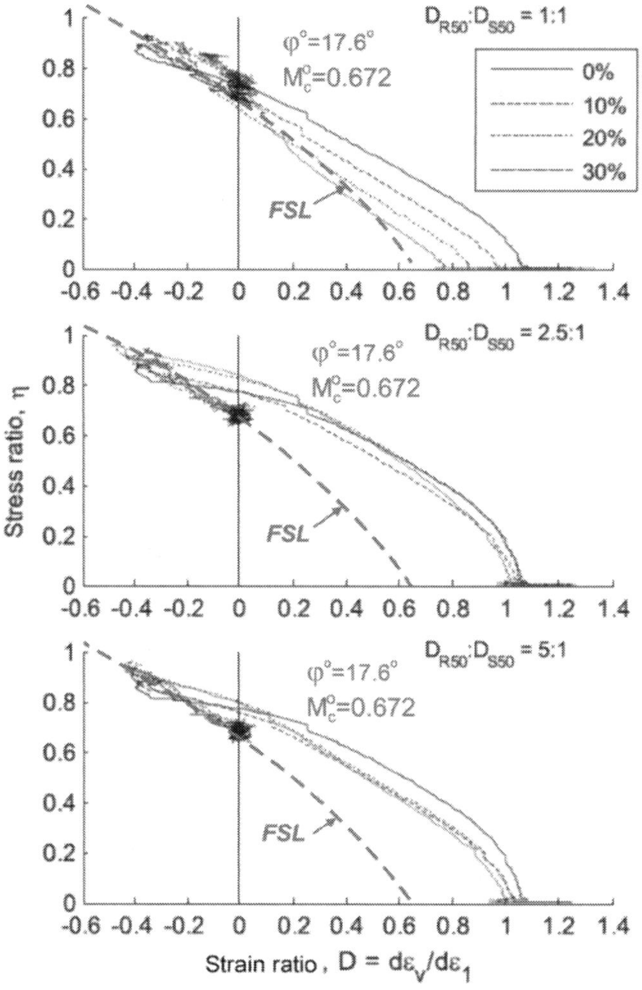

Figure 5. Stress ratio-strain increment ratio for sand-rubber mixtures: a) $D_{R50}:D_{S50}=1:1$, b) $D_{R50}:D_{S50}=2.5:1$, c) $D_{R50}:D_{S50}=5:1$ (adopted from [14]).

5. Conclusion

General Sand-rubber mixtures shear behaviour is similar to the behaviour of the original sand.

The strength and stiffness of sand-rubber mixtures are significantly dependent on rubber size and content.

The Frictional State concept is useful to describe the stress-dilatancy relationships of sand-rubber mixtures.

The DEM simulations give critical frictional state angle and stress-dilatancy relationships differ from the ones obtained in experimental tests.

The stress-dilatancy plane is very useful to describe the complex behaviour of various granular materials.

Acknowledgments

This work, conducted at Bialystok University of Technology, Poland, was supported by the Polish Financial Resources on Science (Project No. WZ/WB-IIL/7/2019 and WZ/WB-IIL/2/2022 and WZ/WB-IIL/2/2022).

References

[1] Mashiri M S 2014 Monotonic and Cyclic Behaviour of Sand-Tyre Chip (STCh) Mixtures *PhD Thesis of University of Wollongong*

[2] Bazibaganye G, Backer A, Vrettos G 2019 Static and cyclic triaxial tests on medium sand and tyre chips mixture *Proc. of the 17th African Regional Conf. on Soil Mech. and Geotech. Eng.* 7-9 Oct. Cape Town pp 163-8

[3] Rouhanifar S, Ibraim E 2015 Laboratory investigation on the mechanics of soft-rigid soil mixtures. In VA Rinaldi, M E Zeballos & JJ Clara (Eds) Deformations Characteristics of Geomaterials pp1073-1080 (Advances in Soil Mechanics and Goetechnical Engineering Vol 6) IOS Press

[4] Promputthangkoon P, Karnchanachetance B 2013 Geomaterial prepared from waste tyres, soil and cement*Procedia Social and Behavioural Sciences* **91** pp 421-8

[5] Pasalkar A A, Bajaj Y M, Wagh A A, Dalvi JD 2015 Comprehensive Literature Review on use of Waste Tyres Rubber in Flexible Road Pavement *Int. J. of Eng. Res. & Technology* **4** pp 685-9

[6] Ling T, Lu Y, Zhang Z, Li C, Oeser M 2019 Value-Added Application of Waste Rubber and Waste Plastic in Asphalt Binder as a Multifunctional Additive *Materials* **12** n 1280

[7] Mashiri M S, Vinod J S, Neaz Sheik M, Tsang H-H 2015 Shear strength and dilatancy behaviour of sand-tyre chip mixtures *Soils and Found.* **55** pp 517-28

[8] Gotteland P, Lambert S, Balachowski L 2005 Strength characteristics of tyre chips-sand mixtures. *Studia Geotech. et Mech.* **27** pp55-66

[9] Bałachowski L, Gotteland P 2007 Characteristics of Tyre Chips-Sand Mixtures from Triaxial Tests*Archives of Hydro-Engineering and Environmental Mech.* **54** pp 25-36

[10] Fu R, Coop MR, Li XQ 2017 The influence of particle type on the mechanics of sand-rubber mixtures *J. Geotech. Geo. Eng.* **143** pp 1-15

[11] Zornberg J G., Cabral AR, Viratjandr C 2004 Behaviour of tyre shred-sand mixtures. *Can. Geotech. J.* **41** pp 227-41

[12] Jastrzębska M 2019 Strength Characteristics of Clay-Rubber Waste Mixtures in UU Triaxial Tests*Geosciences* **9** p 352

[13] Youwai S, Bergado D T 2003 Strength and deformation characteristics of shredded rubber tire-sand mixtures. *Can. Geotech. J.* **40** pp 254-64

[14] Lopera Perez J C, Kwok C Y, Senetakis K 2017 Micromechanical analysis of the effect of rubber size and content on sand-rubber mixtures at the critical state. *Geotex. Geomemb.* **45** pp 81-97

[15] Szypcio Z 2016 Stress-dilatancy for soils. Part I: The frictional state theory. *Studia Geotech. et Mech.* **38** pp 51-7

[16] Szypcio Z 2016 Stress-dilatancy for soils. Part II: Experimental validation for triaxial tests. *Studia Geotech. et Mech.* **38** pp 58-65

[17] Dołżyk-Szypcio K 2018 Natural State Parameter for Sand. Eds: Wu W. and Yu H.-S. *Proc. of China-Europe Conf. on Geotech. Eng.* SSSG, pp 140-3

[18] Bolton M D 1986 The strength and dilatancy of sands*Geotechnique* **36** pp 65-78

Recent Advances in Laterally-Loaded Piles within Mechanically-Stabilized Earth Walls

SJawad[1] and Jie HanF Asce[2]

[1]Lecturer, Dept. of Reconstruction and Projects, Univ. of Baghdad, Baghdad, Iraq.

Email: saif.m@uobaghdad.edu.iq

[2]Glenn L. Parker Professor, Dept. of Civil, Environmental, and Architectural Engineering, Univ. of Kansas, 1530 West 15th St., Lawrence, KS 66045-760 (corresponding author).

Email: jiehan@ku.edu

Abstract. Pile foundations have been increasingly installed within mechanically-stabilized earth walls (MSE) on rocks or compressible soils to support structural loads. Integral abutment bridges, sound walls, and traffic signs are common examples of the structures supported by pile foundations installed within or behind MSE walls. These structures are often subjected to lateral loads resulting from earthquakes, traffic loads, wind loads, and/or thermal expansion and contraction of bridge girders. Typical approaches available to design MSE walls are limited to MSE walls without pile foundations present within reinforced fill. To minimize the interaction between piles and the MSE wall, designers often suggest to place the piles at a distance of six to eight times the pile diameter and/or isolate the piles from the MSE wall by using corrugated pipes. These suggested designs lead to increasing the cost of projects by increasing the length of bridge spans or the diameter of piles and requiring pile embedment. Recently, several studies have been conducted to understand the interaction between piles and MSE walls when the piles are constructed within and close to wall facing. This paper provides a literature review on full-scale tests, reduced-scale tests, and numerical studies of piles constructed within or behind MSE walls. The review will focus on the effects of three key influence factors on lateral capacities of laterally-loaded piles and facing deflections of MSE walls including the plie location behind the wall facing, the ratio of the reinforcement length to the wall height, and the stiffness of the reinforcement layer. This paper will also discuss the available approaches to estimate the forces in the reinforcement layers and the lateral earth pressures behind the MSE wall facing due to the laterally-loaded piles.

1. Introduction

Pile foundations are often subjected to lateral loads in geotechnical applications. For design purposes, analyzing the responses of laterally loaded piles is required to estimate the ultimate lateral load capacities and displacements of the piles. Several methods are available in the literature to estimate the ultimate lateral load capacity and the displacement of a pile in soil. Among these methods are Prescription Value methods [1,2], Limit Equilibrium (LE) methods [3,4,5,6], and Elasticity Linear and Nonlinear methods [5,6,7,8,9,10] However, all the analytical solutions in these methods were derived for piles in a homogenous medium or in layered soils without the presence of MSE wall facing close to the piles. Robertson et al. [11] observed the displacements induced by the laterally loaded pile within the range of six to eight times the pile diameter (D) and any structure within this range could affect the lateral capacity of piles and be also influenced by the piles. Figure 1 shows an example of a laterally-loaded pile within an MSE wall.

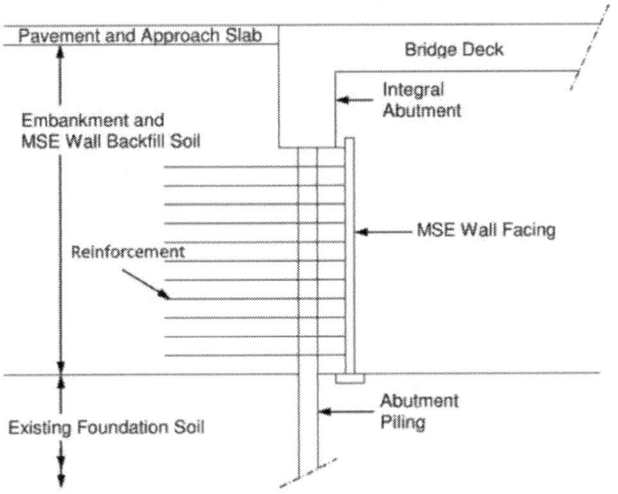

Figure 1. Intergral abutment bridge (after Arenas, 2010)

To minimize the interaction between the piles and the MSE walls, designers suggested using a corrugated pipe (sleeve) around the pile to isolate the effect of the pile from the MSE wall. Ng and Chung [12] developed a three-dimensional numerical model to investigate the effect of sleeving on load transfer from a laterally loaded pile onto wall displacements, lateral earth pressures behind the wall, and the factors of safety of the wall against sliding and overturning. The numerical results showed that the lateral earth pressures induced by an unsleeved pile were higher than those by a sleeved pile. For example, at the lateral load of 5000 kN, the lateral earth pressures from the unsleeved pile at the shallow depths were between K_p and $3K_p$ (K_p is the Rankine passive earth pressure coefficient) multiplied by the vertical stresses with a maximum lateral pressure of 145 kPa at the depth of 2.5 m below the ground surface. After placing a sleeve around the pile, the maximum lateral earth pressure dropped to approximately 90 kPa. Khodair and Hassiotis [13] also developed a three-dimensional model of an integral bridge abutment to investigate the thermal loading effect on the pile as well as the soil surrounding the pile. The numerical model was verified with the results of a fully instrumented abutment and then a parametric study was conducted to investigate the influence of the diameter of the sleeve surrounding the pile. Their results revealed that the axial stresses in the pile decreased with increasing the sleeve diameter and the earth pressures at the perimeter of the sleeve were almost zero regardless the size of the sleeve. Another numerical study conducted by Arenas [14]to evaluate the benefit of the corrugated steel pipes around the abutment piles showed that the use of steel pipe sleeves had not reduced shear forces and moments of the piles. Even though the use of sleeve might significantly reduce the stresses from the piles, the overall cost will increase due to the increase of their diameters and embedded lengths to provide enough lateral resistance against the lateral forces on the piles.

To avoid the use of sleeving pipes, a series of full-scale and reduced-scale tests as well as numerical analyses were conducted to investigate the interactions between the piles and the MSE walls and the influence factors on the lateral load capacity of piles and the performance of MSE walls. Pierson [15]) conducted the first field load testing of laterally-loaded single and group drilled shafts within the reinforced zone of a modular-block MSE wall. In his test, two types of uniaxial geogrid were used to reinforce the backfill material and the shaft length was equal to the wall height. This full-scale test was followed by a series of reduced-scale tests with similar test conditions in term of wall type, reinforcement type, and pile toe depth (i.e., at the base of the MSE wall). Mohammed [16] and Mohammed and Han [17] investigated the factors influencing the performance of laterally loaded piles within MSE walls by conducting 18 reduced-scale models. These factors included pile offset, wall height, reinforcement length and spacing, and geogrid-wall facing connection. Jawad [18] conducted seven reduced-scale model tests and investigated four influence factors: (1) the offset distance of the pile within the MSE wall, (2) the effect of group piles, and (3) the effect of loading type, and (4) the effect of boundary condition. Based on the test results, a simplified additional lateral earth pressure distribution behind an MSE wall facing due to pile loading was proposed. Another series of full-scale tests of piles behind MSE walls have been conducted to study the effect of pile offset from the back face of the MSE wall on both the lateral resistance of the piles and the force resisting inextensible reinforcement [19,20,21,22,23] . These tests consisted of the concrete panel walls and the piles extended to the depth of 15 m below the base of the MSE wall to provide lateral support to the piles.

Additionally, several three-dimensional numerical studies were carried out to simulate the response of the piles behind the MSE wall and investigate the influence of the pile offset, the length of reinforcement (L) to the wall height (H), the reinforcement stiffness, the embedded length of pile, and the backfill properties [14,24,25,26,27,28].

After a literature review of the past studies about the laterally loaded piles within MSE walls, this paper will discuss the effects of the plie location behind the wall, the ratio of the reinforcement length to the wall height, and the stiffness of the reinforcement layer. This paper will also discuss the available approaches to estimate the forces in the reinforcement layers and the lateral earth pressures behind the MSE wall due to the laterally-loaded piles.

2. Lateral load capacity

2.1. Pile offset distance behind wall facing

Berg et al. [29] suggested a minimum offset from the back of the wall facing and the front of the pile as 0.5 m (driven piles) and 1.0 m (drilled shafts) to provide necessary space to achieve proper compaction of backfill material. However, placing the piles within this range might significantly reduce the lateral capacity of the piles. Therefore, several studies have been conducted to examine the effect of pile location behind the wall facing on its lateral capacity. Pierson [15] tested eight drilled shafts within the reinforced zone of a 42 m long, 6 m high MSE wall. The lengths and diameters of the shafts were 6.0 and 0.9 m, respectively, and the spacing between the shafts was 4.6 m. The single shafts were placed at the offset distances of one, two, three, and four times the shaft diameter (i.e., 1D, 2D, 3D, and 4D, D is the shaft diameter) from the center of the shafts to the back of the wall facing, while the group shafts were placed at 2D offset behind the wall facing. In this study, uniaxial geogrids were used as reinforcement. The measured load-displacement curves showed that the lateral load capacity of the shaft located at the 2D offset decreased by 50% as compared with that located at the 4D offset. Also, the individual shaft in the group test showed 20% reduction in the lateral load capacity of the shaft as compared with the shaft in the single shaft test [24].

This percentage of the reduction in the lateral capacity of piles in uniaxial geogrid-reinforced MSE walls decreased in the reduced-scale models. For example, in the reduced-scale model test conducted by Mohammed [16], at the pile head displacement equal to 20% the pile diameter, the lateral capacities of the pile located at the 2D offset behind the wall facing decreased by 10% and 26% as compared with those located at the pile offsets of 4D and 6D, respectively. While in the reduced-scale model test conducted by Jawad et al. [30], the reduction in the lateral capacity of the pile located at the 2D offset was 28% as compared with that at the 4D offset.

In the full-scale tests of piles behind the facing of the concrete-panel MSE wall with steel reinforcement (i.e., steel rib and welded wire grid), the lateral capacity of piles increased with increasing the pile offset behind the wall facing. Price [19] tested five piles in two different sites to study the effect of pile offset from the back face of the MSE wall on both the lateral resistance of the piles and the forces resisted by the MSE steel reinforcement. Two hollow steel pipe piles of 0.32 m in diameter were tested in the first site. These pipe piles were placed at the offsets of 3.75 D and 7.5 D within a 6-m high MSE wall and extended to the depth of 15 m below the base of the wall to reach the hard layer. The other three piles used in the second site were hollow steel pipes of 0.4 m in diameter and were driven approximately 18 m below the base of the wall. One of the piles was located at the offset of 2.9 D behind the MSE wall face, while the other two piles were located at the offsets of 1.6 D and 5.2 D behind the MSE wall wing. Welded wire grid reinforcements were used in both sites with the reinforcement to wall height ratios (L/H) of 1.6 in the first site and 1.1 and 1.7 in the second site. The test results showed that the piles located at 7.5 and 3.8 times pile diameter from the wall facing had identical load-displacement curves. This result was attributed to the longer reinforcement length of 1.6H (H is the wall height), the locations of the piles relatively far from the wall face, and the reinforcement overlap for the pile due to the geometry of the wall near the corner. However, the other test result in Site 2 showed that the lateral load capacity of the pile decreased with the reduction of the offset distance from the back face of the MSE wall to the center of the pile. This result is consistent with that of Pierson [15].

Nelson [20] performed lateral loading tests similar to those conducted by Price [19] but with ribbed strip reinforcement. These tests were carried out on four piles located at 1.3D, 2.7D, 6.3D, and 7.7D offsets behind the facing of a 6.8-m high concrete-panel MSE wall. The test results showed that the final lateral load capacities of the piles located at the offsets of 1.3D and 2.7D decreased by 60% and 30% as compared with that located at 6.3D. Although the piles used in this test were similar to those in the Price [19] test, the pile lateral load capacities were much lower than those in the Price [19] test. For example, at the pile head displacement of 60 mm, the lateral load capacities of the piles located at the 2.9D offset in the Price [19] test and 2.7D in this test were 200 and 75 kN, respectively. The main reasons for this difference in the lateral load capacity are attributed to the different type of the wall and the different reinforcement used in these two tests.

Hatch [21] tested four steel pipe piles of 0.32 m in diameter as placed at the offsets of 1.9D, 3.2D, 4.3D, and 5.3D behind the MSE wall face. The test results showed that the lateral load capacities of the piles located at the offsets of 4.3D and 3.2D were almost similar. Hatch [21] attributed this result to the special location of the pile with the 4.3D offset (i.e., in front of the joint in the wall), while the pile with 3.2D was in front of the panel center. The test results also showed that the behavior of the pile with the 5.3D offset was similar to the reaction pile up to the pile head displacement of 44.5 mm, indicating that the wall had no effect on the lateral load capacity of the pile with the 5.3D offset. The load-displacement curves for the piles located at the offsets of 3.2D, 4.3D, and 5.3D showed a hyperbolic shape, indicating the progressive failure of the surrounding soil with increasing the lateral load. On the other hand, the load-displacement curve for the pile with the 1.9D offset was almost linear, indicating that the soil had a limited contribution in resisting the lateral load and the majority of the load was carried by the pile itself.

Han [22] carried out a series of tests similar to those conducted by Nelson [20] in terms of the pile size, the type of reinforcement, the embedded length of pile, and the type of soil. The piles were placed at the offsets of 1.7D, 2.8D, 3.1D, and 3.9D behind the facing of the 4.5-m high MSE wall. The ratio of the reinforcement length to the wall height was 0.9 in the test. The comparison between the lateral load capacities of piles at the 25 mm pile head displacement showed 50% and 20% reduction in the lateral load capacities of piles located at the offsets of 1.7D and 2.8D, respectively, relative to those of the piles located at the offset of 3.1D or 3.9D.

Besendorfer [23] tested four piles, which were located behind the wall face at the offsets of 1.7D, 2.8D, 2.9D, and 3.9D and extended to 4.5 m below the base of the wall. The reinforcement length to wall height ratio of 0.72 was used in this study, which was close to the typical ratio of 0.7 used in walls that are subjected to static loading. The test results showed approximately equal lateral load capacities of the piles located at the offsets of 2.8D, 2.9D, and 3.9D behind the wall face and approximately 40% reduction in the lateral load capacity of the pile located at the offset of 1.7D as compared with those piles at other offsets. Besendorfer [23] pointed out some reasons that might contribute to this slight difference between the lateral load capacities of the piles located at the offsets of 2.8D and 3.9D behind the wall face. One of the reasons was related to the rainfall that occurred one day before testing the piles with 2.8D and 2.9D offsets, which might add some apparent cohesion to the soil and increased the unit weight of the soil. The other reasons were related to the panel configuration and compaction effort around the piles. Besendorfer [23] also developed a p-multiplier curve as a function of a normalized distance using the LPILE program including his own results and those from Price [19], Nelson [20], Hatch [21], and Han [22] as shown in Figure 2. The regression line showed that the piles with a normalized distance greater than four were not affected by the presence of the wall. Table 1 summarizes the field and laboratory tests available in the literature.

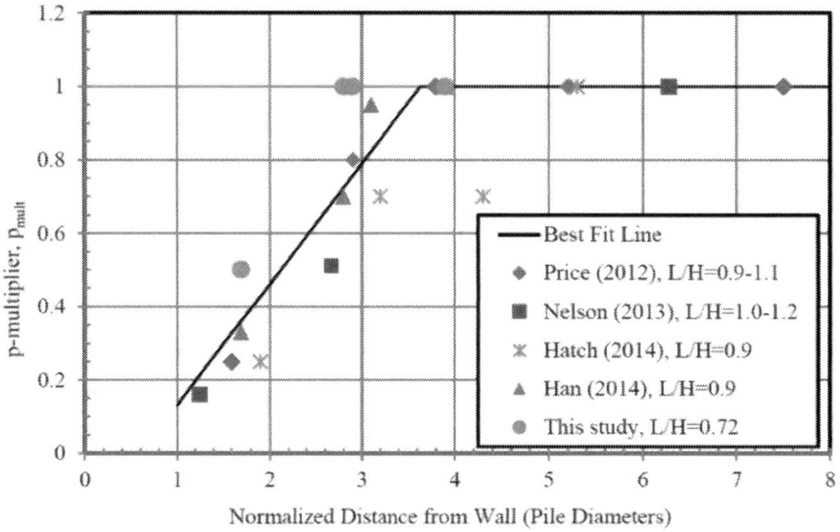

Figure 2. p-multiplier curves (after Besendorfer, 2015)

Table 1. A summary of field and laboratory tests

Test type	Wall and reinforcement	Pile	Key findings	References
Field	H = 6 m, modular block MSE wall, uniaxial geogrid	8 concrete drilled shafts, h = 6.0 m, D = 0.9 m, x = 1D, 2D, 3D, and 4D	The test results showed that the lateral capacity of the shaft located at x = 2D was 50% lower than that of the shaft at x = 4D. The results also showed a 20% reduction of the lateral capacity of each shaft in the three shaft group as compared with that of the single shaft.	Pierson [15]
Field	H = 6 m high, concrete panel MSE wall, welded wire	Five hollow steel pipe piles, h = 21 m to a hard layer, D = 0.32	The test results showed that the piles at x = 7.5D and 3.8D had identical load-displacement curves. This result was attributed to large L = 1.6H and large x	Price [19]

	grid, L/H = 1.1 and 1.6.	m and 0.4 m, x = 3.8D, 5.2D, and 7.5D	from the wall facing. The effect of the wall facing could be eliminated at x = 3.8D and 5.2D for L/H = 1.6 and 1.1, respectively.	
Field	H = 6.8 m, concrete panel MSE wall, galvanized strips, L/H = 1.2	Four hollow steel piles, D = 0.32 m and 0.4 m, h = 12.8 m, x = 1.3D, 2.7D, 6.3D, and 7.7D	The pile at the farthest distance behind the wall facing had the highest lateral resistance. The pile p-y curve showed that the influence of wall facing on the pile lateral resistance vanished at x = 4.5D from the back face of the wall.	Nelson [20]
Field	H = 6.0 m concrete panel MSE wall, welded wire grids, L/H = 0.9	Four hollow steel piles, h =11.5 m, D = 0.32 m, x = 1.9D, 3.2D, 4.3D, and 5.3D	The pile p-y curve showed that the presence of the wall facing had no influence on the lateral resistance of the pile located at x = 4D.	Hatch [21]
Field	H = 4.5 m concrete panel MSE wall, galvanized strips, L/H = 0.9	Four hollow steel piles, h =10 m, D = 0.32 m, x = 1.7D, 2.8D, 3.1D, and 3.9D	At the 25 mm pile head displacement, the lateral load capacities of piles located at x= 1.7D and 2.8D were 50% and 20%, respectively, lower than those located at x = 3.1D or 3.9D	Han [22]
Field	H = 6.0 m concrete panel MSE wall, galvanized strips, L/H = 0.72	Four hollow steel piles, h = 12 m, D = 0.32 m, x = 1.7D, 2.8D, 2.9D, and 3.9D	The piles located at x = 2.8D, 2.9D, and 3.9D had almost the same lateral capacities. This was attributed to the weather and the wall configuration	Besendorfer [23]
Lab	H = 0.45 m and 0.72 m modular block MSE wall, uniaxial geogrid, L/H = 0.7 and 1.25	Aluminum pipe pile, D = 0.063 m, h = 0.45 m and 0.72 m, x = 2 D, 4D, and 6D	The type of geogrid facing connection had little effect on the lateral deflection of wall facing at x ≥ 4D. Mechanical connection reduced the lateral deflection of wall facing as compared with frictional connection at x = 2D. Mechanical connection increased the pile capacity for the wall with long reinforcement as compared with frictional connection.	Mohammed [16]; Mohammed and Han [17] .
Lab	H = 1.0 m modular block MSE wall, uniaxial geogrid, L/H = 0.7	Aluminum pipe pile, D = 0.76 m, h = 1 m, x = 2D and 4D	The lateral load capacity of the pile located at x = 2D was 28% lower than that located at x = 4D. The results of the group pile test showed a 40% reduction of the lateral capacity of piles as compared with that of the single pile.	Jawad [18]; Jawad et al [28] ; Jawad et al [30])

Note: H = wall height, L = reinforcement length, h = pile length, D = pile diameter, x = pile offset.

2.2. Reinforcement stiffness

Several studies have been conducted to examine the effect of reinforcement stiffness on the performance of MSE walls. Leshchinsky and Han [31] investigated the effect of reinforcement stiffness on the global stability of geosynthetic-reinforced multitiered walls by conducting a two-dimensional numerical analysis. The results showed a negligible difference in the computed factor of safety by increasing the reinforcement stiffness from 1000 kN/m to 100000 kN/m.. Zheng and Fox [32] investigated four different values of reinforcement stiffness J= 500, 1000, 1500, and 2000 kN/m and found that the use of reinforcement with the stiffness of 2000 kN/m had approximately 45% reduction in the maximum wall displacement as compared with that with 500 kN/m.

The aforementioned studies investigated the effect of reinforcement stiffness on the wall facing deflection without any presence of laterally-loaded piles within the reinforced zone of the MSE wall. Pierson [24], by conducting a three-dimensional numerical analysis, investigated the effect of reinforcement stiffness in the weak and strong directions on the response of laterally-loaded piles and the performance of the MSE wall. His results revealed that the increase of the stiffness in both directions increased the pile lateral load capacity and reduced the facing deflection. The numerical study conducted by Jawad and Han [33] investigated the effect of reinforcement stiffness on the lateral load capacities of various pile diameters. In his study, the piles were placed at the offset distance equal to two times the pile diameter behind the wall facing. The numerical results showed that the increase of the reinforcement stiffness from 1000 to 3000 kN/m increased the ultimate lateral load capacities of the piles of 0.5 m in diameter and 0.7 m in diameter by 57% and reduced the wall facing deflections at the top by 5.5% to 11%.

2.3. Length of reinforcement

To prevent the pullout failure of the reinforcement layer from the soil and minimize the MSE wall deformation, the required minimum reinforcement length (L) to wall height (H) ratio is typically 0.7 under static loading and 0.9 under seismic loading [29]. For construction purposes, however, designers sometimes use a reinforcement length to wall height ratio (L/H) more than one, for example, 1.6 in Price [19] and 1.2 in Nelson [20] and Hatch [21] at the time of their tests. An increase of the reinforcement length leads to increase the interaction area between the soil and the reinforcement, resulting in an increase in the lateral capacity of piles and reduce the wall facing displacement. Price [19], after conducting full-scale tests, found that the piles located at the offsets of 3.8 and 7.5 times pile diameter had almost the same lateral capacity. One of the reasons for this result was the longer reinforcement length of 1.6H used in the test. Mohammed [16] investigated three different reinforcement length to wall height ratios (i.e., 0.7, 1.0, and 1.3) by conducting reduced-scale model tests of piles behind the MSE wall facing. The test results showed that the lateral load capacity of the pile increased and the wall facing displacement decreased as the reinforcement length to wall height ratio increased, maybe due to the reduced model tests with lower stress levels. However, some studies showed that an increase of the reinforcement length to wall height ratio more than one had an insignificant effect on the lateral load capacity of the pile and the wall facing displacement. Jawad et al. [30] conducted a three-dimensional numerical analysis and investigated the effect of the reinforcement length to wall height ratio (e.g., 0.7, 1.0, and 1.3) on the lateral load capacity of the pile, the wall facing displacement, and the lateral earth pressure behind the wall facing. Their numerical results showed that the load-displacement curves, the wall facing displacements, and the lateral earth pressures behind the wall facing for the piles with the reinforcement length to wall height ratios of 1.0 and 1.3 were almost the same, indicating that the increase of the reinforcement length to wall height ratio more than 1.0 had an insignificant effect on the lateral load capacity of the piles placed at the 2D offset.

3. Wall facing displacement

Since the wall facing is one visible part of the structure, the deformation of the wall facing should be within an allowable range so that it does not affect the aesthetics of the MSE wall. Berg et al. [29] suggested that the allowable lateral displacement for MSE walls under extreme events be in the range of 50 to 100 mm. Pierson [15] measured the lateral displacement of the modular-block MSE wall, which was induced by laterally-loaded piles located at different offset distances behind the wall facing. The vertical profile of the wall facing displacement showed the maximum displacement occurring at the top of the wall, and then the displacement decreased with the depth toward the base of the wall. However, the wall facing displacement profile in the horizontal direction showed that the wall moved more along the wall centerline, and the movement decreased with an increase of the distance from the wall centerline as shown in Figure 3. This figure also shows an increase in the maximum wall displacement and a decrease of the influenced width with a decrease of the pile offset behind the MSE wall facing.

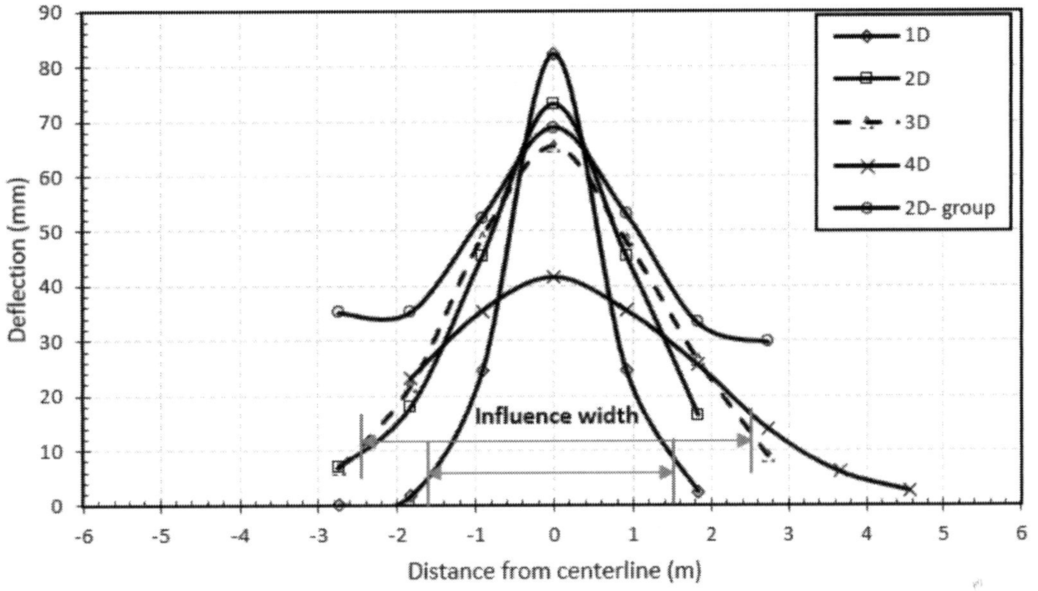

Figure 3. Horizontal profiles of wall facing displacements (after Jawad, 2019)

Price [19] measured the displacements of a concrete-panel MSE wall using three different ways: (1) Linear Variable Differential Transformers (LVDTs), (2) Shape Arrays (SA), and (3) String Potentiometers (SP). Similar to the Pierson [15] test, the pile closer to the wall facing induced more wall deformation with the maximum displacement at the top of the wall. The results of these three measurement methods showed that the wall displacements in front of the pile measured by the Shape Arrays method were larger than those measured by the LVDTs or the String Potentiometers. The reason was that the plastic pipes used in the Shape Arrays method were placed within a relatively un-compacted soil behind the wall. Hatch [21] used the Digital Image Correlation (DIC) system beside SA and SP to measure the wall facing displacements. The results from the SA were not accurate; however, they were used to check the results from DIC and SP. The data from the DIC system showed that the maximum wall displacement occurred at the top of the wall in the middle of the joint. The results from Hatch [21] also showed that the wall facing displacement was limited to one wall panel on either side of the pile and to 1.5 wall panels down from the top. Han [22] showed the displacement results of four different wall configurations induced by the piles located at the offset distances of 1.7D, 2.8D, 3.1D, and 3.9D behind the wall facing. The comparison exhibited that the wall facing displacement induced by the pile located at the offset of 2.8D was larger than others. This phenomenon happened because there was only one reinforcement layer supporting the top panel. Jawad et al. [30], after conducting reduced-scale model tests, found a relationship between the pile head displacement and the maximum wall facing displacement by including the test results from Pierson [15]. The reason for selecting these field tests instead of other tests mentioned before is that these tests had similar test conditions to the model tests in terms of wall type, pile stiffness, reinforcement type, and pile toe depth (i.e., at the base of the MSE wall). From the suggested relationship, the maximum wall displacement can be estimated based on the pile displacement, the pile offset, and the pile diameter.

4. Lateral earth pressure behind the wall

Lateral earth pressure distribution behind the wall facing is one of the key information that can be used for designing MSE or other type of walls. Therefore, several full-scale tests, reduced-scale tests, and numerical analyses have been conducted to estimate the lateral earth pressures behind the MSE wall facing induced by laterally-loaded piles. Pierson [15] and Huang et al. [26] found that the maximum lateral earth pressure, due to the lateral shaft movement, was located within the lower part of the MSE wall, not within the upper part while the FHWA NHI-10-024 guideline [29]. suggested that the lateral pressure distribution from the pile foundation on the MSE wall face decreases linearly or nonlinearly with the wall depth. Jawad et al. [30] measured the additional lateral earth pressure along the back face of the wall using earth pressure cells. The test results from Jawad et al. [30] showed that the additional lateral earth pressure distribution with the wall elevation was non-linear. The maximum lateral earth pressure was located at an elevation of 80% the wall height, while the minimum lateral pressures were observed at the bottom and top of the wall. Based on the measured lateral earth pressures, Jawad [18] proposed a simplified method to design an MSE wall supporting a pile foundation within its reinforced zone. The additional lateral earth pressure calculated in the proposed design method corresponds to that induced by a laterally loaded pile at the pile head displacement of 20% of its diameter. The distribution of the additional lateral earth pressures behind the wall facing was assumed to have a three-dimensional shape and the maximum lateral pressure was assumed to be a percentage of the passive earth pressure, which mainly depended on the pile location within the MSE wall. Jawad [18] also developed an equation to estimate the maximum additional lateral earth pressure based on the proposed shape of pressure distribution.

5. Force in the reinforcement layer

Another key parameter for designing the MSE wall is the force generated in the reinforcement layer during lateral pile loading. Pierson [15] monitored the strains in the uniaxial geogrid layers placed at different elevations to estimate the forces generated by the shaft movement. The test results from Pierson [15] showed that the maximum strain happened near the pile location. Price [19] investigated the relationship between the maximum force in the reinforcement (i.e., welded grid) induced by pile movement and the normalized distance (lateral distance from the center of the pile to the center of a grid divided by the distance from the back face of the wall to the center of the pile). This relationship showed that the induced force in the reinforcement decreased with an increase of the normalized offset distance. Nelson [20] plotted the normalized force (the maximum induced force in the reinforcement divided by the lateral load on the pile) in the strip against the normalized offset distance. The test results from Nelson [20] showed significant scatter of the data about the best-fit curve; therefore, a tentative envelope was plotted to provide a conservative estimate of the additional force in the reinforcement induced by the laterally loaded pile. Hatch [21] also developed a plot of the normalized load versus normalized offset distance including the data, the best-fit line, and the envelope from Price [19] , who also used welded wire grids. The induced forces in the reinforcement layers were higher than those on the Price [19] envelope, showing the data from Price [19] were not adequate for design. Similar to Hatch [21], Han [22] developed a plot of the normalized induced force in the reinforcement versus the normalized offset distance by including the data, the best fit line, and the envelope from Nelson [20] (2013), who also used ribbed steel strips. The induced forces were also higher than the design envelope proposed by Nelson [20]. After conducting a statistical analysis on the data from his own test and Nelson's test, Besendorf [23] created two different regression models to predict the induced forces in the reinforcement layers. In one model, the depth below the ground surface was used as a variable while in the other model the vertical stress was the variable. Jawad [18]calculated the maximum induced tensile forces in the geogrid rib (i.e., the one beside the pile) at the pile head displacement of 20% of its diameter using the proposed additional lateral earth pressure distribution behind the wall facing. The calculated tensile forces from the proposed method showed reasonable match with those measured from the model tests based on the measured strains in the geogrid.

6. Conclusions

Laterally-loaded piles within the reinforced zone of MSE walls are a complicated geotechnical problem due to the interaction between the piles and the MSE walls. Understanding such a problem requires the investigation of the key influencing factors on both the piles and the MSE walls. Therefore, this study summarizes these key factors that affect the response of the pile and the performance of the MSE wall. Based on the field tests, lab tests, and numerical analyses, the following conclusions can be drawn:

1. When the pile offset distance within the MSE wall (i.e., modular blocks or concrete panels) decreased from 4D to 2D, the maximum reduction in the lateral pile capacity was approximately 50%.
2. The location of the maximum induced force in the reinforcement (inextensible and extensible) by the pile movement was within the upper portion of the wall close to the pile location.
3. In the concrete-panel MSE wall, the maximum wall displacement was in the upper concrete panel at the joint, and the wall deformation was limited to one wall panel on either side of the pile and to 1.5 wall panels down. While in the modular-block MSE wall, the maximum wall displacement was also within the upper part of the wall and the influenced width of the wall increased with the increase of the pile offset behind the MSE wall facing.
4. Based on the numerical analysis, an increase of the reinforcement stiffness had a significant effect on the increase of the lateral load capacities of the piles and less effect on the decrease of the wall displacements.
5. When the reinforcement length to wall height ratio (L/H) was more than 1.0, the reinforcement length had an insignificant effect on the lateral load resistance of the piles and the facing displacement of the wall.

References

[1] Feagin L B 1937 Lateral pile-loading tests *Transactions* **102** pp 236-88
[2] McNulty J 1956 Thrust loading on piles *Journal of the Soil Mechanics and Foundations Division* **82** pp 1-25
[3] Broms B B 1964 Lateral resistance of piles in cohesive soils *Journal of the Soil Mechanics and Foundations Division* **90** pp 27-64
[4] Broms B B 1964 Lateral resistance of piles in cohesionless soils *Journal of the Soil Mechanics and Foundations Division* **90** pp 123-58
[5] Meyerhof G G, Mathur S K and Valsangkar A J 1981 Lateral Resistance and Deflection of Rigid Walls and Piles in Layered Soils *Canadian Geotechnical Journal* **18** pp 159-70
[6] Reese L C, Cox W R and Koch F D 1974 Analysis of Laterally Loaded Piles in Sand *Proceedings,* Offshore Technology Conference, Houston **2** pp 473–84
[7] Hetényi M 1946 Beams on Elastic Foundation *The University of Michigan press. Arbor* p 255
[8] Matlock H 1970 Correlations for Design of Laterally Loaded Piles in Soft Clay *Proceedings, Offshore Technology Conference* Houston, Texas, Paper 1204 pp 577–94
[9] Poulos H G 1971 Behavior of laterally loaded piles I. single piles." *Journal of Soil Mechanics & Foundations Div* **58** pp 1245-59

[10] Reese L C and Welch R C 1975 Lateral Loading of Deep Foundations in Stiff Clay Proceedings, ASCE **101**, pp 633–49

[11] Robertson P K, Hughes J M O, Campanella R G, Brown P and McKeown S 1986 Design of Laterally Loaded Piles Using the Pressuremeter *The Pressuremeter and its Marine Applications: Second International Symposium,* ASTM STP 950, J-L Briaud and J.M.E. Audibert, Eds., American Society for Testing and Materials pp 443-57

[12] Ng C W W and Chung I Y M 2005 Three dimensional Numerical Investigations of the Influence of Sleeved Piles on the Stability of a Retaining Wall *Geotechnical Special, Publications* n 132 pp 1-9

[13] Khodair Y A, and Hassiotis S 2005Analysis of soil–pile interaction in integral abutment *Computers and Geotechnics* **32** pp 201-9

[14] Arenas A E 2010 Thermal Response of Integral Abutment Bridges With MSE Walls: Numerical Analyses and a Practical Analysis Tool *Ph.D. dissertation* Dept. of Civil and Environmental Engineering,Virginia Tech

[15] Pierson M 2008 Behavior of laterally loaded shafts constructed behind the face of a mechanically stabilized earth block wall *M.Sc. Thesis* Dept. of Civil, Environmental, and Architectural Engineering, the University of Kansas

[16] Mohammed W K 2016 Factors Influencing performance of a laterally loaded pile with an MSE wall system *M.Sc. Thesis* Dept. of Civil, Environmental, and Architectural Engineering, the University of Kansas

[17] Mohammed W and Han J 2018 Effect of geogrid-facing connection on MSE wall deformations induced by laterally loaded piles *Proceedings of the 11th International Conference on Geosynthetics*, Seoul, Korea, September 16-21

[18] Jawad S 2019 Evaluating Behavior of Laterally Loaded Piles within MSE Walls under Static and Cyclic Loading *Ph.D. Dissertation*, Dept. of Civil, Environmental, and Architectural Engineering, the University of Kansas.

[19] Price J S 2012 Lateral Resistance of Piles Near Vertical MSE Abutment Walls *M.Sc. Thesis*, Dept. of Civil and Environmental Engineering, Brigham Young University

[20] Nelson K R 2013 Lateral Resistance of Piles Near Vertical MSE Abutment Walls at Provo Center Street *M.Sc. Thesis*, Dept. of Civil and Environmental Engineering, Brigham Young University.

[21] Hatch C 2014 Lateral resistance of piles near vertical MSE abutment walls *M.Sc. thesis*, Dept. of Civil and Environmental Engineering, Brigham Young University

[22] Han J 2014 Lateral Resistance of Piles Near 15 Foot Vertical MSE Abutment Walls Reinforced with Ribbed Steel Strips *M.Sc. Thesis*, Dept. of Civil and Environmental Brigham Young University

[23] Besendorfer J J 2015 Lateral Resistance of Pipe Piles Near 20-ft Tall MSE Abutment Wall with Strip Reinforcements *M.Sc. Thesis*, Dept. of Civil and Environmental Engineering, Brigham Young University

[24] Pierson M C 2010 Modeling Drilled Shafts in MSE Block Walls *Ph.D. dissertation*, Dept. of Civil, Environmental, and Architectural Engineering, the University of Kansas

[25] Huang J, Parsons R L, Han J and Pierson M 2011 Numerical analysis of a laterally loaded shaft constructed within an MSE wall *Geotextiles and Geomembranes* **29** pp 233-241

[26] Huang J, Han J, Parsons R L and Pierson M C 2013 Refined numerical modeling of a laterally-loaded drilled shaft in an MSE wall *Geotextiles and Geomembranes* **37** pp 61-73

[27] Rahman M S, Huang J and Bin-Shafique S 2015 Numerical Simulation of a Drilled Shaft in an MSE Wall under Cyclic Loading *IFCEE* pp 929-938

[28] Jawad S, J Han G, Abdulrasool and M Al-Naddaf 2021 Responses of single and group piles within MSE walls under static and cyclic lateral loading *Geotext. Geomembr.*

[29] Berg R R, Christopher B R and Samtani N C 2009 Design of Mechanically Stabilized Earth Walls and Reinforced Soil Slopes–Volume I. FHWA-NHI-10-024. Washington, DC:Federal Highway Administration

[30] Jawad S, J Han, M Al-Naddaf and Abdulrasool G 2020 Responses of laterally loaded single piles within MSE walls *J Geotech. Geoenviron. Eng*. **46** 04020128

[31] Leshchinsky D and Han J 2004 Geosynthetic reinforced multitiered walls *Journal of Geotechnical and Geoenvironmental Engineering* **130** 1225-35

[32] Zheng Y and Fox P J 2016 Numerical investigation of geosynthetic-reinforced soil bridge abutments under static loading *Journal of Geotechnical and Geoenvironmental Engineering* **142** 04016004

[33] Jawad S & Han J 2021 Numerical Analysis of Laterally Loaded Single Free-Headed Piles within Mechanically Stabilized Earth Walls *International Journal of Geomechanics* **21** 04021038

Investigation of force transmission during indirect activation of a geogrid in transparent soil

J Derksen[1] and M Ziegler[2]

[1]RWTH Aachen University,
Mies-van-der- Rohe-Str. 1, 52074 Aachen, Germany
[2]ZAI Ziegler und Aulbach Ingenieurgesellschaft mbH,
Stengerstraße 9, 63741 Aschaffenburg, Germany

E-mail: derksen@gut.rwth-aachen.de

Abstract. The composite material geogrid reinforced soil has a wide range of applications. In most cases the geogrid reinforcement is indirectly activated by an applied load via the surrounding soil, for example in base layers of embankments, foundation pads and fill structures. Within these geogrid reinforced structures, different areas can occur in which the direction of force transmission might differ from soil to geogrid and vice versa. In particular, the force transmission in the soil-geogrid-interface involves a fundamental need for research. Therefore, this paper firstly introduces a conceptional model for indirect geogrid activation, secondly presents a test rig in which the interaction mechanisms can be investigated on micro scale and thirdly determines various areas with different directions of force transmission. For this purpose, transparent geogrid reinforced specimens are biaxially loaded, whereby an area-wide and undisturbed insight into the soil-geogrid-interface is enabled. The transparent soil is created as a two-phase medium consisting of a crushed fused quartz and a white oil with a corresponding refractive index. The displacement fields of the geogrid and the surrounding transparent soil were determined using a digital image correlation (DIC) method based on images taken continuously during the test. Then, strain profiles were derived for the longitudinal tensile members, which were converted into a force distribution along the geogrid. As a result, the conceptional model can be verified.

1. Introduction

In geotechnical practice, the composite material geogrid reinforced soil has successfully performed in a wide range of applications. However, the compound behaviour cannot be derived from the combination of the material properties of both components, but requires consideration of the mutual interaction of soil and geogrid. Furthermore, the interaction behaviour depends on soil properties (density, grain size), characteristics of the geogrid (geometry, type, stiffness) and given boundary conditions (stress level, load characteristics).

In general, the compound behaviour can be described both by a macroscopic view of the stress-strain behaviour of a reinforced soil body and by a microscopic view of the force transmission mechanisms between soil grains and single geogrid members.

Moreover, it is crucial for the interaction behaviour of geogrid reinforced soil whether forces are transferred from the geogrid into the soil or are induced from the soil into the geogrid. Therefore, the direction of force transmission between soil and geogrid has to be considered within the characterisation of the interaction behaviour. In summary, an overarching consideration of the geogrid activation is conceivable, in which a distinction is made between geogrids with a directly applied tensile force and geogrids which are indirectly activated via the surrounding soil.

In case of a geogrid pull out e.g. in anchoring trenches, the soil is directly activated by the applied geogrid displacement whereby the tensile force in the geogrid is transferred into the soil. Several researchers [1,2,3,4,5] determined the micro-mechanical mechanisms during direct activation as friction on geogrid members and earth resistance in front of the transverse members. The geogrid displacement is at least greater than the displacement of the surrounding soil. As a result, there exclusively occurs a load transfer from the geogrid into the surrounding soil during direct geogrid activation. The relative displacement between geogrid and soil decreases with increasing activated length of the grid, until the tensile force is completely transferred to the soil in case of a sufficient anchorage length. Finally, the micro-mechanical mechanisms lead to the total pull out resistance of an embedded geogrid in terms of a macroscopic view. In conclusion, the interaction mechanisms during direct activation are well understood on both macro and micro scale.

In contrast to that, geogrids e.g. in base layers of embankments, foundation pads and fill structures are indirectly stressed by processes in the surrounding soil. The activation of the reinforcement can result from the soil weight of the overburden, external loads or settlement differences in the subsoil. Based on bi- and triaxial compression tests, the global effect of indirect activated geogrids was described as an additional confining pressure or as an increased strength. In the "cohesion concept" the reinforcement effect is considered with an increased cohesion by parallel shifting of the failure criterion of the unreinforced soil in the τ-σ-diagram [6]. In contrast to that, the effect of the geogrid reinforcement is represented in the "confining effect concept" by an additional confining pressure $\Delta\sigma_3$ in the direction of the smaller principal stress without shifting the failure criterion of the soil. The "confining effect" is illustrated in Figure 1 (a) as an alteration of the unfavourable deviatoric stress path in the unreinforced case (continuous arrow) to a path shifted in an isotropic direction (dashed arrow) according to various authors [7,8]. As a result, the favourable deflected stress path of the reinforced soil body reaches the failure criterion at a higher stress state.

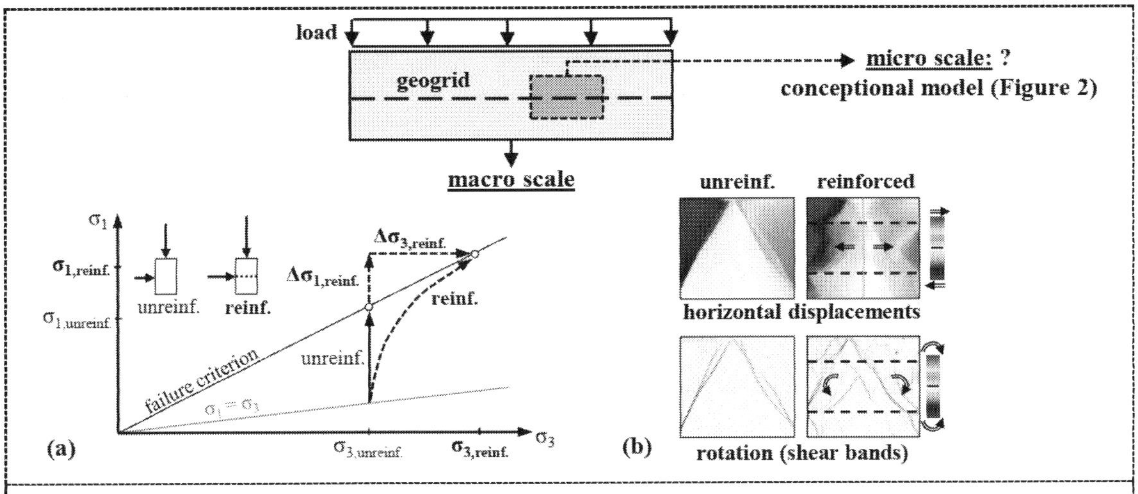

Figure 1. Indirect geogrid activation on macro scale: confining effect concept (left) and results of biaxial compression tests (right) according to Ruiken [9]

Ruiken [9] investigated the stress-strain behaviour of geogrid reinforced soil with biaxial compression tests in plane strain conditions and illustrated the positive effect of the reinforcement. Based on a Digital Image Correlation (DIC) method, soil displacements and slip planes of specimens with and without geogrid reinforcement were evaluated (Figure 1, (b)). In the unreinforced specimen, the known triangular and rigid failure bodies can be identified with a lateral direction of movement. In case of the geogrid reinforced specimen, the reinforcement creates kinematic constraints, which lead to much more subdivided failure bodies. Thus, more uniform deformations of the specimen at a higher load were observed.

While the global effect of indirect activated geogrids is well understood on macro scale, there is a limited understanding of the microscopic soil-geogrid-interaction. Consequently, the micro-mechanical interaction behaviour of geogrid reinforced soil during indirect activation contains a fundamental need of research in terms of deformations and forces that occur in the interface.

2. Conceptional model for indirect geogrid activation

A conceptional micro-mechanical model for force transmission mechanisms during indirect activation of the geogrid by the surrounding soil has been developed as a working hypothesis (Figure 2) and is also presented in [10,11]. Hereby, a uniform distributed load is considered that decreases towards the free end of the geogrid in analogy to a base reinforcement of an embankment.

In the edge area (I), the soil is pushed over the geogrid due to a lack of lateral support, so that shear (spreading) forces are transferred into the reinforcement via longitudinal and transverse tension members. In this "pushout" area the soil displacement is greater than or equal to the geogrid displacement. In contrast to that, in the core area (III) of the uniformly loaded reinforcement plane, there is an equal support from both sides, whereby no shear forces are generated due to the lack of relative movement between grid and soil. Instead, the soil grains interlock in the geogrid openings ("interlocking" mechanism), so that the geogrid stabilizes the grain structure of the soil. Due to the interlock of coarse soil particles in the geogrid openings, higher stress states can be reached without significant increase in deformations [12,13,14]. In the conceptional model it is assumed that there has to be a transition area (II) between the areas (I) and (III) in which the geogrid tensile force is anchored back into the soil by the "pullout" mechanism. As Figure 2 shows, it is currently not known whether the geogrid tensile force is completely (a), partially (b) or even not (c) transferred back into the soil in transition area (II).

Moreover, the stiffness of the soil below the reinforcement must also be considered. If the subsoil is sensitive to settlements, the geogrid will sag in this area due to a lack of vertical support. Then, the horizontal reinforcement is activated by vertical deformations and membrane forces will also occur in the geogrid in addition to the mechanisms described above.

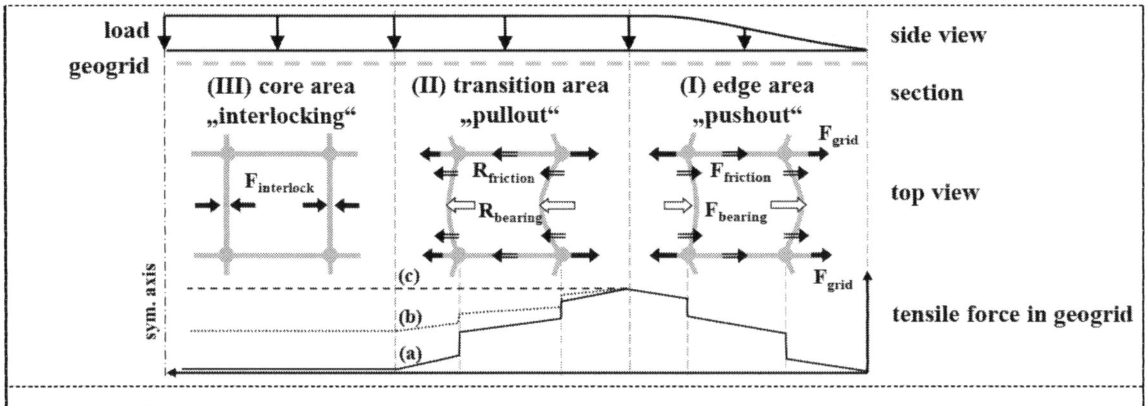

Figure 2. Conceptional model for indirect geogrid activation on micro scale

In conclusion, the authors propose the conceptional mechanical model in order to describe the indirect geogrid activation on micro scale. In contrast to the direct activation, the indirect activation does not only involve one direction of force transmission, but a differentiation must be made between three areas in which the "pushout", "pullout" and "interlocking" mechanisms occur. This model could not be verified so far, because the soil-geogrid-interface is not visible with conventional investigation methods.

However, non-invasive optical methods for stress and deformation measurements are now available to identify the occurring mechanisms.

3. Interaction tests

3.1. Investigation method

The investigation method has to be selected in such a way that the soil-mechanical interaction between geogrid and soil can be recorded undisturbed (non-invasive) and over a wide area. In contrast to conventional measuring methods in which e.g. geogrid strains are recorded with strain gauges and thus the soil-geogrid-interface is influenced in terms of its stiffness or friction properties, the interaction is not disturbed by using optical recording methods.

Already in 1985 Dyer [15] recorded soil stresses during pull out of a geogrid from a transparent medium using the method of photo elasticity and was able to visualize the earth resistance in front of the transverse tensile members. Due to the cost- and time-intensive experiments with polarized light or other methods such as magnetic resonance imaging or computer tomography, these methods were only used occasionally. In recent years, technological advances in optical imaging and computer technology have favoured the development of optical methods in geotechnical engineering [16]. Since then, the Digital Image Correlation (DIC) method is often used in geotechnical research to record soil deformations directly at a transparent boundary of a test box.

Next, transparent soil specimens have been used to investigate soil-structure interaction. Transparent soil is created by combining a translucent solid, usually fused quartz for representing a granular soil, with a suitable pore fluid [17,18]. A fluid or fluid mixture with a matching refractive index must be found for the refractive index of the fused quartz (n = 1.459). The dimensionless refractive index relates the velocity of light in vacuum to the velocity of light in the corresponding medium. Light refraction takes place at the interface between two media with different refractive indices and results in an opaque sample. The refraction of light is minimized if the refractive indices of the two media are as similar as possible, which results in a transparent specimen and makes an embedded geogrid visible (Figure 3). In conclusion, by using a transparent soil, the optical recording methods are not limited to the boundaries of a test box, but can also be applied to planes inside the sample.

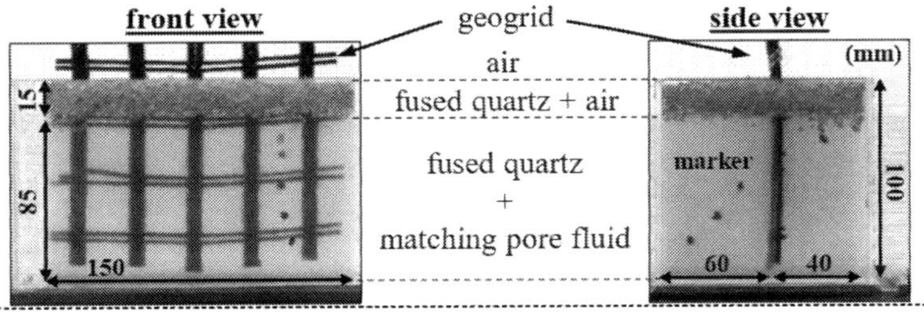

Figure 3. Geogrid in transparent soil (modified acc. to [11])

The good comparability of the mechanical and geotechnical properties of transparent soil with natural non-cohesive soils has been demonstrated several times based on sieve analyses, compression tests, direct shear tests and triaxial compression tests [19,20]. Moreover, no significant grain breakage of the quartz glass occurs and the influence of the high viscosity of the white oil can be neglected as long as the sample is slowly loaded [19]. The test method with transparent soil was recently used to understand the mode of action during direct geogrid activation in real-scale pull out tests [20,21,22].

In conclusion, the experimental technique with transparent soil fulfils the requirements of the investigation method to be used in this study. Therefore, transparent reinforced specimens are used in order to get an insight into the interface during indirect activation. Based on the results to be achieved, the presented conceptional model (Figure 2) can either be verified, has to be adapted accordingly to the new findings or has to be rejected. More detailed information about the materials, test box, sample preparation, experimental set-up and test procedure are given in [10,11].

3.2. Materials

A transparent soil was used that consists of a fused quartz (amorphous silica) and a white oil. First, all optical defective grains with air inclusions were removed from the granulate in order to optimize the optical quality of the transparent soil. Next, the fused quartz was washed, dried and then assembled to the particle size distribution with a mean grain diameter of $d_{50} \approx 7$ mm that is shown and related to other transparent soils used by other researchers in Figure 4 (a). The white oil Marcol™ 52, a colourless, odourless and tasteless purified mixture of liquid saturated hydrocarbons, was used as a matching pore fluid. Due to the refractive index of $n = 1.459$ at 20 °C it was sufficient to use only one pore fluid instead of a mixture of several oils as described by other researchers [19,20,21,22].

Direct shear tests were carried out in order to determine the failure envelops for initial normal stresses up to 400 kN/m² (Figure 4 (b)). The failure lines of the envelops were evaluated firstly with the best regression and secondly with an intercept at a shear stress of zero in the origin. Based on the best regression, the shear strength is determined to $c = 15.4$ kN/m² and $\varphi = 45.0°$, while for $c = 0$ a friction angle of $\varphi_{c=0} = 46.5°$ is obtained.

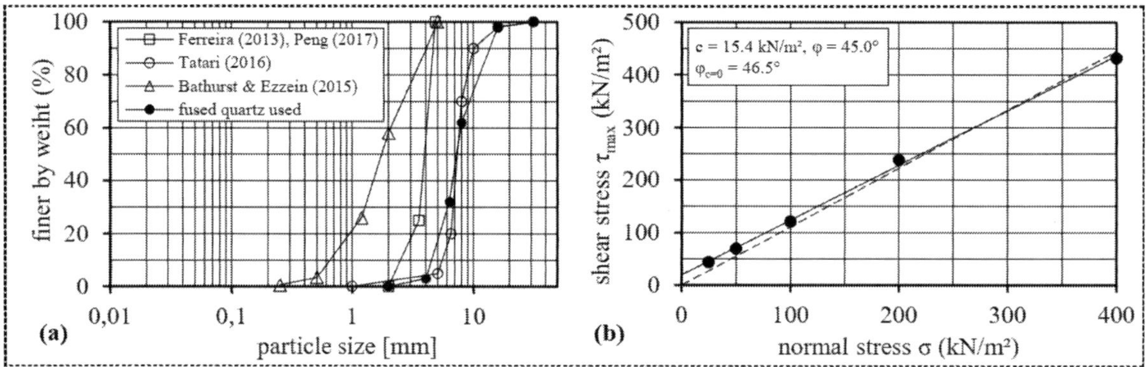

Figure 4. Particle size distribution (a) and direct shear tests (b) for the fused quartz used (modified acc. to [10])

The nominal specifications of the geogrid product used in this research are presented in Table 1. The originally biaxial geogrid was modified manually for the experiments. In the regular geogrid, each transverse tension member consists of two tension elements. For the test shown in this paper, one of the two elements per transverse member was removed in order to show their deflection more distinctly. As a result, the tensile strength in cross machine direction was halved by the modification.

Table 1. Material properties of the geogrid used (modified acc. to [10])

Characteristics	Geogrid orientation	
	md[a]	cd[a] / cd[b]
Polymer	Polyester (PET)	
Structure and coating	woven, polymer coated	
Mass per unit area (g/m²)	≈ 200[a]	
Aperture size (mm)	25	25 / 30
Ultimate tensile strength (kN/m)	≥ 30	≥ 30 / ≥ 15

[a] Property of regular geogrid (md: machine direction, cd: cross machine direction).
[b] Assumed properties of modified geogrid.

3.3. *Experimental setup and test specimen*

The experiments were carried out in a coated aluminum box with acrylic glass windows on the front and bottom (Figure 5 (b)) in which test specimens with dimensions up to 500 x 220 x 135 mm in width, depth and height can be installed. First, the unsaturated quartz granulate was placed in the box to a height of 80 mm. Then the geogrid was installed and covered with 40 mm of quartz glass. The geogrid was fixed on the left side with a vertically movable clamp. On the right side of the test box, a pressure pad was arranged so that a constant lateral pressure can be applied and the soil body can deform laterally at the same time. Afterwards the sample was saturated with the white oil under vacuum to avoid air bubbles in the specimen. The final experimental setup is shown in Figure 5 (a). Above the box there is a compression die connected to a linear variable differential transformer (LVDT) and a load cell arranged. The bottom camera and the lighting system that consist of two line-lasers and LED lights respectively, are located below the box, while the front camera is mounted on a rack in front of the box.

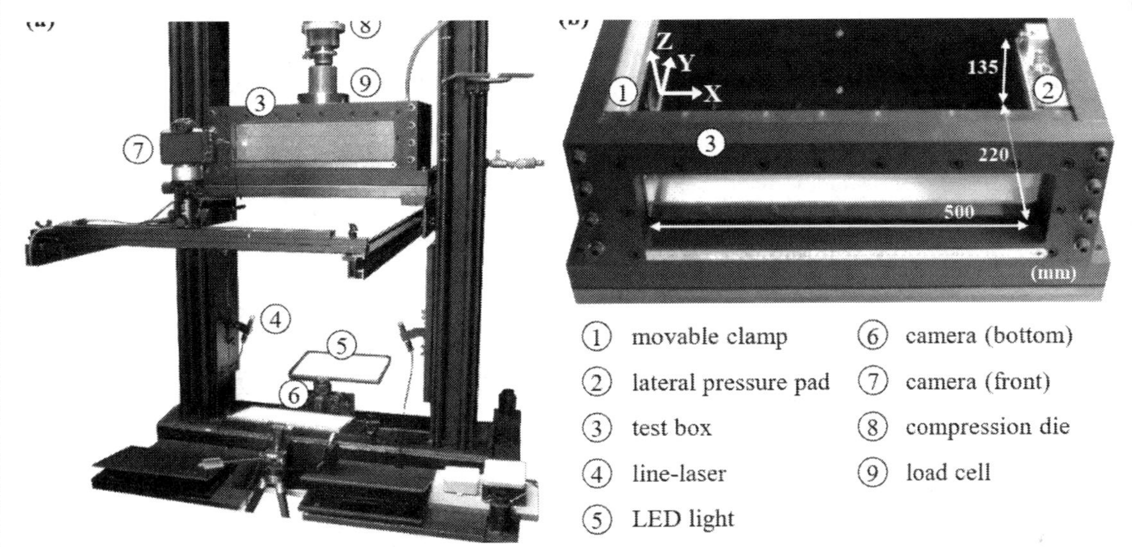

Figure 5. Experimental setup (a) and test box (b) (modified acc. to [10])

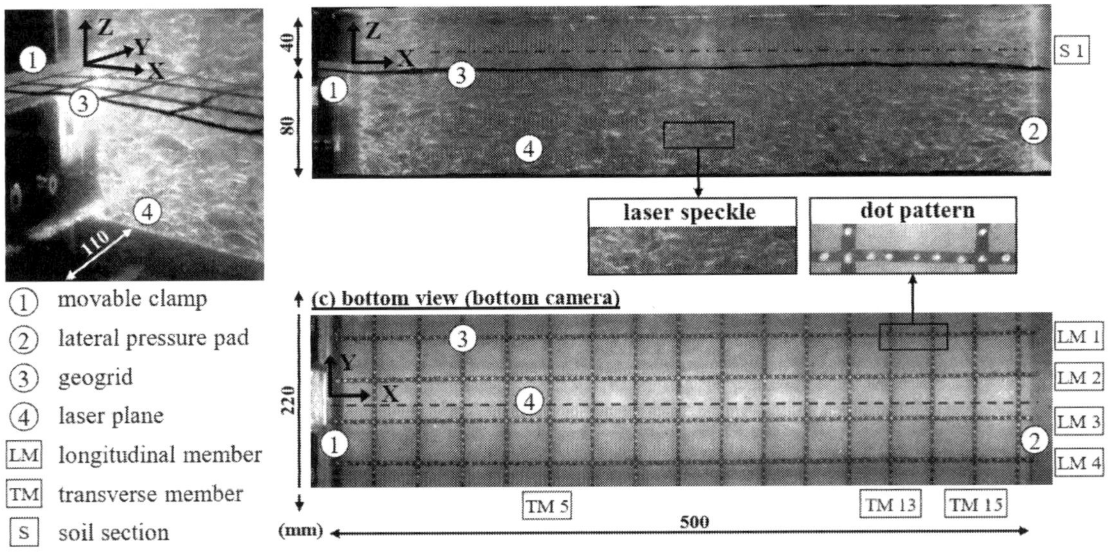

Figure 6. Test specimen in perspective (a), side view (b) and bottom view (c) (modified acc. to [10])

Finally, the specimen was loaded vertically with another pressure pad on top of the sample in a path-controlled manner with a speed of 1 mm/min, while the lateral pressure was kept constant at $\sigma_3 = 3.1$ kN/m² in the test of this study. Meanwhile, photos were taken continuously from the front and the bottom of the box with two digital mirrorless interchangeable-lens cameras. The soil deformations were recorded with the front camera using the speckle patterns generated with line lasers in the centre of the specimen (Figure 6 (a) and (b)). The bottom camera was used to capture the geogrid deformations during LED lightening. A dot pattern has been applied to the geogrid in order to optimize the recording of geogrid deformations due to higher contrast (Figure 6 (b)). The initial states of both laser plane in soil (b) and geogrid (c) are shown in Figure 6.

3.4. Test results and conclusion

During the experiment, the geogrid moves closer towards the bottom camera due to the load. If the recorded images would be evaluated by image correlation, besides the deformations actually occurring in the geogrid, perspective deformations would also result from the changing distance between the geogrid and the bottom camera. In a previous experiment, correction functions were determined in order to exclude these perspective deformations from the recorded images. For this purpose, a white plate with a black dot pattern was moved at a defined load speed in the direction of the bottom camera, while photos were taken. The deformation field of the dot pattern was used to back calculate correction functions for each pixel position in the x-y plane for any deformation state in z-direction. During the interaction test the change of vertical geogrid position is known from the lateral images (front camera) and according to the geogrid displacement in z-direction the corresponding correction function is applied on the image from the bottom camera.

Finally, a Digital Image Correlation (DIC) method was used to calculate the geogrid displacements based on the corrected images of the bottom camera, while the soil deformations were evaluated in the laser plane based on the taken pictures of the front camera.

First, three selected transverse tension members TM 5, 13 and 15 are considered which are marked in Figure 6 (b). For each of these transverse members their total deformations of the sections between the longitudinal tension members LM 1 to 4 are shown at a vertical load of 90 kN/m² in Figure 7 (a). The mean value of the largest deflections of the transverse members TM 5, 13 and 15 are plotted as a function of the vertical load on top of the sample in Figure 7 (b). As Figure 7 illustrates, TM 1 shows no deflection, whereas TM 13 and TM 15 are significantly deflected inwards respectively outwards. The mean values of the largest deflections of all 16 transverse members are plotted in Figure 8 (b) at their corresponding x-position.

IOP Conf. Series: Materials Science and Engineering 1260 (2022) 012005 doi:10.1088/1757-899X/1260/1/012005

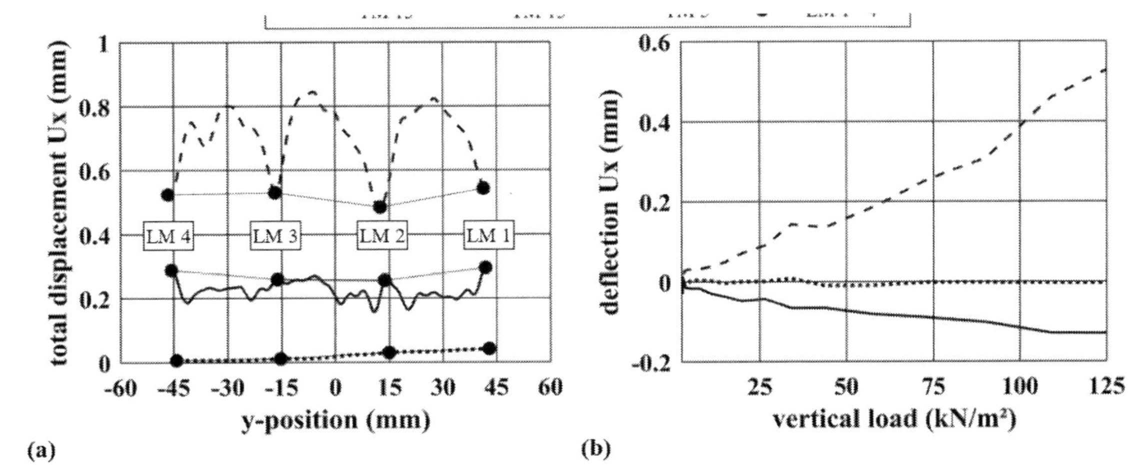

Figure 7. Displacement of transverse members (TM), (a) total and (b) deflection

A section of the displacement field of the transparent soil within the laser plane are shown in Figure 8 (a) for the considered deformation state. The lateral soil displacements along the horizontal section S 1 right above the geogrid (Figure 6 (b)) is shown in Figure 8 (b) as well as the mean displacement of the longitudinal members (LM).

In addition to the deformations, the geogrid forces should also be considered. For this purpose, the mean displacements of the longitudinal members were derived into local strains along the x-direction. Then, the force magnitudes along the geogrid were determined from the calculated local strains in combination with the tensile force-strain curve of the geogrid in air. The resulting force distribution is shown in Figure 8 (c). Moreover, Figure 8 (c) shows the force transmission at each transverse member, where a positive value indicates a force transmission from the soil into the geogrid and a negative value represents a force transmission from the geogrid into the soil.

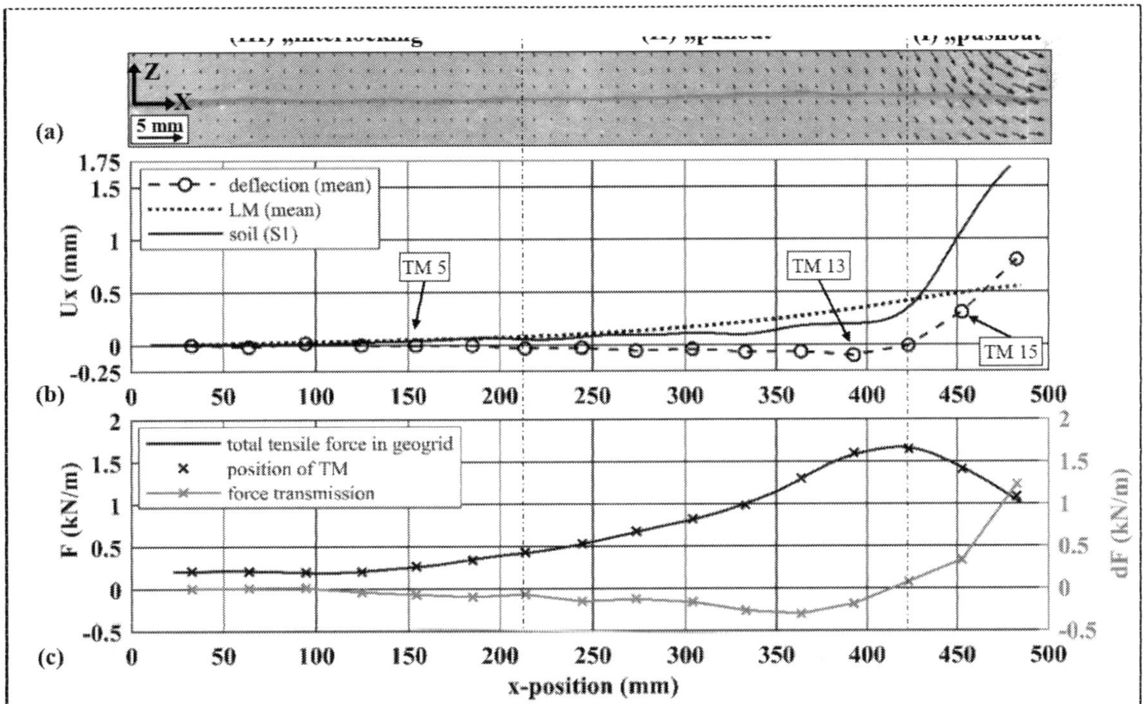

Figure 8. Soil displacement field (a), lateral soil and geogrid displacement (b) and tensile force in geogrid (c) at a vertical load of 90 kN/m²

In conclusion, three areas can be observed along the geogrid based on deformations and forces in x-direction, whereas in each area different interaction mechanisms can be identified. At the edge of the load area, the lateral soil deformation exceeds the geogrid displacement, which means that the soil is pushed over the geogrid, whereby the transverse tensile members are bent outwards. In this "pushout" area spreading forces are transferred into the reinforcement which is confirmed by an increase of geogrid force until a x-position of approximately 425 mm. At this point the relative displacement reverses and the geogrid displacement becomes greater than the soil deformation, leading the transverse tensile members to bend inwards. The decrease of the geogrid tensile force indicates an anchoring of the geogrid force in this "pullout" area. Between x-position 200 mm and the clamp no significant relative displacement between soil and geogrid nor a deflection of the transverse tensile members can be detected. In this "interlocking" area neither a force transmission from the soil into the geogrid nor vice versa can be observed, so that the tensile force in the reinforcement remains approximately constant. Thus, the considered transverse tension members TM 5, 13 and 15 can be assigned to the areas in which the mechanisms "interlocking", "pullout" and "pushout" occur.

The results indicate that the introduced force in the "pushout" area is partially, but not completely, anchored back in the "pullout" area and therefore case (b) of the model in Figure 2 seems to be appropriate. It is assumed that the remaining geogrid force in the "interlocking" area depends for instance on the stiffness of the geogrid. This will be investigated in future experiments. In general, the experimental results confirm the mechanisms described in the conceptional micro-mechanical model (Figure 2), so that the working hypothesis can be verified.

4. Summary and outlook

In the present research a transparent geogrid reinforced specimen was loaded in normal direction to the reinforcement in order to record the soil-geogrid-interface during indirect activation via the surrounding soil.

The displacement fields of the geogrid and of the soil within a laser plane were calculated using a Digital Image Correlation (DIC) method. Hereby, the deflections of transverse members, the axial displacements of longitudinal members and the lateral soil deformations in a section right above the reinforcement were considered. Moreover, the tensile force distribution along the geogrid was determined based on the local strains of the longitudinal members, whereby the force transmission at each transverse member was calculated.

In conclusion, three different areas could be identified along the geogrid in which the interaction mechanisms "pushout", "pullout" and "interlocking" occur when viewed from the edge inwards the reinforced soil. Finally, the previously developed micro-mechanical conceptional model for indirect activated geogrids (Figure 2) could be verified.

The investigated type of geogrid activation is predominant in many applications of geosynthetic reinforcement. The findings can be transferred to application cases and this research has a value for practice. For example, geogrids in base layers, foundation pads or dam base reinforcements are indirectly activated by an applied load via the surrounding soil.

Moreover, it can be stated that interaction experiments with transparent soil are a promising investigation method that give a non-invasive and area-wide insight into the composite material geogrid reinforced soil and visually illustrates how geogrids work.

In a next step, stress and displacement dependent activation functions for the interaction mechanisms "pushout" and "pullout" are derived based on the experimental results. Moreover, further interaction tests will be carried out in order to investigate the effects of geogrid stiffness, ratio of geogrid opening width to grain diameter, different lateral stresses and various load widths on the interaction behaviour.

The overall objective of this research is to review existing soil-geogrid-interface formulations in finite element models and if necessary, to improve their formulations e.g. by using a stress and displacement dependent activation.

5. Acknowledgements

This research is funded by the Deutsche Forschungsgemeinschaft (DFG, German Research Foundation) – 263996107 and supported by the GSI fellowship.

References

[1] Jewell R A, Milligan G W E, Sarsby R W and Dubois D 1984 Interaction between soil and geogrids *Polymer Grid Reinforcement* London UK

[2] Palmeira E M 1987 The study of soil-reinforcement interaction by means of large scale laboratory tests *University of Oxford* Oxford UK

[3] Ziegler M and Timmers V 2003 Neues Bemessungskonzept für die Bemessung der Verankerungslänge von Geogittern *Geotechnik* **8**pp 51-9

[4] Sieira A C C F, Gerscovich D M S and Sayao A S J F 2009 Displacement and load mechanisms of geogrids under pullout condition *Geotextiles and Geomembranes* **27**

[5] Jacobs F 2016 Interaktionsmodell zur Bemessung von Verankerungsgräben mit Geogittern *RWTH Aachen University* Aachen Germany

[6] Hausmann M R and Lee I K 1976 Strength characteristics of reinforced soil *Int. Symposium: New Horizons in Construction Materials* Bethlehem PA USA

[7] Matys M and Baslik R 2004 Study of Interlocking Effect by the Push Test *Tectum Geosynthetic* Bratislava Slovakia

[8] Ziegler M and Ruiken A 2009 Materialverhalten des Verbundbaustoffs „geogitterbewehrter Boden" aus großen triaxialen Druckversuchen *Geotechnik* **32**pp 148-55

[9] Ruiken A 2013 Zum Spannungs-Dehnungsverhalten des Verbundbaustoffs „geogitterbewehrter Boden" *RWTH Aachen University* Germany

[10] Derksen J, Ziegler M. & Fuentes R 2021. Geogrid-soil interaction: A new conceptual model and testing apparatus. Geotextiles and Geomembranes **49**pp 1393-1406

[11] Derksen J, Ziegler M 2019 Interaction behaviour of geogrids in transparent soil. *Proceedings of the 17th African Regional Conference on Soil Mechanics and Geotechnical Engineering*, Cape Town, South Africa pp 267-72

[12] Ižvolt L and Kardoš J 2010 Parameter von Geogittern und ihr Einfluss auf deren Wirksamkeit im Eisenbahnunterbau *Eisenbahntechnische Rundschau* **4**

[13] McDowell G R, Harireche O, Konietzky H, Brown SF and Thom N H 2004 Discrete element modelling of geogrid-reinforced aggregate *Proceedings of the Institution of Civil Engineers Geotechnical Engineering* **159**pp 35-48

[14] Ziegler M 2016 Geosynthetic reinforcement applications *Keynote lecture at 6th European Geosynthetics Congress* Ljubljana Slovenia pp 173-207

[15] Dyer M R 1985 Observation of the Stress Distribution in Crushed Glass with Applications to Soil Reinforcement *University of Oxford* Oxford UK

[16] Black J A and Take W A 2015 Quantification of Optical Clarity of Transparent Soil Using the Modulation Transfer Function *Geotechnical Testing Journal* **38**pp 588-602

[17] Iskander M 2010 Modelling with Transparent Soils: Visualizing Soil Structure Interaction and Multi Phase Flow, Non-Intrusively *Springer* Berlin Germany

[18] Ezzein F M and Bathurst R J 2011 A transparent soil for geotechnical laboratory modelling *Geotechnical Testing Journal* **34**

[19] Ferreira J A Z and Zornberg J G 2015 A Transparent Soil Pullout Testing Device for 3D Evaluation of Soil-Geogrid Interaction *Geotechnical Testing Journal* **38**

[20] Ezzein F M and Bathurst R J 2014 A new approach to evaluate soil-geosynthetic interaction using

a novel pullout test apparatus and transparent granular soil *Geotextiles and Geomembranes* **42** pp 246-55
[21] Tatari A 2016 Interaction behaviour of soil and geosynthetic reinforcement *University of Sheffield* Sheffield UK
[22] Peng X and Zornberg J G 2017 Evaluation of load transfer in geogrids for base stabilization using transparent soil *Transportation Geotechnics and Geoecology* Saint Petersburg Russia

IOP Conf. Series: Materials Science and Engineering 1260 (2022) 012006 doi:10.1088/1757-899X/1260/1/012006

Sodium Carboxymethyl Cellulose-Enhanced GCLs in Freshwater Reservoirs

H O Ozhan[1]

[1] Civil Engineering Department, Yeditepe University, Inonu Mah., Kayisdagi Cad., 326A, 26 Agustos Yerlesimi, Atasehir-Istanbul 34755, Turkey
Email: hakki.ozhan@yeditepe.edu.tr

Abstract. Biopolymers that are environmentally friendly organic products are increasingly used in geotechnical applications instead of synthetic polymers and other chemical additives. Cellulose is a biopolymer that is obtained from the cell walls of plants. In order to evaluate its effect on internal erosion, triaxial permeability tests were performed on a sodium carboxymethyl cellulose-enhanced geosynthetic clay liner (GCL) with Ca bentonite that was placed over a coarse gravel with an average grain diameter of 2.5 cm and permeated with tap water. Furthermore, unconfined compression tests were conducted on the cellulose-added bentonite component of the GCL in order to evaluate the mechanical effect of sodium carboxymethyl cellulose addition to the GCL on internal erosion. Test results indicated that 2% cellulose addition by dry weight maintained sufficient hydraulic and mechanical performances with almost 2.5 orders of magnitude decrease in permeability and 10 times increase in unconfined compressive strength. However, cellulose addition had only a slight enhancement against internal erosion. All the tested GCLs resisted against internal erosion up to a hydraulic head of 25 m. But 30 m hydraulic head resulted in internal erosion for the GCLs enhanced with up to a content of 2% cellulose.

1. Introduction

The main purposes to use a geosynthyetic clay liner (GCL) in a freshwater reservoir are to collect water for agricultural activities (growth of fruits and vegetables), for industrial activities (production of paper and chemicals) or firefighting helicopters that try to stop fires in forests [1,2]. The water that is collected in the reservoir generally do not exceed 30 m [2]. However, in very rare cases, the water level can be up to 50 m [3]. As the water level in the reservoir increases, the hydraulic load that acts on the GCL also increases. Due to this high hydraulic load, there is the possibility for the bentonite component of the GCL to be extruded out [4]. When this bentonite loss is at an extremely high level, the hydraulic conductivity of the GCL suddenly increases 3-5 orders of magnitude. This phenomenon is defined as internal erosion [1,5,6]. Internal erosion of the GCL is mainly affected by the granularity of the subgrade soil over which the GCL is placed. As the grain size diameter of the subgrade soil increases, the risk for the internal erosion also increases [6,7]. Coarse-grained gravel subgrades or cover soils generally caused internal erosion for the GCLs [5-8].

Due to its very low hydraulic conductivity, Na bentonite is usually preferred to be used as the clay component in GCLs for barrier applications; however, Ca bentonite is also used in GCLs with its higher shear strength although it has higher hydraulic conductivity than Na bentonite [9,10]. In order to improve the hydraulic performance of a Ca bentonite, polymeric additives are mixed with the bentonite and GCLs with polymer-enhanced Ca bentonite are also preferred to be placed in landfills or freshwater reservoirs. The polymers that are obtained from living organisms are named as biopolymers. Biopolymers are increasingly preferred as additives instead of synthetic polymers or other chemical additives such as lime and cement due to the fact that the biopolymers do not consist of toxic substances and do not pollute the environment [11]. Biopolymers are also used in geotechnical engineering by increasing the bearing capacity of soils, preventing erosion, enhancing the barrier capability of the lining material in waste containment facilities and purifying groundwater [12].

Cellulose is one of the most abundant biopolymers that is found in the cell walls of plants and functions as a reinforcement material for the plant. Carboxymethyl cellulose is an anionic biopolymer that has a high viscosity. [13] added carboxymethyl cellulose to clay with contents of 2%-16% by dry weight respectively. They investigated some hydraulic properties of the biopolymer-added clay that was in contact with both a $CaCl_2$ solution and seawater. They concluded that both the swell index and the liquid limit of the clay increased as the cellulose content increased. Furthermore, cellulose addition resulted in a decrease in the hydraulic conductivity of the clay. [14] conducted hydraulic conductivity tests on carboxymethyl cellulose-added Na bentonite that was permeated with seawater. Test results indicated that 2% biopolymer addition by dry weight caused the hydraulic conductivity to decrease from 2.93×10^{-7} m/s to 3.5×10^{-10} m/s. 5% carboxymethyl addition also caused a slight decrease in hydraulic conductivity down to 9.11×10^{-11} m/s.

The effect of polymer addition on the down-slope erosion of bentonite from various GCLs was also investigated [15,16]. According to the results, polymer-added GCL did not experience down-slope erosion whereas erosion caused the hydraulic conductivity of the GCLs without polymer enhancement to increase drastically.

The main objective of this paper was to investigate the effect of sodium carboxymethyl cellulose (biopolymer) addition to a GCL on the internal erosion. The GCL was composed of a Ca bentonite between a woven and a nonwoven geotextile and permeated with tap water. The hydraulic head on the GCL was increased up to 30 m and the GCL was placed over a coarse-grained gravel. The biopolymer content in the Ca bentonite was chosen as 0.5, 1, 2, 4 and 8% by dry weight respectively. The relationships between the hydraulic and mechanical properties of the sodium carboxymethyl cellulose-enhanced GCL and internal erosion were tried to be established by conducting triaxial permeability and unconfined compression tests.

2. Materials

The GCL was composed of a powdered Ca bentonite layer sandwiched between a woven polypropylene geotextile with mass/area of 108 g/m^2 and a nonwoven polypropylene geotextile that had mass/area of 203 g/m^2 without reinforcement. The apparent opening size (AOS) values of the woven and the nonwoven geotextile components of the GCL were 0.425 mm and 0.212 mm, respectively. The engineering properties of the Ca bentonite are listed in Table 1.

Table 1. Engineering properties of the bentonite

Property	Value
Specific gravity (-)	2.66
Swell Index (ml/2g)	20
Calcium Montmorillonite Content (%)	85
Liquid Limit (%)	141
Plastic Limit (%)	41
Plasticity Index (%)	100

Sodium carboxymethyl cellulose is a water-soluble, anionic biopolymer that is mainly preferred to be used in food, cosmetics, paper, textiles and oil. It is known as a stabilizer, thickener and viscosity enhancer in many applications [17]. In construction industry, it is generally used as a concrete enhancer and mortar thickener. Sodium carboxymethyl cellulose has a molecular weight of 263.2 g/mol and a density of 1.6 g/cm^3. Its molecular formula is $[C_6H_7O_2(OH)_2CH_2COONa]_n$ [17].

3. Methods

3.1. Triaxial Permeability Tests

The sodium carboxymethyl cellulose with a content of 0.5, 1, 2, 4 and 8% by dry weight was added to the oven-dried bentonite respectively and the mixture was poured into a polyethylene cup. Then, the cup was closed and shaken by hand in order to provide a homogeneous mixture. The mixture remained in the cup for 24 hours [18].

The geotextile components with a diameter of 100 mm were cut from the geotextile rolls with a scissor. Then, the cellulose-bentonite mixture was placed on top of the nonwoven geotextile. The mixture was wetted with tap water homogeneously in order to provide sufficient bonding to the geotextiles. The, the woven geotextile was attached on top of the cellulose-bentonite mixture [10].

The GCL was placed over a rounded, coarse gravel with an average grain diameter of 2.5 cm. The woven geotextile component of the GCL was in contact with the gravel subgrade. The test configuration in the permeameter cell is shown in Figure 1. The test set-up from top to bottom was composed of the top cap, porous stone, filter paper, GCL specimen (nonwoven geotextile component at the top and woven geotextile component at the bottom), gravel subgrade, nonwoven geotextile used as a filter and base pedestal. All of these components would be wrapped around with a latex membrane during the triaxial permeability tests. The aim to use a needle-punched polypropylene nonwoven geotextile beneath the gravel subgrade was to maintain the water flow without clogging. During internal erosion, the bentonite particles migrate from the GCL and might be collected on the filter paper and might clog it. For this reason, the nonwoven geotextile with an apparent opening size of 0.08 mm was used instead of the filter paper with a smaller apparent opening size of 0.03 mm [10]. As the hydraulic gradient was increased to initiate internal erosion, the gravel subgrade might penetrate into the latex membrane and damage it. Moreover, sidewall leakage might occur between the gravel subgrade and the latex membrane. Because of these reasons, the gravel subgrade was wrapped around with another nonwoven geotextile in order to protect the flexible latex membrane against puncture at high hydraulic loads. This nonwoven geotextile had a tensile strength, mass/area and thickness of 32 kN/m, 500 g/m^2 and 3.6 mm respectively. Finally, the test set-up was confined with a flexible latex membrane that was attached with O-rings [10].

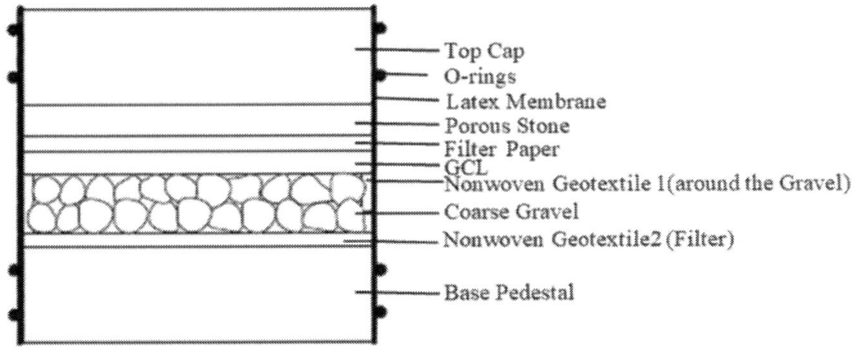

Figure 1. Triaxial Permeability Test Configuration

After the GCL specimen was placed in the flexible-wall permeameter cell as shown in Figure 1, constant head triaxial permeability tests were performed as outlined in [19]. First, the permeameter cell was filled with tap water, then, both the cell pressure and the back pressure were increased up to 550 kPa and 515 kPa and these pressures were kept constant for a period of almost 48 hours in order to provide saturation and consolidation. Afterwards, downward flow was initiated by increasing the pressure on top of the GCL to 530 kPa. As a result, the difference between the top and bottom of the GCl became 15 kPa [19]. The hydraulic head acting on the GCL was increased to 5 m just after the flow became constant as indicated by [10]. The incremental increase of 5 m continued till a hydraulic head of 30 m was obtained. After each hydraulic head increase, the hydraulic pressure acting on the GCL was kept constant for a period of 12 days. This period was chosen because a sudden increase in permittivity and then, the initiation of internal erosion for the tested GCLs was obtained in almost 12 days in previous studies [2,7,10]. Hydraulic head of 30 m was maintained by taking the cell pressure, influent pressure at the top of the GCL and effluent pressure at the bottom of the GCL as 550, 530 and 235.7 kPa respectively [19]. If internal erosion had not been observed during 12 days at 30 m hydraulic head, then, the test would have terminated and the final coefficient of permeability values of the tested GCLs were measured by using Equation 1 as follows:

$$k = \Psi \cdot L \tag{1}$$

Where k (cm/s) is the final coefficient of permeability, ψ (1/s) is the permittivity and L (cm) is the final thickness of the GCL specimen.
From the beginning till the end of each test, permittivity values of the GCLs were calculated by using Equation 2. Permeability tests were terminated for the GCLs that experienced internal erosion until 4-5 orders of magnitude increase in permittivity was measured.

$$\Psi = \frac{\Delta Q}{\Delta h \cdot \Delta t \cdot A} \tag{2}$$

Where ΔQ (cm^3) is the average of water volume that enters into the GCL and leaves the GCL for the taken time interval, Δh (cm) is the water level that acts on the GCL, Δt (s) is the time interval over which the water is permeated through the GCL and A (cm^2) is the cross-sectional area of the GCL.

3.2. Unconfined Compression Tests

First, the sodium carboxymethyl cellulose-enhanced bentonites with a cellulose content of 0.5, 1, 2, 4 and 8% by dry weight were compacted as outlined in [20] respectively. Then, cylindrical steel molds were used to take specimens with a height of 70 mm and a diameter of 35 mm from the cellulose-bentonite mixtures that were compacted at optimum water content (w_{opt}). Afterwards, the compacted specimens were taken into the loading equipment of the unconfined compression test. Then, the specimens were loaded axially at a strain rate of 1%/min. As the load increased, the deformations also increased and the tests were terminated when the applied load became steady and then, decreased. The unconfined compressive strength (q_u) of the compacted cellulose-bentonite specimens at w_{opt} was calculated by dividing the maximum axial load (kN) or the axial load recorded at 0.015 axial strain by the cross-sectional area of the specimen (m^2) as outlined in [21].

4. Test Results and Discussion

4.1. Triaxial Permeability Tests

According to the triaxial permeability test results, all of the tested GCLs did not experience internal erosion at 25 m hydraulic head; however, 30 m water level on top of the GCL without cellulose enhancement and 0.5%, 1% and 2% cellulose-enhanced GCLs resulted in internal erosion. The only GCLs that did not experience internal erosion at 30 m hydraulic head were 4% and 8% cellulose-enhanced GCLs as can be seen in Figure 2. Permittivity values for 4% and 8% cellulose-enhanced GCLs were calculated as $4.8x10^{-12}$ and $3.3x10^{-12}$ 1/s respectively after a permeation period of 13-14 days. However, the other GCLs with lower cellulose contents experienced internal erosion with 4-5 orders of increase in permittivity after a permeation period of 6-8 days as shown in Figure 2. For all the tested GCLs that experienced internal erosion, permittivity increased suddenly in a couple of hours and during the last stages of internal erosion, the rate of increase in permittivity was even higher. For example, permittivity of 2% biopolymer-added GCL increased from $9.6x10^{-12}$ to $9.5x10^{-10}$ 1/s drastically in 10 hours during internal erosion. Then, permittivity increase became 3 orders of magnitude in a few seconds and the final calculated permittivity was $9.4x10^{-7}$ 1/s. A GCL specimen before placed in the permeameter cell is shown in Figure 3a. As the bentonite particles escaped from the GCL during internal erosion, voids were formed in the bentonite structure and there was not a homogeneous bentonite distribution in the GCL as can be seen in Figure 3b.

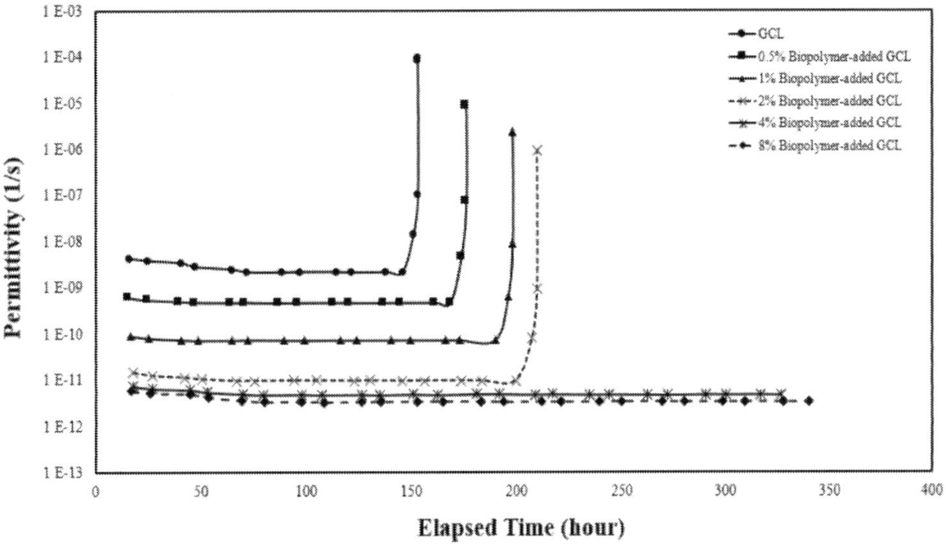

Figure 2. Permittivity-Elapsed Time Graph related to Internal Erosion

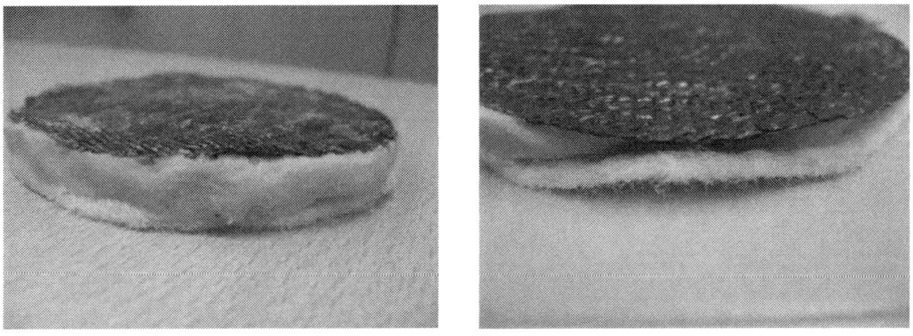

a b

Figure 3. **a** GCL Specimen before Triaxial Permeability Test, **b** GCL Specimen after Triaxial Permeability Test (experienced Internal Erosion)

By measuring the final thickness values of the GCLs that were taken out from the permeameter cells, the final coefficient of permeability values could be calculated. However, these permeability values were calculated just before internal erosion began for the GCLs that experienced internal erosion or for the GCLs that did not experience internal erosion. The final permeabilities versus the biopolymer contents in the GCLs are shown in Figure 4. As the biopolymer content was increased, the permeability decreased. However, the decrease in permeability was considered to be not significant after a biopolymer content of 2% as can be seen in Figure 4. The GCL without the cellulose enhancement had coefficient of permeability of $1.5x10^{-9}$ cm/s whereas the coefficient of permeability of 2% cellulose-added GCL decreased to $7.5x10^{-12}$ cm/s. This decrease was almost 2.5 orders of magnitude. However, further increase, up to 8% in the biopolymer content resulted in only 0.5 additional decrease in permeability. The anionic charge of the sodium carboxymethyl cellulose caused a flocculated structure when the biopolymer was mixed with the bentonite and water. The flocculated cellulose-bentonite particles resulted in a decrease in the pore size of the mixture and the permeability [22,23]. As can be seen in Figure 4, almost 2% sodium carboxymethyl cellulose addition to the bentonite component of the GCL by dry weight can be considered to be enough to fill the pores among the bentonite particles. Additional increase in the cellulose content caused only a slight decrease in the permeability.

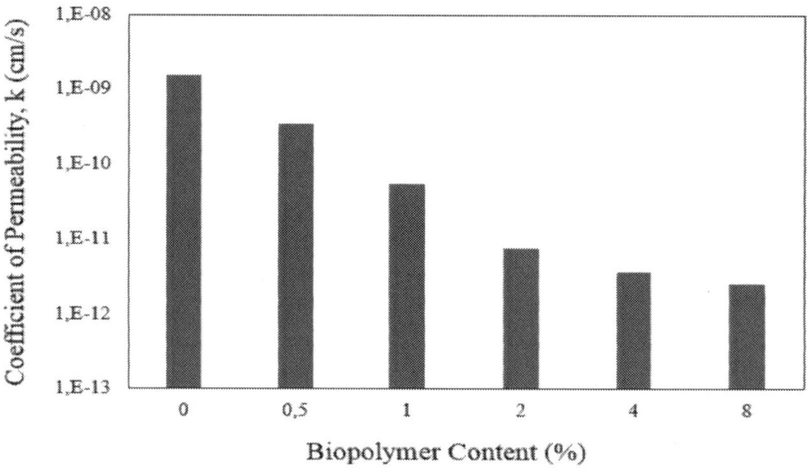

Figure 4. Effect of Biopolymer Content change on Permeability

4.2. *Unconfined Compression Tests*

Unconfined compressive strength values of the bentonite that was used in the GCLs versus the biopolymer contents are given in Figure 5. As the biopolymer content was increased, the unconfined compressive strength also increased. The change in the unconfined compressive strength was found to be similar with the change in the final coefficient of permeability values. The rate of increase was higher up to the biopolymer content of 2%. While the unconfined compressive strength of the bentonite without the cellulose enhancement was calculated as 322 kPa, 2% cellulose-added bentonite had unconfined compressive strength of 3641 kPa. This value was almost 10 times the strength of the bentonite that had no cellulose. However, the unconfined compressive strength of 8% cellulose-added bentonite increased up to only 3909 kPa. As a result, the rate of increase in the unconfined compressive strength was found to be not significant when the cellulose content was further increased from 2%. The sodium carboxymethyl cellulose has the capability to form hydrogels when it gets in contact with water and these hydrogel formations lead to a high binding capacity. As the cellulose content was increased, more hydrogels were formed due to a higher amount of water absorption. As a result, more pores were filled with the cellulose and the capacity of the cellulose-bentonite mixture to hold the particles strongly and closely was enhanced. This could be the main reason why the unconfined compressive strength of the GCLs increased due to an increase in the biopolymer content [17,24].

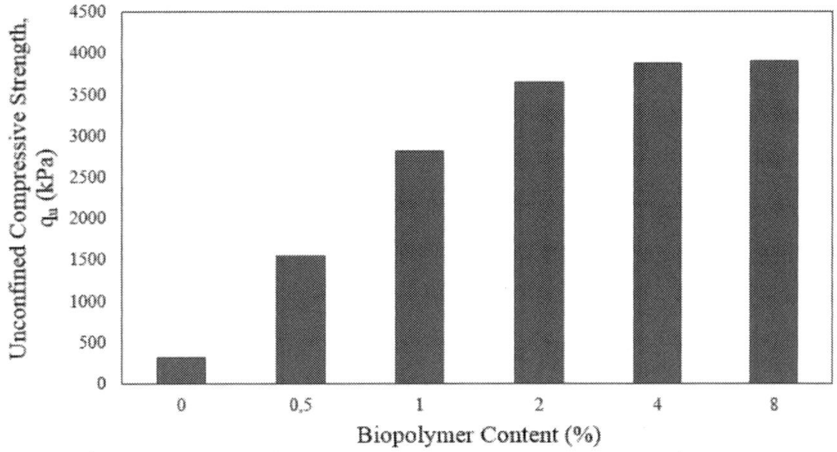

Figure 5. Effect of Biopolymer Content change on Unconfined Compressive Strength

The results of some previous studies showed that internal erosion of the GCL was mainly related to the grain size of the subgrade soil beneath the GCL and the hydraulic load acting on the GCL [2,5-8]. As the grain size and the hydraulic load increase, the internal erosion risk also increases. In this study, the correlation between the addition of a biopolymer, sodium carboxymethyl cellulose, to a GCL with Ca bentonite and internal erosion was found to be not very strong. However, 4% and 8% cellulose-added GCLs did not experience internal erosion at 30 m hydraulic head while the GCLs enhanced with the cellulose up to 2% content experienced. By the way, all of the tested GCLs did not fail at a hydraulic head of 25 m. The addition of the cellulose biopolymer to the bentonite component of the GCL caused higher bonding among the bentonite particles and caused the hydrogels to clog the pores. As a result, the permeability of the GCL decreased whereas its compressive strength increased. The enhanced compressive strength of the bentonite component of the GCL might resist against high hydraulic heads and prevent the bentonite particles to be eroded out.

5. Conclusions

An increase in the sodium carboxymethyl cellulose content resulted in a decrease in the permeability of the GCL. The optimum cellulose content was found to be 2%. When the cellulose amount exceeded 2%, there was almost no change in permeability. 2% sodium carboxymethyl cellulose addition caused a decrease of almost 2.5 orders of magnitude in the permeability of the GCL with Ca bentonite.

An increase in the sodium carboxymethyl cellulose content leaded to an increase in the unconfined compressive strength of the GCL. The optimum cellulose content was measured as 2%. Additional increase in the cellulose content caused only a slight increase in the unconfined compressive strength.

2% sodium carboxymethyl cellulose addition resulted in more than 10 times increase in the unconfined compressive strength of the GCL with Ca bentonite.

The GCL without sodium carboxymethyl cellulose enhancement and 0.5, 1, 2, 4 and 8% sodium carboxymethyl cellulose-enhanced GCLs did not experience internal erosion at hydraulic heads of up to 25 m when the GCLs were placed over a coarse gravel with an average grain diameter of 2.5 cm. However, only 4% and 8% sodium carboxymethyl cellulose-enhanced GCLs did not experience internal erosion at 30 hydraulic head.

In conclusion, both the triaxial permeability and the unconfined compressive strength test results indicated that the sodium carboxymethyl cellulose addition to the Ca bentonite component of the tested GCL improved the hydraulic performance by decreasing the permeability and enhanced the mechanical performance by increasing the unconfined compressive strength.

According to the results of this study, 2% sodium carboxymethyl cellulose was found to be the optimum biopolymer content to be added to the Ca bentonite of the tested GCL that could be placed in freshwater reservoirs. However, for specific cases where water levels higher than 25 m are needed and the subgrade beneath the GCL was a coarse gravel with 2.5 cm grain diameter, a minimum of 4% sodium carboxymethyl cellulose has to be used in the GCL in order to provide a protection against internal erosion.

References

[1] Bouazza A 2002 *Geotext. Geomembr.* **20** pp 3–17
[2] Ozhan H O and Guler E 2013 *Geotech. Test. J.* **36** pp 418–28
[3] Zohary T and Ostrovsky I 2010 *Inland Waters* **1** pp 47–59
[4] Hosney M S H 2014 Performance of Geosynthetic Clay Liners in Cover, Subsurface BarrierandBasal Liner Applications *PhD Thesis*, Queen's University, Ontario, Canada p 865
[5] Rowe R K and Orsini C 2003 *Geotext. Geomembr.* **21** pp 1–24
[6] Shan H Y and Chen R H 2003 *Geotext. Geomembr.* **21** pp 339–354
[7] Ozhan H O and Guler E 2016 *Env. & Eng. Geosci.* **22** pp 157-69
[8] Fox P J, De Battista D J and Mast D G 2000 *Geotext. Geomembr.* **18** pp 179–201
[9] Gleason M H, Daniel D H and Eykholt G R 1997 J. *Geotech. Geoenviron. Eng.* **123** pp 438–45
[10] Ozhan H O 2011 Internal Erosion of Geosynthetic Clay Liners Under High Hydraulic Heads *PhD. Thesis*, Bogazici University, Istanbul, Turkey p 222
[11] Aminpour M and O'Kelly B C 2015 *Proc. of the Second Int. Dam World Conf.* Lisbon, Portugal **1** pp 937 – 46
[12] Mitchell J K and Santamarina J C 2005 *J. Geotech. Geoenviron. Eng.* **131** pp 1222–33,
[13] Di Emidio G, Mazzieri F, Verastegui-Flores R D, Van Impe W and Bezuijen A 2015 *Geosynth. Int.* **22** pp 125-37
[14] De Camillis M, Di Emidio G, Bezuijen A and Verastegui-Flores R D 2016 *Geotext. Geomembr.* **44** pp 739-47
[15] Brachman R W I, Rentz A, Rowe R K and Take W A 2014 *Can. Geotech. J.* **52** pp 395-412
[16] Rowe R K, Ashe L E, Take W A and Brachman R W I 2014 *Geotext. Geomembr.* **42** pp 445-56
[17] Chen H 2015 Lignocellulose Biorefinery Engineering Principles and Applications Woodhead Publishing 1st ed p 274
[18] Razakamanantsoa A R, Barast G and Djeran-maigre I 2012 *Appl. Clay Sci.* **59–60** pp 103–14
[19] ASTM D5887 2009 Standard Test Method for Measurement of Index Flux through Saturated Geosynthetic Clay Liner Specimens Using a Flexible Wall Permeameter (ASTM International, West Conshohocken, PA, USA)
[20] ASTM D698 2012 Standard Test Methods for Laboratory Compaction Characteristics of Soil using Standard Effort (ASTM International, West Conshohocken, PA, USA)
[21] ASTM D2166 2013 Standard Test Method for Unconfined Compressive Strength of Cohesive Soil (ASTM International, West Conshohocken, PA, USA)

[22] Li X and Zhang L M 2009 *Can. Geotech. J.* **46** pp 129–41
[23] Haase H and Schanz T 2016 *Appl. Clay Sci.* **130** pp 62-75
[24] Geng W, Likos W J and Benson C H 2016 *Proc. Int. Conf. Geo-Chicago,* USA GSP 271 pp 498-507

Stepped Reinforced Soil Retaining Walls at Railway Corridor Rokycany – Plzeň

P Hubík[1], M Kašpar[1] and O Vodáček[1]

[1]GEOMAT s.r.o., Pražákova 1008/69, 639 00, Brno, Czech Republic

E-mail: technika@geomat.eu

Abstract. As a part of a modernization of railway corridor Rokycany - Plzeň, a support wall structure was designed on both sides of the slopes. All local conditions have been taken into decision-making what technology to use, in particular, the speed of construction, geotechnical conditions and minimization of the land engagement. Reinforced soil technology was selected to meet all the constraints and requirements and finally the project was completed. As soil cladding, self-standing high concrete panels were used, making the 12 m total height structure unique. Many pictures declaring construction sequence are attached.

1. Introduction

Railway corridor „Rokycany – Plzeň" is located in southwest Bohemia as a part of III. interstate transit railway corridor of Czech Republic connecting Prague with Pilsen. Main objective of the vast modernization of III. interstate was a reduction of a travel time Prague - Pilsen to one hour maximum. Therefore, the railway corridor „Rokycany – Plzeň" on a line section „Ejpovice – Doubravka" needed to be cut short, see figure 1. To reach the goal, a line relocation „Homolka" (km 94,068 – 95,850) was designed, giving approximately 9 minutes time savings.

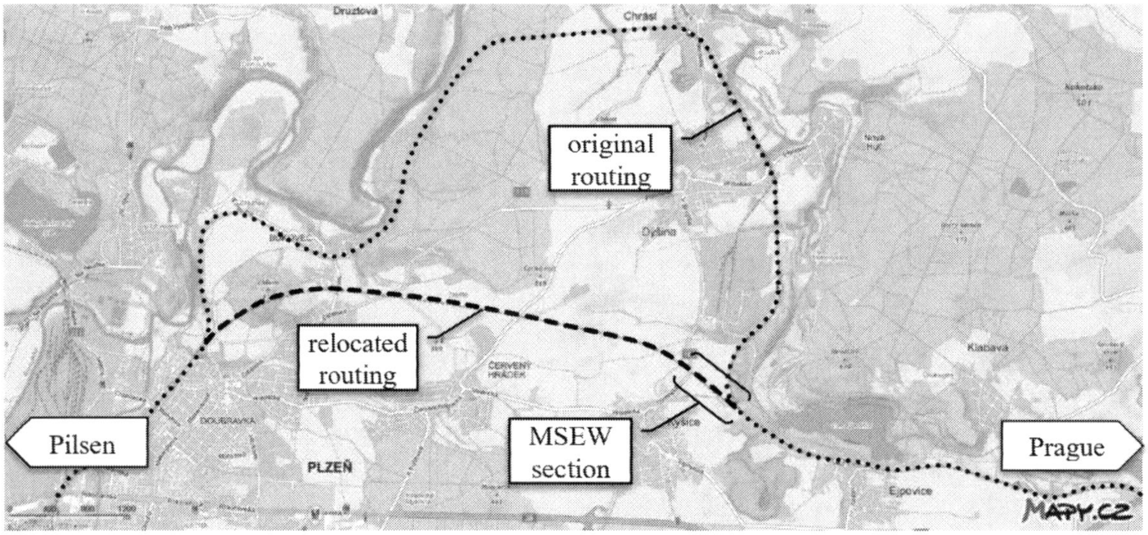

Figure 1. Line relocation; (···) original routing; (- - -) relocated routing

However, relocated line routing went through a 700 m long earthen cut, up to 14 m deep, and it were the slopes of the cut that needed to be stabilized permanently.

Optimization of the project put forth several major requirements, among which following were of an upmost importance: to minimize both temporary and permanent land takes, to choose a quick-to-build technical solution and a maintenance-free construction.

Mechanically stabilized earth walls (also MSEW thereinafter) of a step-like shape with full-height concrete panel facing met all the requirements. Moreover, conventional systems such as concrete retaining walls either precast or monolithic proved difficult to design due to geological conditions, the MSEW therefore became a final solution.

The walls were designed on both sides of the cut, stabilizing it permanently. Construction works began in 2015, each of the walls is up to 12 m high step-like structure divided into three levels, which makes it one of the biggest structures on a railway corridor of Czech Republic.

2. Geological conditions

Quaternary sediments consist primarily of diluvial and fluvial soils, secondarily of anthropogenic sediments, thickness varies in a range of 1,7 – 6,0 m along the 700 m long section. Sediments are represented by silty sands and silty gravels of various plasticity limits, kaolinic clay type soils are detected locally, slates are of an occasional appearance. Tertiary period is represented by volcanic rocks.

Quality of soils varies significantly due to lithological composition combined with shallow ground water head, moreover, ground water yield is influenced by rainfalls and clay isolating layers. Water springs were expected during construction process (figure 2), geological conditions were complicated (figure 3), hence the structure, complicated itself, was allocated to 3rd Geotechnical category.

Figure 2. Groundwater conditions

Figure 3. Geological conditions

3. Design philosophy

It is important to realize that there are no standards for MSEW design in Czech Republic, so whereas some failure mechanisms and limit states could be assessed in accordance with EN 1997-1, as befits any geotechnical structure, some of the important ones could not. In such cases, BS 8006-1:2010 procedures were adopted. Moreover, experience had a major importance, gained in the years past from similar MSEW structures, bridge abutments near Jánovce, Slovakia for one. Experience-based corrections were performed, failure mechanisms determined and calculation methods chosen accordingly.

Along with calculation-related obstructions, technological point of view needed to be taken into account as well, because one of the main goals was to enable a fast and fluent construction process. Therefore, an original retaining system with a concrete panel facing was invented (see figure 4 and figure 5) and all of the important structural details with it.

Due to technological limits, transportation limits, construction speed and with respect to the structure geometry, full-height concrete panels were chosen as a facing, with a limit height of 4,2 m.

Figure 4. Storing of panels on site

Figure 5. Storing of panels on site (detailed)

Three sets of soil properties needed to be known for computations, the first and second one being properties of infill and retained fill respectively. The third one described subsoil properties as per geological layers and consisted of various subsets.

Both retained fill and subsoil properties are given by the nature of soils, but infill was prescribed by a shear strength properties and a grain distribution curve.

From a long-term perspective, no actions were expected on top of the structure, apart from construction and maintenance vehicles during both construction and lifespan period respectively, hence simple surcharge stood for a dominant action, neglecting local effects.

Pilsen region is not exposed to any seismic activity.

4. Calculation

Since there are no local standards for designing reinforced soil in Czech Republic, foreign standards take the place. Generally, three groups of failure mechanisms for ultimate limit state analysis required a check to prove the structure satisfactory. The first one, a global stability represented by a slip plane analysis, was performed in accordance with Bishop's comprehensive method. External stability failure mechanisms, the second group, contained check of bearing capacity in base plane, sliding in base plane and eccentricity (tilt) in base plane. Both global stability and external stability analyses are common for almost every retaining structure, therefore EN 1997-1 procedures could be adopted. In this case, Design approach 3 proved normative.

Above described groups of failure mechanisms however do not affect reinforcement strength nor quantities directly, quite the contrary, it is assumed that reinforcement is strong and numerous enough so the reinforced soil behaves as a homogeneous body.

For this assumption to be justified an internal stability failure mechanisms, the third group, needed to be assessed. The reinforcement strength and interaction with an infill have a major impact on the design, described via failure mechanisms of reinforcement rupture, direct sliding and pull-out resistance. Calculation procedures of EN 1997-1 could no longer be applied, because reinforced soil is not included in the standard. Among the foreign standards, BS 8006-1:2010 is rather conservative and its Tieback Wedge method was therefore chosen for calculations.

Serviceability limit state related matters such as horizontal and shear deformations or intrinsic settlement of infill are as important as ultimate limit states, but seldom computable accurately. That applies twice as much in reinforced soil. Still, serviceability limit state needed to be dealt with. In a first place, limit geogrid extension of 0,5 % has been adopted in Tieback Wedge method, which gave a rough estimate of deformations. This deformation however was generally not likely to occur, due to fluctuating shear interaction along the reinforcement, shear resistance in a base plane and also with respect to a fact that concrete panels are a rigid facing. The panels were likely to lean out, as already monitored few years back at bridge abutments near Jánovce, Slovakia.

Therefore, after the deformation analyses had been performed, calculation results have been corrected empirically and the panels were to be erected with an inclination of 80:1 – 40:1, depending on a height of the panel and its position in the structure.

5. Technical solution

The structure was designed as a step-like MSEW with a maximum height of 12 m, divided into 3 levels. Rigid facing of each level consisted of full-height concrete panels of width of 1,5 m and various height, with a height of 4,2 m as a general upper limit, hence the stepping.

Facing prevents a loss of soil from the soil body and gives the structure its appearance. From a static point of view however, facing is of no importance, except for connection strength.

Essence of stability, especially the internal one, lies in the reinforcement, or rather reinforcement and it's interaction with the soil. Tensile strength, embedded length and vertical spacing of reinforcement are the main driving parameters obtained from the static calculations. All of the reinforcing geogrids were of a uniform length of 6 m, but varied in both ultimate tensile strength and vertical spacing, the deeper the structure went the stronger and denser the reinforcement needed to be, see table 1.

Table 1. Parameters of reinforcement

MSEW level	Embedded length (m)	Short-term strength (kN/m)	Design strength (kN/m)	Connection strength (%)	Vertical spacing (m)
Lower	6	110	60.5	100	300
Middle	6	80	36.2	100	600
Upper	6	55	23.2	100	600

Geological conditions, hydrological conditions and height of the MSEW varied quite significantly across the 700 m long section, also the structure was founded in various depths. All of the mentioned has a major impact on deformations, both vertical and horizontal, and since conditions vary along the MSEW, deformations were expected to evolve accordingly.

Structure was therefore divided into a longitudinal sections, just as it was divided into height levels. Between each of the 15 m long sections (10 panels) a gabion insertion of 0,5 m width took place as to mitigate longitudinal unevenness. Again, the solution was experience based, bridge abutments near Jánovce, Slovakia were dealt with the same way.

Gabions jutted 130 mm out of the face, due to architectural reasons, and were connected to the panels via steel bars with thermoplastic coating, see figure 6 and figure 7.

Gabion retaining structures and reinforced soil slopes were also designed at the ends of the section, where shape of the surrounding terrain turned to simple slopes or to bridge abutments respectively.

Figure 6. Gabion insertion

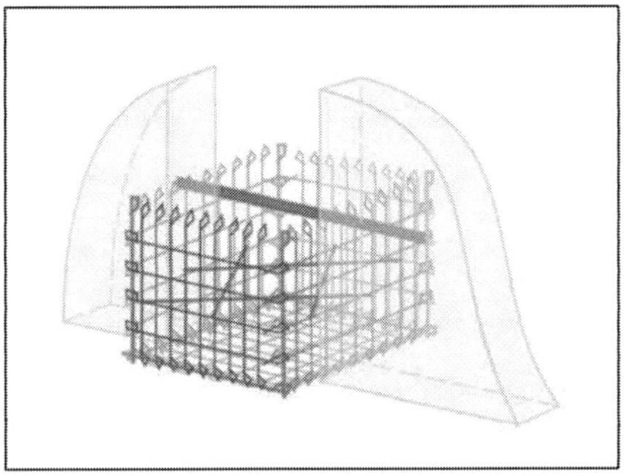

Figure 7. Gabion insertion (detailed)

6. Construction

6.1. Temporary stabilisation of earthen cut

Prior to construction of the MSEW, earthen cut slopes were to be stabilized temporarily, thus enabling a space for the structure itself. Construction period was assumed to last 3 years and there was no room for sloping due to land take boundaries, stabilization was achieved by a soil nailing technique. Soil nails were up to 10 m long in a spacing of 1,25 x 1,25 m with a surface protection of shotcrete and two-layer steel mesh reinforcement.

Figure 8. Soil nailing

Figure 9. Soil nailing (detailed)

6.2. Mechanically stabilised earth walls

Every level of MSEW was founded on a strip footing over a granular subbase layer. The footing of reinforced concrete C 25/30 XC3, XF3 and dimensions of 2,5 x 0,15 m helped to mitigate longitudinal settlement and made it possible not only to erect the panels but also to keep them braced during subsequent operations.

Prior to erecting the panels a series of steel mandrels have been anchored into the footing, in front of the toe of the MSEW, behind them a wooden prisms were laid as to set out a precise position of the panels. Lifting, manipulating and erecting the panels required a crane work (see figure 10, next page).

Bracing system was as unique and original as the retaining system itself and actually eliminated many complicated structural details and constructional obstructions. Series of volutes were built in both footing and panels, making it possible to fix bracing system elements via bolts and also to connect the elements one to another, thus creating a lattice truss structure.

There has been a simple rectification put between the footing and the bracing system, so setting an arbitrary inclination of the panels was a matter of fastening a hex nut. Mobilization of tensile forces in the reinforcement was not disrupted thanks to flexible joint at the upper end of bracing. Such a structure performed as self-bearing and could be set with a milimetric accuracy (see figure 11, next page).

Special rubber plugs were inserted in between joints of both panel-strip contact and panel-panel contact, preventing the panels from crushing each other (see figure 12, next page).

During the construction process a bracing of the panels had been assured and once reinforced soil has been finished, bracing no longer hindered the needed deformations.

Connecting the geogrids to facing panels was a next task to solve. Making an end-bearing connection using HDPE monolithic geogrids is usually achieved by cast-in tails connected to the length of embedment via bar type connector. However, such a connection is not possible with PET woven geogrids that were used here and an alternative way needed to be applied. To do so a series of bent steel rods were partially cast in the panels, thus creating eyelets on back side of the panels. Through the eyelets a steel bars were pulled, with geogrids wrapped around them and re-embedded back to infill zone, making an end-bearing connection. Both steel rods and steel bars had a thermoplastic coating as to prevent rust development.

Figure 10. Setting panels into position

Figure 11. Lattice truss with rectification

Horizontal deformations needed to be as low as possible and though reinforcement extension affects the deformations significantly, so does the infill material. Should round grained aggregate be used, large deformations were likely to occur, therefore a crushed stone mixture 0/63 (sharp edged grains) has been prescribed.

Drainage at the back of panels was achieved by a crushed stone 16/32 (figure 13), water then drained away to subsurface drain system of the railway. Back side of the panels has been provided a bituminous water protection layer and dilate joints were then covered with a strips of nonwoven geotextile. At the very top of each MSEW a concrete channel drain was designed to collect and drain surface waters.

Closing of the MSEW in longitudinal direction was achieved by different ways depending on the surrounding terrain. At the end of Rokycany direction the earthen cut changed from MSEW structure to simple slopes, here a gabion retaining wall was designed as to mitigate the jump-like change. In Pilsen direction, the MSEW merged to bridge abutments, indirectly through a reinforced soil slope (RSS).

Figure 12. Rubber plugs

Figure 13. Drainage, eyelets, coating

7. Summary

In 2013, a vast modernization of III. interstate railway corridor Prague – Pilsen begun with a goal to reduce a travel time to one hour maximum. In a railway section of Rokycany – Plzeň a relocation Homolka was designed, saving approximately 9 minutes. Relocated line routing went through a 700 m long and up to 14 m deep earthen cut which needed to be stabilized permanently. Maintenance-free, fast-to-build and land-take-thrifty technical solution was a priority and after evaluating many conventional structures mechanically stabilized earth walls were chosen.

Due to difficult and variable geological conditions and also due to construction process the structure was divided into 3 height levels resulting in its step-like shape, in a longitudinal direction 15 m long sections were created, separated by gabion columns.

Original retaining system was being evolved prior to construction, dealing with many complicated structural details and constructional obstructions in advance, making possible the desired construction speed.

Construction process took over 3 years with a satisfactory result. Today, when travelling by train from Rokycany to Pilsen, a passenger can take a look on magnificent retaining system made of reinforced soil, one of the biggest ever on a railway corridor in Czech Republic.

Figure 14. MSEW on the railway corridor Rokycany – Plzeň

References

[1] Kašpar M and Vodáček O 2017 *Sborník Příspěvků Konference se Zahraniční Účastí Zakládání Staveb (Brno)* vol 45 (Praha: ČGtS ČSSI) pp 91–98
[2] SUDOP PRAHA 2007 *Modernizace Trati Rokycany - Plzeň* (Praha)
[3] KOLEJCONSULT & servis 2014 *Modernizace Trati Rokycany - Plzeň* (Brno)
[4] British Board of Agrément 2007 *Road and Bridges Agrément Certificate No 07/R143* (Watford)

Effect of mechanical damage under repeated loading and abrasion on the tensile behaviour of a reinforcement geocomposite

J R Carneiro[1], F Carvalho[1], F Almeida[1], M de Lurdes Lopes[1]

[1]Construct-Geo, Faculty of Engineering, University of Porto, Rua Dr. Roberto Frias, 4200-465 Porto, Portugal

E-mail: rcarneir@fe.up.pt

Abstract. The installation process and the occurrence of abrasion may cause unwanted changes on the properties of the geosynthetics. When identified as relevant degradation agents for a given application, their effect has to be properly taken into account during the design phase in order to guarantee that the geosynthetics will perform correctly their functions over time. In this work, a reinforcement geocomposite (formed by a nonwoven polypropylene geotextile reinforced with polyethylene terephthalate filaments) was exposed to the isolated and combined effects of two degradation tests: mechanical damage under repeated loading and abrasion (the geocomposite was tested on both sides, which were structurally different). Damage assessment was performed by visual inspection and tensile tests. Based on the changes occurred in tensile strength, reduction factors were determined. The degradation tests provoked extensive damage on the geocomposite, having a negative impact on its tensile behaviour. Contrary to mechanical damage under repeated loading, the effect of abrasion on the geocomposite was influenced by the side that was tested. Finally, some differences were found between the reduction factors determined by the traditional method (multiplication of reduction factors obtained in isolation for each agent) for the combined effect of mechanical damage under repeated loading and abrasion and those resulting from the successive exposure to both degradation agents.

1. Introduction

It is unquestionable that the higher the level of knowledge about the construction materials, the higher is the reliability on the correct behaviour of the structures in which they are applied. Therefore, it is important to characterize the performance of many materials available in the market in order to avoid committing errors at the design phase. Regarding geosynthetics, this task comprises, among other things, carrying out investigations for improving the knowledge about their response when exposed to the action of different degradation agents, which may cause undesirable changes in their properties. Examples of degradation agents of geosynthetics include: installation damage, creep, abrasion, chemical substances like acids or alkalis, high temperatures, atmospheric oxygen, and weathering agents [1]. These agents may have a negative impact on the geosynthetics, consequently, affecting the stability and performance of the engineering structures in which these materials are used.

The installation on-site of geosynthetics involves carrying out different operations that may affect the properties of these construction materials. The placement and handling of the geosynthetics, as well as the placement and compaction of filling materials over them, in which heavy vehicles are often used, are examples of procedures that may induce damage to the geosynthetics. Indeed, in some applications, the geosynthetics may be submitted to stresses during the installation process higher than those expected during service life [2]. The installation procedures may also induce abrasion on the geosynthetics, since frictional forces can be mobilized in the interface between these materials and the filling materials. The possible occurrence of abrasion is not restricted to the installation process. In some applications, and due to the occurrence of cyclic loads (e.g. traffic of vehicles in roadways and railways infrastructures), abrasion may also occur over time.

The installation damage of geosynthetics can be evaluated by field tests or by laboratory tests (which try to simulate the damaging actions occurring under real conditions). The field tests are expensive, time consuming and require heavy equipment, being their use quite limited. For inducing mechanical damage under repeated loading on geosynthetics, a laboratory method (EN ISO 10722 [3]) was developed by the European Committee for Standardization. Many authors have employed this method to estimate the installation damage of geosynthetics [4-6], while others researched correlations between the method and the damage that occurs under real conditions [7,8]. For evaluating the resistance of the geosynthetics against abrasion, a method (EN ISO 13427 [9]) was also developed by the European Committee for Standardization.

When considering the use of geosynthetics to perform a specific function, it is important to know the effect of the degradation agents on their properties. To obtain the design values, partial reduction factors are used to account for the effect of each degradation agent. If it is necessary to consider the effect of two, or more, degradation agents, the partial reduction factors are usually multiplied (method often used at the design phase). For reinforcement applications, the design value of the tensile strength (T_D) of the geosynthetics can be given by [10,11]:

$$T_D = \frac{T}{RF_{CR} RF_{ID} RF_W RF_{CH} f_s} \tag{1}$$

where T is the tensile strength, RF_{CR}, RF_{ID}, RF_W and RF_{CH} are, respectively, the partial reduction factors accounting for the isolated effects of creep, mechanical damage, weathering, and chemical and biological agents, and f_s is a factor of safety (to allow for extrapolation uncertainly). The multiplication of partial reduction factors (each accounting for the isolated effect of an agent) may not always represent correctly the combined effect of the agents, leading to unreliable designs (which may result in a premature failure of the engineering structures). Indeed, previous works shown that the reduction factors obtained by this method for the combined action of the degradation agents may be inaccurate (by underestimating) when interactions occur between those agents [12-15]. Interactions have been found between: (1) damage during installation and creep [16,17], (2) mechanical damage under repeated loading and abrasion [12,13,15], and (3) between chemical agents [14,18].

In this work, a reinforcement geocomposite was submitted to the isolated and combined actions of mechanical damage under repeated loading and abrasion (the material was tested on both sides, which were structurally different). Damage assessment was carried out by visual inspection and by monitoring changes in the tensile behaviour of the geocomposite. Based on the changes occurred in tensile strength, reduction factors were determined. The reduction factors obtained by the traditional method for the combined effect of mechanical damage under repeated loading and abrasion (multiplication of reduction factors accounting for the isolated effect of each agent) were compared with those directly found in the successive exposure to both degradation agents.

2. Materials and methods

2.1. Geocomposite

The experimental campaign developed in this work was carried out with a geocomposite formed by a nonwoven geotextile made from polypropylene fibres, reinforced with polyethylene terephthalate (PET) filaments in both the machine and cross-machine directions of production. The filaments were attached to one side of the geotextile by seams (side with filaments). The other side of the geocomposite consisted of the geotextile (side without filaments) (the structure of the geocomposite is presented in section 3).

The geocomposite had, according to the manufacturer, a tensile strength of 42 kN.m^{-1} (EN ISO 10319 [19]) in both directions. Its mass per unit area and thickness were, respectively, 351 (±16) g.m^{-2} and 2.26 (±0.09) mm. These properties, which are presented with 95% confidence intervals determined according to Montgomery and Runger [20], were characterised in accordance with EN ISO 9864 [21] and EN ISO 9863-1 [22], respectively.

The geocomposite studied in this work can be used for performing the function of reinforcement in many engineering applications, including transportation infrastructures such as roadways or railways (in which abrasion might occur). In these infrastructures, the geocomposite can be applied, for example, in subgrade stabilisation, railtrack reinforcement, basal reinforcement or to steepen soil slopes. In addition to reinforcement, the material may also perform functions of separation, filtration or drainage.

2.2. Degradation tests

Two types of degradation tests were conducted: mechanical damage under repeated loading and abrasion tests (description of the degradation tests in the next two subsections). The geocomposite was exposed in isolation to each test (single exposures to mechanical damage under repeated loading and abrasion), and successively submitted to both tests (successive exposure). The degradation tests were carried out on both sides of the geocomposite (which were structurally different): the side with filaments and the side without filaments.

2.2.1. Mechanical damage under repeated loading. A prototype equipment developed at the Faculty of Engineering of the University of Porto, which is thoroughly described in Lopes and Lopes [23], was used to carry out mechanical damage under repeated loading tests (hereinafter MD tests) in accordance with EN ISO 10722 [3]. The first step of the MD tests was the placement of a sublayer of *corundum* (synthetic aggregate with a particle size distribution in the range of 5-10 mm) with a height of 37.5 mm in the lower box of the equipment, followed by compaction. Then, an additional sublayer of *corundum* of the same height was included in that box and also submitted to compaction (the compaction of both sublayers of aggregate was carried out by applying a pressure of (200 ± 2) kPa for 60 s over the area of the lower box). At the end of these procedures, the geocomposite (specimen with a length of 500 mm and a width of 250 mm) was placed over the compacted layer of *corundum* (formed by the two sublayers of 37.5 mm), and an upper box was installed and filled with a loose layer of the aggregate with a height of 75 mm (both the lower and upper boxes had square bases with a side of 300 mm and a height of 87.5 mm). The geocomposite (installed between the layers of *corundum*) was subjected to dynamic loading between (5.0 ± 0.5) kPa and (500 ± 10) kPa at a frequency of 1 Hz, for 200 cycles. Finished the loading, the test was stopped and the geocomposite carefully recovered in order to avoid additional damage. In total, 5 specimens were tested for each degradation condition.

2.2.2. Abrasion. The abrasion tests were performed in another prototype equipment developed at the Faculty of Engineering of the University of Porto. These tests, which were conducted according to EN ISO 13427 [9], consisted of submitting the geocomposite (attached to a stationary platform) to the action of a P100 abrasive sheet, which was fixed on a sliding platform. In order to promote the rubbing of the geocomposite, it was triggered a back-and-forth linear motion of the sliding platform, under a controlled pressure of 6 kPa, for 750 cycles (each cycle consisted of a double passage of the abrasive through the geocomposite). For each degradation condition, 5 specimens were tested.

2.3. Damage evaluation

The assessment of the damage occurred on the geocomposite during the degradation tests was carried out through the following approaches: qualitatively, by visual inspection, and quantitatively, through tensile tests. The results obtained for undamaged samples, which were also tested, were used as reference for comparison purposes.

The tensile tests were conducted in conformity with EN ISO 10319 [19] on a LR 50K testing machine from *Lloyd Instruments*. The geocomposite (specimens 100 mm long (between grips) and 200 mm wide) was tested in the machine direction of production under a displacement rate of 20 mm.min^{-1}. Tensile strength (T, in kN.m^{-1}) and elongation at maximum load (E_{ML}, in %) were the properties determined in these tests. The latter was assessed using a video-extensometer, which monitored the distance between two reference points separated from each other by 60 mm (30 mm above and 30 mm below the axis of symmetry of the specimen). The average values of the tensile properties were determined based on the results of 5 tested specimens and are presented with 95% confidence intervals according to Montgomery and Runger [20]. It was also calculated the residual tensile strength ($T_{Residual}$, in %), in order to represent by percentage the variations occurred in tensile strength after the degradation tests. This parameter was obtained by the quotient between the tensile strengths of the damaged and undamaged samples of the geocomposite.

2.4. *Reduction factors*

Based on the changes occurred in the tensile strength of the geocomposite (during the degradation tests), reduction factors (RF) for the isolated and combined effects of MD and abrasion were determined as:

$$\mathrm{RF}=\frac{T_{\mathrm{Undamaged}}}{T_{\mathrm{Damaged}}} \tag{2}$$

where $T_{Undamaged}$ and $T_{Damaged}$ correspond, respectively, to the tensile strength of the geocomposite before and after the degradation tests.

According to the traditional method, the reduction factor for the combined effect of two, or more, degradation agents can be obtained by multiplying the reduction factors obtained in isolation for each agent. When considering the combined effect of MD and abrasion, the reduction factor according to the traditional method ($RF_{MD+ABR\ TRAD}$) can be found using the following expression:

$$\mathrm{RF}_{\mathrm{MD+ABR\ TRAD}}=\mathrm{RF}_{\mathrm{MD}}\mathrm{RF}_{\mathrm{ABR}} \tag{3}$$

where RF_{MD} and RF_{ABR} represent, respectively, the reduction factors for the isolated actions of MD and abrasion.

It is important to stress that the reduction factors presented in this work were obtained in particular degradation conditions, which may not represent the *in situ* conditions of the engineering applications, and therefore should not be considered for design purposes. Their determination was performed in order to evaluate possible differences between the reduction factors obtained by the traditional method for the combined effect of MD and abrasion and their counterparts directly found in the successive exposure to both degradation agents.

3. Results and discussion

3.1. *Visual inspection*

3.1.1. Side with filaments. The degradation tests caused different types of damage on the geocomposite. From what it was possible to observe, the single exposure to MD was the least damaging degradation condition. Still, both the longitudinal and the cross-machine direction filaments, as well as the nonwoven geotextile, suffered some degradation (figure 1b). The original arrangement of the filaments suffered some changes (the distance between filaments and their linearity was affected) and some cuts occurred in the filaments (without completely breaking them). By carrying out a close observation, it was also possible to detect the existence of cuts in fibres, punctures and small holes in the nonwoven geotextile (defects not easily observed at the magnification of figure 1b).

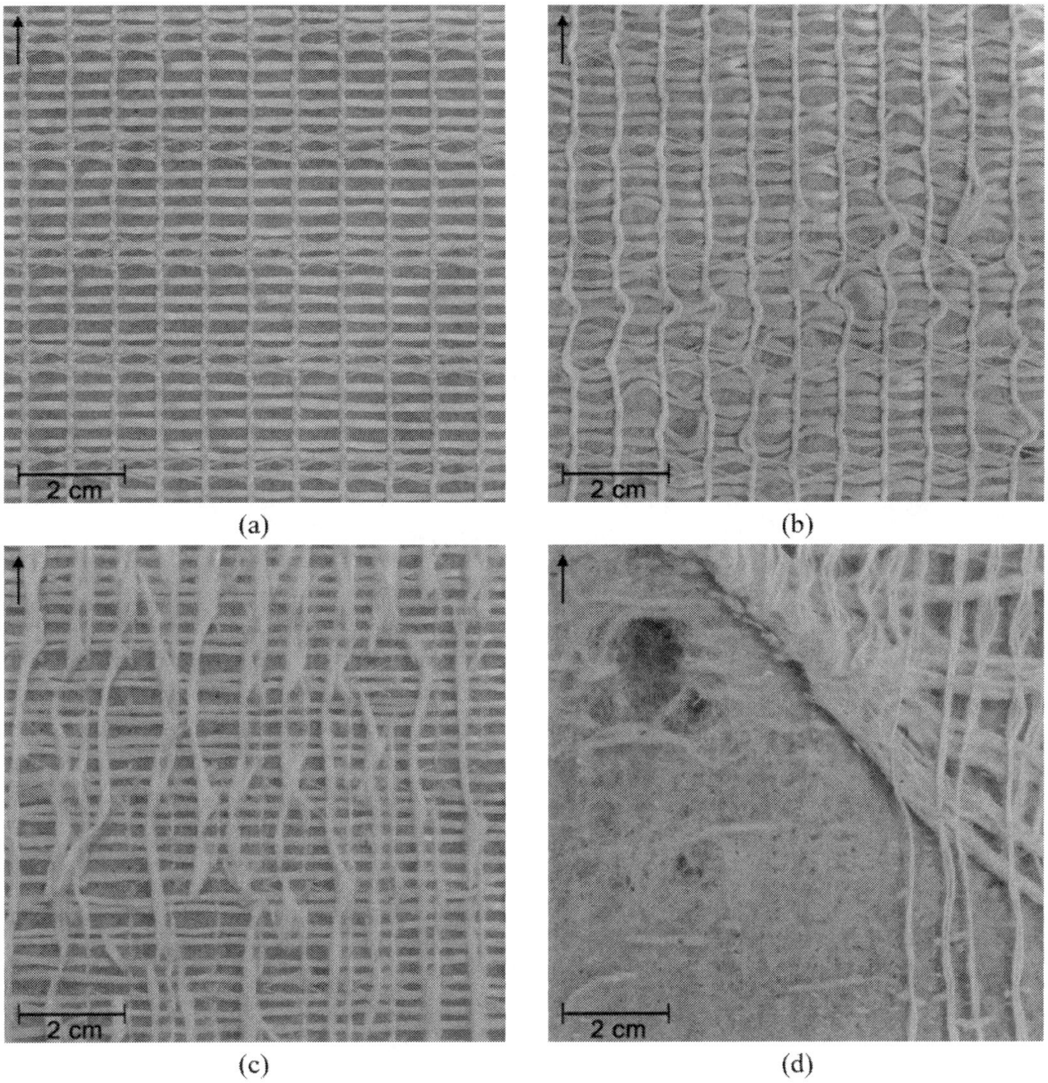

Figure 1. Visual analysis of the geocomposite tested on the side with filaments: (a) undamaged sample; (b) single exposure to MD; (c) single exposure to abrasion; (d) successive exposure to MD and abrasion (the arrow indicates the machine direction of production)

The abrasion tests provoked the detachment and break of the longitudinal filaments (figure 1c), being the cross-machine direction filaments much less affected (since they are placed below the longitudinal filaments, they were not in direct contact with the abrasive). The action of the abrasive: broke the seams attaching the filaments to the nonwoven geotextile, led to a reduction in thickness of the longitudinal filaments, and promoted the formation of clusters of damaged longitudinal filaments above the cross-machine direction filaments. Regarding the nonwoven geotextile, which was not in direct contact with the abrasive, no relevant damage was observed.

The successive exposure to MD and abrasion left the geocomposite highly damaged (figure 1d). The longitudinal and cross-machine direction filaments were cut and detached from the nonwoven geotextile (all seams were cut), leading to the formation of large clusters of filaments (the original structure of the filaments was totally destroyed). In addition, cuts and detachment of fibres and formation of holes with significant diameter were observed in the nonwoven geotextile. The successive exposure to both degradation tests was, unquestionably, the worst scenario to the geocomposite.

3.1.2. Side without filaments. Excepting the single exposure to MD, the side in which the geocomposite was tested had a determinant impact on the degradation occurred. During the MD tests, in addition to cuts in fibres, the particles of *corundum* (with a rough and angular structure) provoked punctures and a large number of small holes on the nonwoven geotextile (defects not easily seen at the magnification of figure 2b). The seams that guaranteed the attachment of the filaments to the nonwoven geotextile have emerged practically unscathed. It is also worth mentioning that very small particles of *corundum*, which resulted from the splintering of particles of higher dimension, were detected imprisoned in the nonwoven structure. Regardless of the side in which the geocomposite was tested, the defects caused by the MD tests were similar (this section described only the damage found in the side without filaments, while the damage occurred in the opposite side was analysed in the previous section). This can be explained by the fact that the geocomposite was placed between two layers of *corundum* (both layers induced damage) independently of the side facing up during the MD tests (the MD tests imposed a cyclic vertical load to the geocomposite, being the side facing up defined as the side faced to the loading mechanism).

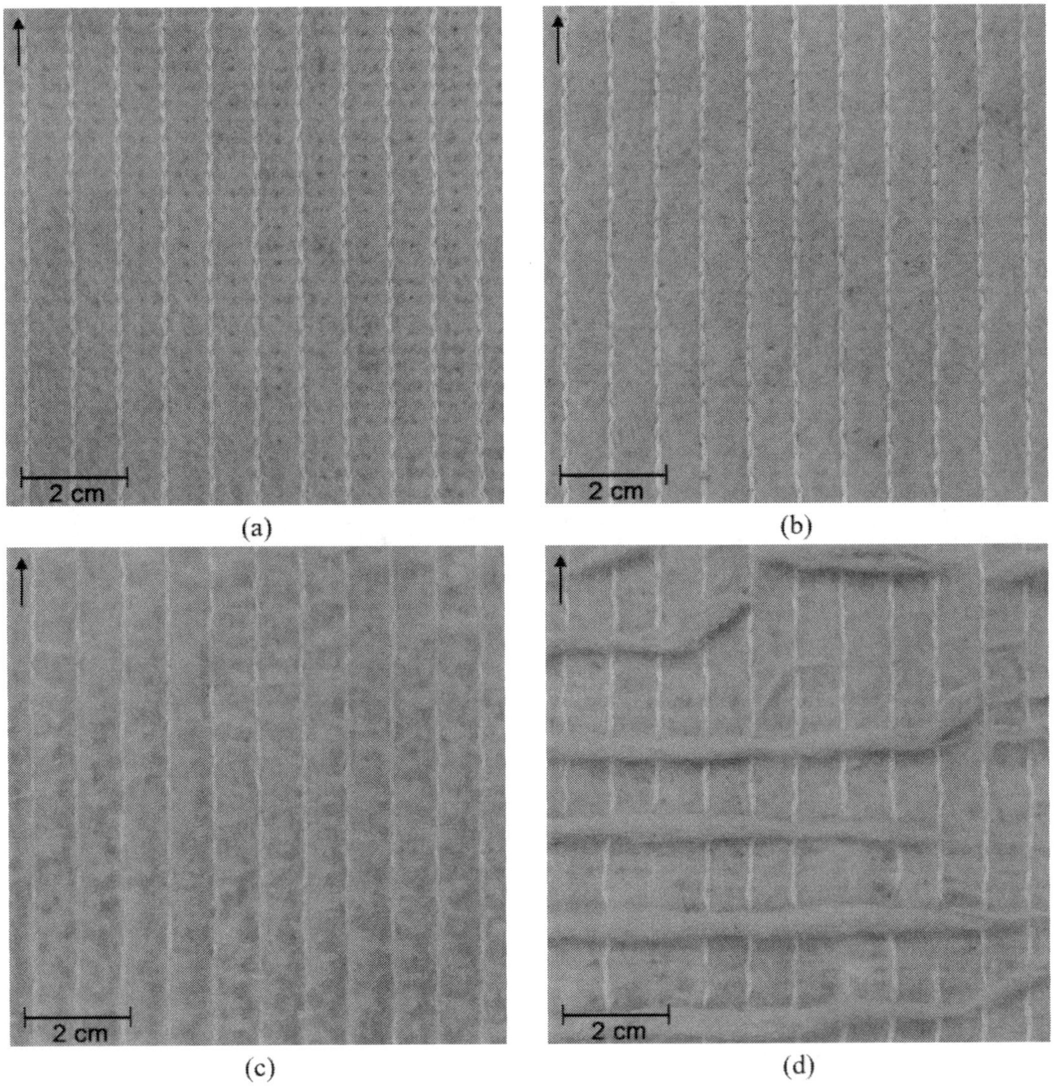

Figure 2. Visual analysis of the geocomposite tested on the side without filaments: (a) undamaged sample; (b) single exposure to MD; (c) single exposure to abrasion; (d) successive exposure to MD and abrasion (the arrow indicates the machine direction of production)

The abrasion tests caused the detachment and cut of fibres, resulting in the formation of small clusters of damaged fibres on the surface of the nonwoven geotextile (figure 2c). Most seams had the ability to withstand the action of the abrasive (only a small number of seams suffered damage). These defects, which were not highly pronounced, cannot be easily observed at the magnification of figure 2c, but a notorious change in texture can be seen compared to the undamaged sample (figure 2a). Regarding the filaments (which were on the other side), no relevant damage was found. This was an expected outcome since there was no contact between the filaments and the abrasive (the nonwoven geotextile, which was the element of the geocomposite directly exposed to the abrasive, protected the filaments from degradation). Indeed, and contrary to the MD tests, the damaging actions in the abrasion tests are mostly induced on one side of the specimens.

The successive exposure to both degradation tests had the most relevant impact on the geocomposite, leading to the formation of small holes and the detachment and cuts of fibres on the nonwoven geotextile (figure 2d). The fibres that were detached formed clusters significantly larger than those observed after the single exposure to abrasion. In addition, some seams were broken, resulting in potential weak points on the arrangement of the filaments. The defects found in the side with filaments were identical to those observed after the single exposure to MD (it is worth remembering that, contrary to abrasion, the MD tests are able to induce damage simultaneously on both sides of the geocomposite).

The different defects found in the geocomposite readily indicated the occurrence of relevant changes in its tensile behaviour, highly affecting its reinforcement function. To quantify those changes, tensile tests were performed, being the results presented in the following section.

3.2. Tensile behaviour

3.2.1. Undamaged samples. The damage suffered by the geocomposite during the degradation tests was quantitatively characterised by comparing the tensile behaviour of damaged and undamaged samples. A typical tensile force-elongation curve of an undamaged specimen can be seen in figure 3.

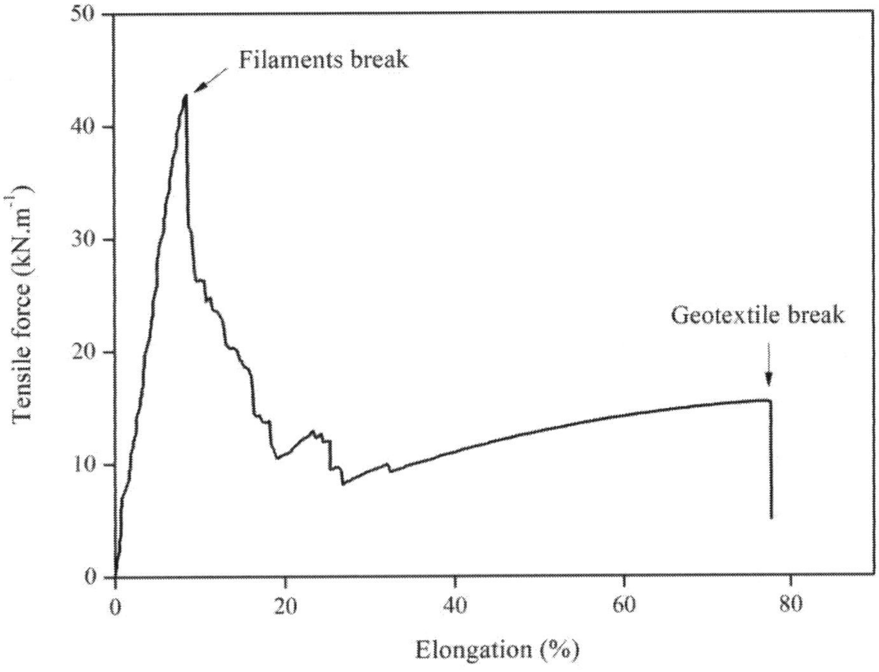

Figure 3. Tensile force-elongation curve of an undamaged specimen of the geocomposite

The tensile force-elongation curve shown in figure 3 reveals that the geocomposite undergoes two different peaks under loading. The first peak (maximum load of the geocomposite) corresponds to the break of the PET filaments (with low deformability) and occurs at a relatively low elongation (≈ 10%). Following the failure of the filaments, the nonwoven polypropylene geotextile (with high deformability and low mechanical strength compared to the filaments) is mobilized. The second load peak, which is observed at high elongations (≈ 70-80%) corresponds to the failure of the nonwoven geotextile. For the undamaged sample, the tensile strength and the elongation at maximum load of the geocomposite were of, respectively, 42.42 (±2.46) $kN.m^{-1}$ and 10.4 (±0.5) % (tensile properties determined in the machine direction of production; 95% confidence intervals shown in brackets).

3.2.2. Side with filaments. The degradation tests caused considerable changes in the tensile behaviour of the geocomposite (when tested on the side with filaments) (table 1). Despite seeming to be much less aggressive than abrasion, the single exposure to MD had a significant impact on the tensile strength of the geocomposite (loss of 45.0%). This reduction in resistance can be associated with the rough nature and angular shape of the particles of *corundum*, which had the capacity to induce cuts on the filaments of the geocomposite, affecting its tensile strength. Despite the lower tensile strength, and considering the 95% confidence intervals, no relevant modifications were found in elongation at maximum load.

Table 1. Tensile properties of the geocomposite tested on the side with filaments.

Degradation test	T ($kN.m^{-1}$)[a]	E_{ML} (%)[a]	$T_{Residual}$ (%)	RF
MD	23.31 (±2.28)	9.2 (±2.2)	55.0	1.82
Abrasion	15.45 (±4.68)	90.3 (±15.1)	36.4	2.75
MD + abrasion	7.23 (±1.85)	28.2 (±14.2)	17.0	5.87

[a] 95% confidence intervals in brackets.

The tensile behaviour of the geocomposite was more affected by the single exposure to abrasion than to MD, which is consistent with the defects visually detected. Indeed, a reduction of 63.6% occurred in tensile strength, being the maximum load achieved at a high elongation (90.3%). This high elongation corresponded to the failure of the nonwoven geotextile, which was less damaged than the filaments and was now the most resistant element of the geocomposite. It is important to refer that the maximum load observed at the break of the filaments (which are responsible for the geocomposite being able to perform the reinforcement function) was 7.53 $kN.m^{-1}$, corresponding to 17.8% of the original tensile strength of the geocomposite. This low value is in agreement with the extensive damage presented by the filaments (figure 1c).

The successive exposure to MD and abrasion provoked the highest reduction in tensile strength (loss of 83.0%), which was not a surprise considering the aspect of the geocomposite noticed during the visual inspection (figure 1d). Indeed, the non-existence of a structure of filaments, as well as the holes found in the nonwoven geotextile, were indicators of such outcome. The maximum force was observed at an elongation of 28.2%, corresponding to the failure of the nonwoven geotextile (which was considerably damaged). The highly damaged filaments had practically no contribution for the tensile behaviour of the geocomposite.

3.2.3. Side without filaments. When tested on the side without filaments, the geocomposite also suffered relevant changes in its tensile properties (table 2). However, some relevant differences compared to the tests performed on the side with filaments were noticed.

The results obtained after the single exposure to MD were relatively similar, regardless of the side in which the geocomposite was tested. Indeed, losses in tensile strength of 45.0% and 41.9% were observed when the geocomposite was, respectively, tested on the sides with filaments and without filaments. The corresponding elongations at maximum load were also very close (respectively, 9.2% and 9.3%). Taking into account the defects visually detected in the geocomposite, this result was somewhat expected. As mentioned before, both layers of *corundum* induced damage to the geocomposite during the MD tests, independently of the side facing up during those tests.

Table 2. Tensile properties of the geocomposite tested on the side without filaments

Degradation test	T (kN.m^{-1})[a]	E_{ML} (%)[a]	$T_{Residual}$ (%)	RF
MD	24.65 (±2.87)	9.3 (±1.0)	58.1	1.72
Abrasion	36.94 (±1.19)	11.1 (±2.7)	87.1	1.15
MD + abrasion	23.31 (±2.86)	9.9 (±2.4)	55.0	1.82

[a] 95% confidence intervals in brackets.

Contrary to what happened after the single exposure to MD, the results obtained after abrasion were highly dependent on which side the geocomposite was tested (an expected outcome taking into account the defects found during the visual analysis). The loss in tensile strength was only 12.9%, a significantly lower reduction compared to the case in which the side with filaments was directly exposed to the action of the abrasive (reduction of 63.6%). In addition, the elongation at maximum load was 11.1% (having no relevant changes compared to the undamaged sample), in contrast with the higher elongation found when the geocomposite was tested on the side with filaments (90.3%). This lower deterioration of the tensile behaviour of the geocomposite can be ascribed to the fact that the filaments (elements with higher resistance) have not been directly exposed to the abrasive (suffering only minor damage as indicated by the relatively small loss in tensile strength), being protected by the nonwoven geotextile.

The impact in the tensile behaviour of the geocomposite induced by the successive exposure to MD and abrasion was not much different from the single exposure to MD. Indeed, the tensile strength and elongation at maximum load of the geocomposite were quite similar after those tests. This shows that the exposure to abrasion (carried out after the exposure to MD) had no relevant impact in the tensile behaviour of the geocomposite. Despite the defects found in the nonwoven geotextile (figure 2d), the filaments had no additional damage compared to the single exposure to MD (as visually detected and corroborated by the tensile tests). Similar to what happened in the single exposures to abrasion, the exposed side had a relevant influence on the degradation suffered by the geocomposite. In fact, the loss occurred in its tensile strength was of 45.0% when exposed on the side without filaments, contrasting with the loss of 83.0% when exposed on the side with filaments. In addition, elongation at maximum load had no considerable differences compared to the undamaged sample when the tests were carried out on the side without filaments (increase from 10.4% to 28.2% after the tests performed on the side with filaments). The lower degradation when exposed on the side without filaments can be explained, once again, by the inexistence of contact between the filaments and the abrasive during the exposure to abrasion (the nonwoven geotextile was the element exposed to the abrasive and, consequently, the one which suffered more damage).

3.3. Reduction factors

The results presented in the previous section shown that, in some cases, the tensile behaviour of the geocomposite was different depending on the side (with or without filaments) that was directly exposed to the degradation agents. With the results obtained for tensile strength, reduction factors were calculated for the isolated and combined effects of MD and abrasion (tables 1 and 2). The reduction factors obtained directly from the successive exposure to MD and abrasion were compared to those determined by the traditional method (multiplication of the reduction factors obtained for the isolated effects of MD and abrasion) for the combined effect of those agents (figure 4).

When tested on the side with filaments, the reduction factor calculated through the traditional method for the combined effect of MD and abrasion was 14.7% lower than its counterpart directly found in the successive exposure to those agents. In this case, the traditional method was unable to account properly (by underestimating) the combined effect of the degradation agents. By contrast, a similar comparison when the geocomposite was tested on the side without filaments shown that the traditional method was slightly overestimating the combined effect of MD and abrasion. Indeed, the reduction factor determined by the traditional method was 8.8% higher than that found in the successive exposure to the degradation agents, setting a conservative approach.

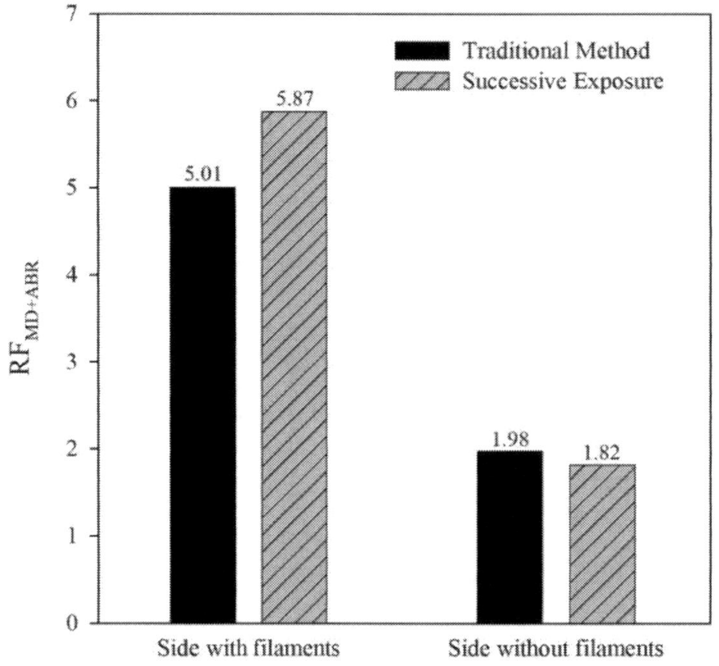

Figure 4. Comparison of the RF_{MD+ABR} obtained by the traditional method and from the successive exposure to mechanical damage under repeated loading and abrasion

The use of inaccurate reduction factors might result in an incorrect design. If conservative approaches are adopted, there are no strong reasons to question the normal behaviour of the structures in which the geosynthetics are installed. The same cannot prevail if the effect of the degradation agents is not properly accounted for, by underestimating. The reduction factors must represent correctly the combined effect of the degradation agents, taking into account the interactions (synergisms) that might occur between them.

4. Conclusions

The single and successive exposures to MD and abrasion induced considerable changes on the tensile behaviour of the geocomposite, compromising its ability to perform the reinforcement function. When tested on the side with filaments, abrasion was more damaging (higher impact in tensile behaviour) than MD. By contrast, the opposite occurred when the geocomposite was tested on the side without filaments. However, independently of on which side was tested, the MD tests induced similar changes on the tensile behaviour of the geocomposite. Therefore, the previous differences were due to the degradation caused by the abrasion tests, which was higher, or lower, depending on which side the geocomposite was tested. This can be ascribed to the characteristics of the MD and abrasion tests. Indeed, the MD tests can induce degradation simultaneously on both sides of the geocomposite (which was installed between two layers of *corundum*), while the abrasion tests can mostly induce degradation on the side directly exposed to the abrasive. This way, the PET filaments (most resistant elements of the geocomposite) were much more damaged when exposed to the abrasive than when protected by the nonwoven geotextile. It is worth referring that no relation has been researched between the standardized laboratory abrasion tests (EN ISO 13427 [9]) and the effects of abrasion under real degradation conditions (which might be different).

The comparison between the reduction factors determined by the traditional method for the combined effect of mechanical damage under repeated loading and abrasion with those obtained directly from the successive exposure to both degradation agents shown the existence of some differences, although not pronounced. It appears that, as found in similar works for other geosynthetics [12,13,15], the traditional method may not always be able to represent accurately (by underestimating) the combined effect of MD and abrasion, since interactions (synergisms) may occur between those agents. This was observed when the geocomposite was tested on the side with filaments, but not when tested on the side without filaments (in which the abrasion tests had a low impact on the PET filaments and, therefore, a limited influence in the tensile strength of the geocomposite). Finally, it is important to underline that the degradation tests were performed under particular conditions that might not be representative of the field conditions and, thereby, the reduction factors determined in this work should not be considered for design purposes.

Acknowledgments

The authors would like to thank TenCate Geosynthetics Iberia for providing the geocomposite. This work was financially supported by: (1) project PTDC/ECI-EGC/28862/2017 – POCI-01-0145-FEDER-028862, funded by FEDER funds through COMPETE 2020 – *Programa Operacional Competitividade e Internacionalização* (POCI) and by national funds (PIDDAC) through FCT/MCTES; (2) Base Funding – UIDB/04708/2020 of the CONSTRUCT – Instituto de I&D em Estruturas e Construções – funded by national funds through the FCT/MCTES (PIDDAC).

References

[1] Ingold TS 1994 *The Geotextiles and Geomembranes Manual* 1st ed (Oxford, United Kingdom: Elsevier Advanced Technology) p 610

[2] Shukla S K and Yin J-H 2006 *Fundamentals of Geosynthetic Engineering* 1st ed (Leiden, The Netherlands: Taylor & Francis/Balkema) p 432

[3] CEN 2007 Geosynthetics - Index test procedure for the evaluation of mechanical damage under repeated loading - Damage caused by granular material (Brussels, Belgium: European Committee for Standardization) EN ISO 10722

[4] Carneiro J R, Morais L M, Moreira S P and Lopes M L 2013 Evaluation of the Damages Occurred During the Installation of Non-Woven Geotextiles *Mater Sci Forum* **730-732** pp 439-44

[5] Huang C-C and Chiou S-L 2006 Investigation of installation damage of some geogrids using laboratory tests *Geosynth Int* **13** pp 23-35

[6] Huang C-C 2006 Laboratory simulation of installation damage of a geogrid *Geosynth Int* **13** pp 120-32

[7] Huang C-C and Wang Z-H 2007 Installation damage of geogrids: influence of load intensity *Geosynth Int* **14** pp 65-75

[8] Pinho-Lopes M and Lopes ML 2014 Tensile properties of geosynthetics after installation damage *Environ Geotech* **1** pp 161-78

[9] CEN 2014 Geosynthetics - Abrasion damage simulation (sliding block test) (Brussels, Belgium: European Committee for Standardization) EN ISO 13427

[10] ISO 2007 Guidelines for the determination of the long-term strength of geosynthetics for soil reinforcement (Geneva, Switzerland: International Organization for Standardization) ISO/TR 20432

[11] Carlos D M, Carneiro J R, Pinho-Lopes M and Lopes M L 2015 Effect of Soil Grain Size Distribution on the Mechanical Damage of Nonwoven Geotextiles Under Repeated Loading *Int J of Geosynth and Ground Eng* **1** p 9

[12] Rosete A, Lopes P M, Pinho-Lopes M and Lopes M L 2013 Tensile and hydraulic properties of geosynthetics after mechanical damage and abrasion laboratory tests *Geosynth Int* **20** pp 358-74

[13] Dias M, Carneiro J R and Lopes M L 2017 Resistance of a nonwoven geotextile against mechanical damage and abrasion *Ciência e Tecnologia dos Materiais* **29** pp 177-81

[14] Carneiro J R, Almeida P J and Lopes M L 2018 Laboratory Evaluation of Interactions in the Degradation of a Polypropylene Geotextile in Marine Environments *Adv Mater Sci Eng* **2018** p 10

[15] Almeida F, Carlos D M, Carneiro J R and Lopes M L 2019 Resistance of Geosynthetics against the Isolated and Combined Effect of Mechanical Damage under Repeated Loading and Abrasion *Mater* **12** p 15

[16] Allen T M and Bathurst R J 1994 Characterization of Geosynthetic Load-Strain Behavior After Installation Damage *Geosynth Int* **1** pp 181-99

[17] Greenwood J H 2002 The Effect of Installation Damage on the Long-Term Design Strength of a Reinforcing Geosynthetic *Geosynth Int* **9** pp 247-58

[18] Carneiro J R, Almeida P J and Lopes M L 2014 Some synergisms in the laboratory degradation of a polypropylene geotextile *Constr Build Mater* **73** pp 586-91

[19] CEN 2015 Geosynthetics - Wide-width tensile test (Brussels, Belgium: European Committee for Standardization) EN ISO 10319

[20] Montgomery DC and Runger GC 2010 *Applied Statistics and Probability for Engineers* 5th ed (New York, United States of America: John Wiley & Sons, Inc.) p 784

[21] CEN 2005 Geosynthetics - Test method for the determination of mass per unit area of geotextiles and geotextile-related products (Brussels, Belgium: European Committee for Standardization) EN ISO 9864

[22] CEN 2016 Geosynthetics - Determination of thickness at specified pressures - Part 1: Single layers (Brussels, Belgium: European Committee for Standardization) EN ISO 9863-1

[23] Lopes M P and Lopes M L 2003 Equipment to carry out laboratory damage during installation tests on geosynthetics *Geotecnia (J Port Geotechical Soc)* **98** pp 7-24

Semipermeable membrane behavior of geosynthetic clay liners

C D Shackelford[1,2] and J Scalia IV[1]

[1]Department of Civil and Environmental Engineering, Colorado State University, 1372 Campus Delivery, Fort Collins, CO 80523-1372, USA

E-mail: shackel@engr.colostate.edu

Abstract. The focus of this paper is on providing an overview of the results of the nearly 20 years of research related to the discovery, examination, and potential benefits of semipermeable membrane behavior of bentonite-based geosynthetic clay liners (GCLs) commonly used as barriers or components of barriers in a variety of hydraulic and chemical containment applications, such as water and chemical retention ponds, tailings and coal ash impoundments, and solid waste landfills. First, the concept of solute restriction underlying semipermeable membrane behavior and the development of a solute-restriction or membrane efficiency coefficient (ω) and chemico-osmosis (q_π) are described. Second, previously measured values of ω are presented for clays and GCLs for the purpose of illustrating the physical and chemical factors affecting ω. Third, the potential benefits of both ω and q_π in terms of hydraulic and chemical containment applications are illustrated, and the relative importance of ω versus q_π is described. Finally, some of the remaining issues related to the semipermeable membrane behavior of GCLs, such as the uniqueness of the measured ω and the persistence of ω at high salt concentrations, are discussed.

1. Introduction

Semipermeable membrane behavior refers to the ability of porous media to restrict the migration of aqueous miscible chemical species (solutes), while allowing unimpeded migration of the solvent, i.e., molecular water (H_2O). The restriction of solutes gives rise to the development of a chemical concentration difference, ΔC, which results in chemico-osmosis whereby the liquid phase (i.e., water) migrates from a zone of lower solute concentration to a zone of higher solute concentration. The resulting flux of the liquid phase (water) is referred to as the chemico-osmotic liquid flux, q_π.

The ability of clays to restrict salt migration and exhibit semipermeable membrane has been recognized for at least 60 years (e.g., [1]). In particular, high activity clays, such as bentonites, with a significant fraction of clay-sized particles comprising some montmorillonite (smectite) clay mineral, are known to exhibit semipermeable membrane behavior (e.g., [2]). This knowledge, together with the knowledge that solute restriction and chemico-osmosis can potentially enhance the performance of bentonite-based chemical containment barriers, such as geosynthetic clay liners (GCLs), compacted mixtures of sand or clay amended with bentonite, and soil-bentonite (SB) backfills for SB vertical cutoff walls, has stimulated extensive research over the past approximate 20 years focused on determining the existence and magnitude of semipermeable membrane behavior of these barriers [3-14].

The purpose of this paper is to explain the concepts of semipermeable membrane behavior and chemico-osmosis for porous membranes, summarize the primary findings of research conducted over the past approximate 20 years to determine the existence and magnitude of semipermeable membrane behavior in GCLs, and discuss the potential significance of the results obtained to date. In addition, some of the more significant remaining issues with respect to the semipermeable membrane behavior of GCLs that still need to be resolved are identified.

2. Semipermeable membrane behavior in clays

First, conceptual explanations for the existence of membrane behavior in uncharged, ideal (perfect) and non-ideal (leaky) membranes are provided. Second, the conceptual basis for the existence of membrane behavior in charged membranes, such as clays, is illustrated, and the mineralogical, physical, mechanical, and chemical factors affecting the existence and magnitude of membrane behavior in clays are illustrated conceptually. Finally, results from the literature are shown to provide supporting evidence for the conceptual framework.

2.1. *Ideal semipermeable membranes*

Ideal or perfect semipermeable membranes are those in which all solutes are restricted from migrating through the membrane. The behavior that occurs in the presence of ideal membranes can be illustrated with the assistance of the schematic cross-sectional scenarios of the reservoir system shown in Figure 1. As shown in Figure 1a, both the left (L) and right (R) reservoirs initially contain deionized or pure (molecular) water (H_2O), such that the water levels in both reservoirs are the same. Under this condition, the activity of the water, a_w, in both reservoirs is unity (i.e., $a_w = 1$), and since the chemical potential of water, μ_w, is a function of the activity of the water [i.e., $\mu_w = f(a_w)$], there is no difference in chemical potential of the water across the membrane (i.e., $\Delta\mu_w = 0$) and, therefore, no gradient for flow of water through the membrane. However, since the activity of water decreases with increasing solute concentration [15], instantaneously replacing the water in the left reservoir with a salt solution (e.g., $KCl \rightarrow K^+ + Cl^-$) at a known initial salt concentration, C_o, reduces the activity of the water in the left reservoir, a_{wL}, to a value less than unity (i.e., $a_{wL} < 1$), while the activity of the water in the right reservoir, a_{wR}, remains the same (i.e., $a_{wR} = 1$) [15]. The resulting difference in activity of the water establishes a chemical potential difference across the membrane (i.e., $\Delta\mu_w > 0$) from the right reservoir to the left reservoir, resulting in a chemico-osmotic water flux, q_π, in the same direction (Figure 1b).

Chemico-osmosis will continue until equilibrium in activity is achieved, at which time $q_\pi = 0$. At equilibrium, the water level in the left reservoir will have increased by a magnitude of $H_\pi/2$ relative to the initial condition, whereas the water level in the right reservoir will have decreased by the same magnitude (i.e., $H_\pi/2$). In other words, there is an increase in water pressure in the left reservoir, and a decrease in water pressure in the right reservoir, i.e., relative to the initial condition. The overall difference between the water levels in the two reservoirs at equilibrium is H_π, which represents the maximum chemico-osmotic pressure head for the system. Note that H_π typically is represented as the maximum chemico-osmotic pressure difference across the membrane, $\Delta\pi$ ($= H_\pi \cdot \gamma_w$, where γ_w = the unit weight of water, i.e., 9.81 kNm^{-3}). The magnitude of H_π or $\Delta\pi$ for a given system is a function of the type of salt (e.g., NaCl vs. KCl vs. $CaCl_2$) in the system, and the difference in the water activities or salt concentrations between the two reservoirs [15, 16].

2.2. *Non-ideal semipermeable membranes*

The situation for uncharged, non-ideal membranes is similar to that for ideal membranes, except some of the pores in the membrane are sufficiently large to allow passage of solutes, whereas other pores are solute restricting. For this reason, non-ideal membranes also are referred to as leaky membranes.

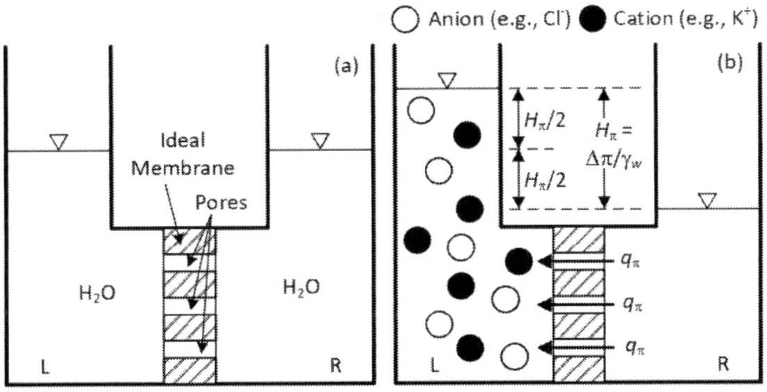

Figure 1. Schematic illustrations of the concept of semipermeable membrane behavior for uncharged, ideal (perfect) membranes: (a) initial condition where only water (H_2O) is contained in the left (L) and right (R) reservoirs bounding the membrane; (b) system at equilibrium after the water in the left reservoir has been replaced with a salt (e.g., KCl) solution.

As an example, consider the schematic cross-sectional scenarios of the reservoir system for a non-ideal membrane shown in Figure 2. The initial condition with respect to the water levels in the two reservoirs illustrated in Figure 2a is the same as that shown in Figure 1a, because both reservoirs contain pure water and, therefore, the same activity and chemical potential. However, the membrane is non-ideal in that there is a single pore that is larger than the salt ions and, therefore, not solute restricting. Thus, as shown in Figure 2b, when the water in the left reservoir is replaced instantaneously with the salt solution, a concentration gradient is established across the membrane, from the higher salt concentration in the left reservoir to the lower salt concentration in the right reservoir, such that the salt ions diffuse from the left reservoir into the right reservoir through the large pore, with the resulting diffusive mass flux of solute, J_d, through the membrane being described by Fick's first law [17]. Although water still will flow from the right reservoir to the left reservoir via chemico-osmosis, the inability of the membrane to provide complete solute restriction results in a chemico-osmotic pressure head, h_π, or chemico-osmotic pressure difference, ΔP_π (= $h_\pi \cdot \gamma_w$), that is lower than the respective values attained in the case of the ideal membrane, i.e., $h_\pi < H_\pi$ and $\Delta P_\pi < \Delta\pi$ (Figure 2b).

Note that the values of h_π and ΔP_π will represent equilibrium values only if the initial salt concentrations in each reservoir are maintained constant (i.e., $C_L = C_o$ and $C_R = 0$), e.g., by continuously replenishing the liquid in the left reservoir with fresh salt solution, and continuously flushing the liquid in the right reservoir with pure water. Otherwise, the process of diffusion will eventually equilibrate the salt concentrations in both reservoirs, i.e., $C_R = C_L = C_o/2$, resulting in $a_{wR} = a_{wL}$, $q_\pi = J_d = h_\pi = 0$, and water levels in each reservoir that are the same as those for the initial condition (Figure 2a).

In the case where the boundary concentrations in the reservoirs are maintained constant, then steady state values of h_π and ΔP_π will be established at equilibrium based on the type and concentration of salt used in the system and degree of solute restriction exhibited by the non-ideal membrane. In this case, the membrane efficiency for the relative degree of solute restriction is represented as follows:

$$\omega = \frac{h_\pi}{H_\pi} = \frac{\Delta P_\pi}{\Delta\pi} \quad (1)$$

where ω is the chemico-osmotic or membrane efficiency coefficient, which ranges from zero for no solute restriction, i.e., no membrane behavior, to unity for ideal membrane behavior with complete solute restriction (i.e., $0 \leq \omega \leq 1$).

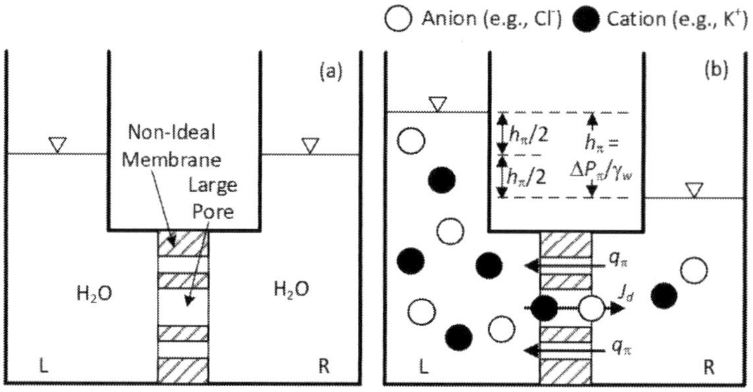

Figure 2. Schematic illustrations of the concept of semipermeable membrane behavior for uncharged, non-ideal (leaky) membranes: (a) initial condition where only water (H_2O) is contained in the left (L) and right (R) reservoirs bounding the membrane; (b) system at equilibrium after the water in the left reservoir has been replaced with a salt (e.g., KCl) solution.

2.3. Solute restriction in charged clay membranes

The membrane behavior observed in clays is conceptually the same as that illustrated in Figure 2, except that clays act as charged membranes. In this case, the mechanism resulting in solute restriction often is referred to as anion exclusion, as illustrated conceptually in Figure 3.

Clay platelets (particles) comprising aluminum phyllosilicate clay minerals (e.g., illite, kaolinite, montmorillonite) typically carry negative surface charges due to isomorphic substitution. The negative charge deficiency within the clay platelets leads to a distribution in negative electrical potential, Ψ, that diminishes with distance away from the platelet surface (note that the Ψ is distributed on both sides of the surfaces shown in Figure 3, but only one side is shown for simplicity). The thickness of this distribution is designated as d_Ψ in Figure 3. Within this electrical potential distribution, cations (counter ions) are readily attracted to the negative particle surface, whereas anions (co-ions) are repelled, resulting in a net excess of cations with an overall net positive (counter) charge that is sufficient to balance the negative particle charge to satisfy electroneutrality for the particle. The cations that are attracted to the particle surface can readily exchange with an equivalent charge of other cations in the free pore water, i.e., the pore water beyond the influence of Ψ (i.e., beyond d_Ψ). For example, one Ca^{2+} ion in the free pore water can exchange with two Na^+ counter ions. For this reason, these counter ions are referred to as exchangeable cations, and the total capacity for a clay to hold exchangeable cations on the basis of equivalents of charge per unit mass of clay is the cation exchange capacity.

As illustrated conceptually in Figure 3, exclusion of anions in the free pore water occurs when adjacent clay particles are sufficiently close such that the effective pore distance, d_{ep}, representing the free pore water thickness between the two adjacent platelets through which anions could potentially migrate, is less than the diameter of the migrating, hydrated anion in the free pore water, D_h, i.e., $d_{ep} < D_h$. In this case, any cation associated with the migrating anion also will be excluded from entering the pore due to the requirement to maintain electroneutrality in the free pore water. Thus, the ability of a clay to restrict solute migration will depend on the magnitude of d_{ep} relative to D_h, and the magnitude of d_{ep} is, in turn, a function of the relative magnitudes of the total pore width, d_p, and d_Ψ, i.e., $d_{ep} = d_p - 2d_\Psi$ (Figure 3). The resulting magnitudes of d_p and/or d_Ψ, and therefore, d_{ep}, are a function of mineralogy and particle sizes of the clay, the applied stress on the clay, the chemistry of the pore water, and the degree of saturation.

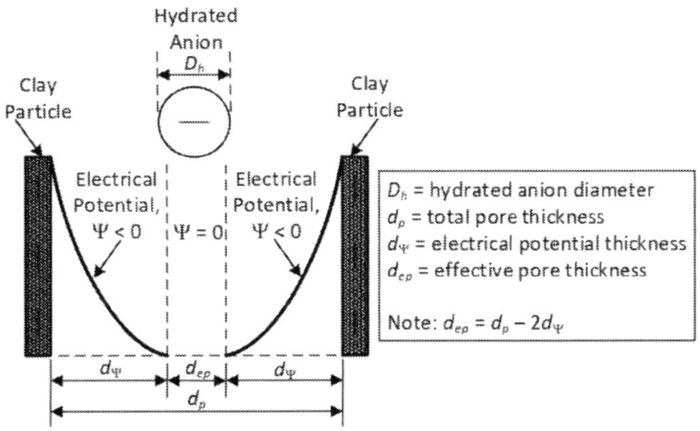

Figure 3. Simplified cross-sectional view of a single pore between two adjacent and parallel, charged clay platelets (particles) illustrating the mechanism by which the distributions in negative electrical potentials result in solute restriction (anion exclusion).

2.3.1. Effects of mineralogy and particle size. The effects of mineralogy and particle size can be illustrated with the aid of Figure 4, wherein the relative differences between clay particles comprising kaolinite versus those comprising montmorillonite are illustrated. As shown, kaolinite clay platelets typically are substantially larger than montmorillonite clay platelets. Since pore size decreases with decreasing particle size, pore sizes for clay (d_p) comprising montmorillonite (e.g., bentonite) generally are much smaller than those for clay comprising kaolinite (i.e., kaolin). In addition, the extent of d_Ψ, which is represented by the double layer water in Figure 4, relative to size of the solid (mineral) particle is substantially greater for clay platelets comprising montmorillonite relative to those comprising kaolinite. As a result, the influence of the negative electrical potential distribution on the interaction between montmorillonite-based clay particles is substantially greater than that between kaolinite-based clay particles. For these reasons, solute restriction (anion exclusion) typically is significantly greater in bentonite than in kaolin.

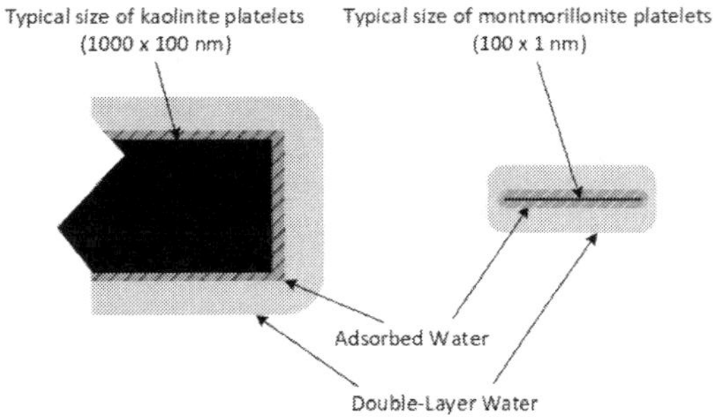

Figure 4. Relative difference between kaolinite platelets and montmorillonite clay platelets.

2.3.2. Effect of applied stress. With respect to Figure 3, differences in applied stress, σ, are reflected by differences in the relative magnitudes of d_p and d_{ep} for the same d_Ψ. At lower σ, the adjacent clay platelets are relatively far apart, such that d_p and d_{ep} are relatively large, and D_h tends to be smaller than d_{ep}, reducing the tendency for solute restriction. In contrast, at higher σ, the adjacent clay platelets are forced closer together, reducing both d_p and d_{ep}, such that eventually d_p is decreased to the extent that solute restriction occurs (i.e., $D_h > d_{ep}$). Thus, the likelihood that clay exhibits membrane behavior increases with increasing σ.

2.3.3. Effect of pore-water chemistry. For a given d_p, decreasing the concentrations, C, of solutes (ions) and/or decreasing the charges, z, of the counter ions (e.g., Na^+ vs. Ca^{2+}) extends the influence of d_Ψ relative to d_p and, therefore, increases the likelihood of solute restriction ($D_h < d_{ep}$). In contrast, at high C and/or high z, d_Ψ is decreased relative to d_p, which decreases the likelihood of solute restriction.

2.3.4. Effect of degree of saturation. Sample-Lord and Shackelford [18] postulated that the membrane efficiency in clays under unsaturated conditions would be enhanced relative to that of saturated conditions to varying extents depending on the relative degree of saturation, all other factors being the same. At relatively high degrees of saturation, the existence of occluded air bubbles within the free pore solution zone, i.e., the zone where $\Psi = 0$ corresponding to the effective pore thickness (d_{ep}), would lead to partial solute restriction in this zone because anions would be forced to migrate only through the aqueous phase existing between the air bubble and the outer edge of the clay mineral electrical potential (d_Ψ). This effect essentially reduces the effective pore thickness through which the anions can migrate. As the degree of saturation reduces further, the air phase eventually becomes continuous, and solute restriction is further enhanced as the aqueous phase in the free solution zone is further reduced. Of course, complete solute restriction would be expected in the limit when free pore water within the clay becomes discontinuous, which would occur at relatively low degrees of saturation.

2.4. Results for clays

2.4.1. Effects of pore-water chemistry, porosity, and mineralogy. The effects of pore-water chemistry, porosity, and mineralogy on ω are illustrated in Figure 5. Based on data from [2], ω is shown in Figure 5a as a function of the average salt concentration in normality (N) across saturated montmorillonite clay specimens, C_{avg}, for both NaCl and $CaCl_2$ salts and several values of specimen porosity, n. Based on data from [19], ω for saturated kaolinite clay specimens at an n of 0.53 is shown as a function of C_{avg} of NaCl solutions in molarity (M) in Figure 5b.

As shown in Figure 5a, for a given salt and n, ω decreases with increasing C_{avg}, which reflects the effect of solute concentration on membrane behavior, i.e., for a given d_p, increasing C compresses d_Ψ, increases d_{ep} and reduces solute restriction. Also, for a given C_{avg} and n, ω for NaCl is higher than ω for $CaCl_2$, which reflects the effect of counter ion charge on membrane behavior, i.e., for a given d_p, increasing z from +1 (Na^+) to +2 (Ca^{2+}) compresses d_Ψ, increases d_{ep} and reduces solute restriction. Finally, for a given salt and C_{avg}, ω decreases with increasing n as expected on the basis that increasing n correlates with an increase in d_p and d_{ep} for a given d_Ψ, and a decrease in solute restriction.

The effect of mineralogy is illustrated by comparing the results for kaolinite clay shown in Figure 5b versus those for montmorillonite clay shown in Figure 5a. Although the trend of ω decreasing with increasing C_{avg} is still apparent in the data for the kaolinite clay (Figure 5b), the ω values in Figure 5b for kaolinite clay are substantially lower than the ω values in Figure 5a for montmorillonite clay at the same C_{avg} for the same salt (NaCl). For example, ω is 0.0235 for the kaolinite clay at an n of 0.53 for $C_{avg} = 0.002$ M NaCl, whereas ω is 0.68, or 29 times greater, for the montmorillonite clay at an n of 0.91 for C_{avg} of 0.0019 M NaCl (= 0.0019 N NaCl). Note that the slightly negative ω values shown for the kaolinite clay at higher C_{avg} are the result of a process referred to as 'diffusion-osmosis' [20].

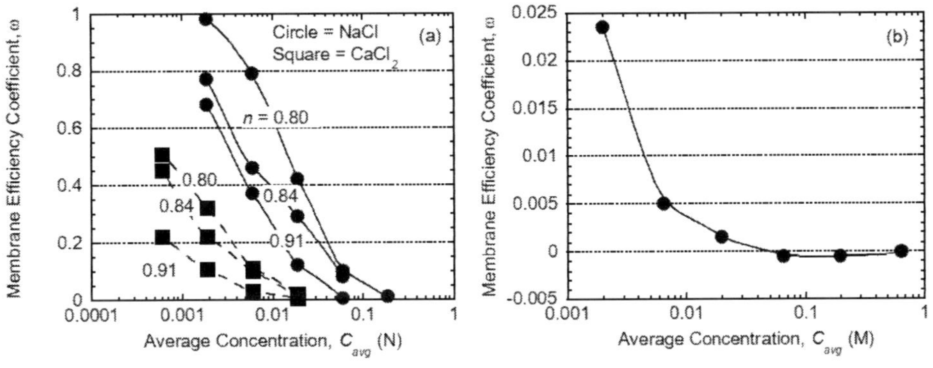

Figure 5. Membrane efficiency of clay specimens as a function of the average salt concentration across the specimen: (a) results for montmorillonite clay as a function of porosity (n) for NaCl and $CaCl_2$ salt solutions (data from [2]); (b) results for kaolinite clay for NaCl salt solutions and an n of 0.53 (data from [19]).

2.4.1. Effect of applied stress. The effect of applied stress, σ, on clay membrane behavior is shown in Figure 6 based on the measured results reported by Olsen [21] for kaolinite clay tested with NaCl solutions. As shown in Figure 6a, an increase in σ from 7 to 1400 kPa results in a decrease in n from 0.590 to 0.114, after which a subsequent decrease in σ to 28 kPa results in an increase in n to 0.213. As shown in Fig. 8b, the increase in σ from 7 to 1400 kPa results in an increase in ω from 0.014 to 0.453, after which the subsequent decrease in σ to 28 kPa results in a decrease in ω to 0.445. Thus, ω is directly related to σ and indirectly related to n. These results also illustrate that increasing σ can significantly enhance the membrane behavior of clays comprising minerals that are not expected to impart significant membrane behavior such as kaolinite.

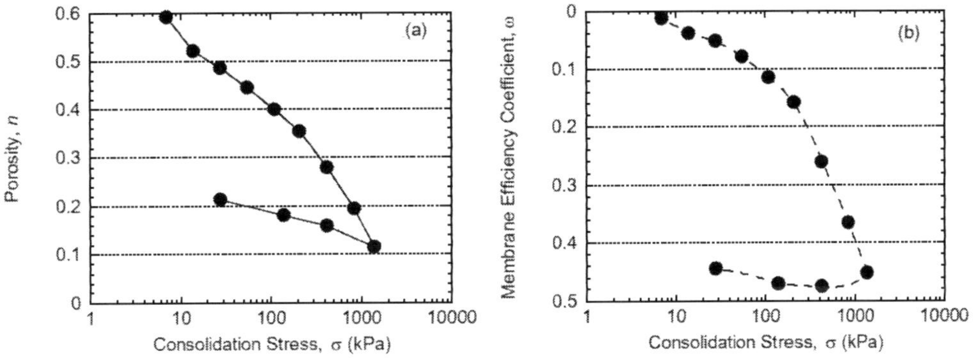

Figure 6. Membrane behavior of kaolinite clay as a function of applied effective stress based on NaCl salt solutions: (a) porosity versus consolidation stress; (b) membrane efficiency versus consolidation stress (data from [21]).

2.4.2. Effect of saturation. The effect of the degree of saturation, S, on the measured membrane behavior of sodium bentonite for three different source concentrations, C_o (= $2C_{avg}$), of KCl as reported by Sample-Lord and Shackelford [14] is shown in Figure 7. The tests were conducted using a special testing apparatus that controlled the degree of saturation of the bentonite specimens via the use of axis translation and high air entry disks at the specimen boundaries (see [22]). The previously noted, hypothesized trend of increasing ω with decreasing S is apparent in the results shown in Figure 7, although the range in S was limited due to the lengthy durations of the tests, which ranged from 232 to 335 days, and the availability of only one testing apparatus. Also, for a given S, ω decreased with increasing C_o, as expected based on the effect of pore-water chemistry (e.g., Figure 4).

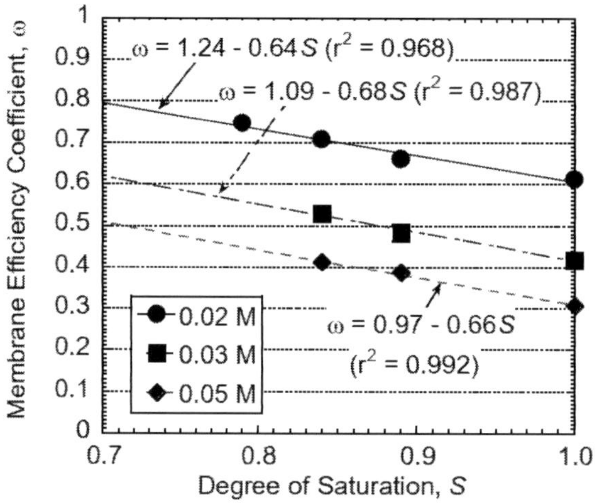

Figure 7. Measured membrane efficiency coefficients of unsaturated sodium bentonite as a function of degree of saturation and source KCl concentration (data from [14])

3. Membrane behavior in GCLs

3.1. Effect of pore-water chemistry

Malusis et al. [3] developed a laboratory testing apparatus to measure ω of clay soils, whereby a salt concentration difference, ΔC, was established across a clay specimen contained within a rigid-wall cell, by continuously circulating an electrolyte solution containing a salt at a constant concentration, C_o, across one side of the specimen while simultaneously circulating deionized water ($C = 0$) across the other side of the specimen at the same rate. The liquids were circulated in separate, closed loops, and the circulation rates were sufficiently fast to maintain essentially constant concentration boundary conditions. Malusis and Shackelford [4] then used the testing apparatus to evaluate the semipermeable membrane behavior of a needle-punched GCL containing sodium bentonite and subjected to KCl solutions with C_o ranging from 0.0039 to 0.047 M KCl, and compared their measured ω values with those from [2,19] shown in Figure 5a. The resulting values of ω are shown as a function of C_{avg} (= $C_o/2$) in Figure 8.

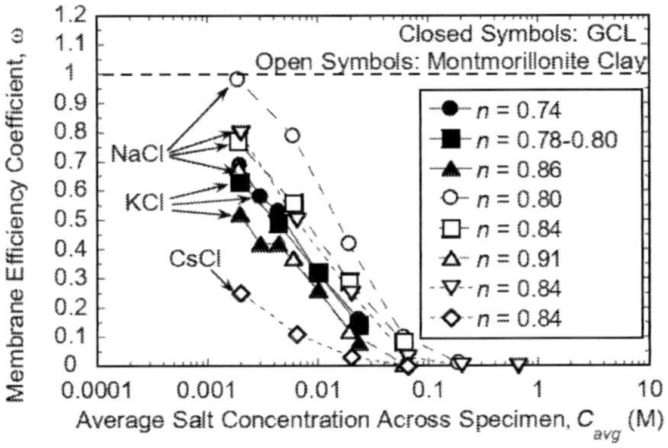

Figure 8. Comparison of the measured membrane efficiency coefficients for specimens of a GCL from [4] versus those for montmorillonite clay specimens from [2] and [19] as a function of the average salt concentration across the specimen for different salt solutions and different specimen porosity (*n*).

As indicated in Figure 8, the results for the GCL are consistent with those for the montmorillonite clay and indicate significant membrane behavior for the GCL over the range of concentrations evaluated. Differences in the results between the GCL and montmorillonite clay can be attributed to the different salt solutions used in the studies, differences in the mineralogy of the bentonite in the GCL and the montmorillonite clay, and the different porosities of the specimens. Regardless of any differences in ω values shown in Figure 8, the results reported in [4] suggest that bentonite-based GCLs can behave as semipermeable membranes, even at relatively high *n*. Knowledge that GCLs can behave as semipermeable membranes is important from the viewpoint that such membrane behavior can enhance the performance of GCLs used as chemical containment barriers by restricting aqueous miscible contaminant migration.

3.2. *Effect of applied stress*

Kang and Shackelford [23] adopted the testing procedure developed by Malusis et al. [3] for a flexible-wall cell, which provided for the ability to back-pressure saturate the specimen prior to measurement of ω, and to evaluate the effect of isotropically applied stress, σ, on the measured ω. Kang and Shackelford [9] then used the apparatus to evaluate the effect of σ on the ω of a GCL for the same C_o of KCl as used by Malusis and Shackelford [4], as shown in Figure 9. As expected, for a given C_o, the ω of the GCL increased with increasing σ, and for a given σ, ω increased with decreasing C_o. These results suggest that GCLs that are under high applied stresses, such as those used in bottom liner systems for solid waste containment, are likely to behave as semipermeable membranes, especially when exposed to liquids with low concentrations of solutes.

3.3 *Effect of degree of saturation*

The effect of *S* on ω of unsaturated GCLs has not been evaluated explicitly. However, given the aforementioned similarity between the membrane behavior of a GCL and montmorillonite clay (i.e., bentonite) as shown in Figure 8, the ω of GCLs is expected to increase with decreasing *S*.

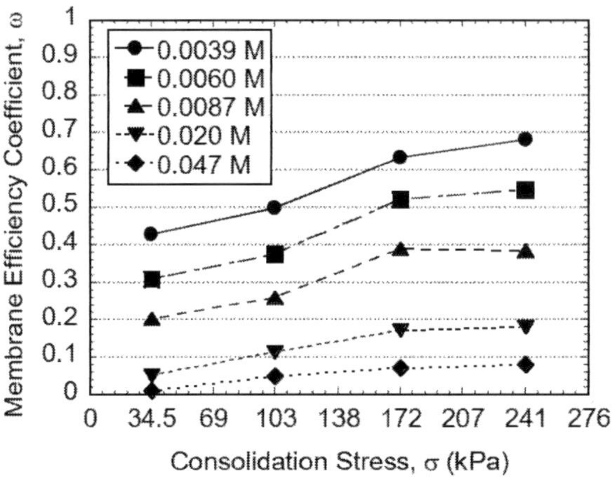

Figure 9. Measured membrane efficiency coefficients for specimens of a GCL as a function of consolidation effective stress and the source KCl concentration (data from [9]).

4. Relative importance of ω and q_π

A conceptual schematic of one-dimensional liquid and solute fluxes for a chemical containment scenario is shown in Figure 10. Although liquid and solute fluxes through a barrier can occur at any orientation based on the application (e.g., horizontal in the case of cutoff walls or vertical in the case of a bottom waste containment liner), a simplified vertical transport scenario through a liner is shown in Figure 10 to illustrate the relative contributions of q_π and the chemico-osmotic solute flux, J_π, i.e., the advective drag of solutes as a result of q_π in the direction opposite of J_d. The total flux of solute, J, comprises J_d and the hyperfiltrated advective solute flux, J_{ha}, in the direction from within the containment system to the surrounding environment, and J_π in the opposite direction into the waste containment system (i.e., $J = J_{ha} + J_\pi + J_d$, where $J_\pi < 0$).

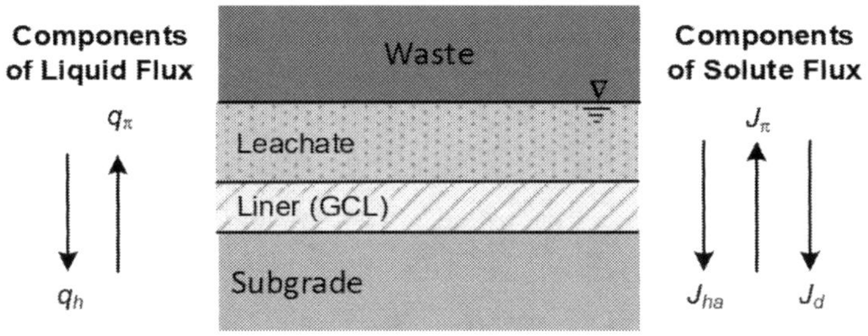

Figure 10. Schematic illustration of one-dimensional liquid and solute flux in a chemical containment scenario where the concentration of solutes within the leachate are greater than those at the base of the liner: q_h = hydraulic liquid flux; q_π = chemico-osmotic liquid flux; J_{ha} = hyperfiltrated advective flux; J_π = chemico-osmotic counter advection; and J_d = restricted diffusion (after [24]).

Malusis et al. [25] evaluated the percent reduction in J for clay barriers exhibiting membrane behavior through the use of coupled solute transport model scenarios parameterized with experimental data for bentonite-based barriers in contact with monovalent salt solutions. For diffusion dominated systems, as is the case for many chemical containment barriers with low hydraulic conductivity and low hydraulic gradient, the percent reduction in J was calculated to be substantial, ranging from 3.8 to 74 % when the effect of ω was included in the transport simulation. However, the percent reduction in solute flux attributed solely to J_π was minor, ranging from 0.5 to 10 % for the diffusion dominated systems.

Unfortunately, given limited datasets available from the experimental literature, additional data are needed to fully elucidate if greater percent reductions in J may be achieved by some semipermeable clay barriers.

Also, there is the remaining question regarding the uniqueness of ω [26]. That is, the values of ω shown in Figs. 5 and 8 were plotted as a function of the average salt concentration across the specimen C_{avg}, where $C_{avg} = (C_{top} + C_{bottom})/2$, and C_{top} and C_{bottom} are the salt concentrations at the top and bottom boundaries of the specimen, respectively. In the case of the ω values in Fig. 8, $C_{top} = C_o$ and $C_{bottom} = 0$, such that $C_{avg} = C_o/2$. However, for the data from [2] and [19] shown in Fig. 5, $C_{top} < C_o$ and $C_{bottom} > 0$. So, with respect to the results for the GCL specimens shown in Fig. 8, are the values of ω uniquely related to C_{avg}? For example, if $C_{top} = 3C_o/4$ and $C_{bottom} = C_o/4$ such that $C_{avg} = C_o/2$, would the value of ω be the same as that when $C_{top} = C_o$ and $C_{bottom} = 0$? The authors hypothesize that the answer to this question is no, but this hypothesis has not been evaluated explicitly to date.

5. Conclusions

The existence of membrane behavior in clays leads to solute restriction via anion exclusion and chemico-osmosis, which can be beneficial in terms of the chemical containment function of clay barriers such as GCLs. Accordingly, this paper has provided a review of the concept of membrane behavior in porous media, with particular emphasis on the existence and magnitude of the membrane efficiency coefficient, ω, in clays and GCLs.

The results of several studies conducted over the past approximate 20 yr have indicated that GCLs exhibit significant membrane behavior over a range of salt concentrations. The results indicate that the measured ω increases with decreasing concentration of the salt in the pore water, decreasing valence (charge) of the salt cation, decreasing porosity or increasing applied stress, σ, and decreasing degree of water saturation, S. The results imply that GCLs subjected to high σ in solid waste chemical containment applications, e.g., municipal and hazardous waste landfills, tailings impoundments, etc., likely behave as semipermeable membranes, thereby enhancing the chemical containment performance of the GCLs. However, additional research pertaining to evaluating the significance and persistence of the membrane behavior of GCLs under high salt concentrations as well as the uniqueness of ω with respect to different boundary salt concentrations at different σ is warranted.

Acknowledgments

The results for GCLs reported in this paper are based on research funded by the U.S. National Science Foundation (NSF) under grants CMS-9634649, CMS-0099430, CMS-0624104, and CMMI-0926205. The opinions expressed in this paper are solely those of the authors and are not necessarily consistent with the policies or opinions of the NSF.

References

[1] McKelvey J and Milne I 1960 The flow of salt solutions through compacted clay *Proc. 9th Nation. Conf. Clay. Clay Miner.*, Pergamon, New York, pp 248–59

[2] Kemper W and Rollins J 1966 Osmotic efficiency coefficients across compacted clays *Proc. Soil Sci. Soc. Am.* **30** pp 529–34

[3] Malusis M, Shackelford C and Olsen H 2001 A laboratory apparatus to measure chemico-osmotic efficiency coefficients for clay soils *Geotech. Test. J.*, **24** pp 229–42

[4] Malusis M and Shackelford C 2002 Chemico-osmotic efficiency of a geosynthetic clay liner *J. Geotech. Geoenviron. Eng.*, **128** pp 97–106

[5] Yeo S, Shackelford C and Evans J 2005 Membrane behavior of model soil-bentonite backfills *J. Geotech. Geoenviron. Eng.*, **131** pp 418–29

[6] Henning J, Evans J and Shackelford C 2006 Membrane behavior of two backfills from field-constructed soil-bentonite cutoff walls *J. Geotech. Geoenviron. Eng.*, **132** pp 1243–49

[7] Kang J and Shackelford C 2010 Membrane behavior of compacted clay liners *J. Geotech.*

Geoenviron. Eng., **136** pp 1368–82
[8] Mazzieri F, Di Emidio G and Van Impe P 2010 Diffusion of calcium chloride in a modified bentonite: Impact on osmotic efficiency and hydraulic conductivity *Clay. Clay Minerals*, **58** pp 351–63
[9] Kang J and Shackelford C 2011 Consolidation enhanced membrane behavior of a geosynthetic clay liner *Geotext. Geomembr.* **29** pp 544–56
[10] Bohnhoff G and Shackelford C 2013 Improving membrane performance via bentonite polymer nanocomposite *Appl. Clay Sci.*, **86** pp 83–98
[11] Dominijanni A, Manassero M and Puma S 2013 Coupled chemical-hydraulic-mechanical behavior of bentonites *Géotechnique*, **63** pp 191–205
[12] Meier A and Shackelford C 2017 Membrane behavior of compacted sand-bentonite mixture *Can. Geotech. J.*, **54** pp 1284–99
[13] Dominijanni A, Guarena N and Manassero M 2018 Laboratory assessment of the semipermeable properties of a natural sodium bentonite *Can. Geotech. J.*, **55** pp 1611–31
[14] Sample-Lord K and Shackelford C 2018 Membrane behavior of unsaturated sodium bentonite *J. Geotech. .Geoenviron. Eng.*, **144** 04017102
[15] Fritz C, Scalia IV J, Shackelford C and Malusis M 2020 Determining maximum chemico-osmotic pressure difference across clay membranes *J. Geotech. Geoenviron. Eng.*, **146** 06019018
[16] Fritz J 1986 Ideality of clay membranes in osmotic processes: a review *Clay. Clay Miner.* **34** pp 214–23
[17] Shackelford C and Daniel D 1991 Diffusion in saturated soil: I. Background *J. Geotech. Eng.*, **117** pp 467–84
[18] Sample-Lord K and Shackelford C 2014 Membrane behavior of unsaturated bentonite barriers *Geo-Congress 2014: Geo-Characterization and Modeling for Sustainability* ed M Abu-Farsakh, X Yu and L Hoyos (Reston, Virginia, USA: American Society of Civil Engineers) pp 1900–09
[19] Kemper W and Quirk J 1972 Ion mobilities and electric charge of external clay surfaces inferred from potential differences and osmotic flow *Proc. Soil Sci. Soc. Am.* **36** pp 426–33
[20] Olsen H, Yearsley N and Nelson K 1990 Chemico-osmosis versus diffusion-osmosis *Trans. Res. Rec.* vol 1288 (Washington DC: Transportation Research Board) pp 15–22
[21] Olsen H 1969 Simultaneous fluxes of liquid and charge in saturated kaolinite *Proc. Soil Sci. Soc. Am.*, **33** pp338–44
[22] Sample-Lord K and Shackelford C 2017 Apparatus for measuring coupled membrane and diffusion behavior of unsaturated sodium bentonite *Vadose Zone J.*, **16** http:// doi:10.2136/vzj2016.12.0140
[23] Kang J and Shackelford C 2009 Clay membrane testing using a flexible-wall cell under closed-system boundary conditions *Appl. Clay Sci.*, **44** pp 43–58
[24] Malusis M, Shackelford C and Olsen H 2003 Flow and transport through clay membrane barriers *Eng. Geol.*, **70** pp 235–48
[25] Malusis M, Scalia J, Norris A and Shackelford C 2017 Quantifying the significance of chemico-osmotic counter-advection on solute transport through semipermeable clay barriers *Int. Symp. Coupled Phenomena in Environ. Geotech. (CPEG2)*, (Leeds U.K.: University of Leeds) paper 51
[26] Shackelford C, Malusis M and Olsen H 2003 Clay membrane behavior for geoenvironmental containment *Soil and Rock America Conf. 2003* ed P Culligan, H Einstein and A Whittle (Essen, Germany: Verlag Glückauf GMBH) Vol 1, pp 767–74
[27] Meier A, Sample-Lord K, Castelbaum D, Kallase S, Moran B, Ray T and Shackelford C 2014 Persistence of semipermeable membrane behavior for a geosynthetic clay liner *Proc., 7th Int. Conf. Environ. Geotech.* ed A Bouazza, S Yuen and B Brown Nov. 10-14, 2014, Melbourne, Australia (Barton, Australia: Engineers Australia) (ISBN 978-1-922107-23-7) pp 496–503
[28] Shackelford C, Meier A and Sample-Lord K 2016 Limiting membrane and diffusion behavior of a geosynthetic clay liner *Geotext. Geomembr.*, **44** pp 707–18

IOP Conf. Series: Materials Science and Engineering 1260 (2022) 012010 doi:10.1088/1757-899X/1260/1/012010

The application of the observational procedure for construction of athletic stadium in Bytów using geosynthetic reinforcement in complicated ground conditions

A Łopatka[1], M Pilch [1]

Przedsiębiorstwo Realizacyjne INORA Sp. z o.o., Poland

E-mail: tech@inora.pl

Abtract. Bytów is a city in northern Poland, in the Pomeranian Voivodeship. The characteristics of the region are peatlands with thick layer of soft peat deposites. In one of these areas an athletic stadium was designed with a static consolidation as a soil improvement method. The applied method included prefabricated vertical drains and surcharge embankment. To ensure slope stability, in base of embankment was used geosynthetic reinforcement.

The selected method of soil improvement required to use of observational procedure. It was the reason for the development of the complex automatic monitoring system. Developed system was a source of valuable information for designer, cotractor and investor. Monitoring system ensured the safety at the every of construction stages and allowed to evaluate the behaviour of rainforced soil mass. This case of study discusses about practical use of monitoring system in hard ground conditions.

1. Ground conditions and soil improvement

The subsoil under designed stadium consist of peat. Thickness of soft layer was between 3 – 11m, and degree of humification according to Von Post scale H3 – H7. Performed geological investigation, [3], [4], showed that peat layer is very compressible, (E_{oed}=100 – 250 kN/m^2) and very soft (S_u=6,7 – 11,2 kN/m^2, Φ'=14 – 23 °. As an improvement method a static consolidation was chosen with prefabricated vertical drains (PVD). Drains were installed in triangular pattern with axial space 1,5m between them. To reduce settlement in exploatation period a surcharge embankment was used. Due to the variable thickness of peat layer, height of embankment was between 3 and 6m. To ensure slope stability, embankment was constructed in stages and in embankment base a high strenght geosynthetic reinforcement was used with characteristic short time strenght between 600 to 1600 kN/m.

Figure 1 view of the construction site

2. Monitoring system

The chosen soil improvement method in such hard ground conditions was the reason for adopting observational procedure, according to eurocode EC-7. Observational procedure is recommended in cases where it is difficult to get good prediction of soil – structure interaction using only static analysis. Therefore the complex automatic monitoring system was developed which was a source of valuable information. The major function of monitoring system was to verify the property of assumptions which were made in the static calculations [5]. Especially it was important to controll consolidation process for ensuring embankment stability in all construction stages. Moreover, in the vicinity of the stadium there was an existing skatepark. There was a fear, that extruded peat layer would damage it, so it was necessary to evaluate influence of the work on the existing structure.

Developed monitoring system consisted of:

- water preasure gauges
- vertical inclinometers
- piezometers
- benchmarks

Benchmarks,water pressure gauges and piezometers were located uniformly in whole improved area. Vertical inclinometers have been positioned by embankment. Inclinometers IW1 and IW2 were in place in where there was the highest thickness of peat layer. Inclinometer IE1 was located between embankment and existing skate park construction. (Figure 2). Monitored area equaled over 15 000 m^2.

At a crucial period of construction of embankment, parallel readings of benchmark and inclinometer took place once a week. Water pressure and piezometer readings were once a hour. All data were automatically processed to provide the full information of the existing situation.

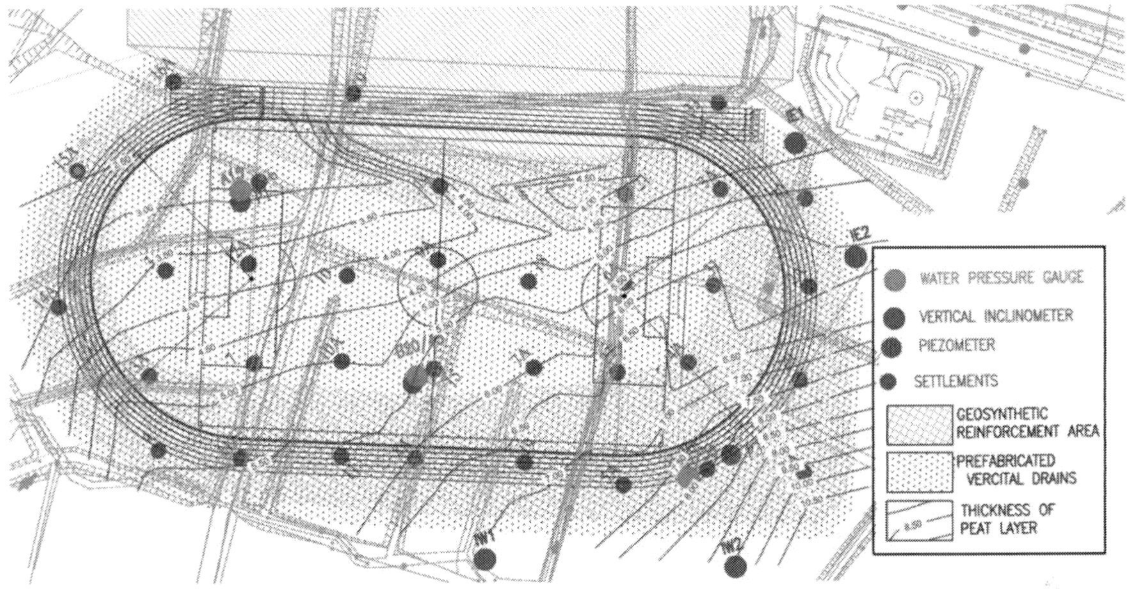

Figure 2 location of elements of monitoring system

3. Observational procedure

In executable project [5], it was assumed that the embankment would be built in six stages, with periods of consolidation between each stage, (initially - 20 days). At the static calculation, it was assumed that each stage will be performed after the primary consolidation from the previous stage is completed. This assumption allowed to determine the increase of undrained shear strength of soil which was the basis for stability calculations for the next stage.

Because a loaded peats are characterized by a large decrease in the coefficient of permeability, there was a danger that the initially assumed period of 20 days of consolidation would be too short, (even using PVD). Therefore the consolidation had to be monitored and the results of the observations were the basis for modifying the construction rate and consolidation period between the stages

Due to high water level and large settlements, the constructed embankment, (with all installed monitoring elements), was gradually submerging into the water. Therefore interpretation of the consolidation process, using only method based on water disspiation, could lead into incorect results. That is why a combination of two metods were used. The consolidation process was analyzed recording the growth of settlements in time, (e.g. according to ASAOKA). An additional elements to evaluate consolidation process were based on analysis of water dissipation in soft layer. Additionally, the stability of the embankment in the place where the largest thickness of peat was deposited, diagram for construction control of embankment on soft ground according to Matuso Kawarmura was used [6]. This method requires observation of both vertical and horizontal displacements near embankment base, (Figure 3).

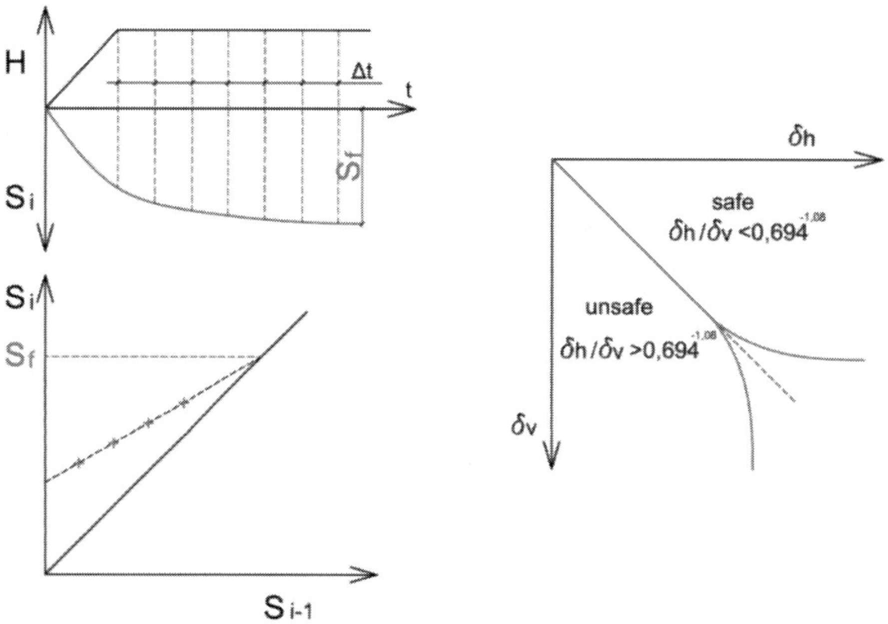

Figure 3 Asaoka' method for final settlement determination (left), Matuso – Kawamura diagram (right)

4. Readings

Monitoring has shown that during the construction time, value of settlements was between 0,4 and 2,5m. The largest value was in the place with the highest thickness of peat layer and with the highest embankment height (Figure 4).

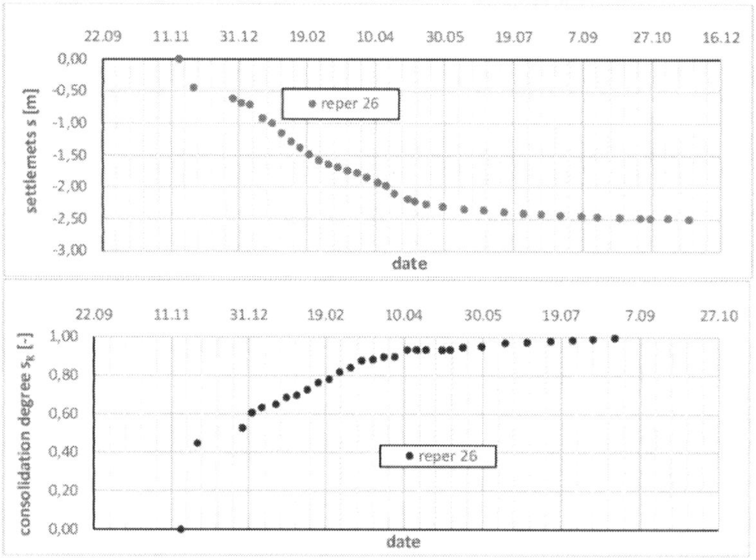

Figure 4 Settlement and degree of consolidation for benchmark 26 (Asaoka method)

Lateral extrusion of peat showed by inclinometer IE1 equals 0,03 m, so there was no negative influence of the work on the existing structure of skatepark. Value of horizontal displacement was negligible and the lateral extrusion did not cause any negative effects such as lift / settling or cracks in skatepark structure. The highest extrusion was observed in inclinometer IW2, (0,87m, figure 5). It was in the same area in where the highest settlements was observed. According to Matuso Kawarmura method, displacements of the embankment are in "unsafe" area in all stages above the third one, (Figure 6). However, the accurate stability analyses have shown safety of embankment. This is due to reinforcement in embankment base in which high strenght geosynthetic was used, (length of geosynthetic insert equals 50m and characteristic short-term strenght 1600 kN)

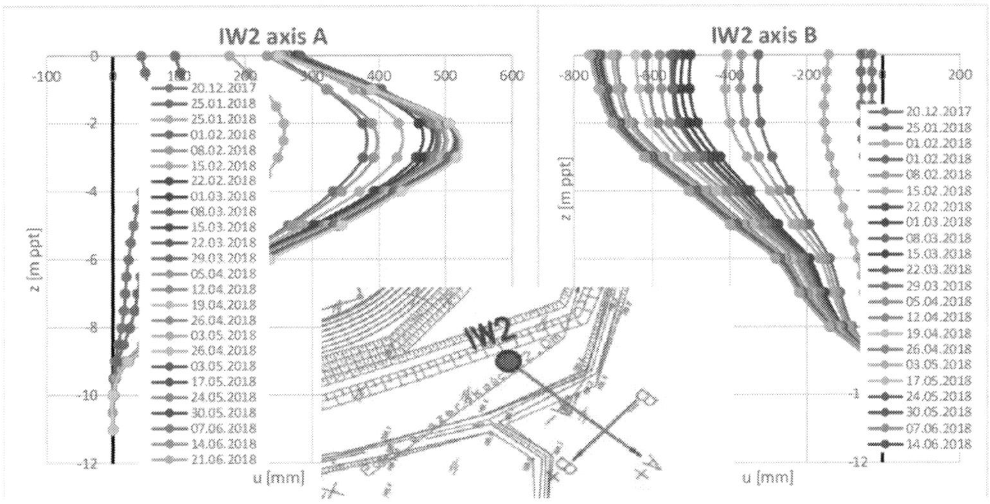

Figure 5 lateral extrusion – inclinometer IW2

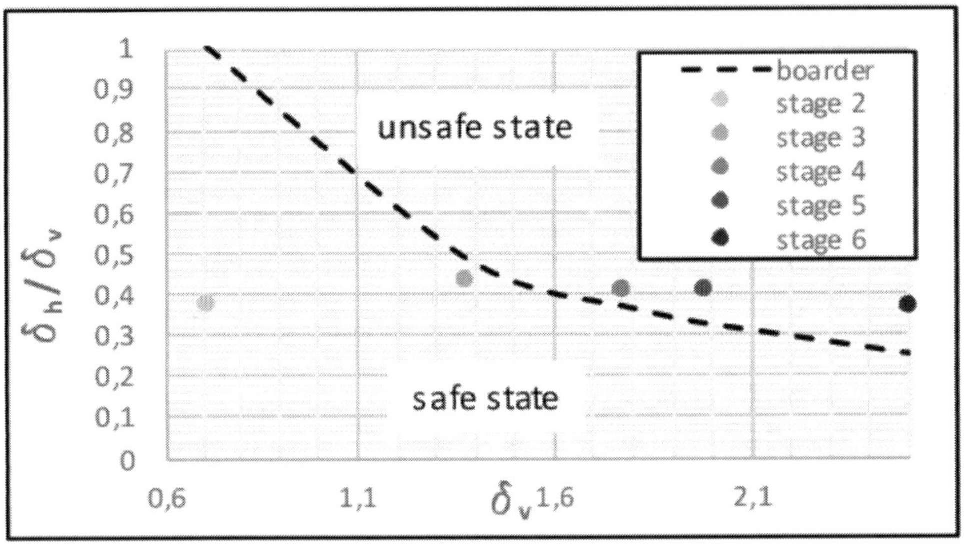

Figure 6 Dependency δh/δv – δv, (Matuso Kawarmura method), for benchmark 26 and inclinometr IW2

The consolidation process went slower than forecast in the calculations, (Figure 7, 8). Therefore, the consolidation period between stages 4 and 5 was extended into 50 days. It was necessary for sufficient soil strengthening to ensure embankment stability in further stages.

Due to high requirement about behaviour of improved soil, the investor has established a special acceptance conditions for it at the end of construction period. The maximum difference in settlement for all benchmark during the two months of inspection should not be greater than 1 cm. Undrained shear strength of soil should be greater than 150% of *in situ* value. Observations of benchmarks in inspection time and additional field investigations showed that both conditions were fulfilled, (Δs=9mm, S_u=240 – 760% of *in situ* value).

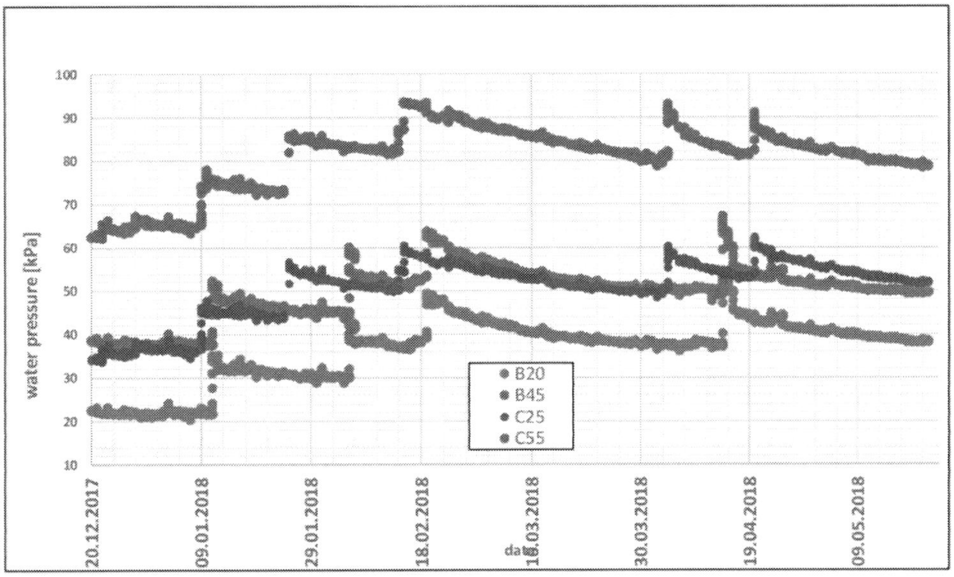

Figure 7 Water pressure dstribution

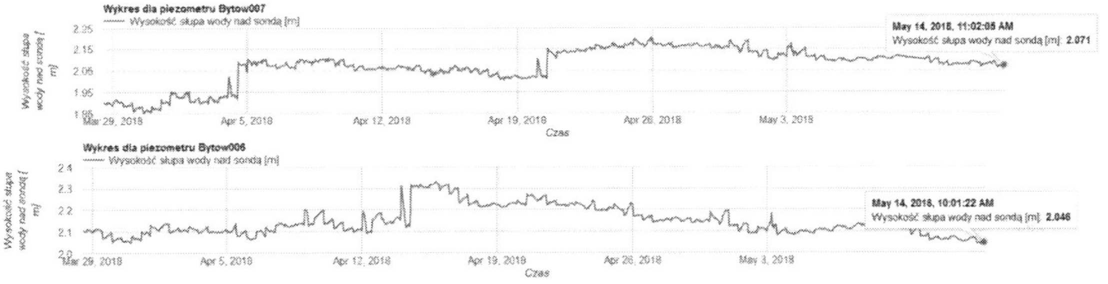

Figure 8 Water level distribution (from piezometers)

5. Conclusions

The article discusses the practical use of monitoring system in hard ground conditions in which relying only on the forecasts from static analysis is very risky. We should also mention that this type of systems are nowadays installed in a large scale, giving very valuable information about the condition of the structure both during the construction and exploatation of the object.

6. References

[1] EN 1997-1:2008 Eurocode 7: Geotechnical design – Part 1: general rules

[2] Von Post L 1924 Das genetische System der organogenen Bildungen Schwedens. Internationale bodenkundliche Konferenz. Rom

[3] Badania Geotechniczne określające warunki gruntowo – wodne na obszarze sąsiadującym od strony południowo – zachodniej z terenem MOSIR Bytów (działki: 180, 181, 182, 185 – obręb *Świątkowo) przeznaczonym pod planowaną budowę stadionu, kortów tenisowych, dróg dojazdowych i parkingów*, MS.GEO, Bytów, październik 2014.

[4] Sprawozdanie z badań geotechnicznych wykonanych dla projektu posadowienia stadionu w Bytowie, GEOTIM, lipiec 2017

[5] Projekt Wykonawczy – Budowa Stadionu Lekkoatletycznego w Bytowie – Wzmocnienie słabonośnego podłoża, INORA, wrzesień 2017
[6] Matsuo M and Kawamura K 1977 Diagram for Construction Control of Embankment on Soft Ground. Soils and Foundations 17pp 37-52

IOP Conf. Series: Materials Science and Engineering 1260 (2022) 012011 doi:10.1088/1757-899X/1260/1/012011

Resilient modulus of crushed stone material reinforced with geogrids

J Kaluder[1], S Lenart[2], M Mulabdic[1] and K Minazek[1]

[1] Faculty of Civil Engineering and Architecture Osijek, Josip Juraj Strossmayer University of Osijek, Vladimira Preloga 3, 31000 Osijek, Croatia
[2] Slovenian National Building and Civil Engineering Institute, Dimičeva ulica 12, 1000 Ljubljana, Slovenia

Email: jkaluder@gfos.hr

Abstract. Unbound base layers can be stabilized with geogrids that reduce lateral movement of granular material thus improving its stiffness through particle interlocking inside geogrid openings. Good particles-geogrid interlocking depends significantly on the ratio of geogrid aperture size to the average soil particle size. This paper describes testing of a resilient modulus and a permanent deformation of geogrid reinforced granular material. Biaxial and triaxial geogrids made with 3D printer were used in which geogrids were of different aperture sizes and rib thickness. It was thus possible to test parametrically the effect of geogrid geometry and rib stiffness on the resilient behaviour and permanent deformations. Experiments were conducted on cylindrical specimens of 160 mm diameter and 320 mm height and were reinforced with one or several layers of geogrids. The results of cyclic triaxial tests for unreinforced and reinforced specimens with created geogrids are presented in the paper, as well as the results of the previous research on commercially available geogrids. It is observed from the presented results that resilient behaviour of the tested granular material was not improved with the use of geogrids, while geogrids gave certain improvement with respect to permanent deformations. Presented research is a part of the ongoing project at the University of J.J. Strossmayer in Osijek and the Slovenian National Building and Civil Engineering Institute.

1. Introduction

The function of road base layers is to improve traffic load distribution and its transfer to the subbase layers while ensuring adequate load-bearing capacity. If their strength and stiffness are inadequate, rutting and cracks on road surface can develop. Geogrids are considered to improve the mechanical properties of unbound soil layers. The contribution of the geogrid to the reduction of permanent deformation can be determined by cyclic load triaxial test with standardized method. Perkins et al. [1] conducted tests on aggregates with a single layer of geosynthetic (geogrid, geotextile and geocomposite). In this research the use of geosynthetics did not improve the resilient modulus while a decrease in permanent deformations was observed on the specimens reinforced with geosynthetics. Similar conclusions could be drawn from the research [2-4].

The interaction of non-cohesive (unbound) soils and geogrid is influenced by geogrid rigidity, position and number of geogrid layers in the soil specimen, geogrid geometry and the ratio of geogrid aperture size (A) with the average grain size (D_{50}). If a geogrid aperture is too large or too small with respect to the soil grain size, good interlocking may not occur. Sarsby [5], according to [6] suggested the optimal ratio of geogrid aperture (A) and mean grain size of at least 3.5. Holtz et al. [7] state that the size of the geogrid aperture should be ≥ D_{50} but ≤ of 2 x D_{85}. Minazek and Mulabdic [8] conducted pull-out tests on biaxial geogrids and showed that the optimal A/D_{50} ratio is around 2. Authors [9] have come to the similar conclusion suggesting the A/D_{50} ratio of 1.8. Brown et al. [10] state that the optimal ratio between the aperture size and the nominal size of the ballast (50 mm) is 1.2-1.6, while Hussaini [11] proposes geogrid aperture size from $0.95D_{50}$ to $2.5D_{50}$ with an optimum of 1.2. Han et al. [12] give the optimal A/D ratio 1.30-1.71 for biaxial geogrids and 1.08-1.43 for triaxial geogrids.

The aim of the study presented in this paper was to evaluate the contribution of geometry and size of the geogrid aperture to the reduction of permanent deformation of a selected well-graded non-cohesive soil under cyclic loading. For this purpose, cyclic load triaxial tests were conducted on reinforced specimens with four different geogrids produced by a 3D printer.

2. Materials and testing

2.1. Materials

The soil material used in the test was crushed dolomite with grading curve (shown at Figure 1) which is formed in accordance with the requirements for unbound base layers according to Croatian Roads Regulations [13]. According to [14] the soil is classified as well-graded gravel (GrW).

In the research it was planned to adjust the geogrid aperture size (A) to some extent to soil grain size (D_{50}). This was achieved by 3D printing of the geogrid samples. A used filament was Z-ULTRAT. This filament, which is a mixture of ABS filament (acrylonitrile-butadiene-styrene copolymer; 90-100%), polycarbonate (0-3%) and additives and colorants (0-10%), is characterized by durability and impact resistance [15]. Two different geogrid aperture shapes were used in the tests: square (biaxial geogrids) and triangular (triaxial geogrids). The A/D_{50} ratios for biaxial geogrids were 2 and 5. The thickness and width of geogrid ribs were also varied. Biaxial grids had the same thickness and width of ribs, and they were 1 mm (for $A/D_{50} = 2$, length of rib 11 mm, here named B12) and 2 mm (for $A/D_{50} = 5$, length of rib 27.5 mm, here named B25). Triaxial geogrids were printed in two ways: with ribs of 10 mm in length and rib thickness and width of 1 mm (referred as $A/D_{50} = 2$, named T12) and with ribs of 24 mm in length, and rib thickness and width of 2 mm (referred as $A/D_{50} = 5$, here named T25). The biaxial and triaxial geogrids are shown in Figure 1.

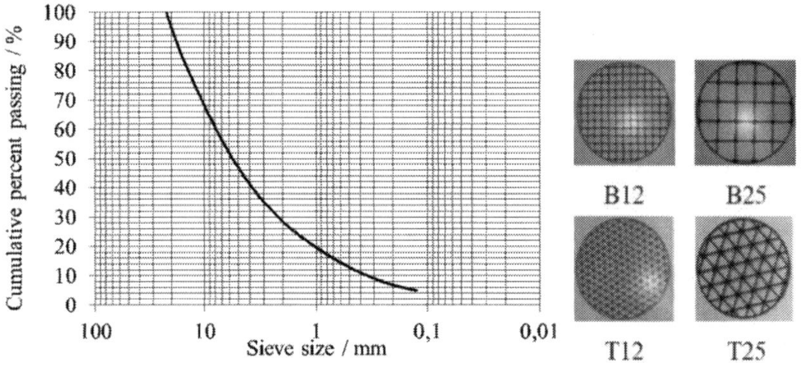

Figure 1. Left: Grading curve of soil specimen material; right: 3D printed specimens of biaxial and triaxial geogrids.

2.2. Specimen preparation

Tests were conducted on cylindrical specimens of 32 cm in height and 16 cm in diameter. Specimens were created by compacting soil in five layers of equal height. Vibrating hammer was used at compaction effort corresponding to Modified Proctor compaction. Specimens were compacted to a maximum dry density of 2.32 g/cm^3 determined by Modified Proctor test (according to [16]) at a moisture content of 2.7%. Prior to installing the geogrid in pre-determined position, the surface of a lower compacted soil layer was loosened a bit in order to achieve a better interaction with the geogrid. Geogrid positions in the specimen is shown in Figure 2.

After the removal of a split mould, a protective membrane was placed, and axial linear variable displacement transducers were fixed on the previously installed studs. There was a total of three axial linear variable displacement transducers positioned at 120° angle from each other and within a middle third of specimen height, and 60° from radial linear variable displacement transducers (according to [17] Annex A - A.1). The position of the axial transducers and studs for radial transducers is given in Figure 2.

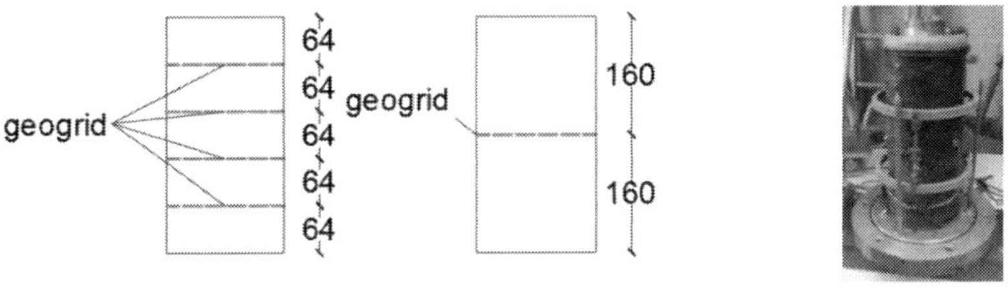

Figure 2. Left and middle: arrangement of geogrids in the specimen; right: arrangement of transducers (dimensions in mm).

2.3. Cyclic load triaxial test

Testing of reinforced and unreinforced specimens with cyclic triaxial test was performed according to [18].The specimens were subjected to a conditioning phase of 20000 cycles and at a cell pressure of 100 kPa and a deviator stress of 600 kPa. After conditioning, a resilient modulus testing was conducted at a small number of stress cycles for different stress combinations defined by [18]. Lastly, the specimens were subjected to a permanent deformation testing at 80000 load cycles with a cell pressure of 250 kPa and a deviator stress of 375 kPa. The duration of one cycle was 3 seconds with load duration of 1.4 seconds. One unreinforced and eight reinforced specimens (with four different geogrids) were tested (Table 1).

Table 1. Tested specimens and type of reinforcement

Specimen	Reinforcement
NG	no geogrid
1B12	1 layer, biaxial geogrid, 1 mm rib, $A/D_{50} = 2$
1T12	1 layer, triaxial geogrid, 1 mm rib, $A/D_{50} = 2$
1B25	1 layer, biaxial geogrid, 2 mm rib, $A/D_{50} = 5$
1T25	1 layer, triaxial geogrid, 2 mm rib, $A/D_{50} = 5$
4B12	4 layers, biaxial geogrid, 1 mm rib, $A/D_{50} = 2$
4T12	4 layers, triaxial geogrid, 1 mm rib, $A/D_{50} = 2$
4B25	4 layers, biaxial geogrid, 2 mm rib, $A/D_{50} = 5$
4T25	4 layers, triaxial geogrid, 2 mm rib, $A/D_{50} = 5$

3. Results and discussion

Resilient moduli were determined for 18 stress combinations according to [18] where the specimen was subjected to 100 load cycles for each stress combination. The values of the resilient modulus with the bulk stress (k-θ model) are given in Figure 3. These results can be characterised by the following findings: a) resilient moduli of reinforced specimens showed lower values compared to the resilient modulus of unreinforced specimen, b) specimens reinforced with one geogrid of larger apertures had higher values of the resilient moduli compared to specimens reinforced with geogrids of smaller apertures for both biaxial and triaxial geogrids, c) in tests where four geogrids were used, specimen reinforced with triangular geogrid of larger apertures produced higher values of resilient modulus compared to specimen reinforced with triaxial geogrid of smaller aperture, d) tests with four biaxial geogrids of larger apertures showed lower values of resilient modulus compared to tests with biaxial geogrid of smaller aperture, d) geogrids with triangular apertures generally produced higher values of resilient modulus than grids with square apertures, e) generally, there is no significant difference between specimens reinforced with one geogrid or four geogrids in terms of the resilient modulus value.

In terms of permanent volume deformations (Figures 4 and 5), geogrids showed positive influence on the specimen behaviour in the way that permanent deformations were smaller for reinforced specimens compared to permanent deformations for unreinforced soil. The following conclusions could be drawn from the results shown in Figure 4: a) the highest permanent volume deformation within the first 1000 cycles was associated with the unreinforced specimen and the specimen with biaxial geogrid of a smaller aperture size, c) the smallest volumetric strains are associated with the specimen reinforced with large aperture triangular geogrid, d) after very high number of cycles, reinforced specimens showed two times lower cumulative permanent volume deformation than unreinforced specimen, and again, larger apertures geogrids performed better, this time the square aperture better than the triangular aperture.

Results regarding permanent deformation for specimens reinforced with four geogrids are shown in Figure 5. It can be concluded that: a) within the first 1000 cycles the highest permanent volume deformations are associated with the unreinforced specimen and the specimen with the larger biaxial geogrid, b) all reinforced specimens showed more stable behaviour and lower permanent volume deformation than unreinforced specimen, c) specimens with geogrids of $A/D_{50} = 5$ ratio had the lowest accumulation of permanent deformation, d) along with the unreinforced specimen, the largest permanent deformations were recorded on the specimen reinforced with one biaxial geogrid of smaller aperture size. The obtained results indicate that with more load cycles, reinforced specimens would show a significant reduction of accumulated permanent deformation compared to the unreinforced sample.

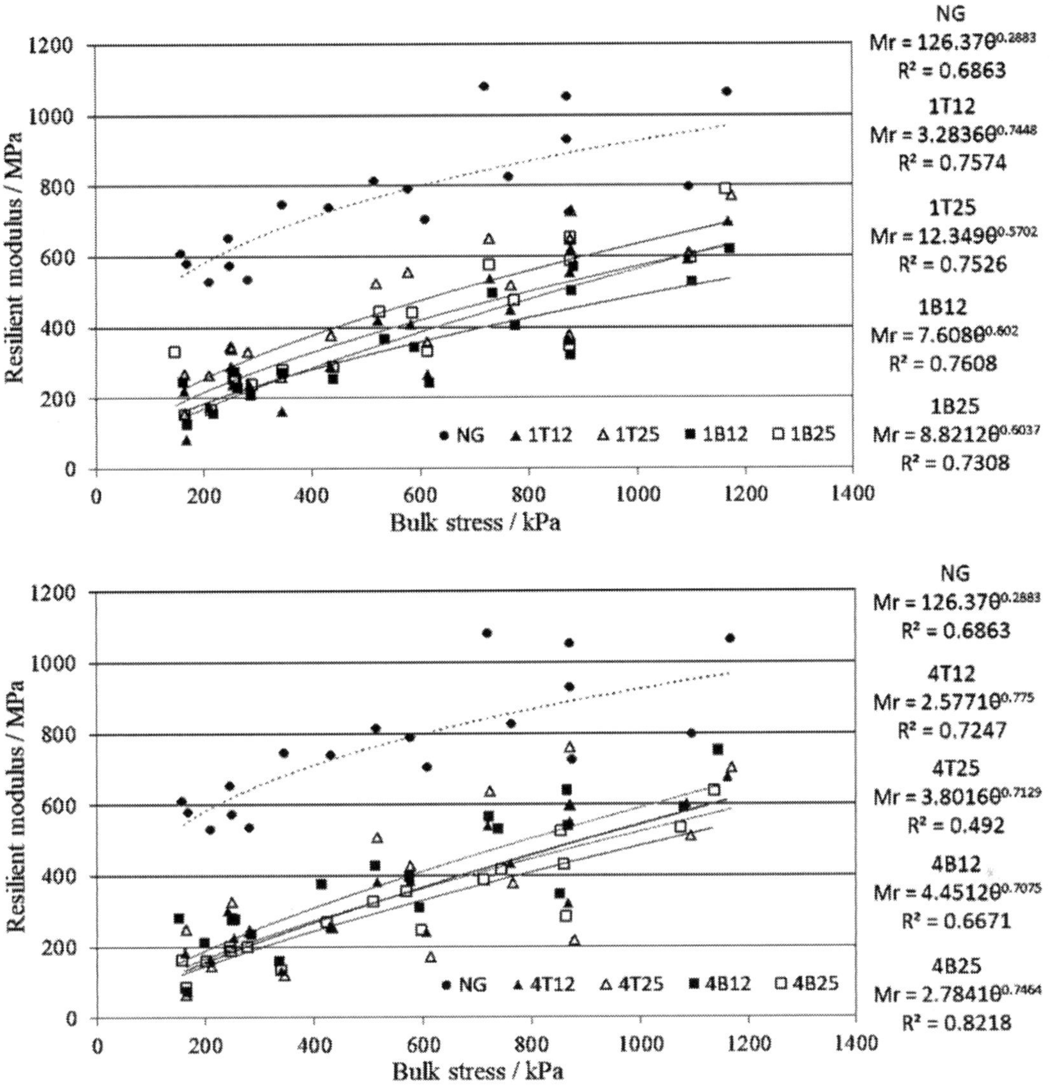

Figure 3. Values of resilient moduli of the unreinforced and reinforced specimens, top: with 1 geogrid in the middle of specimen height; bottom: with 4 geogrids

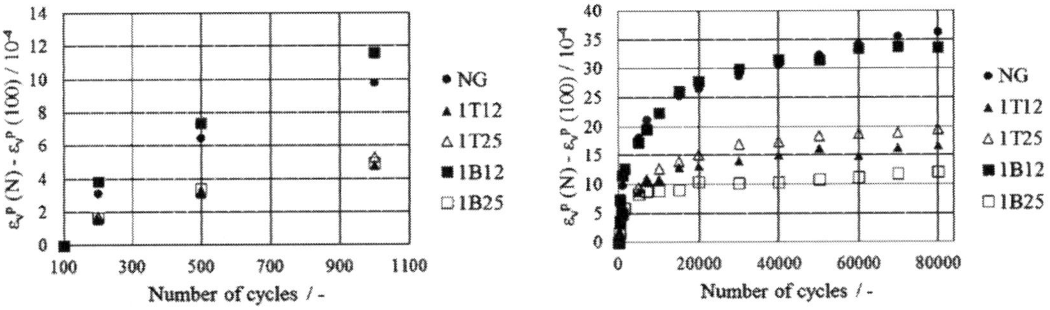

Figure 4. Permanent volume deformations of unreinforced and reinforced specimens with 1 geogrid in the middle, left: results for first 1000 load cycles; right: results for all 80000 load cycles

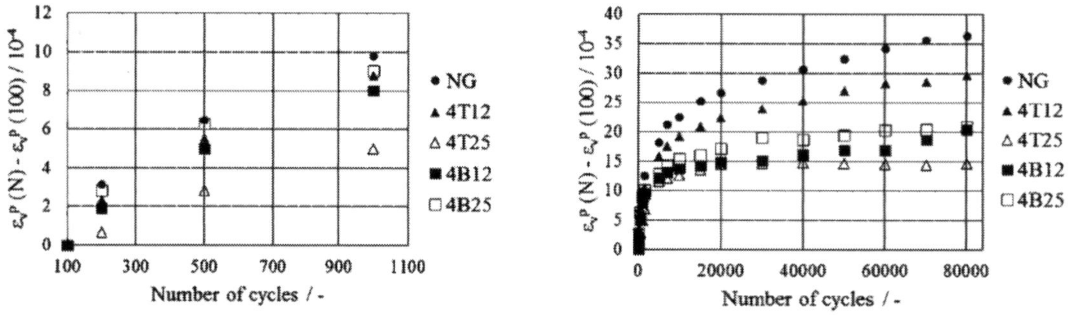

Figure 5. Permanent volume deformations of unreinforced and reinforced specimens with 4 geogrids, left: results for first 1000 load cycles; right: results for all 80000 load cycles

4. Conclusion

Cyclic triaxial shear tests were conducted on unreinforced and reinforced specimens of well-graded non-cohesive soil in order to determine the influence of position and number of geogrids, their geometry and size of geogrid aperture on resilient modulus improvement and permanent deformation reduction. Four different geogrids were made with 3D printing, having triangular and square openings, 1 and 2 mm thick ribs, and different aperture size ratios with D_{50} of 2 and 5. Specimens with one and four geogrid layers were tested. The obtained results show improvement of the non-cohesive soil properties when reinforced in the form of a reduction of permanent deformations. However, improvements of the resilient modulus of reinforced specimens were not observed. Reinforced specimens with geogrids of larger aperture size ($A/D_{50} = 5$) and rib thickness of 2 mm performed better than specimens with geogrids of smaller aperture size ($A/D_{50} = 2$) and rib thickness of 1 mm. It should be emphasised that in case of four geogrids in the specimen, their distances are too small compared to what is used in practice. On the one hand, short geogrid distance deteriorates specimen modulus, but on the other hand, it helps in better lateral restraining of the specimen. Further testing of reinforced specimens with geogrids of different rib dimensions (width, thickness and length) is needed and planned.

Acknowledgments

The authors would like to thank the Slovenian National Building and Civil Engineering Institute for the help in carrying out the research on cyclic load triaxial equipment.

5. References

[1] Perkins S W, Christopher B R, Cuelho E L, Eiksund G R, Hoff I, Schwartz C W, Swano G and Want A 2004 Development of design methods for geosynthetic reinforced flexible pavements (FHWA Report DTFH61-01-X-00068) p 263

[2] Moghaddas Nejad F and Small J 2003 Resilient and permanent characteristics of reinforced granular materials by repeated load triaxial tests *Geotech. Testing J.* **26** pp 152-66

[3] Abu-Farsakh M, Nazzal M and Mohammad L 2008 Characterizing the behaviour of geogrid-reinforced crushed limestone base materials using monotonic and cyclic load tests *Proc. 4th Int. Gulf Conf. on Roads* 10-13 November 2008 (Doha) ed I L Al-Qadi *et al* (London, Taylor&Francis Group) pp 753-9

[4] Abu-Farsakh M, Souci G, Voyiadjis G and Chen Q 2012 Evaluation of factors affecting the performance of geogrid-reinforced granular base material using repeated load triaxial tests *J. Mat. Civil Eng.* **24** pp72-83

IOP Conf. Series: Materials Science and Engineering 1260 (2022) 012011 doi:10.1088/1757-899X/1260/1/012011

[5] Sarsby R 1985 The influence of aperture size/particle size on the efficiency of grid reinforcement *Proc. 2nd Canadian Symp. Geotext. Geomembr.* (Edmonton) pp 7-12

[6] Koerner R M 2005 *Designing with geosynthetics* 5th Edition (New Jersey, Prentice Hall) p 816

[7] Holtz R D, Christopher B R and Berg R R 2008 *Geosynthetic design and construction guidelines, Participant notebook NHI Course No 13213* (FHWA Publication FHWA HI-95-038 (revised) Federal Highway Administration Washington DC) p 484

[8] Minazek K and Mulabdic M 2013 A review of soil and reinforcement interaction testing in reinforced soil by pullout test *Građevinar* **65** pp 235-50

[9] Palmeira E and Gongora I 2016 Assessing the influence of some soil-reinforcement interaction parameters on the performance of a low fill on compressible subgrade. Part I: Fill performance and relevance of interaction parameters *Int. J. Geosynth. Ground Eng.* **2** p 17

[10] Brown S F, Kwan J and Thom N H 2007 Identifying the key parameters that influence geogrid reinforcement of railway ballast *Geotext. Geomembr.* **25** pp 326–35

[11] Hussaini S K K 2013 An experimental study on the deformation behaviour of geosynthetically reinforced ballast *PhD thesis* (University of Wollongong) p 211

[12] Han B, Ling J, Shu X, Gong H and Huang B 2018 Laboratory investigation of particle size effects on the shear behaviour of aggregate-geogrid interface *Constr. Build. Mater.* **158** pp 1015-25

[13] Opći tehnički uvjeti za radove na cestama knjiga III – Kolnička konstrukcija 2011 (Zagreb IGH) p 145

[14] EN ISO 14688-2:2018 Geotechincal investigation and testing – Identifiacation and classification of soil – Part 2: Principles for a classification (ISO 14688-2:2017)

[15] Zortrax 2018 Z-ULTRAT Safety data sheet 2.00 (Poland Zortrax S A) p 8

[16] EN 13286-2:2010 Unbound and hydraulically bound mixtures – Part 2: Test methods for laboratory dry density and water content – Proctor compaction

[17] EN 13286-7:2004 Unbound and hydraulically bound mixtures – Part 7: Cyclic load triaxial test for unbound mixtures

[18] AFNOR 1995 NF P98-235-1 – Test relating to pavements. Unbound granular materials. Part 1: Repeated Loading Triaxial Test

The influence of thermally and mechanically induced short-term strains on smooth HDPE geomembrane properties

M Mulabdić[1], A Jukić[2], F Faraguna[2] and K Minažek[1]

[1] Faculty of Civil Engineering and Architecture Osijek, Josip Juraj Strossmayer University of Osijek, Vladimira Preloga 3, 31000 Osijek, Croatia
[2] Faculty of Chemical Engineering and Technology, University of Zagreb, Marulićev trg 19, 10000 Zagreb, Croatia

Email: leta@zg.t-com.hr

Abstract. HDPE geomembranes are usually used in bottom liners in landfills. Their integrity is required during installation and during the lifetime of the landfill. It is commonly accepted by researchers and suggested by different technical specifications that there should be no strain present during the installation of a geomembrane, although some of them propose that the strain during the installation should be no more than 3%. This is not easy to satisfy in some situations. Strains in the geomembranes can be induced by overnight temperature changes, before covering of the geomembrane (cold weather causes stretching of the geomembrane), which in turn may create trampoline effect in the veneer of sloped foundation soil. Covering of the geomembrane can wait higher temperatures to bring geomembrane into good close contact with the substructure profile or can be forced in order to bring the geomembrane into that position, which means accumulation of strains in the geomembranes. The question arises: What are the changes that occur in geomembrane properties for thermally and/or mechanically induced strains in the geomembranes, are they different, and how the level of induced strain and aging of strained geomembrane can change its properties? This paper presents a large project related to laboratory study of this phenomenon, and the first results regarding its short-time behaviour. Mechanical and structural changes of the geomembrane properties are tested and monitored in the first few days after inducing thermal and/or mechanical strains of 0%, 3% and 6%. Specially developed frames were used for the preparation and conditioning of the geomembrane samples. Test results suggest that strains induced by temperature drop and/or mechanical extension do not affect short-term geomembrane properties. Research continues on samples that will be monitored for four years.

1. Introduction

Geomembrane in landfill liners (usually HDPE) is subjected to severe stresses, from installation through the landfill lifetime, which can compromise its performance and/or its duration. Researchers agree that tensile strain should be kept at a very low level. Strains in geomembrane (and that means stresses too) can be induced in the geomembrane during installation and before the filling of waste starts. Induced strains (stresses) together with the temperature that will be developed by waste degradation, and which includes influence of filtrate, can significantly shorten lifetime of the geomembrane [1,2,3]. Geomembrane should be in intimate contact with its bedding which should be carefully prepared [3-9], When geomembrane is installed and not covered, decrease of temperature will cause stretching of the geomembrane in veneer of slopes and around dividing embankments, where bedding planes change (Figure 1). This will cause lifting of the geomembranes which is known as trampoline effect or bridging

[3]. That situation makes construction work difficult and increases the risk for damage of the geomembrane [3,4,6,10,11,12,13,14,15,16]. At the bridging location the geomembrane can be either cut out and replaced or pressed by the weight of drainage material that will bring it in contact with bedding soil, taking in account induced tensile strain in the geomembrane. However, limits of allowable strain in geomembrane are not clearly stated. Some researchers propose 3% as a good upper limit of tensile strain in geomembrane [17], and some documents [9] suggest a very strong strain upper limit of 0.25%.

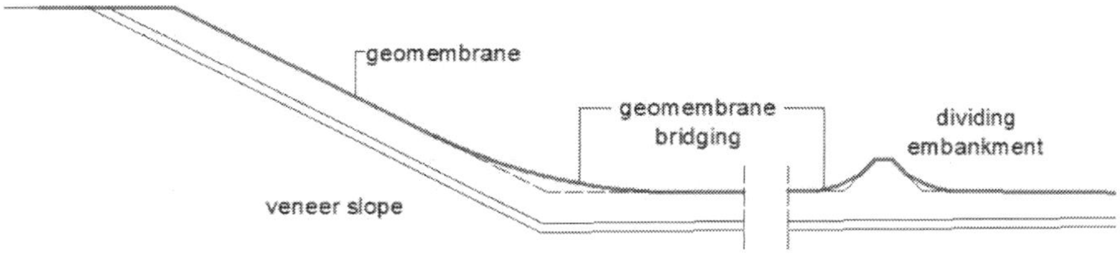

Figure 1. Trampoline effect at veneer of slopes

This paper presents the first short-term results of the research that will examine mechanical and structural changes of the smooth HDPE geomembrane in time after tensile strains are induced in geomembrane by temperature changes (trampoline effect) and mechanically (loading of lifted geomembrane by drainage material) in the period of four years.

Six types of samples in total were prepared (sample here denotes a piece of geomembrane cut from the role in order to fit metal frame used for stretching it) under different conditions as described in Table 1 for both MD (machine direction) and CMD (cross machine direction) geomembrane samples' orientation. Specially constructed metal frame was used to hold the sample under the required strain and at the required temperature (see Figure 2). Five specimens were formed from every sample for testing.

Mechanical strain produced in a frame was 3% and 6%, and thermal strain was a consequence of exposing a sample to temperature of -10°C in a special chamber. Sample was prepared as a sheet fixed in a frame which was capable of imposing extension on it and/or was exposed to specified temperature conditions, and then cut into five specimens further testing – see Table 1.

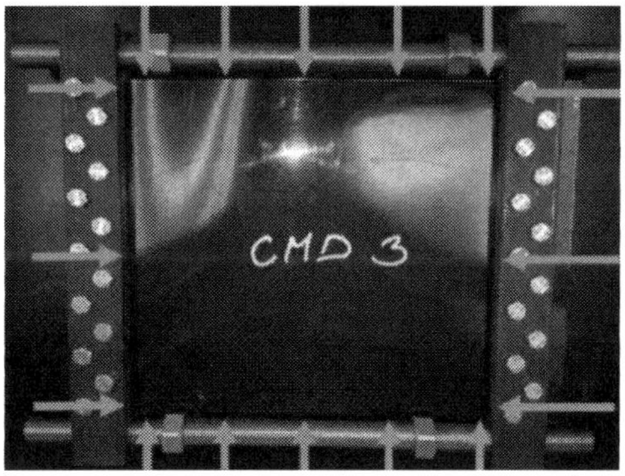

Figure 2. Geomebrane with arrows pointing at the direction of sample cutting: 1 – MD samples, 2 – CMD samples (left): geomembrane sample in metal frame with arrows pointing at the position of length measurements (right)

Table 1. Types of samples and their conditioning, HDPE smooth geomembrane

Sample type[a]	Sample conditioning
A	non-stretched, temperature 20°C at test
B	non-stretched, cooled at temperature -10 °C for 48 hours then tested at 20°C
C3	stretched to strain of 3% for 48 hours, released and tested at 20°C
C6	stretched to strain of 6% for 48 hours, released and tested at 20°C
D3	in frame cooled at -10°C for 48 hours, then stretched to 3% for 48 hours, released and tested at 20°C
D6	in frame cooled at -10°C for 48 hours, then stretched to 6% for 48 hours, released and tested at 20°C

[a] Separate samples were formed for MD and CMD direction. Testing of mechanical properties within 1 hour after conditioning.

2. Testing of mechanical properties

Five specimens were tested from every group (A-D6, table 1), for MD and CMD direction, for their yielding stress and strain according to the standard EN ISO 527-03 [18]. Figure 3 gives the results of tensile test for all the samples.

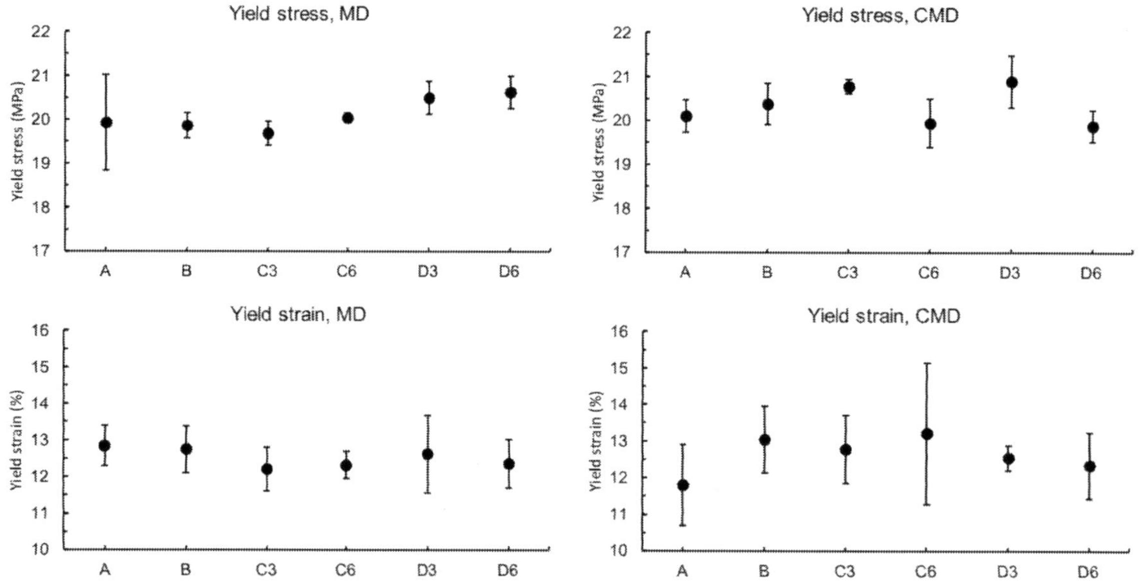

Figure 3. Yielding stress and strain for MD and CMD direction

It can be seen that different conditioning of the samples does not produce significant changes in stress and strain at yield point. For MD direction yield stresses are almost the same as for non-conditioned sample, except for B and C3 that are somewhat lower. For CMD direction yield stresses are somewhat higher for C3 and D3 samples compared to all others.

Yielding strain is almost the same for all the samples, for MD direction, while for CMD direction A, D3 and D6 samples have similar values and B, C3 and C6 have somewhat higher values. Generally, stresses and strains at yield do not change significantly, with no severe change of sample conditioning for the type of conditioning used in this research and for a short time of conditioning (several days).

For each group (A-D6) in MD direction E modulus was calculated for five specimens under the same conditions at strains of 1%, 2% and 3%, based on the tensile test record, as a tangential value around those strains. Average E value for five specimens (for each group) having the same conditioning is presented in Figure 4. It seems that E value drops to around a half of its value for virgin specimen at

1% to 2% strain and drops to 30% at 2% to 3% strain. Values of E are not sensitive to the applied short-time conditioning type, but initial values (at 1% strain) for conditioned samples are somewhat lower than for virgin sample.

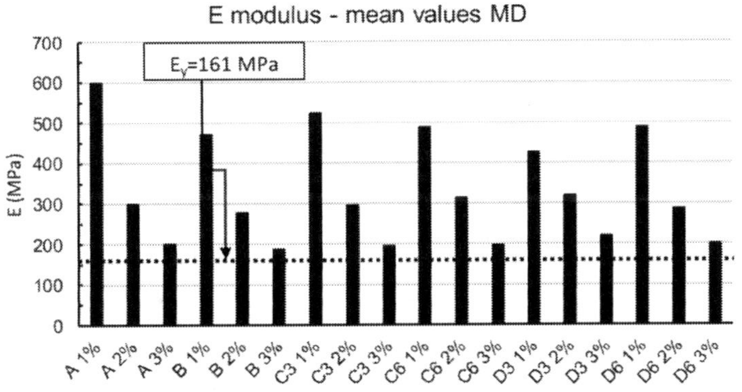

Figure 4. Average values of five tangential E modulus for every group at strains of 1%, 2% and 3% for MD direction, E_y – secant E modulus for stress ranges from 0-yield stress

Specimen dimensions were measured when they were put in the frame and at the end of the conditioning while still in the frame. As expected, in all specimens that were stretched, a reduction of width was observed (Tables 2 and 3). Taking into account possible measurement error of 0.4 %, MD samples (direction 1) and CMD samples (direction 2) show almost the same reduction of width at the same experimental conditions. It can be concluded that geomembrane stretching direction (MD versus CMD) has no influence on width reduction of the specimens. Specimens from C3 and D3 show width reduction of up to -1.24%, while specimens from C6 and D6 show higher width reduction of up to -2.94%.

Table 2. Width change of MD specimens at the end of conditioning.

	w_1 / %	w_2/ %	w_3/ %	w_4/ %	w_5/ %
1B	0.00	0.41	0.41	0.41	0.00
1C3	0.00	-0.41	-0.82	-0.41	0.00
1C6	-1.22	-2.04	-2.04	-2.04	-1.63
1D3	-0.41	-1.22	-0.82	-0.41	0.00
1D6	-1.22	-2.45	-2.46	-2.05	-1.63

Table 3. Width change of CMD specimens at the end of conditioning

	w_1 / %	w_2/ %	w_3/ %	w_4/ %	w_5/ %
2B	0.41	0.41	0.41	0.41	0.41
2C3	-0.41	-0.82	-0.82	-0.82	0.00
2C6	-1.24	-1.67	-1.67	-1.67	-1.24
2D3	-0.41	-0.83	-1.24	-1.23	-0.82
2D6	-1.24	-2.51	-2.94	-2.51	-1.66

3. Structural properties

Two different types of monitoring were used for detection of structural changes in geomembrane which occurred as a consequence of different conditioning:

a) Differential scanning calorimetry (DSC) analysis was performed by the DSC823e Mettler Toledo instrument. Masses of examined specimens were 10–20 mg. One heating run from 25 to 180°C was performed at the heating rate of 10°C min^{-1}.

b) Scanning electron microscopy (SEM) recording was carried out by Vega 3 Tescan microscope with a detector of secondary electrons. All specimens were previously sputter coated with Au or Pd in the argon plasma to enhance their conductivity.

3.1. Influence of testing conditions on crystallinity and melting temperature

After performing tensile tests, specimens were taken near the middle of the elongated part for DSC measurements. Specimens from virgin sample 1A were taken and analyzed 10 times to determine standard deviation of melting temperature and enthalpy of melting due to inhomogeneity, while all the other samples were analyzed in triplets. Results will be given as averages with corresponding standard deviations. During the heating in range from 25 to180°C one endothermic signal was observed which is compatible with the melting of the crystalline domains of polyethylene specimen (Figure 5).

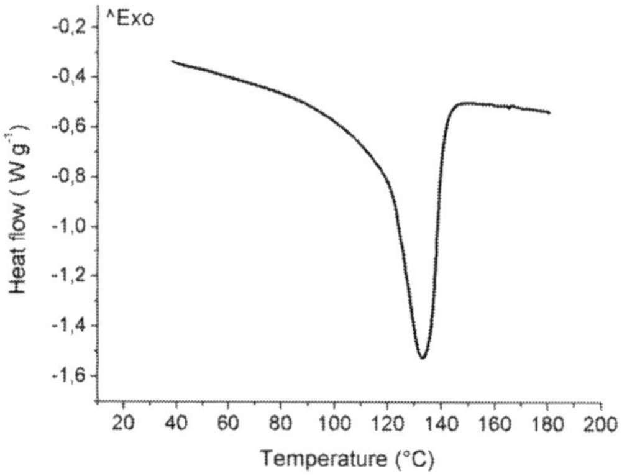

Figure 5. Typical DSC thermogram of the geomembrane with endothermic signal of melting (Specimen 1A)

Average sample melting temperature in both sets, independent of testing conditions, changes within ±1.5°C with respect to the melting temperature of virgin sample (figure 6). It can be concluded that, in the tested range of temperatures and strains, different conditions do not affect the melting temperature of the geomembrane significantly.

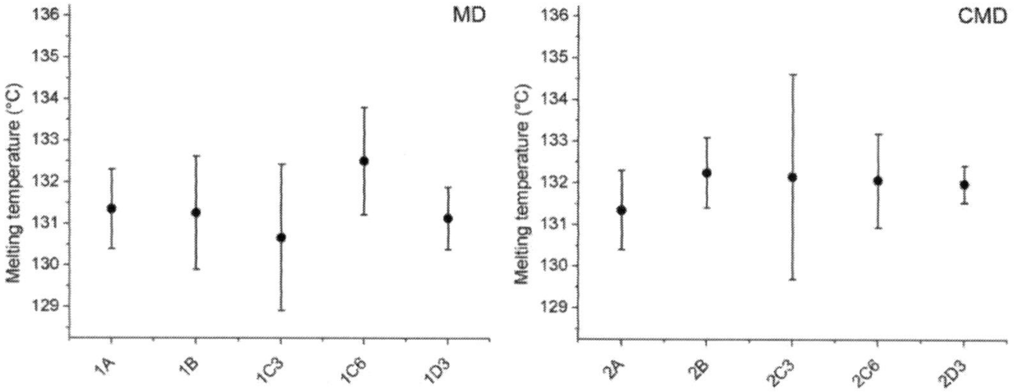

Figure 6. Melting temperatures of samples kept at different conditions: 1st set (left), 2nd set (right)

The average melting enthalpies of samples for MD direction (direction 1) show a decrease compared to the virgin (non-conditioned), except for 1B. In case of experiments 1C3 and 1C6 that are stretched at the room temperature, there is no difference between those stretched for 3% and 6%, while for those that are stretched and kept at -10°C, the one that is more stretched has a lower average melting enthalpy (figure 7).

All CMD specimens (direction 2) have lower melting enthalpies. Here, independent of the temperature, the specimens are stretched and kept at that position (experiments 2C3, 2C6, 2D3, 2D6). More stretched specimens tend to show lower average melting enthalpy. The decrease of average melting enthalpy is more evident in case of specimens 2D3 and 2D6 (stretched and kept at -10 °C).

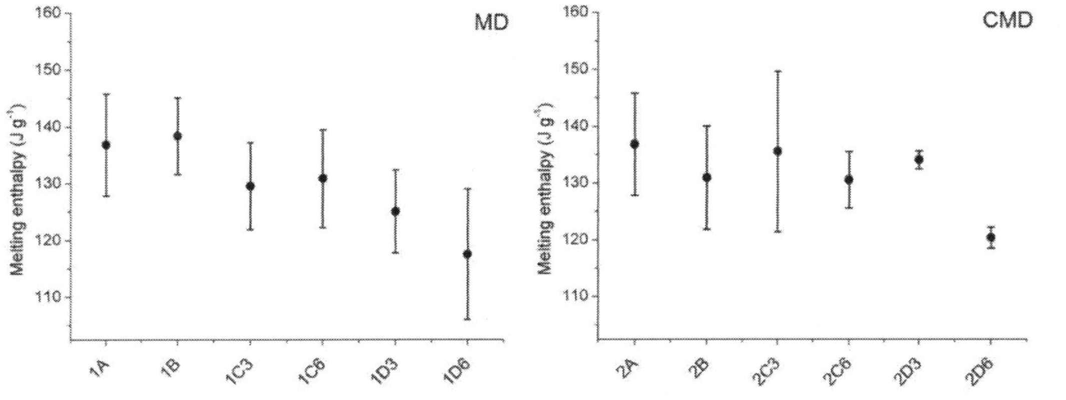

Figure 7. Melting enthalpies of specimens kept at different conditions: 1st set (left), 2nd set (right)

3.2. Influence of testing conditions on specimen morphology

The specimens were cut and the cross section was analyzed by scanning electron microscopy with the aim to analyze the specimen's morphology after tests at different conditions. Also, specimens were analyzed after breakage caused by the testing machine. No major/distinct morphology difference was observed between different cross sections of specimens after testing conditions (Figures 8-12).

Figure 8. SEM images of the specimens 1A, 1C6, 1D6 (upper row) and 2A, 2C6, 2D6 (lower row)

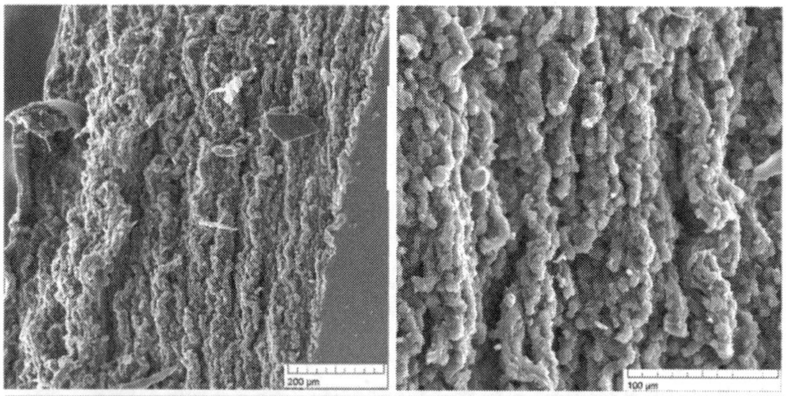

Figure 9. SEM images of the specimen 1A

Figure 10. SEM images of the specimen 1C3

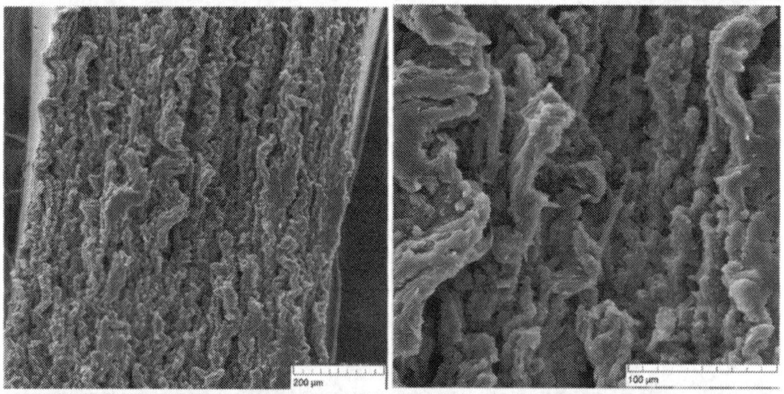

Figure 11. SEM images of the specimen 2A

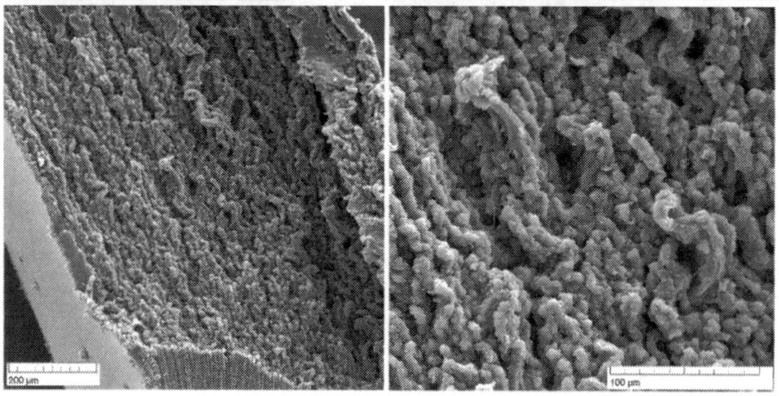

Figure12. SEM images of the specimen 2C3

4. Conclusion

The results of testing short-term changes of mechanical and structural properties of the smooth HDPE geomembrane after being pre-strained and/or cooled are presented. Several types of conditioning of the geomembrane specimens was adopted to simulate conditions that cause HDPE geomembrane bridging in landfill liner. Cooled specimens, stretched specimens and first cooled and then stretched specimens were subjected to the tensile test and afterwards their morphology and structure were examined. It seems that relatively dramatic changes in temperature (20 to -10°C) and the strains of 3-6%, or their combination, do not change the yield stress and yield strain significantly, nor the modulus of elasticity in short-time conditioning (several days). Long-term behaviour may be different, and the influence of leachate action can contribute to larger changes of the geomembrane properties, which needs to be further investigated. The memory of geomembrane preconditioning, even for a short-time period, may become significant for the long-term geomembrane properties. This research is a part of the four-year program that will monitor changes in pre-strained and cooled geomembranes in a longer period.

References

[1] Peggs I D 2003 Geomembrane Liner Durability Contributing Factors and the Status Quo *1st United Kingdom Geosynthetics Symp.* UK Chapter of IGS, Invited Keynote Speaker

[2] Rowe R K and Islam M Z 2009 Impact of Landfill Liner Time-Temperature History on the Service Life of HDPE Geomembranes *Waste Management* **29** pp 2689-99

[3] Scheirs J 2009 A Guide to Polymeric Geomembranes: A Practical Approach, West Sussex : Wiley Series in Polymer Science, John Wiley and Sons Ltd.

[4] Jaros D L 1991 CERCLA Landfill Closures: Construction Considerations (Landfill closures - Geosynthetics, Interface Friction and New Developments, editor R.M. Körner, Reprinted from *Geotextiles and Geomembranes* **10**, Elsevier Science Publishers Ltd, England)

[5] Daniel D E 1993 *Landfills and Impoundments, Geotechnical Practice for Waste Disposal,* (Springer Science + Business Media, B.V.)

[6] Müller W W 2006 *HDPE Geomembranes in Geotechnics* (Berlin Heidelberg Springer-Verlag)

[7] International Association of geosynthetic installers – IAGI 2007 *HDPE and LLDPE Geomembrane Installation Specification*, Revision

[8] Koerner R M and Koerner G R 2013 *The Intimate Contact Issue of Field Placed Geomembranes with Respect to Wave (or Wrinkle) Managemen* (GSI White Paper #27, Geosynthetic Insitute, USA)

[9] BAM 2016 *Guidelines for the cCrtification of Geomembranes Used to Line Landfills*, (issued by Division 4.3 "Contaminant Transfer and Environmental Technologies, 7th edition)

[10] Bonaparte R, Daniel E D and Koerner R M 2002 *Assessment and Recommendations for Improving the Performance of Waste Containment Systems* (Washington DC: U.S. Environmental Protection Agency, EPA/600/R-02/099 NTIS PB2003-103048)

[11] Elton D J and Peggs I D 2002 Geomembrane Research Needs *Geosynthetics International* **9**, pp 283-300

[12] Erickson R B, Thiel R S and Peters J 2008 The Ongoing Quality Issues Regarding Polyethylene Geomembrane Material Manufacturing & Installation *The First Pan American Geosynthetics Con & Exhibition*, Cancun, Mexico

[13] Hsuan Y G, Koerner R M and Comer A I 2013 Cold Temperature and Freeze-Thaw Cycling Behavior of Geomembranes and Their Seams, (GSI White Paper #28, Geosynth. Insitute, USA)

[14] Thiel R 2013 *Proc. of the 25th Annual GRI Conf., Geosynthetics* 2013 *A 25-Year Perspective on Waste Containment Liner and Cover System Designs using Geosynthetics* (in coordination with IFA, Long Beach, CA)

[15] Rowe R K 2015 Protecting the Environment with Geosynthetics *International Conf. on Geotechnical Engineering*, Colombo, Sri Lanka

[16] United Facilities Guide Specifications 2015 Full Impermeable Liner System, UFGS- 33 56 19

[17] Peggs I D, Schmucker B and Carey P 2005 Assessment of Maximum Allowable Strains In Polyethylene and Polypropylene Geomembranes *Geo Frontiers* 2005(Austin, Texas)
[18] ENISO527-03 *Plastics-Determination of Tensile Properties - Part 3:Test Conditions for Films and Sheets*

Geodynamic Modelling of Geogrid-Stabilised Aggregate on Soft Ground in High Speed Rail Environments Using Field-Derived Shear Wave Data.

B Ogunmakin[1], K Zamara[2], J Kawalec[2], M C Todisco[1]

[1] Coffey Geotechnics Ltd UK

[2] Tensar International Ltd UK

Email: basil.ogunmakin@coffey.com

Abstract. Geogrids have been used for many years to stabilise and enhance the performance of rail ballast and sub-ballast in terms of maintaining both vertical and lateral track alignment. However, the ability of geogrids to improve these properties within a high-speed rail environment at close-to-critical velocity remain uncertain. The potential for geogrid to improve the stiffness and the ability of a geogrid-stabilised aggregate layer to transmit ground waves such as Rayleigh waves has great potential for reducing dig out depth in soft ground in cuttings and reduce the need to use great thicknesses of engineered, high quality aggregate in embankments in order to mitigate the effects of such waves. New research has been conducted on a trial embankment comparing 4 geogrid-stabilised trial sections with a control section. This embankment has been subjected to low strain stiffness testing using geophysical methods to determine if the stabilised sections can offer an increase in ground wave velocity over the control and thereby offer the potential to reduce engineered construction thickness beneath the track on high-speed rail lines. Initial geodynamic modelling using the results of the on-site testing suggests that specific geogrids have a significant potential to reduce overall construction costs by reducing the need to dig-out-and-replace in soft ground in order to mitigate the effects of Rayleigh waves.

1 Introduction

This paper presents the findings of a field test carried out to investigate the use of a mechanically stabilised layer (MSL) to improve the small strain stiffness of earthworks. This is to inform the assessment of the suitability of MSL as a viable solution for improving the geo-dynamic response of track infrastructure (ballast) and sub-grade material under high speed train loads. As the demand for high speed train lines continues to rise, faster travel time, higher than previously achieved on such infrastructure anywhere else in the world is also on the rise. An example of this is the High Speed Train 2 (HS2) in the UK which was is being designed for speed up to 350km/h (97mph). This brings about added complexity to the challenge of understanding the dynamic response of the infrastructure supporting the trackbed and developing cost effective means of mitigating the effects vibration resulting from the speed. High speed train type of loading generates mainly surface waves which result in ground vibrations. When the train speed approaches the Rayleigh wave velocity (Vr) of the track-ground system (at about 320km/hr depending on the properties of the sub-track material), excessive deformation of the track could occur. This is known as the critical speed or Rayleigh waves effects and could result in track safety issues, costly long-term maintenance; and potential damages to adjacent structures.
The key mechanical property required to improve the performance of track bed earthworks under dynamic loading is the shear wave velocity, which is largely controlled by the overall stiffness of the earthworks [1].

One of the traditional methods for solving this problem includes replacement of substandard material in the foundation with high quality granular material. Where the thickness of the substandard soil is substantial the cost of 'dig out and replace' can be significant.

The work of Kaynia et al. [1] investigated two approaches for mitigating critical speed effects from high speed train moving loads on an embankment namely, ground stiffening approach and ground strengthening approaches. Examples of ground strengthening considered in the paper include use of piles (concrete, lime etc) also previously studied by Karlsrud [2]. The use of concrete slab or concrete box girder, if higher stiffness is required, were considered as examples of ground strengthening. Kaynia et al. [1] concluded the ground stiffening is a more efficient countermeasure to mitigate against the critical speed effects of high speed train moving loads.

Several authors (e.g. Alves Costa et al.[3], Ntotsios et al. [4]) have shown that the dispersive nature of a layered soil foundation system results in a much higher resistances to critical speed effects (or geodynamic response) in comparison to homogenous foundation with similar properties. The benefits of layering can therefore be incorporated into developing a cost-effective solution for critical speed effects.

Tensar stabilisation geogrids have been shown to provide improved vertical large strain stiffness in a mechanically stabilised layer (MSL) and have been used for this purpose in railway infrastructure [5]. The use of MSL as a layered track bed foundation system could therefore be a cost effective alternative.

There are two key questions that need to be answered to demonstrate the suitability of MSL in this context:

1. Can the use of geogrid in a mechanically stabilised granular soil (i.e. MSL) provide improvement in the very small strain stiffness of the soil?
2. Is such improvement sufficient to resist dynamic loading at the speed from 'modern' high speed trains?

In view of the above, Continuous Surface Wave (CSW) tests have been carried out on a trial embankment consisting of four test sections (with geogrids) and a control section (without geogrids) to provide information on the stiffness of MSL at small strain. The outcome of this series of tests will be presented and discussed.With regards to the second key question above, 3-dimensional (3D) finite element modelling is carried out to investigate this. The result of this analysis is also presented in this paper.

2 Trial embankment and testing methodology

A large-scale trial embankment is used in this research to investigate if the inclusions of TriAx geogrid in granular material to produce mechanically stabilised layers (MSL) of soils can result in improvements in the dynamic stiffness of the stabilised soil. Geophysical surface wave technology using Continuous Surface Wave (CSW) is used to measure Rayleigh waves velocity of the embankment material. CSW is testing is a non-invasive seismic technique for ground stiffness measurement. This relies on measurement of the ground's response to a vibratory source or 'shaker'. The velocity of surface Rayleigh waves (Vr) generated by the shaker are measured by an array of geophones as they travel through the soil. The velocity of the surface waves generated is related to ground stiffness, with the depth of wave penetration increasing with reducing frequency and increasing wavelength (λ). That is, high frequency waves generated by the source, travel in shallow layers only whilst lower frequencies travel both in shallow and deep layers. Wave velocity will therefore vary with frequency and depth [6]. This phenomenon is commonly known as geometrical dispersion and it is different to the dispersion related to attenuation [7].

Surface waves are formed by the interaction of P (pressure) and S (shear) waves in proximity to the ground surface. The influence of P waves in determining Vr is small compared to the influence of S waves. The S wave velocity (Vs) is the key parameter required to permit determination of ground stiffness as obtained through inversion of the field dispersion curve. This is derived from Equation 1 below:

$$V_R = \frac{0.87+1.12v}{1+v} \times V_S \quad (1)$$

(V_R) and V_S are the Rayleigh Wave and shear velocity respectively for an elastic soil and v is Poisson ratio.

The shear wave velocity profile derived by the inversion can be converted to a small strain shear modulus (G_0) using a simple relationship with soil density (Equation 2). Given the nature of the relationship with, and limited variance of, soil density the derivation of G_0 is relatively insensitive to assumed soil density if not known.

$$G_o = \rho V_s^2 \quad (2)$$

G_o is the small strain stiffness property and ρ is material density.

The stiffness values obtained by CSW testing are small-strain stiffness values relevant to strain levels below approximately 0.002%. The test was carried out using the following two Seismic sources and array geophones:

- Standard Shaker (80kg) Shaker - 10 to 91Hz
- Electromagnetic Shaker – 50 to 400Hz

2.1 *The Trial Embankment Configuration, Test Site and Ground Conditions*

The tests were carried out on a trial embankment in a Limestone Quarry. The test embankment is 2.0m high and 40m long and is divided into 5 zones as shown in Table 1 and Figure 1.

Table 1. Fill material and geogrid to be used in the control and trial embankments

Embankment	Control	Trial Zone 1	Trial Zone 2	Trial Zone 3	Trial Zone 4
Fill Material*	Class 6F1	Class 6F1	Class 6F1	Class 6F1	Class 6F1
Geogrid in MSL	None	TriAx TX150	TriAx TX130	TriAx TX170	TriAx TX190

*Class 6F1 is a granular material as with grading properties as defined in the Specifications for Highway Works. (MCHW Series 600).

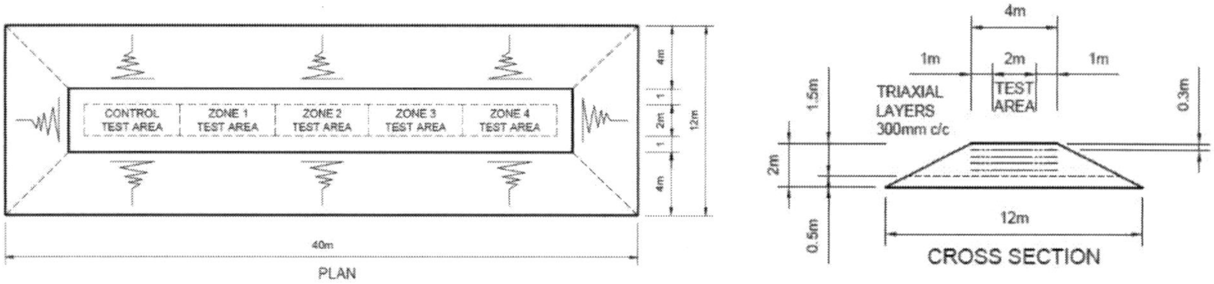

Figure 1. Trial Embankment Geometry: plan view and cross section through zones 1 to 4.

There was no ground investigation information on the material below the test embankment. However, it was understood from site observations and discussions with quarry personal that it is Made Ground which likely consists of quarry waste of varying grading (fine grained soil to boulder size grains) and material loosely compacted.The test was carried out on a rainy day. The soil below the Control Embankment and the TX150 zone was observed to be particularly wet in comparison to the rest of the embankment.

3 Test results and Discussions

As discussed in Section 2, the 'dispersion curve' shows the measured field data, Rayleigh Wave velocity (Vr) against frequency (f). Shear Wave velocity was subsequently obtained using Equation 1, assuming Poisson ratio ($\boldsymbol{\upsilon}$) of 0.26 for the embankment material. A summary of the test results is presented in Section 3.1.

The range of the combined frequencies of the two seismic sources used in the tests is between 10Hz to 400Hz. As discussed in Section 2, penetration depth is directly dependent on source frequency characteristics and predominantly on shear wave velocity of the medium. For example, the average shear velocity of test embankment is about 200m/s, then a 10Hz component of the Rayleigh wave generated would have penetrated to about 7 - 10m below ground level (bgl). Similarly, a 400Hz component of the wave would have penetrated to about 0.2 – 0.3m bgl.

Based on this, the test results show shear wave velocity profiles to 15.0m bgl. However, given that the zone of interest of the research is limited to the depth of the test embankment (i.e. 2.0m bgl), the shear wave velocities for each of the CSW test for the top 2.0m are presented in Figures 2 and 3.

Figure 2. Longitudinal shear wave velocity profiles at 0.02% strain. LONG – Longitudinal test with geophone array running along the length of the embankment. TX130S, 150,170 & 190 – Test embankment with different grades of TriAx geogrid. Control Embankment without geogrid

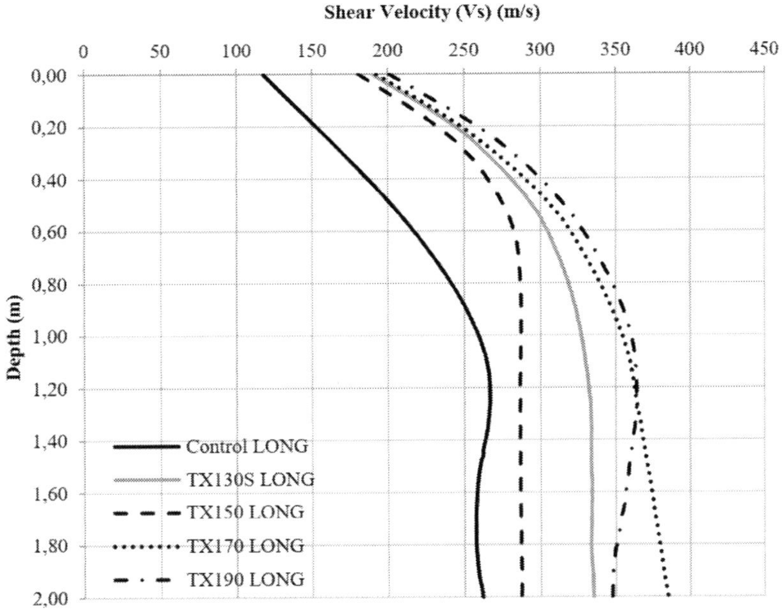

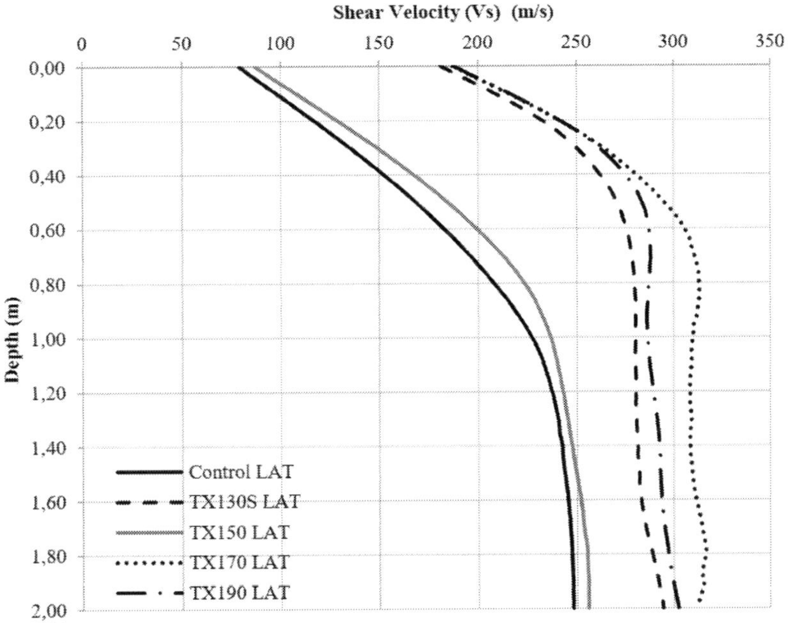

Figure 3. Lateral Shear wave velocity profile at 0.02% strain. LAT - Lateral test taken with geophone array running across the width of the embankment. TX130S, 150,170 & 190 – Test embankment with different grades of TriAx geogrid. Control Embankment without geogrid

A second round of CSW tests were carried out 2 months after the first set to confirm repeatability of the results. The outcome of the test shows very similar shear velocity profiles to those presented in Figures 2 and 3 above from the first round of testing.
To verify that similar degree of compaction was achieved in the embankment, Nuclear Density Meter (NDM) tests were carried out on the embankment. The NDM tests were carried out in the top 200mm of the test embankment only. The results of the NDM tests showed the average bulk density achieved across the length of the test embankment is between 2.25 to 2.29Mg/m^3. This is within acceptable range in terms uniformity of compaction of the material.

The following points summarise the results of the CSW tests:

- In general, the longitudinal stiffness for both the control and the test embankments was greater than the lateral stiffness by about 25%;
- TX190 is the stiffest of the geogrid used in the test embankments. In line with expectation, it appeared to have provided the most improvement in the small strain longitudinal stiffness modulus in comparison to the control embankment by about 70% at the top of the layer reducing to about 30% at the bottom,
- TX170 appeared to have improved the small strain longitudinal stiffness of the test embankment by about 60% at the top of the layer reducing to 40% at the bottom;
- TX130s has similar effect to TX170 at the top of the layer (i.e. 60% improvement) reducing to about 30% at the bottom;
- TX150 provided the least improvement on stiffness of the test embankment at about 60% improvement in the longitudinal strain at the top of the layer reducing to 10% at the bottom. The result for TX150 appears to be an anomaly as TX150 geogrid is stiffer than TX130s, hence it was expected that TX150 would provide similar or better improvement than TX130s.

The response of the TX150 zone appeared to be irregular, as the overall improvement from this geogrid in the MSL is less than what was achieved TX130 which is a softer geogrid. There are two probable reasons for this:

- Variability in the compactive effort applied to TX150 in comparison to the rest of the embankment: The results of NDM tests showed the average bulk density of the embankment in TX150 is similar to the rest of the test embankment. However, there is a possibility that this may be different at depth due to the second reason below.
- Effect of the difference in geogrid aperture size (or aspect ratio) on particle interaction: It is possible that weaker interlock was achieved in the TX150 due to a change in the aspect ratio of the apertures of the geogrid relative to particle sizes. This will need to be investigated further.

4 Numerical Modelling

We have shown in Section 3 that small strain stiffness of granular soil can be improved with geogrid in a mechanically stabilised layer (MSL). The question of whether such an improvement is enough to mitigate against critical speed effects of high-speed train is explored using 3D finite element analysis. The numerical modelling tools used for this assessment is Plaxis 3D, a general purpose commercially available 3D FE software.

The model size was 36m wide (x axis) by 40m long (y-axis) by 17m deep (z-axis). Wave reflection at the boundaries was reduced by selecting standard fixities and absorbent boundaries.

The dynamic loads were applied as moving point loads based on the Swedish X2000 HS train configuration. The rail was modelled using beam elements with properties representing a UIC60 rail. The beam was simply supported with support spaced every 60cm. The sleepers, spaced 60cm centre to centre, were connected to the rail with node-to-node anchor elements. The moving loads were applied adopting the dynamic multiplier method available in Plaxis 3D. This method allows shear forces representing the effects of the moving train to be applied as dynamic multipliers to each point load. A dynamic multiplier is expressed as a time-shear force signal in Plaxis 3D. The time step selected in the model were based on the train speeds and the space of beam supports, therefore given a constant train speed, time step was constant in the simulation. Three train speeds, V of 83, 97 and 111m/s, were used.

This model (Figure 4) was calibrated against COMSOL, a 3D finite element software developed by NGI [8] and validated against a vibration case at Ledsgaard site in Sweden.

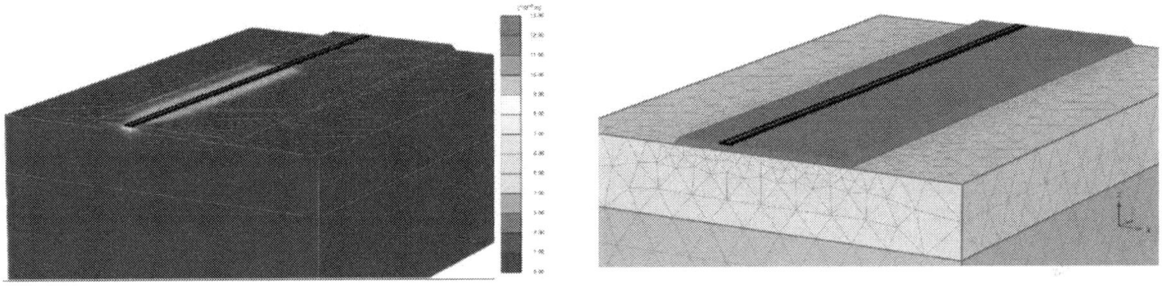

Figure 4. Graphical illustration of Plaxis model used in the study

IOP Conf. Series: Materials Science and Engineering 1260 (2022) 012013 doi:10.1088/1757-899X/1260/1/012013

5 Ground model & Soil Properties

Figure 5 illustrates the ground model and the all the various countermeasures considered in this paper. Three train speeds, V of 83, 97 and 111m/s, were used in order to evaluate the performance of the MSL treatment with respect to sub-critical to super-critical speeds. However, in this paper only the results of the simulation at 111m/s is presented.

As discussed in Section 1, the key properties for the assessment of the geodynamic behaviour of soil are stiffness properties at small strain and shear wave velocity. For the purpose of our analysis we made certain key assumptions to estimate small strain stiffness (E_o) for the soil types considered in our assessment based on the work of Atkinson [9] and Jardine et al. [10].

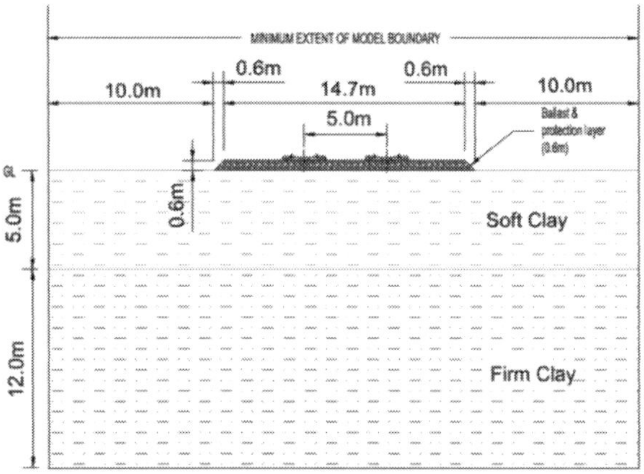

Case 1 (Base Case) : untreated ground

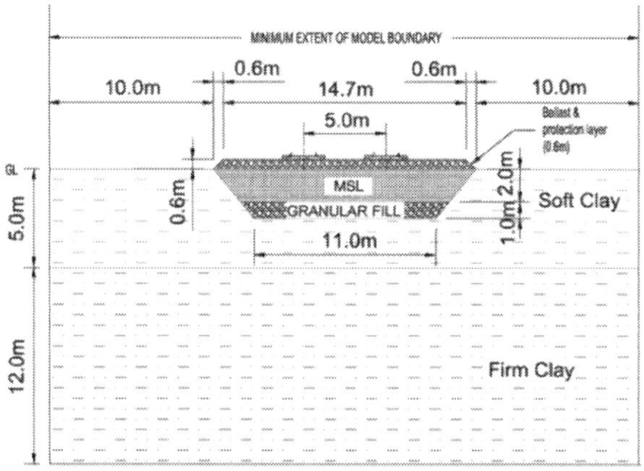

Case 2 – Layering 2.0m MSL with 1.0m granular fill

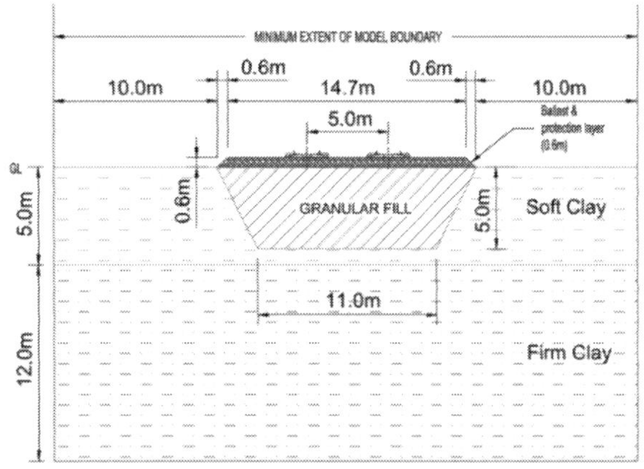

Case 3 – Granular Replacement to 5.0m depth

Figure 5 Assumed ground models and configurations of countermeasures considered in the assessment

The parameters of the granular fill and MSL are derived from the outcome of the field trial described in Section 3. A summary of the soil properties used in the assessment is presented in Table 2.

Table 2. Characteristic Geotechnical Properties

	Units	Soft Clay	Firm Clay	*Granular Fill	**MSL
Bulk Density	Mg/m3	2	2	2	2
PI	%	25	25	-	-
Effective Friction Angle		-	-	36	36
Undrained shear Strength (Cu)	kPa	30	30 + 6.25z	-	-
Youngs modulus at 0.01% strain (Eo)	MPa	*51	*51 + 10.6z	**184	***414
Shear Modulus at 0.01% strain (Go)	MPa	17	17 + 3.63z	80	180
Poisson's ratio	-	0.5	0.5	0.3	0.3

Z = depth from 5m below ground level.
*$E_o = 1700C_u$ (at 0.01% strain), based Jardine et al (1984) and Atkinson (2000)
**Field Trial – Control Embankment. by 12.5% / Eo=2.3Go
*Field Trial –Embankment based (TX170 & TX190) / Eo=2.3Go

5.1 FE Model Calibration

As discussed in Section 5.0, the Plaxis model used in this study is calibrated against COMSOL a 3D FE model developed at NGI and calibrated against a vibration site at Ledsgard in Sweden. The results of these validation exercises are presented in Figures 6 and 7.

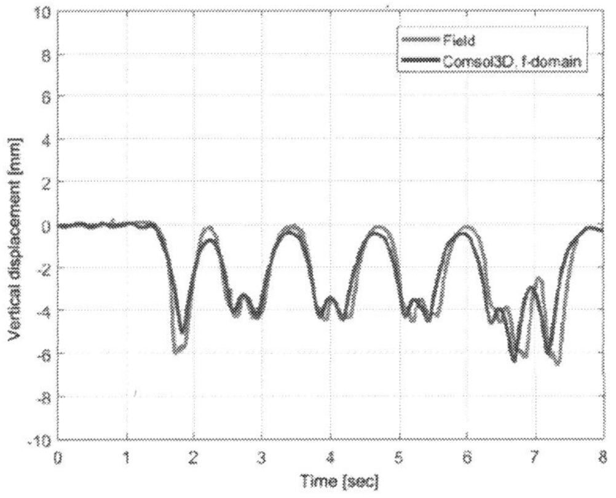

a)

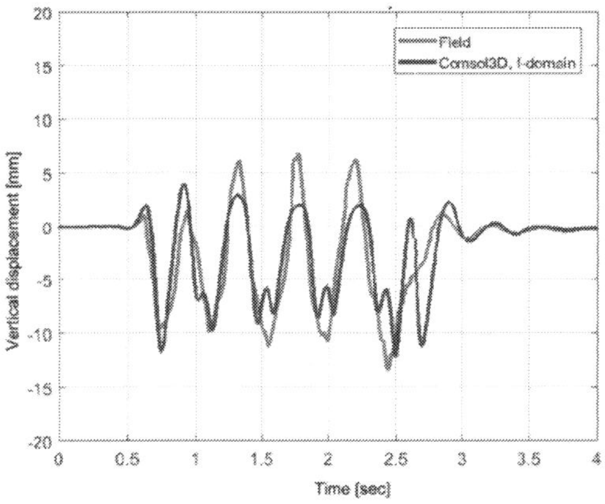

b)

Figure 6. Comparison of filed test data at Ledsgard with COMSOL simulations for train speed 70 km/h (a) and train speed 185 km/h (b)

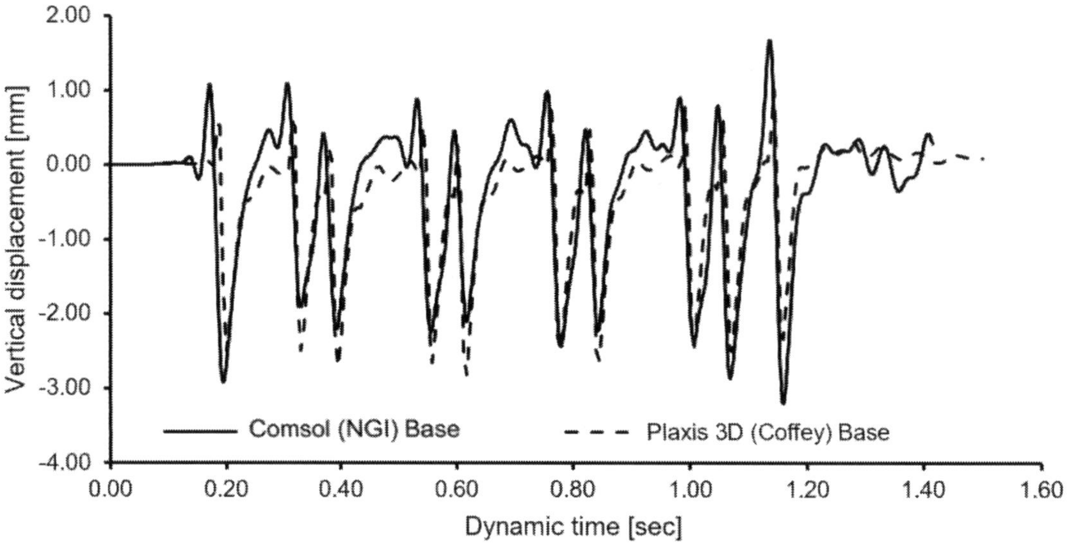

Figure 7 Calibration of Plaxis model against COMSOL using the base case analysis at V=111m/s.

As shown on Figure 7, the output from Plaxis 3D compares very well to the results from COMSOL for the base case analysis. Once a reasonable agreement had been established between the two software programmes, the suitability of MSL in mitigating against the effects of Rayleigh waves from high speed train is investigated. The result of the analysis at the maximum train speed of 400km/hr (i.e. V=111m/s) is shown in Figure 8.

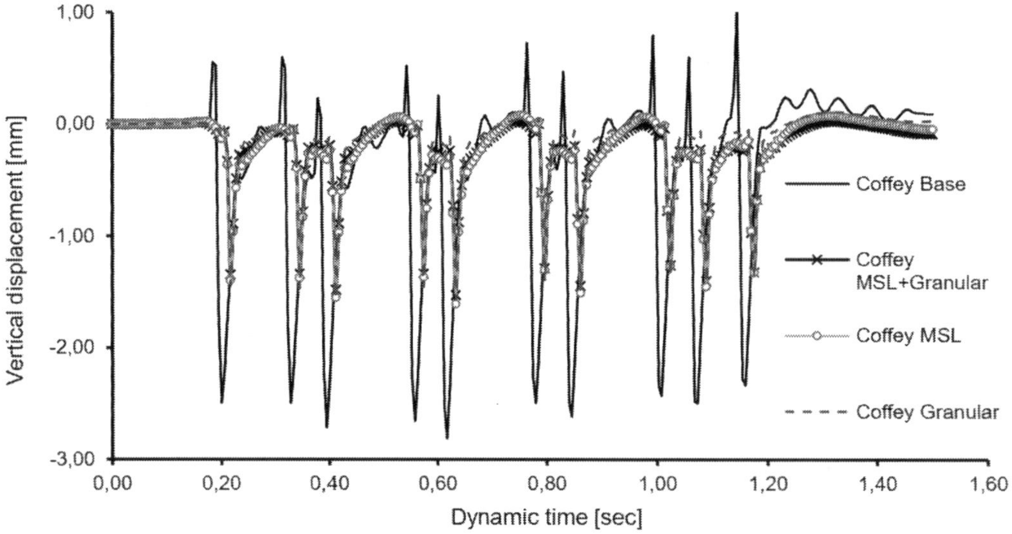

Figure 8 Displacements calculated at middle of railway and comparison among the cases at V= 111m/s

These results show that the use of 2.0m of MSL with 1.0m of granular layer in the layering options can achieve approximately the same degree of improvement as 5.0m of granular replacement. The performance of the MSL solution at lower speed is comparable to the degree of improvement achieve at the maximum speed (i.e. V=111m/s) tested in our analysis.

6 Conclusion

Due to the uniqueness of the problem of Rayleigh waves for high speed train earthworks and supporting structures, innovative solutions are required to proffer suitably robust and cost-effective solutions which satisfy the stringent performance requirements for such high profile infrastructures.
One of the common methods for solving this problem includes replacement of substandard material with high quality granular material to a depth below the track. In this paper, we have explored an alternative solution for solving the Rayleigh wave problem using Tensar geogrid, TriAX, in a mechanically stabilised granular layer (MSL).
Plaxis 3D, a commercially available three-dimensional finite element software, was used for the analysis to simulate the effects of moving train at different speeds (83, 97 and 111m/s).
The results of the analysis show that soil layering (with 2.0m MSL and 1.0m of granular layer) reduced deflection by approximately 40% in comparison to Case 1 (i.e. no treatment). These levels of improvement are comparable to that achieved with the 5.0m granular replacement option.

The outcome of this study demonstrates substantial reduction in the depth of treatment required to resist dynamic load from high speed train can be achieved using MSL. This will not only result in substantial cost savings; it is also a more sustainable solution.

References

[1] Kaynia A M, Madshus C and Zackrisson P 2000 Ground Vibration from High Speed Trains: Prediction and Countermeasure *J. Geotech. & Geoenv. Engrg ASCE* **126** pp 531-7

[2] Karlsrud K 1999 General aspects of transportation infrastructure. Proc. *12th Eur. Conf. Soil Mech. Geotech. Engrg.* **1** pp 17–30 Balkema, Rotterdam

[3] Alves Costa P, Calçada R and Silva Cardoso A 2012 Track-ground vibrations induced by railway traffic: In-situ Measurements and Validation of a 2.5D FEM-BEM Model *Soil Dynamics and Earthquake Engineering* **32** pp 111-28

[4] Ntotsios E, Thompson D J and Hussein M F M 2019 A comparison of ground vibration due to ballasted and slab tracks *Transportation Geotechnics* **21**

[5] Cook J, Buckley J, Dobie M and Bhavsar R 2015 The Use of Multiaxial Geogrids in Rail Trackbed Stabilisation *International Heavy Haul Association Conference* 21-24 June 2015 Perth, Australia.

[6] Heymann G 2007 Ground Stiffness Measurement by the Continuous Surface Wave Test *Journal of the South African Institution of Civil Engineering* **49** pp 25-31

[7] Dong-Soo K and Jin-Sun L 1999 Propagation and Attenuation characteristics of Various Ground Vibrations *Soil Dynamic and Earthquake Engineering* **19** pp 115-26

[8] Kaynia A M 1999 VibTrain A computer code for numerical simulation of train-induced ground vibration *Research Report* 514063-2 Norwegian Geotechnical Institute, Sept. 1999

[9] Atkinson J H 2000 Non-Linear Soil Stiffness in Routine Design *Geotechnique* **50** pp 487-508

[10] Jardine R J, Symes M J and Burland J B 1984 The Measurement of Soil Stiffness in the Triaxial Apparatus *Geotechnique* **34** pp 323-40

A rheological model for georeinforced embankments based on piled foundations

V Mangraviti[1], L Flessati[1] and C di Prisco[1]

[1] Department of Civil and Environmental Engineering (DICA), Politecnico di Milano, P.zza Leonardo da Vinci 32, 20133 Milano (MI), Italy

Email:viviana.mangraviti@polimi.it

Abstract. Concrete columns are commonly used methods to improve the performances of embankments on soft soil strata. To further reduce settlements and stresses on the soft foundation soil, geosynthetic reinforcements are often installed at the base of embankments. Existing design methods for these "geostructures" commonly does not take into consideration both the effect of the embankment construction process and the stiffness of the system components (embankment, soft soil, column and geosynthetics) as design parameters. As a consequence, these approaches do not allow the estimation of the settlements at the top of the embankment. In this paper a new rheological model capable of assessing both tensile forces in the geosynthetic and settlements at the embankment top induced by the construction process is presented. In this model, the variables defining the system performance (tensile forces in geosynthetics, stresses acting on piles and settlements) are explicitly related to the embankment height, interpreted as a generalized loading variable. The model has been validated on the results of 3D numerical analyses.

1. Introduction

In recent years, the employment of deep foundations as settlement reducers has become increasingly popular in the design of earth embankments on soft soil strata. Geosynthetic layers are also often positioned at the embankment base to further improve the system response. Owing to the presence of both piles and geosynthetics, interaction mechanisms, reducing stresses on the soft foundation soil and transferring them towards the piles, take place. These complex interaction mechanisms are not fully understood, despite the well-documented effectiveness of this technique. The approaches suggested in the most used design standards [1,2] are based on the concept of "arching effect" [3-7] allow to estimate the stresses applied to both piles and geosynthetic layers. Nevertheless, since these approaches completely disregard the influence of the stiffness of the elements composing the system, the assessment of settlements is not possible.

In di Prisco et al. [8] a new model to interpret the mechanical response of an unreinforced embankment on piled foundations was proposed. This model puts in relation the increment in the embankment height (employed as a generalized loading variable) with the increment of both average and differential settlements at the embankment top. In Mangraviti et al. [5-8], the authors extended the model to the case of reinforced embankments by discussing the influence of geosynthetic stiffness (J).

In this paper, the capability of the model to reproduce the evolution of both tensile force in the geosynthetic and displacements at the embankment top is presented. The model employs geometrical and mechanical properties as constitutive parameters and explicitly puts in relation the tensile force to the embankment height. Thus, the model allows the geotechnical engineer to adopt a direct displacement-based design approach.

2. Numerical model and results

In the most general case, in principle, the problem is three-dimensional. Nevertheless, when the embankment transversal width is significantly larger than its height, flank effects may be disregarded and only one central axisymmetric cell can be considered as representative of the overall behaviour of the system (Figure 1a). The cell, whose diameter is equal to the spacing between piles (s), includes: (i) the pile of length l and diameter d, (ii) its surrounding soil, (iii) the embankment h thick (iv) the geosynthetic layer. The origin of radial coordinate r is in the centre of the pile top (Figure 1a).

The problem has been numerically modelled by means of the finite difference numerical code FLAC3D (Itasca Consulting Group). Due to axisymmetry, only one quarter of the representative cell has been considered (Figure 1b).

The concrete pile, here considered reaching the bedrock, has been modelled as an elastic element. The mechanical behaviour of both the embankment and the foundation soil is modelled by means of an elastic-perfectly plastic constitutive relationship with a Mohr-Coulomb failure criterion and a non-associated flow rule. An elastic membrane element, characterized by axial stiffness J, has been used to model the reinforcement. Smooth interface elements are introduced between the pile and the foundation soil. Normal displacements are not allowed on the lateral boundary of the domain and at the base. In order to reproduce the construction process of the embankment, the numerical analysis has been subdivided in several stages. Each stage corresponds to the construction of an embankment layer of thickness 25cm. As a consequence, the geometry of the spatial domain progressively evolves, adding at each stage a new layer of elements at the top of the model. In this case the effect of the consolidation processes has been disregarded and drained conditions have been considered.

Table 1. Mechanical properties of the system

	Unit weight, γ (kN/m^3)	Young's modulus (MPa)	Poisson's ratio (-)	Friction angle (°)	Dilatancy angle (°)
Foundation soil	18	10	0.3	30	0
Embankment	18	1	0.3	40	0
Pile	25	30000	0.3	-	-

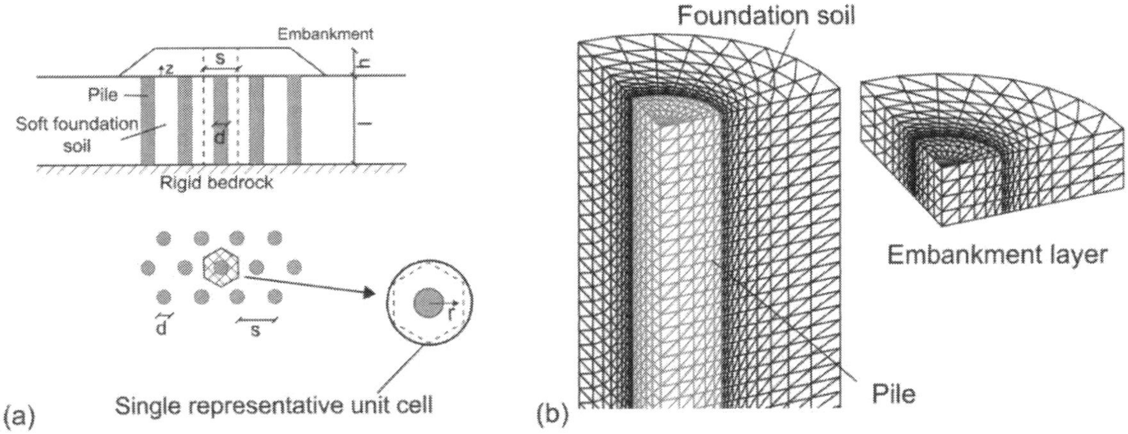

Figure 1 a Problem geometry and representative axisymmetric cell. **b** Numerical model

For the sake of brevity, the results concerning only one reference geometry (s=1.5m, d=0.5m and l=5m) and one value of geo-reinforcement stiffness are hereafter illustrated (J=1000kN/m). All the other employed mechanical parameters are reported in Table 1. For the sake of completeness, the authors also considered the same problem for γ = 10° and 15°. Nevertheless, the mechanical processes are not affected by γ, although a slightly decrease in both average and differential settlements is observed (results here omitted for the sake of brevity).

According to di Prisco et al. [8], the mechanical response of the system can be described by using the following non-dimensional variables:

$$H = h/d \tag{1a}$$

$$\Sigma_c = \frac{\sigma_c}{\gamma d} \text{ and } \Sigma_f = \frac{\sigma_f}{\gamma d} \tag{1b}$$

$$U_{b,c} = \frac{u_{b,c}}{d}\frac{E_{oed,f}/l}{\gamma} \text{ and } U_{b,f} = \frac{u_{b,f}}{d}\frac{E_{oed,f}/l}{\gamma} \tag{1c}$$

$$U_{t,c} = \frac{u_{t,c}}{d}\frac{E_{oed,f}/l}{\gamma} \text{ and } U_{t,f} = \frac{u_{t,f}}{d}\frac{E_{oed,f}/l}{\gamma} \tag{1d}$$

$$U_{t,diff} = U_{t,f} - U_{t,c} \text{ and } U_{t,av} = \frac{U_{t,f}(S^2-1)+U_{t,c}}{S^2} \tag{1e}$$

where subscript c stands for column ($0<r<d/2$), subscript f for foundation ($d/2<r<s/2$); $\sigma_{\cdot}$ and $u_{b,\cdot}$ are the vertical average stresses and displacements at the base of the embankment; $u_{t,\cdot}$ the average displacement at the top of the embankment; $U_{t,diff}$ and $U_{t,av}$ are the non-dimensional differential and average settlements at the top; $S=s/d$ is the non-dimensional spacing; $E_{oed,f}$ is the foundation soil oedometric modulus.

The numerical results (Figure 2) clearly show the role played by the pile: (i) the stresses Σ_c are much larger than Σ_f (Figure 2a); (ii) the vertical displacements at the top of the pile are negligible (Figure 2b) whereas the settlements at the top of the foundation soil follow a non-linear trend; (iii) when H is sufficiently large, differential settlements at the top of the embankment stop increasing reaching an asymptotic value (Figure 2c).

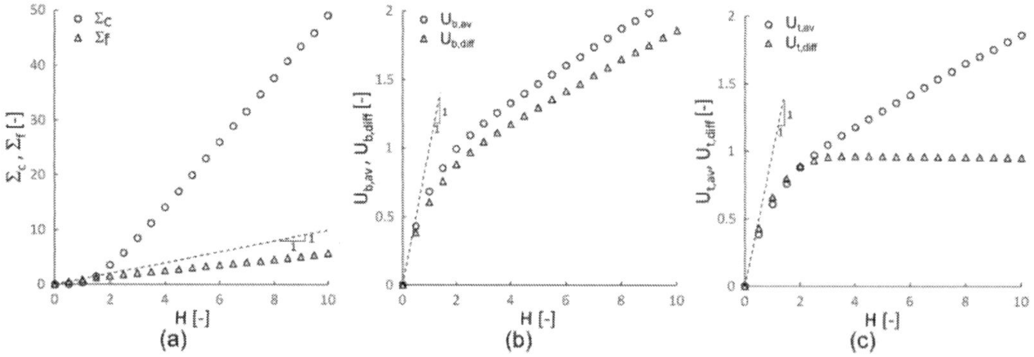

Figure 2. Numerical results during construction stages: **a** non-dimensional stresses **b** non-dimensional displacements at the base of the embankment **c** non-dimensional average and differential displacements at the top of the embankment

To emphasize the role of piles, in Figure 2 dotted lines inclined 1:1 are also plotted: they are obtained by imposing a column stiffness coincident with the foundation soil stiffness.

On the top of the embankment (Figure 2c) the average and differential displacements are initially coincident. In this initial phase, plastic strains develop in a narrow zone close to the top corner of the pile (Figure 3). The evolution of this yielded zone is described by $H_p = h_p/d$. When $H_p=H^*$ the plastic zone stops evolving and the differential displacements stop increasing, although the average displacements continuously increase. H^* is related to the plane of equal settlements [5], defined as the locus where the increment of differential displacements is negligible.

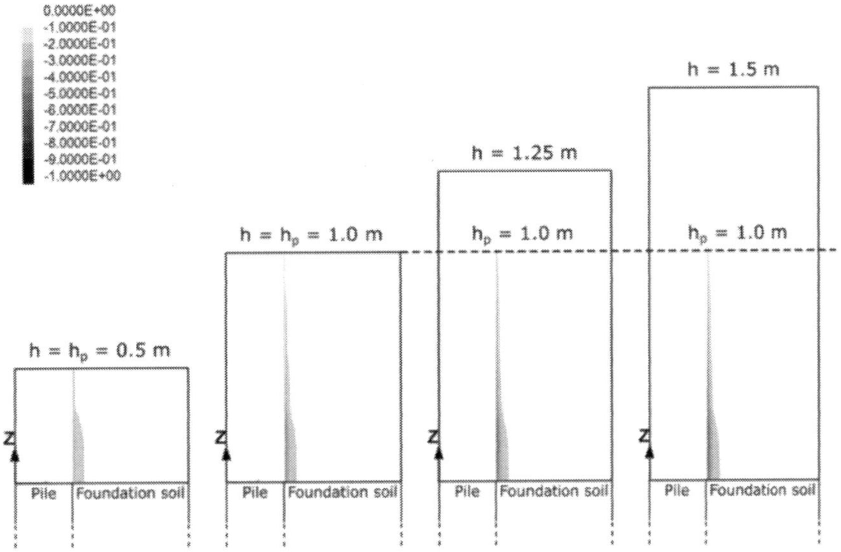

Figure 3. Evolution of plastic strain during the construction process

3. Mathematical model

According to di Prisco et al. [8], the behaviour of a non-reinforced piled embankment can be described by means of a rheological model (Figure 4). To evaluate settlements at the top of the embankment, an incremental relationship between the generalized loading variable $\dot{H}$ and the displacements at the top of the embankment $\dot{U}_{t,diff}$ and $\dot{U}_{t,av}$ is used:

$$\begin{bmatrix} \dot{U}_{t,diff} \\ \dot{U}_{t,av} \end{bmatrix} = \begin{bmatrix} C_{diff} \\ C_{av} \end{bmatrix} \dot{H} \tag{2}$$

C_{diff} and C_{av} represent the non-dimensional compliances of the overall system, analytically evaluated by employing the rheological scheme illustrated in Figure 4, where C_1^r and C_2^r represent the compliances related to the column and to the foundation soil, whereas C_1^{irr} and C_2^{irr} are the compliances related to the portion of embankment H_p thick. The superscript *irr* underlines the irreversibility due to both the material yielding and to the geometrical non-linearity associated with the evolution of H_p with the loading function $\dot{H}$. C_3 is the compliance of the soil stratum $(H - H_p)$ thick. The plastic slider in Figure 4 describes the flux of shear stresses throughout the previously mentioned yielded zone.

In order to take into account the effect of the geosynthetic layer below the embankment base, according to di Prisco et al. [8] only the elastic spring C_2^r needs to be suitably modified with respect to the non-reinforced case [8]. In particular, the authors considered a spring whose stiffness $(1/C_2^r)$ progressively increases with the non-dimensional displacements $U_{b,f}$. The spring stiffness is obtained by integrating the differential equation describing the mechanical behaviour of an asymmetrical membrane resting on an elastic soil and by interpolating the results. The trend of the reinforced soil stiffness for different values of J is reported in Figure 5. For the sake of brevity only the results obtained for J=1000kN/m will be discussed here in the following.

The obtained constitutive model describing the behaviour of the reinforced system (Eq. 2) depends on: (i) the mechanical parameters of column, soil (Table 1) and geosynthetic (J); (ii) the geometrical variables (s, d, l) and (iii) the ratio between horizontal and vertical stresses (k) in the portion of embankment in which irreversible strains accumulated. According to di Prisco et al. [8] k is the only parameter of the model which is not directly related on the system geometry/mechanical properties. Here k is assumed to be not affected by the presence of the geo-reinforcement and to be coincident with the one previously calibrated for non-reinforced systems [8].

The model allows to estimate both the maximum tensile force (T_{max}) in the membrane and the non-dimensional embankment settlements ($U_{t,diff}$, $U_{t,av}$, $U_{b,f}$). The results obtained from the mathematical model are compared with the numerical ones in Figures 6 and 7.

Figure 6 puts in evidence that the proposed model is capable of qualitatively reproducing the variation of T with H. From a quantitative point of view, it always gives safe side estimations. In contrast, in Figure 7a very satisfactory agreement is obtained in terms of displacements at the base (Figure 7a), differential (Figure 7b) and average (Figure 7c) displacements at the top.

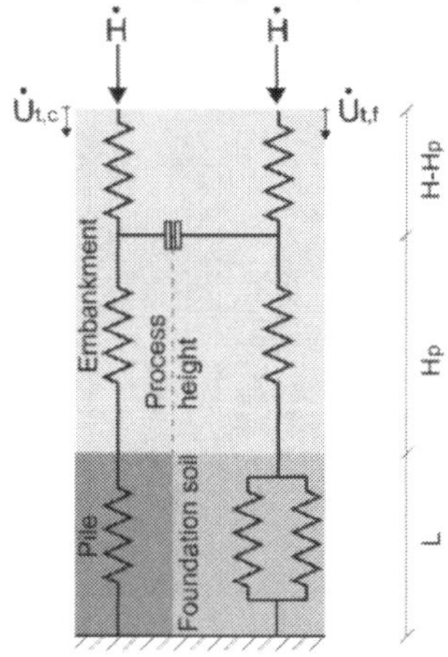

Figure 4. Rheological model for non-reinforced systems

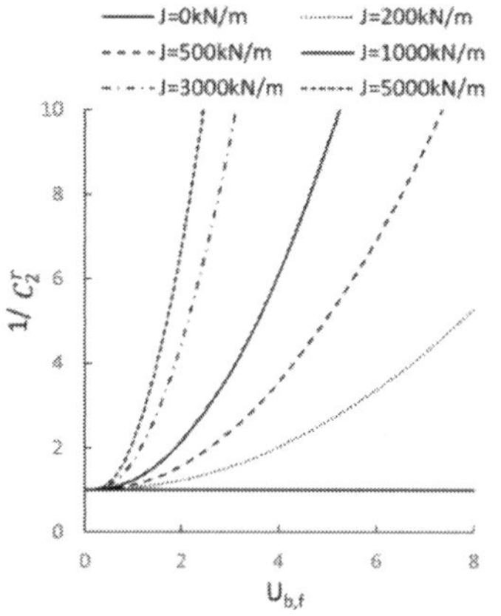

Figure 5. Stiffness of reinforced foundation soil vs embankment base displacement for different J values

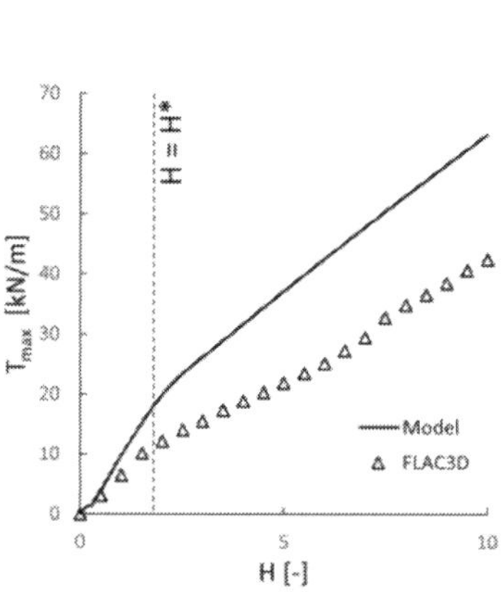

Figure 6. Tensile force in the geosynthetic vs height of the embankment: comparison between numerical data and model prediction

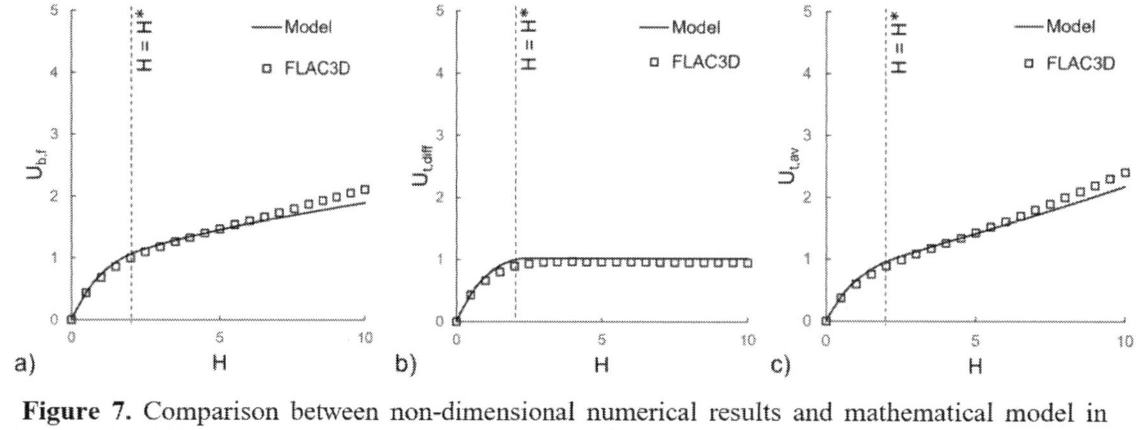

Figure 7. Comparison between non-dimensional numerical results and mathematical model in terms of **a** differential settlements at the top of the embankment, **b** average settlements at the top of the embankment and **c)** displacement at the base of the embankment

4. Conclusions

In this paper a mathematical model capable of reproducing the mechanical response of piled embankments reinforced by means of single geosynthetic layers is presented. The model considers the embankment height as a generalized loading variable and allows to evaluate (i) the tensile force in the membrane and (ii) settlements at both the top and base of the embankment. Regarding the membrane design, the model predictions in terms of tensile forces in the membrane are on the safe side. In conclusion, the mathematical model can be considered as an effective tool to design reinforced piled embankments according to a displacement-based design approach.

Acknowledgement

All the numerical results have been obtained by using the commercial code FLAC3D within the framework of the Itasca Education Partnership (IEP) program. The authors are grateful to Itasca Consulting Group and Harpaceas for the use of the software license.

References

[1] BS8006-1 2010 *Code of Practice for Strengthened/reinforced Soils and Other Fills*. British Standards Institution, ISBN 978-0-580-53842-1
[2] EBGEO 2010 *Empfehlungen für den Entwurf und die Berechnung von Erdkörpern mit Bewehrungen aus Geokunststoffen e EBGEO* German Geotechnical Society, Auflage, ISBN 978-3-433-02950-3 (in German)
[3] Terzaghi K 1943 *Theoretical soil mechanics*. Chapman And Hali, Limited John Wiler And Sons, Inc; New York.
[4] McKelvey III J A 1994 The anatomy of soil arching *Geotextiles and Geomembranes* **13**pp 317-29
[5] Naughton P J 2007 The significance of critical height in the design of piled embankments *Soil Improvement* pp 1-10
[6] McGuire M P 2011 Critical Height and Surface Deformation of Column-Supported Embankments *Ph.D. thesis* Department of Civil Engineering, Virginia Polytechnic Institute and State University, Virginia, U.S.A
[7] Filz G, Sloan J, McGuire M P, Collin J and Smith M 2012 Column-supported embank-ments: settlement and load transfer. *Proceedings of the Geotechnical Engineering State of the Art and Practice* ASCE pp 54-77
[8] di Prisco C, Flessati L, Frigerio G and Galli A 2020 Mathematical modelling of the mechanical response of earth embankments on piled foundations, *Géotechnique* 70(9), 755–773
[9] Mangraviti V 2021 Theoretical modelling of embankments based on piled foundations. Ph.D. thesis, Politecnico di Milano
[10] di Prisco C, Flessati L, Galli A and Mangraviti V 2019 A simplified approach for the estimation of settlements of earth embankments on piled foundations. *Geotechnical Research for Land Protection and Development* In proceedings of CNRIG 2019 pp 640-8
[11] Mangraviti V, Flessati L, di Prisco C 2022 Mathematical modelling of the mechanical response of Geosynthetic-Reinforced and Pile-Supported embankments, *Under Rev*
[12] Mangraviti V, Flessati L, di Prisco C 2022 Geosynthetic-reinforced and pile-supported embankments: theoretical discussion of finite difference numerical analyses results, *Under Rev*

Simulating mechanical response of GRS walls with different construction methods

KM Krishna[1], A Prashant[2]

[1] Ph.D. Scholar, Indian Institute of Technology, Gandhinagar, Gujarat, India
[2] Professor, Indian Institute of Technology, Gandhinagar, Gujarat, India

E-mail: mohan.kolli@iitgn.ac.in, ap@iitgn.ac.in

Abstract. GRS walls are constructed using a variety of fascia such as modular blocks, full height panels, Gabions, etc., which involves different construction sequences. The simulation of construction sequence is important in numerical modeling for understanding the evolution of deformation profiles and reinforcement forces with different fascia and corresponding construction methods. In this study, detailed Finite Element modeling of the GRS wall using OpenSEES as a numerical tool is presented. It includes discussion on the steps involved in simulating construction methods of wall with block and full height panel fascia along with their impact on overall response. A simple staged construction technique is adopted in simulation. The issues and techniques involved in the simulation of staged construction sequence and compaction stresses are explained. The considerable effects of construction sequence and modeling simplifications on deformation profile and reinforcement force distribution are discussed in detail.

1. Introduction

Geosynthetic Reinforced Soil (GRS) Walls offer an edge to designers due to the availability of different wall configurations, reinforcements, and facing types with the major benefit lying in its staged construction method [1]. Also, due to cost benefits, the applications of GRS Walls have extended all over the Transportation (Embankments, Abutments), Water (Dams, waterfront structures), and Public space infrastructures (Retaining walls). The design of GRS walls as per different codes is conventionally based on ultimate limit states, which involves design against damage or failure. The design of conventional GRS walls also requires satisfaction of serviceability limit states, which involves limiting unacceptable/tolerable deformations. By design, GRS walls are often over-conservative with emphasis given only to ultimate limit states and hence, serviceability is generally overlooked. The available tools for estimating deformation are lacking in including the whole physical/load transfer phenomena [1]. There is a need to understand deformation behavior well to address such issues and to develop new tools for deformation estimations. A complete deformation behavior is also needed to improve working stress condition methods like the K-stiffness method [2] as the evolution of deformation dictates the evolution of forces. This article concentrates on one of the phenomena affecting behavior, i.e., construction method effect on evolution of deformations and forces along with critical modeling aspects.
The general understanding of the mechanical response of the GRS wall is that it depends upon the relative stiffness and strength of fill, reinforcement, fascia, and toe [3]. There is a reduction in the lateral fascia deformations, change in reinforcement force distribution and decrease in reinforcement loads with increase in global stiffness of fascia (bending stiffness) and toe resistance [4,8] . The other major aspect

that affects GRS wall behavior is the construction method. GRS walls are constructed with different type of fascia, and these facing systems involves a different type of construction techniques (precast/cast-in-situ facings) and different type of connections between fascia and reinforcement which can have a significant role in the way deformations and forces evolve. A numerical study of two methods of construction of wraparound fascia (full height formwork and layer-by-layer formwork) was performed in the past [9]. It was noticed that there is a significant difference at the end of the construction stage and post-construction stage. It is interesting to see how the construction method affects the deformation behavior of block faced, and the panel faced the wall in addition to fascia and toe rigidity. Many researchers have studied GRS walls using numerical modeling techniques, including [5,10,11,12]]. In this study, the development of a finite element model for GRS walls in OpenSEES (Open System for Earthquake Engineering Simulation) is presented along with validation with full-scale test data. As OpenSEES has no GUI, staged construction technique is developed in house, unlike other commercially available software. Further, some of the modeling issues have been discussed that are involved in the modeling GRS Wall. Numerical models of block faced and panel faced GRS walls are developed to study the effects of construction method, and it is observed that construction techniques should be considered in developing equations, charts, or methods for deformation and force estimations at working stress conditions.

2. Development of a finite element model for GRS wall in OpenSEES

In this study, Finite Element Models of GRS wall is developed in OpenSEES, which is an object-oriented, open-source software framework (http://opensees.berkeley.edu/). As OpenSEES has no user interface and cannot perform pre-processing and post-processing, GiD 14 has been used for that purpose. For validation purposes, block faced wall is chosen, which is tested at the Royal Military College of Canada (Wall-1) reported by [11]. The wall is of 3.6 m height with 8^0 battered block fascia. The backfill soil and fascia are discretized using four-node bilinear iso-parametric Quadrilateral elements. The constitutive behavior of backfill soil is simulated using the Pressure Dependent Multi-Yield material (PDMY) model available in OpenSEES. The plane strain soil strength and modulus parameters of the model (see Table 1) are extracted from [11]. Default parameters given are considered for other parameters like contraction, dilation, and liquefaction parameters of the PDMY model [13]. The fascia blocks are assumed to behave as linear elastic and isotropic material of Young's modulus 1 GPa.

The reinforcements are modeled using truss/bar elements with linear elastic behavior, but with no bending and compressive stiffness. As in the tests, strains are much lower than 2%, and elastic secant modulus is considered good enough to model the reinforcement behavior. The interfaces between soil and reinforcement is considered as bonded. The assumption of no pull-out of geogrid is acceptable [11]. The block-block interface, soil-panel interface, and soil-block interface is modeled using the zeroLengthInterface2D element (Node-to-Segment (NTS) frictional contact element) available in OpenSEES. The NTS contact element follows Mohr-Coulomb's frictional law. Block-Block peak interface shear angle is 57°, with an apparent cohesion of 46 kPa [11]. In the zeroLengthInterface2D element, we can define only the friction angle. Hence, the equivalent friction angle of 62^0 is chosen. The soil-block interface shear angle is chosen as 44°.

The toe stiffness is modeled using a zero-length linear elastic spring element at the toe of block fascia. The toe stiffness is selected as 7000 kN/m appropriately. As the walls are constructed on a rigid structural concrete foundation, a fixed support condition is assumed at the bottom of the model. Roller support condition is given to the side of the model assumed that to be rigid boundary. As FE results are sensitive to mesh size, a mesh convergence study is performed, and a converged uniform mesh size of 0.05 m x 0.075 m is used for both the models. The discretized model, along with the test schematic, is shown in Figure 1.

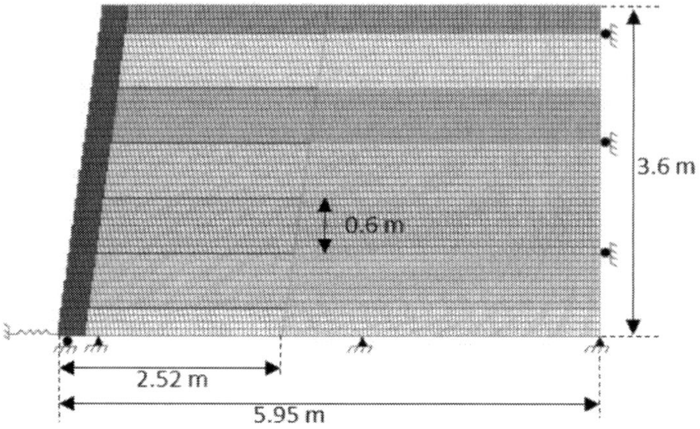

Figure 1. Schematic of the FE model of Wall-1 reported by Hatami and Bathurst [11]

Table1. Parameters used in PDMY model

Parameter	Value
Initial Elastic Modulus	115000 kPa
Pressure Dependency Factor	0.5
Friction Angle	44^0
Reference Pressure	100 kPa
Contraction Parameter	0.03
Dilation Parameters	0.8, 5
Peak Shear Strain	0.04
Phase Transformation Angle	27^0
Density	1.68 gm/cc
Initial Void Ratio	0.45

2.1 Staged construction techniques

The wall is simulated/constructed in stages of 0.15 m lifts. Once each layer is brought into equilibrium under gravity, compaction stresses are simulated. Different commercial software uses different techniques for simulating staged construction like updating upper layer nodes information in compliance with deformation of the present layer, defining ghost elements, etc., which are not directly available in OpenSEES. The following mentioned issues and techniques could be noticed by the user of OpenSEES to perform staged construction (use Figure 2 as reference):

1. Discretization challenges:

While discretization of the domain in GiD, it should be ensured that interfaces at reinforcement and soil layer, nodes are formed at the same location. Similarly, for the interface of block-block and soil-block also corresponding nodes should be formed at the same location.

2. Techniques for staged construction:

The main aim of staged construction techniques is to define a stress-free new layer but with updated coordinates for the deformation of previously analyzed layer. When analyzing the nth stage, exclude nodes of the bottom of the nth layer (defined already with $n-1^{th}$ layer), and including nodes of bottom of $n+1^{th}$ layer nodes have to be defined at once (Refer Figure 2). All the elements, load patterns of the nth layer also to be defined together and can be analyzed. The nodes of the bottom of the $n+1^{th}$ layer and top nodes of the nth layer should be constrained by an equal degree of freedom, which ensures $n+1^{th}$ layer bottom nodes are deformed along with the bottom layer and their nodal coordinates get updated automatically. In $n+1^{th}$ stage analysis, if the user wants to define an interface between two layers, constraints can be removed, and then interfaces can be defined. It is also possible to update the $n+1^{th}$

layer bottom nodes manually [14] by using the information of displacements of top nodes of the nth layer.

Even though the technique used does not maintain lift thickness, but it is relevant to actual construction where fill is placed till the required level compensating bottom deformations. In small deformation problems, the added thickness will be small, and variation in results will be insignificant. However, in OpenSEES, Kotrys [14] tried another technique, where body forces are applied as nodal loads and incremented in steps for the entire model and got satisfactory results. Staged construction allowed to simulate compaction stresses. [15] recommended the Type-2 technique as a more accurate and validated model satisfactorily. Hence, the Type-2 technique is used in this study to simulate compaction induced stresses for two models with 8 kPa loading. Newton-Raphson with a line search algorithm is used for solving the non-linear scheme. As the stiffness matrix formed will be unsymmetrical due to the presence of contact elements, the Umfpack object is used to store and handle the system of equations formed.

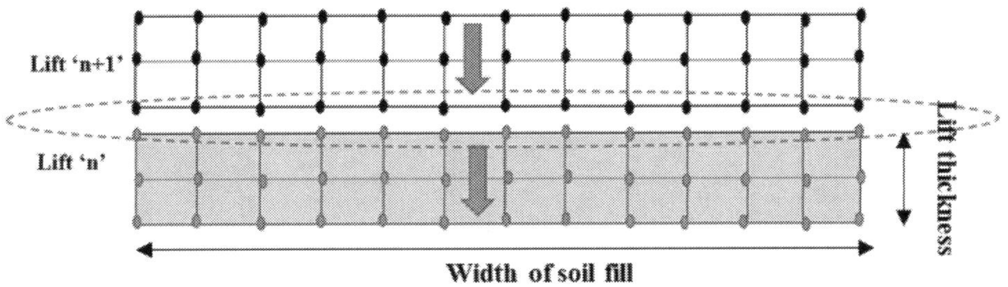

Figure. 2. Schematic of Staged construction

2.2 Validation
The validation of the developed FE model is performed by comparing the FE results with full-scale test measurements [11]. Lateral fascia displacements (Figure 3a) and reinforcement strain distribution (Figure 3b) are used as comparison items. The FE model results showed good agreement with the measured results at the end of construction. As the numerical model of the GRS wall is validated with experimental results, it can be extended further to different types of materials, heights, and facings for further insights on the behavior of GRS walls.

3. Construction stage modeling issues
The FE modeling of the GRS wall involves several challenges, like the availability of calibrated parameters and construction sequence simulation. It is worth discussing some of such issues in particular related to block fascia. Given the parameters, the GRS wall can be simulated in two ways: 1) In stages, and 2) As a full model. In staged construction, each lift is simulated step-by-step following the actual construction sequence. The construction procedure not only involves simulation in stages but also includes simulation of additional compaction stresses after each stage. It is to be noted that simulation of exact compaction (uncompacted thickness of soil compacted to lift thickness) is very challenging, and hence, indirectly compaction stresses are introduced to compacted lifts itself in the literature [15,16] . Neglecting compaction can lead to a significant difference in the estimation of lateral deformation, stresses, and deformation profile at the end of construction. In full model simulation, the whole model is brought into equilibrium at once with facing restrained against lateral deformation. The face restraints are then removed altogether, and again model is brought to equilibrium. In this method, a staged effect of gravity on the deformation profile is difficult to achieve, and even compaction stresses cannot be simulated.
For instance, the model used for validation in section-2 is simulated now by two simulation methods. Figure 4a shows the deformation profile and Figure 4b shows the reinforcement force profile of the block faced wall. There is a clear difference in deformation evolution between the two simulation methods. Total deformations of the wall with full simulation are less than that of in staged simulation.

But, the relative deformations in staged simulation are less than that of full model simulation. In full model simulation, both total and relative deformations are the same as deformations are restrained during gravity and then released. The evolution of forces along height is reflected from relative deformation evolution. As the relative deformations in staged simulation were less, forces mobilized were less. In rigid faced walls like panel fascia, the difference in results can occur mainly due to lack of simulation in compaction stresses than staged gravity loading effect.

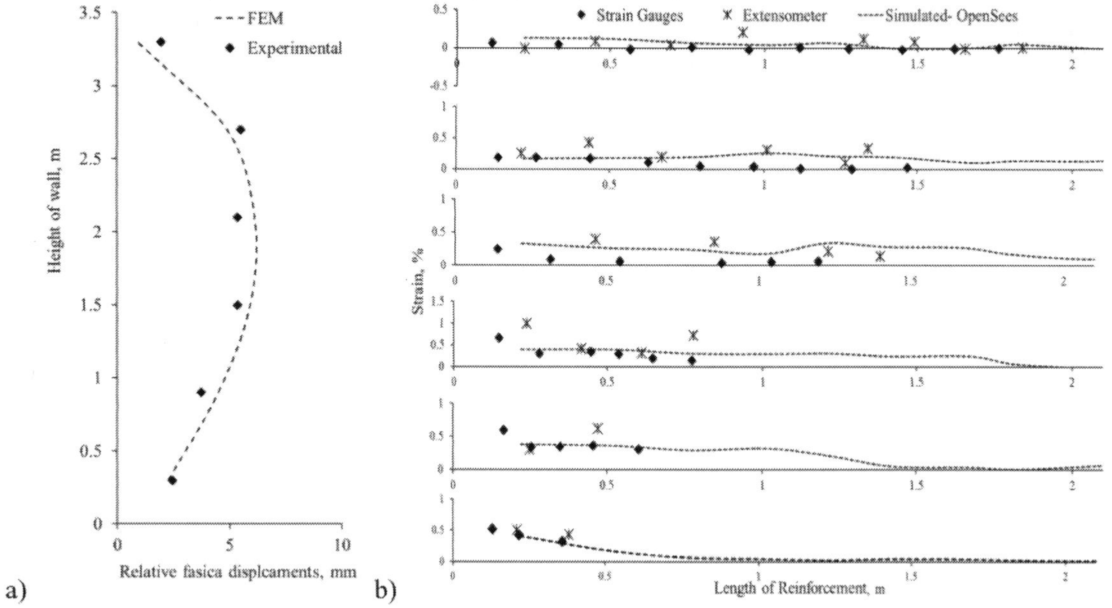

Figure 3. Comparison of Lateral fascia displacement and reinforcement strains from tests and simulations

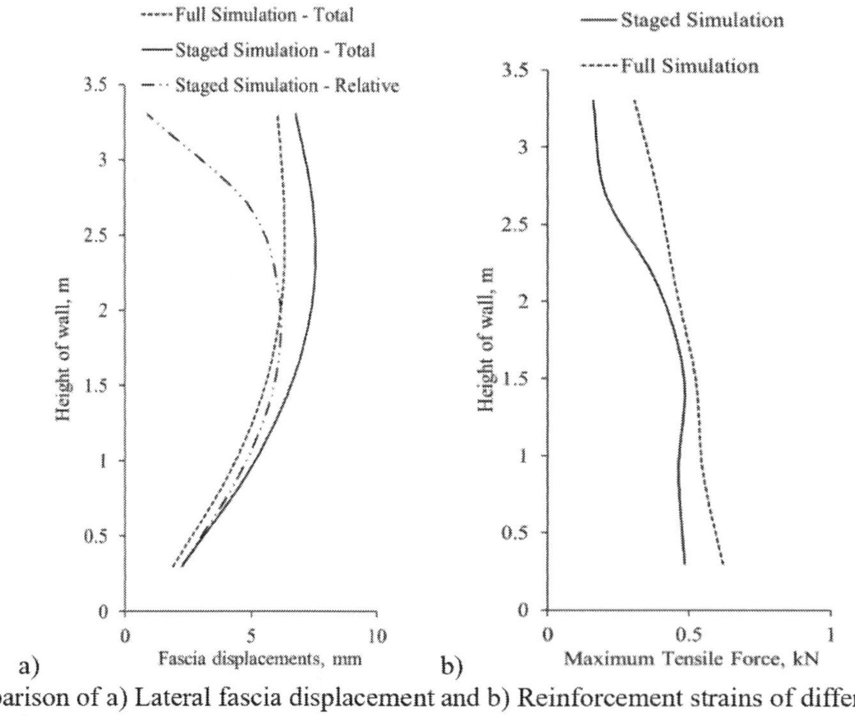

Figure 4. Comparison of a) Lateral fascia displacement and b) Reinforcement strains of different simulation methods

4. Effect of construction method on the evolution of deformation profile and reinforcement forces

Scotland et al.[9] shown that the lateral deformation resulting from layer-by-layer construction of a wraparound fascia (propping in stages only) was greater than that for full height construction (full height propping) of a wraparound fascia. But in contrast, it was observed that full height construction resulted in higher post-construction deformations than layer-by-layer construction. Similarly, block faced and full height panel faced are constructed in different ways will also show differences in mechanical response not only due to fascia and toe effect but also due to differences in construction methodology. Two numerical models were considered to understand such differences. The same block faced wall used for validation is chosen. Another model of same geometry but with full-height panel fascia of 0.3 m thick of the same width as block is considered and with no batter. Full height panel walls are generally constructed without batter. So, the batter was kept vertical and but remaining parameters kept the same in both models. The reinforcement stiffness is chosen as 700 kN/m in both the models with a length of 2.22 m and 0.6 m spacing. The toe condition assumed to be fixed even though unpractical but for qualitative comparison and understandings, it is assumed so. Soil properties according to Hatami et al. [11] is only used in both models. The height of the wall is 3.6 m, and 8 kPa compaction loads are used.

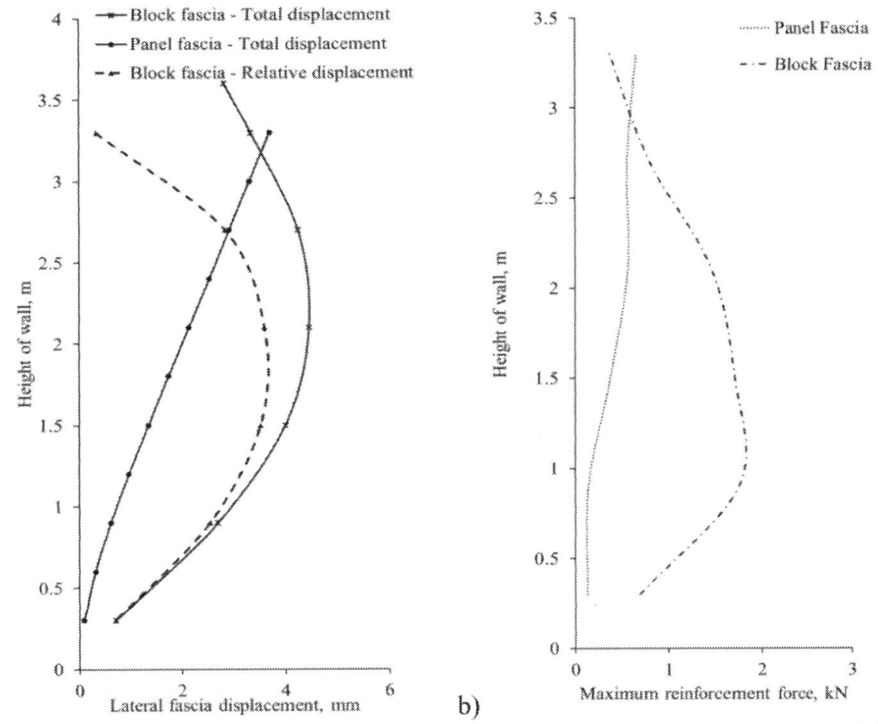

Figure 5. Comparison of a) Lateral fascia displacement and b) Reinforcement forces of block faced and panel faced walls

In the case of block fascia, there is no formwork used to form lateral restraint during backfilling and compaction. Fascia itself helps in contain backfill during compaction. Fascia blocks can move during compaction. In full height panel fascia construction, precast panels are erected and propped till the end of backfilling. After completion of the wall, props were released that allowed the wall to deform laterally. During backfill compaction itself, block fascia can move, while panel fascia cannot. The final deformed profile depends on staged deformations and hence effects the reinforcement force profile. Figure 5a and 5b show the total, and relative deformation profile of block faced wall and panel faced wall, respectively. It is only after the removal of rigid props, deformations and the reinforcement forces are mobilized in the wall. The deformed shape in block faced wall is different from the panel faced wall due to the relative displacement of blocks. Flexible nature of blocks compared to panel resulted in small bulging under staged gravity loading. The major thing to be noticed is the difference in total and relative deformations. At toe due to toe rigidity, they are almost the same but with height, there is a clear distinction. The relative deformations are mobilized in stages with an increase in fill above that level. The top relative deformation is less due to the final placement. The total displacement might be high due to bulging at lower levels which blocks are allowing. In panel, the deformed shape is restricted by the fascia. It is clearly seen in force profiles, that relative deformation profiles are cause of major forces. Due to high deformations at the top of the panel, there are more forces. For block faced wall, forces are high in the middle. It is in compliance with the trapezoidal model according to Miyata et al. [2]. Overall forces are less in panel due to continuous face rigidity with toe fixity.

5. Post-construction behavior under uniform surcharge loading

For major applications, post-construction deformations are major criteria for design rather than deformations during construction. To study the post-construction performance, the same two models in the previous section were used. Panel faced wall showed less deformation than block fascia even under surcharge loading. As observed in Figure 6a, with difference in toe resistance, the block fascia response is not much changed but the panel response shows significant change. It is to be noted that bottom soil nodes are fixed from sliding; hence the deformation profiles represent mostly fill deformations and there is no external sliding. Figure 6b shows maximum reinforcement forces at the end of the surcharge application. There is quite a difference in forces for fixed toe condition among two types of fascia. For higher toe stiffness, the panel faced wall show higher rigidity and resulted in fewer forces in reinforcement. For less stiff toe, the difference is less in both fascia, and at the toe, the panel faced wall showed higher forces than block due to more toe displacement.

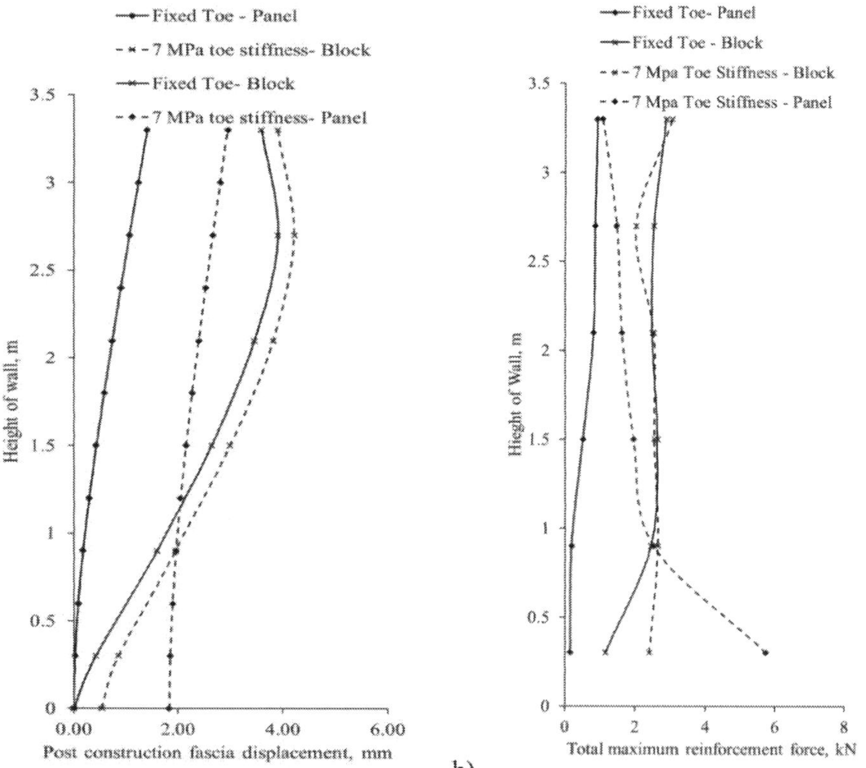

a) b)

Figure 6. Comparison of a) Post construction face displacements and b) Reinforcement forces at end of surcharge of block faced and panel faced walls

6. Conclusions

The staged construction technique that can be adopted in OpenSEES is summarized with the validation of results. While simulating the GRS wall, the results depend upon whether it is a full model simulation or it is staged simulation. The construction method adopted for different fascia has an effect on its deformed shape and force profile in reinforcements. The reinforcement forces are the result of evolved deformations. Post-construction deformations of panel faced wall depends highly on toe stiffness, unlike block faced wall. Tools for estimation of deformation or forces need to account for the construction technique besides the stiffness of the components.

Acknowledgments

The authors are thankful to the suggestions made by Prof. G. Venkatappa Rao and Mr. Venkataraman to improve the quality of the work.

References

[1] Berg R R, Christopher B R, Samtani N C Design and Construction of Mechanically Stabilized Earth Walls and Reinforced Soil Slopes–**I**, US Department of Transportation, Federal Highway Administration, Washington DC, Publication No. FHWA-NHI-10-024 2009

[2] Miyata Y, Bathurst R J 2007 Development of the K-stiffness method for geosynthetic reinforced soil walls constructed with c-φ soils *Canadian Geotechnical Journal***44**pp 1391-416

[3] Lee W F Internal stability analyses of geosynthetic reinforced retaining walls *Doctoral dissertation* University of Washington

[4] Tatsuoka F Permanent geosynthetic-reinforced soil retaining walls used for railway embankments in Japan *InProceedings of the International Symposium on Geosynthetic-Reinforced Soil Retaining Walls* 1992 Balkema

[5] Rowe R K, Ho S 1997 Continuous panel reinforced soil walls on rigid foundations *Journal of Geotechnical and Geoenvironmental engineering* **123**pp 912-20

[6] Bathurst R J, Vlachopoulos N, Walters D L, Burgess P G, Allen T M 2006 The influence of facing stiffness on the performance of two geosynthetic reinforced soil retaining walls *Canadian Geotechnical Journal***43**pp 1225-37

[7] Liu H, Yang G, Hung C 2017 Analyzing reinforcement loads of vertical geosynthetic-reinforced soil walls considering toe restraint *International Journal of Geomechanics* **17**04016140

[8] Mirmoradi S H, Ehrlich M 2015 Numerical evaluation of the behavior of GRS walls with segmental block facing under working stress conditions *Journal of Geotechnical and Geoenvironmental Engineering***141** 04014109

[9] Scotland I, Dixon N, Frost M, Fowmes G, Horgan G 2016 Modelling deformation during the construction of wrapped geogrid-reinforced structures *Geosynthetics International* **23**pp 219-32

[10] Karpurapu R, Bathurst R J 1995 Behaviour of geosynthetic reinforced soil retaining walls using the finite element method *Computers and geotechnics* **17**pp 279-99

[11] Hatami K, Bathurst R J 2005 Development and verification of a numerical model for the analysis of geosynthetic-reinforced soil segmental walls under working stress conditions *Canadian Geotechnical Journal***42**pp 1066-85

[12] Liu H, Wang X, Song E 2009 Long-term behavior of GRS retaining walls with marginal backfill soils *Geotextiles and Geomembranes***27**pp 295-307

[13] Elgamal A, Yang Z, Parra E 2002 Computational modeling of cyclic mobility and post-liquefaction site response *Soil Dynamics and Earthquake Engineering* **22**pp 259-71

[14] Kotrys P E 2004 Finite element modeling of reinforced soil walls *Master dissertation*, University of California, San Diego

[15] Mirmoradi S H, Ehrlich M 2014 Modeling of the compaction-induced stresses in numerical

analyses of GRS walls *International Journal of Computational Methods* 2014 **11**1342002
[16] Mirmoradi S H, Ehrlich M 2018 Numerical simulation of compaction-induced stress for the analysis of RS walls under working conditions *Geotextiles and Geomembranes***46**pp 354-65

The behaviour of containment bund using stacked geotextile tubes filled with lightly cemented soft soil

J W Koh[1], S H Chew[1], Y J E Aw[1], K E Chua[2], H M A Yim[2], D S E Tan[2], Y S Toh[2]

[1] Department of Civil and Environmental Engineering, National University of Singapore, 21 Lower Kent Ridge Road, Singapore 119077
[2] Housing and Development Board, 480 Lor 6 Toa Payoh, Singapore 310480

E-mail: ceekohj@nus.edu.sg

Abstract. This land reclamation project used stacked geotextile tubes to construct a 1 km long containment bund. In the past, sand is the preferred infilling material of geotextile tubes due to its ease of construction. However, due to sand scarcity in Singapore, soft soil becomes a potential alternative infilling material. One of the challenges of using soft soil as infilling material is that there will be an excessive settlement and concerns about the overall stability of the stacked tubes. Hence, the soft soil is lightly cemented before infilling into the geotextile tubes. This paper presents the design consideration and the construction sequence of the construction of a stacked geotextile tubes containment bund. Extensive instrumentation was installed at a section of the containment bund to study the behaviour of this geotextile tube. The instrumentation includes strain gauges, pore pressure transducer and total pressure cell. After two years of construction was completed, Standard Penetration Test (SPT) and laboratory triaxial test were used to evaluate the strength of the infilling material. All the instrument data and strength tests results were used to ascertain the performance of the stacked geotextile tubes containment bund.

1. Introduction

Land reclamation is one of the methods that most high population density countries employ to enlarge their land area, including Singapore [1]. Until today, Singapore has reclaimed approximately 150 km^2 or 25% of the original landmass [2]. In the past, the preferred infilling material for a land reclamation project in Singapore was sand due to its ease of construction and excellent performance. Recently, soft soil such as dredged marine clay and excavated soil from the construction site has become an alternative fill material for land reclamation projects due to the sand scarcity problem [3]. Those soils have high moisture content and behave like slurry in nature when hydraulically filled. Hence, it is a must to construct a containment bund along the edge of the land reclamation area to contain the soft soil fill material. Usually, the containment bund can be made up of merely a sand bund or geotextile tube filled with sand slurry. Then, again, the availability of a large amount of sand is always an issue. It may not be environmentally sustainable in the long term as well. Therefore, the construction of geotextile tubes filled with soft soil is becoming attractive, even though it has some difficulties. The problem with using soft soil as infilling material is it has high moisture content and low strength. Hence, the geotextile tube will have an excessive settlement and significant shape deformation after a long time. It is also noted that the dewatering process of the geotextile tube filled with soft clay may see some initial piping of clay particles through the geotextile sheet. Then, its dewatering rate will be reduced due to the binding of the filter cake at the interior perimeter of the geotextile tube [4].

An innovative way to minimise the outfall of using soft soil-filled material in a geotextile tube is to mix the soft soil with a small amount of cement before infilling it into a geotextile tube. This lightly cemented soft soil serves two purposes: it prevents the initial piping of fine particles and provides the required strength for stability in a shorter time. This concept was tested in a miniature geotextile tube study [5] and a pilot test on a single piece of actual-size geotextile tube filled with lightly cemented soft soil [6]. Both studies concluded that the concept of utilising lightly cemented soft soil as infilling material for geotextile tube is feasible. The geotextile tube in the tests was able to withstand the designed imposed loading with minimal deformation. Hence, lightly cemented soft soil-filled geotextile tubes were subsequently utilised to construct a containment bund in a mega reclamation project in Singapore. The objective of this paper is to examine the performance of this containment bund. Extensive instruments/sensors, including strain gauges, earth pressure cells and pore pressure transducers, were installed at the critical section. The behaviour of the geotextile tubes was analysed using field monitoring data during and post-construction phases. Besides, standard penetration test (SPT) and laboratory triaxial test were employed to evaluate the infilling material strength after two years. The strain development of the geotextile material, the stresses inside the filled geotextile tubes, and the strength development of the infilling material over time will be evaluated and reported in this paper.

2. Design of Containment Bund

At a reclamation site, an approximately 1 km long pyramidal shape containment bund was constructed. It was made up of numerous units of 40.3 m long geotextile tubes. The geotextile tubes were stacked up into a pyramidal shape in section view and placed in a staggered arrangement along the alignment of the bund. The pyramidal shape containment bund was made up of six (6) units of geotextile tubes stacked up in three (3) layers, as shown in figure 1. The first layer has three (3) geotextile tubes (GB1-L, GB1-R and GB1-C) placed side by side. The second layer has two (2) geotextile tubes (GB2-L and GB2-R), which were stacked above and positioned in-between the first layer's geotextile tubes. The third layer has only one geotextile tube (GB3). The height and width of a single filled geotextile tube used in this project are about 2 m and 5.29 m, respectively. The total height of the completed containment bund is roughly 6 m.

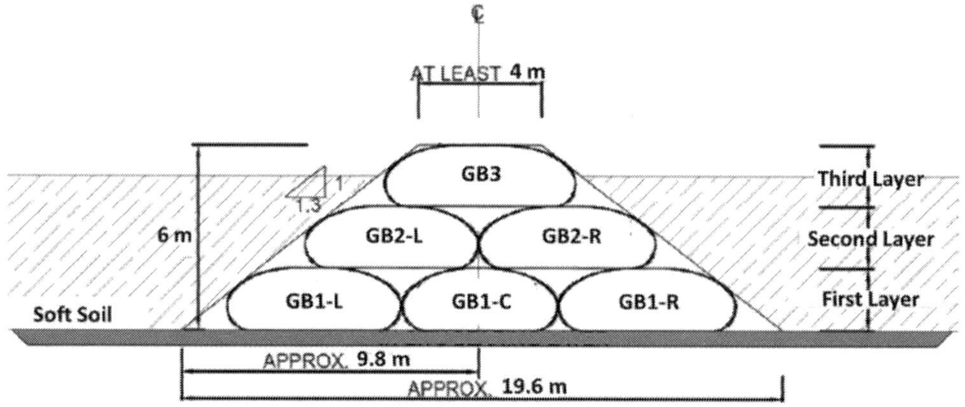

Figure 1. Schematic diagram of a stacked geotextile tubes containment bund in sectional view

2.1. Installation of geotextile tubes

As mentioned, each section of the containment bund was made up of three (3) layers of stacked geotextile tubes, where the first layer was made up of three (3) geotextile tubes placed side-by-side. The first layer's geotextile tubes were installed with a 1.5 m overlapping with each other to close up the gap that may create in between the geotextile tubes after filling. Figure 2 shows the installation sequence of the first layer of geotextile tubes. The construction started with the left geotextile tube (GB1-L), followed by the right geotextile tube (GB1-R). The centre geotextile tube (GB1-C) was installed last to fill the space between the completed left and right geotextile tubes. After completing the whole section with the first layer geotextile tubes, soft soil was backfilled on one side of the bund (i.e. called "contained area") up to the same height as the installed geotextile tubes. The installation work of the second layer geotextile tubes continued only after the whole "contained area" was filled up. This construction sequence avoids any stability issue that may encounter by the geotextile bund during the soft soil backfilling work.

The second layer of geotextile tubes comprises two geotextile tubes (GB2-L and GB2-R), which were then installed from left to right and stacked on top of the first layer's geotextile tubes. Similarly, backfilling of the soft soil on the "contained area" continued up to the same height as the second layer geotextile bunds.

Finally, the topmost layer of the geotextile tube (GB3) was installed, followed by the backfilling of the soft soil up to the final height.

The completed 6 m height containment bund could successfully function as a containment bund for the soft soil backfill in the "contained area" by following the construction sequence above. The whole 1 km long containment bund was constructed successfully.

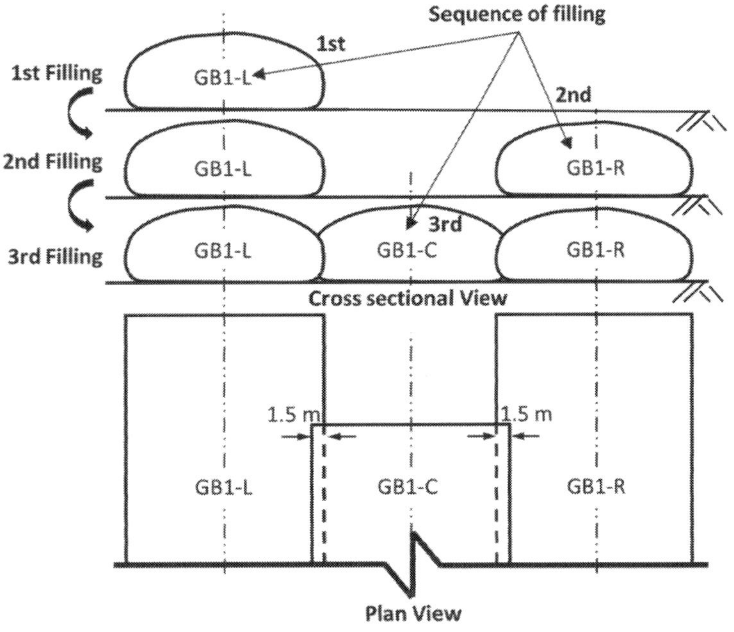

Figure 2. Installation sequence of first layer geotextile tubes

2.2. *Materials of geotextile tube*

The geotextile tubes used were stitched up with Polypropylene (PP) woven fabric. The circumference and longitudinal length of a geotextile tube are 12.6 m and 40.3 m, respectively. There were eight (8) injection ports located at an interval of approximately 5m on the top of the geotextile tube. The geotextile tube was designed to provide adequate strength and the ability to withstand harsh offshore conditions. Table 1 shows the properties of the geotextile material.

2.3. Infilling material properties - Lightly cemented soft soil

The infilling material used in this project consisted of soft soil and a small amount of cement. The soft soil was reused from the excavated soil from local construction sites of the basement excavation and tunnelling project. The soft soil contains mostly silt and clay, and the particle size distribution curve is shown in figure 3. It was mixed with water until a bulk density of 1.25-1.35 Mg/m^3, equivalent to a moisture content of 140 %. Subsequently, the Ordinary Portland Cement (OPC) was added to mix uniformly with the slurry soil. It was then pumped into the geotextile tubes in a slurry state.

Table 1. Material properties of the geotextile tube

Properties	Unit	Values	Test Standard
Wide width tensile strength (MD/CD)	kN/m	120/120	ISO 10319
Strain at max. tensile strength (MD/CD)	%	15/10	ISO 10319
Seam strength (CD)	kN/m	85	ISO 10321
CBR puncture strength	kN	14	ISO 12236
Drop cone perforation diameter	mm	10	ISO 13433
AOS O_{90}	mm	0.40	ISO 12956
Water permeability Q_{50}	$l/m^2/s$	15	ISO 11058
Abrasion resistance (strength retained)	%	75	ASTM D4886
UV resistance at 500 Hours (strength retained)	%	90	ASTM D4355

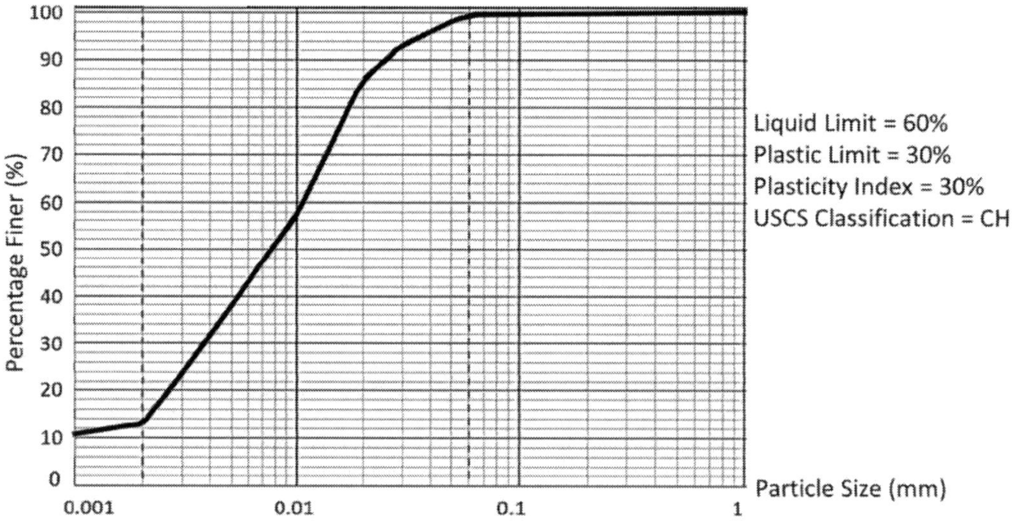

Figure 3. The particle size distribution of the fine-grained soil

Many types of research were conducted on the behaviour of cement mixed clayey soil in Singapore for deep mixing and jet grouting applications [7], [8] and [9]. On the graph of unconfined compressive strength versus cement content, it was proposed that two zones (i.e. "inactive zone" and "active zone") can be observed depending on the amount of cement content [9]. The "inactive zone" is referred to as the soil mixing with a low range of cement content, while the soil mixing of higher cement content is termed the "active zone". It was found that in the "inactive zone", cemented soil exhibited a lower increase rate of the unconfined compressive strength (qu) as compared to that in the active zone [9] &

IOP Conf. Series: Materials Science and Engineering 1260 (2022) 012016 doi:10.1088/1757-899X/1260/1/012016

[10]. With only 5-10% of cement content proposed in this project, it is within the "inactive zone", and only a marginal amount of strength gain is expected. Nevertheless, it is still beneficial to have this cementitious material as it will bind the clayey particle together and reduce the initial piping, as well as provide sufficient strength for the overall bund stability.

The bulk density of this lightly cemented soil used in this project was 1.3 – 1.4 Mg/m^3.

3. Field Monitoring of Geotextile Tubes

Four (4) units of instrumented geotextile tubes were well monitored for their behaviour during and after the infilling process. The instrumented geotextile tubes were the first layer's left (GB1-L) and the first layer's centre tubes (GB1-C), the second layer's left tube (GB2-L), and the top layer tube (GB3). Table 2 shows the installation timeline of the instrumented geotextile tubes for a better overview.

Table 2. Construction timeline of instrumented geotextile tubes

Instrumented Geotextile Tube	Installation Date
GB1-L	11th May 2017
GB1-C	12th May 2017
GB2-L	6th September 2017
GB3	13th November 2017

3.1. Instrumentation of geotextile tubes

Total pressure cell (TPC) and pore pressure transducer (PPT) located near the injection port was employed to measure the total pressure and pore water pressure, respectively. Also, strain gauges were attached, along the circumference, longitudinal and diagonal directions, at the inside surface of the instrumented geotextile tubes to monitor the strain mobilisation. Table 3 shows the instrumentation detail of the geotextile tubes. The instrumentation layout in the respective geotextile tubes is shown in figure 4(a) and 4(b).

Table 3. Construction timeline of instrumented geotextile tubes

Type of Instrumentation		Number of Instrument			
		GB1-L	GB1-C	GB2-L	GB3
Total Pressure Cell (TPC)		1	1	1	1
Pore Pressure Transducer (PPT)		1	1	1	1
Strain Gauge	Circumference Direction (CD)	11	-	8	8
	Longitudinal Direction (LD)	6	-	6	6
	Diagonal Direction (DD)	5	-	5	5

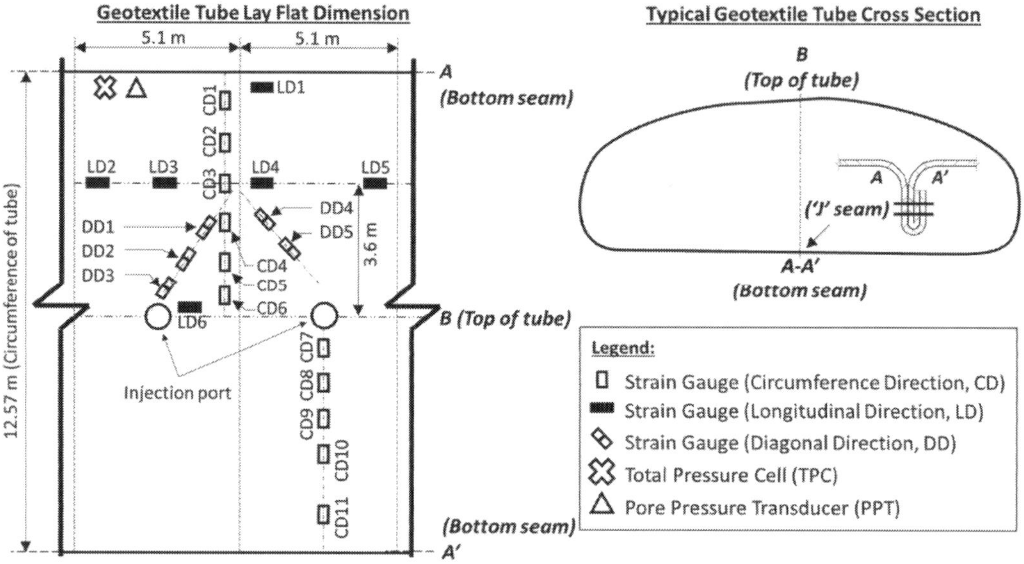

Figure 4(a). Instrumentation layout of GB1-L and GB1-C (TPC, PPT and DSG were only installed on GB1-C)

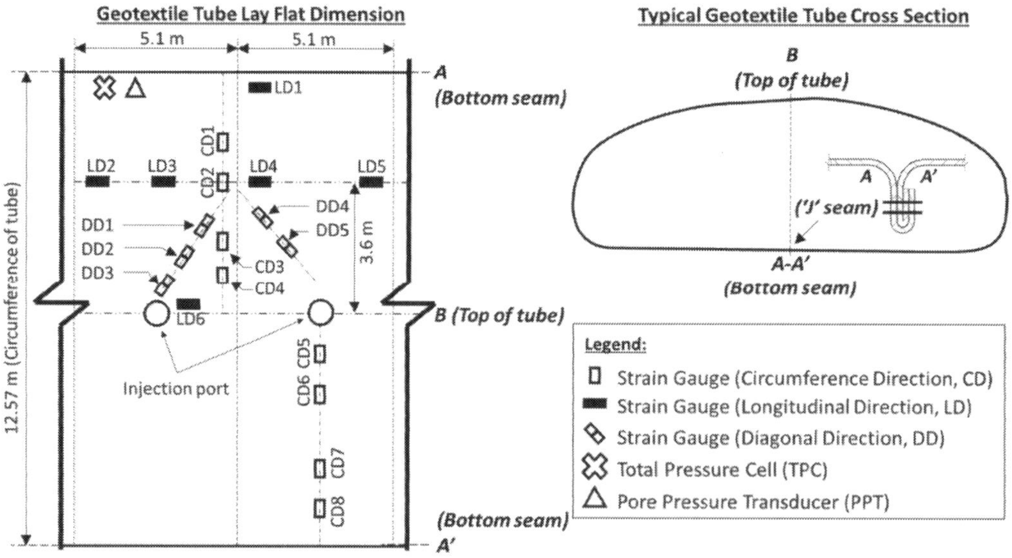

Figure 4(b). Instrumentation layout of GB2-L and GB3

3.2. Data recording and analysis of geotextile tubes

The pore pressure development in the four (4) instrumented geotextile tubes is shown in figure 5. In general, the pore pressure surged up at the very beginning of the infilling of the geotextile tube, which rapidly reduced after that. Subsequently, the pore pressure increased gradually as the backfilling of soft soil in the "contained area" progressed. From figure 5, it can be observed that the maximum pore pressure was recorded on the first layer geotextile tubes (GB1-L and GB1-C), followed by the second layer geotextile tube (GB2-L) and the topmost layer geotextile tube (GB3). This observation reflects well the increasing load from the topmost to the bottom layer geotextile tubes. For all curves, the highest pore pressure occurred after constructing the containment bund and backfilling of soft soil.

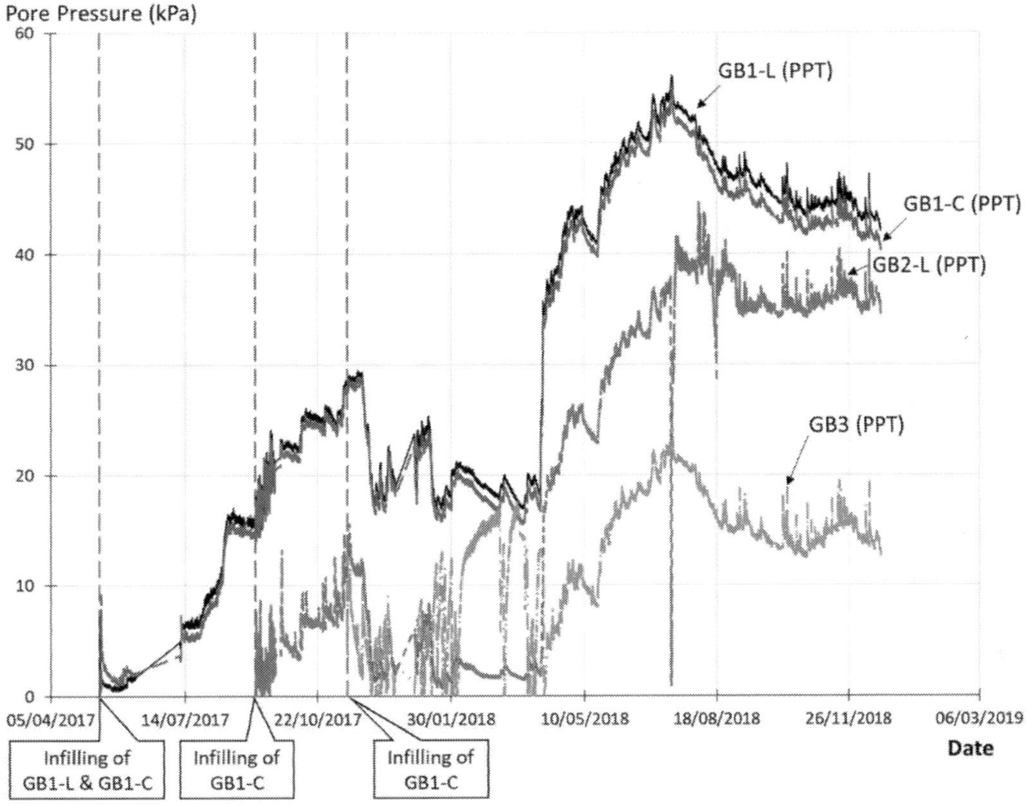

Figure 5. Pore pressure development in the geotextile tubes

Figure 6 shows the total pressure registered in the four (4) instrumented geotextile tubes. GB1-C experienced high pressure, up to ~300 kPa, whereas other geotextile tubes had lower total pressure throughout the entire monitoring period. As expected, the high pressure in GB1-C was due to the direct loading from the upper layer geotextile tubes. Although GB1-L is also located at the bottommost layer, it was not situated at the centre of the whole pyramidal shape bund, and hence, the pressure exerted was lesser than that of GB1-C.

The geotextile tube at the bottommost layer would likely experience a harsher condition from the analysis of these two pressure sensors. Hence, more critical design considerations should be considered for this layer of geotextile tubes.

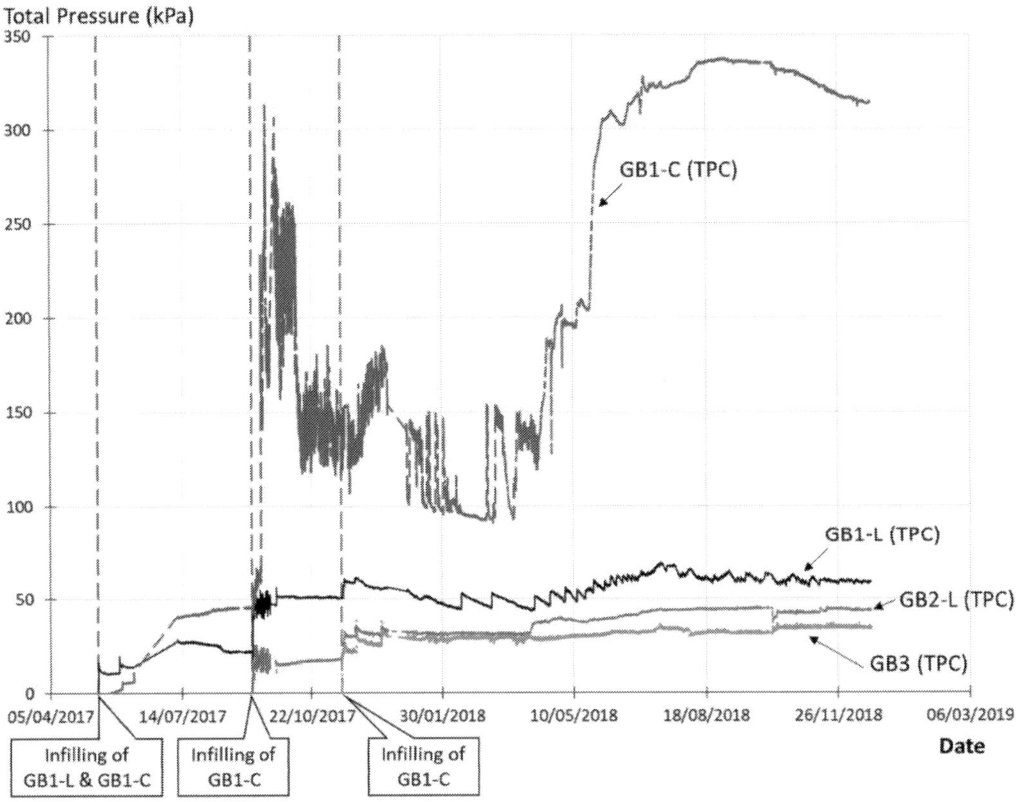

Figure 6. Total pressure development in the geotextile tubes

Figure 7(a), 7(b) and 7(c) show the strain development measured by the "critical strain gauges" of each geotextile tube in circumference direction (CD), longitudinal direction (LD) and diagonal direction (DD), respectively. This "critical strain gauge" is the strain gauge that registered maximum strain amongst all strain gauges in that direction on that instrumented geotextile tube. From Figure 7, it can be seen that during the infilling of each geotextile tube, a sudden surge of strain was observed during the high-pressure pumping process. The highest mobilised strain is about 14 %, measured in the diagonal direction (DD4) of GB2-L. Fortunately, the strain was not continued for an extended period. For circumference direction, the maximum strain, 9 %, was registered in GB3 at the location of CD8, which is attached at the highest curvature point of the circumference direction. It was also shown that the strain level was maintained until the end of the monitoring period. This is thus very critical to include this in the design consideration. On the other hand, the strain readings in the longitudinal direction were significantly lower than that in the other directions.

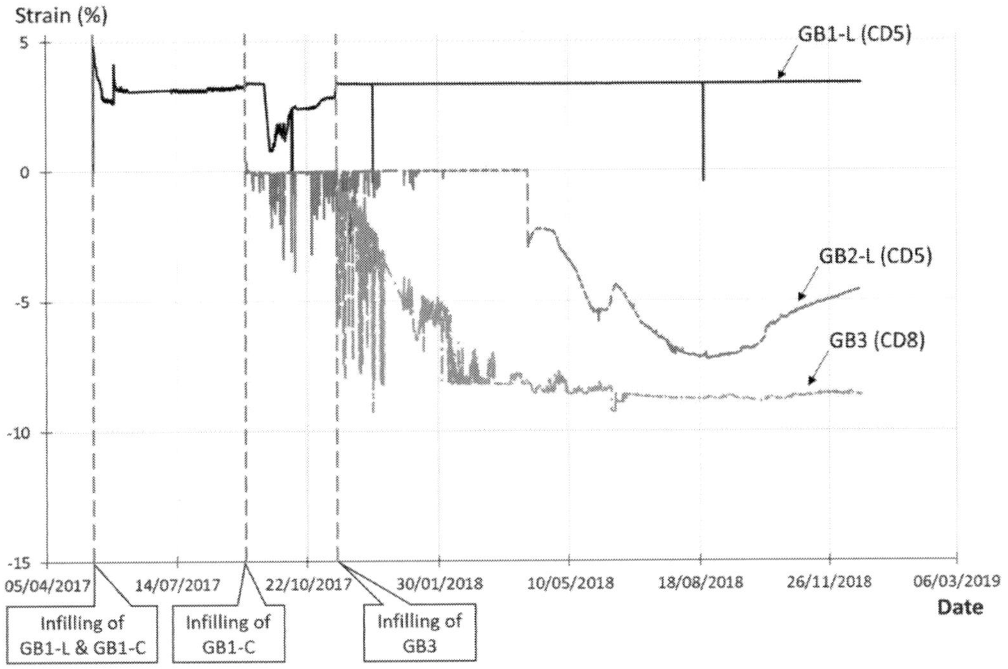

Figure 7(a). Strain reading in the circumference direction

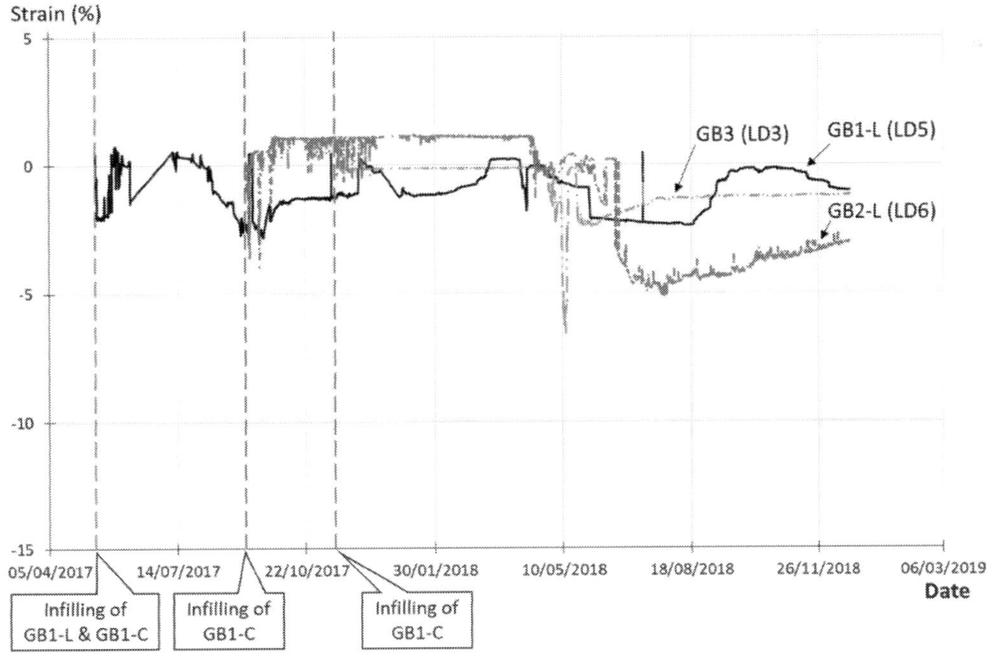

Figure 7(b). Strain reading in the longitudinal direction

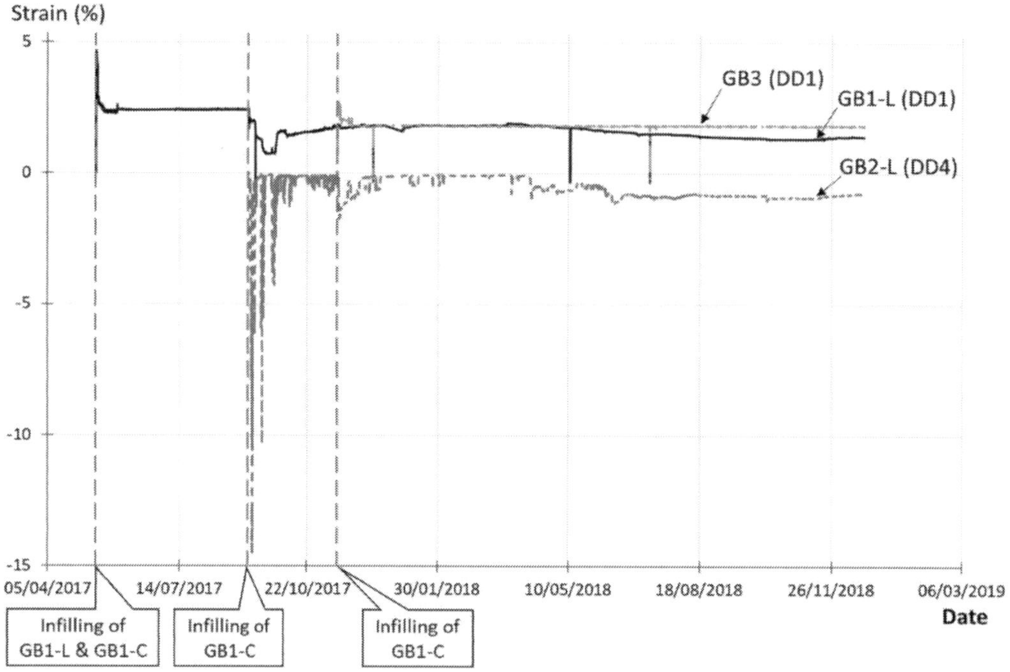

Figure 7(c). Strain reading in the diagonal direction

3.3. Post-construction site investigation analysis of geotextile tubes

After two years of installation of the containment bund, a standard penetration test (SPT) was carried out on the infilling material at 1.225 m, 3.225 m, and 5.225 m depth from the top surface of the GB3. Besides, core samples of 75 mm diameter and 1000 mm length undisturbed samples were extracted at two different depths from each layer of the geotextile tube. These UD samples were sent to the laboratory for strength determination using the Consolidated Undrained Triaxial test (CU). These in-situ and laboratory tests aimed to check the cemented soil strength developed inside the geotextile tube after two years. Figure 8 shows the detailed location of the SPT test and UD sampling locations.

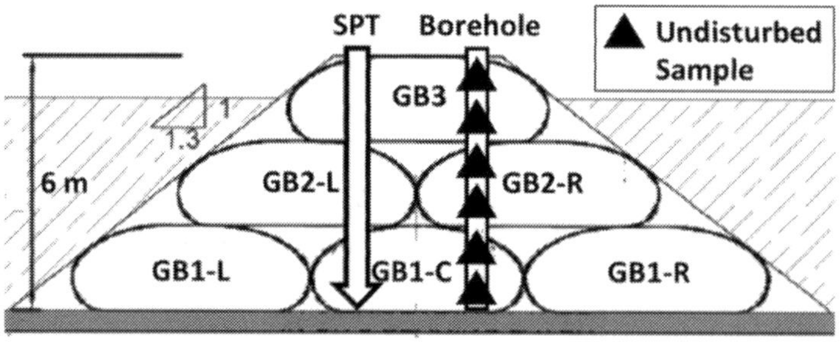

Figure 8. In-situ test and borehole detail location

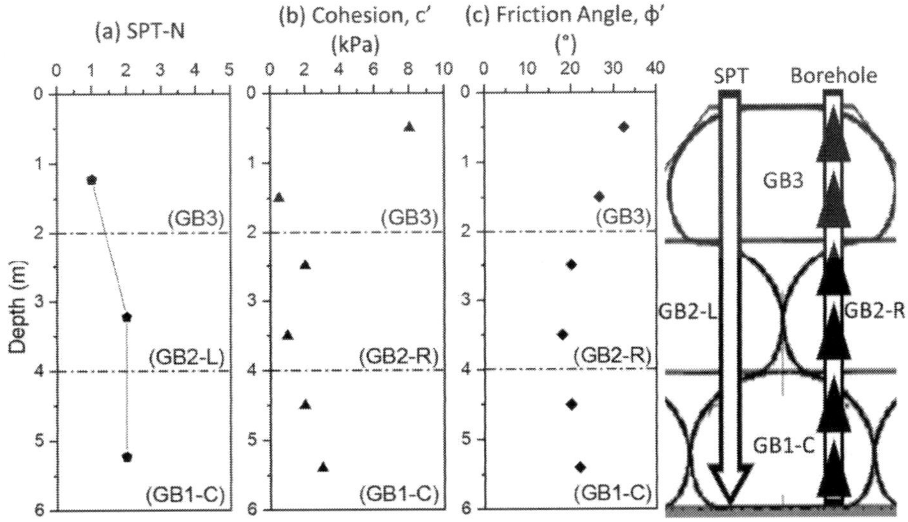

Figure 9(a), (b) & (c). SPT-N value, cohesion and friction angle of the cemented soil with depth

From the SPT results, as shown in figure 9(a), the N-value at the middle of each layer geotextile tube was found to be 1 to 2. The undrained shear strength, Su, of the cemented soil can be estimated using correlation (Su = 6.25N) as presented by [11]. Hence, the undrained shear strength was improved from zero at the slurry state to about 6-12 kPa at the end of 2 years. This value is indeed very close to the in-situ undrained shear strength of soft Singapore Marine clay [7]. It should be noted that, as illustrated in figure 9(a), the three SPT points were conducted in the middle of the respective geotextile tubes. In general, the infilling material in the central portion of a geotextile tube has been subjected to the lowest dewatering rate than the location nearer to the geotextile tube's top and bottom surfaces. Hence, the cemented soil at the other positions should have higher strength than that in the central portion of the geotextile tube.

Figure 9(b) and 9(c) show the consolidated undrained triaxial test results. GB1-C and GB2-R have quite similar effective cohesion, c' and friction angle, ϕ', which is in the range of 2-3 kPa, and 18°-22°, respectively. In contrast, GB3 has very low effective cohesion (c' =0.5 kPa) at the bottom and marginally higher (c' =8 kPa) at the top portion of the geotextile tube. Similarly, GB3's friction angle near the surface is 32 ° compared to that at the bottom part, 26°. The results show that the cemented soil at the topmost surface of the stacked geotextile tubes generally has the highest strength parameter. The observed behaviour may be due to the desiccation of the cemented soil, where the topmost portion is likely to receive lots of heat from sunlight.

The strength development of the infilling soft soil may not be too high, but it is indeed adequate to provide the needed stability of the overall containment bund section.

4. Conclusion

This paper presented the primary consideration and the construction procedure for the stacked geotextile tubes containment bund, of which the geotextile tubes were infilled with lightly cemented soft soil. The adopted design and construction procedure takes into account the effect of the lateral pushing force arising from the backfilling of soft soil by the side of the containment bund. This project adopted a lightly cemented soft soil as the infilling material for geotextile tube because of sand scarcity in Singapore and for environmental sustainability reason. This technique seems to provide sufficient strength for stability and avoid significant deformation of the geotextile tube. A section of this containment bund was instrumented to monitor its behaviour during and after construction.

The pore pressure and total pressure monitoring data show that the bottommost layer of geotextile tubes always experienced a higher loading. The loading is either from the stacked upper layer geotextile tubes or backfills soft soil in the "contained area" by the side. Different design requirements or considerations may be applied to a different layer of the geotextile tube, where the bottommost layer is more critical than the upper layers. The recorded strain readings indicated that the diagonal direction and circumference direction of a geotextile tube would experience a higher level of stretching and tension.

The SPT and CU strength test results show that the lightly cemented soft soil inside the geotextile tube had just gained sufficient strength over the two years. The increment in strength will help in the overall stability of the stacked geotextile tubes.

In conclusion, this project shows that stacked geotextile tubes filled with lightly cemented soft soil can effectively serve as a containment bund in land reclamation projects.

References

[1] Pui SK 1986 100 Years of foreshore reclamation in Singapore *Int. Conf. on Coastal Eng.* **20** 2631 36

[2] Lim SH 2020 Seascape the landscape of Singapore, repurposing land in a land scarce nation *Proc. World Conf. on Floating Solutions 2019* **41** 385-410

[3] Lee SL, Karunaratne GP and Yong KY 1987 Layered clay-sand scheme of land reclamation *J. Geotech Eng.* **113**pp984-95

[4] Leshchinsky D, Leshchinsky O, Ling HI and Gilbert PA 1996 Geosynthetic tube for confining pressurised slurry: Some design aspects *J. Geotech. Eng.* **122**pp682–90

[5] Chew SH, Eng ZX, Chua KE, Goh PL and Tan SE 2017 Performance of pilot test of geotextile tube filled with lightly cement-mixed-soil *GeoAfrica 2017 Conference, Morocco*

[6] Chew SH, Yim HMA, Koh JW, Eng ZX, Chua KE and Tan SED 2018 Performance of pilot test of geotextile tube filled with lightly cemented clay *11th Int. Conf. on Geosynthetics, Korea*

[7] Chew SH, Kamruzzaman AHM and Lee FH 2004 Physicochemical and engineering behavior of cement treated clays *J. Geotech. Geoenviron. Eng* **130**pp 696-706

[8] Lee FH, Lee Y, Chew SH and Yong KY 2005 Strength and modulus of marine clay-cement mixes *J. Geotech. Geoenviron. Eng* **131**pp 178-86

[9] Kamruzzaman AHM, Chew SH and Lee FH 2009 Structuration and destructuration behavior of cement-treated Singapore marine clay *J. Geotech. Geoenviron. Eng.* **135**pp 573-89

[10] Zhang RJ, Santosa AM, Tan TS and Phoon KK 2013 Strength of high water-content marine clay stabilised by low amount of cement *J. Geotech. Geoenviron. Eng.* **139**pp2170-81

[11] Terzaghi K and Peck RB 1967 Soil mechanics in engineering practice *John Wiley, New York* 729

IOP Conf. Series: Materials Science and Engineering 1260 (2022) 012017 doi:10.1088/1757-899X/1260/1/012017

Innovative geocomposite for dredged sediments depollution

B W Forogo[1,2], G Stolz[1], M Coquery[3], S Bonelli[1], N Chaouch[1], N Touze[4]

[1] Aix Marseille Univ, INRAE, Aix-en-Provence, France
[2] AFITEXINOV, Champhol, France
[3] INRAE, UR RiverLy, Villeurbanne, France
[4] Université Paris-Saclay, INRAE, SDAR, Jouy-en-Josas, France
E-mail : wilfried.forogo@inrae.fr\wilibana@yahoo.fr

Abstract. In France, sediments from river dredging are considered as waste. Without any special treatment, they must be stored in landfills. The dredged sediments of our area of study (North of France) are polluted with trace metals (antimony and zinc are above the French regulation threshold). The main objective of this study is to develop a new method of active depollution and dewatering treatment in order to enable the reuse of treated sediments. This treatment is carried out through ponds equipped with innovative geocomposite. This innovative geocomposite has two typical passive functions (filtration and drainage). In addition, it is functionalized by an innovative electrokinetic technology to confer it active roles (depollution and dewatering). Electrokinetic consists in applying a difference of electrical potential to a porous medium. We present here the results of a preliminary laboratory study that simulates this new method on a short-term (3 weeks) experiment. The results show that the electrokinetic treatment has an impact on the chemical and physical properties of sediments. The treatment reduces by 50% the concentration of zinc in the leachate after lixiviation.

1. Introduction

Today, the issue of sustainable development is topical. Thus, several solutions have been developed in different fields such as energy with green energies and construction, with more ecological materials. In the same perspective of practicing a more sustainable management of waterways, a new method of sediment treatment resulting from river dredging is being developed. In fact, in France, sediments from river dredging are considered as waste as soon as shore management is considered. Without any special treatment, the polluted sediment must be stored in landfills. The dredged sediments of our area of study (North of France) are polluted with trace metals (antimony and zinc are above the threshold from the French regulation). These waste storage facilities require more and more space and alter the ecological balance of the areas where they are located. In addition, to avoid groundwater pollution, the landfills must be lined which increases the cost of managing these sediments. The main objective of this study is to develop a new method of active depollution treatment in order to enable the reuse of treated sediments. In France, the inert nature of such sediments is defined by Article R.541-8 of the French Environmental Code [1] as any waste that does not undergo any significant physical, chemical or biological modification and does not deteriorate the materials with which it comes into contact in a manner likely to result in harm to the environment or human health. The thresholds taken into account in this regulation to determine the inertness or not of the sediments are those of the admission thresholds in a landfill defined by the European decision of 19 December 2002 [2] concerning the standards for admission to landfills. These thresholds are measured by a leaching test according to standard NF EN 12457-2 [3].

This treatment is carried out through ponds equipped with innovative geocomposite. The innovative geocomposite has two typical passive functions that are filtration and drainage. In addition, the geocomposite is functionalized by an innovative electrokinetic technology to confer it active roles. Electrokinetic consists in applying a difference of electrical potential to a porous medium. Electromigration and electroosmosis are two of the phenomena that confer an active depollution and dewatering roles to the innovative geocomposites. Electroosmosis is a flow of fluid in the matrix of charged particles that appears when an electrical potential difference is applied. The application of an electrical potential difference in a porous medium causes the appearance of an electroosmotic flow due to the displacement of the charges bound to the surface of the particles. Electroosmosis allows the decontamination of sediments by the transport of solubilized pollutants out of sediments and intervenes for the extraction of sediments by the mobilization of interstitial fluid. Electromigration or migration of ions is the movement of ions due to the application of an electrical potential difference. During the application of an electrical potential difference, the electrodes are polarized. The anode is the positive electrode and the cathode is the negative electrode. Under the effect of the electrical potential gradient, cations (positive sign ions) migrate to the cathode and the anions (negative sign ions) migrate to the anode. Pollutants must be solubilized first to be mobilized by electroosmosis and electromigration [4]. Indeed, during the application of electrokinetic we observe the appearance of a basic front at the cathode and an acidic front at the anode. The acidic front is beneficial for the electrokinetic treatment because it allows a solubilisation of trace metals, but the basic front is detrimental because it causes the precipitation of some trace metals. This heterogeneous spatial distribution of pH causes other heterogeneities such as the development of a heterogeneous flow through the electroosmotic permeability coefficient that is sensitive to pH, and the development of a heterogeneous interstitial pressure that can lead to the appearance of suction near the cathode [5]. Despite the complex phenomena occurring during the treatment, electrokinetic has been used successfully for depollution and dewatering. Indeed, Bourgès-Gastaud *et al.* [6] developed an experimental device to test the ability of an electrokinetic geocomposite to dewater mature mine tailings. Their tests showed that the electrokinetic combined with the mechanical dewatering extraction made it possible to increase the solid content of the mine tailing from 42% to 66% and to increase the shear strength from 0 to 40 kPa on average. Wang *et al.* [7] managed to reduce by 43% (6 days of treatment and a current density of 0.390 mA/cm^2) the concentration of arsenic. This explains why we chose to study electrokinetic for sediment treatment.

The innovative geocomposite is functionalized with the aim of incorporating it an active depollution role and an active dewatering role. In the configuration of ponds, the electrodes are set up horizontally to respect the primary objective of the ponds, which is the gravitational dewatering (see Figure 1). The flow produced by the electrokinetic is in the direction of the gravitational flow and a constant water supply is provided at the surface of the pond (at the anode) to allow a permanent flow during the treatment. The flow crosses the sediments in order to solubilize the pollutants and in order to extract them.

This paper presents the experimental study of the sediment treatment of a dredged material by this innovative geocomposite. An experimental cell has been developed to simulate in laboratory the treatment conditions in ponds.

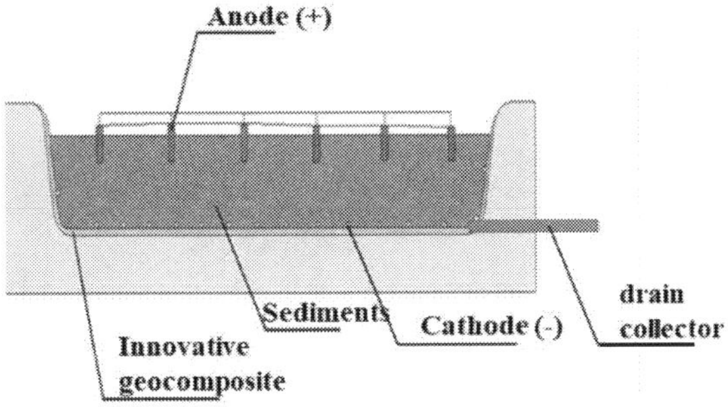

Figure 1. Principle of the treatment pond

2. Material and method

2.1. Material

2.1.1. Sediments. The sediments come from a storage pond belonging to " Voies navigables de France (VNF)". This pond is located in the city of Noyelle-sous-Lens (North of France). The concentrations of the chemical elements present in the sediments are presented in Table 1.

Table 1. Chemical composition of the studied dredged sediment; concentrations of total trace elements and after leaching (D.M: dry matter)

Metal	Unit	Initial	Inert threshold (following [2])
Arsenic (As)[a]	mg/kg D.M.	16.9	
Cadmium (Cd)[a]	mg/kg D.M.	6.1	
Chrome (Cr)[a]	mg/kg D.M.	61.1	
Copper(Cu)[a]	mg/kg D.M.	315	
Lead (Pb)[a]	mg/kg D.M.	246	
Mercury (Hg)[a]	mg/kg D.M.	1.16	
Molybdenum (Mo)[a]	mg/kg D.M.	2.02	
Nickel (Ni)[a]	mg/kg D.M.	31.2	
Zinc (Zn)[a]	mg/kg D.M.	1100	
AFTER LEACHING TEST (NF EN 12457-2)[3]			
Antimony (Sb[b])	mg/kg D.M.	0.078	0.06
As[c]	mg/kg D.M.	<0.2	0.5
Barium (Ba[c])	mg/kg D.M.	0.31	20
Cd[b]	mg/kg D.M.	0.033	0.04
Cr[c]	mg/kg D.M.	<0.1	0.5
Cu[c]	mg/kg D.M.	0.28	2
Pb[c]	mg/kg D.M.	0.17	0.5
Hg[b]	mg/kg D.M.	<0.001	0.01
Mo[b]	mg/kg D.M.	0.241	0.5
Ni[c]	mg/kg D.M.	0.15	0.4
Selenium (Se[b])	mg/kg D.M.	0.031	0.1
Zn[c]	mg/kg D.M.	4.02	4

[a] NF EN ISO 11885[8] – NF EN 13346 Method B[9]
[b] NF EN ISO 17294-2 [10]/ NF EN 16192[11]
[c] NF EN ISO 11885[8] / NF EN 16192[11]

2.1.1. Experimental cell. A new experimental cell was developed to conduct this experiment. The schematic diagram of the experiment is given in Figure 2. The experimental cell has the following dimensions: 0.25 m wide by 0.35 m long and 0.60 m high. It has orifices for inserting electro conductivity (EC) probes. The bottom of the cell is lined up with a drainage geocomposite (filtering and draining nonwoven needle-punched geotextile with mini-drain), functionalized by an electrode to decontaminate the sediment. The surface of the sediment is lined with an electrode (the anode) in titanium alloy coated with mixed oxide of precious metals to enable it to be corrosion-proof. The anode has a shape of a grid. The experimental cell is equipped with the following sensors:

- EC probes follow the evolution of the electrical conductivity inside the sample. These probes are placed at three points: one in the anode area, one in the cathode area and one in the central area of the sample;

- A LabVIEW control panel and an electrical resistance of 3 Ohms follow the evolution of the intensity during the treatment;

- A scale measures continuously the amount of water used for the depollution and the amount of water extracted during the dewatering phase.

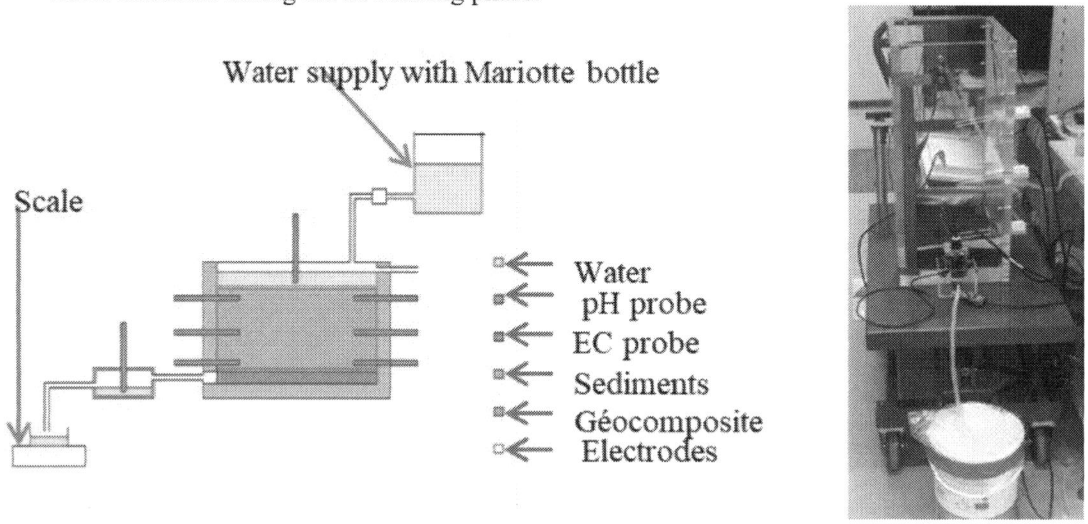

Figure 2. The experimental cell principle

2.1.2. Innovative geocomposite. The innovative geocomposite is a drainage geocomposite. This drainage geocomposite has intrinsic functions which are:

- Filtration, which is the retention of soil or other particles subjected to hydrodynamic forces while allowing the passage of fluids through or in a geosynthetic material [12]. In the geocomposite the filtration function is provided by a geotextile filter. The suitability of geotextile for sludge filtration has been proven and filtration criteria defined [13-17]. The geotextile filter only acts in the first moments of the filtration to allow the formation of the "filter cake", which then ensures the role of filter. The geotextile filter follows the recommendations of Bourgès-Gastaud *et al.* [17] and Delmas *et al.* [13];

- Drainage, which is the collection and transport of rainwater, groundwater and / or other fluids in the plane of a geosynthetic [12]. The drainage function is provided by a needle-punched nonwoven drainage core and mini-drains.

In addition to its intrinsic functions, the geocomposite is functionalized by the electrokinetic technology to give it active roles of depollution and dewatering. The geocomposite is functionalized by the addition of a galvanized steel electrode in 35 mm^2 wire form. The wire-shaped electrode is placed in the same gusset as the mini-drain (see Figure 3). The characteristics of the geocomposite are shown in Table 2.

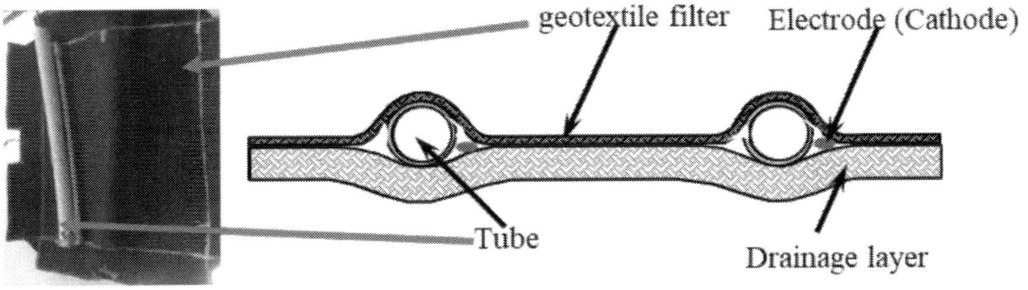

Figure 3. The innovative geocomposite material. At the left, photography of top face of innovative geocomposite (geotextile filter in black) and tube with wire electrode. At the right, schematic representation of the section in the thickness of innovative geocomposite

Table 2. Characteristics of the geocomposite

Properties	Values	Standard Methods
Thickness under 2 kPa normal load	5 mm	NF EN ISO 9863-1[18]
Filtration opening size	110 µm	NF EN ISO 12956[19]

2.2. *Method*

2.2.1. Depollution experiment. The sediments (60 kg with initial height of 44 cm) are introduced into the cell with an initial solid content (SC see equation (1)) of 50% in successive layers.

$$SC= \frac{\text{solid mass}}{\text{solid mass + liquid mass}} \times 100\ \% \qquad (1)$$

Beforehand, the sediments are sieved through a 6 mm sieve to remove coarse particles, and re-wetted with distilled water to a solid content of 50%. The valve allowing the supply of water is opened in order to maintain a constant height of water. It is necessary to keep a constant water level at the anode in order to facilitate the electromigration phenomenon, and to allow electrokinetic continuity over time. Dewatering of the anode could even result in stopping of electrokinetic. The constant water level is maintained for 48 hours to measure the hydraulic conductivity of the sediments (Phase 1). Then the electrokinetic is started with constant water supply (Phase 2). At the end of Phase 2, the electrokinetic is maintained and the constant water supply is stopped. The sediment is dewatered by electrokinetic until electrokinetic stop (no more water extracted at the cathode) (Phase 3). At the end of the dewatering, the cell is disassembled. The final pH and the final mass water content are measured on the thickness of the sediment column. Then the sediment is divided into three equal areas: the anodic area, the central area and the cathodic area and the pollutant concentration are measured in these areas.

2.2.2. Chemical analyses. Different chemical analyses are carried out on the three areas of the sediment:

- On the solid part (sediments): total concentration of trace metals. This analysis is performed after acid digestion (regal water mineralization - heating block) according to standard NF EN ISO 11885[8] – NF EN 13346 Method B[9];

- On the eluate after a leaching test of the sediment performed according to standard NF EN 12457-2 [3]: total concentration of trace metals in eluate after filtration. The leaching test is used to determine the sediment's ability to release trace metals under the effect of a stream of water.

3. Results

The results presented in this section were obtained for an experiment running over a period of 3 weeks. This period is relatively short compared to the height of the sediments column to be treated (i.e. 44 cm).

3.1. Final state of innovative geocomposite after treatment

After the treatment, the innovative geocomposite was disassembled and weighed. The innovative geocomposite was in a good visual state, with impregnation of sediment and precipitate (see Figure 4). Table 3 shows the state of impregnation of the innovative geocomposite and conductive materials. Conductive materials did not lose mass during processing and are resistant to corrosion. The cathode has a mass gain due to the incorporation of precipitate within it.

Table 3. Variation of mass in the geocomposite and conductive material at T0 and after the 3-weeks experiment

	Initial mass (g)	Final mass (g)	Impregnation rate
Filter+ drainage layer	84.11	164.2	542.7 g/m²
Mini-drain	33.12	34.15	2.94 g/m
Anode	44.26	44.26	
Cathode	106.2	107.3	

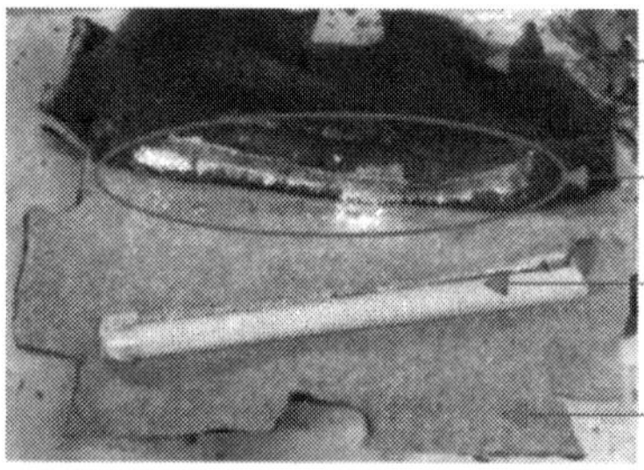

Figure 4. Final state of disassembled innovative geocomposite after the 3-weeks experiment

3.2. Physical parameters

The following caption is used in all section 3.2. and allows to follow the duration of each phase of treatment:

- | (black line): Starting electrokinetic
- | (green line): Stopping the water supply
- | (brown line): No more supernatant at the anode
- | (red line): Stopping electrokinetic

3.2.1. Evolution of the pH. Figure 5 (a) shows the evolution of the pH level of the liquid extracted at the cathode, the supernatant at the level of the anode and the middle of the sediment. The pH level of the fluid extracted at the anode remains stable during the first 20 hours of the electrokinetic treatment before starting to grow rapidly to reach a value of 12 and remains stable during the treatment. The pH in the middle of the sediment remains the same throughout the treatment with a value of about 6.8. The pH sensor at the anode supernatant failed, the pH was measured punctually and is around 1.8. Unfortunately we do not have the kinetics of pH evolution at the anode supernatant level. Figure 5 (b)

IOP Conf. Series: Materials Science and Engineering 1260 (2022) 012017 doi:10.1088/1757-899X/1260/1/012017

shows the evolution of the pH on the thickness at the end of the experiment. There is a heterogeneity in the thickness. The pH rises from the anode to the cathode from 6 to 11.5 with a sharp growth at 5 cm from the cathode. Indeed at 7.5 cm from the cathode the pH is about 6.9 while at 2.5 cm it is 11.5. This heterogeneity is not only observed on the pH but also on the electrical conductivity, as we will see in the section 3.2.5.

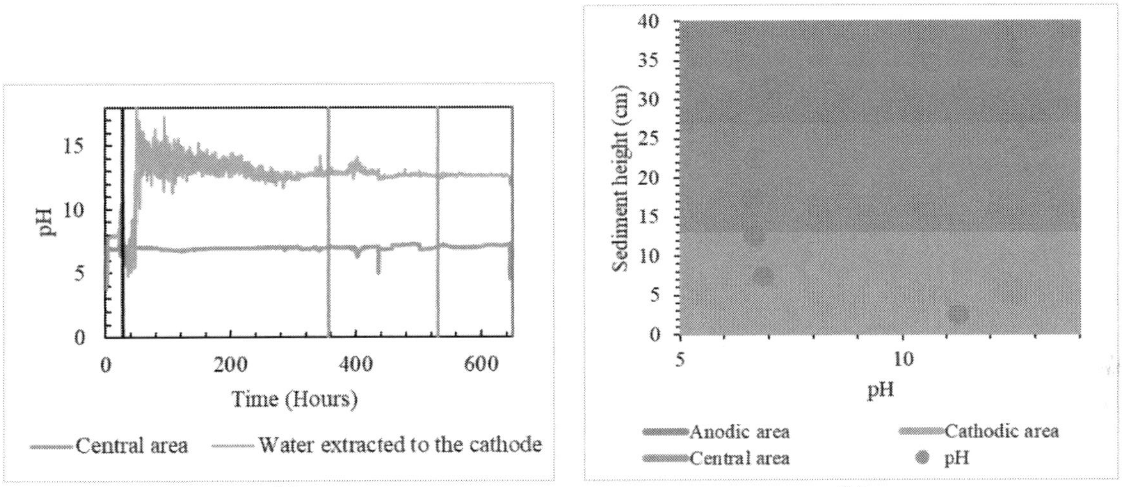

(a) (b)

Figure 5. (a) Evolution of pH as a function of time during the treatment within the cell(b) Spatial profile of pH along the height of the sediment (at the end of experiment).

3.2.2. Settlement. Figure 6 shows the evolution of the height of the sediment during the electrokinetic treatment. The settlement occurs from the first moments and especially during the period of gravity dewatering. Indeed, during the gravity period, the sediment height decreases by 2.5 cm and then the kinetics of settlement lows down. During the first moments of the electrokinetic treatment (phase 2), the sediment height decreases by 1.5 cm and stabilizes at 41.5 cm and remains unchanged throughout the treatment. The settlement during the electrokinetic treatment occurs at the same time as the appearance of the basic pH at the cathode which causes a suction in the cathodic area and therefore a consolidation which could explain this settlement.

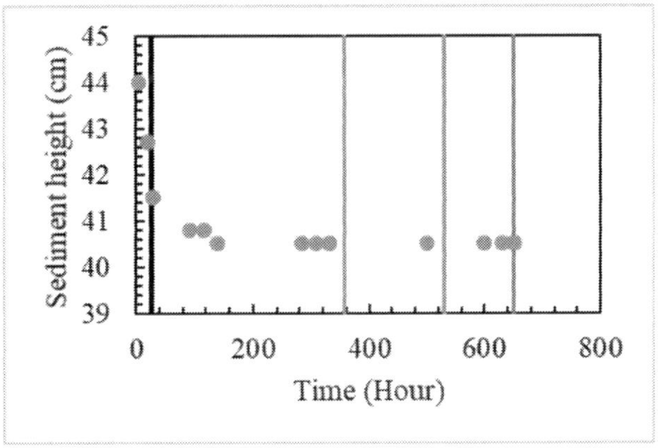

Figure 6. Evolution of the height of sediment in the experimental cell as a function of time during the 3-weeks treatment

3.2.3. Mass of water extracted at the cathode. The mass of water extracted increases almost linearly during the first two phases (see Figure 7 (a)). This trend has been observed by several authors during depollution treatment [20-23]. In the last two phases of the treatment the increase in the mass of extracted water decreases due to the stoppage of constant supply of water which causes a decrease in the hydraulic load. The mass of water extracted reaches a plateau and stagnates during the last phase because there is no more supernatant. By observing Figure 7 (b), the flow rate decreases during phase 1. It can be due to the settlement that the sediment undergoes, which modifies the state of consolidation of the latter and consequently its hydraulic conductivity. During the second phase, the application of electrokinetic increases the flow rate due to the addition of the electroosmotic flow and the hydraulic flow. Then the flow decreases progressively, which may be due to the heterogeneity of pH which leads to a heterogeneous electroosmotic flow by modifying the coefficient of electroosmotic permeability.

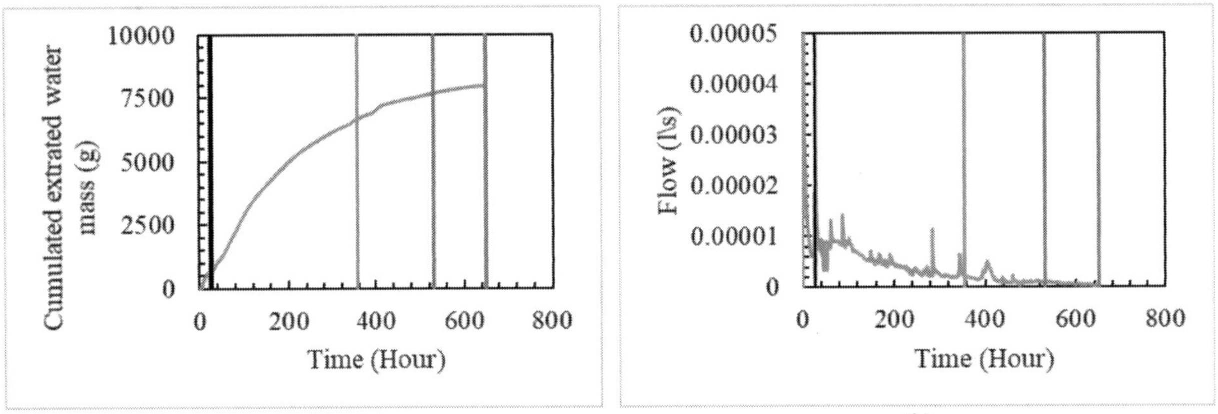

(a) (b)

Figure 7. (b) Evolution of cumulated extracted water to the cathode as a function of time.
(b) Evolution of flow as a function of time

3.2.4. Electrical intensity. Figure 8 shows the evolution of electrical intensity as a function of time. The intensity is at its maximum as soon as the electrokinetic are installed and reaches a value of 450 mA. Then the electric intensity gradually decreases to reach a minimum value around 50 mA. This behaviour has been observed in several depollution or dewatering studies [23-25]. At the beginning of the treatment, the ions are present which explains the high value of the intensity. During the treatment, the ions concentration decreases due to electromigration, absorption and precipitation reactions. This drop in concentration causes a decrease in electrical conductivity and this explains the decrease in intensity during treatment. The basic medium at the anode causes precipitation of trace metals that decreases the conductivity.

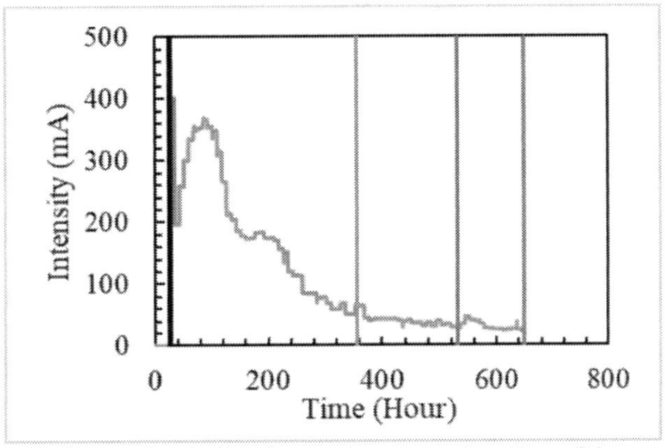

Figure 8. Evolution of electrical intensity during the treatment as a function of time.

3.2.5. Electrical conductivity. In Figure 9 (a), a single value of electrical conductivity is observed throughout the medium in the initial state. Over time, the electrical conductivity of the sediment is reduced at the cathode area. Firstly, this can be explained by the fact that the cathodic area is gradually settled by the weight of the sediments. Secondly, because of the appearance of the basic front, the ions precipitate at the cathode area, which causes a decrease of the electrical conductivity of the medium. For the anode and the middle area, an increase of electrical conductivity is observed at the beginning of the treatment followed by a decrease. The tendency of the electrical conductivity of the sediment at the anode is similar to that of the electrical intensity. The electrical conductivity of the medium is experiencing a sudden drop which appears 40 hours after the start of electrokinetic. This sudden drop is due to the appearance of the basic pH at the cathode that causes suction in the cathode area, a settlement and therefore a decrease in porosity. The electrical conductivity of the liquid extracted at the cathode gradually increases from the start of the electrokinetic to a maximum value of 4 dS/m and remains around this value throughout the treatment (see Figure 9 (b)). This denotes the migration of ions out of the sediment due to electrokinetic and the presence of ions in the liquid extracted at the cathode increases its electrical conductivity. The sensor at the anode level has failed so we cannot interpret the results of this sensor.

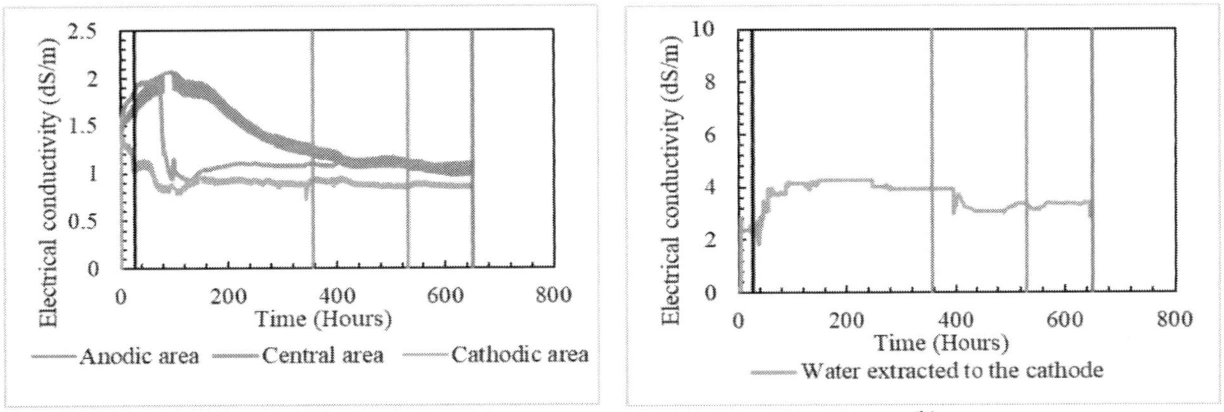

(a) (b)

Figure 9. (a) Evolution of electrical conductivity in the sediment as a function of time during treatment.
(b) Evolution of electrical conductivity in water as a function of time during treatment.

3.2.6. Water content. Figure 10 shows the evolution of the mass water content in the sediment. The final mass water content in the sediment decreases from the anode to the cathode which is in the opposite direction of the flow from the anode to the cathode. Other studies have observed the same pattern of mass water content [26-28]. This behaviour can be explained by the heterogeneity that develops in sediments during electrokinetic treatment.

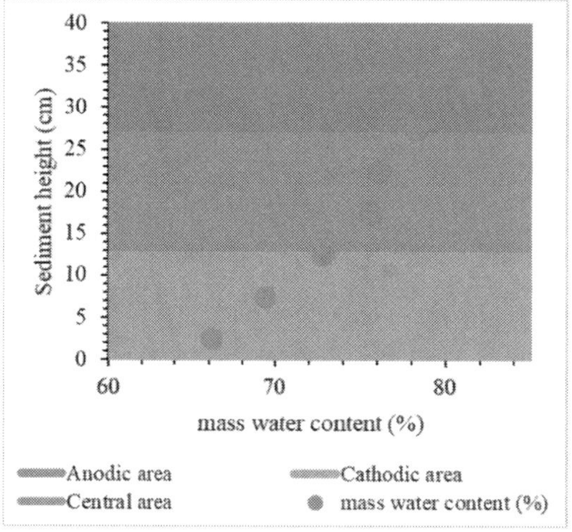

Figure 10. Spatial profile of the mass water content along the height of sediment at the end of the 3-weeks experiment.

3.3. Chemical parameters

In this part we present the results of the chemical analyses. The analyses are made over three specimens taken from the anodic area, the cathode area and the middle of the sample. The comparisons with the initial values are indicative because we did not have the opportunity to have an initial pollutant content of the sediments used for the experiment. Therefore we used the values measured at the sampling site as initial pollutant content values.

3.3.1. Analyses on sediment. Figure 11 shows the normalised concentration of trace metals in the sediment after treatment in the different areas. Cd, Cr, Pb and Zn normalized concentration is equal or less than unity for anodic and central area and superior to unity for cathodic area. It means that the concentrations reduce at the anodic and central areas and increases at cathodic area. For As, Hg and Ni the normalised concentration has opposite trend. The difference of normalised concentration between the different areas is small that could by explain by the short treatment time and the low refinement of the thicknesses of the concentration measurement area. From these observations two behaviours emerge :

- the first is a cationic behaviour, that is to say that the metals are present in a cationic form and move from the anode to the cathode under the effect of electromigration and electroosmosis. These metals are cadmium, chromium, copper, lead and zinc; Concentrations increased at the cathode in comparison to the 2 other areas after the 3-weeks experiment.

- the second is an anionic behaviour, that is to say that the metals are in an anionic form and move from the cathode to the anode under the effect of electromigration but go against the electroosmotic flow. These metals are Hg, As and Ni. Concentrations are slightly higher at the anode in comparison to the 2 other areas after the 3-weeks experiment.

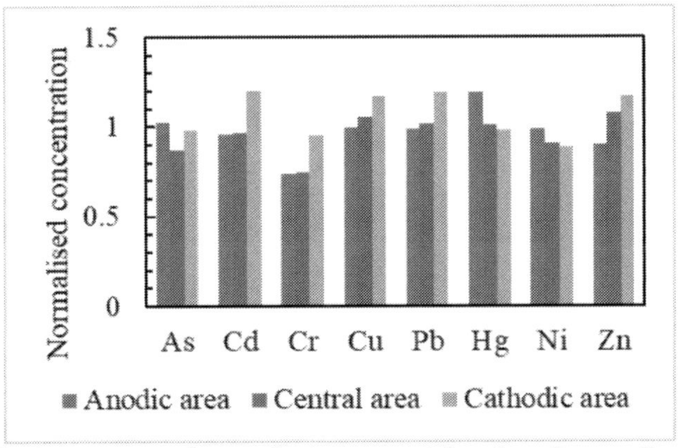

Figure 11. Normalised concentration of trace metals in the sediment after treatment. $\text{Normalised concentration} = \frac{\text{initial concentration}}{\text{final concentration}}$

3.3.2. Analyses on leachate. Figure 12 shows the normalised concentration after standard leaching of the metals in the different areas. We note that some metals such as cadmium have a reverse tendency in leachate in comparison to that of sediments. In fact, Cd has a higher concentration at the anode than at the cathode. This is explained by the fact that Cd precipitates (in a non-soluble form) in the cathodic area because of the basic pH and solubilize in the anodic area because of the acidic pH. Zn and Ni have the same tendency and are below the limit of quantification (of the analytical method) at the cathode area. This implies that electrokinetic treatment can stabilize these metals in forms that do not solubilize so allows sediments to be inert. But other trace metals, like Sb, Ba, Cu, Mo and Se that have a high concentration at the cathode, do not precipitate under the effect of basic pH and thus are not stabilized.

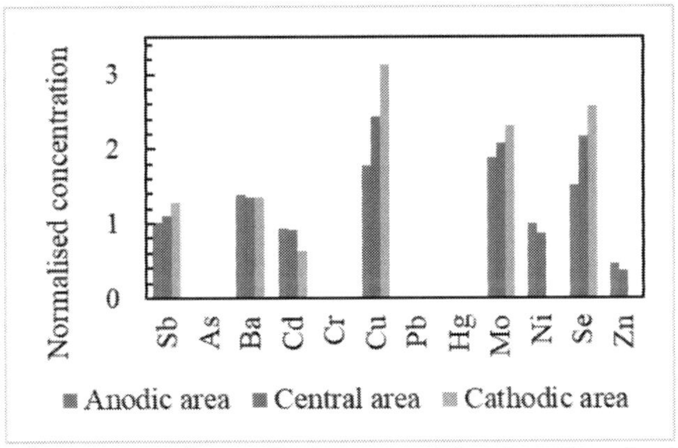

Figure 12. Normalised concentration of trace metals in the leachate after treatment. $\text{Normalised concentration} = \frac{\text{initial concentration}}{\text{final concentration}}$

4. Conclusions

This paper presents an experimental study of the depollution of dredged sediment from waterways with electrokinetic incorporated on innovative geocomposite. A new experiment cell was developed to simulate the condition of the treatment pond. The first results show that the innovative geocomposite accomplished well its intrinsic functions, the anode is corrosion-proof and the electrokinetic treatment is very complex and heterogeneous. Indeed during the treatment, there is a modification of the physical properties of the sediment due to the settlement of the sediment, which causes the reduction of the hydraulic conductivity. The application of electrokinetic produces an electroosmotic flow that increases the total flow. This flow decreases during the treatment due to the heterogeneity of pH that induces heterogeneous electroosmotic flow. An acidic condition is developed at the anodic side and a basic condition is developed at the cathodic side. This heterogeneous pH condition is concomitant with a heterogeneous electro conductivity in the sediment. The electro conductivity decreases during the treatment like the electrical intensity. In spite of its complexity, the electrokinetic treatment enables to lower the concentration of zinc in leachates by 50%, and even to eliminate Zn and Ni after precipitation at basic pH near the cathode. In contrast, the concentrations of antimony after leaching does not change at the end of the experiment, so electrokinetic has no effect on Sb, Ba, Cu, Mo and Se. Additional tests over longer periods will be performed to fully evaluate the performance of this new electrokinetic method to depollute dredged sediments.

Acknowledgments
The authors gratefully acknowledge all the members of the French project VALSSINNOV supported by BPI France (AnteaGroup, Irstea, VNF, Vinci Terrassement and Afitexinov), and Lysiane Dherret (UR RiverLy, LAMA team) for trace metals analyses by ICP-OES.

References
[1] Legifrance 2016 Article R541-8 Modifié par Décret n°2016-288 du 10 mars 2016 - art. 6.
[2] AIDA 2012 Décision n° 2003/33/CE du 19/12/02 établissant des critères et des procédures d'admission des déchets dans les décharges, conformément à l'article 16 et à l'annexe II de la directive 1999/31/CE. https://aida.ineris.fr/consultation_document/1595)
[3] AFNOR 2002 NF EN 12457-2 Characterization of waste - Leaching - Compliance test for leaching of granular waste materials and sludges - Part 2 : one stage batch test at a liquid to solid ratio of 10 l/kg for materials with particle size below 4 mm (without or with size reduction).
[4] Acar Y B and Alshawabkeh A N 1993 Principles of electrokinetic remediation *Environmental Science and Technology* **27** pp 2638-47
[5] Alshawabkeh A N, Sheahan T C and Wu X 2004 Coupling of electrochemical and mechanical processes in soils under DC fields *Mechanics of Materials* **36** pp 453-65
[6] Bourgès-Gastaud S, Stoltz G, Dolez P, Blond É and Touze-Foltz N 2015 Laboratory device to characterize electrokinetic geocomposites for fluid fine tailings dewatering *Canadian Geotechnical Journal* **52** pp 505-14
[7] Wang J Y, Huang X J, Kao J C M and Stabnikova O 2007 Simultaneous removal of organic contaminants and heavy metals from kaolin using an upward electrokinetic soil remediation process *J. Hazard. Mater.* **144** pp 292-9
[8] AFNOR 2009 NF EN ISO 11885 Water quality - Determination of selected elements by inductively coupled plasma optical emission spectrometry (ICP-OES).
[9] AFNOR 2000 NF EN 13346 Characterization of sludges - Determination of trace elements and phosphorus - Aqua regia extraction methods - Caractérisation des boues.
[10] AFNOR 2016 NF EN ISO 17294-2 Water quality — Application of inductively coupled plasma mass spectrometry (ICP-MS) — Part 2: Determination of selected elements including uranium isotopes.
[11] AFNOR 2012 NF EN 16192 Characterization of waste - Analysis of eluates - Caractérisation des déchets.
[12] AFNOR 2015 NF EN ISO 10318-1 Geosynthetics - Part 1 : terms and definitions.
[13] Delmas P, Baral C, Stoltz G, Majerus T and Laidié N 2016 Filtration of clayey sludge by the use of geotextiles : experimental study *EuroGeo 6* 25-28 September 2016

[14] Moo-Young H K and Tucker W R 2002 Evaluation of vacuum filtration testing for geotextile tubes *Geotextiles and Geomembranes* **20** pp 191-212
[15] Kutay M E and Aydilek A H 2005 Filtration performance of two-layer geotextile systems
[16] Muthukumaran A E and Ilamparuthi K 2006 Laboratory studies on geotextile filters as used in geotextile tube dewatering *Geotextiles and Geomembranes* **24** pp210-9
[17] Bourgès-Gastaud S, Stoltz G, Sidjui F and Touze-Foltz N 2014 Nonwoven geotextiles to filter clayey sludge: An experimental study *Geotextiles and Geomembranes* **42** pp 214-23
[18] AFNOR 2016 ISO 9863-1:2016 Geosynthetics - Determination of thickness at specified pressures - Part 1 : single layers
[19] AFNOR 2010 NF EN ISO 12956 Geotextiles and geotextile-related products - Determination of the characteristic opening size
[20] Ben Hassine A, Souli H, Dubujet P and Trabelsi-Ayadi M 2014 On the use of electrokinetic in the kaolinite soil decontamination *Journal of Materials and Environmental Science* **5** pp 2555-9
[21] Shaplro A P and Probstein R F 1993 Removal of contaminants from saturated clay by electroosmosis *Environmental Science and Technology* **27** pp 283-91
[22] Iannelli R, Masi M, Ceccarini A, Ostuni M B, Lageman R, Muntoni A, Spiga D, Polettini A, Marini A and Pomi R 2015 Electrokinetic remediation of metal-polluted marine sediments: experimental investigation for plant design *Electrochim. Acta* **181** pp 146-59
[23] Song Y, Ammami M T, Benamar A, Mezazigh S and Wang H Q 2016 Effect of EDTA, EDDS, NTA and citric acid on electrokinetic remediation of As, Cd, Cr, Cu, Ni, Pb and Zn contaminated dredged marine sediment *Environ. Sci. Pollut. Res.* **23** pp 10577-86
[24] Cherifi M, Hazourli S and Ziati M 2009 *Proceedings of the Jmsm 2008 Conference,* ed A Cheikhrouhou (Amsterdam: Elsevier Science Bv) pp 1021-30
[25] Darmawan and Wada S I 2002 Effect of clay mineralogy on the feasibility of electrokinetic soil decontamination technology *Applied Clay Science* **20** pp 283-93
[26] Reddy K R, Saichek R E, Maturi K and Ala P 2002 Effects of soil moisture and heavy metal concentrations on electrokinetic remediation *Indian Geotechnical Journal* **32** pp 258-88
[27] Souli H 2016 Evolution du comportement mécanique et de la structure des géomatériaux sous l'effet du transport de particules. Habilitation à Diriger des Recherches. ed U J M d S Etienne
[28] Al-Hamdan A Z and Reddy K R 2008 Geochemical assessment of metal transport in glacial till during electrokinetic remediation *Environmental monitoring and assessment* **139** pp 137-49

IOP Conf. Series: Materials Science and Engineering 1260 (2022) 012018 doi:10.1088/1757-899X/1260/1/012018

Evaluation of the performance of a drainage geocomposite in a simple cover system based on 10 year measurements

E Birle[1], M Melsbach[1], D Heyer[1]

[1] Zentrum Geotechnik, TU Munich, Baumbachstr. 7, 81245 Munich
Email: e.birle@tum.de

Abstract. In the course of the geotechnical investigations for a new runway at Munich Airport the hydro-mechanical behaviour of organogenic clays was studied. With the aim of an efficient use of materials the organogenic clay shall be used for noise and view barriers. As the organogenic clays have increased arsenic contents the percolation of such structures should be minimized. For the detailed investigation of the hydro-mechanical behaviour and for the determination of seepage quantities, a large scale, 5 m high, 30 m long and 25 m wide test fill was constructed. Due to the low permeability of the organogenic clays in compacted condition a cover design with a drainage mat and a top layer without an additional sealing layer was chosen. Measurements of the water balance were collected since the installation in 2008. The results show very low leachate quantities, although humid climatic conditions with average annual precipitation of approx. 745 mm were present. The low leachate rates can be mainly attributed to the capillary-breaking effect of the drainage geocomposite and the high water retention capacity of the topsoil. Based on the results the presented simple cover system can be considered an efficient and effective solution for minimizing infiltration water into core materials of embankments and noise barriers in road-way design.

1. Introduction

In the course of the geotechnical investigations for a new runway at Munich Airport the hydro-mechanical behaviour of organogenic clays was studied. The organogenic clays, which form the topsoil in the area of the Munich Airport, have geogenically increased arsenic contents. For the planned expansion of Munich Airport (amongst others, the construction of a new runway is planned) large amounts of these topsoils will have to be removed.

With respect to material recovery, the organogenic clay shall be used for noise and view barriers. For reasons of groundwater protection, the barriers have to be designed with the aim of minimizing the elution of arsenic from the clays. This can principally be achieved by covering the organogenic clays with sealing layers, e.g. clay liners or geomembranes. However, due to the low permeability of the organogenic clays in compacted condition, a cover design with a drainage mat and a top layer without an additional sealing layer was chosen.

For the detailed investigation of the hydro-mechanical behaviour and for the determination of seepage quantities, a large scale, 5 m high, 30 m long and 25 m wide test fill was constructed. Extensive monitoring systems were installed for the determination of the water balance of the fill, including seepage flow through the compacted organogenic clay. Data of the water balance were collected since the installation in 2008. Results of the measurements were reported by Birle et al. [1] and Birle and

Heyer [2][3]. In the following the design of the fill and the measuring instrumentation is briefly described. Thereafter, the seepage flow through the simple cover system, which consists of the topsoil and a drainage geocomposite only is analysed and the hydraulic interaction between the topsoil and the geotextile is described.

2. Materials and design of the test fill

2.1. Core Material

For the construction of the test fill, topsoil was excavated from a site near Munich Airport. According to DIN 18196 and the Unified Soil Classification System (USCS), this soil is classified as an organogenic clay (OH) with a liquid limit in the range of 94 % to 133 % (average value of approx. 124 %) and a plastic limit in the range of 55 % to 87 % (average value of approx. 44 %). The organic content determined by the ignition method is between 25 % and 30 %. The compaction behaviour was studied by Standard Proctor tests. Other than in mineral soils, for the organogenic clay no distinct Proctor Optimum could be determined. A maximum dry density of approximately 0.8 g/cm^3 with corresponding water contents, ranging between 55 % und 75 %, was noted. For the determination of the water permeability laboratory tests were performed. According to the results, the coefficient of permeability depends strongly on the compaction water content: for a water content of 55 % a coefficient of permeability of $5 \cdot 10^{-8}$ m/s and for a higher water content of 75 % a lower value of $5 \cdot 10^{-9}$ m/s was determined [4]. To distinguish this soil from the material which was used for the top layer, it is also referred to as the "core material" in the following.

2.2. Drainage mat

For the drainage mat, which was placed between the core material and the top layer, a drainage geocomposite consisting of a wave-extruded monofilament drainage core and nonwoven geotextiles on the top and bottom for filtration and separation was used. According to the product specifications the effective opening width of the nonwoven layers was determined in line with DIN EN ISO 12956 and is 0.12 mm. The lateral drainage capacity is $1.5 \cdot 10^0$ L/(m·s) for a hydraulic gradient of 1 under vertical stresses of 20 kN/m^2 (determined according to DIN EN ISO 12958).

2.3. Topsoil

Another organogenic clay, which was also taken from a site near Munich Airport, was used as topsoil of the test fill. With average values of the liquid limit of 93 %, the plastic limit of 56 % and the organic content (determined by the ignition method) of 25 %, this organogenic clay is similar to the core material. The topsoil was placed with an excavator shovel and lightly pressed on, but not compacted with a roller. The dry density of the topsoil was determined directly after placement to approx. 0.77 g/cm^3. But only one year later due to consolidation effects a significantly higher dry density of 1.09 g/cm^3 was measured. Based on permeability tests, an average value of the coefficient of permeability of $5 \cdot 10^{-8}$ m/s can be assumed for the topsoil [5].

2.4. Design of the test fill

The design of the fill is shown in figure 1. The test fill is 5 m high, 30 m long and 25 m wide with a slope inclination of 1:2. The compacted organogenic clay is covered by a drainage mat and the 0.6 m thick topsoil layer. As the hydraulic properties of the organogenic clay depend strongly on the compaction water content, the fill was divided in two parts. In the western side the soil was compacted with an initial water content of approximately 55 % ("dry" compacted) and in the eastern side with an initial water content of approx. 75 % ("wet" compacted). The compaction was performed by means of a heavy roller (25 tons) with polygonal drums. Like that, dry densities of the dry and wet compacted material between 0.75 g/cm^3 and 0.85 g/cm^3 were achieved. After the compaction of the organogenic clay the drainage mat was installed. Finally the topsoil layer was spread with a thickness of approx. 60 cm and grass was sown.

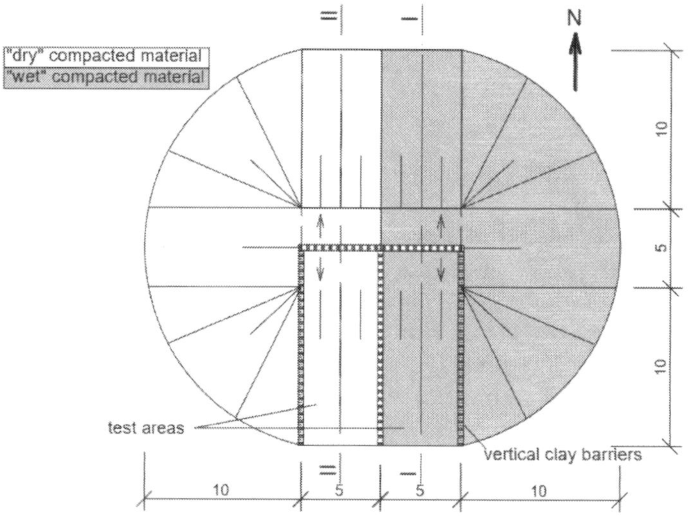

Figure 1. Plan view of the test embankment

2.5. *Monitoring system*

Two areas of 5 m width and 12.5 m length were separated from the remaining part of the fill by vertical clay barriers of very low permeability in the southern area of the fill (see Figure 1). For the assessment of the water balance extensive measuring systems were installed in these test areas, in which two-dimensional flow conditions can be assumed. Leachate is collected at the base by a drainage mat on top of a geomembrane. A tube connected to the geomembrane discharges the water to a gauge well situated on the south side of the fill, where the leachate discharge is measured by electronic tipping buckets. The interflow in the drainage mat between core and topsoil is collected at the bottom of the test areas and quantified similarly. The flow of water in the compacted organic soil can be derived from variations of suction and water content, measured by tensiometers and TDR-probes. The positions of these sensors after installation are shown in Figure 2.

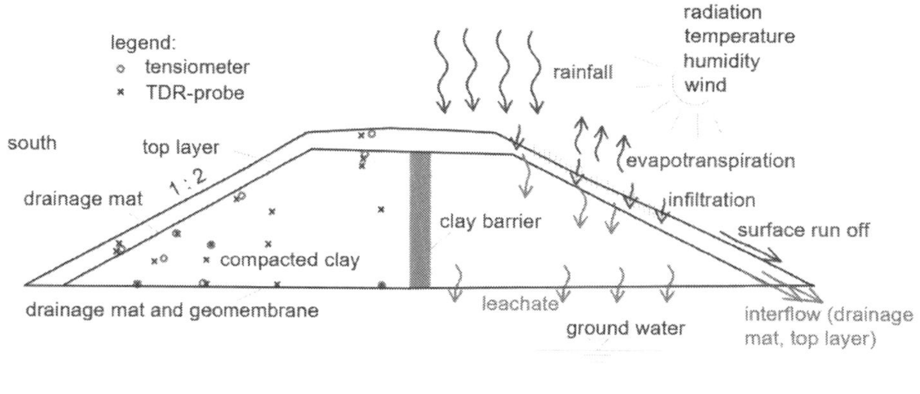

Figure 2. Cross section II-II of the fill with the final arrangement of the tensiometers and the TDR-probes (left section) and the description of the water balance (right section)

For the determination of the volumetric water content by means of the signals of the TDR-probes, a calibration of the TDR-probes had been performed. For the simulation of the water balance of the fill the climate conditions have to be known. The climate is measured by a meteorological station close to the fill. It consists of several sensors measuring the air temperature and humidity, the effective radiation, the wind velocity and direction as well as the amount of precipitation. The water balance of the fill is shown schematically in Figure 2 (right section).

3. Measurements

3.1. Precipitation

The precipitation recorded at the fill between November 2008 and November 2018 is shown in figure 3. The data presented refer mainly to the values measured directly on the test fill, unless no data were available due to problems with the local measuring device. In this case, the precipitation values from the German Meteorological Service (DWD) for Munich Airport were used.

The annual precipitation pattern is typical for the south of Germany showing a maximum in the summer months. This is caused by the accumulating effect of the Alps that is particularly significant during the summer months. The accumulated precipitation reaches 7475 mm, which corresponds to an average annual precipitation of 747 mm. Especially, the years 2013, 2014, 2015 and 2016 were slightly drier and the years 2010 and 2012 significantly wetter than the average long-time annual precipitation of approx. 750 mm given by the German Meteorological Service for Munich Airport.

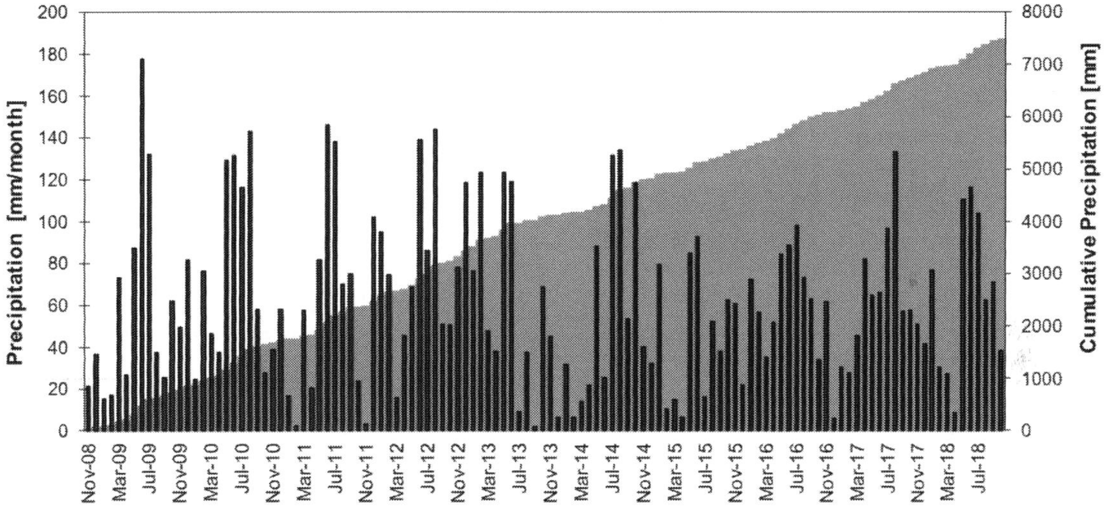

Figure 3. Precipitation measured at the test fill or taken from the German Meteorological Service (DWD) for Munich Airport

3.2. Leachate and interflow in the top drainage geocomposite

Figure 4 shows cumulative data of precipitation, leachate at the base and interflow from the top drainage mat, which were measured for the "wet" and "dry" side of the test fill. For the wet side 99 mm were collected at the base of the test area in the entire observation period and 112 mm from the drainage mat placed above the compacted organic clay. In comparison to the total precipitation of 7475 mm these are very small fluxes (approx. 1.3 % and 1.5 %). For the dry compacted clay even smaller fluxes for the leachate at the base (45 mm) as well as for the drainage mat (58 mm) were recorded. This is caused by the low water content of the "dry" compacted clay leading to the main portion of seepage water being retained by it.

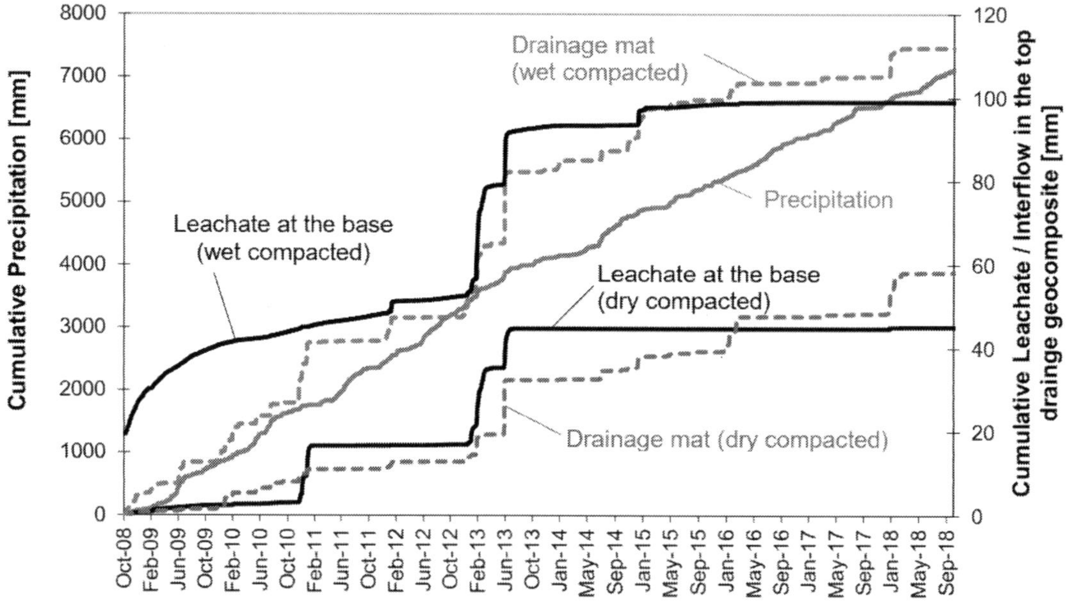

Figure 4. Cumulative fluxes of precipitation, leachate and interflow in the top drainage mat for the wet and dry compacted part of the test fill

Measurements of the TDR-probes located in the topsoil (Figure 5a), show that the maximum water contents are usually reached in the winter months. This is reasonable as the transpiration by the vegetation is low during this time. The interflows from the drainage mats are shown in figure 5b. A comparison of the water content measurements with the interflows indicates that significant discharge from the drainage mats only occurs when the topsoil has reached values of the volumetric water content in the range of 45 % to 55 %, which means that the topsoil is almost saturated. This occurs usually in winter, when no evapotranspiration takes place or in summer after a wet period with high precipitation (e.g. in summer 2013). As the wet compacted clay was almost saturated after compaction, less water is absorbed by the core material and a slightly higher discharge from the drainage mat is noticeable in comparison to the dry compacted core material. However, similar to the dry side, an interflow in the drainage mat takes place only occasionally, when the topsoil has reached water contents above the water content at field capacity and thus infiltrating water cannot be retained further by the topsoil.

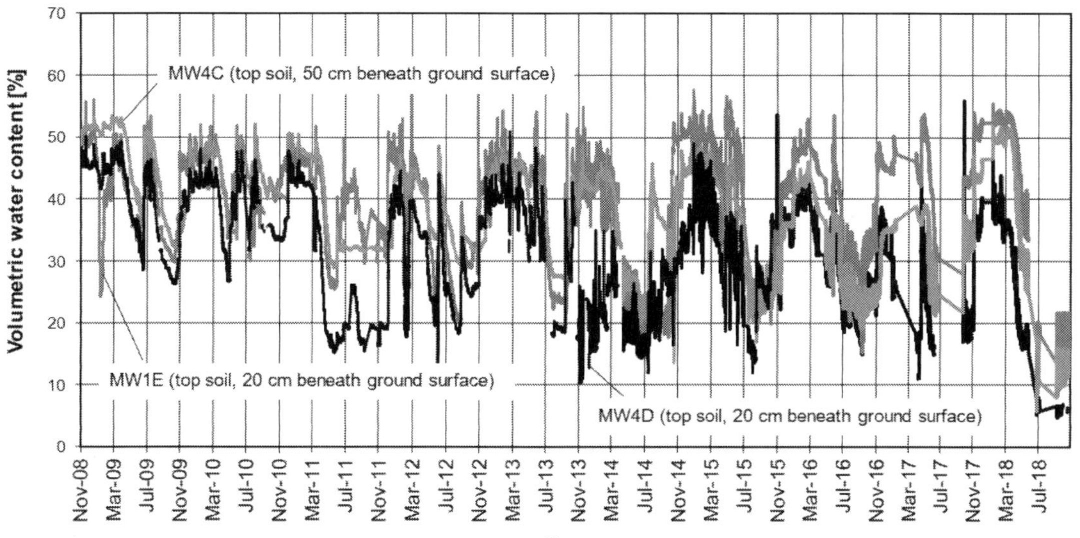

a

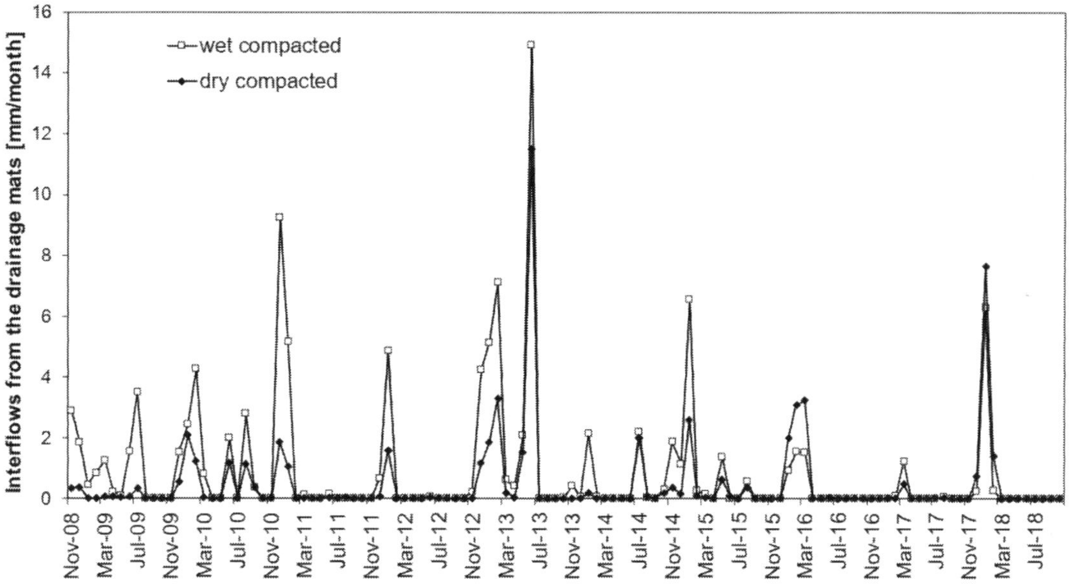

b

Figure 5. Volumetric water contents measured in the topsoil **a** and interflows from the top drainage mats **b**

4. Analysis of the water balance

According to the water balance equation the precipitation (N) must be equal to the sum of evapotranspiration (ET), surface runoff (SR), seepage flow (seepage at the base S_B and interflow from the top drainage mat S_I) and storage (W):

$$N = ET + SR + S_B + S_I + W \quad (1)$$

For the wet compacted clay Birle&Heyer [3] concluded based on the measurements of the water content sensors which were installed in the wet compacted core material that it was almost saturated after compaction and that no significant storage had occurred since 2008. For the dry compacted material in contrast, Birle&Heyer [3] described that it has the potential to retain a significant portion of the infiltrating water and thus the seepage at the base of the dry compacted organogenic clay and the interflow from the drainage mat of the dry compacted clay is only half of that collected from the wet compacted clay.

The actual evapotranspiration was analysed by Birle&Heyer [3] for the period between November 2008 and June 2014. Based on the climate data, they determined the potential evapotranspiration by the Penman-Monteith equation as described by Allen et al. [6]. To get the actual evapotranspiration they used the approach of Disse [7] referring to the measured data of the water content in the topsoil. With this approach, they determined an actual evaporation rate of 84% based on precipitation. Assuming that the evaporation rate did not change significantly since 2014 the actual evapotranspiration for the period between 2008 and 2018 is calculated to 6279 mm.

As the water storage can be neglected for the wet compacted organic clay, equation (1) can be reformulated and the surface runoff, which is not measured directly, can be calculated from the precipitation, the evapotranspiration and the seepage flow as follows:

$$SR = N - ET - S_B - S_I \tag{2}$$

Considering the cumulative fluxes of the leachate at the base and interflow from the drainage mat for the wet part as shown in figure 4 the accumulated surface runoff between 2008 and 2018 is determined to 985 mm, which corresponds to 13 % of the precipitation. As the same material was used for the topsoil above the wet and the dry compacted part, a similar surface runoff is expected for the dry part. Accordingly, for the dry part, the stored water quantity can be determined from the water balance equation (1). Table 1 summarizes the components of the water balance of the dry and wet compacted part.

Table 1. Components of the water balance for the period from November 2008 to November 2018. Percentages in brackets are related to the cumulative precipitation of 7475 mm

	Evapotranspiration ET [mm]	Surface runoff SR [mm]	Seepage at the base S_B [mm]	Interflow from the drainage mat S_I [mm]	Water storage W [mm]
Wet compacted part	6279 (84 %)	985 (13.2 %)	99 (1.3 %)	112 (1.5 %)	0 (0 %)
Dry compacted part	6279 (84 %)	985 (13.2 %)	45 (0.6 %)	58 (0.8 %)	108 (1.4 %)

5. Performance of the top drainage geocomposite

According to the analysis of the water balance in chapter 4 it is evident that the amount of seepage water infiltrating the compacted clay (core material) is very low. This can be attributed mainly to the performance of the topsoil in combination with the drainage geocomposite installed between the topsoil and the core material. As the data in Table 1 show, only 2.8 % ($S_B + S_I$) of the precipitation seep into the top drainage mat. Due to the high water retention capacity of the topsoil a large amount of seepage water can be stored before the topsoil is fully saturated. The drainage geocomposite acts as a capillary break and does not draw water from the topsoil. Thus, the percolation from the topsoil into the underlying drainage geocomposite is only gravity driven, which means that it occurs, when the water content at field capacity of the topsoil is exceeded. This corresponds to almost saturated conditions as explained below. Additionally the saturated hydraulic conductivity of the topsoil and the inclination of the slope of the test fill are of importance. Based on flexible wall permeameter tests acc. to Birle et al. [5] the saturated hydraulic conductivity of the topsoil is in the range of $2 \cdot 10^{-8}$ m/s through $2 \cdot 10^{-7}$ m/s. That comparatively low hydraulic conductivity limits the maximum water infiltration rate under saturated conditions and increases a surface runoff. The latter is also favoured by the steep slope inclination.

The capillary break effect of the nonwoven geotextile is illustrated by figures 6 and 7. They show the soil-water retention curves (SWRC) and the unsaturated hydraulic conductivities of the topsoil and a nonwoven geotextile. The SWRC of the topsoil was experimentally determined by use of a suction-controlled oedometer cell (ATT) for suctions up to 500 kPa and a dew-point hygrometer for the high suction range. Additionally measurements with tensiometers were performed for the determination of the main wetting path in the low suction range (below 100 kPa). Sample preparation and test procedures are described by Birle et al. [5]. Details about the functionality of the suction-controlled oedometer cell (ATT) are given by Birle [1]. The data for the non-woven geotextile were taken from McCartney et al. [8]. In Figure 6 additionally to the measurements the best-fit SWRCs defined using the van Genuchten-Mualem model [9] are shown. The hydraulic conductivity functions in Figure 7 were defined using the van Genuchten-Mualem model under consideration of the saturated hydraulic conductivity values. For the topsoil a saturated hydraulic conductivity of $5 \cdot 10^{-8}$ m/s was considered.

Due to its high porosity the non-woven geotextile shows a very low air-entry suction value of approx. 0.6 kPa and a very flat slope of the SWRC. In contrast, the topsoil has a much higher air-entry value and a high water retention capacity. Under consideration of the data given in Figure 6 a field capacity of 63 % and an effective field capacity of 40 % can be assumed for the tested sample of the topsoil.

The capillary break effect of the nonwoven geotextile against the topsoil is clearly visible from figure 7. As soil suction at the interface between the topsoil and the nonwoven geotextile is continuous, for suction values larger than approx. 1 kPa the hydraulic conductivity of the nonwoven geotextile is much smaller than the hydraulic conductivity of the topsoil. Thus, a significant water uptake by the nonwoven geotextile takes place only for suction values lower than 1 kPa. Similar results about the capillary-breaking effect of nonwoven geotextiles were reported by McCartney et al. [8], Krisdani et al. [10], Bathurst et al. [11] and Zornberg et al. [12].

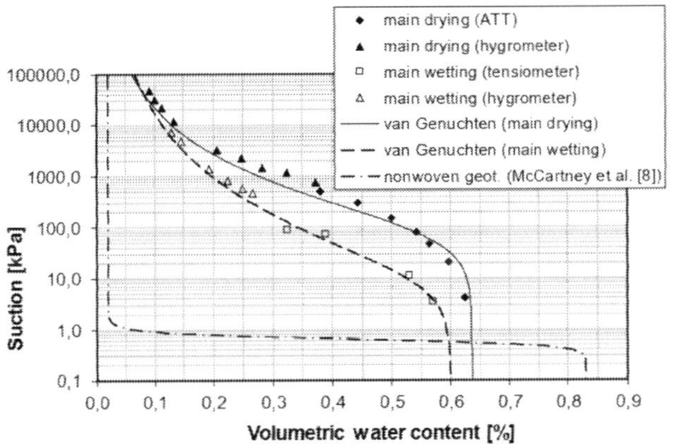

Figure 6. SWRC of topsoil according to Birle et al. [5] and nonwoven geotextile according to McCartney et al. [8]

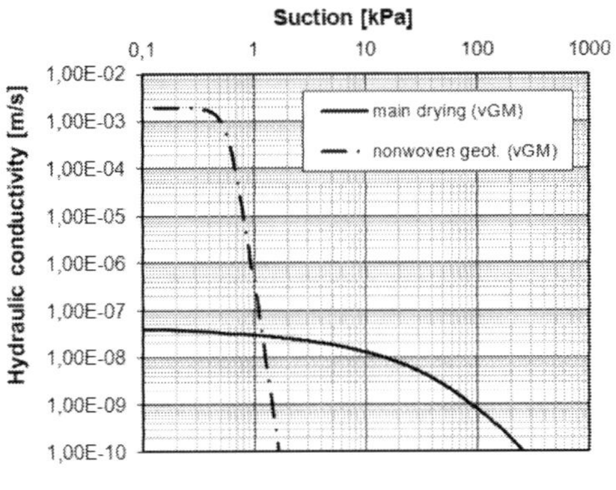

Figure 7. Hydraulic conductivity of topsoil and nonwoven geotextile with van Genuchten&Mualem model (vGM)

The comparison of seepage water into the dry and wet compacted clay shows that also the hydraulic properties of the core material have a certain impact on the leachate. Due to its low hydraulic conductivity and low soil suction the water uptake by the wet compacted clay is low, which favours the interflow in the drainage geocomposite. In contrast, the dry compacted clay takes up considerably more water, which is why a similar seepage rate as in the wet part can be expected at some point in the future.

6. Conclusions

The results of 10 years of measurement on a test fill which was covered by a 60 cm thick topsoil and an underlying drainage geocomposite show very low leachate quantities, although humid climatic conditions with average annual precipitation of approx. 745 mm were present. The low leachate rates can be mainly attributed to the capillary-breaking effect of the drainage geocomposite installed between the topsoil and the core material and the high water retention capacity of the topsoil. Similar to results presented by other authors [8][10][11][12] an inflow of water from the top layer into the drainage geocomposite occurs only under almost saturated conditions, which however, rarely arise in the seasonal course. Additionally, the steep slope inclination of the test fill (1:2) was assessed to favour surface runoff and reduce seepage flow.

For embankments and noise barriers in road-way design the presented simple cover system can be an efficient and effective solution for minimizing infiltration water into the core material. Specific attention should be paid to the hydraulic properties of the top layer as it strongly influences the seepage rate. Soils with a high water retention capacity (e.g. as a result of a high fines or organic content) are materials of choice in this context. Compared to the topsoil, the core material has only a minor influence on the amount of seepage water. But with respect to minimising seepage through the core material, the portion of percolate from the topsoil being diverted and discharged by the top drainage geocomposite is essential. For this reason, a core material with a low hydraulic permeability should be chosen.

References

[1] Birle E, Boso M, Heyer D 2010 Investigation of the water flow in an embankment built of organic soil: Experimental results and large scale field test *5th Int. Conf. on Unsaturated Soils*. vol. II pp 1299-304

[2] Birle E, Heyer D 2012 Water balance of an earth fill built of slightly contaminated organic clay *GeoCongress 2012* Oakland

[3] Birle E, Heyer D 2016 Water balance of an earth fill built of organic clay, *E-Unsat 2016 3rd Europ. Conf. on Unsaturated Soils* Paris

[4] Birle E 2012 Geohydraulische Eigenschaften verdichteter Tone unter besonderer Berücksichtigung des ungesättigten Zustandes, Dissertation, TU München, 2012

[5] Birle E, Heyer D, Rentemeister M, Boso M, Schweller G, Stalter S, Gahbauer M, Vogt N 2010: Grundbautechnische Untersuchungen und umweltgerechte Verwendung von organischen Böden in Geländeaufschüttungen, Lehrstuhl und Prüfamt für Grundbau, Bodenmechanik, Felsmechanik und Tunnelbau, Technische Universität München, unpublished report

[6] Allen R G, Pereira L S, Raes D, Smith M 1998 Computing crop water requirements - FAO irrigation and drainage paper 56, http://www.fao.org/docrep/X0490E/x0490e00.htm

[7] Disse M 1999 Physics and Chemistry of the Earth **24** pp. 325-30 Elsevier Science Ltd.

[8] McCartney J S, Kuhn J A, Zornberg J G 2005 Geosynthetic Drainage Layers in Contact with Unsaturated Soils *Proc. of the 16th Int. Conference of Soil Mechanics and Geotechnical Engineering (ISSMGE)* pp 2301-05

[9] van Genuchten M T 1980 A closed-form equation for predicting the hydraulic conductivity of unsaturated soils *Soil Science Society of America Journal* **44** pp 892-8.

[10] Krisdani H, Rahardjo H, Leong E-C 2006 Experimental study of 1-D capillary barrier model using geosynthetic material as the coarse-grained layer *Proceedings Unsaturated Soils* 2006, Carefree AZ, pp 1683–94.

[11] Bathurst R J, Ho A F, Siemens G 2007 A column apparatus for investigation of 1-D unsaturated-saturated response of sand-geotextile systems *ASTM Geotechnical Testing Journal* **30** pp 433–41

[12] Zornberg J G, Bouazza A, McCartney J S 2010 Geosynthetic capillary barriers: current state of knowledge *Geosynthetics International* **17** pp 273–300

Alternative laboratory approach for determination of geogrids performance in mechanically stabilized layers

Zikmund Rakowski [1], Leos Horníček [2], Jacek Kawalec [3]

[1] Consultant, Cesky Tesin , 737 01, Czech Republic
[2] zech Technical University in Prague, Faculty of Civil Engineering, Thakurova 7, Prague, 166 29, Czech Republic
[3] Tensar International sro, Cesky Tesin, 737 01, Czech Republic; Silesian University of Technology, Gliwice, Poland

Email: kawalec@tensar.pl

Abstract. New laboratory plate test method is proposed for the determination of geosynthetics performance and efficiency of the interlocking in mechanically stabilized layers. The laboratory equipment and experimental methodology are described in details. Proposed method was tested in a set of trials which are evaluated in the paper. Resulting methodology is based on matrix type of tests evaluation, the matrix is formed by several measured parameters. Experiments executed in the laboratory confirmed that the principle of the method is correct and the methodology can be applied for geogrid performance evaluation of various geogrids.

1. Introduction

Geosynthetics performance and efficiency of the interlocking in mechanically stabilized layers is still very open scientific issue. There are number of field observations and publications in that matter [1-10] but naturally the site conditions in each case have very big influence on the results and therefore it is rather difficult to arrange such real testing sections which would allow mutual comparisons. There are large scale accelerated pavement tests (APT) which allow to construct almost identical testing sections which could be trafficked in relatively short time with even millions of axle passes. But APT is not the cheap experiment and requires very specialized and expensive testing equipment. Also the problem of repeatability of the same conditions in new series of tests appears.

The paper describes some alternative laboratory approach, in rather minor scale, which allows to characterize the efficiency of interlocking and mechanical stabilization much faster, simpler and with the equipment which is available in many laboratories.

2. Mechanical stabilization

When the aggregate particles penetrate into the apertures, then retains against the ribs of the geogrid. The result is interlocking of the aggregate grain in the aperture under applied load. The stone grain is locked in the aperture when abutting forces are becoming active. The interlocking is defined as the restriction of the movement of the particles of granular layer under applied load. Shear displacements between grains are very restricted, practically not appearing. The system is very stiff and performs like "quasi rock", see Figure 1. Interlocking causes that the grains are immobilized and confined, consequently the displacement of them is very restricted and stabilized in the given position. This is on the contrary to not stabilized grains which can move even in macro scale. Micro deformations are ones of the important features of mechanical stabilization. Interlocked grains together with geogrid form a kind of a composite material manifesting new quality, which is not a simple sum of the properties of granular material and geogrid. Strongly reduced deformability leads to increased stiffness of the composite. The confinement is strongest in the level of the geogrid and relatively thin layer above it. Above that the transition zone appears which transits in to the zone not influenced by the geogrid layer, see Figure 2.

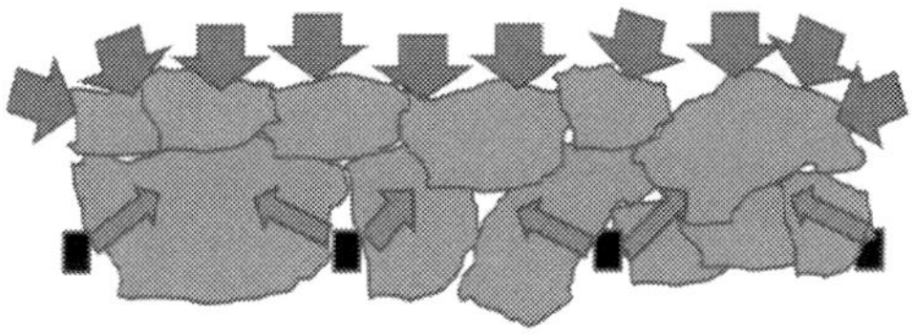

Figure1 Interlocking of grains into geogrid

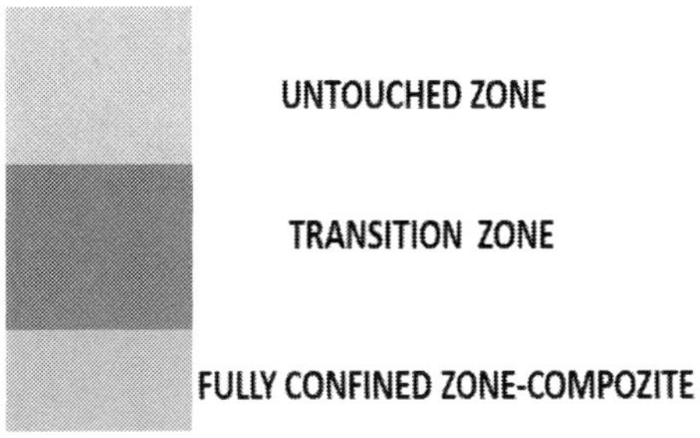

Figure2 Typical zonality of mechanically stabilized layer

3. Small box plate test

3.1. Experimental equipment

The box with dimensions of 800 x 400 x 300 mm (l x w x h) was built for the purpose of testing, Figure 3. The size of the box enables to execute physical experiments in 2D boundary conditions. The box is rather small and easy to manipulate but in the same time dimensions are big enough from the point of view of the scale effect related to typical geogrid and granular material dimensions.

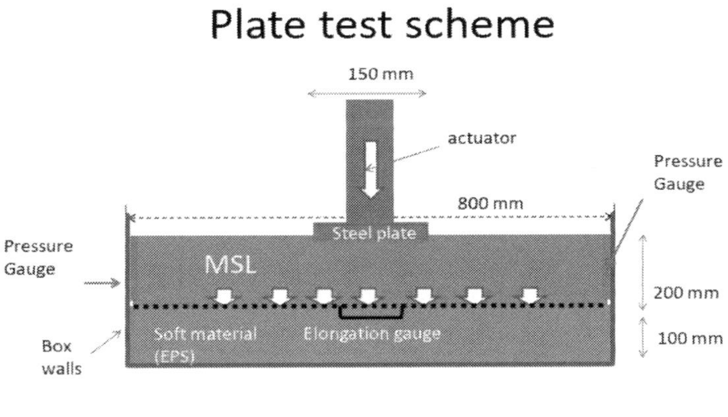

Figure.3 Plate test scheme

Loading was applied by the steel plate with ground plan dimensions of 300 x 150 mm connected to the actuator rig. Static and dynamic regimes of loading were arranged.

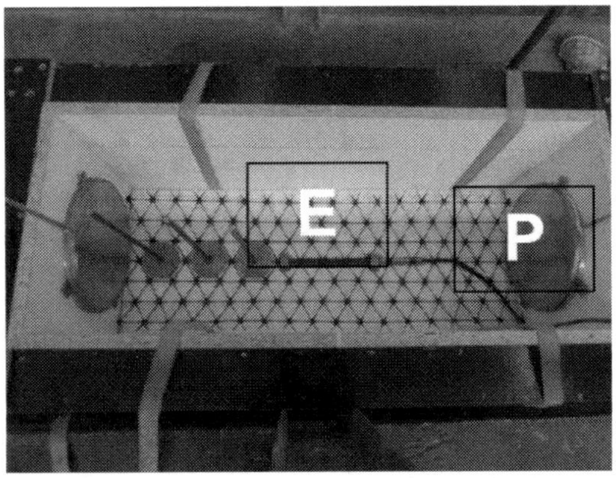

Figure 4 The box with geogrid before filling

Figure 5. Testing box completed

E - elongation gauge (strainmeter), P - pressure gauge, S - settlement measurement equipment

The base of the box is formed by 100 mm thick block of expanded polystyrene (EPS), which allows proper penetration of the grains into the geogrid apertures and guarantees repeatable conditions for all experiments. Vibration-wire gauge (base 185 mm) was mounted on the geogrid rib to measure its elongation during loading. Pressure gauges which are placed on front side walls to measure lateral pressure there (Figure.4) work on the same principle. Settlement of the actuator rig completed the set of parameters, Figure 5. In the first approach, geogrid settlement gauges were installed (Figure5) but due to collision with the proper compaction were omitted later. The proper compaction for effective interlocking is very crucial. Two stages of the compaction were executed in the experiment preparation: preliminary - handy mechanical compactor by two layers of stone by 100mm each, final - with the actuator, the load, 1-19kN, frequency 3Hz, 3k cycles in total.

In static regime the load applied was in a range of 0-10 kN in 1 kN steps, followed by load release to 0 kN. In dynamic regime the load was in a range of 1-12kN, frequency 3Hz, number of cycles up to 5k. Following parameters were measured: vertical position of the actuator, lateral pressures acting on the contact pressure cells and the deformation of the geogrid's longitudinal rib detected by the strainmeter. The settlements of the geosynthetic is an alternative measurement if applied. After the loading and after the removal of the fill from the box, the dimensions of the trough deformation on the EPS block surface were recorded – the length, depth and the volume of the trough by filling it by fine silica sand and measuring its volume then.

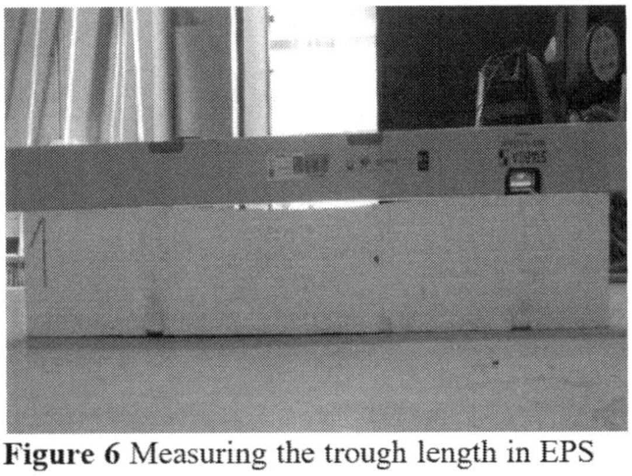

Figure 6 Measuring the trough length in EPS

Figure7 Measuring the trough volume in EPS

3.2. *Experimental results*

In the first experimental stage 3 types of geogrids (GG) and one type of geotextile (GTX) were tested. Two differently graded aggregate mixtures were used: 0-8 mm and 8-31.5 mm.

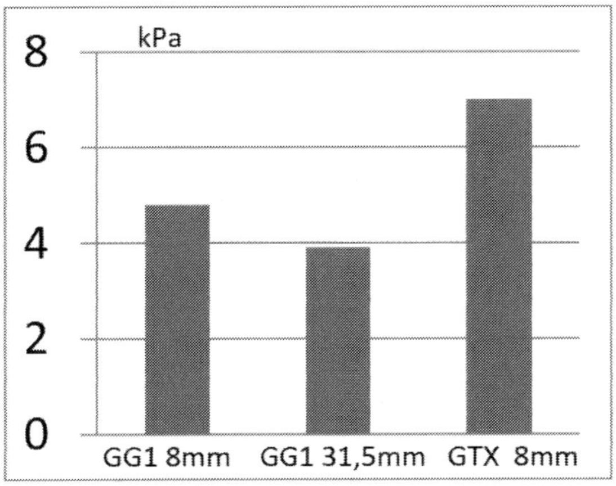

Figure 8 Horizontal pressure – static tests

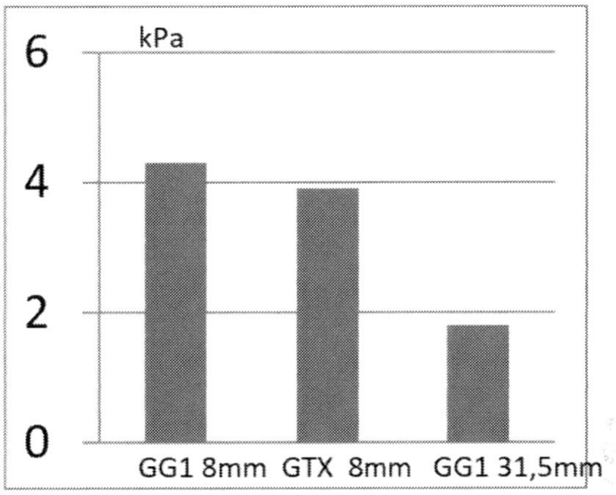

Figure9 Horizontal pressure – dynamic tests

Horizontal pressure records in static (Figure8) and dynamic (Figure 9) tests showed visible difference between geotextile and geogrids and also between tests of the same geogrid but with different aggregate fraction. The best confinement (lowest horizontal pressure) was recorded in case of GG1 geogrid with 0-31.5 mm stone mixture.

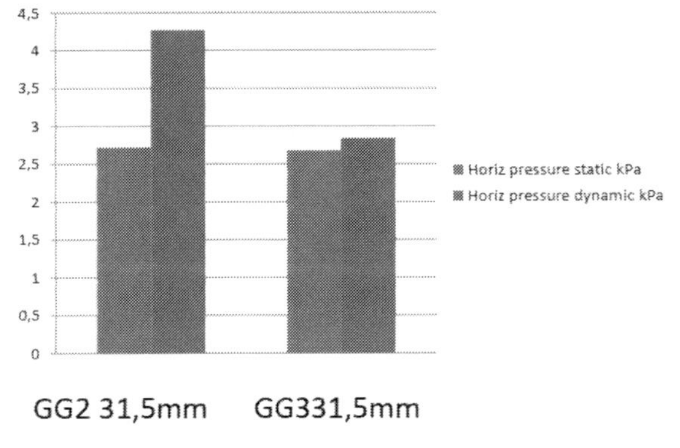

Figure10 Horizontal pressure GG1-GG2

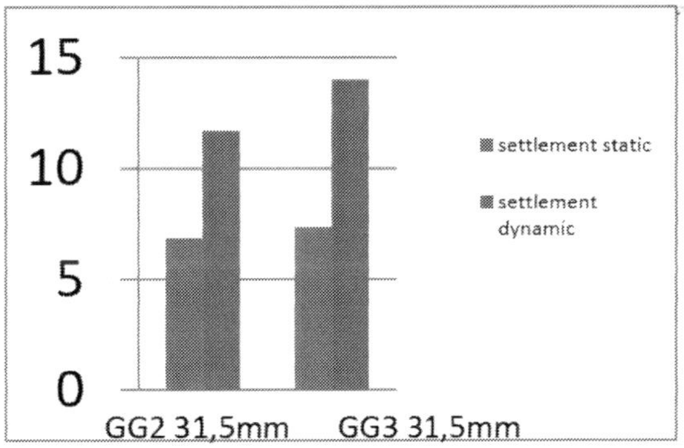

Figure 11 Plate settlement GG1-GG2

Two very different types of geogrids were compared in further tests (Figure10, 11). It is visible that cyclic loading brings better difference between geogrids. GG2 manifests better performance. Those two tests are presented separately. In this test much effective dynamic compaction was applied, while in previous one only manual compactor was used.

Permanent depression in EPS caused by loading both in static and dynamic regimes is very good indicator of the load distribution, Figure12, 13. Geogrid GG1 with 0-31.5 mm stone provided the best results, it means widest load distribution. In case of GG2 and GG3 the comparison was done by depression in EPS volume, Figure 14, with the same conclusion. Interesting observation is that the width/volume of the trough print in the EPS is greater in case of cycling loading.

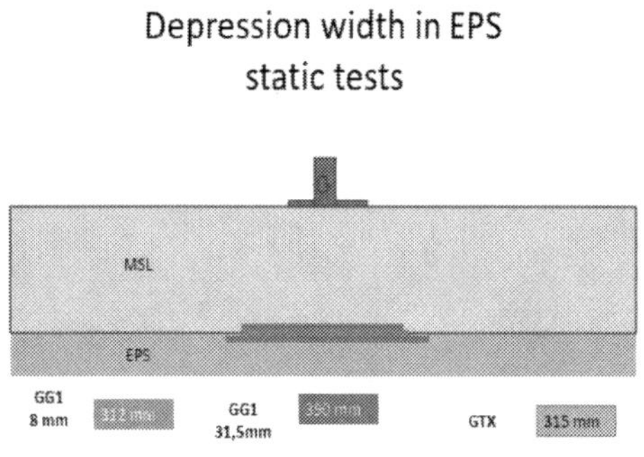

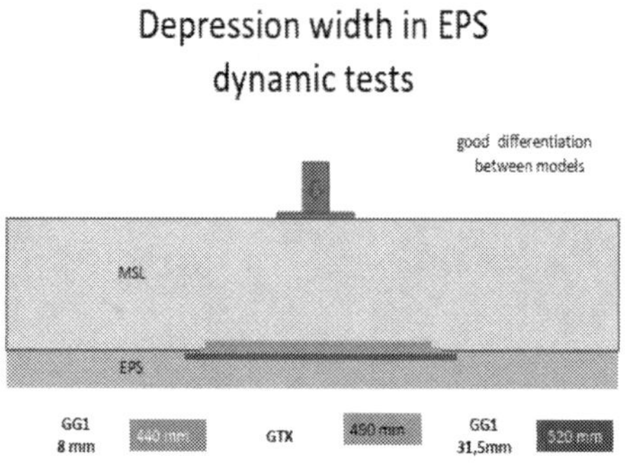

Figure12 Load distribution – static test **Figure13** Load distribution – dynamic test

.

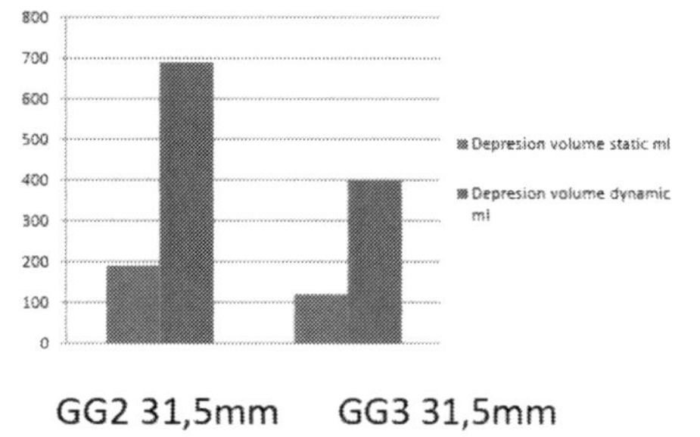

Figure 14 Load distribution GG2/GG3

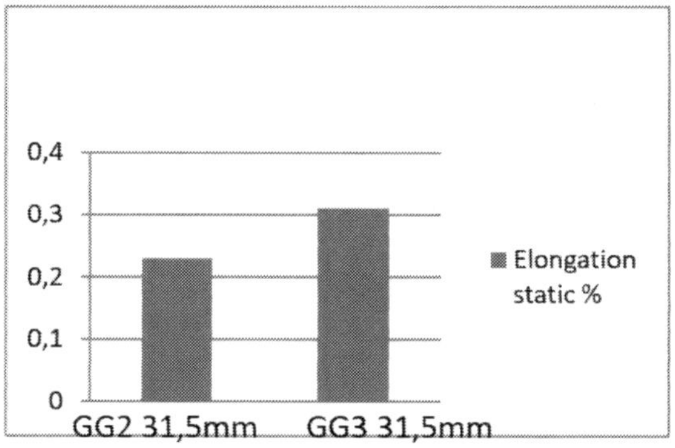

Figure15 Geogrids total elongation

An elongation of geogrids at the end of the trial is the further parameter measured during experiments. Figure 15 shows the comparison of two very different type of geogrids and Figure16 between models with the same geogrid but different aggregate.

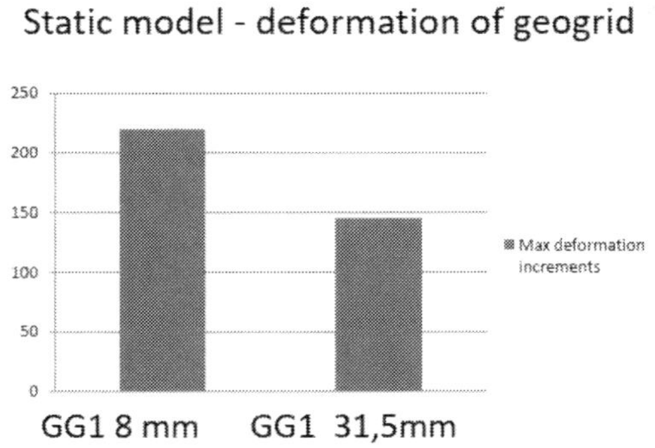

Figure 16 Geogrid elongation with different fill

4. Analysis of the resulsts

The tests and parameters mentioned above can be arranged into the kind of matrix, Tab.1. Red font indicates dynamic compaction, green marked values are the best performance concerning specific parameter.

Table1 The matrix of tests/parameters values

Test		Product / Parameter	GG1 31.5 mm	GG1 8mm	GG2 31.5mm	GG4 31.5 mm	GTX	GG3 31.5mm
Plate test	static	settlement /mm/	7.30	15.48	6.85	N/A	10.63	7.34
		side pressure /kPa/	1.18	3.68	2.72	N/A	3.71	2.68
		geogrid elongation /mm/	0.925	0.788	0.428	N/A	N/A	0.569
		EPS trough width /mm/	390	312	245	N/A	315	250
	dynamic	settlement /mm/	9.82	7.32	11.70	N/A	7.83	13.95
		side pressure /kPa/	5.93	1.55	4.27	N/A	3.27	2.84
		geogrid elongation /mm/	0.393	0.502	0.410	N/A	N/A	0.314
		EPS trough width /mm/	520	440	450	N/A	490	330

The next possibility of the evaluation is so called index matrix, Table 2. In that case the best result has index 1 and other values are related to that. Finally, the average for all parameters is calculated.

Table 2. The matrix of tests/parameters indexes

Test		Product / Parameter	G G1 31.5 mm	G G1 8mm	GG 2 31.5mm	G TX	G G3 31.5mm
Plate test	static	settlement /mm/	1.06	2.12	1.00	1.46	1.01
		side pressure /kPa/	1.00	3.11	2.30	3.14	2.27

		geogrid elongation /mm/	**2.16**	**1.84**	**1.00**	**X**	**1.33**
		EPS trough width /mm/	**1.00**	**1.20**	**1.37**	**1.20**	**1.36**
	dynamic	**settlement /mm/**	**1.34**	**1.00**	**1.48**	**1.07**	**1.90**
		side pressure /kPa/	**4.38**	**1.55**	**2.75**	**2.10**	**1. 83**
		geogrid elongation /mm/	**1.25**	**1.60**	**1.30**	**X**	**1.00**
		EPS trough width /mm/	**1.00**	**1.15**	**1.14**	**1.05**	**1.36**
			1.28	**1.50**	**1.42**	**1.53**	**1.46**

It is visible in Table1 that the matrix approach is the proper one as relaying on one single parameter would not be sufficient and correct for the evaluation of the geosynthetic performance. The index matrix (Table .2) is very useful because enables the quantification of the performance differences between geosynthetics and various models.

Table 2 indicates that the GG1 in combination with 8-31.5mm stone has the best performance index. In fact, it can be used as a kind of reference model for next trials.

5. Conclusions

Proposed method and methodology of interpretation of the measured data can find good application in the evaluation of geosynthetics performance, especially geogrids, in mechanically stabilized layers. The approach enables comparative analyses between various geosynthetic products and also from the point of view of geosynthetic/aggregate combination, i.e. composite layer evaluation. The experiments are neither costly nor time consuming and can be done with rather common laboratory equipment.

References

[1] Giroud J P, Han J, Tutumluer E, & Dobie M J D 2022 The use of geosynthetics in roads *Geosynthetics International*, 1-34.

[2] Robinson W J, Wayne M H, & Tamrakar P 2022 Evaluation of Hexagonal Multi-Shape, Multi-Axial Geogrids in Unsurfaced Road Applications. *Transportation Research Record*, 03611981211069941.

[3] Jeremy Robinson W, Tingle J S, Wayne M H, Kwon J, & Norwood G 2020. Instrumentation Response of Full-Scale Multi-axial Geogrid Stabilized Flexible Pavements. In *Accelerated Pavement Testing to Transport Infrastructure Innovation* (pp. 564-573). Springer, Cham.

[4] Sun X, Han J, Fei J, Guo J, & Zhang W 2020 Geosynthetic-stabilized flexible pavements: Solution derivation and mechanistic-empirical analysis. *Geotextiles and Geomembranes*, ***48***, 468-478.

[5] Sweta K, & Hussaini S K K 2018 Performance Evaluation of Ballast-Subballast Interface Stabilized with Geogrids. In *Proceedings of China-Europe Conference on Geotechnical Engineering* pp. 1738-41 Springer, Cham.
[6] Cook J, Dobie M, & Blackman D 2016 The development of APT methodology in the application and derivation of geosynthetic benefits in roadway design In *The Roles of Accelerated Pavement Testing in Pavement Sustainability* pp 257-75 Springer, Cham.
[7] Rimoldi P, Shamrock J, Kawalec J, & Touze N 2021 Sustainable use of geosynthetics in dykes *Sustainability*, **13**, 4445.
[8] Rakowski Z, & Kawalec J 2020 The Technology of Mechanically Stabilized Layers for Road Structures in Cold Regions. In *Transportation Soil Engineering in Cold Regions,* **2** pp 63-70. Springer, Singapore.
[9] Kawalec J 2019. Stabilisation with geogrids for transport applications–selected issues. In *MATEC Web of Conferences*, **265** p 01001. EDP Sciences.
[10] Rakowski Z, & Kawalec J 2009 Mechanically stabilized layers in road construction. In *XXVII Baltic Road Conference, Ryga.*

Effects of geogrid reinforcement on the seismic performance of lightweight embankments

A Edinçliler* and Y S Toksoy

Boğaziçi University Kandilli Observatory and Earthquake Research Institute, Earthquake Engineering Department, Istanbul, TurkeyEmail: aedinc@boun.edu.tr

Abstract. Highway embankments have experienced earthquake induced damages from minor to catastrophic scales during the past few decades despite the fact that highways are considered as major lifelines that should be in continuous operation under any circumstances. Recent advances in highway engineering suggest that geosynthetic reinforcement can be considered as a novel approach to mitigate the earthquake hazards on highway embankments successfully. The use EPS geofoam inclusions is one of the prominent improvement methods available to increase the seismic performance of lightweight embankments. It's a well-known method to reduce the magnitude of earthquake-induced dynamic forces. The aim of this numerical study is to determine the effects of using geogrid reinforcement on the seismic performance of lightweight embankments. Non-linear dynamic analyses were performed using the PLAXIS software using real earthquake records. Evaluation of the results reveal that inclusion of geogrid reinforcement layer beneath the lightweight embankment model successfully decreases earthquake-induced dynamic forces and increases the seismic performance of the embankment model in comparison of the nominally identical structure with no geogrid reinforcement. Numerical results obtained from this novel technique suggest that the implementation of a geogrid layer beneath such lightweight embankments can successfully and effectively increase the seismic performance of such structures.

1. Introduction

Dynamic performance of roads or highways, which are basically embankments, slope-or dam-like structures, is crucial because highways are vital lifelines which must be constantly in service even after disastrous events to provide required safety and emergency needs. Thus, the stability and durability of slopes and embankments is a serious problem under static and dynamic loads that should be solved from the early construction process [1-3].

Construction of highway embankments over soft soil sites using expanded polystyrene (EPS) block to mitigate consolidation settlement is a mature technology and has been implemented by various countries around the world. In addition, geofoam technology provides a unique solution for embankments constructed over buried utility corridors. Utilizing geofoam technology can help prevent possible structural damage of buried utilities when compared to embankments constructed using conventional compacted earthen fill [4].

Geofoam is a geosynthetic product composed of foam. This lightweight product takes increasing attention of researches due to its unique static and dynamic properties. Both experimental and numerical studies have been performed to investigate the effectiveness of expanded polystyrene (EPS) geofoam buffers in reducing the earthquake induced seismic loads on geotechnical structures such as retaining walls. A smart extension of this application is to use vertical compressible inclusions of EPS geofoam to attenuate earthquake-induced dynamic earth pressures against rigid walls. It is known that EPS geofoam has superior cushion properties due to the individual air bubble body which is capable of reducing the impact and vibration effects [5]. The results from the parametric analyses have been compiled into design charts that quantify the efficiency of geofoam as a function of geofoam buffer thickness and density, wall height, dynamic stress-strain properties of the retained soil mass, and characteristics of the base input excitation [6-13]. Recent comparative numerical experiments reveal that inclusion of EPS geofoam as a seismic buffer successfully decreases earthquake-induced dynamic earth pressures and transmission of accelerations, which prove that it can be possible to mitigate earthquake hazards on quay walls by simply implementing geofoam cushions [14-16].

Geosynthetics are synthetic materials that are commonly used to solve civil and geotechnical engineering problems. The integration of geosynthetic materials to the civil, geotechnical and earthquake engineering applications is an advantageous and cost-effective way to achieve the required stability conditions of related structures. Geosynthetics are capable of absorbing dynamic forces and transmitting less dynamic forces to engineering structures. The use of geosynthetics under foundations or beneath embankments can absorb seismic energy and mitigate dynamic excitations that transmit to upper layers of soils and foundation of overlying structures. This phenomenon occurs due to the tensile strength of the geosynthetic material and the soil-structure interaction between soil and geosynthetic material which increases the shear resistance of soil under dynamic conditions [17].

Reinforced soil foundations can successfully be used in various geotechnical engineering applications, including bridge approach slabs, bridge abutment, building footings, and embankments. Inclusion of geosynthetics by means of reinforcement in soil foundations is a cost-effective solution to increase the bearing capacity and to decrease the settlement of footings compared to the conventional methods such as replacing natural soils or increasing the dimensions of the footings. The product of choice is geogrid, which is the most commonly used geosynthetic materials to reinforce the foundation soil. Understanding the dynamic interface shear properties of geosynthetic materials and geosynthetic reinforced soils have gained attention of many researchers for the past few decades [18-19]. Energy dissipation along structures and geosynthetic interfaces was also studied in literature [20-22].

A study on the bearing capacity of footings on geogrid reinforced sand and indicated that there is an optimum reinforcement spacing for multi-layer reinforced sand. The optimum depth would be larger for settlement ratios greater than 6 and the highest bearing capacity occurs at embedment depth of approximately 0.3B [23]. Another study related focused on the effects of reinforcement parameters of foundations on carrying load and settlement. The reinforcement parameters such as vertical space between reinforcement layers, dimension and number of reinforcement layers and degree of compactness were selected as variables. Experiments revealed that when the number of reinforcement is N = 3, maximum carrying capacity was obtained and it was determined that soil improvement was not only dependent on number of layers but also varied with total reinforcement depth and vertical space between reinforcements [24]. A study related with the bearing capacity of circular foundations settled on geogrid reinforced sand was performed by using finite element method. It was seen that the bearing capacity was increased when the first reinforcement layer was selected as 0.3D and the number of reinforcement layers were selected as N= 4 [25]. Another study proved that the inclusion of geogrid layers increases bearing capacity of foundation and decreases the settlement [26]. This statement was declared by a study which was performed on design factors affecting the settlement of strip foundation on geogrid-reinforced sand. Findings of their study include that the settlement increases until the number of reinforcement is N = 3. The amount of settlement increases until 2nd level of reinforcement depth rate (u = 0.5B) and it decreases after this level [27]. A similar study on the investigation of the bearing capacity of a strip foundation on geogrid reinforced sand suggests that increasing number of geogrid layers has a positive effect in bearing capacity. A comparative study on the effects of geogrid reinforcement on bearing capacity of a soil under static loading reveals that bearing capacity increases and settlement decreases with respect to the increment of reinforcement layers [28]. A numerical study focused on the seismic performance of mid-rise buildings overlying the geogrid reinforced sand. It was revealed that the reinforcement depth has an important effect for the seismic performance [29]. Recent extensive parametric studies on seismic performance of low and mid-rise buildings on geogrid reinforced sand showed how the use of geosynthetics for soil reinforcement purposes can significantly and effectively increase the stability of the foundation soil by providing an additional tensile strength to the soil mass and unique energy absorption properties [30-31].

In light of the previous experience and expertise, the aim of this unique study is to determine the effects of geogrid reinforcement on the seismic performance of lightweight highway embankments. The study is parametric in a way that the seismic performance of the highway embankments is investigated by performing seismic performance analyses using different real earthquake records both for geogrid-reinforced and unreinforced sand foundation cases. The careful evaluation of results obtained from each model and dynamic excitation of interest prove that the proposed soil improvement method of geogrid-sand foundation is capable of increasing the seismic performance of lightweight highway embankments considerably.

2. Numerical study

Model design and highway embankment material selection criteria are carefully defined to meet the 1993 AASHTO regulations [32]. The thickness of asphalt concrete (AC) layer and the granular aggregate base (GAB) layer have been selected as 15cm and 30cm, respectively. GAB layer is located right beneath the AC layer and above the subgrade layer. For the lightweight EPS Geofoam embankments, geofoam blocks are the subgrade layer itself. The aggregate base and subgrade materials are classified as Type-A for 1993 AASHTO.

The geogrid used in the numerical analysis has an ultimate tensile strength of 40kN/m2. However, PLAXIS use axial stiffness (EA) parameter instead of T_{ult} value. Provided from the design charts, the axial stiffness value of the proposed geogrid material under 5% strain level has been calculated as 2000kN/m and used as an input value. One layer of geogrid layer which is placed 1m beneath the lightweight embankment model has been previously found to be enough with respect to the preliminary finite element analysis and hand calculations. A strength reduction factor (R_{inter}) 0.85 has been selected for the interface between geogrid and soil as suggested. The geogrid material meets the requirements as given in FHWA NHI-07-092 [33]. The model mesh has been generated as "very fine" for the credibility of the results. A prescribed displacement has been defined at the bottom of the model for the dynamic excitations The geogrid reinforced embankment model and its dimensioning is represented in Figure 1 and Figure 2.

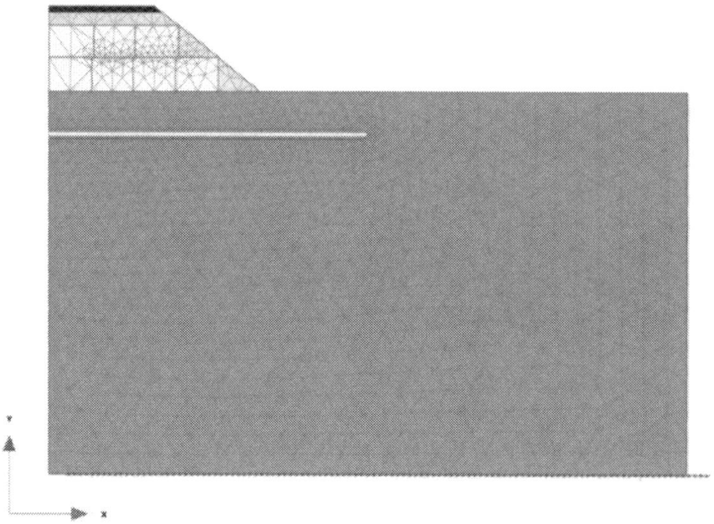

Figure 1. Numerical model of reinforced lightweight EPS geofoam embankment model

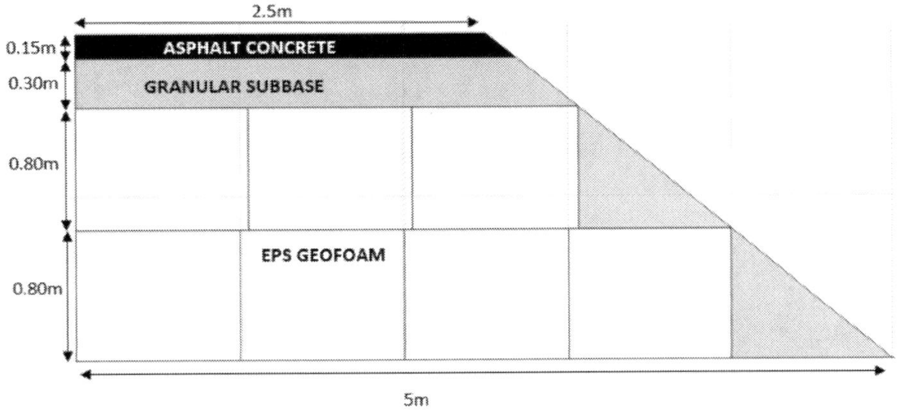

Figure 2. Dimensioning of lightweight EPS geofoam embankment model

Selection of the EPS Geofoam material has been done in accordance with 2004 NCHRP Guideline and Recommended Standard for Geofoam Applications in Highway Embankments [34]. Based on the requirements, selected EPS geofoam material has a density of 20kg/m3 and as proposed in AASHTO material designation, EPS50. Each block has the identical dimensioning of 100x100x80cm. Two layers of EPS blocks has been used as recommended. Design values of engineering parameters for the EPS geofoam and other materials are given in Table 1.

Table 1. Material properties used in finite element modelling

	Asphalt Concrete	Granular Subbase	Subgrade Material	Very Firm Silty Clay	EPS Geofoam
Material Model	Linear Elastic	Mohr-Coulomb	Mohr-Coulomb	Mohr-Coulomb	Linear Elastic
γ_{unsat} (kN/m3)	24	22	18	18	0.2
E' (kN/m2)	2.1E6	100E3	25E3	50E3	5000
Poisson Ratio	0.4	-	0.34	-	0.01
Cohesion (kN/m2)	-	0.01	40	17	-
Ø	-	45°	30°	12°	-

Numerical simulations have been performed using the commercially available PLAXIS 2D 2018 software with finite element modelling technique. As the embankment model is symmetric, only the right half has been modelled and subjected to dynamic excitations. Dynamic response analysis requires special boundary conditions to avoid reflection of seismic wave which may possibly cause distortions and unrealistic results. Thus, viscous boundaries are defined to the edges of the mesh.

During the study, two different real earthquake records have been used which are the 2014 Gökçeada Earthquake and the 1999 Kocaeli Earthquake. Earthquake records have been obtained from BU-KOERI database. The 2014 Gökçeada Earthquake record (Meteoroloji Station) has PGA value of 0.18g and 2.49Hz of pre-dominant frequency (M_w:6.5) and the 1999 Kocaeli Earthquake (Yarımca Station) has a PGA value of 0.32g and a pre-dominant frequency of 0.29Hz (M_w:7.6). Related accelerograms can be seen in Figure 3 and Figure 4.

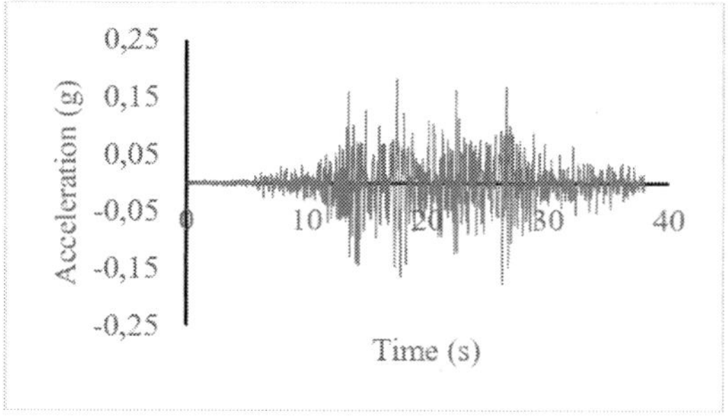

Figure 3. Gökçeada Earthquake accelerogram (Meteoroloji Station)

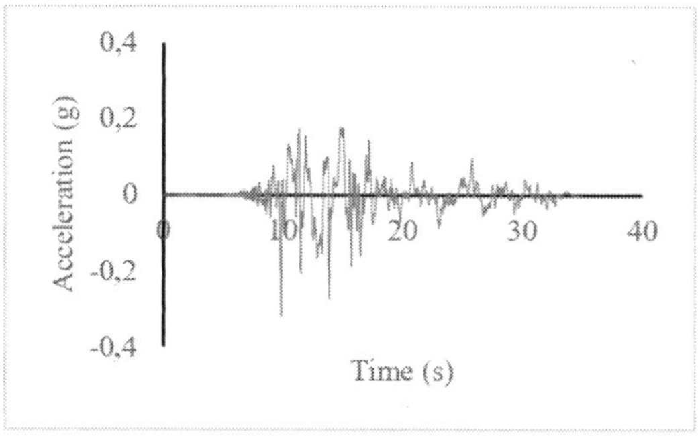

Figure 4. Kocaeli Earthquake accelerogram (Yarımca Station)

3. Results

Numerical results obtained from seismic performance analyses for Unreinforced and Reinforced models are represented by means of total displacement and transmitted acceleration values for Gokceada and Kocaeli Earthquakes. In addition, direct measurements are taken from pre-defined locations located on critical parts of the highway embankment models as can be seen in Figure 5.

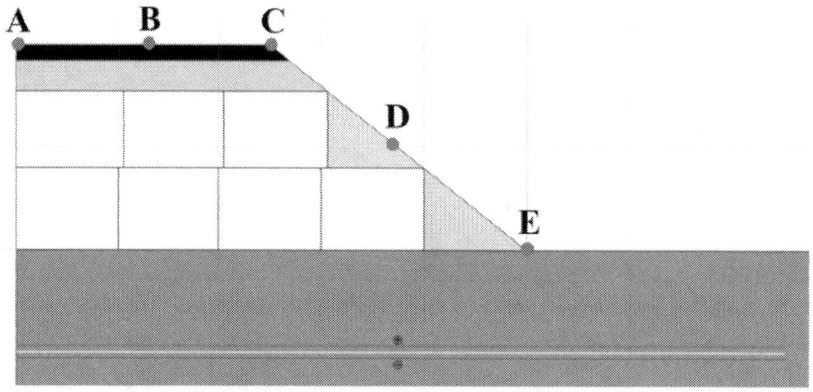

Figure 5. Pre-defined critical points on highway embankment Models

3.1. Obtained results from Gökçeada Earthquake

Numerical simulations with Gökçeada Earthquake record show that soil reinforcement using geogrids can successfully decrease the total displacements and diminish the transmitted accelerations on lightweight highway embankment models. Total displacement distributions for both models under Gökçeada Earthquake excitations are shown in Figure 6-7.

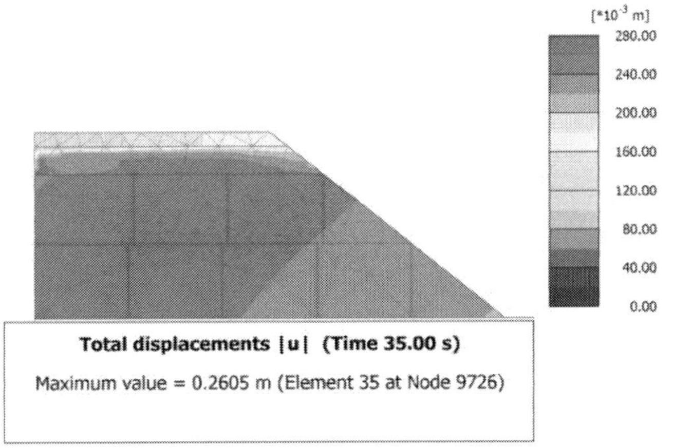

Figure 6. Total displacement distributions of reinforced model

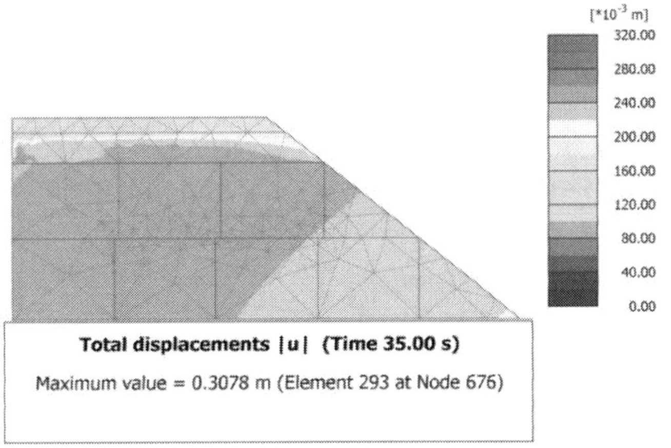

Figure 7. Total displacement distributions of unreinforced model

As can be inferred from the Figures, soil reinforcement can effectively lessen the degree of displacements along the reinforced model. Obtained total displacements from pre-defined critical points are tabulated in Table 2.

Table 2. Total displacement values under Gökçeada Earthquake

	A	B	C	D	E
Reinforced Model (m)	0.11	0.14	0.17	0.22	0.21
Unreinforced Model (m)	0.12	0.15	0.18	0.24	0.22

As can be inferred from Table 2, total displacements are less on asphalt concrete layer at crest level and gradually increases toward to edges and the toe of the models. At crest level, total displacements in unreinforced model are 0.12m, 0.15m and 0.18m for A, B and C locations, respectively. Relatedly, measurements taken from D and E are 0.24m and 0.22m, respectively. Inclusion of a geogrid layer to the foundation soil successfully decreases the total displacements as crest level measurements in reinforced model are 0.11m, 0.14m and 0.17m for A, B and C locations; 0.22m at D and 0.21m at toe of the model. Transmitted Acceleration values under Gökçeada Earthquake are given in Table 3.

Table 3. Transmitted acceleration values under Gökçeada Earthquake

	A	B	C	D	E
Reinforced Model (g)	0.05	0.12	0.28	0.05	0.02
Unreinforced Model (g)	0.09	0.16	0.32	0.13	0.06

As can be seen in Table 3, the geogrid reinforcement in the foundation soil absorbs a portion of the seismic energy which results in decreased transmitted accelerations for all critical locations. In the unreinforced model, the transmitted acceleration values at crest level are 0.09g, 0.16g and 0.32g for A, B and C, respectively. The measurements at D and E are 0.13g and 0.06g, respectively. However, due to the reinforcement effect, crest level transmitted accelerations are effectively decreased to 0.05g, 0.12g and 0.28g. Toe level acceleration measurement is 0.02g in this case.

3.2. Obtained results from Kocaeli Earthquake

FEM simulations with Kocaeli Earthquake record also indicate that soil reinforcement using geogrids is quite efficient to decrease the total displacements and diminish the transmitted accelerations on lightweight highway embankment models. Total displacement distributions for both models under Kocaeli Earthquake excitations are shown in Figure 8-9.

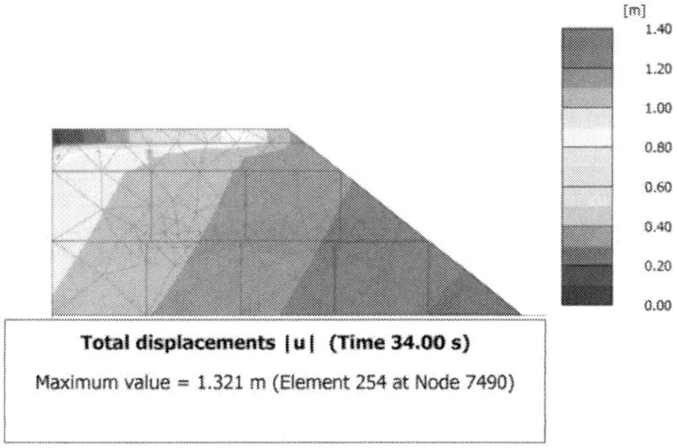

Figure 8. Total displacement distributions of reinforced model

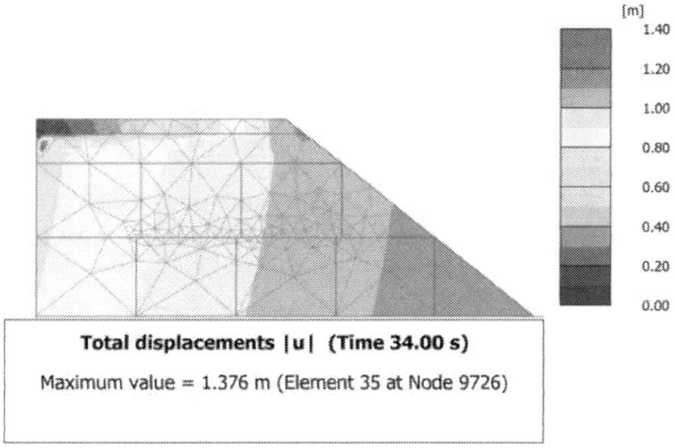

Figure 9. Total displacement distributions of unreinforced model.

As seen in total displacement distributions, gegrid reinforcement absorbs a portion of seismic energy and lessens the degree of displacements along the reinforced model. Obtained total displacements from pre-defined critical points are tabulated in Table 4.

Table 4. Total displacement values under Kocaeli Earthquake

	A	B	C	D	E
Reinforced Model (m)	0.10	0.57	1.07	1.11	1.19
Unreinforced Model (m)	0.14	0.58	1.09	1.25	1.32

Given in Table 4, total displacements are less on asphalt concrete layer at crest level than other critical parts and gradually increases toward to edges and the toe of the models. At crest level, total displacements in unreinforced model are 0.14m, 0.58m and 1.09m for A, B and C locations, respectively. Relatedly, measurements taken from D and E are 1.25m and 1.32m, respectively. Inclusion of a geogrid layer to the foundation soil decreases the total displacements as crest level measurements in reinforced model are 0.10m, 0.57m and 1.07m for A, B and C locations; 1.11m at D and 1.19m at toe of the model. Transmitted Acceleration values under Kocaeli Earthquake are given in Table 5.

Table 5. Transmitted acceleration values under Kocaeli Earthquake

	A	B	C	D	E
Reinforced Model (g)	0.01	0.13	0.44	0.04	0.03
Unreinforced Model (g)	0.06	0.26	0.55	0.26	0.21

The geogrid reinforcement in the foundation soil absorbs a portion of the seismic energy which results in decreased transmitted accelerations for all critical locations. As can be seen in Table 5, in the unreinforced model, the transmitted acceleration values at crest level are 0.06g, 0.26g and 0.55g for A, B and C, respectively. The measurements at D and E are 0.26g and 0.21g, respectively. However, due to the reinforcement effect, crest level transmitted accelerations are effectively decreased to 0.01g, 0.13g and 0.44g. Toe level acceleration measurement is 0.03g in this case.

4. Evaluation of the results and conclusions

This is the first study in the literature with the aim of investigating the seismic performance of lightweight highway embankments on geogrid-reinforced sand foundation. A dimensionally identical embankment model placed over the unreinforced and single layer geogrid reinforced sand was subjected to real earthquake records of Gökçeada and Kocaeli Earthquakes, which vary in amplitude and frequency domain. Performance indicators have been selected as total displacements and transmitted accelerations at pre-defined critical locations.

The careful evaluation of results obtained from each model and dynamic excitation of interest clearly indicate that the proposed soil improvement method of geogrid-sand foundation is capable of increasing the seismic performance of lightweight highway embankments. In all cases the embankment model on a geogrid-reinforced sand foundation performs much better than the unreinforced case.

Total displacements are reduced by up to 9% under Kocaeli Earthquake and to 29% under Gökçeada Earthquake excitations in reinforced cases. Crest level displacement measurements (A, B and C) are much better improved than any other pre-defined critical locations specifically the point A, in reinforced cases regardless of the dynamic excitation. Toe level (E) total displacement measurements are 5% and 10% less under Gökçeada and Kocaeli Earthquakes, respectively.

A similar situation applies to the transmitted acceleration values as obtained results from reinforced models are 66% and 86% less when compared to unreinforced models for Kocaeli and Gökçeada Earthquakes, respectively. On the contrary to the displacement measurements, geogrid reinforced foundation application especially improves the toe level transmitted acceleration values.

Considering the higher reduction ratios under Kocaeli Earthquake record, it is seen that the proposed soil improvement method is more effective under high-amplitude, low-frequency ground motion of this study due to the tensile strength of the material and the friction along the interfaces. Authors of this study suggest that further experimentation and investigation are needed as obtained results are valid for the indicated parameters of this preliminary study.

IOP Conf. Series: Materials Science and Engineering 1260 (2022) 012020 doi:10.1088/1757-899X/1260/1/012020

References

[1] Edinçliler A 2007 Using waste tire–soil mixtures for embankment construction *International Workshop on Scrap Tire Derived Geomaterials "Opportunities and Challenges"* Kanto Branch of Japanese Geotechnical Society pp 319–28

[2] Edinçliler A, Baykal G and Saygılı A 2010 Influence of different processing techniques on the mechanical properties of used tires in embankment construction *Waste Management* **30** pp 1073–80

[3] Edinçliler A and Toksoy Y S 2017 Physical Model Study of the Seismic Performance of Highway Embankments with and without Geotextile *Journal of Earthquake and Tsunami* **11**

[4] Ozer A T, Akinay E 2019 First Geofoam Roadway Embankment Application in Turkey *Proceedings of 5th International Conference on Geofoam Blocks in Construction Applications EPS 2018* Springer Switzerland

[5] Lin L K, Chen L H and Chen R H L 2010 Evaluation of geofoam as a geotechnical construction material *Journal of Materials in Civil Engineering* **22** pp 160–70

[6] Pelekis P C, Xenaki V C and Athanasopoulos G A 2000 Use of EPS Geofoam for Seismic Isolation of Earth Retaining Structures: Results of an FEM Study *Proceedings of the Second European Geosynthetics Conference*, Bologna, Italy pp 843-6

[7] Hazarika H 2001 Mitigation of Seismic Hazard on Retaining Structures - A Numerical Experiment *Proceedings of the 11th International Offshore and Polar Engineering Conference* pp 459-64

[8] Hazarika H and Okuzono S 2002 An Analysis Model for a Hybrid Interactive System Involving Compressible Buffer Material *Proceedings of the 12th International Offshore and Polar Engineering Conference* pp 622-9

[9] Hazarika H and Okuzono S 2004 Modeling the Behaviour of a Hybrid Interactive System Involving Soil, Structure and EPS Geofoam *Soils and Foundations* **44** pp 149-62

[10] Zarnani S and Bathurst R J 2005 Numerical Investigation of Geofoam Seismic Buffers using FLAC *Proceedings of the North American Geosynthetics Society (NAGS)/GRI19 Conference*Las Vegas, Nev., 14-16 pp 5.10a

[11] Zarnani S and Bathurst R J 2006 Application of EPS Geofoam as a Seismic Buffer: Numerical Study using FLAC *Proceedings of the 59th Canadian Geotechnical Conference*, Vancouver, B.C

[12] Athanasopoulos G A, Nikolopoulou C P, Xenaki V C and Stathopoulou V D 2007 Reducing the Seismic Earth Pressure on Retaining Walls by EPS Geofoam Buffers - Numerical Parametric Study *Proceedings of the Geosynthetics Conference*, Washington, DC

[13] Zarnani S and Bathurst R J 2009 Numerical Parametric Study of Expanded Polystyrene (EPS) Geofoam Seismic Buffers *Canadian Geotechnical Journal* **46** pp 318-38

[14] Edinçliler A and Toksoy Y S 2016 Numerical Study of Improving Seismic Performance of Quay Walls *12th International Congress on Advances in Civil Engineering ACE* 2016, Boğaziçi University, Istanbul, Turkey

[15] Edinçliler A and Toksoy Y S 2016 Effects of Two Different Seismic Cushions on Seismic Performance of Quay Walls: Numerical Study *6th European Geosynthetics Congress EuroGeo6*, Ljubljana, Slovenia

[16] Edinçliler A Toksoy Y S 2017 The influence of Geofoam Material Parameters on the Seismic Performance of Quay Walls *G7 7th National Geosynthetics Conference*, Boğaziçi University, Istanbul, Turkey (in Turkish)

[17] Edinçliler A and Toksoy Y S 2017 Shake Table Tests to Measure the Dynamic Performance of Geotextile-Reinforced Embankment *Periodica Polytechnica Civil Engineering* **61** pp 803-14

[18] Yegian M K, Yee Z Y and Harb J N 1995 Response of Geosynthetics under Earthquake Excitations *Proceedings of Geosynthetics* pp 677-89

[19] Yegian M K, Yee Z Y and Harb J N 1995 Seismic Response of Geosynthetic/Soil systems

Geotechnical Special Publication **46** pp 1113-25
[20] Kavazanjian E, Hushmand B and Martin G 1991 Frictional base isolation using layered soil-geosynthetic liner system *Proceedings of the Third U.S. Conf. on Lifeline Earthquake Engineering* pp 1140-51
[21] Yegian M K and Lahlaf A M 1992 Dynamic interface shear properties of geomembranes and geotextiles *Journal of the Geotechnical Engineering* **118** 760-79
[22] Zimmie T F, De A and Mahmud M B 1994 Centrifuge modeling to study dynamic friction at geosynthetic interfaces *Proc. Fifth International Conference on Geotextiles, Geomembranes and Related Products* pp 415-8
[23] Yetimoglu, Wu J T H and Saglamer A 1994 Bearing Capacity of Rectangular Footings on Geogrid-Reinforced *Sand Journal of Geotechnical Engineering* **120**pp 2083-99
[24] Adams M T and Collin J G 1997 Large model spread footing load tests on geosynthetic reinforced soil foundations *Journal of Geotechnical and Geoenvironmental Engineering* **123**pp 66-72
[25] Yıldız A A, Laman M, Örnek M and Demir A 2006 Numerical an Analysis Circular Foundations Supported by Sand Reinforced with Geogrid *Second National Conference on Geosynthetics*pp 75-84 (in Turkish)
[26] Alamshahi S and Hataf N 2009 Bearing capacity of strip footings on sand slopes reinforced with geogrid and grid-anchor *Geotextiles and Geomembranes* **27** pp 217-26
[27] Demiröz A and Tan Ö 2010 An Experimental study for settlement of strip foundation on Geogrid-Reinforced Sand *Scientific Research and Essays* **5**pp 3306-12
[28] Marto A, Oghabi M and Eisazadeh A 2013 The Effect of Geogrid Reinforcement on Bearing Capacity Properties of Soil under Static Load; A Review *Electronic Journal of Geotechnical Engineering* **18** pp 1881-98
[29] Edinçliler A and Yildiz O 2016 Seismic performance of mid-rise buildings on geogrid reinforced sand *6th European Geosynthetics Conference* pp 1346-55
[30] Edinçliler A and Toksoy Y S 2017 Geogrid-Donatılı Kum Üzerine İnşa Edilen Orta Katlı Binaların Sismik Performansının Değerlendirilmesi *3. Zemin-Yapı Etkileşimi Konferansı*, İzmir, Türkiye (in Turkish)
[31] Edinçliler A and Toksoy Y S 2017 Parametric Study on Seismic Performance of Low and Mid-Rise Buildings on Geogrid Reinforced Sand *GeoAfrica 2017 Conference*, Marrakech, Morocco
[32] AASHTO 1993 *AASHTO Guide for Design of Pavement Structures*, American Association of State Highway and Transportation Officials, Washington, D.C
[33] FHWA NHI-07-092 2008 *Geosynthetic Design & Construction Guidelines*, U.S. Department of Transportation Federal Highway Administration, Washington, USA
[34] Stark T D, Arellano D, Horvath J S and Leshchinsky D 2004 *Guideline and Recommended Standard for Geofoam Applications in Highway Embankments* NCHRP Report 529, Washington, USA

Seismic design of geosynthetic-reinforced earth retaining walls following a pseudo-static approach

D Gaudio[1, 2], L Masini[1] and S Rampello[1]

[1] Department of Structural and Geotechnical Engineering, Sapienza University of Rome, via Eudossiana 18, 00184 Rome, Italy
[2] present address: Department of Engineering, University of Cambridge, Trumpington Street, Cambridge, CB21PZ, UK
E-mail: dg564@cam.ac.uk

Abstract. Geosynthetic-reinforced earth (GRE) retaining walls show a better performance than conventionally-designed walls during destructive earthquakes, due to their capability of redistributing seismic-induced deformations within the reinforced zone. In this paper, a recently-proposed method to design GRE walls is first recalled, where the wall is designed to trigger an internal plastic mechanism in the presence of strong earthquakes. Following a pseudo-static approach, the seismic coefficient k is therefore assumed equal to the internal seismic resistance of the wall k_c^{int}. The seismic coefficient is then calibrated against given seismic wall performance, expressed in terms of limit values of earthquake-induced displacements. Permanent displacements are evaluated through empirical relationships that were previously developed on the basis of a parametric integration of an updated Italian seismic database. Effectiveness of the proposed procedure is then demonstrated by assessing, through Finite Difference nonlinear dynamic analyses, the seismic performance of two walls, namely a GRE and a conventional gravity wall, characterised by the same seismic resistance but triggering an internal and external plastic mechanism, respectively. They are both subjected to a real strong motion, capable of activating a plastic mechanism. Results showed that lower permanent displacements are accumulated in the GRE wall where internal mechanisms are triggered.

1. Introduction

Geotechnical systems subjected to strong earthquakes usually go beyond the elastic regime, strongly involving nonlinear and inelastic soil behaviour. Instead of trying to avoid this occurrence, one can profitably take advantage of plastic mechanisms triggered during the seismic event, provided that they are not followed by a sudden drop of system resistance, but are rather characterised by ductile stress-strain relationships. This design paradigm, the so-called *Capacity Design*, was firstly conceived for structural engineering [1] and now is increasingly adopted in the geotechnical design as well. This approach for seismic design typically leads to strongly reduce the inertial forces transmitted to the structure though accumulating permanent displacements till the end of the seismic event. In the framework of *Performance-Based Design* (PBD), safety check is then verified if these displacements are lower than given threshold values, the latter related to the considered limit state and to the design working life of the structure at hand.

The recalled concepts have been recently applied for the seismic design of geosynthetic-reinforced earth (GRE) retaining walls [2, 3], as their good performance under strong seismic loading is widely recognised from both experimental [4] and numerical analyses [5, 6], as well as from post-earthquake surveys [7]. The main reason why GRE walls behave better than conventionally-design gravity walls is to be attributed to their capability of redistributing seismic-induced deformations into the reinforced zone, provided that reinforcement layers are adequately ductile [8], as geosynthetics usually are.

In this paper, a recently-proposed procedure to design GRE walls following a pseudo-static approach is first summarised. In this design strategy, the walls are designed to activate internal plastic mechanisms during strong ground motions, so as to involve the ductile behaviour of geosynthetic reinforcing layers. To this end, the seismic coefficient k to be used in Limit Equilibrium (LE) computations is assumed to be equal to the critical seismic coefficient related to an internal plastic mechanism, k_c^{int}, that in turn is to be lower than the ones resulting from external plastic mechanisms (e.g. sliding, bearing capacity, overturning), k_c^{ext}, so that an internal mechanism is actually involved. In order to implicitly meet the requirements of PBD in terms of displacements through LE calculations, the seismic coefficient k is then calibrated against given levels of seismic performance, here expressed in terms of threshold displacements d_y, where permanent displacements come from a parametric integration of an updated version of the Italian seismic database via the rigid-block Newmark method [9].

Validity of the proposed design procedure is demonstrated through an illustrative calculation example carried out by performing Finite Difference (FD) nonlinear dynamic analyses, through which the seismic performance of a GRE wall is compared with that of a conventional gravity wall. The two walls are characterised by the same seismic resistance (i.e. the same critical seismic coefficient k_c) but the GRE wall is designed on purpose to trigger an internal plastic mechanism, while the conventional gravity wall can activate external mechanisms only. In the analyses, the walls are subjected to the same real horizontal acceleration time history, representative of a seismic event intense enough to trigger plastic mechanisms. This calculation example validates the proposed design procedure that aims at designing GRE walls for internal plastic mechanisms.

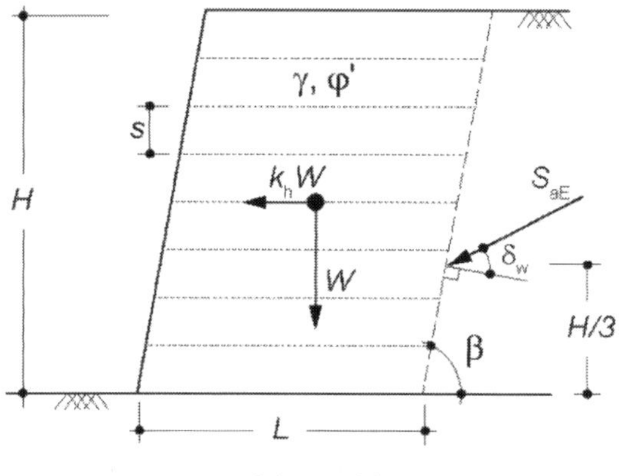

Figure 1. Layout of the problem considered in the parametric study.

Table 1. Non-dimensional parameters used in th parametric study

Parameter	Value
L/H	0.7 – 0.8 – 0.9 – 1.0
β	70° – 80°
φ'	32° – 35° – 38°
k_c	0.005 ÷ 0.300

2. Proposed design method

Problem layout considered in developing the proposed design procedure is given in figure 1. A fill of height H is retained by an earth structure with slope angle β and is reinforced by n geogrid layers, all characterised by uniform spacing s, same length L and tensile strength T_T. At the soil-geogrid contact a purely frictional strength is assumed with an angle of shearing resistance $\varphi'_{s/GSY} = \tan^{-1}(f_{s/GSY} \cdot \tan\varphi')$, where $f_{s/GSY}$ is the interface friction factor.

Following a pseudo-static approach to represent seismic-induced effects, forces acting onto the retaining wall are the self-weight W, the active earth thrust in pseudo-static conditions S_{aE} and the inertial force $k \cdot W$ acting at the centre of gravity of the wall. Preliminary studies showed that the vertical component of inertial forces does not have a remarkable influence on the semi-empirical relationships that will be used in the following to calibrate the seismic coefficient k [2]: therefore, this component is neglected in the sequel ($k_v = 0$).

Internal plastic mechanisms were first detected through the upper-bound theorem of limit analysis, finding for each system the plastic mechanism that maximises the tensile strength demand T_T to reinforcing layers (i.e. the critical plastic mechanism). A Matlab routine was implemented to this purpose, so as to find the values of geometrical parameters defining the critical mechanism (e.g. angle of inclination of a sliding planar surface). Values of the non-dimensional quantities adopted in the dimensionless parametric study are given in table 1. For the non-dimensional parameters that are not listed in table 1, the following constant values were adopted, as their influence on the strength demand turned out to be negligible: $s/H = 1/16$ and $f_{s/GSY} = 0.8$. The latter is clearly an approximation as the interface friction factor actually depends on several factors [10]. More details on the plastic mechanisms can be found in [2].

In addition to the strength demand required to geosynthetic reinforcing layers to bring the system into equilibrium, the detected plastic mechanisms provided the shape coefficient $C = d/d_0$ as well, that is defined as the ratio between the earthquake-induced horizontal permanent displacement at the toe of the wall, d, and the displacement resulting by double integrating the relative motion of a rigid block sliding on a horizontal plane, d_0 [9]. By definition, the shape coefficient takes into account the shape of the critical mechanism developing in the GRE wall for a given set of values of the non-dimensional parameters given in table 1.

Shape coefficients were then applied to obtain the corrected displacements d starting from Newmark displacement d_0, as schematically shown in figure 2. Permanent displacements d_0 were first computed through a parametric integration of an updated Italian seismic database [11], where acceleration time histories recorded during earthquakes occurred in the time frame 1972 ÷ 2017 on the Italian national territory are collected. These time traces were split up into five groups, corresponding to the five subsoil classes defined by the Italian Building Code [12]. For each group, acceleration time histories were scaled either up or down to match predefined peak acceleration values, namely $a_{max} = 0.05$, 0.15, 0.25 and $0.35g$, removing those requiring scaling factors F less than 0.5 or more than 2. Permanent displacements d_0 were then calculated through the double integration of the equation of relative motion considering both signs of accelerations and values of the ratio $k_c/k_{max} = 0.1 \div 0.8$, where $k_{max} = a_{max}/g$ is the maximum seismic coefficient of the seismic input. These displacements were therefore multiplied by the shape coefficient C to compute the corrected displacement, $d = C \cdot d_0$, and relevant upper-bound (95th percentile) semi-empirical relationships expressed through the following functional form were obtained:

$$d = B_{1c} \cdot e^{-A_c \frac{k_c}{k_{max}}} \tag{1}$$

where coefficient B_{1c} and A_c are the intercept and the slope of the curve in a semi-logarithmic plane. Coefficient B_{1c} was computed adopting, for a given value of the ratio k_c/k_{max}, a log-normal distribution of displacement around their mean value, whereas A_c was assumed to be equal to the slope of the median (50th-percentile) curve.

The upper-bound semi-empirical relationships were finally used to establish an equivalence between the corrected permanent displacements d and the seismic coefficient k to be used in LE pseudo-static analyses, so that k was calibrated against given threshold values of permanent displacements, d_y. In particular, equation (1) can be inverted to obtain, for given subsoil class, acceleration level and threshold displacement d_y, the ratio $\eta = k_c/k_{max}$ (figure 3): therefore, if the seismic coefficient $k = \eta \cdot k_{max}$ is used in a LE pseudo-static analysis of the wall and a safety factor $F_S = 1$ is obtained (i.e. limit conditions are attained), then a maximum permanent displacement $d = d_y$ can be expected. If, instead, a safety factor $F_S > 1$ is computed, then the critical seismic coefficient of the wall k_c is by definition greater than the adopted seismic coefficient k, and a maximum permanent displacement lower than the threshold value can be predicted, $d < d_y$. Different values of d_y can be adopted, depending on soil behaviour (e.g. ductile or brittle) and on the structures that can be involved by wall movements. Computed values of coefficient η are listed in tables 2 and 3 for threshold displacements for GRE walls corresponding to $d_y = 2$ and 5 cm, respectively. For $\eta < 0.10$ the minimum value $\eta = 0.10$ was assumed ($a_{max} = 0.35g$, subsoil category E). As expected, coefficient η, that reduces a_{max}, increases for decreasing threshold displacements d_y, that is for a better required seismic performance, and for increasing peak acceleration a_{max}, that is for more intense earthquakes. The highest values of η are mostly computed for subsoil category C, while the lowest ones are obtained for subsoil category E.

In the proposed design procedure, this calibrated seismic coefficient is to be used to design the tensile strength of geosynthetic reinforcing layers T_T, imposing $k = k_c^{int}$. Then, the length of the reinforcing layers L can be selected so as to obtain a critical seismic coefficient related to external plastic mechanisms higher than the internal one, this meaning $k_c^{ext} > k_c^{int}$, in order to promote the activation of internal mechanisms during the seismic event. Imposing internal mechanisms to be triggered will strongly improve the performance of GRE walls, as already showed in previous studies and discussed in the following through the results of numerical analyses.

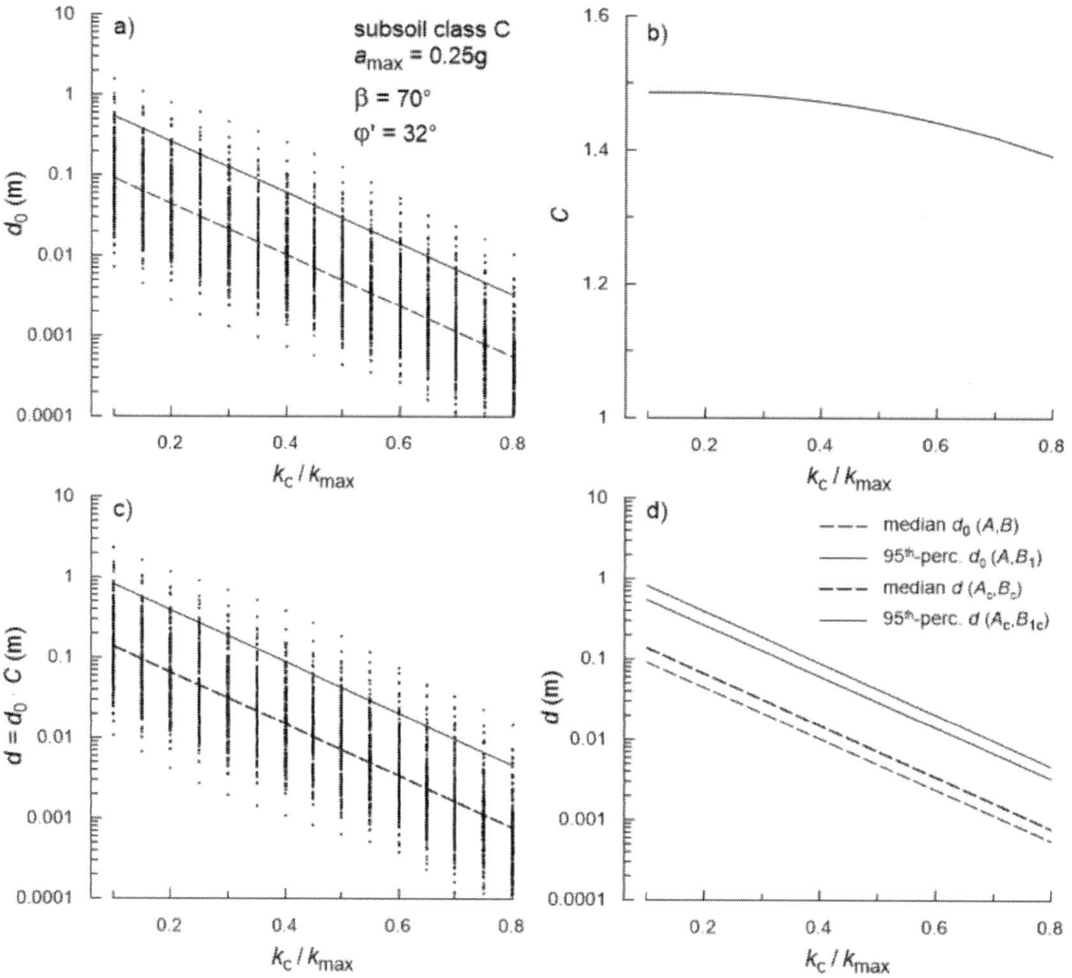

Figure 2. Evaluation of corrected permanent displacement d (subsoil category C, $a_{max} = 0.25g$) and comparison with displacement d_0.

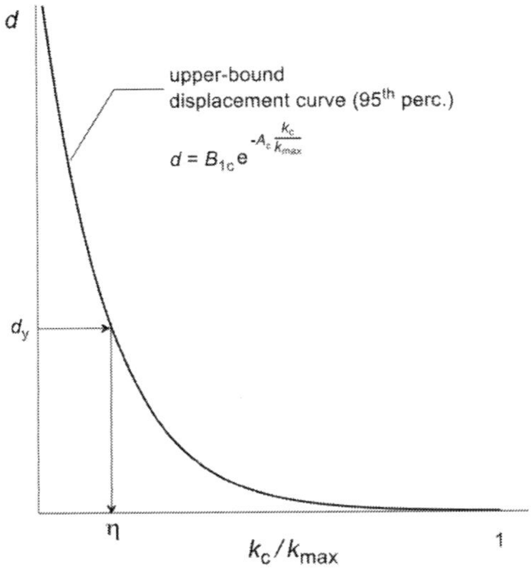

Figure 3. Scheme for calibrating seismic coefficient k as a function of the desired seismic performance.

Table 2. Values of η for d_y = 2 cm

a_{max}	Soil category				
(g)	A	B	C	D	E
0.35	0.61	0.58	0.67	0.57	0.50
0.25	0.52	0.53	0.60	0.55	0.43
0.15	0.43	0.45	0.52	0.52	0.33
0.05	0.28	0.32	0.38	0.49	0.20

Table 3. Values of η for d_y = 5 cm

a_{max}	Soil category				
(g)	A	B	C	D	E
0.35	0.50	0.45	0.55	0.46	0.38
0.25	0.40	0.40	0.48	0.44	0.31
0.15	0.31	0.33	0.40	0.41	0.20
0.05	0.16	0.20	0.26	0.38	0.10

3. Validation of the proposed design procedure

Effectiveness of designing GRE walls to trigger internal plastic mechanisms during intense seismic shakings is demonstrated in this paragraph, where the results of plane-strain nonlinear dynamic analyses, performed using the FD code FLAC v.5 [13] are discussed. Two different walls are considered in the analyses: the first represents a GRE wall for which an internal plastic mechanism is activated during earthquake loading, while the second is a conventional gravity wall that can only mobilise external plastic mechanisms during the seismic event. Characteristics and numerical modelling of these walls are presented first, together with the detection of plastic mechanisms: then, results of FD nonlinear dynamic analyses are discussed, showing the better seismic performance of the GRE wall.

3.1. Numerical modelling and detection of plastic mechanisms

Two idealised walls are considered, retaining the same backfill and founded on the same soil deposit, but activating internal and external plastic mechanisms, respectively (figure 4). The backfill is characterised by a height H = 15 m and a slope angle β = 80° relative to the horizontal.

The first wall (case 1 in figure 4) is a GRE wall designed, through the procedure previously described, to activate on purpose plastic mechanisms involving the geosynthetic reinforcing layers, this meaning to obtain an internal seismic coefficient, k_c^{int}, lower than the one related to external mechanisms, k_c^{ext}. This wall is characterised by 25 uniformly spaced geogrid layers (s = 0.6 m) and a length-to-height ratio L/H = 0.75 (L = 11.25 m). The fill is made by a coarse-grained soil with an angle of shearing resistance φ' = 35°, while the foundation soil has φ'_f = 28° and an effective cohesion c'_f = 10 kPa: both soils were characterised by a unit weight γ = 20 kN/m^3. Geogrid reinforcing layers are assigned a constant tensile strength T_T = 25 kN/m, a yield strain in the axial direction ε_y = 2 % and an infinite extensional ductility: it is worth mentioning that the latter is a primary hypothesis to simulate reinforcing layers ductile enough to provide a good seismic performance, as recently discussed in [8]. Resistance of soil-reinforcement contact was assumed to be purely attritive with a Mohr-Coulomb failure criterion and an angle $\varphi'_{s/GSY}$ = φ' = 35°. For this wall, the following values of the internal and external critical coefficients are computed through the kinematic (upper-bound) theorem of limit analysis: k_c^{int} = 0.101 and k_c^{ext} = 0.196.

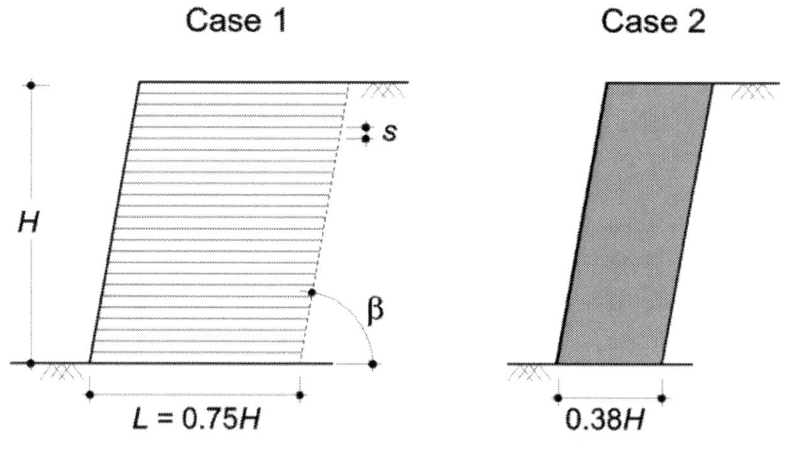

Figure 4. Layout of the two considered retaining walls.

The critical seismic coefficient k_c^{int} was also obtained by performing nonlinear pushover analyses using the code FLAC v.5 [13]. In these analyses, the horizontal component of volume forces, expressed as a fraction k_h of gravity g, was increased until numerical convergence could not be attained anymore. In these conditions well-defined internal plastic mechanisms were detected in the numerical model and the horizontal seismic coefficient k_h became, by definition, the critical seismic coefficient of the wall, $k_c = k_c^{int}$. In the numerical analyses soil behaviour was described using a linear elastic-perfectly plastic model with Mohr-Coulomb failure criterion and a zero dilatancy ($\psi = 0$). Geogrid reinforcing layers were modelled via FLAC *strip* elements, characterised by a linear elastic-perfectly plastic $T - \varepsilon_a$ relationship: therefore, axial (extensional) stiffness was only considered, as usually done for geosynthetic reinforcement. This would represent a polyester (PET) geogrid. Further details on the numerical model can be found in [8]. The value $k_c = k_c^{int} = 0.060$ was computed, that is lower than the one obtained through limit analysis. This result can be ascribed to two main reasons: the upper-bound solutions provided by the kinematic theorem of limit analysis and the different assumption made on the flow rule, associative in the upper-bound theorem ($\psi = \varphi'$) and non-associative in the FD analyses ($\psi = 0$).

The second wall (case 2) simulates a conventional gravity wall, characterised by the same seismic resistance of the GRE wall but related to an external plastic mechanism ($k_c^{ext}{}_{case\ 2} = k_c^{int}{}_{case\ 1}$). To this end the width of this wall, $L = 5.7$ m $= 0.38 \cdot H$, was selected to provide $k_c = k_c^{ext} = 0.060$, equal to the one already computed through the pseudo-static analyses for the GRE wall (case 1). In the FD numerical model, the gravity wall was simulated through a linear elastic material so as to inhibit any internal plastic mechanism.

Figure 5 shows the contours of shear strains at the end of the pseudo-static analyses, these identifying the plastic mechanisms of the two walls at hand. It is worth recalling that the values of the shear strains are not of interest when a failure mechanism is developed. For the GRE wall (case 1) two concurrent mechanisms are observed, both involving the reinforcing layers: the first one can be fairly approximated by a log-spiral rotational mechanism starting from the toe of the wall, while the second one is a two-block mechanisms passing through most of reinforcing layers and then approaching the backfill close to the top of the wall. Conversely, for the conventional gravity wall (case 2) any possible internal mechanism was prevented on purpose and therefore a fully external collapse mechanism is developed, this involving both foundation soil and backfill. Although characterised by the same seismic resistance ($k_c = 0.060$), the different deformation patterns occurring for the two walls will give rise to different seismic performance when subjected to the same seismic input, as it will be shown in the following.

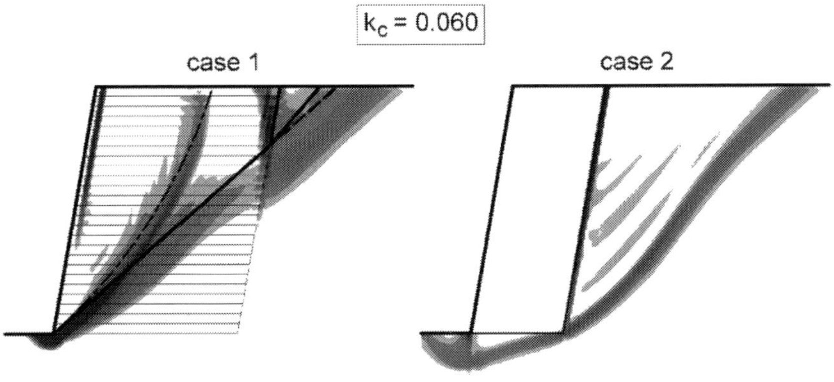

Figure 5. Plastic mechanisms of the walls (modified from [8]).

3.2. FD nonlinear dynamic analyses

Seismic performance of the walls was assessed by performing FD nonlinear dynamic analyses, adopting the same grid as the one used for the pushover analyses, just replacing the static boundary conditions with the FLAC free-field boundaries along the vertical sides while vertical displacements were still constrained at the base of the model, to simulate an infinitely rigid (i.e. fully-reflecting) bedrock. The seismic input was applied at the base of the grid through an acceleration time history.

In the dynamic analyses, soil behaviour was described via the hysteretic model available in the library of the FD code, coupled with Mohr-Coulomb failure criterion already recalled for the pseudo-static analyses. In the hysteretic model, the small-strain stiffness and the backbone curve are to be defined. To this end, parameters defining the dependency of small-strain shear modulus G_0 on the mean effective stress p' were selected to reproduce a typical small-strain stiffness of a medium plasticity sandy-silt for the foundation soils and a dense sand for the backfill: more details are given in [8]. Similarly, the backbone τ-γ curve, where τ and γ are the shear stress and strain respectively, was set to simulate shear modulus decay (G/G_0) and damping ratio increase (ξ) curves typically adopted for coarse-grained soils. These curves are plotted in figure 6: the obtained curves fairly resemble the ones provided by Seed and Idriss [14] and Vucetic and Dobry [15], except for the values of the hysteretic damping ratio obtained for high shear strains, $\gamma > 0.1$ %. A viscous damping ratio $D = 1$ % was added to reduce the numerical noise resulting from the hysteretic model at very low shear strains, where the hysteretic damping is close to zero: to this end, the Rayleigh formulation was adopted, selecting as central frequency the fundamental frequency of the soil deposit, $f = 1.02$ Hz.

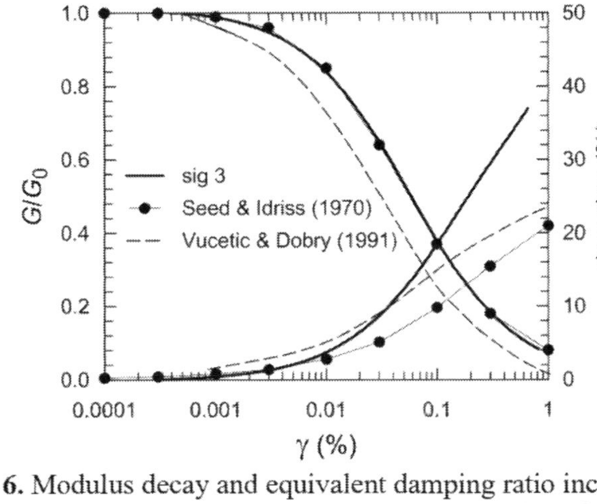

Figure 6. Modulus decay and equivalent damping ratio increase curves adopted in the numerical analyses (modified from [8]).

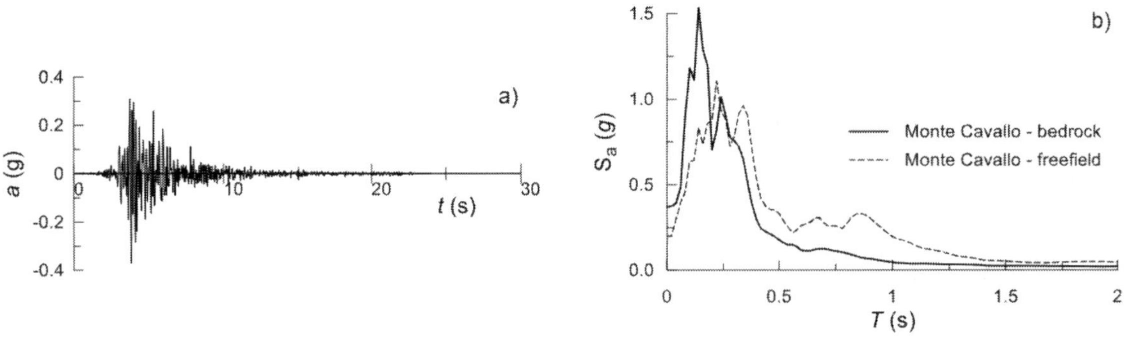

Figure 7. (a) Seismic input adopted in the analyses; (b) input and free-field acceleration spectra.

Both walls were subjected to the scaled (reduced in amplitude by a factor $F = 1.25$) horizontal acceleration time history recorded in Monte Cavallo during the October 2016 Central Italy earthquake (figure 7a), extracted from the updated version of the Italian seismic database [16].

This input motion is characterised by a peak acceleration $a_g = 0.37$g, Arias intensity $I_A = 0.75$ m/s, mean period $T_m = 0.20$ s [17] and a significant duration $T_D = 4.23$ s [18]. To comply with the adopted FD grid, the input was low-pass filtered ($f_{max} = 15$ Hz) and subjected to a quadratic baseline correction to obtain zero velocities and displacements at the end of the record. In figure 7b the 5 %-damped elastic acceleration spectrum obtained at ground surface in free-field conditions (that is, far enough from the wall) is compared with the one of the input motion: a noticeable reduction of spectral acceleration is observed at low periods ($T < 0.25$ s), this inducing a remarkable reduction of the peak acceleration $PGA = a_{max} = Sa(T = 0) = 0.20g$, while a strong amplification of motion occurs at higher periods ($T = 0.25 \div 2$ s).

Plastic mechanisms were actually triggered during the seismic event for both walls, as shown in figure 8, where the contours of shear strains at the end of the dynamic calculation phase are plotted. It is evident that the same plastic mechanisms as the ones obtained from the pseudo-static analyses (see figure 5) are mobilised. The deformation patterns associated to the different plastic mechanisms resulted in different permanent displacements developed at the end of the seismic event, that is in a different seismic performance of the walls.

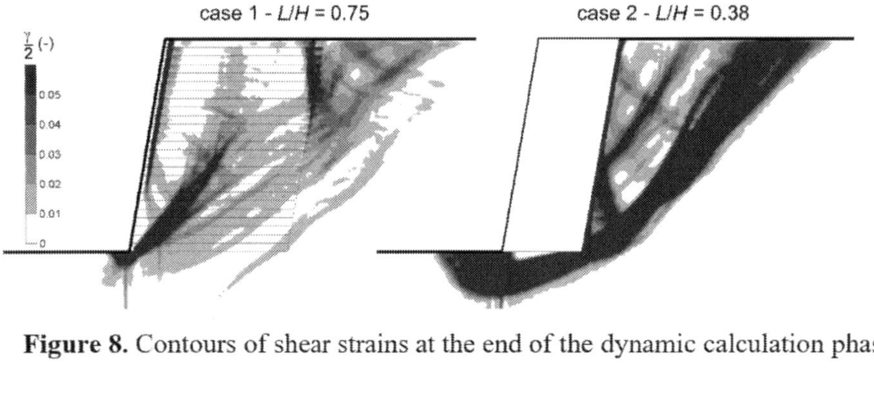

Figure 8. Contours of shear strains at the end of the dynamic calculation phase.

Figure 9 shows the time histories of the horizontal displacements at the base (B) and the top (T) of the wall façade, for both the walls under consideration. Displacements relative to that computed in free-field conditions, at the same relevant depth, are plotted in the figure. Values of $d_B = 0.26$ m ($d_B/H \approx 1.7$ %) and $d_B = 0.82$ m ($d_B/H \approx 5.5$ %) were computed at the base of the GRE and the conventional gravity wall (CGW), respectively, with a ratio $d_{B,\,CGW}/d_{B,\,GRE} \approx 3.2$. Similarly, at the top of the wall façade displacements $d_T = 0.31$ m ($d_B/H \approx 2.0$ %) and $d_T = 0.56$ m ($d_B/H \approx 3.7$ %) are computed, this providing a ratio $d_{B,\,CGW}/d_{B,\,GRE} \approx 1.8$. Wu and Prakash [19] indicated values of $d_B/H \leq 2.0$ % as permissibile non-dimensional horizontal displacements: therefore, according to this criterion seismic performance of the GRE wall can be still deemed acceptable, while the one of the conventional gravity wall cannot.

Relative displacements computed at the base of the wall façade were also compared with those resulting from the semi-empirical relationships provided in [2] (figure 10). Indeed, these relationships can also be used to obtain a fair estimate of the order of magnitude of the expected displacements, as well as to calibrate the seismic coefficient k. For the case at hand, semi-empirical curves proposed for $a_{max} = 0.25$ g and subsoil class C are deemed the most appropriate to be used, with a ratio $k_c/k_{max} = 0.06/0.20 = 0.3$. It can be seen that the horizontal permanent displacement calculated for case 1 (GRE) wall plots close to the 95 % upper-bound curve, whereas the displacement computed for case 2 (conventional gravity) wall is much further from the curve and cannot be properly estimated with the proposed semi-empirical relationship. Moreover, the displacement resulting from the FD nonlinear dynamic analysis of the GRE wall falls into the computed range of displacements (grey-shaded area), computed through the parametric integration of the Italian seismic database using the rigid-block Newmark method, whereas the same does not hold for the conventional gravity wall.

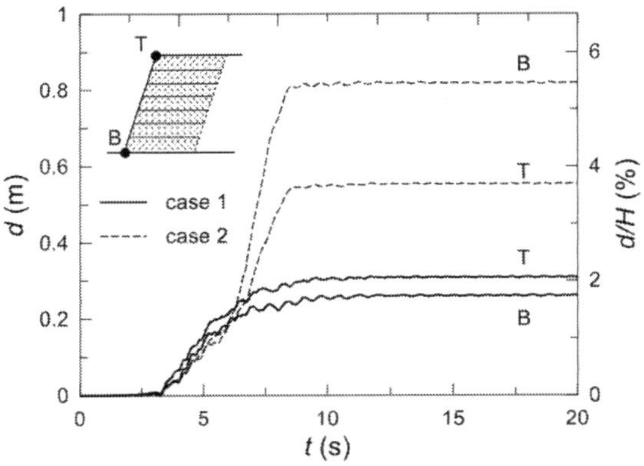

Figure 9. Time histories of horizontal displacements relative to free-field.

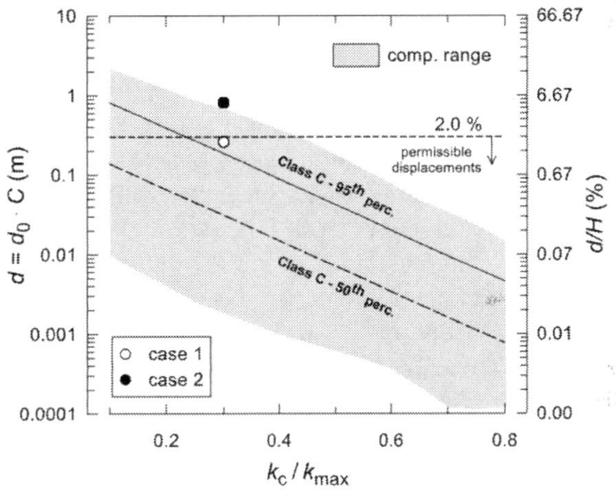

Figure 10. Comparison of permanent displacements obtained from dynamic analyses with those estimated with proposed semi-empirical relationship for $a_{max} = 0.25g$.

4. Concluding remarks

Seismic performance of geosynthetic-reinforced earth (GRE) retaining walls recently proved better than conventionally-designed retaining walls, thanks to the presence of reinforcing layers that redistribute seismic-induced strains into the reinforced zone. In this paper, a straightforward procedure recently developed to design GRE walls through the pseudo-static approach is firstly recalled. On the basis of upper-bound semi-empirical relationships linking permanent displacements to the ratio k_c/k_{max}, different values of the seismic coefficient k are provided as a function of the desired seismic performance of the wall, expressed by threshold displacement d_y (= 2 and 5 cm), earthquake intensity, quantified by the peak acceleration a_{max} (= 0.05, 0.15, 0.25 and 0.35g) and subsoil category (A, B, C, D, and E). In the proposed design procedure, it is suggested to design the GRE wall imposing the seismic coefficient k to be equal to the critical seismic coefficient related to an internal plastic mechanism, k_c^{int}, this being linked to the tensile strength of the reinforcing layers, T_T. The length of the reinforcing layers is then evaluated so that the critical seismic coefficient associated to the external plastic mechanisms, k_c^{ext}, is higher than k_c^{int}: this to trigger internal plastic mechanisms during strong seismic events.

Effectiveness of the proposed procedure has been demonstrated showing that a GRE wall designed to activate internal plastic mechanisms during a severe earthquake actually behaves better than a conventionally-designed gravity wall, the latter triggering external plastic mechanisms only. To this end, FD nonlinear dynamic analyses were carried out, where the two walls were subjected to the same seismic record: the results showed that permanent displacements cumulated at wall façade for the GRE wall are much lower (up to about 70 %) than the ones computed for by the conventional gravity wall, this leading to a much better seismic performance.

It is useful recalling that conclusions drawn in this paper result from some underlying assumptions, these narrowing the range of applicability of the study. First, geosynthetic reinforcement layers are modelled as infinitely ductile: this is clearly a strong assumption, though quite realistic if bearing in mind the highly-ductile behaviour of geosynthetics compared to other materials (e.g. steel, glass, etc…). Second, the implicit hypothesis of infinitely flexible wall façade is made in this study, this being in principle suitable for wrap-around facing only, but neither for full-height rigid facing nor for modular blocks: nonetheless, concepts related to the better seismic performance of GRE walls still maintain.

References

[1] Paulay P and Priestley MJN 1992 *Seismic Design of Reinforced Concrete and Masonry Buildings* (New York: Wiley)

[2] Gaudio D, Masini L and Rampello S 2018 A performance-based approach to design reinforced-earth retaining walls *Geotext. Geomembr.* **46** pp 470 – 85

[3] Gaudio D, Masini L, Rampello S 2021 A Procedure to Design Geosynthetic-Reinforced Earth-Retaining Walls Under Seismic Loadings. In: Barla, M., Di Donna, A., Sterpi, D. (eds) Challenges and Innovations in Geomechanics. IACMAG 2021. Lecture Notes in Civil Engineering, **126** pp 300-8, Springer, Cham https://doi.org/10.1007/978-3-030-64518-2_36

[4] Yazdandoust M 2018 Seismic performance of soil-nailed using a 1g shaking table *Can. Geotech. J.* **55** **pp** 1 – 18

[5] Gaudio D, Masini L and Rampello S 2018 Seismic performance of geosythetic-reinforced earth retaining walls subjected to strong ground motions *Proc. China-Europe Conf. Geotech. Eng. (Vienna)* (Springer Series in Geomechanics and Geoengineering 216849) pp 1474 – 78

[6] di Filippo G, Biondi G and Moraci N 2019 Seismic performance of geosynthetic-reinforced retaining walls: experimental tests vs numerical predictions *Proc. 7th Int. Conf. Earthq. Geotech. Eng. (Rome)* pp 2574 – 82

[7] Koseki J, Nakajima S, Tateyama M, Watanabe K, Shinoda M 2009 Seismic performance of geosynthetic reinforced soil retaining walls and their performance-based design in Japan *Proc. 7th Int. Conf. on Perf. based Design in Earthq. Eng. (Tokio)* pp 149 – 61

[8] Masini L, Callisto L and Rampello S 2015 An interpretation of the seismic behaviour of reinforced-earth retaining structures *Géotechnique* **65** pp 349 – 58

[9] Newmark NM 1965 Effects of earthquakes on dams and embankments *Géotechnique* **15** pp 139 – 93

[10] Cardile G, Pisano M and Moraci N 2019 The influence of a cyclic loading history on soil-geogrid interaction under pullout condition *Geotext. Geomembr.* **47** pp 552 – 65

[11] Gaudio D, Rauseo R, Masini L and Rampello S 2020. Semi-empirical relationships to assess the seismic performance of slopes from an updated version of the Italian seismic database *Bull Earthquake Eng* **18** pp 6245–81 https://doi.org/10.1007/s10518-020-00937-6

[12] Ministero delle Infrastrutture 2018 Norme Tecniche per le Costruzioni *Gazzetta Ufficiale della Repubblica Italiana* **42** (Rome: Decreto Ministero Infrastrutture 17.01.2018)

[13] Itasca 2005 *fast Lagrangian analysis of Continua v. 5.0. User's manual* (Minneapolis: Itasca Consulting Group)

[14] Seed H B and Idriss I M 1970 *Soil moduli and damping factors for dynamic response analyses* Report No. EERC70-10 (Berkeley, California: Earthquake Engineering Reaearch Centre)
[15] Vucetic M and Dobry R 1991 Effect of soil plasticity on cyclic response *J. Geotech. Eng.* **117** (1) 89 – 107
[16] Luzi L, Puglia R, Russo E and ORFEUS WG5 2016 *Engineering strong motion database, version 1.0* Istituto di Geofisica e Vulcanologia, Observatories & Research Facilities for European Seismology https://doi.org/10.13127/ESM
[17] Rathje E M, Abrahamson N A and Bray J D 1998 Simplified frequency content estimates of earthquake ground motions *J. Geotech. Geoenviron. Eng.* **124** pp 150 – 9
[18] Trifunac M D and Brady A G 1975 A study on the duration of strong earthquake ground motion *Bull. Seismol. Soc. Am.* **65 pp** 581 – 626
[19] Wu Y and Prakash S 1996 On the seismic displacements of rigid retaining walls *ASCE Geotech. Special Publ.: Analysis and Design of Retaining Structures against Earthquakes ASCE* pp 21 – 37

IOP Conf. Series: Materials Science and Engineering 1260 (2022) 012022 doi:10.1088/1757-899X/1260/1/012022

Plastic in the ocean and global warming: New challenges for geosynthetics

H Hoyme[1], L Vollmert[2], H Ehrenberg[1]

[1] NAUE GmbH & Co. KG, Gewerbestr. 2, 32339 Espelkamp, Germany
[2] BBG Bauberatung Geokunststoffe GmbH & Co. KG, Gewerbestr. 2, 32339 Espelkamp, Germany
E-mail: hhoyme@naue.com

Abstract. The application of geosynthetics is state of the art onshore, at the coast and offshore. Geosynthetics complement conventional construction methods, or even replace them and generate advantages for the client, the designer and even more for the environment, e.g. by reducing carbon footprint. Decades of research and experience have resulted in standards and products, which are optimised for long-term performance and reliable application. Geosynthetics, by definition, are products covered by soil and thus protected against impacts or abrasion. As they are often used in contact to water, they are tested to be groundwater neutral. On the other hand, however, concerns about the use of plastic are omnipresent and must be taken into account. In some very limited but not negligible applications, temporary or periodical environmental influences have to be discussed, e.g. permanently or temporarily uncovered geotextile applications such as scour protection measures, coastal protection or riverbed stabilisation. As the geosynthetics, beside their technical beneficial use, can be subject to abrasion in this limited range of applications, further options for optimisation shall be discussed. One possible alternative is to design geosynthetics from or with the use of biodegradable raw materials. The subject of this article is the presentation of geosynthetic products made from alternative raw materials which biodegrade after their planned period of use. Other applications may require a defined degradation process over time, for example erosion control products. Making use of biodegradable materials, the potential risk of causing damage to the flora and fauna in case of an undesirable entry into the environment can be limited.

1. Introduction

Climate change, the need to reduce CO_2 emissions, plastic in the ocean and many other issues related to ecology have been on almost everybody's lips since the youth movement "Fridays for Future" has appeared strongly. But what does this have to do with the special area of geosynthetics in the construction industry?

First of all, geosynthetics, as the name implies, usually consist of plastics which are made mainly from fossil fuels such as petroleum. Geosynthetics are used in many different areas of the construction industry for many decades. They are used particularly frequently due to their technical advantages in the areas of infrastructure and hydraulic engineering. Environmental protection measures such as landfill lining, groundwater protection in industry and in infrastructural applications, remediation of brownfields and contaminated sites as well as water supply using ponds and irrigation channels are developed by making use of membranes, drainage products and protection layers. Significant benefit can be drawn from the use of geosynthetics in coastal protection measures [6], [7], helping to solve one of the biggest environmental issues in the next decades: coastal erosion. Filter layers under armour layers and sand containers of different shapes and sizes provide robust and flexible construction elements in one. Further common applications are found in road and railway constructions, reducing the carbon footprint of the construction.

Depending on the application, geosynthetics have different functions, such as draining, filtering, protecting, packaging, reinforcing or combinations thereof. In some areas of application, they complement conventional construction methods, in others they replace them and generate great benefits for the client or from a planning point of view. But there are also particular advantages for ecology if, for example, one compares the CO_2 consumption of a classic geotechnical solution made primarily of mineral building materials with a construction method using geosynthetics. Economically, solutions with geosynthetics have a major advantage over classic solutions that often need a lot of mineral construction material.

Geosynthetics have found their way into numerous regulations over the past decades. The focus is usually on durability and long-term performance as well as purity of the raw materials. Following the regulations, the todays products are optimised in this sense using purity of the raw materials (1A quality). The amount of recycled raw material to be used is strongly limited, but industry is working on improved cyclic economy concepts in this regard.

Considering the aforementioned development and use, plastics in the ocean is, in general, not a topic of classical geosynthetics. On the other hand, in some limited hydraulic applications, geosynthetic constructions are only for temporary use or are used in situations, where exceptional stress can result in local damages, coverage and protection by soil can be reduced or abrasion has to be considered as regular stress. In this special situation, the design of current products offers potential for optimisation of the characteristics. By the use of biodegradable materials, protection layers, wearing layers or entire products can provide benefit for these situations and shall be discussed in detail.

2. Classical geosynthetics

2.1. *Ecological footprint*

Especially when it comes to the ecological footprint, e.g. measured in terms of CO_2 emissions, the alternative use of geosynthetics offers significant advantages. In recent years, some comparative life cycle analyses (LCA) with different construction methods were carried out using project examples. A comparison between soil stabilisation with fine lime and a solution with geogrids using the example of the district road K34 near Aachen, Germany [1], [3], [4] is presented here.

The results (Figure 1) show a significant saving of the greenhouse gas CO_2 when using a geogrid instead of a traditional soil improvement with lime stabilisation. Of course, such a result cannot be generalized for all projects, it always depends on the individual case and the respective boundary conditions. However, in the vast majority of the project examples examined, the solutions using geosynthetics provide a reduction in the order of a magnitude compared to the classical method.

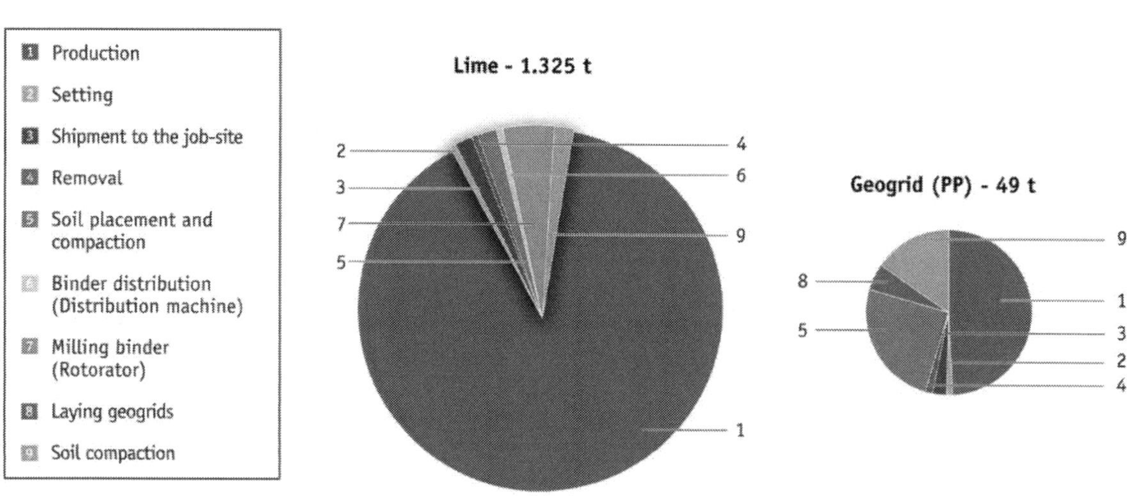

Figure 1. Cumulative CO_2 emissions when using hydraulic binders for soil stabilisation in comparison to geogrids used for example for the new construction of district road K34, district of Aachen, 2008

2.2. *Durability*

Another aspect that should not be neglected is the durability of geosynthetics. In public eye, plastic as waste is often equated with plastic as a product. However, today it is impossible to live and work without plastics; as examples may be mentioned their use in means of transport packaging and protection, in industrial industry such as technical devices or in medical technology. But what can be improved is our handling of the products after the end of their lifetime. Correct recycling and an associated closed material cycle are certainly desirable, but are still only implemented in some areas today. Compared to other applications, geosynthetics are produced for a usability of decades and in landfill applications even for hundreds of years (storing the implemented energy) and therefore they can be considered to be sustainable products per definition. As most of the products are produced by single-origin polymers, a high recycling quote can be expected in future realistically. Concepts are currently under verification to ensure practical and efficient solutions and can be expected in the near future.

2.3. *Direct impact on the environment*

The use of geosynthetics is designed, by definition, in such a way that the geosynthetic material itself is covered and encapsulated with soils. Thus, no abrasion or other undesired entry of particles into the environment can occur until the end of the desired service life. An example of this is the use as a concrete nonwoven for roads (Figure 2). The nonwoven is encapsulated between the different layers of concrete and is only uncovered when the road is removed. During the course of its use, however, it significantly increases the lifespan of the road by the factor of approx. two, providing drainage and filtration.

Figure 2. Concrete road separating nonwoven, completely encapsulated in the operating phase

In Germany, since many years a system for evaluating the impact of geosynthetics (raw material, stabilisers, avivage etc.) is in place. All kind of geosynthetics are tested following the German Regulation M Geok E 16, chapter 6.29, and will be checked regarding the requirements of chapter 7.7 [2]. The test method itself is a leaching test and the chemical analysis is done regarding requested elements and classes from the German federal soil protection ordinance (BBodSchV). Only geosynthetics fulfilling these requirements are allowed to be installed.

There are also applications in which the geosynthetic material is in direct contact with environmental influences, e.g. for revetments, coastal protection measures or in the area of scour protection. Products that have been developed for these purposes have a predefined lifespan. During this time, as little as possible or ideally no particles of the geosynthetic should be released into the environment as abrasion or other loss. However, since the introduction of particles into the environment cannot always be excluded with 100% certainty, there is an increasing tendency in some applications and markets to use alternative solutions and this often leads to a step back to conventional designs. The advantages which the use of geosynthetics offer from an ecological as well as from an economic perspective are then lost to the tendering authorities or private investors.

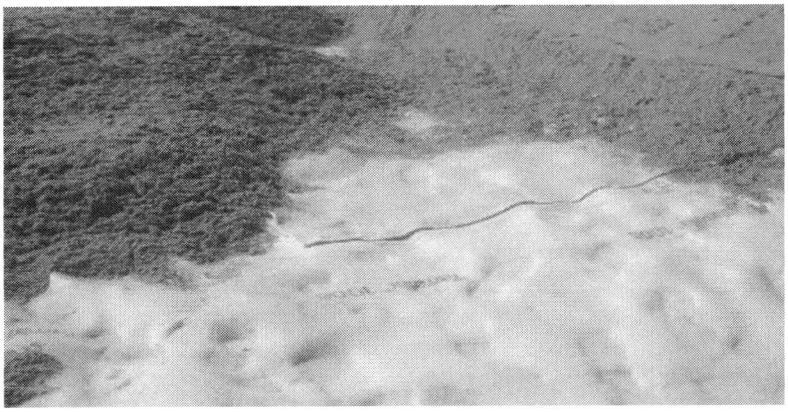

Figure 3. Nonwoven filter fabric in direct contact with the environment, also during the operating phase

3. Biodegradable geosynthetics

For some limited applications as for example uncovered geotextiles exposed to abrasion, geosynthetics from biodegradable raw materials are an option to make use of. In these cases, the property to degenerate is of major importance in case abrasion takes place and particles can migrate to the environment. Other applications may require degradation over time, such as erosion control products. A third group, temporary applications where the product serves for a limited period in covered and protected environment, but is removed shortly, can benefit from these characteristics.

The characteristic of the biodegradable product is that even an unwanted entry into the environment, for example from abrasion or damage or removal, does not lead to any harm for flora and fauna. Nevertheless, they achieve their service life required for the application and fulfil their function during this time when the design is tailored to the respective project boundary conditions.

To achieve this goal, extensive tests were carried out with different materials. The best results in terms of mechanical properties with simultaneous biodegradability could be achieved with industrially manufactured staple fibres. The fibres used are made from 100 percent renewable raw materials. It should be noted that the raw materials for the manufacture of the fibres only come from certified sources. In the production of the fibres, particular emphasis is put on recycling the process water. Another advantage of the raw material used is that the water consumption during manufacture, for example compared to the production of cotton fibres, is around 95% lower. Due to the industrial production of the fibres, comparable to the production of synthetic fibres from fossil fuels, a constant material characteristic and industrially controlled quality can be achieved, unlike with natural fibres such as flax, jute or coconut (coir).

Figure 4. Biodegradable staple fibres, raw material for the 100% biodegradable geotextile Secutex® Green

3.1. Applications

In many applications, permanent preservation of the product properties is not necessary. For filter applications it can be accepted that a tensile strength, which is important for the installation, decreases in the installed state. In other cases, it is even intended that product properties change with a time delay. There are also applications in which the conditions for biodegradation are practically non-existent.

Examples of the criteria mentioned can be found in applications in the following areas:

- separation, filtration and protection layers,
- gardening and landscaping,
- temporary pathway and road construction,
- facing at temporary reinforced slope and retaining wall systems,

- road construction in particularly sensitive areas such as Flora-Fauna-Habitat (FFH), in alpine areas, in highly sensitive water protection areas, etc.,
- bank protection on waters / nature-oriented hydraulic engineering
- protective layers on embankments of waters and as dune and coastal protection,
- sand container as scour protection / bank protection / artificial reef,
- beach path and sand trap fence,
- filtration and separation layer for alpine pathways,
- covering of snow fields and glaciers.

Figure 5. The newly developed geotextile nonwoven, which is similar in appearance and feel to the classic nonwoven

3.2. German geotextile robustness classes

Due to the constant quality of the fibres and the nonwoven produced, the new biodegradable geotextile can be divided into German geotextile robustness classes. This classification refers to the properties of weight and puncture force, which are also decisive for the classic nonwoven made of plastic.

Table 1. German geotextile robustness classes (GRK) of multifunctional separating and filtering nonwovens in a biodegradable variant

	[GRK 2]	GRK 3	GRK 4	GRK 5
Weight per unit area	300 g/m²	400 g/m²	600 g/m²	900 g/m²
Thickness	3,0 mm	3,3 mm	4,0 mm	4,5 mm
Puncture force	≥1.000 N	≥1.500 N	≥2.500 N	≥3.500 N

3.3. Certificates

There is a large selection of international certifications for a wide variety of product properties. The property of "biodegradability" is now fairly well standardised, there are different labels, but these are mostly based on the results of similar, comparable or even identical tests.

An often-criticised property of biodegradable products is that they can only decompose at high temperatures. In many cases this means, that the biodegradable products in the recycling plants have to be sorted out and then incinerated. The nonwoven made from renewable organic raw materials presented in this article meets the requirements according to European standards with regard to industrial compostability and in particular also with regard to compostability in domestic compost. However, the focus of the desired biodegradability is not compostability, but rather the expanded environmental safety in applications in which a release by abrasion cannot be ruled out constructively, as can be ensured in most conventional applications.

3.4. Further investigations

Further research is currently underway with various alternative raw materials on their processability, service life, biodegradability and metabolisability.

The service life of biological materials largely depends on the boundary conditions to which they are exposed during the duration of their use. The ambient temperature, humidity and biological activity are particularly important factors influencing the service life of biodegradable materials. Long-term test series with different raw materials, masses per unit area and environmental conditions such as type of soil, temperature and humidity have been planned in cooperation with scientific institutions and are currently being implemented.

Another property of materials is the so-called metabolisability or metabolism. It describes the conversion of materials in bodies of different organisms to intermediate products, the metabolites. Further metabolic processes consist of many individual series of enzymatic reactions that deliver specific products and which in turn are further broken down.

With regard to nonwovens in geotechnical engineering, this is of particular interest in the context of hydraulic engineering. During installation, operation and removal of geosynthetics like nonwovens, loosening of fibres from the nonwoven surface may occur. Since the nonwoven is made of metabolisable material, fibres and nonwoven residues accidentally released will be harmless for flora and / or fauna.

Scientifically supported investigations are being carried out on this topic, the first results of which are presented here partly. Figure 6 shows an image of a nonwoven fabric with a scanning electron microscope (SEM) after incubation with certain constituents of gastric acid from aquatic organisms. The fibres show a clear attack by enzymes on the fibre surface. This clearly indicates the degradability of the organic fibres e.g. in the stomachs of marine animals.

Further studies regarding the metabolism of different raw materials by aquatic organisms are currently being implemented. Studies on the micro-robust metabolisability of the pure product without environmental substances, i.e. unlike tests for biodegradability in soils or compost, have been started in the first quarter of 2020.

Figure 6. Image by a scanning electron microscope of fibres of a biodegradable geotextile attacked by enzymes, 100 μm ≙ 0.1 mm

4. Summary

This article shows the advantages of using geosynthetics compared to classic construction methods with purely mineral building materials. An example shows the reduction in CO_2 consumption when using modern geosynthetic solutions.

There are questions in civil engineering that do not depend on the extremely long durability inherent in the plastics. For these questions biodegradable nonwovens offer benefit for the environment by application of protection and wearing layers as well as entire biodegradable products. In most of such applications, they show a sufficient service life, since in the majority of such applications, the boundary conditions with regard to temperature and surrounding biological activity are low enough that there is no immediate decomposition of the basically biodegradable product. This new product group represents an expanded option for the use of geosynthetics, particularly in applications that require a temporary function, or in applications in which abrasion cannot be ruled out structurally.

References

[1] Egloffstein T 2009 Bauverfahren mit mineralischen Baustoffen und Bindemitteln im ökologischen Vergleich mit dem Einsatz von Geokunststoffen. 6. Geokunststoff-Kolloquium der NAUE GmbH & Co. KG

[2] FGSV 2016 Merkblatt über die Anwendung von Geokunststoffen im Erdbau des Straßenbaus (M Geok E 16)

[3] Frischknecht R, Büsser-Knöpfel S, Itten R, Stucki M, Wallbaum H 2013 Comparative Life Cycle Assessment of Geosynthetics versus Conventional filter layer *Proceedings of the 18th International Conference on Soil Mechanics and Geotechnical Engineering*, Paris

[4] Frischknecht R, Büsser-Knöpfel S, Itten R, Stucki M, Wallbaum H 2013 Comparative Life Cycle Assessment of Geosynthetics versus Concrete Retaining Wall *Proceedings of the 18th International Conference on Soil Mechanics and Geotechnical Engineering*, Paris

[5] Hoyme H, Su J H, Kono J & Wallbaum H 2019 Nonwoven geotextile scour protection at offshore wind parks, application and life cycle assessment *Proceedings and Monographs in Engineering, Water and Earth Sciences* pp 315-21

[6] Prambauer M, Wendeler C, Weitzenböck J, Burgstaller C 2019 Biodegradable geotextiles – An overview of existing and potential materials *Geotextiles and Geomembranes* **47** pp 48-59

[7] Heerten G, Heibaum M, Haselsteiner R, Werth K 2009 Danger of flooding - new safety measures in dike construction by using geosynthetics *Proceedings of the 17th International Conference on Soil Mechanics and Geotechnical Engineering* Kairo, Egypt

[8] Dassanayake D T, Oumeraci H, Werth K, Heerten G 2011 Setting up monitoring plans to assess the performance and the durability of exposed geotextile encapsulated sand elements in coastal engineering applications *2nd International Workshop on Geosynthetics and Modern Materials in Coastal Protection and Related Applications* 2-5 March2013 Madras, India

IOP Conf. Series: Materials Science and Engineering 1260 (2022) 012023 doi:10.1088/1757-899X/1260/1/012023

Mechanically stabilised layer by geogrids in road construction as an environmental aspect of sustainable development

M Gołos[1]

[1] Tensar Polska Sp. z o.o, Poland

Polish Chapter of PSG-IGS, Poland

Email: mgolos@tensar.pl

Abstract. The purpose of this paper is to focus on the advantages of environmental aspects of mechanically stabilised layer by using geogrids (MSL). MSL is utilised in road construction as a sub-base layer or capping layer for a week sub-soil improvement. The concept of a MSL says that unbound granular layer is stabilized by the non-reinforcing hexagonal geogrid by the way of interlock with the aggregate. The mechanism of interlock means that both geogrid and aggregate interact under applied external load. The beneficial consequence on the serviceability of that layer by the resisting of the movement of the particles is defined as stabilisation. Mechanically stabilised layer by geogrid allows to save both money and time of construction compared to traditional technologies. Moreover, it also substantially supports sustainable development being a successful alternative to traditional solutions in road construction, such as: soil replacement, cement and other hydraulically bound mixtures or granular unbound layers which are non-stabilised by geogrids. MSL gives massive benefits for the environment, allowing for significant reduction of energy usage and greenhouse gases emission (mainly CO_2). The paper shows the methodology of carbon footprint approach, which is an equivalent amount of CO_2 emission, resulting from all component processes related to a specific technology used in road construction.This paper describes also the basic rules of how to calculate the amount of this reduction by using a special tool, so called Carbon Calculator (CC) designed by one of the leading geogrid manufacturer. CC allows to determine the reduction of CO_2 as a result of geogrid utilisation in MSL compared to other well-known and long-standing solutions, mainly traditional granular unbound and cement or other hydraulically bound mixtures. A number of calculations of carbon footprint with MSL and other traditional solutions used as a sub-base or capping layer in the road construction have been done and the results of them have been presented in the further part of this article.

1. Introduction

Geogrids were invented in late 1970s and have been considerably reducing aggregate usage and minimize environmental impact, specifically in road pavement constructions. Geogrids used in mechanically stabilised systems can reduce the need for aggregate or fill quantities in those constructions. Mechanically Stabilised Layer by using geogrids relates to the stabilising function which has been quite recently identified and characterized by an ISO committee in 2018 [2]. Since that time stabilizing function is a separate one which differentiates stabilization from reinforcement. Mechanically stabilised layer by hexagonal geogrids refers to stabilise unbound granular aggregate, mainly treated as sub-base, frost resistance and capping layers placed on the week and soft subgrade to achieve a required strength and compaction level before placing the principle road pavement (base course and asphalt package of layers). It's been described and widely detailed in several studies so far. However, the environmental aspect has been analysed in this article by comparing MSL option to other widely used and long-standing solutions in pavement engineering, i.e. granular unbound layers non-stabilised by geogrids and hydraulically bound mixtures (aggregate) or soils.

It is worth to know that the idea of MSL has developed over the recent years and it has been still progressing. It has been presented some benefits of installing hexagonal geogrid in a granular layers used in working platforms and beneath foundations to improve load spread and bearing capacity on weaker clay soils [8,9. Hexagonal geogrids can be also utilised for stabilizing aggregate of a base course layer on firm subgrades. In such approach the positive effect of stabilization can result in reductions in upper layers thickness (both asphalt and base aggregate course) for a given design life [7,10,15] . There is also a potential to utilize the aggregate recycled from the crushed concrete which is screened into the pretty regular particle fractions [5]. It can be reused to form new fills and aggregates to be stabilized with the hexagonal geogrids in the equivalent way as the unbound granular aggregate layers are enhanced by the geogrids. All above approaches allow thinner granular layers to be installed and they bring cost, time and carbon savings associated with the smaller volumes of material. However, as they need to be considered quite extensively and comprehensively it has not been exposed on calculations and analyse in this paper.

2. Mechanically stabilized layer (MSL) of an unbound aggregate with geogrid

MSL takes place when unbound aggregate is placed and compacted on a layer of stiff non-reinforcing hexagonal geogrid made of polypropylene (PP) and then a composite of geogrid and unbound aggregate creates. Aggregate particles interlock and confined within the stiff geogrid matrix. Composite system created in such way is called exactly MSL. Lateral restraint provided by the stiff ribs and junctions of a geogrid reduces strains and in this way increases the stiffness of the MSL, when comparing to the layer of the same thickness with no geogrid. The mechanism of interlock means that both stiff non-reinforcing hexagonal geogrid and aggregate interact under applied cyclic external load. During the relatively short period of time associated with the passing of the transient load, unbound granular layer is subjected to stress (both its vertical and components) whereas the associated horizontal strains have a radial distribution (Fig. 1).

This mechanism prevents the layer deformation. The main and most desirable effect of stabilisation of aggregate with a stiff non-reinforcing hexagonal geogrid is an increase in a modulus of MSL. Following to e.g. AASHTO 1993 *Guide for Design of Pavement Structures* [14] this modulus of MSL is a function of the comprehensive stress condition (Θ), presented in the formula (1).

$$E = k_1 \Theta^{k_2} \quad (1)$$

where:

Θ - a sum of principal stresses ($\sigma_1 + \sigma_2 + \sigma_3$)

k_1, k_2 – constants depending on aggregate type

Above means that if the external traffic load is applied to a MSL, the stiff structure of a geogrid prevents aggregate particles from moving, mainly in a horizontal direction. This allows to growth the horizontal stresses (σ_2 and σ_3), and as a result also the comprehensive stress state Θincreases. Therefore the modulus of MSL increases and – as a result - its bearing capacity increases finally [4].

Figure 1. General idea of the radial distribution of traffic load in all directions of the unbound granular layer

The issue of MSL has been confirmed in both laboratory and in-situ tests [4]. The tests showed e.g. that the modulus of a MSL samples increases by 5% to 20% compared to non-stabilized at laboratory sections [6] and the resilient modulus of MSL samples increases by 5% to 30% compared to non-stabilized at in-situ sections [12,13]. Herein calculations on MSL thickness reduction have been carried out following to:

- linear elastic analysis of multi-layers foundation structure (mainly two or three-layered structures) [3],
- increase the elastic modulus of unbound granular layers (capping and sub-base or frost resistance) thanks to its stabilisation with hexagonal geogrids and then
- *Boussinesque Equation* which allows for calculation of surface modulus, e.g. E2=100MPa if the deflection (displacement) value under given load is known (thanks to a number of available software) [3].

3. General concept of Carbon footprint

Carbon footprint is an amount of carbon dioxide released into the atmosphere as a result of the activities of a particular individual, organization or community. In relation to civil engineering it is a certain assessed value informing about the grade of CO_2 released into the atmosphere as a result of taking place of all construction component processes. Carbon footprint is measured in kilograms kCO_2e or tones tCO_2e of carbon dioxide equivalent [22]. To assess the total carbon footprint for certain civil engineering technology, e.g. roadway application which is a base of this paper, it is needed to consider and summarise CO_2 emission for all activities related to its main technology process, mainly [4]:

- quarrying source of natural aggregate fill,
- manufacturing all required materials, e.g. cement and other hydraulically bound mixtures, chemical binders and concrete, bitumen and asphalt, geosynthetics,
- transporting and delivering of all above materials from quarried aggregate sources and/or manufacturing plants to the construction site,
- installing, placing, compacting the aggregate and asphalt layers,

and in terms of using any geosynthetics, also:

- tensioning (if required), placing and/or assembling and overlapping them into road construction.

4. The concept of a bespoke Carbon Calculator tool for using MSL compared with associated technologies

When comes to MSL one can achieve the substantially saving in percentage of carbon emissions. This saving is due to both: a reduced thickness of aggregate to be used in the construction when compared with a non-stabilised granular layer (1) and/or a lack of usage of cement or other hydraulically binders that generate a high demand of carbon footprint (2).

In order to evaluate the carbon emissions savings and quantify the environmental benefits of using a MSL, an independent consultants [16] commissioned by the hexagonal geogrid manufacturer and geosynthetic technology leader has developed a bespoke tool, called Carbon Calculator (CC). CC follows an approach contained in [21] and has been mainly launched to estimate the percentage carbon saving available by using an MSL with hexagonal geogrid/s for stabilisation when compared with the construction of a non-stabilised granular layer (1) of an equivalent performance. Recently the CC tool has been updated and modified for estimating the percentage carbon saving available by using a MSL when compared also with a certain type of cement bound mixtures layer (2) of an equivalent performance.

It is also worth to add that CC tool is available to calculate and provide both construction cost and time benefits. To add up, when properly providing the input data concerning initial traffic, its annual growth, targeted design life as well as all needed information about pavement maintenance, rehabilitation and salvage, the outcome of the life cycle cost benefit can be given.

The calculation of carbon savings considers MSL and the associated:

- non-stabilised layer (without the inclusion of geogrid) and/or
- cement and other hydraulically bound mixtures layer

for a given set of input parameters. The assessment process is divided into three main following sections [16]:

1. Determining the embodied carbon within construction materials at their respective "factory gates" [1,17,23,4].
2. Assessing the carbon emissions during the delivery of these materials to the site [4].
3. Assessing the carbon emissions for the completion of construction of the comparable roadway constructions (stabilised, non-stabilised and hydraulically bound) [16,23,4].

There are also three other critical assumptions which carbon footprint calculator is based on. They are taken from the general [11], construction [18] and materials [1,17,20,21,23] point of view [16].

All needed data referring to embodied carbon for the bound granular materials, aggregate, bitumen and grades of asphalt could be provided by [1]. The exampled construction materials – some of them have been used for the calculations related to this paper - are presented in Table 1. What is more, information included in [17] allows compaction operations to be included in the resulting evaluation of carbon footprint for asphalt layers of manufactured and installed asphalt layers (in $kgCO_2/kg$) [4]. In terms of the layers of cement bound mixtures following assumptions had to be done:

- There are three type of cement bound layers taken into consideration in present analysis: sub-base layer of cement bound mixture (C5/6), frost resistance (or sub-base) layer of cement bound mixture (C1.5/2) and capping layer of cement stabilised soils (C0.4/0.5);
- As no embodied carbon values available for cement bound aggregate (mixture) following to [1] then all of above layers have been assumed as cement stabilised soil (in situ soils);
- As [1] (see Table 1) presents only the values at 5% and 8% of cement stabilised soils then cement stabilised soil at 8% has been used as a base for the 2% alternative and a value of 0.04 $kgCO_2/kg$ has been used for both C1.5/2 and C0.4/0.5;
- Following to Table 1 value of 0.06 $kgCO_2/kg$ has been used for C5/6.

Table 1. Inventory of Carbon main summary [1]

Lp.	Materials	*EC - Embodied Carbon* [$kgCO_2/kg$]
1.	Bitumen, general	0,38-0,43
2.	Asphalt, 4% (bitumen) binder content (by mass)	0,059
3.	Asphalt, 8% (bitumen) binder content (by mass)	0,076
7.	Cement stabilised soil @ 5%	0,06
8.	Cement stabilised soil @ 8%	0,082
9.	Fly ash stabilised soil	0,039
10.	Aggregate (gravel or crushed stone)	0,0048
11.	Granite aggregate	0,64
12.	Lime aggregate	0,087
14.	Sand	0,0048
15.	Polypropylene (PP)	4,98

5. Examples of calculating carbon footprint for some variants of a pavement design Mechanically stabilised layer by geogrid (MSL) allows to save both money and time of construction compared to traditional technologies. The incorporation of a MSL in a pavement design provides also a substantial benefits for the environment, allowing for a significant reduction of energy usage and CO_2 emission. To prove and to quantify above well know thesis, one pavement structure with four alternatives of an equivalent performance have been subjected to the analysis. All presented pavement variants were loaded with the same heavy-medium traffic, i.e. category KR4 - according to the *Design Catalogue of Flexible and Semi-rigid Pavement* [19]. The four analysed versions of pavement structures were differentiated only in terms of type and thickness of lower construction strata, i.e. capping and sub-base (or frost resistance, if needed) layers. The set-up of the lower layers has been determined to achieve the required compaction and bearing capacity estimated as deformation modulus E2 (in this case E2≥100 MPa) below the principle pavement structure. This is called as the upper construction layers (which are unbound granular base overlaid by the package of 3 layers of asphalt). This follows again to assumptions and requirements mentioned in [19].

It needs to be noticed that in the case presented below the whole upper structure layers of the pavement, so both unbound granular base layer and asphalt layers were not optimised through the usage of hexagonal geogrid in aggregate base (Pavement Optimisation). The set-up of upper layers of pavement structure has been then adopted as invariable following to the recommendation of [19]. Pavement Optimisation treated as the increasing pavement design life or decreasing its thickness, both asphalt and aggregate base course layers, for a given design life was not subjected to this analyse. So no additional hexagonal geogrid was incorporated in unbound granular base course layer for pavement optimisation process.

The four analysed versions of pavement differentiated by the lower structure layers, i.e.:

1. pavement structure with the lower layers of both non-stabilised unbound granular capping and sub-base layers,
2. pavement structure with the lower layers of non-stabilised unbound granular capping layer and hydraulically bound sub-base layer (cement),
3. pavement structure with the lower layers of both cement bound capping and frost resistance (sub-base) layers, and finally
4. pavement with the lower layers of decreased thicknesses due to stabilised (MSL) both unbound granular capping and sub-base layers,

have been designed with an equivalent performance as to reach the same level of required traffic life. All other assumptions done to the pavement calculations are presented below:

- Heavy-medium traffic: KR4, where the total standard axel load (ESALs100kN) is determined as 7,3mln >N100>2,5mln,
- Subsoil strength: E2≥25 MPa (G4 - according to [19]),
- Length of road section: 1km,
- Width at road level: 6m,
- Road level: +0,58m
- Thickness of top soil to be excavated in all cases: ~0,30m,
- Maximum excavation for the pavement structure no. 1 with both non-stabilised unbound granular capping and sub-base layers: -0,46m=0,58-1,04m,
- Maximum excavation for the pavement structure no. 2 with non-stabilised unbound granular capping layer and cement and other hydraulically bound sub-base layer: -0,40m=0,58m-0,98m,
- Maximum excavation for the pavement structure no. 3 with both cement and other hydraulically bound capping layer and frost resistance layers: -0,29m=0,58m-0,87m,
- Maximum excavation for the pavement structure no. 4 with decreased thicknesses due to stabilised (MSL) both unbound granular capping and sub-base layers: -0,30m=0,58-0,88,
- Width at deepest applicable formation level: ~9,75m,
- Width at deepest applicable formation level due to MSL reduction: ~9,25m,
- Construction site: assumed: Zielona Gora, Lubuskie voivodeship, West Poland
- Distances between have been assumed as: asphalt plant and construction site: 10km; quarry (aggregate/fill location) and construction site: 25km; cement bound mixture and construction site, the same as above, i.e.: 25 km; construction site and the place of disposal (landfill) of

unacceptable existing soils and top soil: 10km; distribution point (geogrid supply) and construction site: 320km.

To sum up, all of the variants of the whole pavement structure have been designed and assessed as follows:

Pavement structure variant no. 1: with both non-stabilised unbound granular capping and sub-base layers (total thickness: 104cm)

- 4cm: wearing course of AC
- 6cm: binder course of AC
- 10cm: asphalt base course of AC
- 20cm: base course of unbound granular crushed aggregate mixture C90/3
  E2=100MPa
- 24cm: sub-base layer of an unbound granular aggregate mixture of CBR≥60%
- 40cm: capping layer of an unbound granular aggregate mixture of CBR≥20%
- Geotextile as separation and filtration
  Existing subgrade: E2=25MPa

Pavement structure variant no. 2: with non-stabilised unbound granular capping layer and cement bound sub-base layer (total thickness: 98cm)

- 4cm: wearing course of AC
- 6cm: binder course of AC
- 10cm: asphalt base course of AC
- 20cm: base course of unbound granular crushed aggregate mixture C90/3
  E2=100MPa
- 18cm: sub-base layer of cement bound mixture: C5/6
- 40cm: capping layer of an unbound granular aggregate mixture of CBR≥20%
- Geotextile as separation and filtration
  Existing subgrade: E2=25MPa

Pavement structure variant no. 3: with both cement bound capping and frost resistance layers (total thickness: 87cm)

- 4cm: wearing course of AC
- 6cm: binder course of AC
- 10cm: asphalt base course of AC
- 20cm: base course of unbound granular crushed aggregate mixture C90/3
  E2=100MPa
- 22cm: frost resistance (or sub-base) layer of cement bound mixture: C1.5/2
- 25cm: capping layer of cement (or lime) stabilised soils: C0.4/0.5
  Existing subgrade: E2=25MPa

Pavement structure variant no. 4: with decreased thicknesses of both unbound granular capping and sub-base layer due to MSL implementation (total thickness: 88cm)

- 4cm: wearing course of AC
- 6cm: binder course of AC
- 10cm: asphalt base course of AC
- 20cm: base course of unbound granular crushed aggregate mixture C90/3
  E2=100MPa
- 18cm: sub-base layer of an unbound granular aggregate mixture of CBR≥60% stabilised by hexagonal geogrid (MSL)
- 30cm capping layer of an unbound granular aggregate mixture of CBR≥20% stabilised by hexagonal geogrid (MSL)
  Existing subgrade: E2=25MPa

The final analysis has been done only in terms of carbon footprint usage followed to carbon calculator (CC) called *Spectra Value Tool v.6.03* and is presented hereinafter. Due to diverse and variable costs of construction materials depending on very local factors and circumstances, as: country location, distance from the quarried aggregate source, manufacturing plant of cement bound mixtures and asphalt plant to the project site, local material and operational prices, etc., the calculations in terms of construction cost and time savings have not been carried out within this analysis. There has not been done the analyse due to the life cycle cost benefit either.

6. Results

As a result of the carried out calculations by using carbon calculator (CC) tool following amounts of carbon usage [kgCO_2e] have been achieved (Table 2) and juxtaposed with all four analysed pavement structures (Figure 2).

Table 2. Calculation results of carbon emissions

Measured parameter	**Pavement structure no. 1 (PS1):** with two non-stabilised unbound granular layers	**Pavement structure no. 2 (PS2):** with one non-stabilised unbound granular layer & one cement bound layer	**Pavement structure no. 3 (PS3) :** with two cement bound layers	**Pavement structure no. 4 (PS4):** with two stabilised unbound granular layers by hexagonal geogrid (MSL)
Carbon emission during the whole construction time [kgCO_2e]	1.354.964	1.399.560	1.451.593	1.114.199

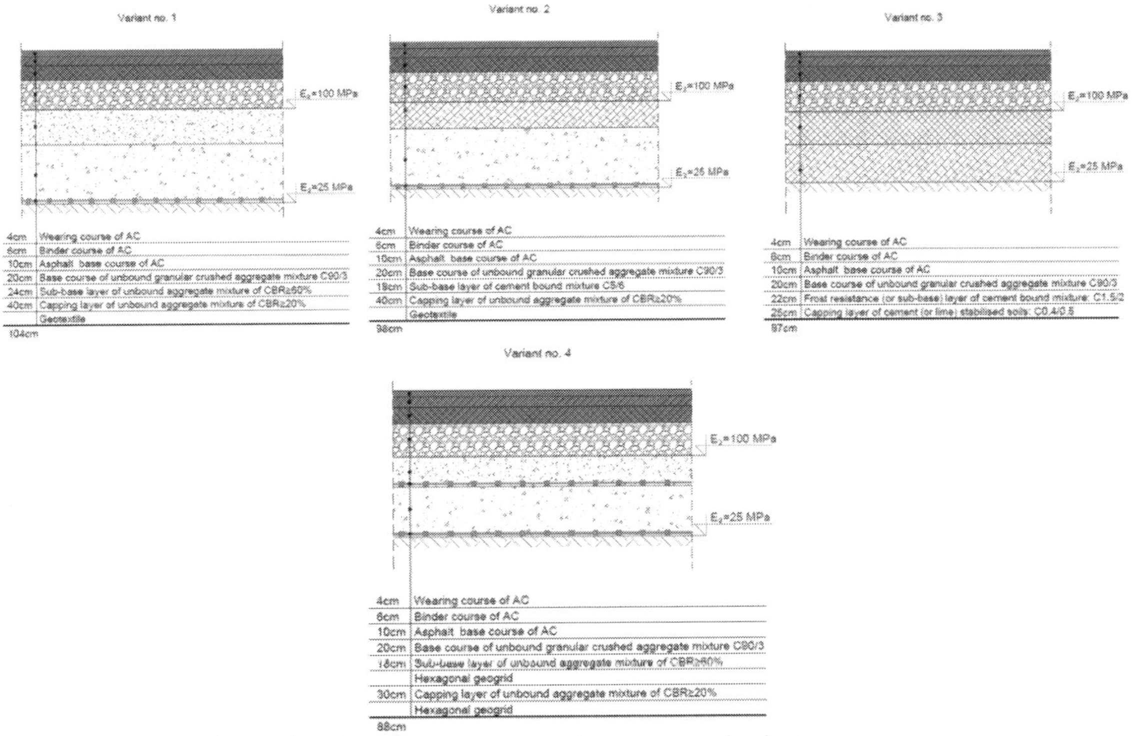

Figure 2. Four variants of analysed pavement structures summarised

The savings of total carbon emission of pavement structure variant no. 4 (PS4) - with decreased thicknesses due to MSL implementation - compared to 3 other traditional pavements used in road construction have been then achieved (Table 3) as 17.8% to PS1, 20,4% to PS2 and 23,2% to PS3.

Table 3. Calculation results of carbon emission savings

Measured parameter	**Pavement structure no. 1 (PS1):** with two non-stabilised unbound granular layers	**Pavement structure no. 2 (PS2):** with one non-stabilised unbound granular layer & one cement bound layer	**Pavement structure no. 3 (PS3):** with two cement bound layers
Carbon emission savings by using **pavement structure no. 4 (PS4)** compared to others [$kgCO_2e$]	240.765	285.361	336.526
Carbon emission savings by using **pavement structure no. 4 (PS4)** compared to others [%]	17,8	20,4	23,2

7. Conclusions and discussion on further development

Mechanically stabilised layer (MSL) by geogrid allows to save cost and time of construction (economical aspect) in comparison to traditional technologies in road construction, such as: granular unbound layers which are non-stabilised by geogrids or cement and other hydraulically bound mixtures. In addition, MSL also considerably endorses an environmental aspect being a significant and successful alternative to above traditional solutions in common use in lower layers of road construction, i.e.: sub-base, frost resistance and/or capping layers. Thereby MSL has a positive impact on the environment, allowing for a significant reduction of energy usage and carbon footprint. Calculations and the analyse of four different versions of typical pavement structure trafficked up to 7,3 mln ESALs 100kN (differentiated in terms of type and thickness of lower construction layers), clearly showed the visible savings of total carbon emission for using a PS4 (MSL) version compared to others.

Presented comparison of total carbon emission shows a significant benefit of MSL against cement bound layers (CBL) - savings of 23.2% in favor of MSL vs. CBL have been achieved. Furthermore, due to excluding the quarrying of aggregate/soil in the cement-bound layers in the calculations carried out by Carbon Calculator (CC), presented results in terms of embodied carbon seem to be slightly underestimated. So it can be expected to achieve higher amount of carbon emission savings obtained by structure incorporating MSL compared to version with cement bound layers.

Some plastics have been polluting our environment and they cause ecological problems for future generations. Only about 20% of the total plastic produced worldwide is being recycled or incinerated. However geosynthetics, specifically geogrids, which are the family of the plastic products can be used towards the protection of environment. Correctly and properly manufactured and then utilised, they can support a sustainable development. Taking into account the presented results and considering the further sustainable development in roadway applications, carbon reduction calculation should become a standard check procedure at design stage of any kind of pavement applications, specifically when considered in the lower layers of pavement.

References

[1] Hammond G and Jones C 2011 *Inventory of Carbon & Energy (ICE)* Version 2.0 Sustainable Energy Research Team, Dept. of Mechanical Engineering, University of Bath,

www.bath.ac.uk/mech-eng/sert/embodied
[2] ISO 10318-1:2015+A1:2018 Geosynthetics – Part 1: Terms and definitions
[3] Judycki J 2014 *Analysis and designing of flexible and semi-rigid pavement* (*Analizy i projektowanie konstrukcji nawierzchni podatnych i półsztywnych)*, Wydawnictwo Komunikacji i Łączności, Warszawa, Poland
[4] Kawalec J, Golos M, Mazurowski P 2018 Environmental aspects of the implementation of geogrids for pavement optimisation, Resilient and Safe Road Infrastructure *IOP Conference Series: Materials Science and Engineering* **356** 012018
[5] Kawalec J, Kwiecień S, Pilipienko A, Rybak J 2017 Application of crushed concrete in geotechnical engineering – selected issues World Multidisciplinary Earth Sciences Symposium *IOP Conference Series: Earth and Environmental Science* **95** 022057
[6] Kwon J, Wayne M, Norwood G J and Tingle J S 2012). The Implementation of Findings From Accelerated Pavement Testing in Pavement Design And Construction Practice, Advances in Pavement Design through Full-scale Accelerated Pavement Testing – Jones, Harvey, Mateos & Al-Quadi (Eds.), Taylor&Francis Group, London, UK
[7] Lee H S 2017 Recommended Practice for Incorporating Geogrids in ME Pavement Design *Applied Research Associates*, Champaign, USA
[8] Lees A 2019 The bearing capacity of a granular layer on clay *Proceedings of the Institution of Civil Engineers, Geotechnical Engineering* In press
[9] Lees A, Matthias P 2019 Bearing capacity of a geogrid-stabilised granular layer on clay *Technical papers of Ground Engineering GE* (geplus.co.uk)
[10] Mazurowski P, Buckley J, Kawalec J 2019 Modification of mechanistic-empirical pavement design method to incorporate the influence of hexagonal stabilisation geogrids *AAPA International Flexible Pavements Conference*, 18th, 2019, Sydney, Australia
[11] Powell W D, Potter J F, Mayhew H C and Nunn M E 1984 The structural design of bituminous roads (No. LR 1132 Monograph)
[12] White J D 2014 *In Situ Performance Verifiaction of Geogrid-Stabilized Aggregate Layers*: Dewitt County, Cuero, Texas – Schlinke Rd., Geomatters Technologies Inc., Ames, USA
[13] White J D2014 *Preliminary APLT Comparision of Ev Strain Modulus and In-situ Resilient Modulus/Permanent Deformation*, Geomatters Technologies Inc., Ames, USA
[14] American Association of State Highway and Transportation Officials *AASHTO* 1993 *Guide for Design of Pavement Structures*, Washington, USA
[15] American Association of State Highway and Transportation Officials *AASHTO* 2009 *Standard Practice for Geosynthetic Reinforcement of the Aggregate Base Course of Flexible Pavement Structures* AASHTO R 50-09, Washington, USA
[16] Coffey Geotechnics Ltd. *Tensar Carbon Calculator Assumptions – based on International Carbon Calculator V4.0*, 701.2AB_M_001B_Assumptions for Carbon CalculatorV4.0
[17] DEFRA Conversion Factors 2012 UK Government Department for Environment, Food and Rural Affairs, https://www.gov.uk/government/publications/greenhouse-gas-reporting-conversion-factors-2017
[18] Manual of Contract Documents for Highway Works 2016 Vol. Specification for Highway Works Series 600 Earthworks, Amendment – February 2016; Table 6/1 and Table 6/4
[19] Design Catalogue of Flexible and Semi-rigid Pavement, Poland 2014 (*Katalog Typowych Konstrukcji Nawierzchni Podatnych i Półsztywnych*), Załącznik do zarządzenia nr 31 GDDKiA z dn. 16.06.2014r
[20] Plastics Europe Eco-profiles (2005) http://www.plasticseurope.org/application/files/7415/1747/5136/eco-profile_methodology_version2-0_April2011.pdf.
[21] Publically Available Specification PAS 2050 2011 *Specification for the assessment of the life cycle greenhouse gas emissions of goods and services*, http://shop.bsigroup.com/upload/shop/download/pas/pas2050.pdf

[22] The Cut Costs and Carbon Calculator, https://www.carbontrust.com/resources/tools/cut-costs-and-carbon-calculator-catering/
[23] Waste and Resources Action Programme (WRAP) CO2 Emission Estimator Tool 2010 http://www.wrap.org.uk/search/gss/co2%20emission%20estimator%20%20

IOP Conf. Series: Materials Science and Engineering 1260 (2022) 012024 doi:10.1088/1757-899X/1260/1/012024

Applications of Microbially Induced Calcium Carbonate Precipitation on Non-Woven Geotextiles

E Duman[1], Y E Sarıçiçek[1], O Pekcan[1], R Gurbanov[2], A G Gözen[3]

[1] Civil Engineering Department, Middle East Technical University, 06800 Çankaya/Ankara, Turkey
[2] Bioengineering Department, Bilecik Seyh Edebali University, Bilecik, Turkey
[3] Biological Sciences Department, Middle East Technical University, 06800 Çankaya/Ankara, Turkey
Email:opekcan@metu.edu.tr

Abstract. Recently, an innovative ground improvement method known as microbially induced calcium precipitation (MICP) has been proposed as an alternative to conventional ground improvement techniques. Having numerous advantages such as being environmentally friendly and promoting sustainable solutions, MICP has been studied by researchers from various civil engineering sub-disciplines to solve problems related to crack propagation of concrete, insufficient soil strength characteristics, soil erosion, asphalt cracks, etc. This study aims at moving MICP one step further by introducing a hybrid approach that combines MICP and geotextiles to produce a good quality product for reliable field use. Within this scope, MICP is applied on non-woven geotextiles. Physical (imaging) and mechanical (puncture resistance) properties of non-woven geotextiles subjected to MICP treatments under soaking conditions are investigated. Puncture resistance tests are performed and the details of laboratory work including specimen preparation and MICP treatments are presented in this paper. This study, together with its implications for future works, also highlights the next steps to be taken for properly using the MICP method when other geosynthetics are used.

1. Introduction

Microbially induced calcite precipitation (MICP) is a recently proposed innovative ground stabilization method that harnesses the bio-geo-chemical processes to plug calcium carbonate within the soil matrix [1]. The method utilizes urea hydrolysis induced by ureolytic bacteria, to precipitate calcium carbonate in the pores of the soil. The emergence of MICP stems from the necessity of more environmentally friendly solutions to replace conventional ground improvement methods such as injection of micro-fine cement, epoxy, and silicates [2] since artificial grouts have toxic constituents that can affect the quality of groundwater and soil. Moreover, the use of chemical grouts is prohibited in some countries [3].

In a sustainability-driven age, the importance of new improvement techniques like MICP have become more apparent for geotechnical engineers. Therefore, MICP is currently the subject of an increasing number of studies in the geotechnical community. To date, researchers have investigated the precipitation observation [4,5,6,7], the effect of MICP on strength and deformation [8,9,10,11], hydraulic [12,13,14], and thermal [15,16,17] characteristics of sands, mitigation of soil erosion [18,19], and liquefaction [20,21] problems. The outcomes in the literature are promising, yet most of them are still limited to laboratory scale. These properties changed by the precipitation of calcium carbonate within the pore medium of the material. Ureolytic bacteria hydrolyses the urea and produces ammonium and carbonate ions (1). In the presence of calcium ions, calcium carbonate precipitates (2) in the pores.

$$CO(NH_2)_2 + 2H_2O \rightarrow 2NH_4^+ + CO_3^{2-} \quad (1)$$

$$Ca^{2+} + CO_3^{2-} \rightarrow CaCO_3 \quad (2)$$

To open new horizons and introduce a possibility of a novel application, the researchers of this study present a hybrid approach that combines MICP and geotextiles. Geotextiles are frequently used in geotechnical works for separation, filtration, drainage, reinforcement and protection purposes [22]. Geotextile applications include but not limited to contact points with drainage layers in waste landfills, subgrade and coarse aggregate layers in paved roads. These applications inherently go hand in hand with deformation and eventually puncture threats for geotextiles. These circumstances compromise the filter, separator, and protection functions as holes will reduce the strength and effectiveness [23]. Therefore, puncture resistance is an important mechanical property of a geotextile.

In this study, the effect of MICP on puncture resistance is investigated; the methodology and results are presented.

As an initial step, the authors herein focus on whether MICP can improve the puncture resistance of geotextiles or not. Since MICP is a relatively new field, not many studies are present, which focus on the combination of MICP and geosynthetics. Recently, bio-cementation has been proven to be effective in improving the pullout resistance of a geogrid reinforced system. However, this study did not focus on the effect of MICP on the mechanical properties of geogrids [24]. It is important to assess this effect since geosynthetics will be buried in the soil for a certain lifetime. Either positive or negative, any change in the mechanical properties of geosynthetics due to the bacteria should be foreseen before any MICP application.

2. Materials and Methods

2.1. Geotextile Specimens

Samples taken from geotextile products were formed into 12 x 12 cm specimens according to ASTM D4833 [25] and placed on plastic containers with pre-drilled three drainage holes, as seen in Figure 1. For comparison, two different commercial non-woven geotextile products were used, one named TP and the other one being AP. Five samples from each geotextile roll were prepared and exposed to treatments. Each sample was paired with control specimens on which no treatments were made. Specimens were labeled as GeotextileType_C# (control specimens) and GeotextileType_SP# (MICP treated specimens). Also, only distilled water treated (GeotextileType_W#) and biological-cementation media treated (GeotextileType_TR#) specimens were tested to assess change in puncture resistance is due to the bacteria.

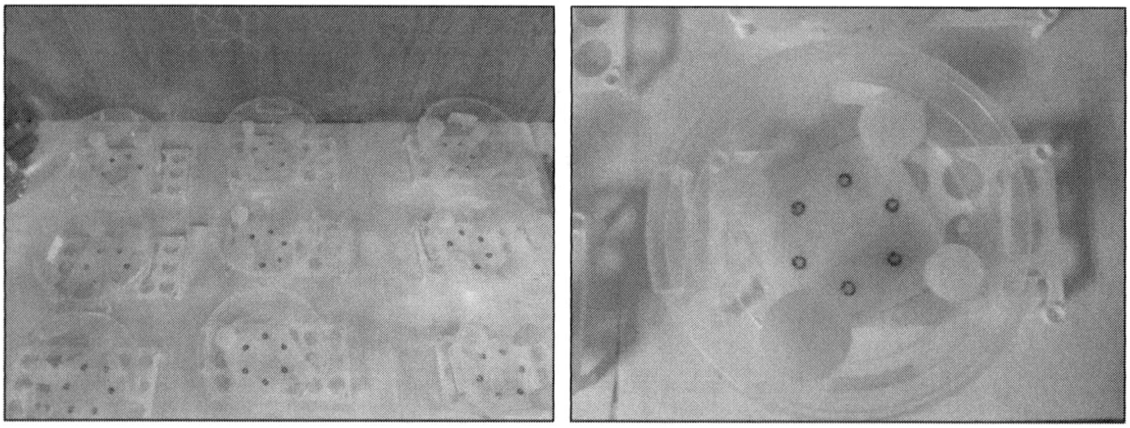

Figure 1. Geotextile specimens in plastic containers prior to treatments

2.2. Bacteria Growth Conditions

Sporosarcina pasteurii (ATCC 11859) was used as urease positive bacteria. The bacteria were initially precultivated in ammonium-yeast extract (NH_4-YE) medium (ATCC 1376) at 30 °C with a shaking incubator at 200 rpm, for approximately 24 hours. The growth medium consists of 20 g/L yeast extract, 10 g/L ammonium sulfate [$(NH_4)_2SO_4$], 0.13 M Tris buffer (pH = 9.0). Precultivated bacteria medium was inoculated in fresh growth media (1% v/v) and then the bacteria were recultivated with the same conditions for 24 hours. The final solution was diluted to the optical density (OD_{600}) of approximately 1.0 and centrifuged at 4000 g for 15 minutes in 15 ml sterile tubes. Following that, the supernatant was removed and pellets were stored at +4 °C for 14 days prior to use.

2.3. Biological and Cementation Media

Biological, representing the medium used for the introduction of bacteria to the samples, and cementation media constituents are shown in table 1. Calcium chloride dihydrate ($CaCl_2 \cdot 2H_2O$) was not included in the biological medium in order to avoid the possibility of early precipitation. Furthermore, sodium bicarbonate ($NaHCO_3$) was not preferred in cementation medium as sodium bicarbonate could react with calcium chloride dihydrate. All the constituents were autoclaved together at 121 °C for 15 min, except urea, as it may be decomposed under high temperatures. Instead, urea was sterilized through a 0.22 μm filter. Following the end of autoclaving and filtering stages, the components were mixed in bottles under aseptic conditions.

Table 1. Chemical components of biological and cementation media

Components	Biological medium concentrations (M)	Cementation medium concentrations (M)
Urea	1.0	1.0
$CaCl_2.2H_2O$	-	0.5
Nutrient Broth	3 g/L	3 g/L
NH_4Cl	0.187	0.187
$NaHCO_3$	0.025	-

2.4. Experimental Procedures

2.4.1. Cementation Treatments

Plastic weights were placed on each corner of the geotextiles to prevent floating when the solutions were added into the container. The overall procedure consisted of two phases. In the first phase, specimens were soaked in 45 ml of biological solution for 16 hours and in the end, the solution was allowed to drain out from the holes. Then, 50 ml of cementation solution was added to container and specimens were soaked in the solution for 24 hours for the first cementation treatment. Starting from the second cementation treatment, the volume of cementation medium for each specimen was determined to be 20 ml which continued initially for four days. At the fifth day, 15 ml of bacteria in biological media were added to each of the specimens in order to maintain the bacterial activity. Cementation medium treatments (20 ml) continued for four more days. After each treatment, the solutions were repeatedly drained out from the container. Finally, specimens were kept at 25 °C for six days prior to testing. The same treatment schedule is also performed with only distilled water and a biological and cementation medium without bacteria injection.

Table 2. Treatment schedule (bs: biological solution, cs: cementation solution)

# Day	Treatment
Day 1	45 mL bs
Day 2	50 mL cs
Day 3	20 mL cs
Day 4	20 mL cs
Day 5	20 mL cs
Day 6	20 mL cs
Day 7	15 mL bs
Day 8	20 mL cs
Day 9	20 mL cs
Day 10	20 mL cs
Day 11	20 mL cs

2.4.2. Puncture Resistance Test

The procedure described in ASTM D4833 [24] was followed. Initially, the specimen was mounted to the holding plates by making sure that the specimen extends from the clamps, as shown in Figures 2, 3, and placed at the center of the testing machine. Finally, it was punctured with a rate of 300 mm/min.

Figure 2. Geotextile specimen clamped at holding plate

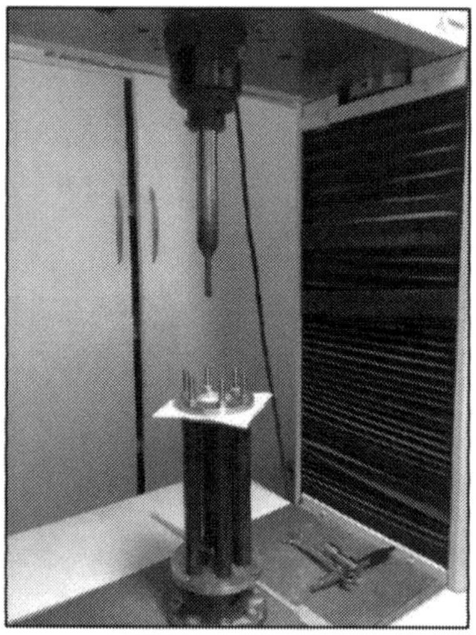

Figure 3. Holding plate placed at testing machine

3. Results and Discussions

Following the end of the treatment procedure, the accumulation of calcium carbonate particles on top of the geotextile specimens, in the form of a crust-like appearance, was clearly visible, as seen in Figure 4. The formation of this layer head started immediately after the first cementation treatment. As the cementation treatment continued, the thicknesses of the crust layers were observed and they increased with time. During the test, the samples were still humid although they were kept at 25 °C for six days. Pairs of treated-nontreated (control) specimens were tested from each geotextile type.

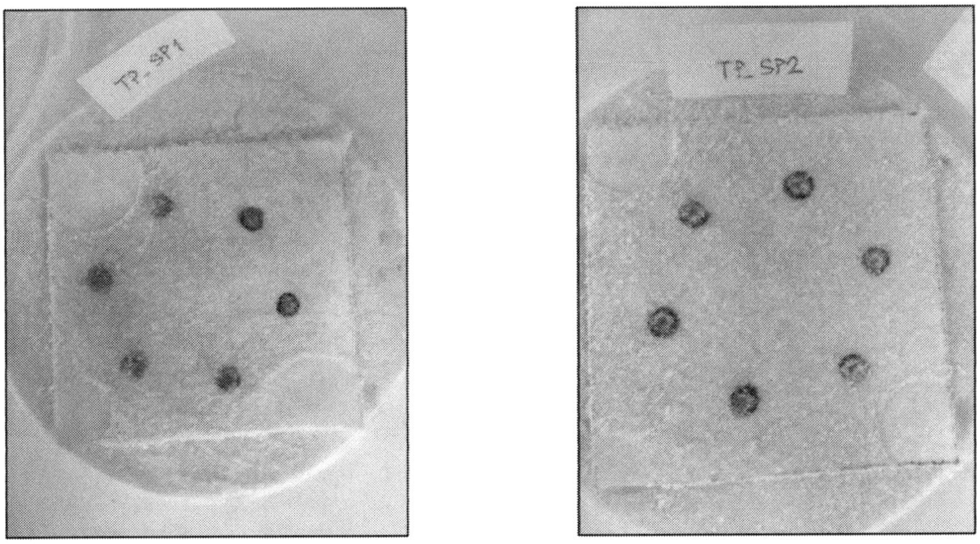

Figure 4. Geotextile specimens after cementation treatments

In both of the geotextile types, the decrease in puncture resistances was observed unexpectedly. In geotextile AC, the average puncture resistance of MICP treated specimens is 455 N; in contrast, it is 581 N for the control group. There is a significant decrease in resistance. The average puncture resistance in TP is 529 N for MICP treated case with respect to 605 N of its control group. At this point, it is important to underlie that in one of the geotextile types (AC), the puncture resistance of two control samples varied as much as 13% (Table 3), although these samples prepared from the same part of the geotextile. Therefore, comparing the treated case of TP to its control group might be deceptive. But even if the only distilled water treated and MICP cases were compared, there is no difference in the resistance. At least for the TP geotextile, MICP might not have a positive effect on puncture resistance.

Table 3. Test results for geotextile AC

Treatment type#	Specimen	Puncture Resistance (N)
Control (No treatment)	AC_C3	587
	AC_C4	614
	AC_C5	544
Distilled water only	AC_W1	587
	AC_W2	593
Chemical treatment only	AC_TR1	691
	AC_TR2	589
MICP treatment	AC_SP1	537
	AC_SP2	502
	AC_SP3	399
	AC_SP4	396
	AC_SP5	443

Table 4. Test results for geotextile TP

Treatment type#	Specimen	Puncture Resistance (N)
Control (No treatment)	TP_C5	605
Distilled water only	TP_W1	511
	TP_W2	520
Chemical treatment only	TP_TR1	637
MICP treatment	TP_SP1	614
	TP_SP2	414
	TP_SP3	543
	TP_SP4	560
	TP_SP5	514

Some of the specimens were prepared from the same location to avoid the heterogeneous nature of the geotextile to a certain extent. The force vs. displacement graph of pairwise prepared specimens is shown in Figure 5. One interesting finding from these graphs is that MICP treated specimens showed lesser resistance and failed at lower displacements than control groups. Only TP_C4 and TP_SP4 were failed at similar vertical displacement. It could be inferred that the crust layer might not have an effect on puncture resistance. Also, after MICP treatments, the material might be lost its ductility, which might have a negative effect on the resistance.

A similar trend was not observed for distilled water only and MICP treated cases. TP_SP3 has higher puncture resistance and failed at larger displacement compared to TP_W1. TP_W2 and TP_SP2, AC_W2, and AC_SP2 were again failed at similar vertical deflections. Unfortunately, chemical treatment only cases do not have any pairs to make any comparison.

Distilled water treated specimens were showed similar puncture resistance compared to the control group in AC type geotextile. However, it is the opposite for TP type geotextile. This difference might be associated with the heterogeneity of the material. It is hard to assess the effect of water treatment and chemical treatment on puncture resistance due to the limited number of tests. Also, not having paired-wise specimens is making any comparison harder.

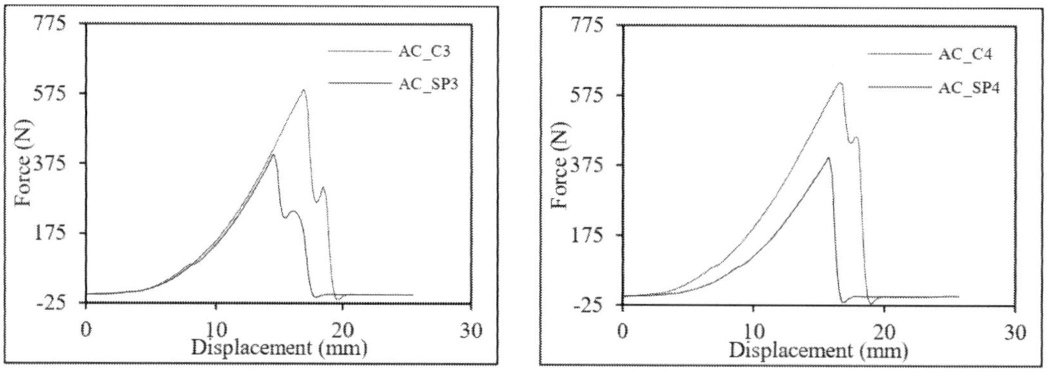

a

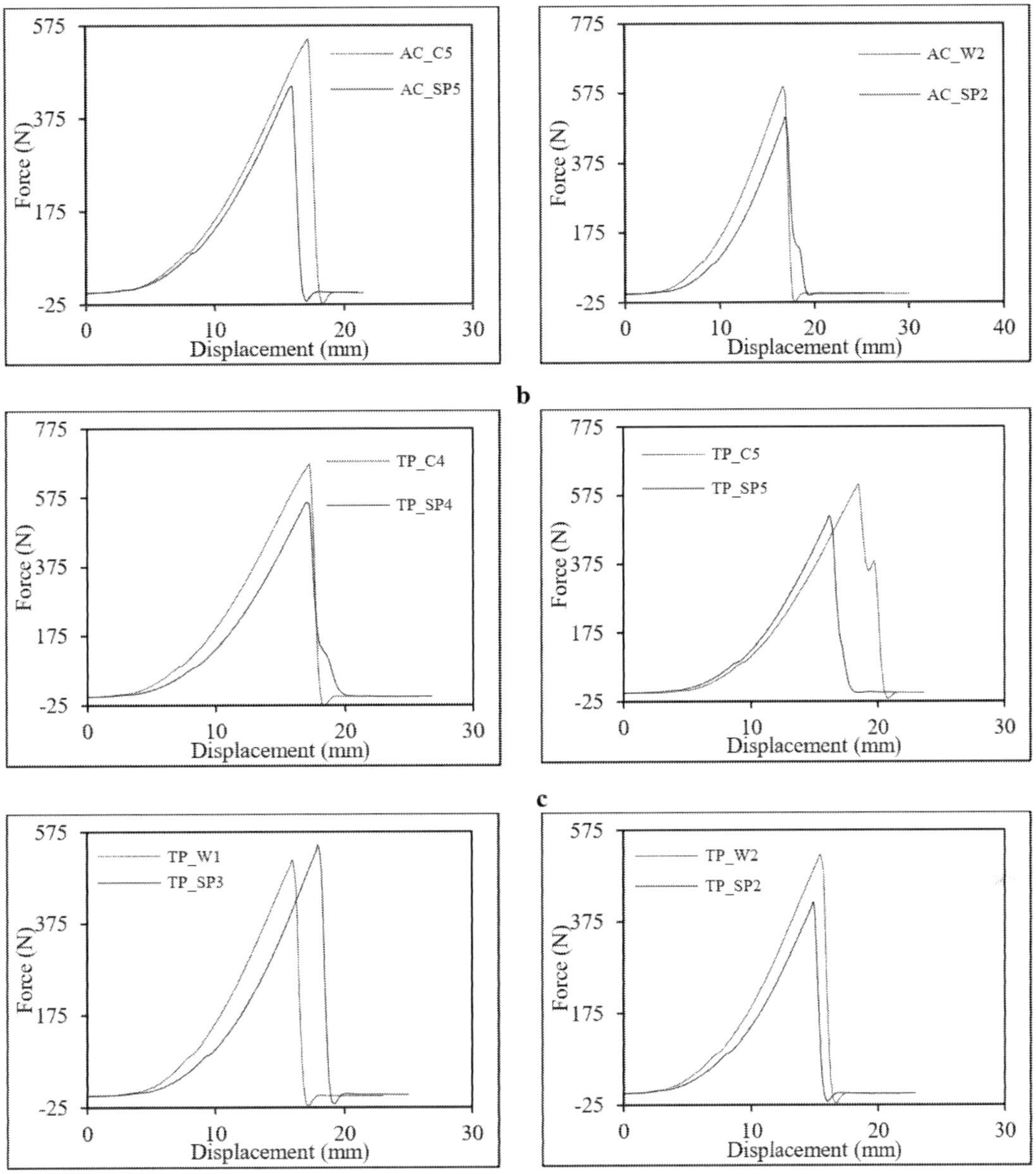

d

Figure 5. Force (N) vs. displacement (mm) graphs for paired wise prepared specimens: **a** AC_C3 and AC_SP3, AC_C4 and AC_SP4 **b** AC_C5 and AC_SP5, AC_W2 and AC_SP2 **c** TP_C4 and TP_SP4, TP_C5 and TP_SP5 **d** TP_W1 and TP_SP3, TP_W2 and TP_SP2

4. Conclusion

This paper herein represents the initial work on how the MICP treatments affect the puncture resistance of two different geotextiles. This study in this sense, is one of the first studies bringing the concept of using MICP and geosynthetics together. Initial findings showed that MICP has no positive effect on the resistance. MICP treated geotextiles tend to have lesser puncture resistance and fail at lower vertical displacements. For assessing any effect of chemical treatment or distilled water treatment, the number of specimens (treated with water only and chemicals) should be increased in further studies.

References

[1] Stocks-Fischer S, Nakamura S, Galinat J K, Bang S S 1999 *Soil Biology and Biochemistry* **31** pp 1563-71
[2] DeJong J T, Mortensen B M, Martinez B C, Nelson D C 2010 *Ecological Engineering* **36** pp 197–210
[3] Karol R H 2003 *Chemical Grouting and Soil Stabilization, Revised and Expanded* (New York: Marcel Dekker)
[4] Gurbuz A, Sarı Y D, Yuksekdag Z N 2015 *Geomicrobiology Journal* **32** pp 853-59
[5] Saricicek Y E, Gurbanov R, Pekcan O, Gozen A G 2019 *Geomicrobiology Journal* **36** pp 42–52
[6] Wang Y, Soga K, DeJong J T, Kabla A J 2019 *Géotechnique* **69** pp 1086–94
[7] Mujah D, Cheng L, Shahin M A 2019 *Journal of Materials in Civil Engineering* **31** pp 1–10
[8] DeJong J T, Fritzges M B, Nüsslein K 2006 *J. Geotech. Geoenviron. Eng.* **132** pp 1381–92
[9] Feng K, Montoya B M 2015 *J. Geotech. Geoenviron. Eng.* **142** pp 1–9
[10] Terzis D, Laloui L 2019 *Acta Geotechnica* **14** pp 639–56
[11] Lin H, Suleiman M T, Brown D G, Kavazanjian E 2015 *J. Geotech. Geoenviron. Eng.* **132** pp 1–13
[12] Rowshanbakht K, Khamehchiyan M, Sajedi R H 2016 *Ecological Engineering* **89** pp 49–55
[13] Hataf N, Baharifard A 2019 *Geomicrobiology Journal* **37** pp 147–58
[14] Li D, Tian K, Wu Y, Nie K, Zhang S 2018 *Construction and Building Materials* pp 251–62
[15] Venuleo S, Laloui L, Terzis D, Hueckel T, Hassan M 2016 *Géotechnique Letters* **6** pp 39–44
[16] Wang Z, Zhang N, Ding J, Li Q, Xu J 2020 *International Journal of Heat and Mass Transfer* **147** 118899
[17] Martinez A, Huang L, Gomez G 2019 *Géotechnique Letters* **9** pp 15–21
[18] Zomorodian S M A, Ghaffar H, O'Kelly B C 2019 *Aeolian Research* **40** pp 34–41
[19] Jiang N, Soga K, Kuo M 2016 *J. Geotech. Geoenviron. Eng.* **143** pp 1–13
[20] Zhiguang H, Xiaohui C, Qiang M 2016 *Earthquake Engineering and Engineering Vibration* **15** pp 673–79
[21] Zhang X, Chen Y, Liu H, Zhang Z, Ding X 2020 *Soil Dynamics and Earthquake Engineering* **129** 105959
[22] Koerner R M 2005 *Designing with Geosynthetics* (New Jersey: Prentice Hall)
[23] Koerner R M 2016 *Geotextiles From Design to Applications* (Duxford: Woodhead Publishing)
[24] Gao Y, Hang L, He J, Zhang F, Van Paassen L *Geotextiles and Geomembranes* **49** pp 646-56
[25] ASTM 2013 *Standard Test Method for Index Puncture Resistance of Geomembranes and Related Products* (Pennsylvania: ASTM International)

IOP Conf. Series: Materials Science and Engineering 1260 (2022) 012025 doi:10.1088/1757-899X/1260/1/012025

Mechanical Improvement of a Reinforced Sand with Açaí (*Euterpe oleracea*) Fibers

M M Lopes[1], M D T Casagrande[1]

[1] Department of Civil and Environmental Engineering-FT, University of Brasilia, Brasília, Brazil.

Email: machadomlopes@gmail.com, mdtcasagrande@unb.br

Abstract. The growing research for environmental sustainability applied in technological engineering works allied to the growth generation of solid waste problematics guided this research. Searching for new locational alternatives for these residues is becoming more and more economically viable and giving new uses to these materials has proved to be an effective way of reusing them. With this in mind, the present work seeks to evaluate the use of natural açaí fibers to reinforce soils in geotechnical works. To carry out this evaluation and investigate the physical-mechanical behavior of a sand reinforced with açai fibers, were proposed direct shear tests with soil-fiber composites in the proportions of 0.5% and 1.0% of fiber in relation to soil dry weight and two percentages of relative density, 50% and 75%. The results demonstrate that the soil-fiber composites have greater peak and post-peak resistance than pure soil which are satisfactory from the geotechnical point of view, fulfilling its role mainly in acting on the shear resistance of the soil, opening possibilities for the study of application in geotechnical works such as landfills under soft soils and with more studies, for paving.

1. Introduction

The growing generation of solid waste, despite being a real challenge for several professionals, has made possible the use of new materials in the most varied areas of knowledge as a way to mitigate the impact generated by conventional known forms of final disposal, such as landfills and incinerators. In light of this problem, several researches have been carried out on the topic of new geotechnical materials and practical cases are already found around the world.

The use of new geotechnical materials in engineering works is based not only on the need to identify new destinations for these materials, but also on the fact that commonly the physical-mechanical properties of the available soils are insufficient for their applications, which makes necessary the reinforcement of these soils.

Is possible to improve the soil properties using different techniques as mechanical compaction, physical-chemical stabilization and soil reinforcement with the inclusion of resistant elements as geotextiles, fibers, grills, strips [1].

The reinforcement of soils technique refers to antiquity which were used by our ancestors mixing soil with roots, logs or wool on civil works [2]. These techniques were forgotten by unknown reasons until the XX century where were rescued on sixties and the elements related to their uses were studied and understood [2]. Soil reinforcement consists of inserting certain material in another in order to increase its properties. The concept of reinforced soil was developed in the 1960s by the French engineer Henri Vidal. The engineer patented one of his discoveries, the "reinforced earth",

where he proved that the soil shear strength improved when introducing a reinforcement element in the soil matrix [3].
The inclusion of fibers can be a great technique of improvement, depending on the interactions between the soil and fibers, due to the possibility to reduce cracking as well as the increase of soil ductility. For the fibers to perform is usually necessary the presence of certain granular fraction [4].
The fibers used for soil reinforcement can be of natural or synthetic origin, where this last group is currently more applied. However, the production of synthetic fibers has a large generation of waste and is linked to the petrochemical industry, which has been showing depletion of its resources. These factors, combined with a growing awareness of the need for sustainable development, bring perspective and motivation for the use of fibers natural.

The vegetable fibers use is relevant when compared to synthetics, those natural fibers has shown to be economically viable, have great availability, is easy to find and to handling [5,6,7,8]. The natural fibers can be obtained from different parts of plants. There are fibers that can be extracted from the stalk as jute, flax and hemp; you can extract fibers of leaves as sisal and curauá, from seeds as cotton and coconut fibers are obtained from the fruit [8,9].

However, plant fibers are materials that in natural environments have dimensional variations that influence the adhesion characteristics of the soil-fiber composite due to changes in temperature and moisture content [5]. In addition, vegetable fibers are biodegradable, unlike synthetic fibers, which are inert. These are the main factors that hinder the diffusion of the use of these materials as soil reinforcement in earth works.

In this context, the present work seeks to evaluate the physical-mechanical properties of an area reinforced with açaí fibers randomly distributed in different relative densities, being able to potentiate the use of soil-vegetable fiber mixtures in engineering works.

2. Experimental Program

The established testing program has as main objective to investigate and identify the effect of the addition of açaí fibers on the mechanical properties of a sandy soil, as well as the influence of the variation in the relative density of the sand being divided into two stages: The first stage investigated the physical properties and characterization of the materials used in the research and in the second stage, the study of the mechanical behavior of the materials was carried out through laboratory tests, encompassing Direct Shear tests with soil-fiber composites.

2.1. Materials

2.1.1. Sandy Soil.

The sandy soil (Figure 1) used in the present study, after being collected, were prepared according to the procedures described in the Brazilian Standard - NBR-6458 [4], being packed in plastic drums of 200 liters capacity.

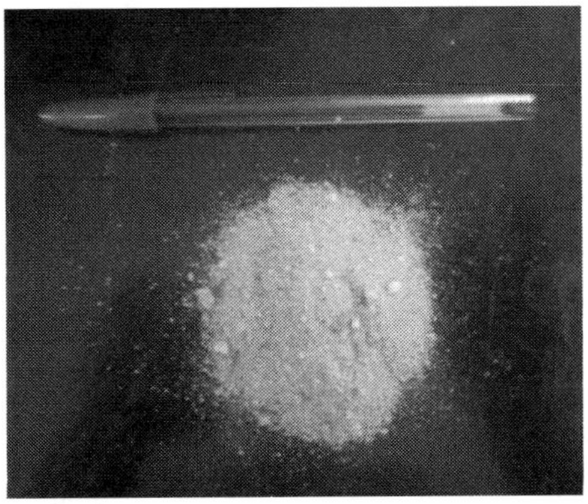

Figure 1. Sandy Soil (Source: Author)

The real specific mass of the grains in the pure soil, obtained in the pentapycnometer, was 2.68 g/cm³. The maximum void index was 0.894 and the minimum 0.586. The uniformity coefficient (CNU) was 2.40 and the curvature coefficient (Cc) was 1.06, with an effective diameter (*D*10) of 0.18 mm and the average diameter (*D*50) of 0.39 mm. The granulometric curve of this sand is presented next (Figure 2).

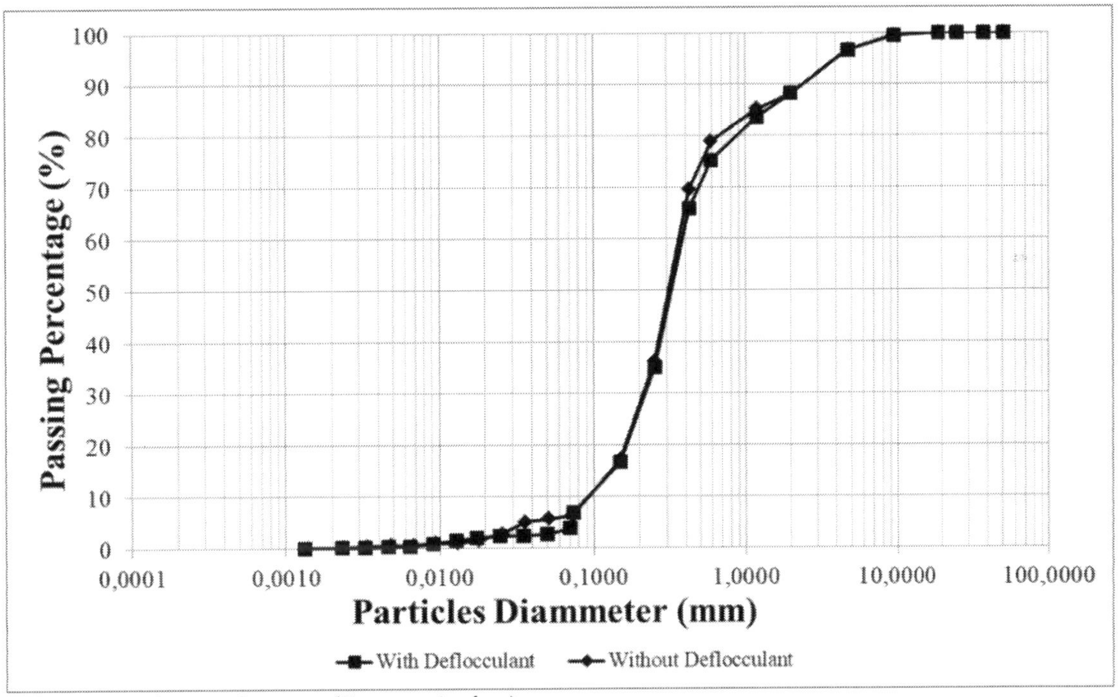

Figure 2. Sand Granulometry (Source: Author).

The granulometric curve shows that the studied soil is composed of 96.2% sandy fraction and 3.8% clay fraction. The soil is classified as medium sand by NBR 6502 [5] and, according to the Unified Soil Classification System - USCS, a well-graded sand (SW), without plastic characteristics.

2.1.2. Açaí Fibers

The fibers used as a reinforcement element were the vegetable fibers from the açaí bagasse (Euterpe oleracea) after the pulp was removed. They were obtained in contact with the Federal University of Pará (UFPA), which has a machine capable of separating the fibers from the core. These fibers are found in the Northeast of Pará, a Brazilian State, without proper destination, which are often found on the streets.

Açaí fibers were chosen because they have great availability and poorly suited destinations, presenting mechanical characteristics compatible with the need for new geotechnical materials, which can open up new perspectives for studies on soil reinforcement with natural fibers and alternatives on the problems of Urban Solid Waste. The following is a more detailed image of the fibers (Figure 3).

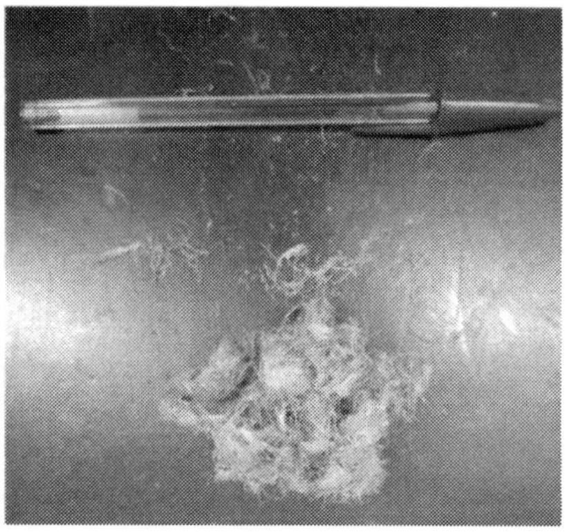

Figure 3. Açaí Fibers (Source: Author).

2.2. Tests and Procedures

Below are the tests and procedures used throughout the experimental program. The tests were carried out in accordance with standards that will be presented accordingly.

2.2.1. Direct Shear Tests.

The direct shear tests were executed according to the standard D3080/D3080M – 11 [6] to soil + 0.5% of fibers and soil + 1.0% of fibers with loads of 50 KPa, 100 KPa and 200 KPa. The first step is to execute the consolidation phase of the experiment and with the consolidation curve define the shear speed by Taylor's method [7] using the time necessary to occur 90% the primary consolidation. Based on that, was determined a speed of 0.5 mm/min and used the same for every sample.

The results show the influence of the fibers content in the shear resistance of the composites soil-fibers, the influence on the friction angle and cohesion as well as the influence of the relative density variation.

3. Results and Discussions

3.1. Relative Density of 50%

The results for sandy soil with a relative density of 50% the composites showed slight improvements in the shear strength parameters as shown in Figure 4, the only one that did not show improvement was S.G._50% _0.5_50, which had a peak strength of 33.82 kPa, a value similar to that of pure soil

(SG_50% _50) which was 33.89 kPa. The composite S.G._50% _1.0_50 had a peak resistance of 38.11 kPa, 12.45% higher than that of S.G._50% _50.

For the confinement of 100 kPa, the composite S.G._50% _0.5_100 had a peak resistance of 69.89 kPa and the composite S.G._50% _1.0_100 had a peak resistance of 76.94 kPa, being 3.11% and 13.52%, respectively, greater than that of pure soil (S.G._50% _100) of 67.78 kPa. The composite S.G._50% _0.5_200 had a peak resistance of 128.39 kPa and the composite S.G._50% _1.0_200 had a peak resistance of 131.28 kPa, being slightly higher than that of pure soil (S.G._50% _200) of 126.67 kPa.

The results show that açaí fibers as a reinforcement of sandy soil with a relative density of 50% work better for the percentage of 1.0% of fibers and at lower confinement stresses, with better performance at stresses of 50 and 100 kPa. Composites with 0.5% açaí fibers had results very close to pure soil, being slightly higher for confinement stresses of 100 and 200 kPa.

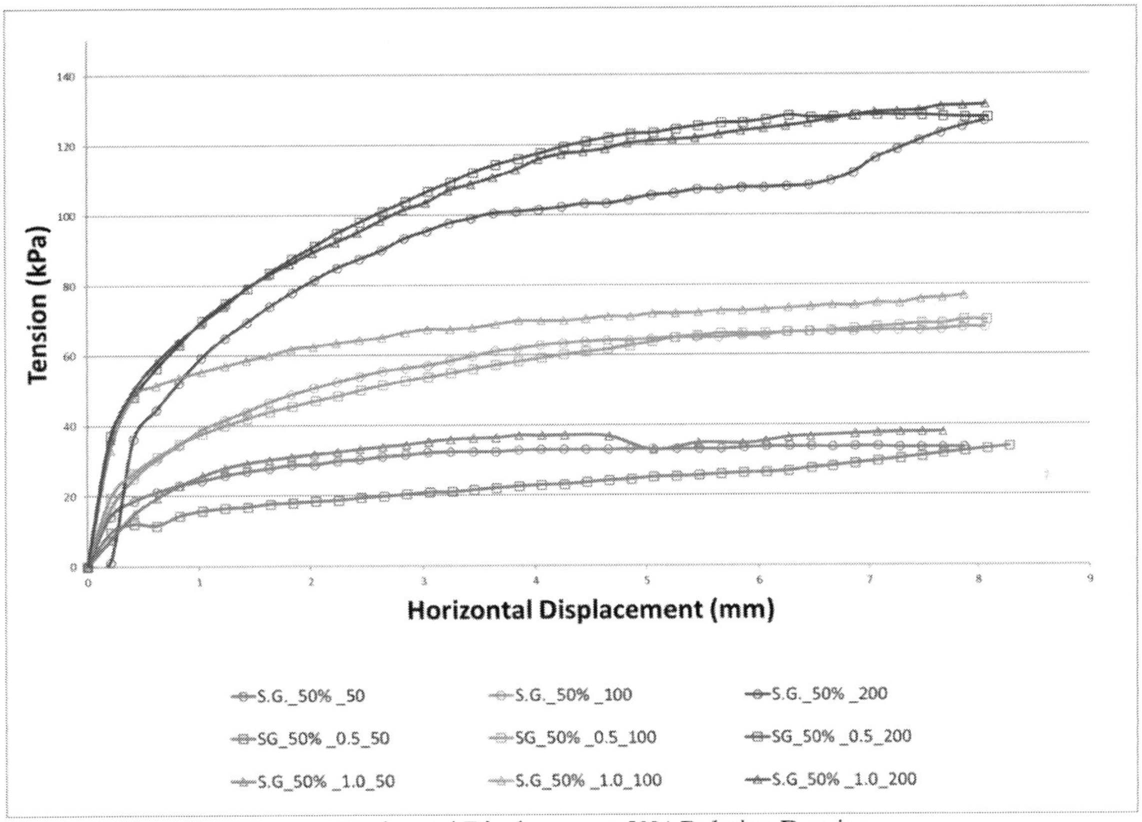

Figure 4. Shear Stress versus Horizontal Displacement 50% Relative Density

In Table 1, it is possible to identify the differences between the friction angle and cohesion parameters between sandy soil with a relative density of 50% and the soil-fiber composites, the equations and the adjustment parameters of the lines to verify the conformity of this adjustment, represented by R^2. The variations in the friction angle are very small, and can be considered negligible. The cohesive intercept has undergone a significant improvement for the composite with 1.0% of açaí fibers (S.G._50% _1.0), 24.64% higher than pure soil (S.G._50%). The increase in cohesion for the composite with 0.5% açaí fibers (S.G._50% _0.5) can be considered negligible.

Table 1. Parameters of the Confinement Stress Lines versus Shear Stress 50% of Relative Density

Composite	Friction Angle (°)	Cohesion (kPa)	Line Equation	R^2

Sandy Soil 50% (S.G._50%)	31.56	4.44	Y = 0.61*X + 4.44	0,998
Sandy Soil 50% + 0,5% of Fibers (S.G._50%_0.5)	31.96	4.57	Y = 0.62*X + 4.57	0,997
Sandy Soil 50% + 1.0% of Fibers (S.G._50%_1.0)	31.38	10.94	Y = 0.61*X + 10.94	0,991

3.2. Relative Density of 75%

In the results for the sandy soil with a relative density of 75%, the composites showed slight improvements in the shear strength parameters for low and medium confinement stresses (50 and 100 kPa) as shown in Figure 5. For the confinement stress of 200 kPa the results of fiber-reinforced composites were lower than those of pure soil.

For the confinement of 50 kPa, the composite S.G._75% _0.5_50 had peak resistance of 40.72 kPa and the composite S.G._75% _1.0_50 had peak resistance of 39.89 kPa, with 11.91% and 9.62%, respectively, greater than that of pure soil (S.G._75% _50) of 36.39 kPa. The S.G._75% _0.5_100 composite had a peak resistance of 77.61 kPa and the S.G._75% _1.0_100 composite had a peak resistance of 72.11 kPa, with 23.63% and 14.87%, respectively, being higher. than that of pure soil (S.G._75% _100) of 62.78 kPa. The composite S.G._75% _0.5_200 had a peak resistance of 113.98 kPa and the composite S.G._75% _1.0_200 had a peak resistance of 115.72 kPa, being lower than that of pure soil (S.G._50% _200) of 125.56 kPa.

The results show that açaí fibers as a reinforcement of sandy soil with a relative density of 75% work well for both percentages of 0.5% and 1.0% of fibers and at lower confinement stresses, with better performance in stresses 50 and 100 kPa.

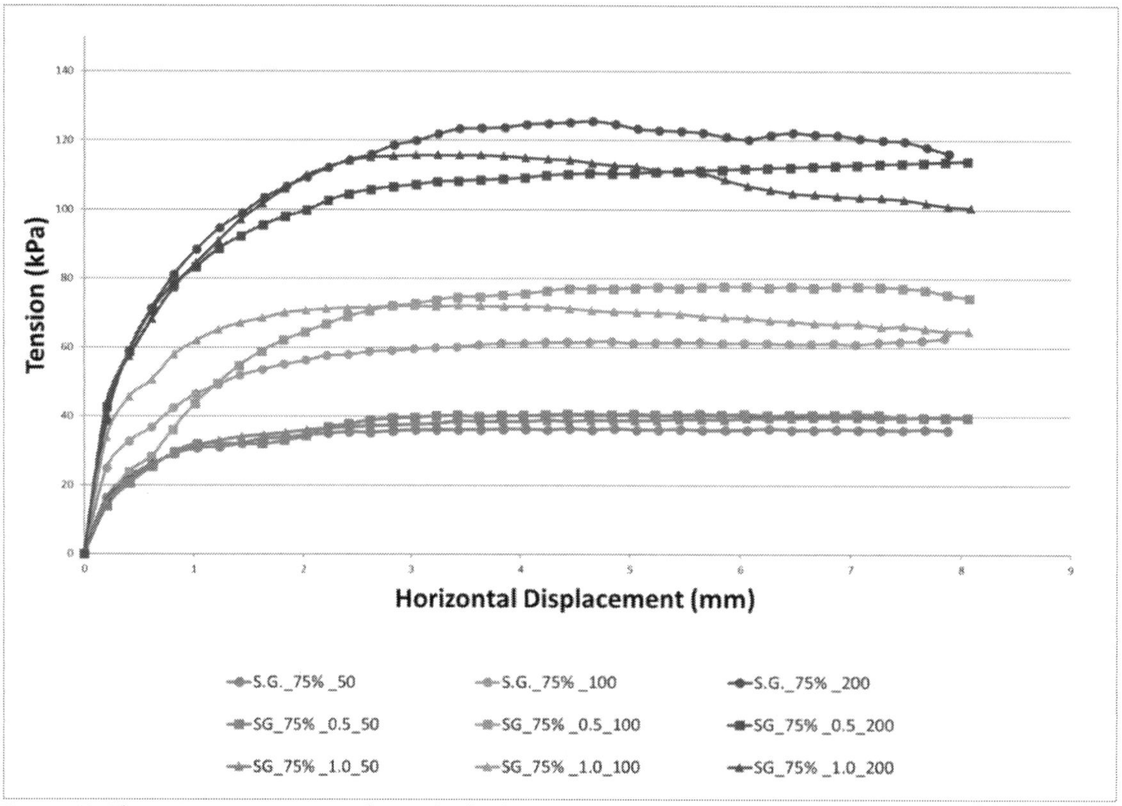

Figure 5. Shear Stress versus Horizontal Displacement 75% Relative Density

In Table 2 it is possible to identify the differences between the friction angle and cohesion parameters between the sandy soil with 75% density and the soil-fiber composites, the equations and the adjustment parameters of the lines to verify the conformity of this adjustment, represented by R^2. There is a small decline in the friction angle for both fiber percentages, while the cohesive intercept has undergone a significant improvement for both composites; the composite with 0.5% of açaí fibers (S.G._75% _0.5) had an increase of 450.8% in relation to the pure soil (S.G._75%) and the composite with 1.0% of açaí fibers (S.G._50% _1.0) had an increase of 361.6% in relation to pure soil.

The improvement in the cohesive intercept of sandy soil with a relative density of 75% was the greatest gain in the addition of açaí fibers to this soil, so that the sands present low cohesion or even null cohesion, this gain was so significant that it was possible remove the specimens from the shear box without breaking them. Thus, depending on the type of application, this matrix-reinforcement arrangement can be interesting from a technical point of view.

Table 2. Parameters of the Confinement Stress Lines versus Shear Stress 75% of Relative Density

Composite	Friction Angle (°)	Cohesion (kPa)	Line Equation	R^2
Sandy Soil 75% (S.G._75)	30.93	5.00	Y = 0.60*X + 5.00	0,998
Sandy Soil 75% + 0,5% of Fibers (S.G._0.5_75)	25.20	22.54	Y = 0.47*X + 22.54	0,963
Sandy Soil 75% + 1.0% of Fibers (S.G._1.0_75)	26.36	18.08	Y = 0.50*X + 18.08	0,989

4. Conclusions

Contrasting the relative densities for the composites with sand it was possible to notice that the sandy soil composites with 50% relative density showed better results for the stresses of 100 and 200 kPa and for the percentage of fibers of 1.0%. For soil-fiber composites with a density of 75%, the gain was greater for confinement stresses of 50 and 100 kPa and for the percentage of fibers of 0.5%. In addition, both had an increase in the cohesive intercept, being much more significant for soil-fiber composites with a relative density of 75%

The reinforcements for sandy soil proved to be effective in several cases, being more effective for sand with a relative density of 75%, which can be explained by the reduction of voids in the soil, which leads to greater contact between the matrix and the reinforcement, that is, a greater soil-fiber interface. An evidence of this phenomenon is the significant gain in the cohesive intercept for these composites, which were the biggest gains among the two sand densities.

The study of soil reinforcement with açaí fibers (*Euterpe oleracea*) proved to be quite positive, where the results showed a real improvement of the two types of soil analyzed, from a geotechnical point of view. For the sandy soil, it is evident that the best application arrangement would be for 75% relative density with the percentage of 0.5% of fibers at lower confinement stresses, between 50 and 100 kPa, which obtained gains in shear strength up to 24% and a significant increase in the cohesive intercept.

The general analysis of the results indicates that the açaí fibers can be used in works where the critical case for the stabilization or functionality of the work is immediately after construction, such as embankment stabilization, temporary landfills, access roads built on saturated soft clay, where the primary function of any reinforcement is to allow the structure to be built. For these cases, açaí fibers can be used as long as they are known to the temporal variation of their mechanical behavior and / or an appropriate treatment is applied to improve their characteristics.

For cases where stabilization needs to be lasting, the existing results leave these possibilities open, being inconclusive, but which can be better evaluated, according to the suggestions for future work to follow.

IOP Conf. Series: Materials Science and Engineering 1260 (2022) 012025 doi:10.1088/1757-899X/1260/1/012025

References

[1] Casagrande M D T 2005 Behavior of fiber reinforced soils subjected to large deformations. *Thesis (Doctorate)* - Federal University of Rio Grande do Sul. School of Engineering. Graduate Program in Civil Engineering. Rio Grande do Sul

[2] Palmeira E M 1993 Soil Stabilization and Strengthening Course: Introduction to the Use of Geosynthetics. Publication No. GAP002B/93 2a. Edition, Master in Geotechnics, University of Brasília. Brasilia

[3] Mitchell J K and Villet W C B 1987 Reinforcement of earth slopes and embankments *NCHRP* Rep. No. 290, Washington, D.C

[4] Sales K C S 2011 Soil Improvements by Inclusion of Natural Fibers p 113 *Dissertation (Masters)* - University of Brasília. Postgraduate program in Geotechnics, Brasilia

[5] Ghavami K, Toledo Filho R D and Barbosa N P 1999 Behaviour of composite soil reinforced with natural fibres *Cement and Concrete Composites* **21**

[6] Dittenber D B and GangaRao H V S 2012 Critical Review of Recent Publications on Use of Natural Composites in Infrastructure *Composites Part A: Applied Science and Manufacturing* **43**pp 1419-29

[7] Martins A P S 2014 Development, mechanical characterization and durability of self-compacting soil-cement composites reinforced with sisal fibers *Thesis (Doctorate)* – UFRJ/COPPE. Rio de Janeiro

[8] Silveira M V 2018 Analysis of mechanical behavior and durability in sand composites reinforced with natural fibers of curauá and sisal *Thesis (Doctorate)* — Department of Civil Engineering. Pontifical Catholic University of Rio de Janeiro

[9] Thomas S, Paul S A, Pothan L A, Deepa B 2011 Natural fibres: structure, properties and applications. In: Cellulose fibers: bio-and nano-polymer composites [S.l.] Springer p 3–42.

[10] ABNT – Associação Brasileira de Normas Técnicas 2016 Gravel grains retained on the 4,8 mm mesh sieve - Determination of the bulk specific gravity, of the apparent specific gravity and of water absorption *NBR 6458*

[11] ABNT – Associação Brasileira de Normas Técnicas 1995 Rocks and soils - Terminology *NBR 6502*

[12] ASTM - American Society for Testing and Materials 2011 *ASTM D3080/ASTM D3080M-11: Standard Test Method for Direct Shear Test of Soils Under Consolidated Drained Conditions*

[13] Taylor D W 1948 *Fundamentals of Soil Mechanics* John Wiley & Sons, New York, USA p 770

IOP Conf. Series: Materials Science and Engineering 1260 (2022) 012026 doi:10.1088/1757-899X/1260/1/012026

Design Method for Geogrid Stabilisation of Railway Ballast and Subballast

P Rimoldi[1], P Pezzano[2], F Trovato[2]

[1]Consultant, Milano, Italy
[2]Officine Maccaferri Spa, Geosynthetic Department, Bologna, Italy

Email: pietro.rimoldi@gmail.com

Abstract. Railways are often built on soft soils. Railways present peculiar characteristics in terms of loads applied to subgrade: a single train may produce hundreds wheels loads in few seconds, that is a train produces rapid, repeated and cyclic loads; loads are first distributed from rails to sleepers, then from sleepers to ballast, and finally from ballast to subgrade; loads are always applied to the same area. The effect of such loads on soft soils generates absolute and differential settlements, which may quickly degrade the quality of the railway segment. Reinforcing geogrids may be used for stabilizing the railway base, in such a way as to decrease settlements and to increase the railway lifetime. The paper aims to present a design method for geogrid stabilisation of railway ballast and subballast, taking into account the wheel loads, the number of passages, the railway geometry, the subgrade characteristics and the geogrid properties. The proposed design method can be considered as an extension of design methods for geogrid stabilisation of road bases and it is based on the same engineering principles.

1. Introduction

Railways, which are often built on soft ground, present special characteristics in terms of loads applied to the soil: a single train can produce hundreds of wheel loads in seconds, so the soil is subjected to fast, repetitive and cyclical loads; loads are distributed first by rails to sleepers, then from sleepers to ballast and possibly subballast; and finally to the foundation soil; loads are always applied to the same area. The effect of these loads on soft soil produces absolute and differential settlements, which can rapidly degrade the quality of the railway. The function of the railway superstructure is to transfer the axial loads of the wheels evenly in the subgrade. If due to the inadequacy of railway superstructure, these loads are not transmitted properly to ballast and subballast, and from these to the subgrade, it is possible that high deformations occur, forcing the trains to slow down the line. The inclusion of layers of geogrid for stabilisation of the ballast and / or subballast layers affords to limit the extent of settlements, ensuring the structure durability and a much greater degree of efficiency. Design methods for geogrid stabilisation have been developed for road bases, but reliable design methods for stabilised

IOP Conf. Series: Materials Science and Engineering 1260 (2022) 012026 doi:10.1088/1757-899X/1260/1/012026

railway ballast and subballast are lacking. Hence this paper presents a design method for geogrid stabilised railway ballast and subballast, which considers the wheel loads, the number of wheel passes, the geometry of the railroad, the characteristics of the subgrade and the properties of geogrids. The proposed design method is an extension of the methods available for the design of stabilised road bases and is based on the same engineering principles. The most recent researches on the behavior and properties of railway ballast have been considered and introduced in the design method.

2. Geometry and materials

The railway superstructure consists of the ballast and possibly the subballast (Figure 1), the sleepers (Figure 2) and the rails.

The ballast is a selected granular material, usually consisting of particles of sharp-edged stones with dimensions of approximately 20-100 mm, which is put on the subballast or on the subgrade to limit lateral and vertical movement of longitudinal rails and sleepers; it transmits and distributes loads to the subballast and / or the subgrade, and it absorbs most of the variations in temperature. The subballast usually consists of granular material (generally sand and gravel mix), which is placed between the ballast and the subgrade to better distribute the load and to help to protect the ballast from the upward pumping of fine particles and from ice intrusion. According to some national regulations, the minimum thickness of ballast shall be 500 mm for railway main lines, and 350 mm for secondary lines, while the subballast generally has a thickness of 200 – 300 mm. The thickness of ballast and subballast may be much greater in the case of soft subgrade or with poor geotechnical characteristics and high level of groundwater.

Figure 1 shows the geometry of a single-track railway line in Italy: the distance between the axes of the rails is 1.435 m. Figure 2 shows the typical geometry of sleepers: the characteristic dimensions L x W x H are 2.60 m x 0.26 m x 0.16 m. The sleepers are usually placed at intervals from 0.60 to 0.80 m. About the geometry of rails: a Vignole type rail, according to EN 13674-1 [1], has a base width of 150 mm.

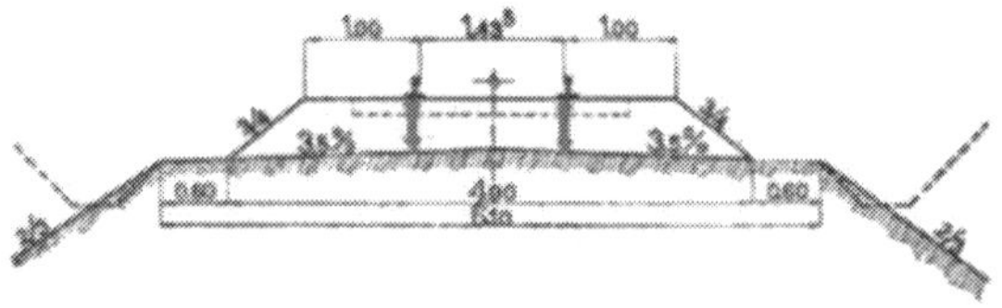

Figure 1. The railway superstructure

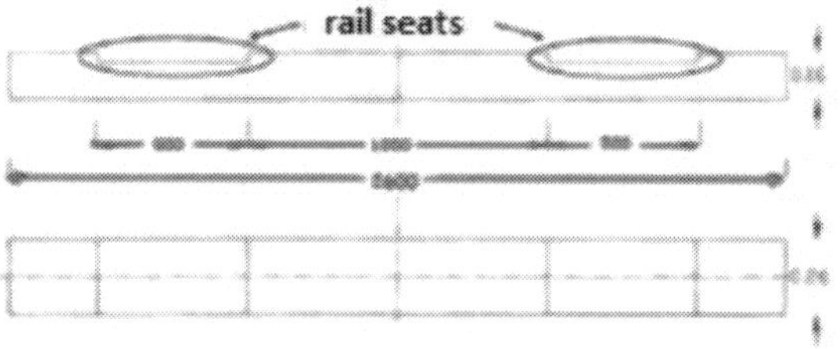

Figure 2. Cross section and plan view of a sleeper

3. Existing design methods for ballasted railway tracks

Design of ballasted railway track foundations requires the accurate estimation of the granular (ballast and subballast) layer thickness that provides protection against track failures induced by train dynamic loads.

Thus, the design of railway track foundations is also referred to as design of the granular layer thickness. The main existing design method is the North American Railway Method.

The manual of the American Railway Engineering Association (AREA) [3,4] specified a minimum thickness of ballast and sub-ballast as 300 mm and 150 mm, respectively (i.e. the minimum granular layer thickness is 450 mm). When designing railway track foundations according to the AREA manual [3], the first step to be taken consists of the determination of the design rail seat load,

P_d. An assumption of P_d = 50 % of the design wheel load is reasonable. The second step is the determination of the sleeper/ballast contact pressure, σ_{sb}. The AREA manual recommends that the maximum value of σ_{sb} should be less than 65 psi (450 kPa) for wood sleepers and less than 85 psi (590 kPa) for concrete sleepers, under the consideration that the track is founded on high-quality ballast. Once the values of P_d and σsb are determined, the AREA Manual recommends calculating the required granular layer thickness, in which the subgrade stress, σ_{bs}, should be equal to the allowable subgrade bearing pressure.

The main objective of designing a minimum thickness of granular layer is to prevent failure of track substructure and limit excessive plastic deformation of ballast. To satisfy this objective, the AREA method [3] recommends using a universal allowable bearing pressure of 138 kPa. As subgrade soils may vary widely in strength, the use of a single bearing capacity value in design is inappropriate. To overcome this limitation, Raymond [5] suggested using the allowable subgrade bearing capacity based on the Casagrande soil classification.

However, the effect of repeated (cyclic) loading on the track cumulative plastic strain and deformation was not considered in the available design methods, except in the British Railways [6] and Li and Selig [7,8] methods.

The design criterion of the British Railways method was based on limiting the deviatoric stress at the subgrade surface to be less than a threshold stress, which is an attractive criterion, as it takes into account the effects of repeated loading on the subgrade cumulative plastic strain. Nonetheless, the threshold stress is not a property of soils and cannot be found in all types of soil. Moreover, this criterion does not reflect the effect of subgrade layer thickness on the cumulative plastic deformation [9].

In order to prevent the progressive shear failure of subgrade, the design criterion of the Li and Selig method [7,8] limits the total cumulative plastic strain at the subgrade surface to be less than the allowable plastic strain for the design tonnage. Furthermore, to prevent the excessive plastic deformation of the track, another design criterion is used whereby the total cumulative plastic deformation of subgrade soil was limited to less than the allowable subgrade plastic deformation for the design tonnage. However, the deformation of ballast layer was ignored completely, although about up to 40 % of the total deformation of railway tracks originates from the ballast layer, as indicated by many researchers [10,11] .

4. Development of a design method

Geogrid stabilisation can be used to stabilize the base of the railway (ballast and / or subballast) in order to reduce settlements and to increase the life cycle. A rational design method for geogrid stabilised railway ballast / subballast has been developed, which considers the wheel loads, the number of wheels passes, the geometry of the railroad, the characteristics of the subgrade and the properties of geogrids. The proposed design method is an extension of Leng and Gabr method [12] [13,14] , for the design of geogrid stabilisation of road bases, and it is based on the same engineering principles. Furthermore, the design method proposed in this paper is a development of the method presented by Rimoldi [15].

4.1 Stabilisation of the ballast

Leng and Gabr [12,13,14] have proposed a method for the design of geosynthetic stabilised unpaved roads, based on an analysis of the simplified distribution of vertical stresses, the correlation between the degree of mobilization of bearing capacity with the depth of rutting and the modulus of geogrids at 2 % strain; the method has been validated with the results of experimental tests.

The adaptation of the method to the case of railways is shown first with reference to the stabilisation of ballast, that is in case of no subballast. Figure 3 shows the scheme and symbols used. Geogrids are placed at the base of the ballast.

The basic assumption of the method of Leng and Gabr [12,13,14] is that the section of a paved road can be assumed as a system of two layers with different elastic moduli.

The theory of Odemark [16] allows to transform a multi-layered system in an equivalent one, where the thickness of the layers is changed so that all layers have the same elastic modulus (Figure 5a shows the scheme for two layers system). Such a system becomes equivalent to an elastic half-space characterized by a single value of the elastic modulus, for which we can apply the well known theory of Boussinesq, which provides the state of stress at any point in the elastic half-space, due to a set distribution of stress applied to the upper surface.

According to the Odemark theory, considering the two layers system ballast – subgrade (Figure 4a), the equivalent thickness he of the ballast, so that it has the same elastic modulus E_{sg} of the subgrade, is given by:

$$h_e = h_b \cdot \left[\frac{E_b \cdot (1-\mu_{sg}^2)}{E_{sg} \cdot (1-\mu_b^2)}\right]^{1/3} = h_b \cdot \chi_1 \qquad (1)$$

where: h_b = thickness of the ballast (m), E_b = elastic modulus of the ballast (kPa), E_{sg} = elastic modulus of subgrade (kPa), μ_b = Poisson's ratio of the ballast (default value = 0.35), μ_{sg} = Poisson's ratio of the subgrade (default value = 0.42); χ_1 = Odemark coefficient ballast – subgrade.

When there is also the subballast (Figure 4b), the thickness h_e, calculated for a single granular layer, has to be increased to the equivalent thickness h_g of the granular layers (ballast + subballast), given by:

$$h_g = h_e \cdot \left[\frac{E_b \cdot (1-\mu_{sb}^2)}{E_{sb} \cdot (1-\mu_b^2)}\right]^{1/3} = h_e \cdot \chi_2 \qquad (2)$$

where: E_{sb} = elastic modulus of subballast, μ_{sb} = Poisson's ratio of the subballast (default value = 0.30); χ_2 = Odemark coefficient ballast – subballast.

The matter is: which modulus should be used? In railway design the initial modulus E_i, the resilient unloading – reloading modulus E_{ur}, and the secant modulus E_{50} are often used.

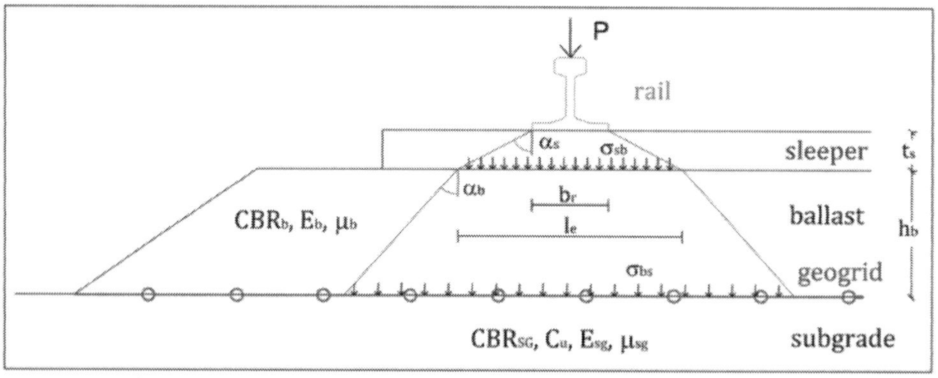

Figure 3. Schematic of the model for the stabilisation of ballast

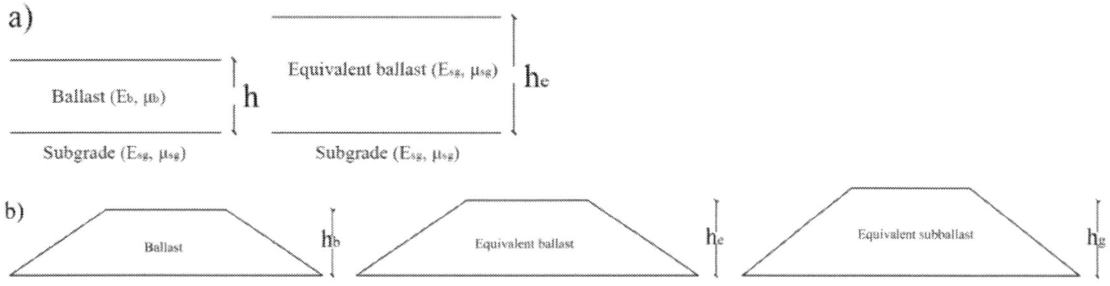

Figure 4. a) Transformation of the actual system in an equivalent one with a modulus equal to that of the subgrade; **b)** Case of ballast and subballast

Since there is no uniformity, in the presently proposed design method the following formulas, commonly used in road engineering, are used to calculate the modulus:

- Modulus of the subgrade E_{sg} (MPa) (Shell formula, from [17]), valid for fine soils with CBR ≤ 10:

$$E_{sg} = 10 \cdot CBR_{sg} \tag{3}$$

- Modulus of the ballast E_b (MPa) and subballast E_{sb} (MPa) (from [18]):

$$E_b = 36 \cdot CBR_b^{0.3} \tag{4}$$

$$E_{sb} = 36 \cdot CBR_{sb}^{0.3} \tag{5}$$

with: CBR_{sg} = CBR index of the subgrade, CBR_b = CBR index of the ballast.

It must be noted that experimental evidence indicates that the modulus of limestone ballast is usually lower than the modulus of basalt and granite ballast; hence it is suggested to assume CBR_b = 90 for limestone ballast, and CBR_b = 100 for basalt and granite ballast.

For suballast typical values are: $CBR_{sb} = 20 \div 30$.

With the Odemark transformation espressed by equation (1), the half space below the sleeper become uniform, elastic and isothropic, that is in the conditions for the validity of the well known Boussinesq theory.

Hence the present method assumes that the vertical stresses are distributed through the equivalent granular layer according to Boussinesq theory for rectangular loaded area [19].

As shown in Figure 5, if we apply a vertical pressure q, evenly distributed on a rectangular area, to a homogeneous elastic half space, since the overburden stress is constant, the application of the pressure q generates equilibrium conditions in every point of the medium. Hence the vertical stress is independent from the medium characteristics. If the load from sleeper to ballast is assimilated to a uniform load on a rectangular area, the Boussinesq theory allows to obtain the value of the vertical stress σ_c at any point along the vertical line passing by one corner of the rectangle. With reference to Figure 5, the following equations apply:

$$R_1 = (L^2 + z^2)^{0.5} \tag{6}$$

$$R_2 = (B^2 + z^2)^{0.5} \tag{7}$$

$$R_3 = (L^2 + B^2 + z^2)^{0.5} \tag{8}$$

$$\sigma_z = \frac{q}{2\pi} \cdot \left[\arctan\left(\frac{L \cdot B}{z \cdot R_3}\right) + \frac{L \cdot B \cdot z}{R_3} \cdot \left(\frac{1}{R_1^2} + \frac{1}{R_2^2}\right)\right] \tag{9}$$

where: σ_z = induced vertical stress at depth z below one corner of the rectangle (kPa), q = uniform pressure on the rectangular area (kPa); z = depth below surface (m); L, B = length and width of the rectangle (m).

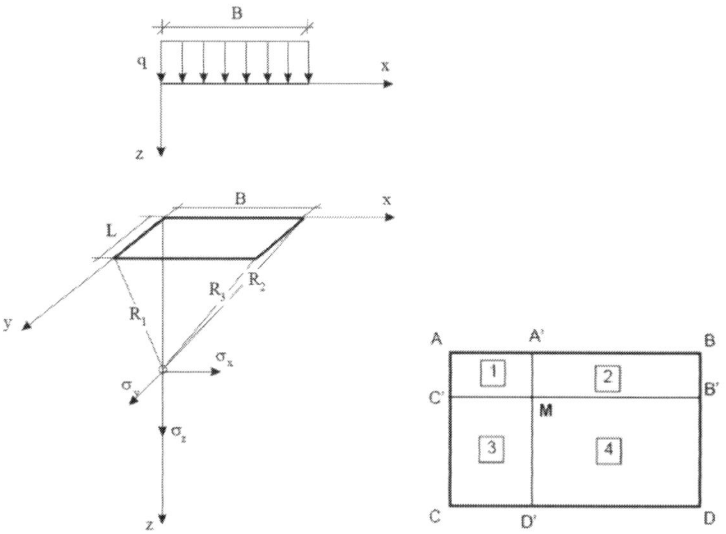

Figure 5. Scheme for Boussinesq equation and for calculating the stress along the vertical line passing by point M inside the rectangular loaded area (modified from [19])

If we want to calculate the stress along a vertical line from a point inside the rectangular loaded area, then it is possible to divide the whole area into 4 rectangles, as shown in Figure 5: by applying the principle of superposition, the total stress at a point will be the sum of the stresses generated by the rectangles 1, 2, 3, 4.

If point M in Figure 5 is exactly in the centre of the rectangle ABCD, then the 4 internal rectangles become all equal and the stress in every point of the vertical line passing by point M will be simply 4 times the stress calculated with Equation (10):

$$\sigma_{ztot} = 4\ \sigma_z \tag{10}$$

In the present condition, with reference to the symbols in Figure 3 and 4, the parameters of the Boussinesq equations are: $L = l_c$; $B = b_s$; $z = h_c$; $q = \sigma_{sb}$; $\sigma_z = \sigma_{bs}$.

Hence the ratio of the vertical stress at top and bottom of ballast is:

$$\frac{\sigma_{sb}}{\sigma_{bs}} = \frac{\pi}{2} \cdot \frac{1}{\left[\arctan\left(\frac{L \cdot B}{z \cdot R_3}\right) + \frac{L \cdot B \cdot z}{R_3}\left(\frac{1}{R_1^2} + \frac{1}{R_2^2}\right)\right]} \tag{11}$$

4.2 Load Transfer

In a ballasted railway track, the rails transfer the wheel loads to the supporting sleepers, which are spaced evenly along the rail length. Similarly, sleepers transfer the load from the rail to the wider ballast area. The ballast and subballast layers transmit the high imposed stress at the sleeper/ballast interface to the subgrade layer at a reduced level through spreading. Figure 6 shows the typical load distribution from the wheel to the rail, sleepers, ballast, subballast and subgrade layers.

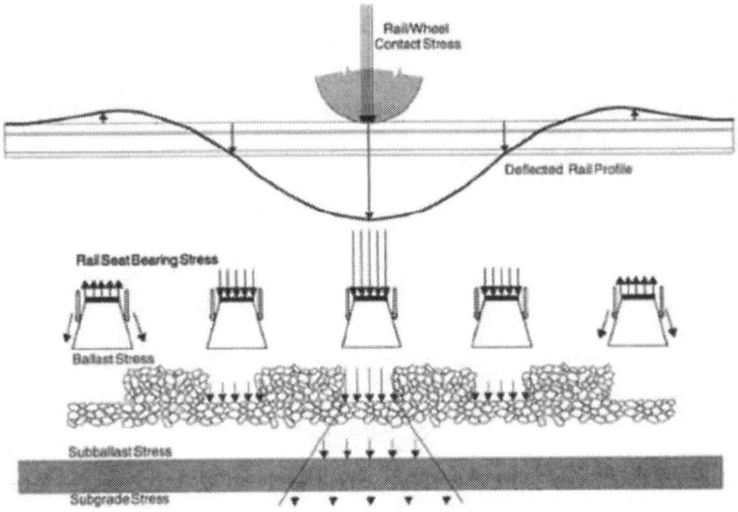

Figure 6. Typical wheel load distribution in track (modified from [22])

For design purposes, the sleeper/ballast contact pressure over the estimated effective sleeper area is generally considered to be uniform; at present the most used methods for its evaluation are by the American Railway Engineering Association (AREA) in USA [3] and the Japanese National Railways (JNR) in Japan [20].

According to the AREA method [3], the sleeper/ballast contact pressure can be expressed by the following equation:

$$\sigma_{sb} = \frac{P}{b_s \cdot l_e} \cdot F_2 = \frac{3 \cdot P}{b_s \cdot l} \cdot F_2 \tag{12}$$

where, σ_{sb} is the average sleeper/ballast contact stress; P is the maximum rail seat load; b_s is the sleeper width; l_e is the length of sleeper supporting the load P; l is the total length of sleeper; and F_2 is a factor depending on the sleeper type and track maintenance.

In Equation (12) the effective length l_e of the sleeper supporting the rail seat load is considered to be one-third of the total length l of sleeper, as shown in Figure 7a (from [21]).

In the (JNR) method [20] a similar distribution of contact pressure between the sleeper and ballast was assumed; however, a different effective length of sleeper, l', was considered, as shown in Figure 7b, and σ_{sb} can be calculated as:

$$\sigma_{sb} = \frac{P}{2 \cdot b_s \cdot l'} \cdot F_2 \qquad (13)$$

where l' is the length from the rail centre to the end of sleeper (Figure 7b).

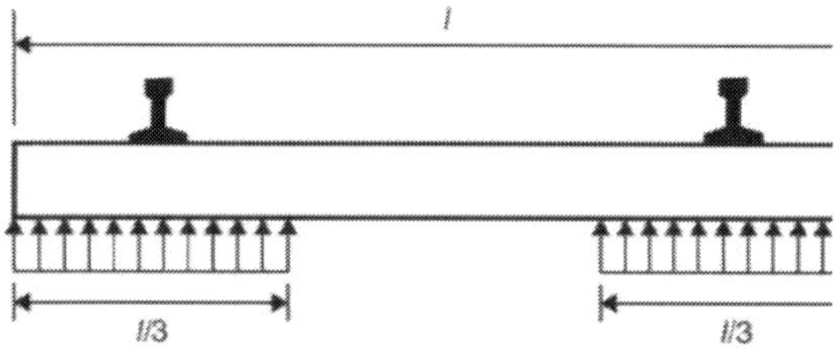

Figure 7. a) Simplified sleeper/ballast contact pressure according to the AREA method (modified from [22])

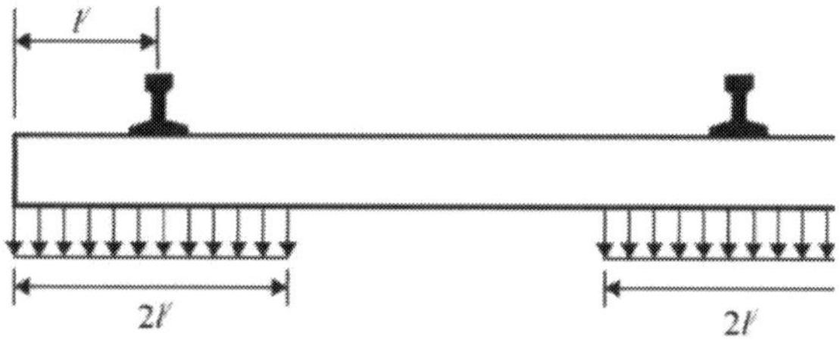

Figure 7. b) Simplified sleeper/ballast contact pressure according to Japanese Standards (modified from [20])

In the presently proposed method, instead, the load from the rail spreads inside the sleeper according to the load spreading angle α_s, that is simmetrically to the rail center (Figure 3).

For standard concrete sleepers the value $\alpha_s = 55°$ provides a value l_e practically equivalent to the value provided by the AREA method [3], while the value $\alpha_s = 60°$ provides a value l_e practically equivalent to the value provided by the JNR method [20]. Therefore, the load spreading angle should be assumed as: $\alpha_s = 55° \div 60°$. The effective length of sleeper l_e becomes:

$$l_e = b_r + 2 \cdot t_s \cdot \tan \alpha_s \qquad (14)$$

where: l_e = effective length of sleeper (m); b_r = rail width (m); t_s = sleeper thickness (m).

The sleeper/ballast contact area A_{sb} (m^2) and contact pressure σ_{sb} (kPa) are therefore:

$$A_{sb} = l_e \cdot b_s \qquad (15)$$

$$\sigma_{sb} = P / (l_e \cdot b_s) \qquad (16)$$

where: P = design rail seat load (kN); b_s = sleeper width (m).

It must be noted that in the above equation (16) for determining the average sleeper/ballast contact stress, the rail seat load, P, is not equal to the wheel load, $P_d = W / 2$, where W is the axle load. In fact, the USACE design manual [23] suggests that the point wheel load of the train be transmitted to the adjacent five sleepers, emphasising the highest load on the sleeper below the wheel (Figure 8). However, the deflection pattern of the track subjected to the wheel load indicated that only three sleepers support the load while the other sleepers remain suspended [22]. Furthermore, [20] reported that 40 % to 60 % of the wheel load is supported by the sleeper that is directly below the wheel.

Hence the rail has the effect of sharing the load of 3 to 5 sleepers (Figure 9): considering the more conservative option of load shared by 3 sleepers, 50 % of the load stresses the sleeper directly under the axle of the train, while the two side sleepers get 25 % each of the load axle. Hence, said W the axle load, it can be considered that the rail seat load P produced by a wheel on the most loaded sleeper is:

$$P = 50\ \% \cdot W / 2 = W / 4 \tag{17}$$

With regard to the loads generated by the wheels of trains, the Eurocode load model LM71 for railway bridge loading from EN 1991-2 (2003) [2] can be considered, which is composed of 4 x 250 kN axles arranged with spacing of 1.60 m (Figure 9). In case of high-speed trains, the heaviest axle load (in driving cars) is typically equal to 170 kN/m.

Hence, for train load like LM71 ([2], see Figure 9): P = 250 / 4 = 62.5 kN

For a high speed train, the highest rail seat load can be assumed as: P = 170 / 4 = 42.5 kN.

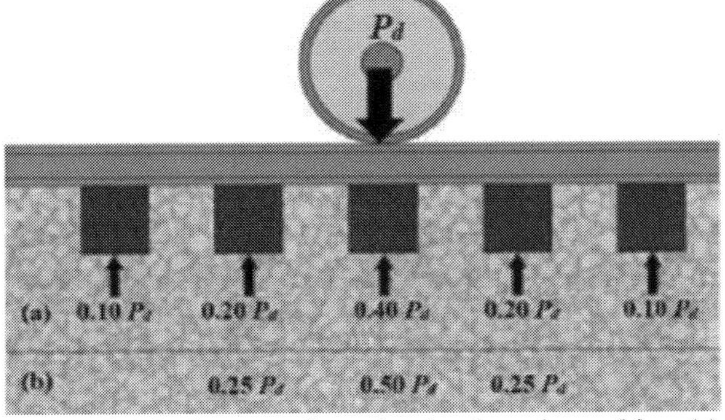

Figure 8. Maximum rail seating loads estimated by: (a) five adjacent sleepers' method; (b) three adjacent sleepers' method (modified from [20,22])

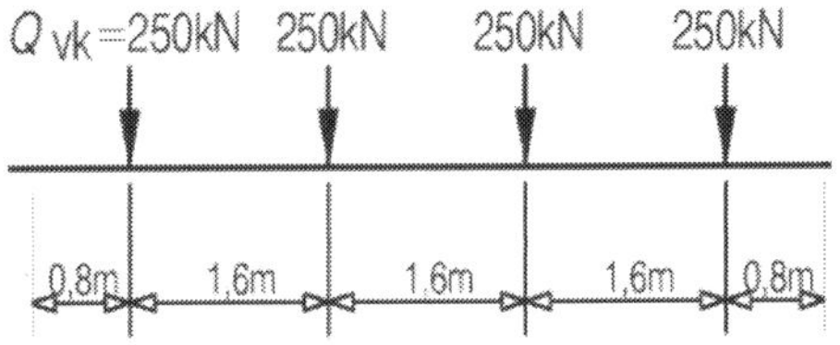

Figure 9. Eurocode load model LM71 for railway bridge loading

4.3 Design rail seat load

To consider the dynamic component of wheel load, the static wheel load shall be multiplied by an influence coefficient generally known as the dynamic amplification factor (DAF). Many factors affect the DAF, including train speed, static wheel load and wheel diameter, unsprung vehicle mass, condition of vehicle and track-ground system, etc. In the existing design methods, a variety of empirical equations can be used for determining the design vertical wheel load. In these methods, the design dynamic wheel load is generally expressed as a function of the static wheel load.

In the presently proposed method, according to [7] [8], the AREA impact factor will be used:

$$P_{dyn} = \varphi \cdot P \tag{18}$$

where: P_{dyn} and P are the dynamic and static wheel load (kN), respectively; and φ is the dimensionless DAF, which is given by:

$$\varphi = 1 + \frac{0.0052\ C}{D_w} \tag{19}$$

where: C is the train speed (km/h), and D_w is the wheel diameter (m).

For cargo trains the following typical values can be assumed: C = 100 km/h, and D_w= 0.90 m; hence a typical value of DAF is: $\varphi = 1.58$.

For high speed trains the following typical values can be assumed: C = 300 km/h, and D_w = 0.92 m; hence a typical value of DAF is: $\varphi = 2.70$.

Hence the design rail seat load should be assumed equal to:

$$P_{dyn} = \varphi \cdot W / 4 \tag{20}$$

4.4 Applied Stress on Subgrade

The load spreading model here assumed is the trapezoidal distribution, as proposed by Indraratna et al. [24], shown in Figure 10. The trapezoidal approximation is a simple approach for estimating the variation in vertical stress with depth. According to this method, it is assumed that the vertical stress diminishes with depth in the form of a trapezoid according to the load spreading angle, α_b, as shown in Figure 10.

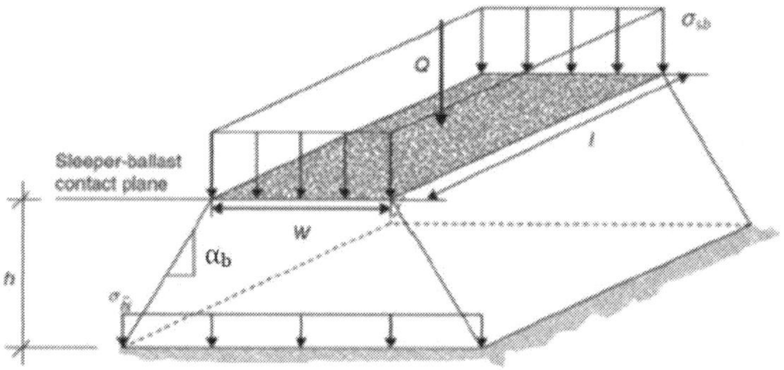

Figure 10. Stress distribution on the subgrade from sleeper/ballast contact stress by trapezoidal distribution (modified from [24])

Subsequently, the stress at the equivalent depth he below the sleeper is:

$$\sigma_{bs} = \sigma_{sb} \cdot \frac{A_{bs}}{A_{sb}} \tag{21}$$

$$A_{sb} = l_e \cdot b_s \tag{22}$$

$$A_{bs} = (l_e + 2 \cdot h_e \cdot \tan\alpha_b) \cdot (b_s + 2 \cdot h_e \cdot \tan\alpha_b) \tag{23}$$

where: σ_{sb} is the average value of vertical stress below the sleeper (i.e. above the ballast), from Equation (16); A_{sb} is the sleeper/ballast contact area; A_{bs} is the ballast / subgrade contact area; and h_e is the equivalent thickness of ballast, from equation (1).

4.5 Allowable settlement

The allowable settlement of the rail tracks is very limited, of the order of 1 to 5 mm; above these values trains are bound to reduced speed. At the base of the ballast or sub-ballast allowable settlements are larger, since only part of the settlements at subgrade interface is transmitted to the sleepers and rails. According to [25], the thresholds for permanent allowable settlements of rails must be considered in accordance with the requirements of railway feasibility, which are of the order of a few millimeters. Such low values can be considered in the presence of a light seismic event, while for heavy seismic events a threshold of tolerance shall be of the order of several centimeters, which has a sense not so much as to ensure the practicability but as to allow a rapid recovery. Hence it can be assumed that the settlement limit at the interface with the subgrade is equal to 50 mm, while the settlement limit at the top of ballast can be assumed equal to 5 mm.

4.6 Accumulated plastic deformation of ballast

For unstabilised ballast, Sayeed [8] has proposed an improved empirical model, which is a modification of a model previously suggested by Shahin [26], for better prediction of the accumulated plastic deformation of ballast.

In the present method, a modification of the Sayeed [8] model is used for the evaluation of the accumulated plastic deformation of unstabilised ballast; the modified model is expressed by the following formula:

$$\varepsilon_{bu} = x \cdot \alpha^{y} \cdot [1 + \ln(N)]^{v} \tag{24}$$

where ε_{bu} is the cumulative plastic strain of unstabilised ballast (%); N is the number of load applications on the ballast; and α, x, y and v are regression parameters depending on the ballast type as summarised in Table 1 (modified from [8]).

At present there are no published method for evaluating the accumulated plastic deformation of stabilised ballast. As reported by Rimoldi [15], Vecchiotti and di Prisco [27] have studied the behavior of railway superstructures by: theoretical analysis of the problem; experimental tests on a small scale; formulation of a numerical model to interpret the experimental results and predict real-scale settlements; development of a method for the estimation of settlements. The model of Vecchiotti and di Prisco [27] allowed to interpret the experimental data and to set the following formula for the estimation of the settlement r:

$$r = \beta \text{ Log } N \tag{25}$$

with: N = number of load cycles; β = coefficient depending on the applied load.

The formulas for calculating the coefficient β are:

- for unstabilised ballast: $\beta_u = (0.0655\ P_{dyn} - 0.8256)$ (26)

- for stabilised ballast: $\beta_s = (0.05163\ P_{dyn} - 0.4245)$ (27)

with: P_{dyn} = design rail seat load (kN), from Equation (20).

Hence, in absence of any other data, in the present method the accumulated plastic deformation of stabilised ballast ε_{bs} is evaluated as:

$$\varepsilon_{bs} = \varepsilon_{bu} \cdot (\beta_s / \beta_u) \tag{28}$$

Table 1. Material parameters for various types of ballast (modified from [8])

Ballast type	x	y	v	α
Basalt	4.82	1.42	0.49	0.30
Granite	1.27	2.41	0.48	0.67
Dolomite	4.23	1.15	0.32	0.40

4.7 Prediction of subgrade deformation

As reported by Sayeed [9], based on experimental data collected from tests, various models were proposed for estimating the cumulative plastic strain of fine-grained soils under repeated loading. Among these models, the most advanced ones that are currently used to predict the cumulative plastic strain and cumulative plastic deformation of track fine-grained subgrade soils are the ones presented by Li [28] and Li and Selig [29], which are modified in the present method as following:

$$\varepsilon_{sp} = \frac{\xi}{100} \cdot N^{\vartheta} \cdot \left(\frac{\sigma_{bs}}{q_c}\right)^{\omega} \quad (29)$$

$$r_u = \varepsilon_{sp} \cdot H_s \quad (30)$$

$$r_u = \sum_{i=1}^{n} \varepsilon_{spi} \cdot H_{si} \quad (30\ \text{bis})$$

where: ε_{sp} is the cumulative plastic strain of track subgrade soil; σ_{bs} is the vertical stress applied at the ballast – subgrade interface; q_c is the bearing capacity of the subgrade soil; N is the number of load repetitions in the subgrade; ξ, ϑ and ω are material parameters given in Table 2; r_u is the rut depth at the ballast – subgrade interface in case of unstabilised ballast / subballast; H_s is the thickness of the soft subgrade layer; ε_{spi} and H_{si} are the cumulative plastic strain and layer thickness for the i-th subsoil layer, in case of stratified subsoil.

For practical calculations, as required for design purposes, the considered thickness of the soft subgrade may be limited to 5.0 m.

At present there are no published method for evaluating the rut depth at the ballast – subgrade interface in case of stabilised ballast / subballast.

Hence, in absence of any other data, as done above, in the present method the rut depth at the ballast – subgrade interface in case of stabilised ballast / subballast is evaluated as:

$$r_s = r_u \cdot (\beta_s / \beta_u) \quad (31)$$

Table 2. Material parameters for various types of soil (modified from [28] and [29]).

Subgrade soil type	ξ	ω	ϑ
Fat clay (CH)	1.20	2.4	0.18
Lean clay (CL)	1.10	1.8	0.16
Elastic silt (MH)	0.84	2.0	0.13
Silt (ML)	0.64	1.7	0.10

4.8 Calculation of the load spreading angle in ballast

We have previously assumed the trapezoidal distribution of the load inside the ballast; now, for simplifying calculations, let's reduce the rectangular loaded area below the sleeper to a square load area of size a_t, having the same area of the rectangular one; hence:

$$a_t = (l_e \cdot b_s)^{0.5} \tag{32}$$

With reference to Figure 10, still considering the trapezoidal distribution according to the load spreading angle α_b, the square loaded area at the ballast – subgrade interface will have size $a_b = a_t \cdot \tan \alpha_b$.

The load from sleeper to ballast shall be the same load that is transferred to the subgrade; hence, considering the equivalent thickness h_e (see Equation (1) and Figure 4.a), it shall be:

$$\sigma_{sb} \cdot a_t^2 = \sigma_{bs} \cdot a_b^2 = \sigma_{bs} \cdot (a_t + 2 \cdot h_e \cdot \tan\alpha_b)^2 \tag{33}$$

Hence, we can calculate the load spreading angle α_b, to be intended as the value at the first application of the train wheel load α_{b1}, as:

$$\tan\alpha_{b1} = \frac{a_t}{h_e} \cdot \left(\sqrt{\frac{\sigma_{sb}}{\sigma_{bs}}} - 1 \right) \tag{34}$$

where $(\sigma_{sb} / \sigma_{bs})$ has to be calculated with equation (11).

4.9 Degradation of the load spreading angle with loading cycles

Based on cyclic plate load tests [12], it has been shown that the spreading angle α_b of the load decreases with the number of load cycles. This corresponds to the intuitive concept that a large number of wheel passages results in a degradation of the ballast, due to the displacement of the grains of soil and the horizontal tensile stresses that are generated because of Poisson's ratio of the ballast, which in turn causes a decrease in the elastic modulus E_b of the ballast: since a less rigid layer spreads the loads worse than a hard layer, it is intuitive that as the number of passages increases the load spreading angle decreases. Therefore Leng and Gabr [12] introduced the coefficient λ_2 of degradation of the angle α, defined as:

$$\lambda_2 = \frac{\tan\alpha_{bN}}{\tan\alpha_{b1}} = \frac{1}{1 + k_2 \cdot \log N} \tag{35}$$

where: $\tan \alpha_{bN}$ = tangent of the angle of load distribution at the N-th loading cycle, $\tan \alpha_{b1}$ = tangent of the initial angle ofload distribution (for N = 1) from Equation (34); N = number of load cycles, k_2 = parameter that defines the degradation of $\tan\alpha_b$.

4.10 Effect of geogrids for stabilisation of ballast

Based on cyclic plate load tests [12] and in situ tests [30], it was shown that the introduction of geogrid stabilisation in the ballast delays and minimizes the degradation of the load spreading angle with the number of passages. In fact, the open structure of geogrids allows the interlocking of the grains of soil, so that the grains are confined by the geogrid, which provides tensile forces at low strain; interlocking and geogrid tensile forces (which produce horizontal compressive stresses on soil granules) prevent lateral movement and expansion of soil under load, produced by the Poisson's ratio of the soil. Thus, the elastic modulus of the ballast is maintained for much longer as the number of passages increases and consequently the degradation of the angle of load diffusion decreases. Based on the above experimental evidence and empirical criteria, the influence of the geogrid is introduced by Leng and Gabr [12] by means of the tensile strength of geogrids $T_{2\%}$ at 2% strain. The coefficient k_2, which defines the degradation of α_b, is related to the tensile strength of geogrids $T_{2\%}$ by the following empirical equation:

$$k_2 = (a_t / h)^{0.81} \cdot \text{Max}\,[(0.58 - 0.000046\, T_{2\%}{}^{4.5}),\, 0.15] \quad (36)$$

The minimum value of k_2 is for N = 1 and is equal to 0.15. Equation (36) contains the intuitive fact that as the thickness h of the ballast increases the degradation of α_b (and then the coefficient k_2) is decreased.

4.11 Bearing capacity mobilization factor

The load bearing capacity of the subgrade, according to the scheme in Figure 3, is defined with the classic formula:

$$q_c = N_c \cdot c_u \quad (37)$$

where: N_c = bearing capacity factor; c_u = undrained shear strength of subgrade (kPa).

Leng and Gabr [12] assumed the following values for the bearing capacity factor N_c:

- N_c = 3.80 for unstabilised unpaved roads (N_c value in the case of saturated cohesive soil conditions at the elastic limit in axially symmetric conditions);
- N_c = 6.04 for unpaved roads stabilised with geogrids (N_c value in the case of saturated cohesive soil with maximum shear stress at the interface).

The same N_c values will be assumed here for unstabilised and stabilised railway ballast and subballast.

The well known empirical relation between CBR_{sg} value and the undrained shear strength of subgrade c_u (kPa) can be used (Giroud and Han [18]):

$$c_u = 30 \cdot CBR_{sg} \quad (38)$$

Leng and Gabr [12] introduced the bearing capacity mobilization factor, m, which allows the expression of mobilized resistance as a function of the bearing capacity and the vertical stress applied on subgrade:

$$m = N_{cm} / N_c = \sigma_{bs} / (N_c \cdot c_u) \quad (39)$$

where: N_c = bearing capacity factor of the subgrade; N_{cm} = mobilized bearing capacity factor of the subgrade; c_u = undrained cohesion of the subgrade.

Leng and Gabr [12], based on experimental evidence, found the following empirical relationship:

$$m = \left(1 - e^{-0.78 \cdot \left(a_t/h\right)}\right) \cdot \frac{N_{cm}}{N_c} \quad (40)$$

The mobilization of bearing capacity of the subgrade occurs at the cost of a settlement on the upper surface of the subgrade itself, i.e. in proportion to the depth of the rut that is created at the ballast – subgrade interface. Leng and Gabr [14] assumed that at the critical value r_{cr} of the rut depth the entire load-bearing capacity of the subgrade is mobilised; the critical rut r_{cr} (m) is related to the number of passes through the following empirical relationship:

$$r_{cr} = 0.025 \cdot (0.125 \cdot \log N + 1.5) \quad (41)$$

By assuming that the mobilization of bearing capacity increases linearly with the depth of rutting r at the ballast – subgrade interface, Equation (39) becomes:

$$m = \left[1 - e^{-0.78 \cdot \left(a_t/h\right)}\right] \cdot \frac{r}{r_{cr}} \quad (42)$$

where the rut depth r has to be calculated with Equation (30) for unstabilised ballast, and with Equation (31) for stabilised ballast.

The mobilized bearing capacity q_{cm} becomes:

$$q_{cm} = m \cdot N_c \cdot c_u \quad (43)$$

4.12 Minimum thickness of ballast

The vertical stress at ballast – subgrade interface σ_{bs} shall be less than the load bearing capacity mobilized in the subgrade, so as not to cause failure of the subgrade due to an excess of localized settlements, that is of rut depth. Still considering the load from sleeper to ballast applied on a square area of size a_t, and the trapezoidal distribution of load in ballast (Figure 10), the following condition shall be verified:

$$\sigma_{bs} = \frac{P_{dyn}}{(a_t + 2 \cdot h \cdot \tan\alpha_{bN})^2} \leq m \cdot N_c \cdot c_u \quad (44)$$

From equation (44) we obtain the minimum thickness h of ballast:

$$h = \frac{1}{2 \cdot \tan\alpha_{bN}} \cdot \left(\sqrt{\frac{P_{dyn}}{m \cdot N_c \cdot c_u}} - a_t\right) \quad (45)$$

The model described above is based on the results of cyclic loading tests performed in the laboratory (Leng and Gabr [12]). To make the method more closely to actual conditions, a model factor, equal to 0.85, is introduced in Equation (45); then, assuming that the degradation of the load spreading angle in ballast with the number of passes is given by Equation (35), Equation (45) becomes:

$$h = \frac{0.85\cdot(1+k_2\cdot\log N)}{2\cdot\tan\alpha_{b1}} \cdot \left(\sqrt{\frac{P_{dyn}}{m\cdot N_c\cdot c_u}} - a_t\right) \tag{46}$$

Since (P_{dyn} / a_t^2) shall be equal to the vertical stress transmitted from sleeper to ballast, σ_{sb}, Equation (46) can be rewritten as:

$$h = \frac{0.85\cdot a_t\cdot(1+k_2\cdot\log N)}{2\cdot\tan\alpha_{b1}} \cdot \left(\sqrt{\frac{\sigma_{sb}}{m\cdot N_c\cdot c_u}} - 1\right) \tag{47}$$

Considering that the thickness h of the base appears in both members of Equation (46) or Equation (47), they should be solved by an iterative process.

By solving Equation (46) with respect to the load P_{dyn}, we obtain the following equation:

$$P_{dyn} = m \cdot N_c \cdot c_u \cdot \left(\frac{2\cdot h\cdot\tan\alpha_{b1}}{0.85\cdot(1+k_2\cdot\log N)} + a_t\right)^2 \tag{48}$$

By solving Equation (47) with respect to the vertical stress σ_{sb} we obtain the following equation:

$$\sigma_{sb} = m \cdot N_c \cdot c_u \cdot \left(\frac{2\cdot h\cdot\tan\alpha_{b1}}{0.85\cdot a_t\cdot(1+k_2\cdot\log N)} + 1\right)^2 \tag{49}$$

Equation (48) or Equation (49) can be used to calculate h when the P_{dyn} or σ_{sb} value is known: starting with a minimum value of h (e.g. h = 0.15 m), h is increased in small steps until the value calculated with Equation (48) or Equation (49) becomes equal to the known value P_{dyn} or σ_{sb}.

Equations (46), (47), (48), and (49) afford to calculate the minimum thickness both for unstabilised ballast, h_u, and for stabilised ballast, h_s: for unstabilised ballast, the parameter k_2 has to be calculated from Equation (36) with $T_{2\%} = 0$, and the parameter m has to be calculated using the rut depth ru from Equation (30); for stabilised ballast the coefficient k_2 has to be calculated from Equation (36) with the $T_{2\%}$ value for the selected geogrid, and the parameter m has to be calculated using the rut depth rs from Equation (31). The value of h used at each iteration influences the parameters k_2, tan α_{b1}, and m.

4.13 Stabilisation of the subballast

When the total granular thickness h'_g is composed of the ballast thickness h'_b and subballast thickness h'_{sb} (Figure 4b), from Eqs. (1) and (2) we get:

$$h'_g = h_e \cdot \chi_2 \tag{50}$$

$$h'_b = \frac{h_g}{\chi_1\cdot\chi_2} \tag{51}$$

$$h'_{sb} = h_g - h'_b \tag{52}$$

The above formulas can be used to calculate the thickness of ballast and subballast both in the unstabilised and stabilised cases.

If a lower stabilized ballast thickness than h'_b is desired, it is possible to set the stabilized ballast thickness h''_{bs}.

In this case the stabilized subballast thickness has to be recalculated, which can be done by considering the decrease in equivalent ballast thickness:

$$\Delta h_{bs} = (h'_{bs} - h''_{bs}) \cdot \chi_1 \tag{53}$$

The granular thickness becomes:

$$h''_g = h'_g + \Delta h_{bs} \tag{54}$$

The stabilized subballast thickness, with the set h''_{bs} stabilized ballast thickness, becomes:

$$h''_{sbs} = h''_g - h''_{bs} \tag{55}$$

Note that, for the stabilised design with ballast and subballast, the above formulas are valid only if both ballast and subballast are stabilised with geogrids having the same tensile strength at 2 % strain, $T_{2\%}$; anyway the opening size of the geogrid for ballast stabilisation and for subballast stabilisation need to be different and suited to the granular size distribution of the two materials.

Compared to the case of design with ballast only, the design with ballast and subballast requires a higher total granular thickness, yet the thickness of ballast gets reduced.

4.14 Calibration and limits of the method

The method of Leng and Gabr [12,13,14] was validated by in situ tests on existing roads: for unstabilised road this design method predicts a base thickness greater than that resulting from in situ tests (i.e. the method in this case is conservative); in the case of geogrid stabilised roads the results of calculations by the method of Leng and Gabr are in good agreement with the results of in situ tests. The design method proposed here for the stabilisation of railway structures is an extension of the method of Leng and Gabr, based on the results of available research, reported in the previous paragraphs. The calibration and validation of the method requires measurements of existing railway lines with ballast and / or subballast stabilised with geogrids. This research has not yet been performed, so it is not possible yet to define the precise limits of the present design method. The subballast and ballast thickness calculated with the proposed method (see example below) are still in good agreement with the results of the researches mentioned above.

4.15 Design example

The design method is applied to the case of a railway line on soft soil.

The input data are shown in Figure 11.

Results of calculations according to the design method are shown in Figure 12.

INPUT DATA	SYMBOL	CARGO TRAIN		PASSENGERS TRAIN	
		VALUE	UNIT	VALUE	UNIT
Design axle load	W	250	kN	170	kN
Train speed	C	100.00	km/h	300.00	km/h
Wheel diameter	D_w	0.90	m	0.92	m
Rail width	br	0.15	m	0.15	m
sleeper thickness	t_s	0.16	m	0.16	m
sleeper width	b_s	0.26	m	0.26	m
Load spreading angle in sleepers	α_s	60.00	deg	60.00	deg
Number of trains/day	N_t	5	kN	15	kN
Number of cars/train	N_w	50	-	8	-
Number of axis/car	N_p	4	-	4	-
design life	T	20	years	20	years
Thickness of soft subgrade	H_s	5			m
CBR of ballast	CBR_b	100.0			-
CBR of subballast	CBR_{sb}	30.0			-
CBR of subgrade	CBR_{sg}	1.50			-
Poisson ratio of ballast	μ_b	0.35			-
Poisson ratio of subballast	μ_{sb}	0.42			-
Poisson ratio of subgrade	μ_{sg}	0.30			-
Ballast type (1 = Basalt, 2 = Granite, 3 = Dolomite)		2			-
Subgrade soil type (1 = Fat clay; 2 = Lean clay; 3 = Elastic silt; 4 = Silt)		1			-
Tensile strength of geogrid reinforcement at 2 % strain	$T_{2\%}$	10.50			kN/m
Set stabilised ballast thickness with subballast below (= 0 if automatic calculation is required)	h_b^*	0.35			m

Figure 11. Input data for the design example

DESIGN SUMMARY	UNSTABILISED			STABILISED		
	SYMBOL	VALUE	UNIT	SYMBOL	VALUE	UNIT
Ballast thickness without subballast	h_u	2.41	m	h_s	0.88	m
Ballast vertical deformation without subballast	d_{vu}	0.110	m	d_{vs}	0.040	m
Ballast thickness with subballast below	h'_{bu}	1.02	m	h'_{bs}	0.41	m
Subballast thickness	h'_{sbu}	1.65	m	h'_{sbs}	0.56	m
Set stabilised ballast thickness with subballast below				h''_{bs}	0.35	0
Subballast thickness with h''_b ballast thickness				h''_{sbs}	0.76	0
Ballast vertical deformation with subballast below	d'_{vu}	0.046	m	d'_{vs}	0.016	m
Rut depth at the ballast – subgrade interface	r_u	0.012	m	r_s	0.003	m

Figure 12. Summary of results for the design example

From the example it is clear that geogrid stabilisation affords a remarkable reduction in the thickness of ballast and subballast.

It has to be noted that also the vertical deformation of ballast and the rut depth at the ballast – subgrade interface are remarkably reduced: hence, in many cases of railway on soft soil, only with geogrid stabilisation it would be possible to keep deformations within acceptable limits.

5. Conclusions

The paper presents a design method for geogrid stabilised railway ballast and subballast, which considers the wheel loads, the number of wheel passes, the geometry of the railroad, the characteristics of the subgrade and the properties of geogrids.

The proposed design method is an extension of Leng and Gabr method [12,13,14] , for the design of geogrid stabilisation of road bases, and it is based on the same engineering principles.

For the development of the design method the latest research on the behavior of unstabilised ballast, subballast, and subgrade under railway loading has been considered.

The authors believe that the proposed design method for geogrid stabilised ballast and subballast is an important advancement, which will allow important savings in the use of selected granular materials in railway projects.

References

[1] EN 13674-1:2011+A1:2017. Railway applications - Track - Rail - Part 1: Vignole railway rails 46 kg/m and above. *CEN*, Bruxelles, Belgium

[2] EN 1991-2 2003 Eurocode 1: Actions on structures - Part 2: Traffic loads on bridges. *CEN*, Bruxelles, Belgium

[3] AREA 1996 Manual for Railway Engineering **1** *American Railway Engineering Association (AREA)*, Washington, DC, USA

[4] AREMA 2003 Practical Guide to Railway Engineering, American Railway Engineering and Maintenance of Way Association. *Simmons-Boardman Publishing Corporation*. Maryland, USA

[5] Raymond G P 1978 Design for railroad ballast and subgrade support. *Journal of the Geotechnical Engineering Division* **104** pp 45-60

[6] Heath D L, Shenton M J, Sparrow RW, and Waters JM 1972 Design of conventional rail track foundations. *Proceedings of the Institution of Civil Engineers* **51** pp 251-67

[7] Li D and Selig E T 1998a Method for railroad track foundation design. I: Development. *Journal of Geotechnical and Geoenvironmental Engineering* **124** p 316

[8] Li D and Selig E T 1998b Method for railroad track foundation design. II: Applications. *Journal of Geotechnical and Geoenvironmental Engineering* **124** p 323

[9] Sayeed A 2016 Design of Ballasted Railway Track Foundations using Numerical Modelling with Special Reference to High Speed Trains. *PhD thesis, Curtin University*. Perth, Australia

[10] Li D, Hyslip J, Sussmann T, and Chrismer S 2002 *Railway Geotechnics*. CRC Press, Tailor & Francis Group, Broken Sound Parkway, NW, USA

[11] Stewart H E 1982 The prediction of track performance under dynamic traffic loading. *PhD Thesis, University of Massachusetts*, Amherst, Massachusetts, USA

[12] Leng J and Gabr M A 2002 Characteristics of geogrid-stabilised aggregate under cyclic load. *Journal of Transportation Research Board*, No. 1786, National Research Council, Washington, DC, USA, pp 29-35, November 2002

[13] Gabr M A 2005 Numerical analysis of stress-deformation response in stabilised unpaved road sections *Geosynthetics International* **12** pp 111-9

[14] Leng J and Gabr M A 2006 Deformation-Resistance Model for Geogrid-Stabilised Unpaved Road. *Journal of Transportation Research Board*, No. 1975, National Research Council, Washington, DC, USA pp 146-54, November 2006
[15] Rimoldi P 2012 Design method for railway bases reinforced with geogrid. *Proc. 2nd International Conference on Transportation Geotechnics*. Sapporo, Japan
[16] Odemark N 1949 Investigations as to the Elastic Properties of Soils and Design of Pavements According to the Theory of Elasticity. *Statens Vag Institute Meddeland* **77** Stockholm, Sweden
[17] Heukelom W and Foster C R 1960 Dynamic testing of pavements. *Journal of Soil Mechanics and Foundations Division*. American Society of Civil Engineers **86** No. SM1.
[18] Giroud J P and Han J 2004 Design method for geogrid-stabilised unpaved roads – Part I: theoretical development. *ASCE Journal of Geotechnical and Geoenvironmental Engineering* **130**pp 776-86
[19] Facciorusso J, Madiai C, Vannucchi G 2011 Pressioni di contatto e diffusione delle tensioni in un semispazio elastic. *Dispense di Geotecnica*. University of Firenze, Italy (In Italian)
[20] Atalar C, Das B M, Shin E C, and Kim D H 2001 Settlement of geogrid reinforced railroad bed due to cyclic load. *Proceedings of the International Conference on Soil Mechanics and Geotechnical Engineering*, Istanbul pp 2045-8
[21] Jeffs T and Tew G P 1991 A review of track design procedures, **2** Sleepers and Ballast. *BHP Research - Melbourne Laboratories*. Railways of Australia, Victoria, Australia
[22] Selig E T and Waters J M 1994 *Track Geotechnology and Substructure Management* Thomas Telford Press
[23] USACE 2008 Engineering Guidance Design Manual. 7th edition. *U.S. Army Engineering Support Center*. Huntsville, AL, USA
[24] Indraratna B, Salim M W, and Rujikiatkamjorn C 2011a *Advanced Rail Geotechnology - Ballasted Track*. Taylor & Francis Group, London, UK
[25] Pagano L and Sica S 2009 Vulnerabilità delle reti viarie. In "*Costruzioni in terra*", Gruppo Nazionale per la Difesa dai Terremoti, Progetto VIA. Roma, Italy. (in Italian)
[26] Shahin M A 2009 Design of ballasted railway track foundations under cyclic loading. *Proceedings of the 2009 GeoHunan International Conference – Slope Stability, Retaining Walls, and Foundations*, Changsha, Hunan, China pp 68-73
[27] Vecchiotti M and di Prisco C 2006 Comportamento Meccanico di Rilevati Ferroviari rinforzati sottoposti a carichi ciclici. *Proc. XVIII Convegno Nazionale Geosintetici. L'Ingegnere e l'Architetto*, n.1 (in Italian)
[28] Li D 1994 Railway track granular layer thickness design based on subgrade performance under repeated loading. *PhD Thesis, University of Massachusetts*, Amherst, Massachusetts, USA
[29] Li D and Selig E T 1996 Cumulative plastic deformation for fine-grained subgrade soils. *Journal of Geotechnical Engineering*, **122**pp 1006-13
[30] Fannin R J and Sigurdsson O 1996 Field Observations on Stabilization of Unpaved Roads with Geosynthetics. *ASCE Journal of Geotechnical Engineering*, **122** pp 544-53

Experimental study to evaluate soil water retention curve of HYPER clay geosynthetic clay liner

M K Khan[1], M D Camillis, G D Emidio and A Bezuijen

[1]Laboratory of Geotechnics, Ghent University, Technologiepark 905, 9052 Zwijnaarde, Belgium

E-mail:muhammadkhizar.khan@ugent.be

Abstract. Geosynthetic clay liners are widely used as hydraulic barrier due to their low hydraulic conductivity but bentonite in the liners loses its effectiveness due to significant thermal fluxes by both diurnal and seasonal heating and cooling cycles. Modified sodium carboxy methyl cellulose-based bentonite clay (HYPER clay) has shown better hydraulic performance in both situations. A possible reason for this improved performance of HYPER clay based geosynthetic clay liner is the improvement in the suction under changing thermal conditions. Thus, the relationship between soil suction and moisture content, also called the soil water retention curve, needs to be estimated. Therefore, we investigated the soil-water retention curve of the HYPER clay based geosynthetic clay liner and compared it with the untreated clay based geosynthetic clay liner. The article presents the suction test results on wetting path conducted on geosynthetic clay liner prototypes containing HYPER and untreated clay assessed by the contact filter paper method and the relative humidity sensor. The results showed that the geosynthetic clay liner containing HYPER clay has a high volumetric water content and thus, high water retention compared to untreated bentonite at a given suction value. In other words, the HYPER clay can be considered as a potential alternative to conventional bentonite due to its improved water retention capacity.

1. Introduction

Geosynthetic clay liners (GCL) have demonstrated the potential to act as a hydraulic barrier owing to their low hydraulic conductivity to water, which leads to many applications such as landfill liners, tailing ponds, dams, and railway liners [1]. To efficiently act as a hydraulic barrier the GCL must be sufficiently hydrated. Also, for long term use, environmental factors that can influence the performance of GCL must be considered. The most significant environmental factor controlling the durability is thermal changes due to both diurnal and seasonal heating and cooling cycles, which results in shrinkage and consequently, the loss of GCL panel overlap [2, 3].

The key constituent of GCL is bentonite, which controls their sealing properties. The bentonite showed very good performance in restricting the flow of water, however, in case of regular wet and dry cycles, its sealing performance declines with each cycle. Hence, Sodium Carboxy Methyl Cellulose (Na-CMC) treated bentonite clay, known as HYPER clay (HC), has been introduced which displayed better hydraulic properties than untreated bentonite clay (UC) [4]. Besides, the effect of wet and dry cycles on the hydraulic conductivity of HC and UC has recently been investigated [5]. The authors demonstrated that the HC barriers exhibited much lower hydraulic conductivity (three orders of magnitude) as compared to UC, after the fourth wet-dry cycle, which proved the improved behaviour of HC. The whole process of wetting and drying cycles involves the change in the moisture content of the GCL, thereby leading to unsaturated condition.

To characterize the unsaturated hydraulic performance of GCL, the relationship between soil suction and moisture content, also called water retention curve (WRC), should be measured [6]. Soil suction is one of the most important parameters describing the conduct of unsaturated soils, which is a measure of the free energy of the pore-water in the soil. Indeed, soil suction describes the susceptibility of the soil to retain water, which is important to explain the long-term performance of the bentonite. Since the water retention of bentonite under changing suction determines its behaviour during wet and dry cycles. This underlies the need to investigate the WRC of the HC, which according to the authors' knowledge, has not been investigated yet.

This paper presents the preliminary study of the wetting WRC curve of HC and its comparison with that of UC, to stimulate the unsaturated condition during the heat waves and seasonal rainfalls, by using contact filter paper method. Furthermore, the HC and UC total suction measured by filter paper method and relative humidity sensor are also compared.

2. Materials and methods

The following sections describe the detailed procedure adopted to prepare and subsequently test the specimens of both HC and UC for WRC. The procedure used was based on a combination of the (American Society for Testing and Materials standard, D5298), ASTM D5298, Acikel et al., 2015; Risken, 2014 [7-9].

2.1. Specimen preparation

Extensive preparation of bentonite specimens is required to minimize disturbance, simulate field conditions, and maintain consistency with previous studies [7, 10]. GCL prototypes and powder samples of UC and HC were used to measure the water retention (i.e., wetting curve) by using filter paper method.

Samples were hydrated to target moisture contents before suction measurements under a 1.5 kilo Pascals overburden pressure. This load was meant to allow near free swell conditions yet promoting adequate contact between the specimens and the filter paper. The specimens were then sealed in two plastic bags. Some water was sprayed between the two plastic bags to further reduce moisture losses from the specimen due to diffusion [8, 10]. The samples were stored for 30 days in a temperature-controlled room of 20°C ± 2°C before suction measurements.

2.1.1. Powder sample. UC and HC were evenly spread with a density of 4.5 kilogram per square meter (kg/m^2) area, typical of GCL, in a 50-millimetre diameter cell. Figure 1 shows the stepwise procedure followed during the specimen preparation. The initial water content of the samples was determined to measure the exact dry mass and to calculate the water necessary to reach target water contents. An acrylic plate and ring were attached with the help of tape, to avoid loss of bentonite during handling. A stack of three dry filter papers Whatman number 42 were placed on the base of the cell. The outer two filter papers were used to protect the inner filter paper from bentonite. This method differs from previous research [*8*] as the filter papers were placed in contact with the soil from the hydration phase, as we used powder bentonite instead of GCL specimens. As can be seen in figure 1, the bentonite was carefully spread on the filter papers and two drainage layers were used to distribute the hydrating liquid and to hold the applied load. Deionized water was slowly poured on the first drainage layer in contact with the bentonite before applying the second drainage layer and the load.

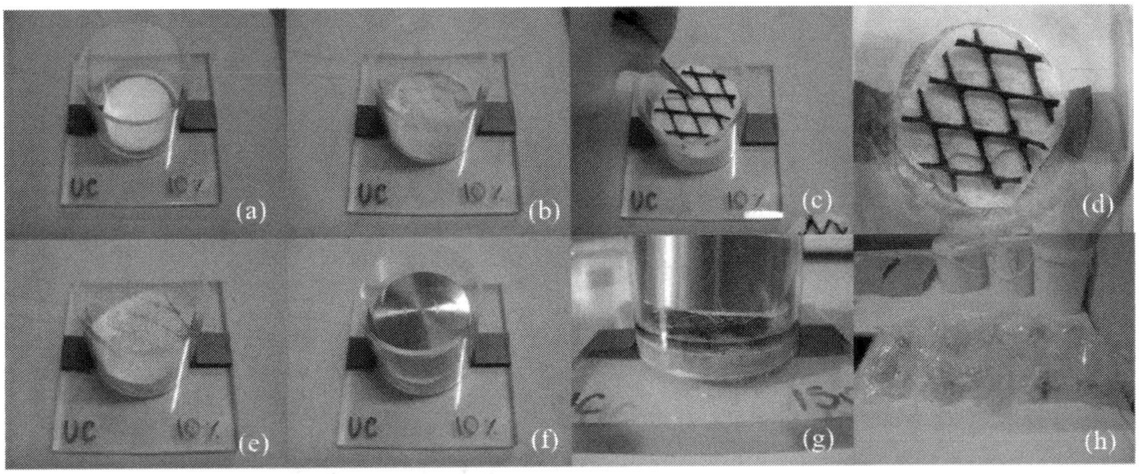

Figure 1. Stepwise procedure (from (a) to (h)) for sample preparation for WRC wetting curve

2.1.2. GCL prototypes. The needle punched GCL prototypes specimens composed of a layer of UC or HC (abbreviated as GCL_UC and GCL_HC, respectively) sandwiched between a woven carrier and a non-woven cover geotextile was used. GCL specimens of 50-millimetre diameters were cut from the prototype rolls with scissor. Before cutting, the perimeter of the specimen was first wetted with a small amount of deionized water to minimize the loss of the bentonite. Half of the mass of water used was then accounted for the water content, considering that the other half of that water mass was absorbed by the GCL panel [*8*]. The initial water content of the bentonite present in the GCL sample was determined to predict the amount of water necessary to hydrate the GCL specimens to their target moisture contents. The perimeter of each GCL specimen was covered with weighted tape, thus reducing preferential drying on the edges. A single GCL specimen was used for each test. For instance, Barroso et al., 2006 [11] showed that using a single GCL specimen, it is possible to eliminate the moisture variation between dual GCL specimens. The water was distributed on both sides of the GCL. The GCL were placed on the acrylic plate with the carrier geotextile oriented downwards and the load was then applied. The setup was then sealed in two plastic bags and some water was sprayed between them to further reduce moisture losses from the specimen due to diffusion. The GCL specimens were hydrated for a period of 30 days in a temperature-controlled room (20°C ± 2°C).

2.2. Measurement of soil potential using filter paper

The principle of the filter paper test method is based on the moisture uptake by filter paper until there is a balance in potential between the filter paper and the soil [12]. The technique involved the use of filter papers in direct contact with the specimens, prepared according to Section 2.1 until suction equilibrium was attained. The filter papers were placed in contact with the specimens from the beginning of the hydration period (30 days) for the tests with powder bentonites. This choice was due to the difficulty of placing the filter papers between the dry powder bentonite and the acrylic plate. On the contrary, in the case of GCL samples, a stack of three filter papers was placed after hydration period. Subsequently, the specimens were kept for further 7 days for equilibrium after placing the filter papers, whereas, ASTM D5298 requires just 7 days of equilibrium period. Whatman Grade 42 filter papers were used throughout the experimental program. Besides, the total suction was measured on just GCL samples at 10% gravimetric water content. In this test, two filter papers were placed on GCL specimen, but isolated with the help of wire mesh, for 7 days after 30 days of hydration period (Figure 2). After the hydration and equilibrium period, the moisture content of the filter paper was measured according to the ASTM D5298, and the suction was calculated from the filter paper's moisture content using the pre-determined calibration curve specified in ASTM D5298.

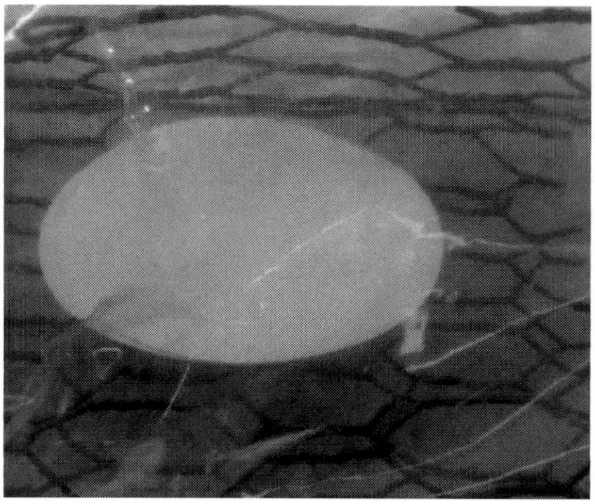

Figure 2. Total suction setup after hydration period

2.3. *Relative humidity sensor test*

The relative humidity test was performed, for high total suction potentials, by relative humidity device "Model-Lascar EL-USB-2" equipped with relative humidity and temperature sensor. The sensor was inserted in the nested bags containing the specimens and activated on a computer. The sensor end of the relative humidity device was centered over the specimen surface (Figure 3). The sensor recorded relative humidity and temperature measurements every 5 minutes for a period of 12 hours which was typically enough time to attain humidity equilibrium in the plastic bag apparatus [8]. The suction was given by the following equation

$$h = -\frac{RT\rho}{M}.\ln(R_h) \quad (1)$$

where,

h is suction in kilo Pascals (kPa), R is the universal gas constant, which is equals to 8.31 Joule per mole kelvin ($J.mol^{-1}.K^{-1}$), T is the temperature in Kelvin (K), R_h is the relative humidity fraction, ρ is the density of water in kilogram per cubic meter (kg/m^3) at temperature T and M is the molecular weight of the water, which is equals to 0.018 kilogram per mole (kg/mol).

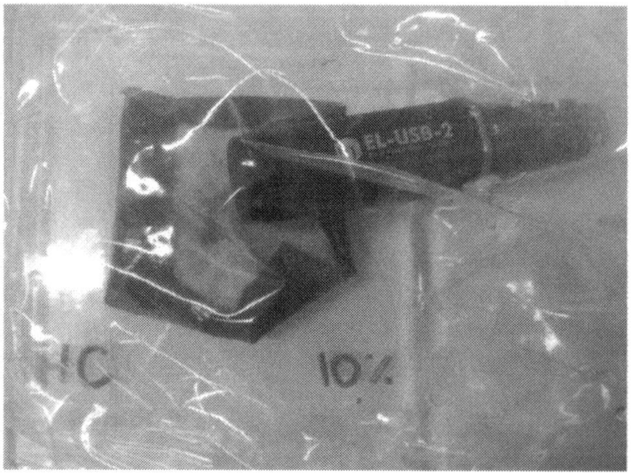

Figure 3. Relative Humidity Sensor during data recording

3. Results and discussion

The section is divided into two parts. The first part highlights the powder WRC whereas, the second part explains the GCL WRC.

3.1. Powder soil water retention curve

The WRC of powder UC and HC relating to gravimetric water content are shown in figure 4. The volumetric water content was not calculated as the initial thickness was not measured to avoid sample disturbance. Both powder UC and HC presented increasing suction with decreasing water content. In general, high water content leads to relatively low suction. Besides, the moisture content is highly variable upon small changes in suction in both the samples. On the contrary, low water contents provide high suction values due to the small amount of water, which remained tightly adsorbed in the intra-aggregate voids and clay particle surfaces [13].

The figure shows comparable suctions at high water contents for both samples. This can be due to the diffuse double layer of both bentonites which is sufficiently open with the high amount of deionized water leading to osmotic swelling. The gravimetric water content sharply decreased increasing the suction for UC. On the contrary, the decrease of the gravimetric water content of HC was more gradual. Thus, for the same water content, the suction is greater in HC as compared to UC and hence more retained water potential. The more retention potential signifies the durability of healing and better control of shrinkage cracking. However, at low water contents, the water retention of the polymer treated clay and UC was approximately the same.

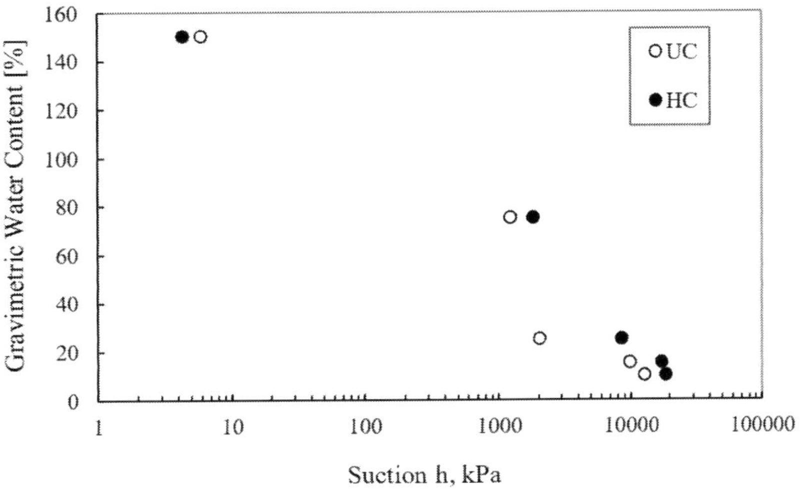

Figure 4. Suction versus gravimetric water content of UC and HC

3.2. Geosynthetic Clay liners soil water retention curve

The WRC of needle punched GCL prototypes are presented in figure 5, which illustrates the volumetric water contents versus GCL_UC and GCL_HC suction. The volumetric water content was calculated to consider the volume change after hydration. Indeed, it provided more reliable results compared to the gravimetric water content as it considered the change in thickness of the samples. Strong evidence of the improved performance of the GCL_HC compared to GCL_UC can be seen, which shows that the water retention trend is shifted towards higher volumetric water contents. The volumetric water content of GCL_HC was higher, or similar, compared to the one of GCL_UC along with the all range of suction values investigated. In other words, HC can retain more water compared to UC. The continuous curve was derived by fitting the available data with Fredlund and Xing equation [14]. It is evident from the figure that the curve breakpoint of GCL_HC corresponds to a suction value higher compared to that of

GCL_UC. The suction related to the curve breakpoint might be identified as the air-entry value of the wetting path WRC, after which the water started to be replaced by air. Furthermore, it can also be observed that the residual suction of conventional GCL is lower compared to the one containing the HC. Therefore, in the presence of high suctions and similar environmental conditions, GCL_HC is more capable of retaining water in its structure, leading to better rehydration and lower hydraulic conductivity. The irreversible adsorption of the polymer in the bentonite structure enhanced water adsorption likely increasing the diffuse double layer thickness. These findings provided further support to the hypothesis that the polymer treatment had a beneficial effect on the swelling and hydraulic performance of bentonite liners [4]. Undoubtedly, this study demonstrated the improved adsorption and retention capacity yielded by the HC technology. A comparison with the water retention behaviour, reported by [15], is presented in figure 6 in terms of volumetric moisture content versus suction, which highlights the published values of the drying and wetting curves of a needle-punched GCL_UC.

In general, this comparison indicated that the results obtained in the present study for GCL_UC were lower than the ones obtained by Beddoe et al. (2011) [15]. The relationship between volumetric water content and suction is likely similar at high suctions. The different performance might be related to the test procedure and the possibility of sample disturbance in picking and placing the filter papers for the next targeted water content which might have disturbed the distribution of the pores and affected the results. Therefore, additional tests with more controlled conditions are required to validate these results.

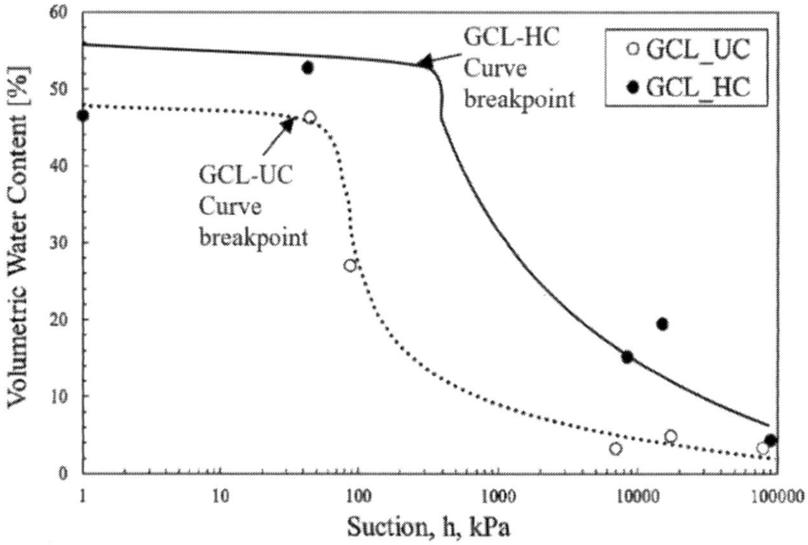

Figure 5. Volumetric water content versus suction of GCL_UC and GCL_HC

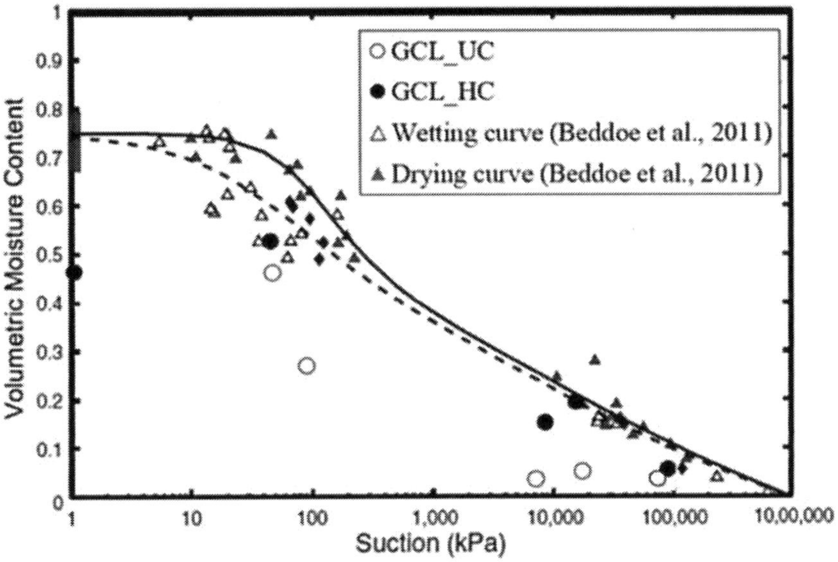

Figure 6. Comparison of the results obtained in this study with the results published by Beddoe et al. (2011) [15]

3.3. GCL total suction

The total suction represents the combine negative potential of osmotic and matric suction. The total suction of GCL_UC and GCL_HC at 10% gravimetric water content is displayed in figure 7. It can be observed from the figure that the total suction of GCL_HC is more than GCL_UC attributed to higher matric suction since the water of same electrical conductivity was used in both the samples depicting same osmotic potential. Furthermore, the suction measured by the humidity sensor is higher than that of filter paper due to high sensitivity of measurements involved in filter paper test. Besides, with increasing suction beyond 100,000 kilo Pascals, the humidity sensor was preferred to use due to increasing accuracy as compared to filter paper test, which needs further investigation [16].

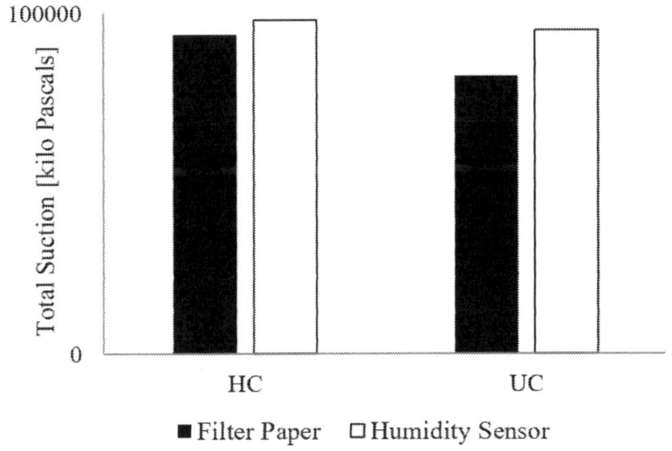

Figure 7. Total suction of GCL_UC and GCL_HC at 10% gravimetric water content

4. Conclusion

This study has several practical implications. Firstly, it demonstrated the enhanced retention capacity of the HC. Secondly, the residual volumetric water content of HC was higher compared to UC at a given suction. Finally, the current data provided insight into the understanding of the hydraulic performance of HC. Therefore, HC might be considered as a potential alternative to conventional bentonites due to its improved water retention capacity. This feature is of importance especially in the presence of unsaturated conditions and the occurrence of wet and dry cycles. Indeed, bentonite liners must act as a barrier against the passage of solutes and being able to withstand environmental changes, such as wet and dry cycles, for a longer period.

This research sets the basis for future studies on the moisture-suction behaviour of HC. In this framework, the drying path, the influence of electrolyte solutions, modification/changes in methodology, and the impact of wet and dry cycles on the constitutive relationship between moisture and suction need to be investigated.

References

[1] Bouazza A 2002 Geosynthetic clay liners *Geotextiles and Geomembranes* **20** 3-17

[2] Thiel R, Giroud J, Erickson R, Criley K and Bryk J 2006 Laboratory measurements of GCL shrinkage under cyclic changes in temperature and hydration conditions. In: *Proceedings of the 8th International Conference on Geosynthetics, Yokohama, Japan* pp 21-44

[3] Koerner G and Koerner R 2005 *Geosynthetics Research and Development in Progress* pp 1-6

[4] Di Emidio G 2010 Hydraulic and chemico-osmotic performance of polymer treated clays. Ghent University

[5] De Camillis M, Di Emidio G, Bezuijen A and Verástegui-Flores R D 2016 Hydraulic conductivity and swelling ability of a polymer modified bentonite subjected to wet–dry cycles in seawater *Geotextiles and Geomembranes* **44** pp 739-47

[6] Likos W J and Lu N 2002 Filter paper technique for measuring total soil suction *Transportation research record* **1786** pp 120-8

[7] Acikel A S, Singh R M, Bouazza A, Gates W P and Rowe R K 2015 Applicability and accuracy of the initially dry and initially wet contact filter paper tests for matric suction measurement of geosynthetic clay liners *Géotechnique* **65** pp 780-7

[8] Risken J L 2014 Development and Use of Moisture-Suction Relationships for Geosynthetic Clay Liners

[9] Standard A 1997 D 5298-94: Standard test method for the measurement of soil potential (Suction) Using Filter Paper *Annual Book of ASTM Standards* **4** pp 157-62

[10] Beddoe R, Take W and Rowe R 2010 Development of suction measurement techniques to quantify the water retention behaviour of GCLs *Geosynthetics International* **17** pp 301-12

[11] Barroso M, Touze-Foltz N and Saidi F 2006 Validation of the use of filter paper suction measurements for the determination of GCL water retention curves. In: *Proceedings of the 8th International Conference on Geosynthetics, Yokohama,* pp 171-4

[12] Almeida E L d, Teixeira A d S, Silva Filho F C d, Assis Júnior R N d and Leão R A d O 2015 Filter paper method for the determination of the soil water retention curve *Revista Brasileira de Ciência do Solo* **39** pp 1344-52

[13] Lu N and Likos W 2004 Rate of capillary rise in soil *Journal of geotechnical and Geoenvironmental engineering* **130** pp 646-50

[14] Fredlund D G and Xing A 1994 Equations for the soil-water characteristic curve *Canadian geotechnical journal* **31** pp 521-32

[15] Beddoe R A, Take W A and Rowe R K 2011 Water-retention behavior of geosynthetic clay liners *Journal of Geotechnical and Geoenvironmental Engineering* **137** pp 1028-38

[16] Hanson J, Risken J and Yesiller N 2013 Moisture-suction relationships for geosynthetic clay liners. In: *Proceedings of the 18 th International Conference on Soil Mechanics and Geotechnical Engineering, Paris 2013,* pp 3025-8

Scour protection effects on the dynamic response of jacket structures under extreme load events

A Khansari[1]**, H Hoyme**[2] **and H Oumeraci**[3]

[1] BBG Bauberatung Geokunststoffe GmbH & Co. KG, Espelkamp, Germany
[2] NAUE GmbH & Co. KG, Espelkamp, Germany
[3] Leichtweiss-Institute for Hydraulic Engineering and Water Resources, Technische Universität Braunschweig, Germany

E-mail: akhansari@bbgeo.com

Abstract. A well-validated Finite Element (FE) model of a jacket structure (OC4 jacket) with pile foundation together with a Computational Fluid Dynamic (CFD) model (OpenFOAM-based) set-up for extreme waves on the structure are applied to perform systematic comparative analyses of the dynamic response of the structure to extreme wave loads with and without scour effects. The pile foundation model is considered by implementing a non-linear soil model to the foundation using load-deflection (p-y) approach to capture the non-linear behaviour of the soil for large pile foundation displacements. The results show that, the scour protection improves the foundation bearing capacity and consequently, decreases the failure probability in terms of both serviceability and ultimate limit states. Moreover, it enhances the dynamic response of the structure under extreme load events caused by breaking waves. Accordingly, an innovative solution for scour protection of offshore-fixed bottom structures has been introduced by "Biodegradable Soft Rock" (sand filled containers produced from bio-degradable geotextile product) which is the newest development in this section.

1. Introduction

The dynamic response of jacket structures to non-breaking waves has been addressed in a large number of diverse studies. However, less attention has yet been paid to calculate their dynamic response under storm wave conditions which may generate breaking waves on the structure and may cause scour effects around the pile foundation of the structure due to the severe current and wave kinematics. Currently in practice, jacket structures are designed under the most extreme non-breaking wave, which may occur once in 100-years. In addition, the negative effects of scour on the dynamic response are neglected by the current design process of multi-member offshore structures in practice. The application of the current approach might result in considerable uncertainties in the estimation of extreme wave loads and consequently also in the dynamic response of the structure since the significant impact force caused by breaking waves on the structure is neglected. Moreover, it neglects the excessive hydrodynamic loads on the structure due to the scour effects. The implications of such extreme wave load events and the associated uncertainties for the dynamic response of the entire jacket structure, particularly including the response of the pile foundation, are still not fully satisfactorily addressed. This latter issue is even more crucial as the foundation piles of jacket structures for wind turbines are significantly shorter (i.e.

with a basically different behaviour) than those commonly found in jacket structures supporting oil and gas offshore platforms for which a large experience is available.

Seabed scour has generally two main effects on the dynamic response of offshore structures: (i) as the soil in the area close to the mudline is removed, the geodynamic mass contribution to inertia is decreased which may lead to the changes in the stiffness and mass matrixes and accordingly the natural frequency of the structure; (ii) scour may increase the hydrodynamic forces on the entire structure.

Although the occurrence of scour around piles of fixed-bottom offshore structures is a well-known phenomenon, only a very limited knowledge on the scour effects on the dynamic response of these structures is available. In this paper a CSD model of a jacket structure with pile foundation (section 2) has been set up together with a CFD model for the simulation of breaking waves on the jacket (section 3). The models are then weakly coupled to calculate the dynamic response of jacket structure and to model the scour effects around the pile foundation and the mudline (section 4). The set-up models are finally applied to perform a comparative analysis to identify the scour induced effects on the dynamic response (section 5).

2. Computational Solid Dynamic (CSD) model of the SP17 jacket structure

The jacket structure as modelled in this study was previously designed by Rambøll A/S ([1] and [2]) in the framework of UpWind project, which was the largest R&D wind energy project Europe.

The four-legged OC4 jacket structure has four levels of X-braces and is founded on four piles with a penetration depth of 45 m into the soil. The piles are grouted to the jacket legs. The transition piece (TP) between the jacket and the tower is a block of concrete that is penetrated by the upper parts of the four jacket legs (see Figure 1). The total height of the jacket from the mudline to the transition piece (TP) is 70.15 m. The water depth is 50 m. The conical tower has a total length of 68m leading to a realistic hub height over the mean sea level (MSL) of 90.55 m.

The dimension of the tubular members is illustrated in Figure 1. Each dimension set is illustrated with a specific colour describing the geometry of the members as well as outer diameter and wall thickness.

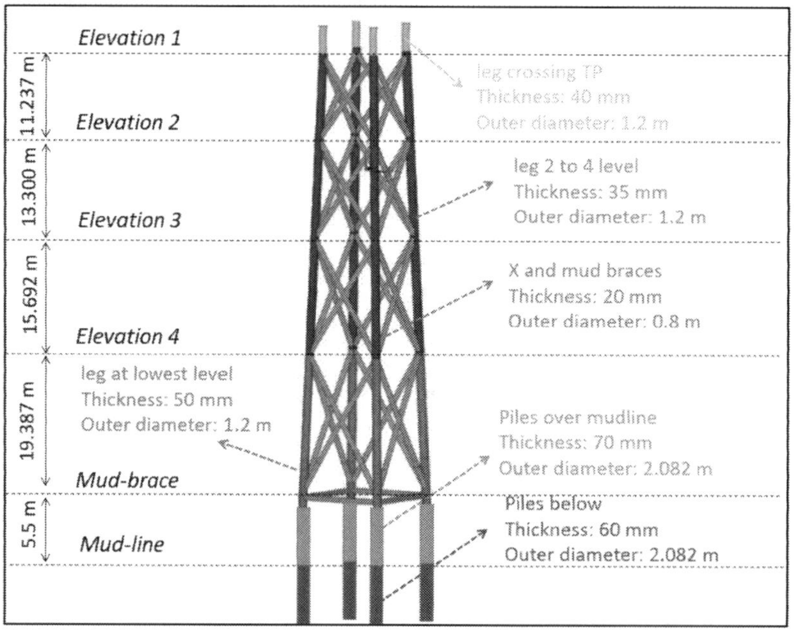

Figure 1. Geometry of the OC4 jacket structure (excluding the transition piece and tower)

2.1. OC4 jacket structure without pile foundation model

In the basic model of the OC4 jacket structure called hereafter ''clamped structure'', which has mainly been modelled by the available tools in previous research studies, the structure is cantilevered at the mudline, which means that all six degrees of freedom (DOF) are set to zero at the mudline. Therefore, only a short length of the piles over the mudline is considered in the model (4.5 meter). In this study, after completion of the evaluation of the performance of the modelled clamped structure, a particular focus is paid to the interactions between pile and soil and as a result, the basic model is extended with the pile foundation and a soil model (see section 2.2). The actual model of a jacket structures is very complicated and without proper idealization and simplifications, an effective modelling would be hardly possible. The simplifications considered in this study to model the jacket structure and the wind turbine can be briefly summarized as follows:

a) The wind turbine components such as rotors, nacelle and assembly (RNA) are modelled as a lumped mass on the top of the wind tower;

b) The rigid concrete block as a transition piece, which connects the wind turbine to the jacket substructure, is modelled by four horizontal stiff beams. The density of the beams surrounding the replaced concrete block is set so that they represent the equivalent mass of the concrete block;

c) In the actual model, the piles and the legs are overlapped in their connection region and the gap between them is filled with grout material. This piece is replaced with an equivalent tubular member representing the same mass and stiffness.

The performance of the numerical model in this study is evaluated in several steps by comparing the results with those of other numerical models developed for the clamped OC4 jacket structure under non-breaking wave loads. The following steps are performed to validate the numerical model and addressed in detailed in [3]:

(i) A *verification of the structural masses* of the OC4 jacket structure including the body of the jacket structure, the transition piece, the tower and the Rotor Nacelle Assembly is conducted through comparison of different studies;

(ii) A *modal analysis* is performed of the *clamped OC4 structure* in the absence of water and soil in order to compare the results obtained in this study with other studies (see Table 1);

(iii) The *total wave loads* induced by non-breaking waves calculated by superposing of quasi-static wave forces on members of the structure, is obtained. The total force response (TFR) at the base of the structure on the direction of the incident wave (fore-aft force) is computed and plotted together with the results from the other studies.

(iv) *The displacement of the clamped OC4 structure* (without pile foundation) under non-breaking waves is calculated and compared with other numerical models developed for the same structure.

The examination of all abovementioned analyses has shown satisfactory results in terms of static and dynamic response of the jacket structure. While the detailed validation results can be seen in [3], the well-validated model has been applied in this study to investigate scour effects on the dynamic response.

2.2. OC4 jacket structure with pile foundation model

The CSD model of the clamped OC4 jacket structure is modified by introducing a pile foundation with an appropriate soil model. The performance of the extended model is examined by comparison of the natural frequencies with the model developed by [4], which represents the only study performed so far on the pile and soil foundation of the OC4 jacket structure.

The natural frequency of the structure is significantly affected by the pile-soil interaction. As a result, the dynamic response of a jacket structure to environmental loads is strongly influenced by the soil behaviour and shall be based on an integrated analysis of the jacket structure, the foundation piles and soil. Diverse models may be utilized to determine pile-soil interaction for a jacket structure. However, only very few can account for the nonlinear response behaviour of the soil. In this study, the implemented pile-soil model for the foundation of the OC4 jacket structure is based on p-y curves developed in a research study (sponsored by a group of five oil companies) on laterally loaded piles for offshore structures including laboratory model testing and field tests with an instrument pile (according to [5] and [6]). These nonlinear springs connected to the pile foundation in different soil layers, are recommended also in several guidelines such as [7] and [8]. Such a modelling approach, called p-y method, has indeed provided relatively good results for offshore platforms. Figure 2 shows a schematic configuration of these springs in different soil layers. These springs work independently; i.e. the deformation of each spring doesn't affect the other springs behaviour. Springs introduced along a pile are classified in three groups: (i) springs resisting against lateral loads to describe the lateral pile-soil interaction (p-y); (ii) springs considering axial pile resistance due to internal and external pile skin friction (t-z); (iii) springs considering the tip resistance (Q-z).

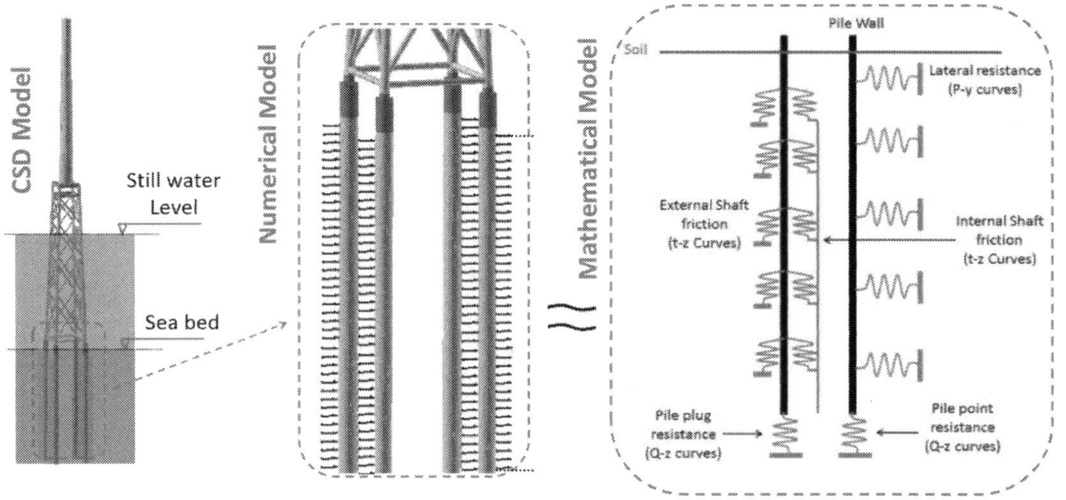

Figure 2. Pile-soil interaction modelling by different types of springs uniformly distributed along the pile

The FE model of the OC4 jacket structure is extended by introducing four foundation piles with a penetration depth of 45 m into the soil. The soil is modelled using several lateral springs connected each meter along the piles by considering the aforementioned non-linear p-y approach (Figure 2). The CSD model of the clamped OC4 jacket structure is modified to consider the effect of pile foundation and soil parameters by performing the following steps:

Step1: The CSD model of the OC4 jacket structure is extended by adding four piles with the diameter of 2.086m, the thickness of 7mm and the length of 45m.

Step2: The pile-soil interaction is modelled by implementing linear springs along the piles (soil plasticity and non-linearity not considered). The plausibility of the extended model is examined by comparing the natural frequencies of the model with the study by [4].

Step3: The pile foundation model is improved by implementing a non-linear soil model to the foundation using a p-y approach in order to consider the non-linear behaviour of the soil for higher displacements of the pile foundation.

2.3. *Performance of the CSD model with pile foundation*

The pile foundation model of the OC4 jacket is set-up for different soil types. In order to evaluate the performance of the extended CSD model for the structure with pile foundation, the natural frequencies obtained for the clamped structure (without pile foundation) and for the structure with pile foundation in clayey and sandy soil types are tentatively compared with the results of [4]. In the latter, the effect of the soil-structure interaction is considered with a particular focus on pile group effect on the response of a jacket support structure to non-breaking waves. The modal analysis is performed for the fully flexible jacket substructure and tower with a lumped mass atop, which represents a rigid RNA at the top of the structure, and the results are summarized in Table 1. As can be seen, the fore-aft natural frequency of the structure with piles in clay is 6% lower than that of the structure with piles in sand, showing a more flexible behaviour of the structure with piles founded in clay. Given the same jacket substructure and piles, the natural frequency of the entire structure increases with higher soil stiffness.

Table 1. Frequencies (Hz) of the OC4 jacket structure only clamped and with pile foundation in sandy and clayey soil types as obtained in this study and by Shi et al. [4]

Model	1st fore-aft	2nd fore-aft
Clamped, i.e. no pile foundation (This study)	0.3219	1.1869
With pile foundation in soft clay (This study)	0.2607	0.6510
With pile foundation in uniform sand (This study)	0.2748	0.8499
With pile foundation in layered sand (This study)	*0.2749*	*0.8512*
With pile foundation in layered sand (Shi et al, 2015)	*0.2771*	*0.8931*
With pile foundation in hard clay (This study)	0.2684	0.7541

It should be stressed that the OC4 jacket structure modelled in this study is symmetric as the turbine's blades, rotor, nacelle and assembly are modelled as a lumped mass, and springs representing soil are connected from both fore-aft and side-side directions. As a result, the fore-aft natural frequencies are equal to side-to-side. Therefore, the comparison of natural frequencies is performed only in fore-aft direction, showing a difference of less than 5% between this study and [4] for pile foundation in layered sand.

Comparison of the natural frequencies obtained from the modal analysis of the OC4 jacket structure with and without pile foundation (Table 1) indicates higher values of the natural frequency for the former case up to 20% and 45% for 1st and 2nd mode shapes, respectively. Therefore, the consideration of the pile foundation with a proper soil model is crucial as it may significantly affect the dynamic response.

3. Wave loading of the jacket structure

In this study the focus was laid on the dynamic response of the structure under breaking waves. Breaking wave induced forces on the entire OC4 jacket structure are calculated through superposition of: (i) the quasi-static force component obtained by means of the Morison equation [9] using the flow parameters calculated by a well-validated CFD model and; (ii) the slamming force component calculated using the prediction formulae as proposed by Khansari [3] for the estimation of breaking wave impact loads on the front, rear and side braces of the jacket structure.

3.1. Quasi-static component of the load

The quasi-static induced force by non-breaking wave on each member of the OC4 jacket structure is calculated using Morison equation. For this purpose, the flow velocity and acceleration obtained from a validated CFD model as set up in OpenFOAM are considered, together with the corresponding drag and inertia coefficients for the calculation of wave forces on the jacket structure. The total wave force is obtained by superposition of the quasi-static forces on each member of the jacket structure. The time series of the calculated forces are then applied on respective members of the jacket structure in the CSD model. Figure 3, shows a simulated non-breaking wave over its wave period on the OC4 jacket structure.

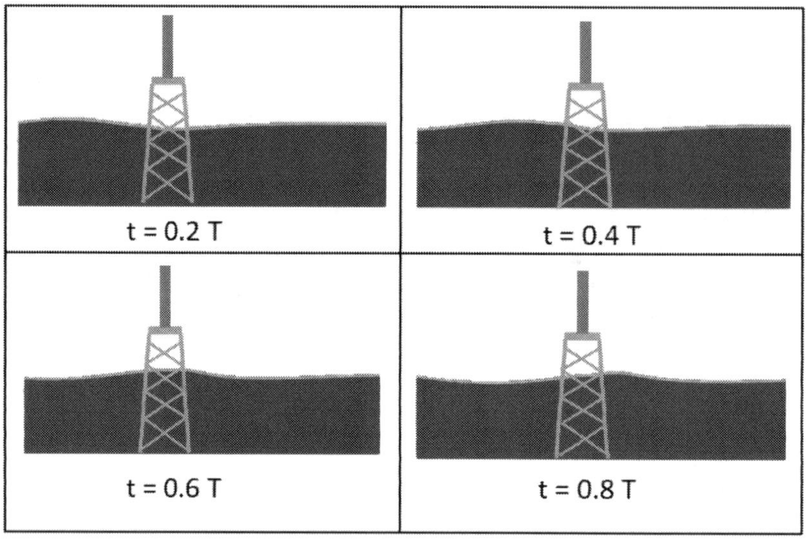

Figure 3. The simulation of quasi-static loads on the OC4 jacket structure by OpenFOAM

3.2. Slamming component of the load

For any prospective formulae for wave impact forces on structures, it should be stressed that both magnitude and duration of the impact force are not only strongly linked, but also equally important for the dynamic response of the structure. Khansari [3] proposed a new approach by combining the frequency response function (FRF) and the Duhamel integral to calculate both magnitude and duration of breaking wave impact forces on truss-type structures using the insight of physical processes which were gained from large scale wave tests. The slamming formulae proposed by [3] has been implemented as follows to calculate magnitude and duration of the impact force on the different faces of the OC4 jacket structure:

(i) *Front face of the jacket structure:* The breaking wave impact force and duration on the front face of the jacket is calculated according to the set of formulae by [3] as originally developed by combining the Frequency Response Function (FRF) approach and the Duhamel's integral method.

(ii) *Side braces of the jacket*: The study by [3] revealed that the wave impact force on the side braces of the structure are not high compared to those impact forces on the front and the rear faces (less than 10%). Therefore, the impact force on side braces is neglected in this study.

(iii) *Rear face of the jacket*: The wave slamming force on the rear face of a truss structure is affected by two main processes called *"sheltering effect"* and *"wave dropping effect"*. Sheltering effect for jacket structures occurs on the structure members behind the front face of the structure (downstream face) thus affecting the breaking wave, which might reach the sheltered members as a broken wave. Dropping effect is reduction of the wave crest height by losing its energy through the propagation from the front face to the rear face of the structure. Both sheltering and dropping effects are taken into account separately by means of the reduction factors as introduced in [3] to calculate the total impact force on the rear face.

The OC4 jacket structure was subject to different wave loads caused by breaking waves (Five load cases: LC1 to LC5 as defined in Figure 4). In each simulation, the characteristics of the incident breaking wave are kept constant, and only the distance between the incipient wave breaking location and the jacket structure is varied to generate different types of breaking waves on the jacket structure as follows:

(i) Load case 1 (LC-B1): the wave breaks far in front of the structure (10-20m) and the breaking wave force on the rear face of the structure is much less than the force on the front face.

(ii) Load case 2 (LC-B2): the wave breaks in front of the structure (3-10m) and induces an extreme impact force on the front face of the jacket structure;

(iii) Load case 3 (LC-B3): the waves breaks directly at the front face and the relative values of the breaking wave force on the rear face to those on the front face of the truss structure are higher than previous load cases (LC-B1 & LC-B2);

(iv) Load case 4 (LC-B4): the wave breaks between the front and rear faces of the structure. Although the incident wave breaks within the structure, there is a significant dynamic force on the front face of the truss structure;

(v) Load case 5 (LC-B5): the wave breaks behind the structure and only quasi-static forces are induced on the jacket structure.

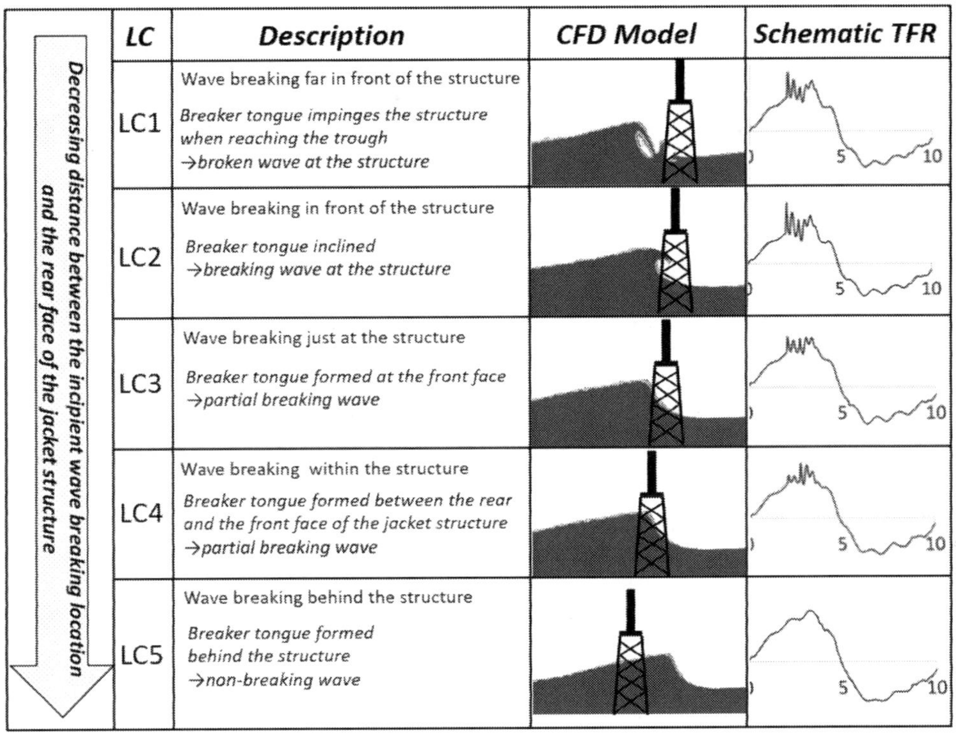

Figure 4. Load cases induced by breaking waves on the OC4 jacket structure and the total force response of the structure

The calculated impact forces are then introduced in the CSD model by considering the impact area on the front and rear faces of the OC4 jacket structure. The total wave force caused by the breaking wave on the jacket structure is calculated by linear superposition of the quasi-static and impact force components.

4. Weakly coupled CFD-CSD Models for the comparative study

4.1. Numerical model without scour effect

The numerical model as extended by adding pile foundation with soil models using non-linear p-y springs was used for the model scour protection. The plausibility of the extended model was tentatively examined by comparing the results with the study performed by [4]. The model of the pile foundation is set-up for four different soil types including: (i) uniform soft clay; (ii) uniform sand soil; (iii) layered sand soil; (iv) uniform hard clay. For the comparative study however, the model with the pile foundation embedded in clayey soil has been used. It has been considered that the model is protected using

Geotextile Sand Containers (GSC) against soil erosion around the pile foundation therefore, the mass of the system does not change and the soil stiffness can be considered along the entire pile foundation (no stiffness reduction is taken into account). In addition, due to the installation of GSC, the hydrodynamic wave loads on the structure would be reduced as a part of the drag force would be taken by the GSC at the base foundation. The aforementioned effects are considered in the weakly coupled CFD-CSD model.

The model of the OC4 jacket structure with pile foundation was used to investigate the dynamic response of jacket structure to different breaking waves. The most critical response was computed for the jacket structure under breaking wave LC-B2. Accordingly, this load case has been considered for the comparative analysis in this study.

4.2. Numerical model with scour effect

For the model without scour protection, both local and global scour effects are taken into account. The set-up of the FE model of the OC4 jacket structure with pile foundation is modified based on the obtained global and local scour depths (Figure 5d) in order to perform a comparative study of the dynamic response of the structure to breaking wave loads with and without scour effect. For the global scour, which is characterised by general erosion and removal of soil over a large area (Figure 5a & d), all p-y curves are regenerated on the basis of a modified seabed level which is taken as the initial seabed level lowered by a height equal to the depth $\boldsymbol{S_G}$ of the global scour. For the local scour, which is characterised by erosion and removal of soil only locally around each pile (Figure 5 a & d), the p-y curves are generated with due account for the depth $\boldsymbol{S_T}$ of the total scour hole. Moreover, the mass of the soil oscillating with piles of the structure is removed according to the total scour depth. With this regard, the maximum scour depth around piles of the OC4 jacket structure, which is considered as the lowest probable bottom level over 30 years structure life time (H=10m; T=10m, h=50m), is calculated as follows:

(i) The global scour depth S_G (Figure 5b & d) is defined by $S_G = 0.37\ D_{calc}$ (according to [10]) where D_{calc} is the pile diameter including marine growth. Considering the value of $D_{calc} = 2.082m$ the maximum global scour depth is equal to 0.77m.

(ii) The local scour depth S_L (Figure 5a & d) around piles of the structure might be defined by the expected value $S_L = 1.3\ D_{calc}$ (according to [8]) for sand soil which means $S_L = 2.71m$.

Therefore, the total scour depth is obtained from the following equation $S_T = S_L + S_G$ as 3.48m.

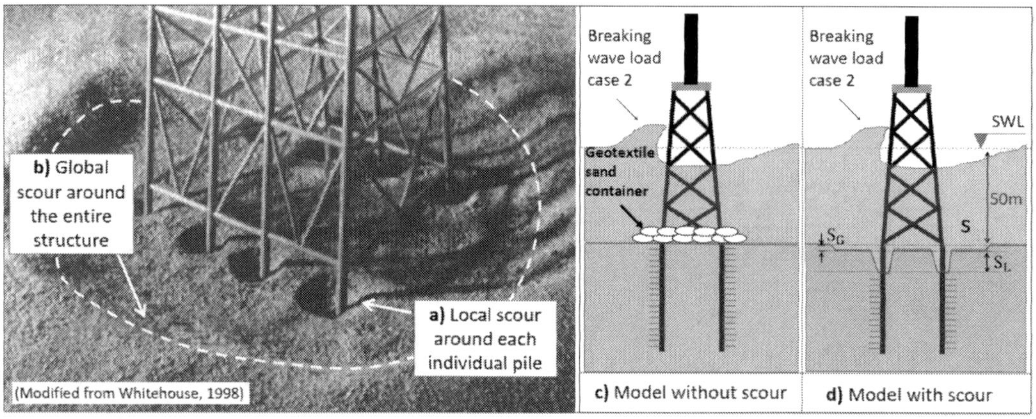

Figure 5. Representation of global and local scour development around jacket structures (a & b modified from [11])

5. Dynamic response of the jacket structure in sever sea state

Figure 6 shows time series of the dynamic response of the OC4 jacket structure with pile foundation in clay to the breaking wave LC-B2 by considering the effect of scour (Figure 5d) and without scour effect (Figure 5c) due to the installation of GSC. The slamming forces are applied at time t=2s & t=2.66s on the front and rear faces of the jacket structure, respectively. The time series of the lateral displacement of the structure is shown in Figure 6 a, b & c. As can be seen, neglecting the scour effect results in an underestimation of the lateral displacement of the structure by up to 25%. The main reasons are the significant reduction of the soil mass and stiffness considering scour effects.

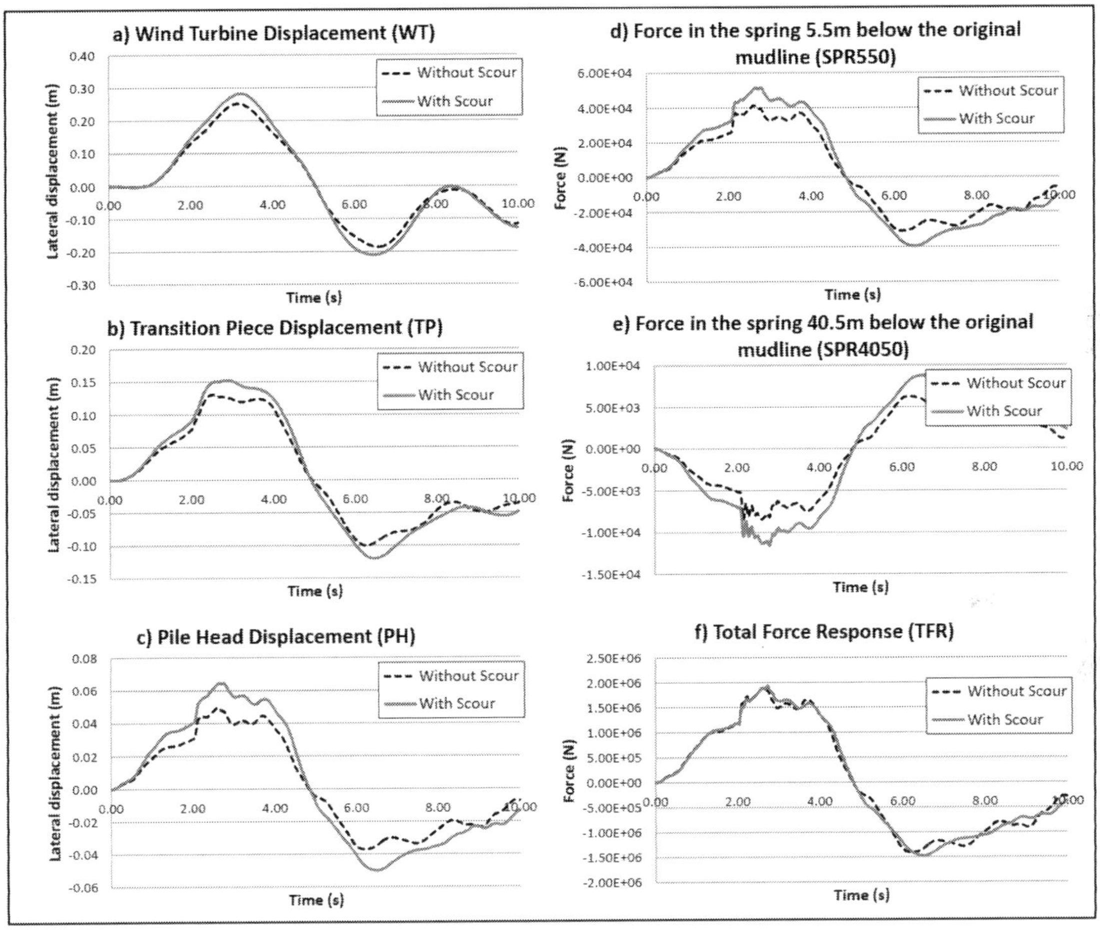

Figure 6. Dynamic response of the jacket structure with pile foundation in clay with and without scour consideration seabed scour (H=10m; T=10m, h=50m)

Basically, the stability of the jacket structure to lateral applied loads is mainly provided by the following structural members (see Figure 7):

(i) Within the braced zone (L_{Br}), the side braces of the structure parallel to the incident wave withstand the lateral forces considering their axial stiffness (EA/l_{br}). In this zone, the jacket structure behaves very stiff (Figure 7).

(ii) Within the unbraced zone (L_{UB}), the piles and legs of the jacket structure withstand the lateral forces considering their bending stiffness (EI/L_{UB}) (Figure 7).

Since the length of the unbraced zone of the jacket structures is relatively short (6m for the OC4 jacket), the bending stiffness of the member in this zone is very high (EI/L) thus ensuring the stability to the structure. However, the scour around the piles may increase the height of the unbraced zone of the structure significantly (from 6m to 9.5m for OC4 jacket) which might reduce the bending stiffness considerably (Figure 7). Therefore, as shown in Figure 6 a, b & c, the displacement of the structure due to scour influences is considerately higher than the model set-up without scour.

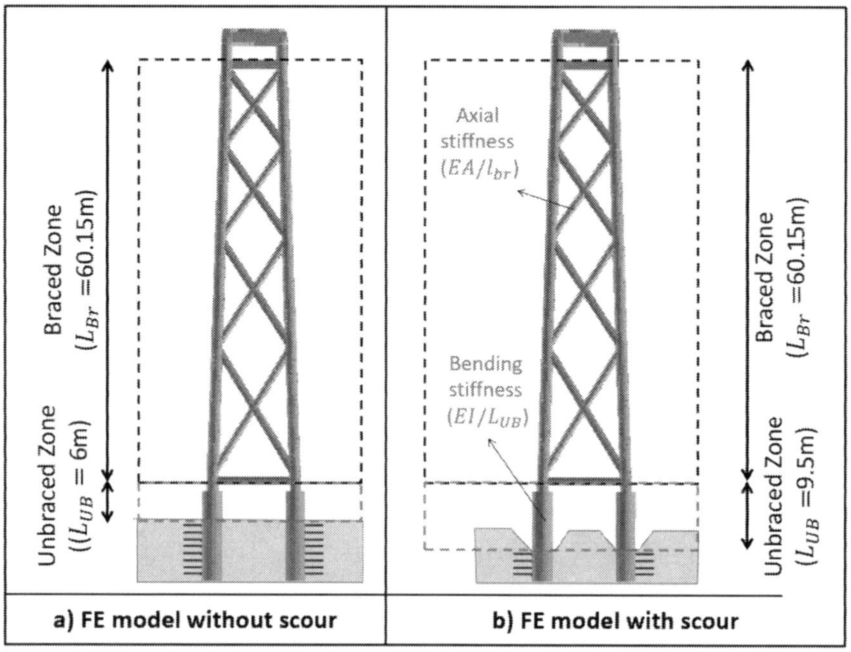

Figure 7. Braced and unbraced zone of the OC4 jacket structure a) without scour; b) with scour effects

In summary, scour (i) reduces the foundation bearing capacity and consequently, might increase the failure probability in terms of both serviceability and ultimate limit state; (ii) reduces the natural frequency of the structure embedded in the soil and therefore, might significantly affect the dynamic response of the structure (Figure 6 d & e). Therefore, the results of this preliminary analysis indicate that the scour effect should be considered in the design process of jacket support structures.

6. An ecological friendly solution to protect the foundation against soil erosion

For scour protection at the base foundation of offshore fixed-bottom structures, one can choose between different solutions regarding design and material [12] [13] [14] [15] [16]. Most common is the classical rock solution, consisting of big stones, normally bedded on a layer of gravel which has to be installed before piling the structure on the sea floor. The rock solution causes a huge impact regarding CO_2 and other emissions for the environment [17]. In order to counter these disadvantages, an innovative solution was examined and the so-called "Soft Rock" was proposed. This means containers made from needle-punched nonwoven are filled with sand and randomly laid on the sea floor around the pile foundation. This can be conducted prior to the piling of the masts of the offshore structure so that the Soft Rock solution can be punched through easily by installing the monopile [13].

Investigations of Soft Rock container made from PP-nonwoven which have been dismantled from an existing wind park show their ability to form a new habitat for flora and fauna at the sea bottom. The sand, penetrating the openings of the nonwoven geotextile in interaction with the staple fibres form a new kind of material which lasts for a life time of an offshore structure[17]. Due to this mixture from nonwoven and sand, the service-life will expand as any abrasion due to waterflow and sand transport close to the seabed will be reduced. Especially in comparison to the classical solution made of gravel and rock, the new solution is in most cases more like a natural seabed. This means that the existing habitat (flora and fauna) in the region of an offshore windfarm will settle inside of this scour protection system. On the other hand, the classic rock solution attracts a rather unusual, non-native flora and fauna that would not live there in an untouched system, since the seabed is sandy in most cases where offshore wind farms are planned. All kinds of scour protection systems have to be monitored by the operating companies. If there occur damages or drifting processes, in all versions of scour protection measures have to be taken into account. These can be done easily by putting additional rocks or additional sand containers to the scour protection.

The newest development for geosynthetic based scour protection solutions are "biodegradable Soft Rock containers". These sand-filled containers have a volume of $1m^3$ similar to the common Secutex® Soft Rock Container produced from Polypropylene nonwoven. The raw material of the biodegradable nonwoven containers, is an industrially manufactured fibre, in a size and dimension which is comparable to the nonwoven fibres made from Polypropylene. Thus, the pore structure, opening size and hydraulic conductivity and therefore the hydraulic and geometrical filter stability against sandy seabed soils and fills can be ensured.

The seams are sewn with a biodegradable thread, which is made from the same raw material as the fibres of the nonwoven fabric and is available on the market. The whole scour protection system is biodegradable and may not harm the environment unlike the classical rock-based solutions for which a large amount of CO_2 is released during the process of extracting the rocks from pits and due to the transportation of this huge amount of stones with large weights by vessels towards the offshore wind parks. Further investigations including the performance of laboratory tests are currently undertaken to assess the durability and the biological behaviour of biodegradable Secutex® Soft Rocks under the boundary conditions at the sea floor. The so far obtained results show promising behaviour of this product achieving a sustainable and environmentally friendly solution which builds the future of scour protection in coastal and offshore problems [18] [19].

7. Key remarks and conclusion

Although the occurrence of scour around piles of a jacket structure is a well-known phenomenon (Figure 5 a & b), to the author knowledge no study has yet addressed the effect of scour on the dynamic response of jacket structures. Previous studies have revealed that scour might generally have two important effects on the dynamic response:

(i) As the soil in the area close to the mudline is removed, the soil resistance may be reduced. Moreover, the geodynamic mass contribution to inertia is also decreased, these changes in the stiffness and mass matrixes, might change the natural frequency of the structure;

(ii) The hydrodynamic forces on the entire structure will be increased as the soil beside the pile foundation will be eroded due to scour effects.

In this study, a weakly coupled CFD-CSD model of a full-scale jacket structure called 'OC4 jacket' was applied to calculate the dynamic response to waves. The model for the dynamic response was then extended to consider pile foundation with appropriate clay and sand soil models. Using the extended model, a systematic parameter and comparative study of the dynamic response of the jacket structure with pile foundation to breaking waves was performed with a particular focus on the effect of scour on the response. The key outcomes might be summarised as follows:

- Pile foundation model significantly affects the dynamic performance of the entire jacket structure (40% for natural frequency), and thus has to be considered in the design process of jacket structures. In general, the lower the soil stiffness in the foundation, the higher are the displacements and deformations of the structure and consequently, the higher are the forces in the members of the jacket structure.
- Scour effect should be considered in the design of jacket structures as it might reduce the foundation bearing capacity and might increase the failure probability in terms of both serviceability and ultimate limit state. Moreover, scour may change the eigen-frequency of the entire jacket structure

(up to 10%) embedded in the soil and as a result, considerably affect the dynamic response. Scour increase the lateral displacement of the jacket structure in this study by up to 25%.

An innovative solution using bio-degradable geotextile sand containers was proposed to overcome the soil erosion around the pile foundation of the jacket structure. This may lead to an improved dynamic response of the jacket structure and simultaneously provides a sustainable ecological friendly solution which may not disturb the existence life at the sea bottom. Currently, the tests are in progress to investigate the durability of bio-degradable geotextile sand containers under different boundary conditions at the seabed.

8. References

[1] Popko W, Vorpahl F, Zuga A, Kohlmeier M, Jonkman J, Robertson Aand Von Waaden H 2012 Offshore Code Comparison Collaboration Continuation (OC4), Phase I–Results of Coupled Simulations of an Offshore Wind Turbine with Jacket Support Structure. *22nd International Society of Offshore and Polar Engineers Conference*. Rhodes, Greece

[2] Vemula N K 2010 Design Solution for the UpWind Reference Offshore Support Structure-Deliverable D4.2.5 (WP4: Offshore Foundations and Support Structures). *Technical report, Rambøll Wind Energy*, Esbjerg

[3] Khansari A 2018 Loading and Dynamic Response of Jacket Structures to Breaking and Non-breaking Waves. Doctoral dissertation, *Leichtweiß Institute for Hydraulic Engineering and Water Resources. Technische Universität Braunschweig*. Braunschweig, Germany

[4] Shi W, Park H C, Chung C W, Shin H K, Kim S H, Lee S S and Kim, C. W 2015 Soil-structure interaction on the response of jacket-type offshore wind turbine. *International Journal of Precision Engineering and Manufacturing-Green Technology* **2**pp 139-48.

[5] Matlock H 1970 Correlations for design of laterally loaded piles in soft clay. *Offshore Technology in Civil Engineering's Hall of Fame Papers from the Early Years*,77-94

[6] O'Neill M W and Murchinson J M 1983 An Evaluation of p-y Relationships in Sands. *A Report to the American Petroleum Institute*.

[7] American Petroleum Institute (API) 2005 2A–WSD Recommended practice for planning, designing and constructing fixed offshore platforms–Working stress design. *In Twenty first.*

[8] *Det Norske Veritas (DNV) 2013 OS-J101*: Design of Offshore Wind Turbine Structures.

[9] Morison J R, O'Brien M P, Johnson J W, Schaaf S A 1950 The force exerted by surface waves on piles. *Petroleum Transactions, AIME*, **2**pp 149-54

[10] Sumer B M and Fredsoe J 2002 *The mechanics of scour in the marine environment*, World Scientific

[11] Whitehouse R 1998 *Scour at marine structures*: A manual for practical applications. Thomas Telford

[12] Heerten, G, 1981 Experience of different scour protection techniques at offshore-structures in the North Sea. *Fifth Australian Conference on coastal and ocean engineering*, Perth pp 22-28

[13] Heerten, G, Peters, K, 2011 Kolkschutz für Offshore-Windenergieanlagen in der Nordsee. geotechnik (34) Heft 1 pp 51-58

[14] Grüne J, Sparboom, U, Schmidt-Koppenhagen R, Wang Z, Oumeraci H, 2006, *Large scale investigations of geotextile sandcontainers used for scour protection of offshore monopiles supporting wind energy turbines*. Proceedings of OMAE2006 25th International Conference on Offshore mechanics and arctic engineering, Hamburg

[15] Mayerle G, Cazzuffi D, Greco P, Sarti L, Vicari M., Usan G, 2006 Laboratory and field tests on a ballasted geocomposite filter for the stabilisation of the seabed in the Venice lagoon inlets, *Proceedings of the 8th International Conference on Geosynthetics*, Yokohama, Millpress pp 793-796

[16] Cazzuffi D, Mainieri V, Sala A. 2012 Application of different ballasted geocomposite filter systems for the stabilisation of the seabed in the Venice lagoon, *Proceedings of ICSE-6 - 6th International Conference on Scour and Erosion*, Paris p 10

[17] Hoyme H, Su J H, Kono J and Wallbaum H 2019 Nonwoven geotextile scour protection at offshore wind parks, application and life cycle assessment, *Proceedings and Monographs in Engineering, Water and Earth Sciences* pp 315-21
[18] Hoyme H, Vollmert L, 2020 *Kunststoffe und Umwelt: Neue Herausforderungen an Geokunststoffe.* Wasserbauliche Mitteilungen, Heft 63, Dresden pp 79-88, ISBN 978-3-86780-627-5
[19] Hoyme H, Vollmert L, Ehrenberg H, 2020 Plastic in the sea: New challenges for geosynthetics. *Journal of Physics: Conference Series,* Euro-Geo7, Warszawa

Performance of drainage geocomposites applied as capillary break layers

T Szatmári[1], **S Fischer**[2]

[1] Low & Bonar, Budaörs, Hungary
[2] Department of Transport Infrastructure and Water Resources Engineering, Széchenyi István University, Győr, Hungary

Email: tamas.szatmari@lowandbonar.com

Abstract. The phenomenon of capillary rise can cause water rise up 2-3m above the high water table within the soil. Due to this effect the shear strength of fine-graded soil is decreased, or it can damage decorative floorings, paved areas, concrete and different manmade structures. Furthermore, in arid regions the ground water is often saline with a high concentration of chloride ions, which can be harmful to vegetation. The traditional methods to solve the above mentioned problems with breaking the capillary rise can be the use of free draining granular soil with very low fines contents, usually with 30 to 50 cm thickness. One of the effective capillary break methods is creating an air void between the soil layers with the help of drainage geocomposites therefore capillary rise of water is prevented. This paper presents the research work of the above described capillary break phenomenon with different types of geocomposite drainage layers in laboratory controlled conditions.

1. Introduction of capillary action

Capillary action happens in unsaturated soils and describes the movement of pore water from lower elevation to the higher driven by the hydraulic gradient acting across the curved pore air/pore water interface [1]. This elevation phenomenon appears in different degrees in case of different type of soils. The height of the altitude in inversely proportionate to the "pipe" diameter, in this way the rough grained gravels, sands the height of the rise is lower than in cohesive soils, see Figure 1. [2].

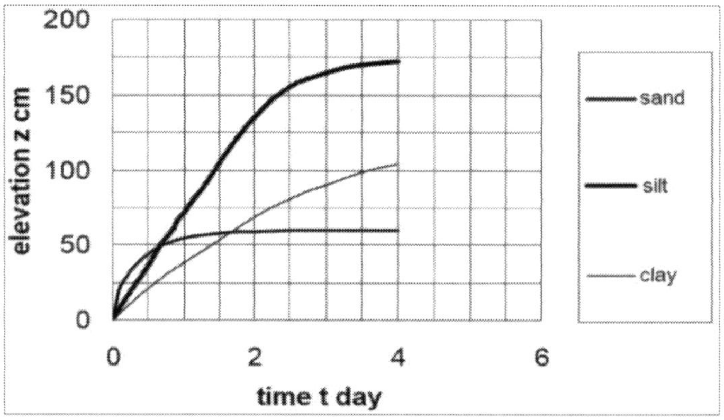

Figure 1. Capillary elevation as function of time, in different soils [2]

1.1. Problems in embankments of infrastructure, soil structures

In case of embankments of infrastructural transportation facilities, foundations of architectural, bridge and hydraulic or other engineering structures, when water table is relatively close to the surface, the phenomenon of capillary rise can cause water rise above the high water table within the embankment body. Due to this effect the shear strength of fine-graded soils is decreased, additional settlements can be arisen that can influence the adequate operation of the facility, in special cases the global stability of the structure can be also affected.

1.2. Problems of saline water

In coastal areas and particularly in arid regions the ground water is often saline with a high concentration of chloride ions. In addition to the problems this poses for provision of drinking water the chloride ions can be harmful to vegetation and manmade structures. Salt concentrations of only 4 or 5 g/l are generally too high for most plant species to survive whilst the salts can damage and discolor decorative marble floorings and paved areas and cause concrete in foundations and basement structures to deteriorate. [3]

2. Capillary break layers

Due to the negative effects of capillary elevation described in 1.1 and 1.2, an effective barrier is advised to use in order to provide a capillary break layer between the saturated and unsaturated zone and prevent the so called capillary fringe zone above the ground water level (see Figure 2).

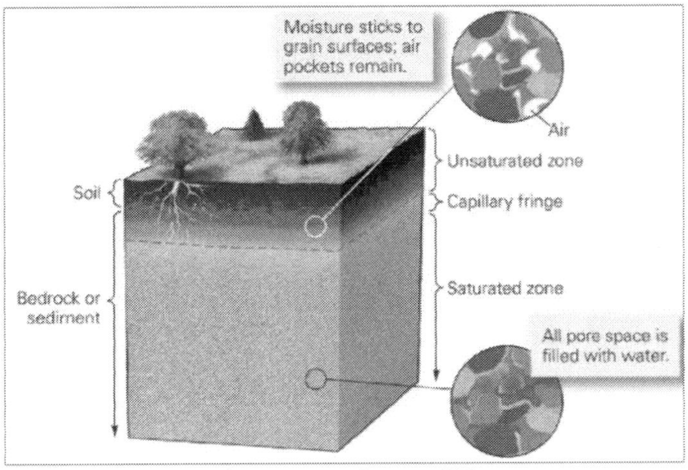

Figure 2. Saturated, capillary fringe, and unsaturated zones [4]

The capillary break layer creates a void across which capillary rise of water moisture is prevented. The traditional methods to solve the above mentioned problems with breaking the capillary rise can be the use of free draining granular layers, usually 30-50 cm thick, with very low fines contents to act as a discontinuous void within the soil strata and prevent the rise of water. However, it has become common to use geosynthetic drainage layers to provide capillary break. [3]

3. Geocomposite drainage layers as capillary break layers

Drainage geocomposites can create an air void between the soil layers with the help of their drainage core, therefore capillary rise of water is prevented. Geocomposites were first developed in the 1970's as the geosynthetic industry began to establish itself and broaden the offering from geotextiles. Drainage was one of the first applications for a geocomposite with the pioneering development beginning with a filament core structure. Currently there are many variations of drainage composites available with widely varying core types all using a geotextile filter on one or both sides. The form of the drainage composites typically consists of two layers of geotextile filter fabric on either side of an open core structure which usually consists of 3D filament structures, cuspated thermoformed sheets, box like structures or geonets formed of two layers of extruded polymer to form a grid. A geocomposite capillary break should be situated above the highest ground water level so it does not become saturated by the ground water.

4. Problem definition

All the various type of drainage geocomposite available function by maintaining an open void that separates soil filters from each other whilst filter fabrics prevent the material becoming clogged with the migration of fine soils. In the present research the answer was searched to the question that in laboratory – controlled – conditions, how and with what kind of manner the different geosynthetic product can ensure the interruption of capillary elevation in fine graded soils. The long term function of drainage geocomposites as capillary break layer is essential to ensure the product is not compromised during the service life. Due to the compressive strength combined with delayed intrusion the decrease of void thickness can happen resulting from the applied stress and the deformation of the geotextile filter, see Figure 3.

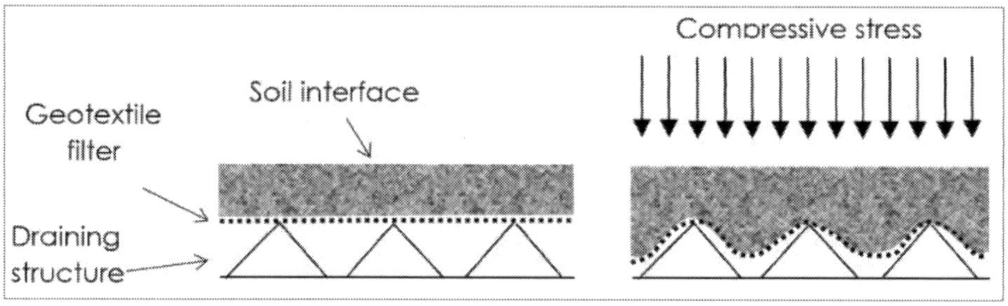

Figure 3. Intrusion phenomenon [5]

5. Testing procedure

During the laboratory test one type of fine-grained cohesive soil material and two different types of geosynthetics were applied, and the vertical stress value was determined as 200 kPa. Measurements were also executed to the case without geosynthetic capillary break layer which was used as reference measurement. Additionally one more measurement was done in order to represent the situation when the geocomposite product is installed in inaccurate way.

5.1. Set up of measurements

For the testing procedure a special measuring system was applied with the dimensions and structural elements visible in Figure 4. A steel pipe with 50cm diameter and 100 cm height was used under the loading frame which was able to constantly load the soil with 200 kPa. A steel loading plate was applied in order to distribute the load equally. Under the loading plate, 65 cm of cohesive soil was built in together with the different geocomposite capillary break samples located 40 cm from the loading plate and insulated with silicone at the edges. The ground water level in the cohesive soil was adjusted just below the geocomposite layer.

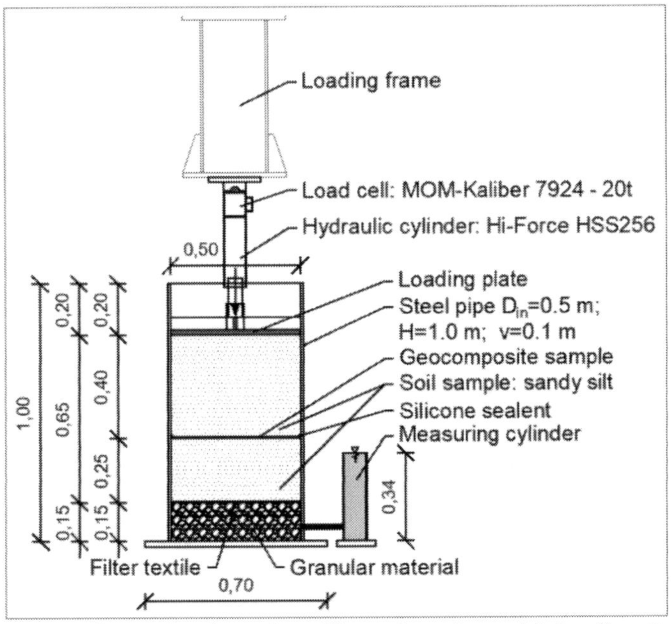

Figure 4. Schematic set-up of examination of capillary elevation, with loading frame up to 20t

5.2. Soil type

For the laboratory test cohesive soil was chosen, with around 70% silt content, and 15-15% sand and clay content (see Figure 5). According pre-tests with the chosen silt soil, the capillary elevation reached the required 50 cm within a day, therefore it could be used for the test series.

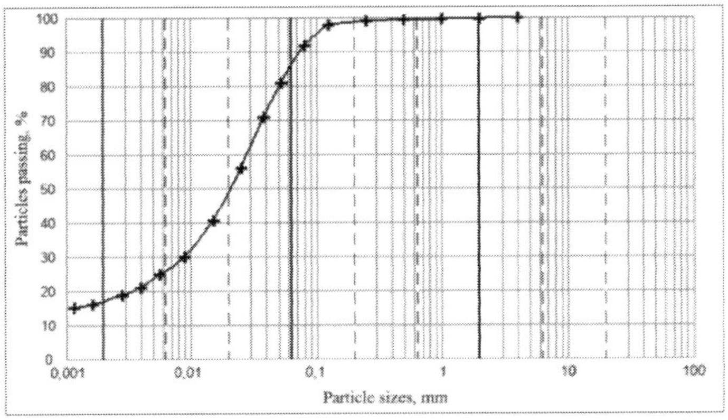

Figure 5. Particle size distribution curve of the applied soil

5.3. Type of tested geocomposite layers

During the laboratory test series two different type of geocomposite products were tested with similar filter and drainage characteristic, however, the formation of drainage cores are remarkably different:

GCO 01 (EnkaDrain 5004C/5-2s/M110PP), visible in Figure 6.

- Semi compressible uniform 3D monofilament "V" shape structure with needlepunched geotextile 110 g/sqm on both sides;
- Thickness under 2 kPa is around 5 mm;
- Flow capacity under 20 kPa and i=1 is around 1.0 l/m*s, measured between foam plates;

GCO 02, visible in Figure 7.

- Rigid, hardly compressible geonet drainage structure with needlepunched geotextile 130 g/sqm on both sides;
- Thickness under 2 kPa is around 7 mm;
- Flow capacity under 20 kPa and i=1 is around 1.8 l/m*s, measured between rigid plates;

Figure 6. GCO 01 product sample

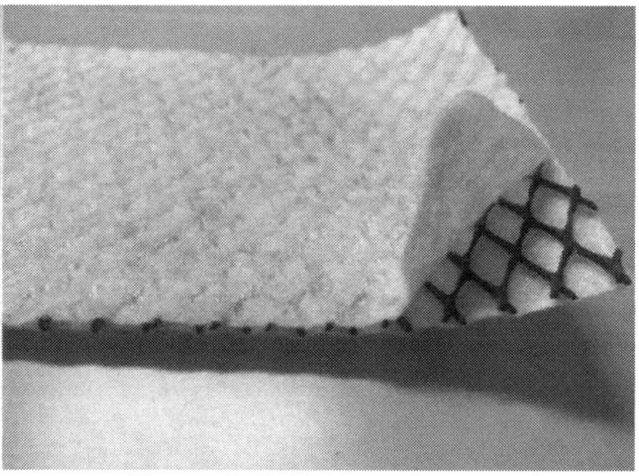

Figure 7. GCO 02 product sample

It is important to mention that the technical data sheets of the two products declare the flow capacity values measured in different ways, for GCO 01 the boundary conditions are compressible foams meanwhile for GCO 02 it is measured between rigid plates.

5.4. Procedure

In all cases the loading period was 5 days with 200 kPa vertical pressure, and water content measurements were executed related to the soil samples excavated from the soil above the installed geocomposite products in each 5 cm height level. The effective measuring process contained the following steps:

- drying of the fine-grained soil until mass constancy;
- insertion of coarse-grained soil as a filter layer into the bottom part of the steel pipe in 15 cm thickness, with one layer of geotextile separation layer;
- installing and compacting of cohesive soil in 25 cm thickness;
- insertion of the geocomposite after a precise preparation to accurate dimensions, together with detailed sealant and insulation at the sides (for the reference measurement obviously this step was skipped);
- loading and compacting of the cohesive soil in 40 cm thickness to reach to maximum height;
- insertion of the full surface loading plate onto the +80 cm height plane;
- application of constant 200 kPa vertical pressure with hydraulic cylinder and controlled by load cell;
- setting of free water surface onto the +34 cm height (~6 cm below the geocomposite layer) and holding it on this value with suppling the water backup;
- holding the process until 5 days;
- disassembling of the measuring system, excavate the soil from the steel pipe in every 5 cm layers, determination of the water content with weigh and drying until mass constancy;

There were four measurements done during the test series. First, a reference measurement was executed in order to examine the capillary elevation in the particular soil type without capillary break layer installed. Then, as second and third assessments the performance of GCO 01 and GCO 02 were analysed. In order to be able to evaluate the performance in irregular built in conditions, a fourth measurement was also done where GCO 01 product was laid to a rough surface instead of well-prepared smooth soil and tried to create an inaccurate installation.

5.5. Process method and results

In Chapter 5.4 detailed manner and process were described about the laboratory tests, the water content in vertical sections (profiles) were determined with weigh and drying until mass constancy of the soil samples excavated from the steel pipe, the water content values were specified according to the 72 hours drying. Water content values were illustrated in graphs for each cases, surface of every layer (soil, geocomposite, water level) were given in them. Obtained results were compared to the reference measurement (without geocomposite) for all cases. The prepared diagrams represent how the geocomposite layers influenced the original water content in vertical profiles, moreover they demonstrate the effectiveness of the capillary interruption in the examined cohesive soil.

Figure 8 shows the measured water content results as a vertical profile using GCO 01, when Figure 9 represents the values in case of GCO 02. In both cases a blue horizontal line displays the sedent water level, a red dashed line presents the level of the geocomposite. The two graphs contain a green line, which indicates the water content during the reference test when there was no geocomposite product installed.

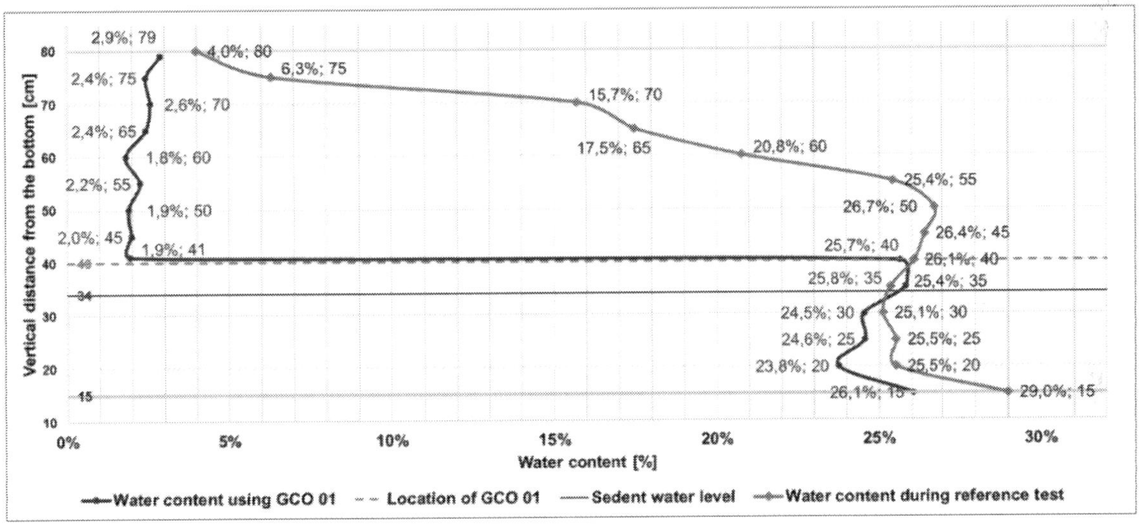

Figure 8. Water content vertical profile related to GCO 01

According to the results in figure 8, it can be stated that the water content value in the +41 cm height plane (just above the GCO 01 product) was decreased from 25-27% to ~2%, as well as above this height the original values were modified to that level. Next to the water content results, visual inspection also confirmed the fact, that there was no water elevation above the level of GCO 01, the soil was completely dry. Based on the above mentioned results, it can be stated that the GCO 01 type can effectively interrupt the capillary elevation in this particular cohesive soil.

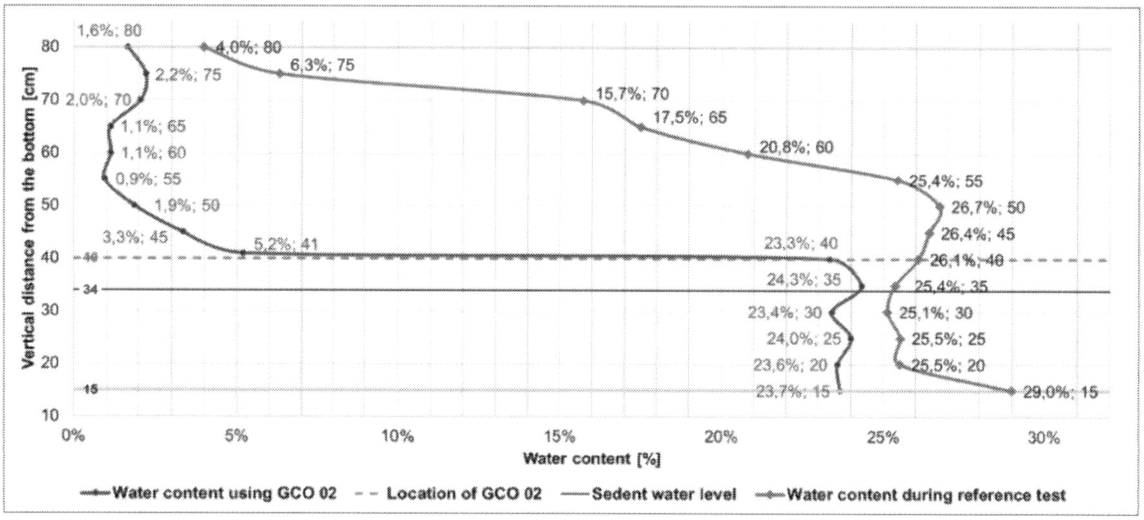

Figure 9. Water content vertical profile related to GCO 02

According to the results in figure 9, it can be seen that the water content value in the +41cm height plane (just above the GCO 02 product) was decreased to 5.2%, as well as above this height the original values were decreased to 0.9-3.3%. Compared to the results seen in Figure 8, a modest water content increase could be examined above the GCO 02 geocomposite product compared to the completely dried condition. This increase is minor, however, during visual inspection it was also visible: the cohesive soil above the product was marginally wet. Based on the above mentioned observations, it can be stated that GCO 02 product cannot effectively interrupt the capillary elevation in the described particular fine-grained soil type.

The cross-sections of the tested product were also checked after excavated them. Regarding GCO 01, clear difference could not be seen compared to the original product sample (Figure 10). However, for GCO 02 a significant intrusion phenomenon was visible on the excavated product sample probably due to its structural characteristics (Figure 11): the inner core of the drainage geocomposite has not been compressed, but the geotextiles on both sides contacts each other which makes a "bridge" between them and probably transmit the moisture through the geocomposite layer. This gives the explanation to the soil samples with higher water content above the geosynthetic layer.

Figure 10. GCO 01 product cross section after the laboratory testafter the laboratory test

Figure 11. GCO 02 product cross section

It was experienced during the laboratory tests, that in case of the irregular built-in case (i.e. the soil plane under the geosynthetic layer is rough, the product is creased) there was no relevant difference compared to the built-in according to regulations on horizontal and smooth soil surface. The water content vertical profiles were almost the same related to both cases.

6. Summary and conclusion

Capillary elevation in fine-graded soils, as one of the emerging problems in civil engineering was introduced in this paper, together with its effects in different fields and areas. During the laboratory tests with drainage geocomposites vertical water profiles were created in order to be able to compare to the results related to reference case. Regarding to the measurement results with the process detailed in chapter 5.4, it can be verified, that the capillary interruption is effective with adequate type of geosynthetic product in the applied cohesive soil material. In this way, the coarse-grained layer as traditional interruption filter layer can be neglected or significantly decreased.

In accordance with the measurement results it can be stated, that those types of geocomposite products where the geotextile filter layers can contact each other beside the inner core due to serious intrusion of the surrounding soil, the capillary interruption effect decreases and it results higher water content in the soil layers above the geosynthetic layer.

References

[1] Lu N and Likos W J 2004 Rate of Capillary Rise in Soil *Journal of Geotechnical and Geoenvironmental Engineering*

[2] Szepesházi R 2008 Geotechnika (Geotechnics, in Hungarian) university book, Széchenyi István Egyetem, Győr p 187

[3] Hérault A and Woods D 2018 The long term performance of geocomposite drainage materials used as capillary break layers *GeoMeast International Congress & Exhibition*

[4] Marshak S Essentials of Geology (Fourth Edition, October 30, 2012) Chapter 16, A hidden reseve: Groundwater, 16.2 Where does groundwater reside? p 474

[5] = Touze-Foltz N , Hérault A, Stoltz G 2014 Evaluation of the decrease in long term water flow capacity of geocomposites due to filter intrusion *7th International Congress on Environmental Geotechnics*

Modified multiaxial tensile tests with geogrids in soil contact

H Zanzinger[1], D Alexiew[2]

[1] SKZ – German Plastics Center, Friedrich-Bergius-Ring 22, 97076 Würzburg, Germany
[2] Alexiew Geoconsulting, Weningshof 5, 48712 Gescher, Germany

Email: h.zanzinger@skz.de

Abstract. For geogrids with the function "stabilization", the (low) deformability is one of the essential parameters. As a punctual wheel load is usually involved, the products are subjected to multiaxial loads. Therefore, one-directional tensile strip tests are not sufficiently meaningful for this application. A modified burst test is more likely to generate a multiaxial tensile load to geogrids. The geogrid is clamped in a circular ring and loaded by a spherically inflating membrane. The geogrid deforms into a spherical segment. This "multiaxial" tensile force-elongation behaviour can be evaluated and compared.
The "multiaxial" test method has been further developed, while applying a gravel layer to the membrane (in a superstructure) and then tensioning the geogrid over it. This means that the geogrid is loaded by the ballast and no longer directly by the smooth membrane. Thus, any interaction effects between geogrid and gravel are also reflected, and the test approaches a so-called performance test. In principle, the evaluation is analogous to the test without gravel, but the measured values and diagrams reflect not only the ratio of load-deformation-stress, but also the interaction effects. Test setup, special features and evaluation are described and illustrated. The experiments and the results for three geogrids are presented and commented.

1. Introduction

There are two main mechanisms by which geosynthetics (GSY) can influence the mechanical behaviour of a mineral base layer, the "lateral restrain" of the aggregate by the GSY and/or the support of the base layer by a "membrane action support" of the GSY. The distinction between these two mechanisms and their relevant applications has been and still is the subject of intensive discussions in a wide variety of committees for several years [1].

The first mechanism allows stabilization by enclosing the fill grains or by providing lateral support, which in both cases counteracts the lateral deformation of the aggregate. The GSY holds the aggregate particles together mainly (but not only) by means of a form-locking bonding. By reducing mainly the lateral movement of the aggregate particles, the deformability of the system is reduced and the load-bearing capacity is increased.

The second mechanism makes use of the supporting membrane effect of a GSY, whereby above all the vertical deformation is reduced. The result is similar to the first mechanism, but the deformations necessary for mobilization are larger.

Newly, the first mechanism is referred to as "stabilization", and the second mechanism is referred to as "base layer reinforcement". Both terms are still controversially discussed both officially and unofficially. One thing has probably become clear: for efficient "stabilization", lower GSY strains and higher interaction (bonding of whatever type) are advantageous. The membrane mechanism requires higher elongation for mobilisation, although sufficient stiffness is also important. According to current knowledge, the transition between "stabilization" and "base layer reinforcement" is fluent. This is partly due to the fact that the actual behaviour of a GSY installed in the soil is very difficult to determine in situ. There are popular, "neutral", recognised calculation methods for designing "base layer reinforcement", such as Giroud & Noiret [2] , Lifa [3], Giroud & Han [4] or EBGEO [5]. This is still not the case with "stabilization": there are only company and product-related recommendations, nomograms etc. Some of the methods are calibrated only for a given product. From the authors' point of view, there is currently a lack of objective, neutral ("generic") and generally accepted design methods based on "generic" GSY parameters. Instead, the calculations are product-specific and thus so far hardly comparable in a product-neutral way.

Ideally, there should be objectively measurable mechanical parameters for GSYs that would allow product-neutral calculation (or at least comparison) and tendering. For research purposes, large-scale field tests were carried out on complete systems. Different systems were compared with each other. The evaluation was carried out after numerous drive-over tests based on the deformation on the surface of the system structure. These tests on the construction site are subject to so many varying influences that ultimately the correct comparison or the establishment of an objective calculation method do not seem plausible. Finally, the systems must then be put out to tenders on a project-related or product-related basis.

Intensive work is being done on draft ISO Technical Report ISO/WD TR 18228-5 [1].

Figure 1. Base layer reinforcement with GGR (Naue)

In this publication, we limit ourselves to GGRs used in road construction under base layers (Figure 1). Other GSYs such as geocells, woven geotextiles or geocomposites are not discussed here, although they are also used for exactly this purpose in a variety of ways. The effects on the subsoil as a result of a wheel load are shown schematically by Wrigley & Zheng [6] on Figures 2 and 3. A GGR is located between the base layer and the subsoil. The compressive stresses spread in the base layer downwards but also to the sides. The radially driving forces S under the base layer are compensated in this model consideration by the (also radially acting) GGR force C.

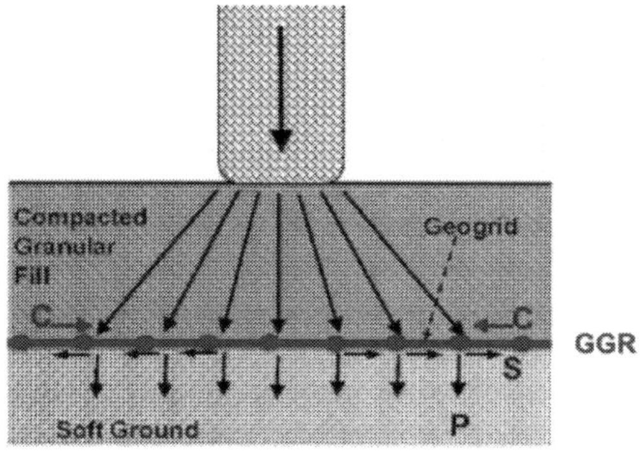

Figure 2. Load transfer due to a wheel load into the GGR [6]

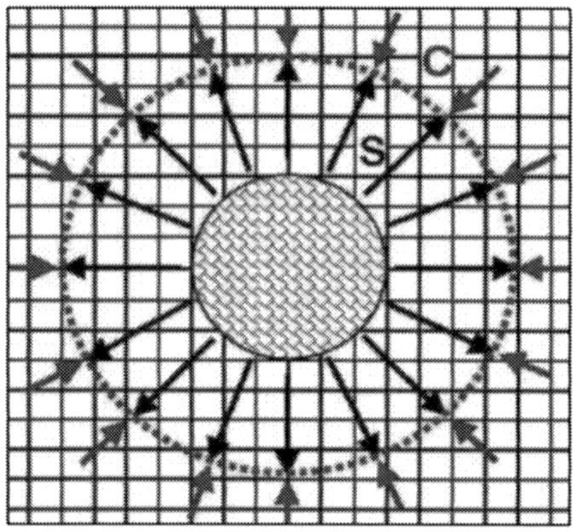

Figure 3. Compensation of the "radial force" S by the "GGR force" C [6]

2. Testing of GGR without soil contact

The task of the GGR is to absorb the "spreading forces" S (Figures 2 and 3) by means of the highest possible interlocking at the lowest possible elongation (i.e. high tensile stiffness). The crux of the matter is this: since it is an axially symmetrical ("radial", axial-isotropic) load acting on the GGR, the corresponding response (counteraction/resistance/confinement behaviour) of the GGR is also axially symmetrical (axial-isotropic, "radial") one. Consequently, this retaining counteraction will depend on an also radial-isotropic tensile stiffness. It is important to understand this, since it is the logic and the focal point of this publication.

Tensile stiffness EBGEO [5] is defined as "secant stiffness", which is the quotient of force per unit width in kN/m and strain (figure 4) and is determined by testing according to EN ISO 10319.
Today, bi- and triaxial GGRs are often used (Figure 5). Recently, there are even four-axial GGR. The biaxial GGR have similar tensile-strain behaviour in both directions, MD (machine direction) and CMD (cross machine direction).

Triaxial GGRs have almost equilateral triangular meshes and "ribs" in three directions. However, the tensile strengths are not identical in MD and CMD [7]. Rather, these GGR have three main axes instead of the two main axes in the biaxial GGR (Figure 5). Thus, a direct comparison of the biaxial GGR with the triaxial GGR based on the uniaxial "strip" tensile test is not directly possible, although EN ISO 10319 now also describes a test method for triaxial GGRs in MD and CMD. More detailed comments on that and first notes on the geogrid burst tests without soil contact described below are given in Zanzinger [8]. To summarize:

(1) With the common uniaxial wide-width tensile test according to EN ISO 10319 a comparative investigation of tensile stiffness as a criterium for the purpose handled here is questionable.

(2) Even more important: ultimately, the tensile behaviour is being tested always in one single direction. This has little to do with the "axially symmetrical radial behaviour" mentioned in the previous chapter.

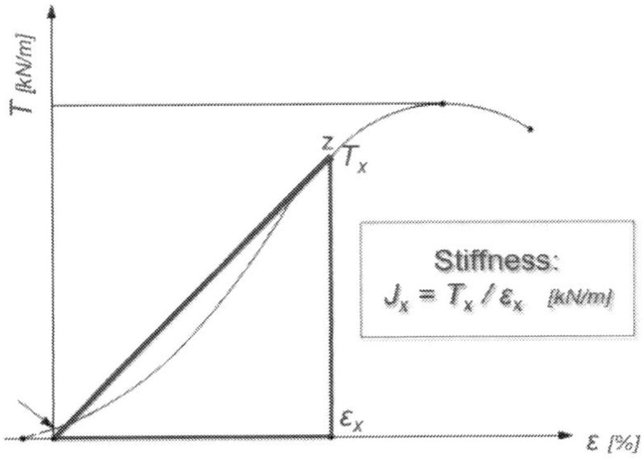

Figure 4. Tensile stiffness

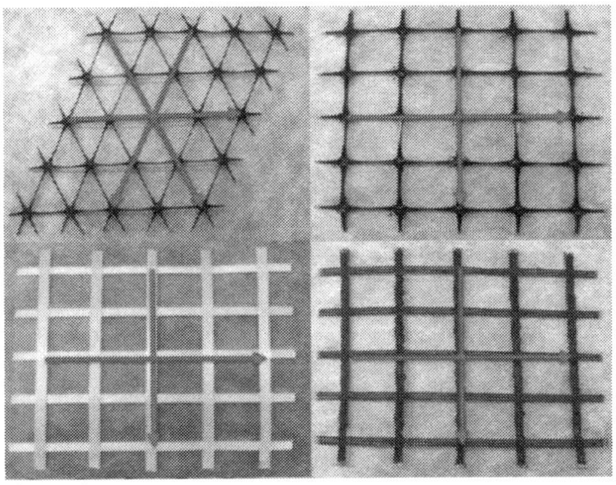

Figure 5. GGRs and their main axes

For this reason, the method for axial-symmetrical testing described below was developed: a radial loading of GGR in a burst test. For this purpose, the procedure for testing geomembranes in accordance with DIN 61551 [9] was modified (figure 6). In this case, a soft elastic membrane made of plasticized polyvinyl chloride (PVC-P) is placed under the GGR and water pressure is applied from below in a pressure vessel. Both the GGR and the PVC-P membrane are radially clamped (Figure 7). The clear diameter of the clamping ring can be 1000 mm or 600 mm.

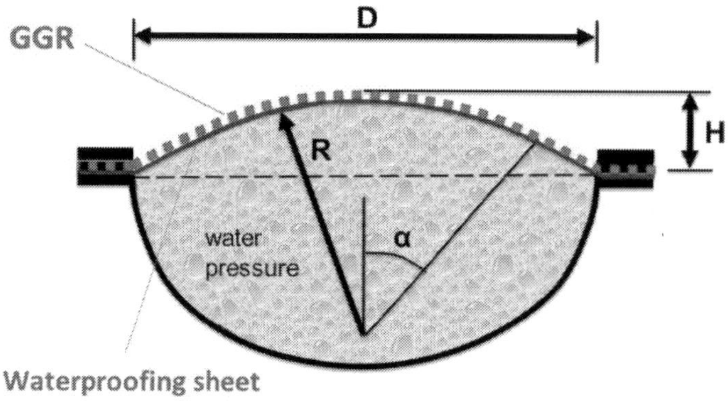

Figure 6. Schematic illustration of the multiaxial burst test with a GGR and the geometrical parameters

Figure 7. Biaxial drawn GGR under load

An isotropic hydrostatic pressure p is applied to the membrane and the arch height H is measured. From this, the multiaxial stress and strain in GGR can be derived. The geometric parameters α (half angle of the circular arc segment) and R (radius of spherical deformation) are required for the calculation. These are determined using equations (1) and (2).

$$\mathrm{R} = \frac{\mathrm{H}}{2} + \frac{\mathrm{D}^2}{8\mathrm{H}} \quad (1)$$

$$\alpha = \arctan\left[\frac{\mathrm{D}}{2(\mathrm{R}-\mathrm{H})}\right] \quad (2)$$

The calculation of strain is based in all standards for burst testing on the assumption that strain is based on the relative change of the diameter ΔD to the initial diameter D_0. One can speak here of "linear strain" or 1D strain (equation (3)). The intention of this calculation approach is probably that one can compare the strain determined in the burst test with the uniaxial strain in the tensile test. This would also be correct if the stress in the burst test was a clamping between two infinitely long clamping strips arranged in parallel. In this case, the outer surface would assume the shape of a circular cylinder segment and one would expect a "linear strain (1D strain)" or cylindrical strain ε.

$$\varepsilon = 100\left(\frac{2\,\mathrm{R} \times \alpha}{\mathrm{D}} - 1\right) \quad (3)$$

In the present case, however, it is the case that due to the radial clamping; the test specimen takes on a spherical segment shape. It is therefore better to determine the spherical strain ε_{2D} (equation (4)), which is based on the change of the surfaces ΔA to the circular initial surface A_0. The "two-dimensional strain

(2D strain)" yields values that are approximately 50% higher than the cylindrical strain. In the following spherical strains will be evaluated.

$$\varepsilon_{2D} = 100 \left(\frac{2\,H}{D}\right)^2 \tag{4}$$

The calculation of the stress σ in the test specimen according to DIN 61551 [9] is based on the vessel formula for a spherical pressure vessel (equation (5)).

$$\sigma = \frac{p \times R}{2\,d} \tag{5}$$

In the case of a GGR, however, one cannot assume a uniform thickness d and operate with stresses. Instead, it is common practice for GGR to state the force per unit width T in kN/m (equation (6)). This is defined as

$$\sigma = \frac{F}{b \times d} \tag{6}$$

with $\sigma = F/A$ with cross section A = b x d, with width b and thickness d and T is defined as

$$T = \frac{F}{b} \tag{7}$$

Here we are talking about the multiaxial tensile force per metre width T_{multi}.

$$T_{multi} = \sigma \times d \tag{8}$$

Including equation (5) we end up with

$$T_{multi} = \frac{p \times R}{2} \tag{9}$$

The relation between the pressure p and the arch height H generated by it (which is further used to determine the strains) can depend on many factors: Geometry, strength, structure, meshes, ortho- or isotropy of the GGR and also the size of the testing equipment.

The diagram from T_{multi} versus the spherical strain ε_{2D} shows a steep increase in the "force-strain curve" for a super-linear rise typical for polypropylene (PP) GGR. For "stabilization", the tensile stiffness is of interest. The multiaxial tensile stiffness J_{multi} (equation (10)) is defined here as the ratio between T_{multi} and ε_{2D} at e.g. 0.5% strain or 2.0% strain.

$$J_{multi} = \frac{T_{multi}}{\varepsilon_{2D}} \tag{10}$$

It is important that the force at which the strain is to be assumed to be zero is also defined. This has been defined by the authors here in accordance with the practice in EN ISO 10319 for these evaluations at 1% T_{max} (tensile strength) uniformly for all tests. The point of intersection in the diagram of T_{multi} is corrected by 1% of the maximum of T_{multi}.

Tests by Antczak et al. [10] on numerous GGRs made of extruded GGRs, laid GGRs and knitted GGRs made of PP or polyester (PET) have shown that these burst tests are very well suited for all biaxial and triaxial GGRs with nominal uniaxial tensile strengths of up to 40 kN/m to determine multiaxial tensile stiffness.

In the burst test, the knot connections are loaded synchronously. This test could also be used for a unified evaluation of the GGRs after installation damage. It is an index test with the same boundary conditions for all GGRs and it allows a direct comparison of any GGR used for base layer stabilization. However, this possible test application is beyond the scope this publication.

3. Testing of GGRs in multiaxial burst tests with soil contact

The multiaxial burst tests described in chapter 2 were performed without soil contact on a smooth membrane. In reality, however, the forces are introduced into the GGR via a coarse-grained base layer. To simulate this, the test was further modified. Instead of placing the GGR directly over the PVC-P membrane, a second ring of identical diameter was placed on top and then a layer of crushed gravel with grain sizes between 8 mm and16 mm (gravel 8/16) was placed over the PVC-P membrane (Figure 8). The GGR is now loaded by the gravel. This allows any interaction effects to be also reflected, the test moves from an index test to a performance test. Note, that generally not only gravel but any other granular soil can be used as "loading medium".

The membrane lies on a stable perforated plate. It is therefore possible to compact the gravel to a specified density. The test and evaluation are carried out analogously to the test without gravel.

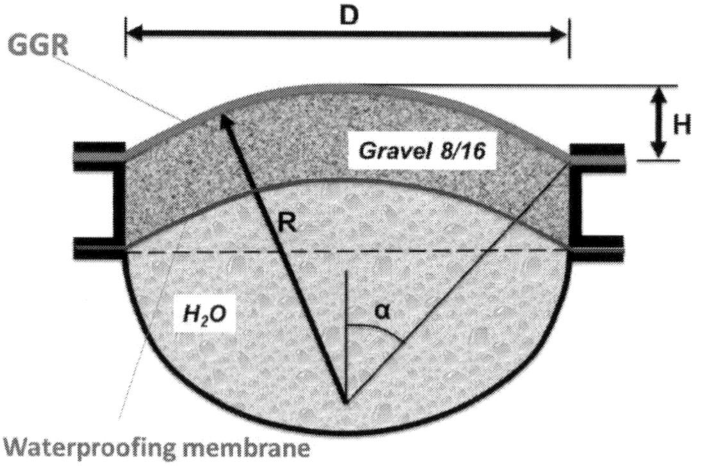

Figure 8. Illustration of the modified multiaxial burst test with GGR with soil contact

For the newly developed test in contact with a ballast layer, pilot tests were carried out with three GGRs of different production methods and different tensile strengths: GGR1 is a biaxial, laid GGR of PP, GGR2 is a biaxial, drawn GGR of PP, GGR3 is a triaxial, drawn GGR of PP. Note that they have different ultimate tensile strengths (UTS) as per EN ISO 10319.

4. Test setup

The inner diameter D of the clamping rings (Figure 8) was 600 mm. The attachment for the gravel is 120 mm high. Crushed, closely graded shell limestone with an installation density of 1.6 t/m³ was used as gravel 8/16 (Figure 9). The GGR was placed on the compacted level gravel surface and clamped in the upper flange. The "inflation pressure" p was generated at a constant approx. 10 kPa/min. To measure the arch (deflection) height H (Figure 8), a mechanical displacement transducer was placed centrally on the GGR (Figure 10).

Figure 9. Testing device with gravel filling before installing the GGR

In the preparation of the tests, the soft elastic membrane made of PVC-P alone was tested in order to "subtract" its influence during the evaluation.

Figure 10. GGR3 at the beginning of the loading

5. Evaluation of the measurements
Evaluation and calculations were carried out analogously to the tests without soil contact. The circumferential wall friction between gravel and steel cylinder, which should reduce the effective pressure load on the GGR, has not yet been considered here. Considerations and tests on this are still ongoing. Since the diameter of the attachment for the gravel is 600 mm and the height only 120 mm, the wall friction loss will probably be kept within limits. In any case, because this influence is the same for all tested GGR, the comparability of the results is maintained. All tests were carried out up to the point of GGR rupture (Figure 11). The subject of local compression and fracture is not discussed in detail here.

Figure 11. GGR1 with gravel after rupture

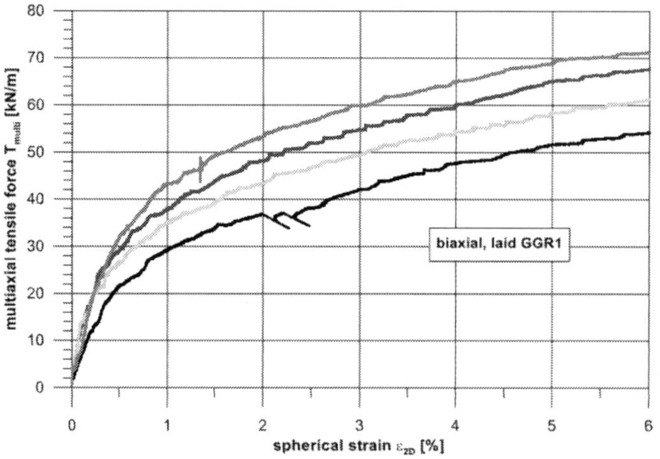

Figure 12. Multiaxial tensile force T_{multi} versus spherical strain ε_{2D} of GGR1 with gravel (showing 4 specimens)

The final relations of multiaxial tensile force T_{multi} to spherical strain ε_{2D} for the three GGRs are shown here in figures 12 to 14. From this the multiaxial tensile stiffnesses J_{multi} for any elongation can be derived. E.g. for ε_{2D} = 0.5% GGR1 shows a tensile stiffness $J_{multi\ 0.5\%}$ of 5428 kN/m, for GGR2 it is 3833 kN/m and for GGR3 is $J_{multi\ 0.5\%}$ = 12390 kN/m.

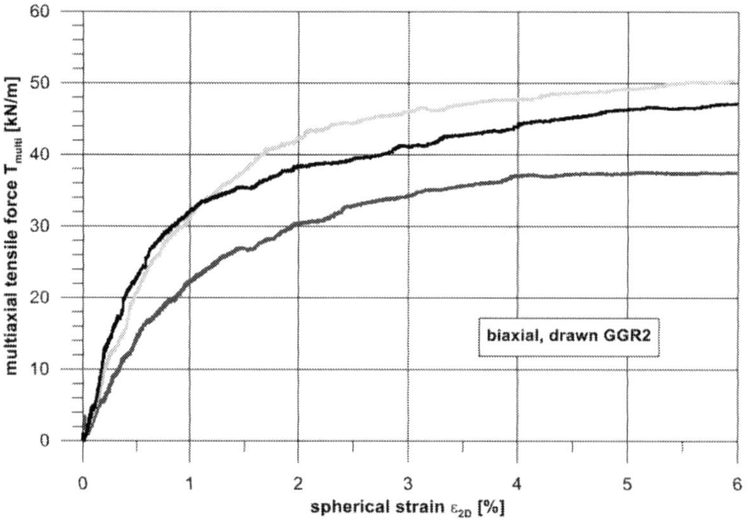

Figure 13. Multiaxial tensile force T_{multi} versus spherical strain ε_{2D} of GGR2 with gravel (showing 3 specimens)

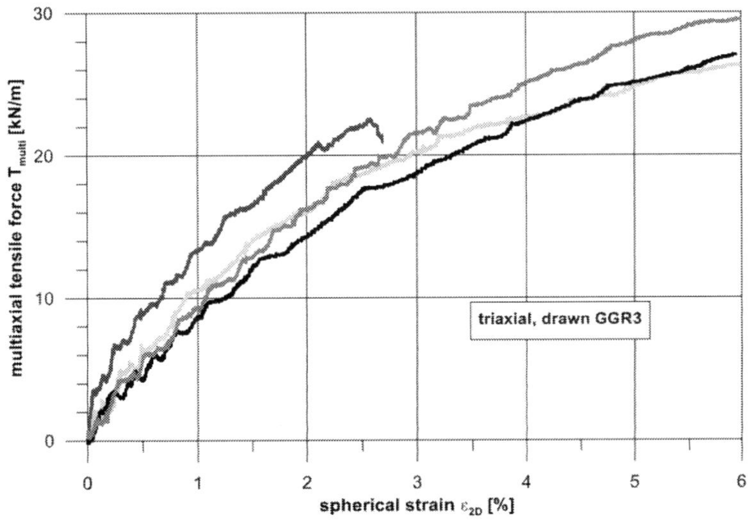

Figure 14. Multiaxial tensile force T_{multi} versus spherical strain ε_{2D} of GGR3 with gravel (showing 4 specimens)

6. Conclusions

This new test provides in our opinion an excellent basis for the evaluation of any GGR, woven geotextiles or geocomposites used for "stabilization". Even the interaction of the aggregates with the GSY will be considered. Therefore, the performance of any stabilization using GSYs in combination with a specific aggregate can be demonstrated with this test method and quantitative properties such as multiaxial tensile stiffness can clearly differentiate these products.

A comparison of tests with and without gravel with identical GGRs, which is not shown here, shows very clear differences between the "naked membrane" testing (chapter 2) and the testing with gravel (chapter 3) in favour of the tests with gravel contact.

In this respect, the idea and the additional effort seem to have been worthwhile.

References

[1] ISO/WD TR 18228-5 Design using geosynthetics - Part 5: Stabilization

[2] Giroud J P and Noiray L 1980 Geotextile-Reinforced Unpaved Road Design. *ASCE Convention*, Portland, September 1981, pp 80-177

[3] Lifa I 2019 *Building with Geosynthetics. A handbook for the geosynthetics user (Bauen mit Geokunststoffen. Ein Handbuch für den Geokunststoff-Anwender)*. Schweizerischer Verband für Geokunststoffe (SVG), in German

[4] Giroud J P and Han J 2004 Design Method for Geogrid-Reinforced Unpaved Roads. Part I: Theoretical development. *ASCE Journal of Geotechnical and Geoenvironmental Engineering*, **130**

[5] EBGEO 2011 *"Recommendations for Design and Analysis of Earth Structures using Geosynthetic Reinforcements – EBGEO"*, Deutsche Gesellschaft für Geotechnik e.V. (ed.), Essen, April 2011, Ernst & Sohn, Berlin, XXII, p 316

[6] Wrigley N E and Zheng H 2008 The confinement effect of different geogrids. *Proceedings of EuroGeo 4*, Edinburgh 243

[7] Aboelwafa M M, Roodi G H and Zornberg J G 2019 In-Isolation Properties of Biaxial and Triangular Geogrids along Various Directions. *Proceedings of Geosynthetics Conference 2019*, Houston, TX, USA, pp 918-28

[8] Zanzinger H 2019 Laboratory Testing of Geogrids and Geocells used for the Function Stabilisation.

8th National Conference on Geosynthetics (Sekizinci Ulusal Geosentetikler Konferansi), 16-17 Mays 2019, Bogazici Üniversitesi Istanbul, Turkey, pp 31-54

[9] DIN 61551 Geosynthetics - Determination of bursting strength (Geokunststoffe - Bestimmung der Berstdruckfestigkeit), in German

[10] Antczak L, Hausmann S and Zanzinger H 2017 Determination of the stiffness of PP geogrids for use under traffic areas using burst tests (Bestimmung der Steifigkeit von PP-Geogittern für die Verwendung unter Verkehrsflächen anhand von Berstversuchen). *DGGT Fachsektionstage Geotechnik 2017*, Würzburg (PowerPoint Präsentation), in German

Micro-structural analysis of cement composites using stone waste: A review

K Devi[1,2,b], B Saini[1] and P Aggarwal[1]

[1]Civil Engineering Department, National Institute of Technology, Kurukshetra-136119

[2]Department of Civil Engineering, Faculty of Engineering and Technology, Shree Guru Gobind Singh Tricentenary University, Gurugram-122505, Haryana

Email: kiranbimbhra@gmail.com, babitasaini6@gmail.com, paratibha@rediffmail.com

Abstract. The direct disposal of industrial by-products to environment is the potential threat to the ecosystem. A possible solution can be utilization of these materials into construction industries to counteract the consumption of cement. Along with the strength and durability properties, study of micro-structural properties of concrete is equally important. The microstructure of concrete largely depends on the type of cement, admixtures, water-cement ratio and curing age. To characterize and quantify the hydration products i.e., calcium silicate hydrate, calcium hydroxide etc. of cement based materials, scanning electron microscopy (SEM), energy dispersive X-ray spectroscopy (EDS), X-ray diffraction (XRD) etc. techniques are being used worldwide. XRD is a powerful tool for the identification of chemical compounds present in the cement-based materials. SEM reveals the information about external morphology, chemical composition, crystalline structure and orientation of the material. EDS is an analytical technique that can be coupled with SEM and used to determinethe elemental composition. A critical review of microstructural analysis using above mentioned techniques for mortar and concrete using stone waste has been done. It was found that these techniques are capable to address the morphological and crystalline structure along with chemical elements and compounds of the materials; and can also be correlated with strength and durability properties. The micrographs showed stone waste, being a filler material aid in forming the dense matrix by filling the voids. Elemental analysis coheres to the chemical compounds (calcium oxide, silica dioxide, etc.) present in the mortar and concrete specimens corresponding to micrographs images.

1. Introduction

Due to continuous depletion and consumption of natural resources during the manufacturing of cement and emission of green house gases to the environment caused to promote research for alternate materials to cement as a partial substitution. On the other hand, wastes from various industries create a nuisance to the ecosystem and demand safe disposal. Industrial wastes can be used in the construction industry as alternate materials to cement or aggregates. The stone powder from the stone industry during the dimensioning process can also be used as a cement replacement to reduce the cement consumption upto a certain extent.The exercise of stone powder (marble, granite, limestone, Kota stone) enhanced the strength and durability properties of mortar and concrete [1-4]. The mechanical and durability properties of mortar and concrete depend on the intrinsic microstructure. The hydration time, water-cement ratio, addition of admixtures and type of cement affect the microstructural properties of mortar/concrete. Low water cement ratio in concrete results into highly dense microstructure. The microstructure of concrete consists of hydrated cement paste, aggregates and interfacial transitional zone (ITZ) [5]. Various hydrous products such as calcium silicate hydrate (CSH), calcium hydroxide (CH), ettringite etc. formed in the matrix during hydration. The improvement and reduction in various hardened properties depend on these hydration products.

The surface characterization of mortar or concrete specimens was conducted to examine its morphology and mineralogy properties and alteration in these microstructural properties with the incorporation of additional materials. The microstructure of cement composites is the microscopical detailing of component of concrete from its macrostructure. The microstructure of concrete is controlled by uniform distribution of cement particles, aggregates and mineral admixtures. The formation of products during hydration reaction in concrete influence the setting or hardening of cement composites, strength, shrinkage permeability etc. The microstructure and performance of cement based material can be controlled by studying the behavior of concrete at the micro level. To characterize the microstructural properties of mortar and concrete, various techniques i.e. SEM, EDS, XRD etc. have been used. EDS is a technique that is used in conjunction with SEM. EDS technique detects X-ray emitted from the sample to characterize the elemental composition of analyzed volume. SEM provides both topographical and compositional analysis of materials. SEM-EDS was used to observe the alteration in the microstructure morphology near the cement particles. EDS was used to detect the contents of mainly existing elements distributed in a certain range of area which indicate microstructure changes in SEM images. The contents of elements were equivalently converted to corresponding oxides in the results of EDS and corresponding oxides are major oxide composition of cement [6].

XRD is one of the most powerful tools for the identification of unknown crystalline phases. By comparing the positions and intensities of diffraction peaks against a library of the unknown of known crystalline phases, target materials can be recognized. In addition, multiple phases in a sample can be identified and quantified. XRD is used to analyze the identification of components of concrete and materials used.

1.1 Research Significance

Since the performance of cement composites depends on the microstructure, thereby study of microstructure of mortar and concrete has become important. The microstructural properties of mortar and concrete were studied by the various characterization techniques. The formation of hydration products and the atomic ratio of various elements/oxides in mortar and concrete were discussed. The stone waste from the stone industries was used to analyze its effect on the microstructural properties of mortar/concrete. In the present study, SEM-EDSand XRD techniques were used to characterize the microstructural properties of mortar and concrete consisting of stone waste i.e. marble waste, granite waste or Kota stone waste either in powder, slurry, dust or other forms. The SEM micrographs are used to investigate the bonding between the cement paste and coarse aggregate by studying interfacial transition zone (ITZ), weaker structure due to porous region and high-water content affects the strength and stiffness. The elemental and chemical composition of mortar and concrete specimens assessed by EDS and XRD analysis was discussed.

2. Methodology

The microstructural characterization was carried out to study morphology and elemental composition of mortar specimens. SEM and EDS were used to analyze the microstructure of cement composites. SEM-EDS and XRD were performed to explore the link between macroscopic and microstructural characteristics of mortar with or without admixtures [7].

2.1 SEM-EDS

SEM scans a focused electron beam over a surface to create image, electron interacts with samples and produces various signals that can be used to obtain information about the surfacetopography and composition. These signals reveal the information about the sample includingexternal morphology (texture), chemical composition and crystalline structure and orientationof materials making up the sample. SEM micrographs provide information about topography, morphology, and composition. By scanning using an electron probe, high-resolution imagesof the morphology of a specimen, with greater depth of field, at very low or high magnification can be obtained. SEM is used to identify phases based on qualitative chemical analysis and crystalline structure. The microstructural analysis of concrete specimens was carried out using SEM which is highly useful to investigate topographical information suchas micro-cracks, voids, different crystalline and amorphous phases present on the surface. SEM has a greater depth of field and high spatial resolution which produces focused images of specimens and analytical data regarding the elemental composition of the features can be obtained from images using EDS. EDS is an analytical technique, used for the elemental analysis or chemical characterization of the specimens. SEM provides a visual answer and EDS provides an elemental answer. To obtain good quality images and avoids introducing cracks or removal of any ettringite nests, cutting of surface of the specimens was avoided, instead pieces were fractured and fractured surfaces were analyzed after coating with gold.To reduce rapid build up of electric charges on non-conducting specimens, gold coating was adopted on concrete surface, before scanning under high energy beam of electron. Measurement of energy and intensity distribution of the characteristic X-ray helps in surface chemical analysis of specimens to confirm the presence of various elements using secondary electron images. The secondary electron images were obtained to analyze the morphology of hydrated. EDS was used to determine the elemental composition of hydrates and confirm the ettringite [8]. EDS is used because it conventionally records the full chosen range of characteristic X-ray which may be processed and quantified. EDS is an analytical technique that is coupled with SEM and used to determine the elemental composition. This technique depicts the plot between X-rays and energy, and this energy peak corresponds to the presence of various elements in the samples. To evaluate the hydrate phase of C-S-H (calcium-silicate hydrate), CH (mineral name portlandite or calcium hydroxide) and AFm (aluminate ferrite monosulphate), the criteria used, has been used which was reported by Mohammed et al. [9]. The presence of C-S-H gel substantially increased the residual strength and reduced the capillary porosity. CO_2 reacts with calcium hydroxide to produce calcium carbonate, known as carbonation and this carbonation reaction enhanced the mechanical strength of concrete and decreased pH value. AFm is crystalline hydrates with formula $3CaO.(Al,Fe)_2O_3.CaSO_4.nH_2O$ and important for the hydration of Portland cement. Table 1 shows the ratio of various elements present in specimens which signifies the formation of hydrous products as suggested by Mohammed et al. [9].

Table 1. Ratio of Ca/Si, (Al+Fe)/Ca and S/Ca

Hydration product	CSH	CH	AFm
Ca/Si	0.8-2.5	$\geqslant$10	$\geqslant$ 4.0
(Al+Fe)/Ca	$\leqslant$ 0.2	$\leqslant$ 0.04	>0.4
S/Ca	-	$\leqslant$ 0.04	>0.15

The broken/fractures sample from mortar/concrete of size 10 mm were taken and sample is soaked in ethanol to stop the hydration process. It is oven dried to remove the moisture. The samples are polished and coated with gold/platinum coating is done to improve the conductivity of electrons. The sample is set in SEM lens and images are taken with different focus. It focused a beam of incident electrons and developed an image by collecting reflected and scattered electron beams [1,10]. EDS is coupled with SEM.

2.2 XRD

X-ray diffraction was used to study the structure of minerals present in the concrete [11]. XRD is one of the most powerful tools for the identification of unknown crystalline phases. By comparing the positions and intensities of diffraction peaks against a library of unknown or known crystalline phases, target materials can be recognized. In addition, multiple phases in sample can be identified and quantified. XRD is used to analyse the identification of components of concrete and materials used [5]. XRD is a non-destructive technique used to determine the elements present in any particular substance. XRD is based on the fact that the measured intensity of a diffraction peak is directly proportional to the content of the substance producing it in the concrete mixture.

XRD involves in directing an incident wave into materials and recording outgoing diffracted wave direction and intensity using a detector. Scattered waves emitted from an atom of different directions. The crystal structure of a material is related to this diffraction pattern. When a sample is scanned in 2θ range all possible diffraction directions can be obtained [8].

To conduct XRD test, fragment of mortar and concrete after destructive compression test were used. The fractured samples were oven dried, grounded with mortal and pestel and sieved through 63 μm aperture size [12].

3. Results and discussion

3.1 SEM

Dense and less porous structure was formed with the use of marble dust blended cement specimens. The paste consists of amorphous particles of CSH and CH crystal that appear in massive layers and ettringite needles are located in pores [13]. The particle of CWP was irregular and angular similar to cement particle as observed from the SEM micrograph [14]. Granite quarry dust (GQD) (60%) had more refine microstructure than GQD at 0% and 100% due to low w/b resulting improvement in compactness and denseness of micro-structure at 150 times magnification of specimens at 28 days curing. GQD (100%) had a large number of pores because of poor compaction of fresh mortar during placing in mould due to less free water in the surrounding matrix. At 3000 times magnification, unhydrated PFA particles were observed for GQD (0%) because the pozzolanic reaction from PFA was not fully hydrated and also number of pores was present. The incorporation of GQD at 60% formed a dense and compact micro-structure matrix. For GQD (100%), poor interlocking micro-structure and number of pores were observed at 28 days. Increase in GQD replacement content reduced w/b that results in improper compaction and also lower content of cement. For 150 times at 90 days, the microstructure improved dueto the formation of more complete hydrated products. Also, number of pores decreased at 90 days in comparison to 28 days. For 3000 times, larger micro-cracks were observed may be because of evaporation of water in the surrounding area. The water absorbed in natural river sand as low in comparison to GQD (60%) which had high water absorption. GQD (100%) has micro-cracks due to high fines content contributed by GQD [15]. SEM images of control mix of autoclaved aerated concrete (AAC) showed the presence of many micro-pores at the pore wall and hydration products with good crystallization leaves i.e. tobermorite. The evaporation of water and chemical reaction is the reason for the formationof pores. For 100% stone sawing mud, the pore distribution was uniform and also tobermorite crystallized leaf structure was present. The main hydration products were CSH and tobermorite; but CH was absent [16]. SEM micrographs of mortar consisting of marble powder showed that structure is not homogenous like control mix. The use of MP increased the intensity of porous area. CSH gel is compact and finely spread over the matrixin case of control mix. Mortar consisting of 10% MP consists of CSH and ettringites; while 30%, 40% and 50% MP showed the large number of voids and crystal of CH. A Needle of ettringite and fibrous crystal of CSH gel were seen in the mortar at higher content of marble powder (MP) at higher magnification. Fibrous crystal of CSH filled the voids which occurred at an early age of hydration. At higher replacement content of MP (40% and 50%), fibrous crystal of CSH gel delayed the hydration process [17]. SEM images of mortar consisting of30% and 40% GP showed that structure was relatively more compact, denser and lesser poreswere present at the interfacial bond between cement and fine aggregates in comparison to control mix [18]. Granite dust (GD) particles were angular in shape due to the cutting and shaping process in their production as observed from SEM images [19]. The combination of copper slag and marble dust formed dense with no pores and cracks. Gel formation was also visible around the sand particles in SEM images [20]. The crack width is visible at micro level at 10% replacement of sandstone slurry as aggregates in comparison to control mix which is responsible for a reduction in strength [10]. No sign of cracking and ASR was observed for mortar consisting 10% and 20% replacement waste marble powder (WMP) due to formation of calcium silicate hydrate and reduction in alkali content while presence of cracks and absence of ASR gel can be seen in case of control mix because ofASR expansion [21]. SEM images showed the lower porosity and more packed structure of mortar with 10% BSS (burnt stone slurry) in comparison to control mix under moist and autoclaved curing conditions because BSS lowered the porosity. Moist cured specimens had ahigher degree of porosity as compared to autoclaved mortar because autoclaved curing produced more packed structure as compared to the former one [22]. SEM image showed the weak bonding between the polished granite aggregates and cement paste those results into reduction in pull off strength [23]. Concrete with 25- 40% GCW (granite cutting waste) had a relatively more compact structure as compared to the control mix. Wide shear cracks and interfacial cracks were observed from SEM micrographs in control mix. Concrete with 25% GCW showed firm contact between aggregates and mortar matrix; and also, denser interface was observed due to the inclusion of GCW. Further addition of GCW in concrete leads to the reoccurrence of interfacial cracks [1]. The small agglomerates and needle-shaped products were observed on the surface of alkali-activated fly ash. Many cracks were observed on the surfaceof alkali activated blast furnace slag specimens due to shrinkage and thermal cracking during high temperature curing. Specimens of control mix had more cracks on the surface of alkali activated slag than S70 mix [24]. The use of Kota stone slurry powder in concrete formed the dense matrix due to its pore filling effect as shown in Figure 1 and Figure 2. It has been observed that more hydrous products were formed at 365 days in comparison to 28 days of water curing due to continued hydration consequently increase in strength was observed. Figure 2 a and b showed dense and compacted matrix which decreased the porosity of matrix and enhanced the strength of the concrete. It has been observed from figure 3 (b) that addition of SSP increased the sulphate resistance of concrete specimens by filling voids and formation of a dense matrix as compared to control mix i.e. Figure 3 a [4,25]. The marble dust on soil not only improved the geotechnical properties by forming compacted matrix consequently voids decreased and improved the pore size distribution and inter particles bonding [26].

a b

Figure 1. SEM micrographs of control mix at **a** 28 days; **b** 365 days

a b

Figure 2. SEM micrographs of SSP in concrete at **a** 28 days; **b** 365 days.

a b

Figure 3. SEM image of **a** plain concrete mix; **b** concrete mixture with SSP under sulphate attack.

3.2 EDS

The major compounds of CWP are SiO_2 and Al_2O_3 as reported from EDS analysis [14]. EDS analysis was conducted to study the change in the chemical composition of control mix and 10% WMP. Control mix had a lower amount of CaO and SiO_2; and higher amount of alkali content in comparison to mix with 10% WMP. Lower amount of alkali controlled the ASR expansion [21]. The Na content in S70 specimens was more than that in SLC specimens i.e. S70 mix had higher proportions of Geopolymer products than SLC and results into a stronger matrix as observed from EDS analysis [24]. The presence of Ca, Si as a major component can be seen in figure 4 and 5. The ratio of Ca to Si indicates the formation of hydration products [4]. From EDS analysis, it was reported that hydration product CSH was formed due to consumption of CH during the pozzolanic reaction of SSP consequently formed dense matrix and enhanced the strength of concrete. On the other hand, lesser amount or absence of sulphur (S) indicates the lesser amount of mono-sulphate and ettringite which was responsible for reduction of strength. The lower values of Ca/Si ratio indicate the improvement in CS by forming CSH and consumption of CH during pozzolanic reaction [25]. Under the exposure to sulphate attack, magnesium sulphate reacted with CSH (calcium silicate hydrate) or CH (calcium hydroxide) and formed MSH (magnesium silicate hydrate) and $Mg(OH)_2$ (magnesium hydroxide). Brucite and gypsum were obtained when magnesium reacts with $Mg(OH)_2$. The mix with SSP (7.5%) had a lower content of magnesium (Mg) which represent a lesser amount of brucite and M-S-H formation that increased the strength in proportion to control mix [4,25].

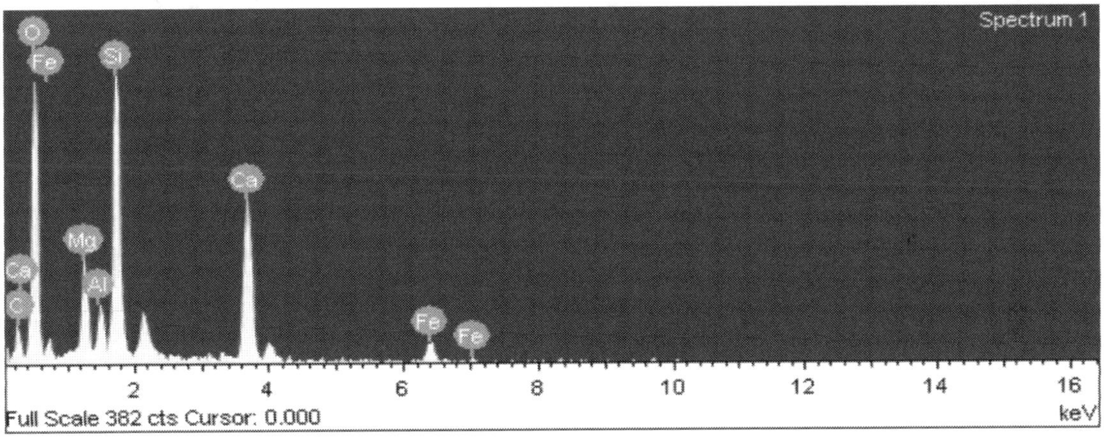

a

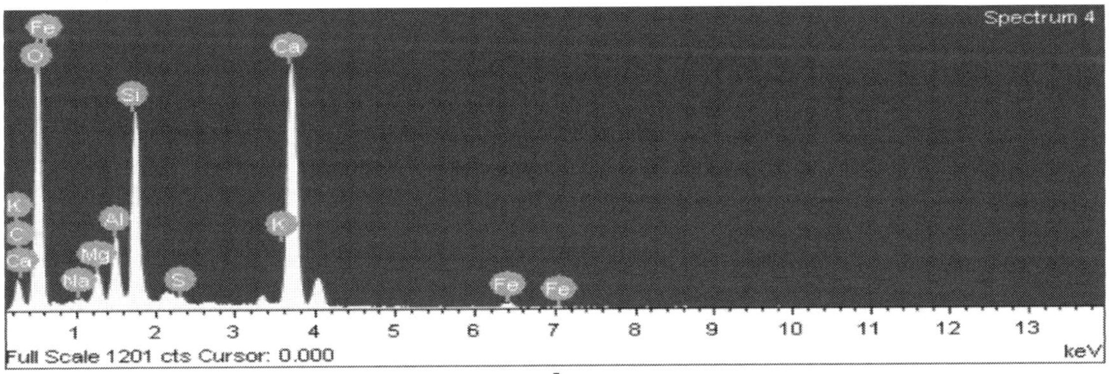

b

Figure 4. EDS images of control mix at a 28 days; b 365 days

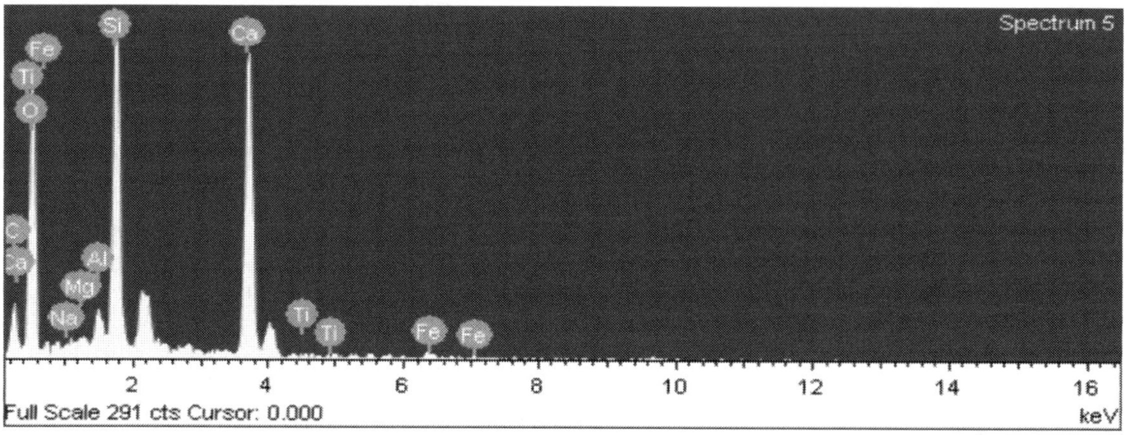

a

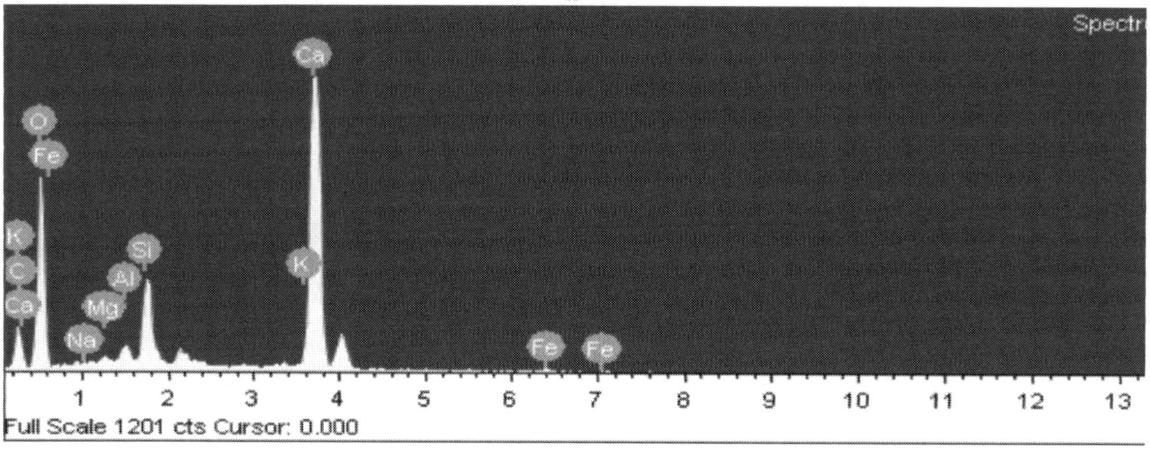

b

Figure 5. EDS images of SSP mix at **a** 28 days; **b** 365 days

3.3 XRD

No observable difference related to CH content for control cement paste and marble dust modified cement paste was found [13]. The presence of CWP deficient the CH amount to react with available silica as confirmed from micro-structural analysis. XRD analysis showed the presence of crystalline SiO2, CSH, CH with no significant difference for 20% and 40% CWP. The degree of hydration was 61%, 64% and 65.7% for cement, 20% CWP and 40% CWP respectively. There was an insignificant difference in the hydration components in mixtures consisting of CWP and control mix [27]. XRD analysis of autoclaved aerated concrete (AAC) with 0% and 100% stone sawing mud as sand replacement showed that C-S-H was the major product that consists of tobermorite and also some amount of quartz which participates in hydration reaction partially [16]. CH, CSH, calcite, calcium silicate and aggregates in the form of quartz are the main components present in the mortar. The presence of MP does not alter the phase composition qualitatively which indicate that MP was inert material [17]. XRD analysis showed the presence of quartz alpha. Peaks of calcium oxide are uniformly seen in C2 mix (75% marble aggregates and 25% conventional aggregates) which results in weakening the bond between cement paste and aggregates that form micro-cracks and cracks [28]. XRD analysis indicate that GD particles had SiO_2, Al_2O_3, $Ca(OH)_2$ and $CaCO_3$ [19]. Broad and diffuse peaks were observed around $2\theta = 35^0$ which implies amorphous or short ordering structure phase. Irrespective of stone powder sludge in alkali activated blast furnace slag, strong peak at $\approx 33.9^0$ was attributed to CSH (I) phase. More hydrotalcite peaks were seen in S70 sample than that of SLC [24]. Marble dust showed dolomite, calcite and quartz whereas soil predominate montmorillonite, aluminium and quartz [26].

The stone waste is used in the cement composites fruitfully with improved performance. Apart from concrete, stone waste can be used in soil stabilization. Soil stabilisation is necessary to alter the geotechnical properties of weak soils by different means. The stabilised soil can be used in construction of pavement and foundation for the structures. The performance of pavement can also be improved with geosynthetics i.e., geogrids, geotextile, geocells etc. as reinforcing material. Geosynthetic materials can be used for strengthening of concrete structures too.

4. Conclusions

The microstructural properties of cement composites have important influence on the strength and stiffness of cement concrete. Concrete is the most unarguable used construction material The strength and durability of cement based material very much depend on the hydration process and formation of hydration products. The microstructural properties of mortar or concrete using stone waste powder have been studied. The addition of stone waste (slurry, powder/dust form) in the mortar and concrete-filled the voids and thus densified the microstructure and reduced the porosity. EDS analysis indicates the elemental composition ofcement composites consisting of stone waste. XRD analysis indicates the presence of a crystalline structure. The presence of MP does not alter the phase composition qualitatively which indicates that MP was inert material as revealed from XRD analysis. Therefore, the utilization of stone waste powder in cement based materials improved the performance due toits filling effects and pozzolanic reactivity. Similarly, it can be used for the soil stabilisation with the same characteristics as used in concrete. Stone waste with synthetic fibre or polymer can also be used to improve the performance of construction materials.

References

[1] Singh S, Nagar R, Agrawal V, Rana A and Tiwari A 2016 J. Clean. Prod. **116** pp 223–35
[2] Singh M, Srivastava A and Bhunia D 2017 Constr Build Mater. **134** pp 471–88
[3] Devi K, Acharya KG and Saini B 2019 *Proc. Int. Conf. on Sustainable Management through Waste Design* Springer Nature Switzerland AG 2019 **21** pp 484–92
[4] Devi K, Saini B and Aggarwal P 2019 *Comput. Concr.* **23** pp 189–201
[5] Singh M and Siddique R 2014 *Constr Build Mater.* **50** pp 246–56
[6] Wu S, Sun K, Zhao H and Zhang F 2018 *Constr Build Mater.* **158** pp 700–6
[7] Senhadji Y, Escadeillas G, Mouli M, Khelafi H and Benosman 2014 *Powder Technol.* **254** pp 314–23
[8] Pillai C S, Santhakumar A R, Poonguzhali A, Pujar M G, Kumar J A, Preetha R and Mudali U K 2016 *Constr Build Mater.* **125** pp 453-65
[9] Mohammed M K, Dawson A R and Thom N H 2013 *Constr Build Mater.* **49** pp 84–92
[10] Arif M, Gupta V, Choudhary H, Kumar S and Basu P 2018 *Constr Build Mater.* **184** pp 432–9
[11] Dave N, Misra A K, Srivastava A, Sharma A K and Kaushik SK 2017 *Constr Build Mater.* **139** pp 447–57
[12] Awoyera P O, Akinmusuru J O, Dawson A R, Ndambuki J M and Thom N H 2018 *Cem Concr Compos* **86** pp 224–37
[13] Aliabdo A A, Abd EMAE and Auda E M 2014 *Constr Build Mater.* **50** pp 28–41
[14] El-Dieb A S and Kanaan D M 2018 SM&T 17
[15] Ban C C, Sern L J and Ramli M B 2019 *Constr Build Mater.* **197** pp 291–306
[16] Wan H, Hu Y, Liu G and Qu Y 2018 *Constr Build Mater.* **184** pp 20–6
[17] Vardhan K, Goyal S, Siddique R and Singh M 2015 *Constr Build Mater.* 96 pp 615–21
[18] Gupta L K and Vyas A K *Constr Build Mater.* **191** pp 155– 64
[19] Li L G, Wang Y M, Tan Y P and Kwan A K H 2019 *Powder Technol.* **342** pp 388–96
[20] Patil M V, Patil Y D and Vesmawala G R 2018 *Int. J. Struct. Eng.* **9** pp 91–100
[21] Munir M J, Kazmi S M S and Wu Y 2017 *Constr Build Mater.* **154** pp 590–9

[22] Al-Akhras N M, Ababneh A and Alaraji W A 2010 *Constr Build Mater.* **24** pp 2658–63
[23] Sharma N K, Kumar P, Kumar S, Thomas B S and Gupta R C 2017 *Constr Build Mater.* **151** pp 158–63
[24] Choi S J, Jun S S, Oh J E and Monteiro P J M 2010 *J. Mater. Cycles Waste Manag.* **12** pp 275–82
[25] Devi K, Saini B and Aggarwal P 2020 *Int. J. Microstruct. Mater. Prop.* **15** pp 58–85
[26] Jain A K, Jha A K, Shivanshi 2020 *Soils Found.* **60** pp 737–51
[27] Kannan D M, Aboubakr S H, EL-Dieb A S and Taha M M R 2017 *Constr Build Mater.* **144** pp 35–41
[28] Kore S D and Vyas A K 2019 *J. King Saud Univ. Eng. Sci.* **31** pp 42–51

Reliability of road pavement constructions reinforced by geosynthetic materials

Gameliak I[1], Nikolaichuk A[2], Zhurba G[3]

[1] Doctor of Engineering Science, National Transport University, Kyiv, Ukraine
[2] Head of the Research Center, Hydrozahist LLC, Kyiv, Ukraine
[3] Director of Euroizol Geosynthetics LLC, Kyiv, Ukraine

Email: gip65n@gmail.com ; anna@euroizol.com

Abstract. Existing experience of unpaved road constructions reinforced by geogrids is described in the article, amended by own practical experience of reinforcement usage in unpaved road constructions. The Object of the research is reinforced and non-reinforced unpaved road constructions. The Purpose of the research is to determine influence of unpaved constructions reinforcement by rigid geogrids on overall bearing capacity and its deviation. The Methods of the research are field bearing capacity dynamic testing and statistic data analysis. The article is a long-term study generalization of unpaved road constructions with grained layers reinforced by rigid geogrids. The study was performed within the period of 2017-2019, and it covers both the construction and operation stages within the period of moisture accumulation and during the drainage period. The results of the study can be used for design and calculation for road constructions by geogrids and geotextile. Forecast assumptions about the object of the research are expansion of the experimental base, taking into account different test methods, different reinforcing materials and different construction conditions.

1. Introduction

Today there are a number of methods, which enable to calculate the required thickness of the base or pavement layer when using geosynthetic materials. The method of J. Giroud and Jie Han is the most well-known calculation method of thickness of granular layers of non-rigid road pavement [1-2], which takes into account the influence of geogrids on the load bearing capacity of structures.

The method was repeatedly improved by J. Giroud and L. Noiray [3], J. Giraud et al. [4], using the results of recent field and laboratory research. J. Giroud and L. Noiray [3] developed an empirical solution for unpaved roads without reinforcement, using the data of field tests and quantitative assessment of the efficiency, reached due to geogrid reinforcement. The solution was based on the limit of equilibrium by the theory of bearing capacity with modifications, in order to take into account the effect on stress due to the membrane effect.

The bearing capacity modular ratio determines the level of reduced bearing capacity, which depends on the deflection in the upper part of the soil base when the surface subsidence reaches the permitted track depth. The stress distribution angle determines the capacity of the base material to redistribute the traffic load. The effect of traffic and geogrids influences on the speed of change of the stress distribution angle, since the condition of the road of the transitional type deteriorates with reloading, which is taken into account in the formula of calculation.

Bearing capacity ratio for unreinforced unpaved roads is known to vary from 2.8 to 3.3. Giroud and Han [1] assumed the ratio of 3.14 (i.e., π), which is the value of elasticity limit for saturated non-draining soil for simple deformations and asymmetric conditions, as well as a zero shear stress interval. For the case when the base layer is separated by geotextile fabric with no interaction, Giroud and Han [1] accepted the value of 5.14 (i.e., $\pi + 2$), as originally proposed by Giroud and Noiray [3], which is the ultimate ratio of bearing capacity for the simple deformation condition and zero shear stress at the boundary of the base and soil of the subgrade.

Thus, grid reinforcement, according to this method, enables to increase the bearing capacity by 5.71/3.14 = 1.8 times when reinforcing with geogrids, and by 5.14/3.15 = 1.6 times when using geotextile fabrics.

Based on the results of our research reported at the IGS International Conference, the method of calculation of reinforcement ratio is proposed, and the value of reinforcement ratio of 1.6 was experimentally confirmed, as well as it is revealed that the use of geotextile fabrics and geogrids significantly reduces the variation. As a result, we obtained the data, which can be used in calculations of road pavement construction in order to estimate increased bearing capacity of reinforced base layer, once reinforced with geosynthetic materials. The obtained reinforcement ratio for GG1 rigid geogrid is at the level of 1.70…1.84 that enables to recommend it for the calculation of reinforced structures. The reinforcement ratio for GG2 varies within 1.07…1.28, and it will increase further during operation in the process of blocking the granular material in the grid cells and putting it into operation. Geotextile materials have the least reinforcing effect and the values of reinforcement ratio of which are 0.85…0.88 and 1.08…1.44 for GT1 and GT2 respectively. For all the cases of geosynthetic materials application a significant decrease in heterogeneity-is observed (the variation ratio decreases from 21.7…35.5% to 10…12% for geogrids and even 4.3…6.1% for geotextile fabrics. Due to increase of homogeneity, the estimated values of modulus of elasticity grow by 1.11…2.24 times.

2. Methodology

At present, calculation methods for road pavement structures of transition type with granular base layers and reinforced by geogrids do not exist in our country. Provision of reliability and durability of road pavement is one of the key tasks in calculating the road pavement at the design stage and ensuring their trouble-free operation during the design lifetime. When calculating the road pavement elastic deflection, the reliability or probability of its trouble-free operation is determined by the formulas 1-4 [6-8]:

$$H= R\ (t,\ N_{sum}) =1 - F\ (t,\ N_{sum}) =P\ (t,\ N_{sum}\leq T), \tag{1}$$

Where $R\ (t)$ is the probability of trouble-free operation, T is the vector of criteria values (T_1, T_2... T_n) of edge condition of the construction; $F\ (t,\ N_{sum})$ is the probability of RPC failure at the moment of time t, or at the value of cumulative effect of the load N_{sum} ; P is the probability of RPC failure.

It is reasonable to calculate the probability of RPC failure (F) not only given to lifetime t, but also considering the cumulative effect of load application N_{sum} separately. We consider this characteristic as the probability of an adverse event within the estimated lifetime t. The cumulative quantity of load application N_{sum} is difficult to predict for estimated life, as it depends on many factors, including the economic conditions of the environment. Thus, it is reasonable to give and calculate the construction reliability separately, according to accepted criteria.

Under standard laws of distribution of *strength* and *stress*, the reliability of the road pavement construction is expressed through the reliability ratio [8]:

$$K_{н}=Prob\ (\beta) =0,5 + \Phi(\beta), \tag{2}$$

Where $\Phi\ (\beta)$ is Laplace normalized function (probability integral), $\Phi_x = \frac{1}{\sqrt{2\pi}}\int_{-x}^{x} e^{\frac{-t^2}{2}}dt$; β is the reliability index, when calculating according to the elastic deflection criteria [1];

$$\beta = \frac{E_{total}-E_{req}}{\sqrt{S^2_{Etotal}+S^2_{Ereq}}} \quad \text{або} \quad \beta = \frac{K_M-1}{\sqrt{C^2_{Etotal}\cdot K^2_M+C^2_{Ereq}}}, \tag{3}$$

Where E_{total} is the overall modulus of elasticity of construction, MPa, which depends on the thickness of the layers h_i, m, their actual modulus of elasticity E_i, MPa and the diameter of the imprint of pneumatics D, m [9],

$E_{req.}$ is the required modulus of elasticity of the construction, MPa, which depends on estimated load and cumulative quantity of load application N_{sum} [9],$S_{E\ total}$ and $S_{E\ req.}$ are mean square deviation of overall (actual) and, accordingly, required modulus of elasticity of the road pavement, MPa,

$C_{E\ total}$ and $C_{E\ req.}$ are coefficient of variation of overall (actual) and respectively required modulus of elasticity of the road pavement layer,

K_M is the capacity margin ratio:

$$K_M = E_{total}/E_{req.} \tag{4}$$

To estimate the overall modulus of elasticity the formula GBN (Industry Standard Norms) B.2.3-37641918-559: 2019 is used [9]:

$$E_{total}^{(i)} = \frac{\left[1{,}05 - 0{,}1\frac{h_i}{D}\left(1 - \sqrt[3]{Etotal_2^{(i+1)}/E_i}\right)\right]E_i}{0{,}71\sqrt[3]{\frac{E_{total}^{(i+1)}}{E_i}arctg\left(\frac{1{,}35eq}{D}\right) + \frac{E_i}{Etotal_3^{(i+1)}}\cdot\frac{2}{\pi}arctg\frac{D}{h_{eq}}}}, \quad \frac{h_{eq}}{D} = \frac{2 \cdot h_i}{D} \cdot \sqrt[3]{\frac{E_i}{6 \cdot Etotal^{(i+1)}}}, \tag{5}$$

Where, i is the number of the considered road pavement layers, counting from the top to the bottom (i = 1, 2, 3 ...),

h_i is the thickness of the ith layer, m,

D is the diameter of the loaded area, m,

$E_{total}\ (i + 1)$ is the overall modulus of elasticity of the half-space underlying the ith layer, MPa;

E_i is the modulus of elasticity of the material of the ith layer, MPa.

The required modulus of elasticity is determined by the formula:

$$E_{req} = (a \cdot \ln(\textstyle\sum N) - b). \tag{6}$$

Where: a=42,843, b=315,68 at Q=115 kN; a=42,843, b= 350,21 at Q=100 kN; a=42,843, b=409,4 at Q= 60 kN.

Taking into account the reinforcement ratio of the granular layer by geosynthetics:

$$E_{reinf} = k_{reinf} \cdot E_1, \tag{7}$$

In case of a three-layer system (asphalt concrete with modulus of elasticity $E_1 = E_{ab}$, a geosynthetic reinforced granular layer $E_2 = E_{total}^{reinf}$ and soil of the roadbed $E_3 = E_{soil}$, it is necessary to solve the system of equations which arise out of the formula (5).

Then, according to the formula (2), one can find the ratio of reliability by taking β – the reliability index (calculation by elastic deflection) [4]:

$$\beta = \frac{Etotal\ \ (k\_reinf) - Ereq}{\sqrt{S_{E_{total}}^2(k\_reinf) + S_{Ereq}^2}} \quad \text{або} \quad \beta = \frac{K_M(k_{reinf}) - 1}{\sqrt{C_{Etotal}^2\ \ (k\_reinf) \cdot K_M^2(k\_reinf) + C_{E_{req}}^2}}, \tag{8}$$

Thus, the reliability of road pavement construction according to the criterion of overall modulus of elasticity on the surface of the layer E_{total}, depends on the actual value of modulus of elasticity of the base reinforced with geosynthetic material as well as on the influence of reinforcing material on the change in the standard deviation of S_E or the ratio of variation C_E of reinforced layer. The value $C_{E\ req.}$ can be taken within 0.15…0.20 depending on the loading conditions for a given category of the road.

For practical application of the above dependences, it is necessary to determine experimentally the reinforcement ratio and the change in the homogeneity of the base layer reinforced with geosynthetics.

3. Main part

The reinforcement ratio was experimentally determined within the period of 2017-2019 when building the road pavement of transitional type to ensure transit of local transport through the section passing partially on the natural soil base, which grades into partially wetlands as well as with the washed up sand of thickness 0.7 - 1.5 m at the base above the peat of thickness 3.0 - 6.0 m.

The main objectives of the research are: to trace the conditions of road pavement layers and their operation, to develop a working method [10] of preparation for testing and its performance, as well as to analyze the results correspondently.

The project of research is located at Iabluneva Street, in the village of Sofiivska Borshchahivka, Kyiv-Sviatoshynsky district, Kyiv region (Figure 1).

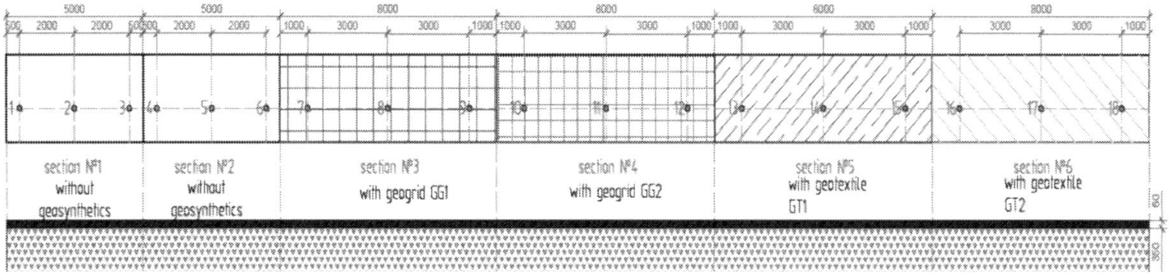

Figure 1. A scheme of the research object in 2014-2019, Yabluneva Street, the village of Sofiivska Borshchahivka, Kyiv-Svyatoshinsky district, Kyiv region

Figure 2 shows the road pavement construction and the tested types of geosynthetic materials at the project No. 1.

- Fine-grained asphalt concrete, type B, grade I, h = 70 mm.

- Stone and sand mixture (SSM) C5, h = 350 mm.

- Together with a) the test area without geosynthetics, 4 similar areas with the reinforcing layer were arranged b) **GG1,** c) **GG2**, d) **GT1,** e) **GT2.**

- Sandy loam, γ=16,5 кN/m^3, φ=24°, c=1 kPa, E=32,0 … 34,1 MPa.

Figure 2. The road pavement construction and tested types of geosynthetic materials at the project No. 1

There were applied most widely used various types and brands of geosynthetic materials in our country. The technical characteristics of rigid geogrids and geotextile fabrics are shown in Table. 1. The estimated characteristics of geosynthetic materials based on the results of tensile strength testing, according to the requirements of DSTU (National Standard of Ukraine) 8607:2015 Geosynthetic Materials. Test Methods [12] were determined by the laboratory.

Table 1 shows the technical characteristics of rigid grids GG1, used as a structural element of the road pavement construction, as well as grids GG2 for comparison.

Table 1. Technical characteristics of geotextile and geogrids

Symbols	GT1	GT2	GG1	GG2
Photo		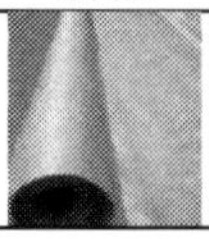	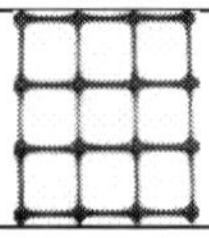	
Type	Needle-punched , nonwoven, thermally bonded geotextile	Nonwoven, thermally bonded geotextile	Biaxial extruded rigid geogrids	Woven flexible geogrids

Material	PP	PP	PP	PET with PVC-coating
Tensile strength, MD/CD, kN/m	16/16	13,1/13,1	40/40	60/60
Strength at elongation 2% MD/CD, kN/m	5/5	4/4	15/15	11,5/10
Strength at elongation 5% MD/CD, kN/m	8/8	6,5/6,5	30/30	19/16
Elongation at break, MD/CD, %	45/45	55/55	11/11	12,5/12
Modulus of elasticity at elongation of 2%, MD/CD, kN/m	250/250	200/200	750/750	575/500
Modulus of elasticity at elongation of 5%, MD/CD, kN/m	160/160	130/130	600/600	380/320

During the testing the device INSPECTOR 4 (Figure 3) was used to determine the dynamic elasticity modulus of FWD that made possible to obtain data samples of required volume within limited period of time (according to the methods [9-10]). The measurements were made in steps, starting with measurements on the surface of the base layer of subgrade just before construction, as well as on the surface of compacted crushed rock after laying and during operation in different periods – periods of moisture accumulation and drying periods. The method provides for three conditional stages of work:
1. Planning stage – preparation for testing, determining the method and setting the scope of testing [10].
2. Testing stage – conducting testing according to the first stage.
3. The stage of statistical processing and analysis of results – the obtained test results are statistically processed, the distribution parameters and the actual values of bearing capacity at the surfaces are determined.

Figure 3. Measurement of elasticity modulus at the surface area within different periods of road structure operation

4. Results
In the process of the above testing, the method was practically tested and can be used to plan and perform testing based on other tools.

Table 2 shows the results of statistical processing of obtained samples of measurements by years. Figure 4 shows an example of the distribution of measured elasticity modulus lengthwise of the test section in relation to its average value.

Table 2. Test results generalization by sites 1-3, 2017-2019

Data	t° C	Section No.1-2 without GS		Section No.3 with GG1		Section No.4 with GG2		Section No.5 with GT1		Section No.6 with GT2	
		E_{v2}, MN/m^2	C_{Ev2}, %	E_{v2}, MN/m^2	C_{Ev2}, %	E_{v2}, MN/m^2	C_{Ev2}, %	E_{v2}, MN/m^2	C_{Ev2}, %	E_{v2}, MN/m^2	C_{Ev2}, %
31.08.17	26	234.82	17.8	286.32	15.3	287.31	13.2	253.50	12.6	254.13	12.0
26.09.17	22	243.71	20.3	305.41	13.6	313.76	11.1	266.28	13.5	270.04	12.9
18.10.17	18	251.31	18.3	308.73	13.4	318.74	11.6	277.07	12.7	282.08	2.9
05.05.18	35	217.19	15.2	270.96	12.1	278.45	10.9	232.13	11.0	235.65	10.9
19.06.18 at 7 a.m.	24	233.65	15.5	284.96	9.7	292.90	9.9	252.68	8.8	254.59	9.3
19.06.18 at 6 p.m.	32	226.33	13.8	282.79	10.1	286.69	4.4	244.45	9.6	251.58	9.6
07.08.18	35	217.81	16.6	267.68	11.2	273.01	7.8	237.14	13.0	241.58	10.6
10.10.18	21	263.78	21.6	330.52	14.3	348.94	12.8	299.17	17.4	302.48	15.8
08.06. 19	29	221.37	11.3	288.16	8.1	290.49	6.6	242.58	6.5	251.78	8.6

The reinforcing effect occurs only after some period of operation of the RPC when blocking the granular material in the cells of geogrids and putting it into operation.

For new GMs, it is necessary to experimentally determine the value of the reinforcement ratio by plate bearing or drop weight testing of unreinforced and reinforced road pavement construction, according to the correspondent methods. In this case, it should take at least 30 days from the arrangement of reinforced layer of the base and its intensive operation.

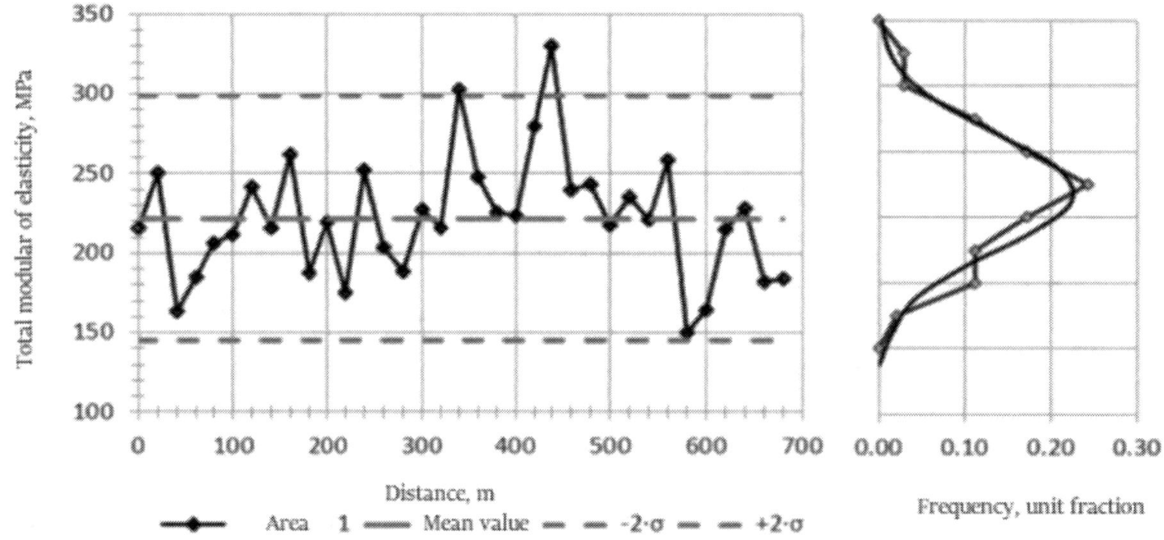

Figure 4. Change and frequency of overall elasticity modulus (a) and its distribution density (b)

To bring the measurement results to the reference temperature + 10°C according to GBN (Industry Construction Norms) B.2.3-37641918-559 the dependence was used

$$K_t = 1,432446 \cdot t^{-0,159946}, \tag{9}$$

Where t is the temperature of the asphalt concrete layer of 70 mm thickness.

Table 3 contains the results of calculation of reliability for specific test areas, taking into account the reinforcement ratio of granular layer, $k_{reinf.}$ and the layer variation ratio, S_V, in these tests according to the formulas (1-7).

Table 3. – Results of the reliability calculation for the test areas

No. of the area	GM type	Elasticity module		Coefficient of Variation (CV)		Safety factor, Km	Mean-square deviation		Reliability index, β	Reliability, K_n
		E_{total}= E_{v2}, MN/m²	$E_{req.}$= $E_{v2,projek}$, MN/m²	C_{Ev2}, %	$C_{Ev2,projekt}$, %		S_{Ev2}, MN/m²	$S_{Ev2,projekt}$, MN/m²		
1	unreinforced	275.66	235.0	16.73	17,00	1.17	46.36	39.95	0.6543	0.74
2	GR1	343.06	235.0	11.97	17,00	1.46	41.18	39.95	1.8819	0.97
3	GR2	351.42	235.0	9.82	17,00	1.50	34.67	39.95	2.1876	0.98
4	GT1	301.03	235.0	11.68	17,00	1.28	35.51	39.95	1.2131	0.89
5	GT2	306.15	235.0	10.28	17,00	1.30	31.65	39.95	1.3786	0.91

Both values, K_n (β) and K_m, characterize the trouble-free operation at a given load application intensity, however, opposed to the safety ratio, the reliability one takes into account not only estimated load bearing capacity but also their distribution variation. As Table 3 shows, the effect at reinforcement, which increases the homogeneity of bearing capacity of the construction, the same way increases its resource in the perspective of intensity grow of load application.

5. Conclusion

The research has confirmed the reinforcement effect published by J. Giraud [1], but it is not less important is to establish the fact of increasing homogeneity of bearing capacity of constructions. Such an effect is well observed by reducing the bearing capacity of the soil base of the construction during the period of spring moisture accumulation, in other words reinforcement levels the influence of soil (natural) base heterogeneity on the overall homogeneity of the construction.

Despite of different areas and years of their arrangement for the same conditions and construction, similar results are obtained. The reinforcement ratio for rigid geogrid GG1 is obtained at the level of 1.41 ... 1.88, which enables to recommend it for the calculations of reinforced construction. Reinforcing effect occurs only after some period of operation of the RPC when blocking gravel into the geogrid cells and putting it into operation.

The unreinforced construction has the lowest safety margin, K_m = 1.17 and the highest C_{E2} = 16.7%. The reliability is K_n = 0.74, which requires reinforcement or increased thickness of the layers of the road pavement.

The GT1 and GT2 geotextile fabrics work almost the same, increasing the safety margin to 1.3 and reducing the coefficient of variation to 10.3… 11.7%. At the same time, the reliability is achieved at the level of 0.89 ... 0.91, which complies with the regulatory requirements.

The greatest effect of the increased reliability up to 0.98 occurs with the reinforcement of the rigid grid GR 2 due to the increased safety factor K_m up to 1.50, and at the same time due to the reduced ratio of the coefficient of variation up to 9.83%.

Based on this it becomes possible to determine what soil types at and what types of geosynthetic materials have a higher reinforcement ratio and are more effective.

It is necessary to continue the study of the reinforcing effect of geogrids with regard to the thickness of the layer, modulus of soil elasticity of the base and the value of the load to verify the theoretical model.

References

[1] Giroud J P, Han J 2004 Design Method for Geogrid-Reinforced unpaved Roads. I. Development of Design Method *Journal of Geotechnical and Geoenvironmental Engineering ASCE* **130** pp 775-86 [in English]

[2] Giroud J P, Han J 2004 Design Method for Geogrid-Reinforced unpaved Roads. II. Calibration and applications *Journal of Geotechnical and Geoenvironmental Engineering ASCE* **130** pp 787-97 [in English]

[3] Giroud J P and Noiray L 1981 Geotexile -reinforced unpaved road design *Proceedings, of the American Society of Civil Engineers Journal of the Geotechnical Engineering Division* **107** pp 1233-54 [in English]

[4] Giroud J P, Ah-Line C and Bonaparte R 1985 Design of unpaved roads and trafficked areas with geogrids *Proc. Symp. On Polymer Grid Reinforcement in Civil engineering* pp 9-12 [in English]

[5] Collin J G, Kinney T C and Fu X 1996 Full-scale highway load test of flexible pavement systems with geogrid reinforced base courses *Geosynthet. Int.* **3–4** pp 537–49 [in English]

[6] Dmitrichenko M F, Dmitriev M M, Gameliak I P, Raikovsky V F, Yakimenko Y M 2012 Nadiynist konstrukciy dorozhniogo odiagu [Road pavements reliability] Navchalnyi posibnyk, NTU – Tutorial, National Transport University p 206 [in Ukrainian]

[7] EN 1990:2002 Eurocode – Basis of structural design. European Committee for Standardization. Brussels: 2003

[8] Rzhanicyn A R 1987 Teoriya rascheta stroitelnych konstrukcyi na nadezhnost [A theory of building constructions design for reliability] Stroyizdat p 239 [in Russian]

[9] GBN (Industry Construction Norms) B.2.3-37641918-559:2019 Dorozhniy odiag nezhorstkiy / Proektuvannia / [Flexible road pavements/ Design] State standard [in Ukrainian]

[10] Gameliak I., Voloshchuk D., Koval T. 2017 Metodyka vstanovlenia zagalnogo modulia pruzhnosti (deformacii) dorozhnogo odiagu za rezultatamy exsperymentalnych vyprobuvan [A methodology of road pavements overall elasticity modulus establishment based on experimental measurements]. Naukoviytvir – Scientific work [in Ukrainian]

[11] Gameliak I, Zhurba G, Nikolaichuk A Reinforcement ratio of road pavement grained layers with application of geosynthetic materials *Proceedings of the 11th International Conference on Geosynthetics. – ICG 11* 16-21 September 2018, Seoul, Korea p.10.

FEM model for analysing the interaction of a geomatrix with a subsoil

Marian Łupieżowiec

Silesian University of Technology, Department of Geotechnics and Roads, Gliwice, Poland

Abstract. The article presents the numerical model which can be applied for the description of the behaviour of a geomatrix mechanically stabilized by a geogrid that reinforces the subsoil. Both stiff loading (geomatrix under building foundation slabs) and susceptible loading (under road embankments) were examined. The selection of parameters was proposed for the analysis based on subsoil tests and data provided by the manufacturers of geosynthetics. The carried out analyses can be used to estimate internal forces in geogrids and their expected deformations. The obtained results can be used as a reference for the development of guidelines for designers of soil reinforcement in terms of boundary problem analysis and design of structures containing geosynthetics.

1. Introduction

Geosysthetics can be excellent structural materials, provided that they are suitably applied [1]. The application of such materials is determined primarily by their mechanical features (strength, stiffness), change of these features over time (durability, aging, rheology) and physical features (weight, geometric dimensions, water absorbability). Not less significant are such features as resistance to external factors and the interaction with other materials - usually with soil medium [2]. The present article is focused on the modeling of the work of geomatrices consisting of compact aggregate stabilized with a geogrid [3]. Such a structure can be used as the reinforcement of subsoil, or it can be treated as the lower part of the structure transmitting stresses to the subsoil [4].
A particularly important feature characterizing a given material in terms of its application as a structural element is its strength. The said quality is defined by stress or, which is often given with respect to geosynthetics, by force per width unit. It guarantees safe operation of the object [5]. Since geosynthetics are almost always embedded in the geomatrix, forming a composite together with the soil, very rarely does the depletion of material strength result in damage to the structure and following it failure [6]. The ductility of such a material must be taken into account, which usually means that when its load-carrying capacity is reached, the given element will stop interacting with soil medium, instead of a sudden and unexpected damage to the entire structure [7]. The geosynthetic can be ruptured, for example, in retaining structures from reinforced soil, while it almost never occurs in geomatrices reinforcing and stabilizing the top part of the subsoil under the object.
Another important feature defining high properties of a geosynthetic material is its stiffness [8], [9], [10]. It is expressed by high force, which can be mobilized in the fibers of the material during its work (stretching) when it is subjected to loading or other external factors. The stiffness of the geosynthetic must be adapted to the properties of the interacting soil, which in the case of geomatrices is most frequently in the form of aggregate (e.g. broken one). A properly built-in geosynthetic guarantees high functional properties of the entire geomatrix [11] [12], [13]. The interaction of the geosynthetic with

aggregate grains is important. In the case of geogrids with a stiff structure, it is done by interlocking of the material with individual grains, while in the case of geomeshes with interlaced connections (flexible ones) it involves the friction force. In the latter case, to ensure proper adhesion, there is no need for high-quality broken aggregates, but the use of inferior material filling the geomatrix results in a significant decrease of the mechanical properties of the structure reinforcing the subsoil. When describing stiffness, it is particularly important to consider the characteristics with respect to small strains (usually 0 - 2%, and in some applications even 0 - 0.5%), which corresponds to the actual strains occurring in a geomatrix [6].

Another very important property of geosynthetics, which distinguishes them significantly from other materials used in construction and geoengineering is their behavior over time (rheological properties). Geosynthetics subjected to constant load undergo creeping, which is a phenomenon consisting in the rise of material strain with the unchanged value of stress [14]. This causes delayed strains and ultimately damage to the material (Fig. 1). If the internal force present in the geosynthetic is close to its strength, the delayed damage can occur almost immediately. The damage (by rupture) of the geosynthetic can occur e.g. in places where it is subjected to stretching for a long time (e.g. in retaining structures). A similar state of stress is reported in such elements in laboratory tests [15], [16]. But in geomatrices, geosynthetics are most often needed at the stage of construction works (they provide a stiff subsoil for foundations or for lower layers of embankments), while their later impact is not very important. At the same time, the values of forces transmitted by geosynthetics are also not very large, and hence we can assume that in this type of structures geomeshes or geogrids will not be ruptured in effect of creeping.

The rheological properties of materials are also characteristic of a phenomenon opposite the to creep – relaxation [14]. It consists in the decrease of stress value over time, while the material is kept under constant strain conditions (Fig. 2). This type of stress state is observed in real conditions when analyzing the geosythetic's performance in various types of structures. Then the passage of time will reduce the stress transmitted by the material, and thus the geosynthetic will not be ruptured, but it will cease to interact. Of course, in some cases (e.g. retaining structures) this may cause failure and damage to this element, while with geomatrices, the stress will be transferred by the aggregate grains. Relaxation can result in the increment of settlement over time, but it usually has nothing to do with the damage done to the built-in geosynthetics. Therefore, the creep safety factors A_1 adopted by engineers (e.g. as proposed by the EBGEO guideline [17], which reduce tensile strength) are in fact trivializing a very difficult and complex engineering problem. In advanced numerical analyses, various types of material models (e.g. Maxwell and Voight models) are used, which are capable of describing rheological properties of materials (changes of stress-strain state over time), including polymeric ones, from which geosynthetics are made [5], [18], [19], [20], [21], [22], [23]. However, in practice, the values of the parameters of these models are not estimated, and moreover, a typical engineer is not prepared in terms of knowledge of mechanics to solve such problems.

Frequently, quite an important load acting on structures in the field of construction and geoengineering are cyclical loads [9], [24], [25], [26]. The problem is particularly important in road communication projects, where the value of load is almost never large enough for the element to be damaged immediately, but a very large number of cycles may result in their inability to be used after some time. The above aspect is taken into account when road pavements are designed, but the geosynthetics used in road construction also tend to lose mechanical properties due to cyclic interactions. In practice, the said problem has not been recognized sufficiently well and is often overlooked when designing structures with the use of geosynthetics [6], [10].

The last of the discussed properties characterizing geosynthetics involves their resistance to external factors such as solar radiation, temperature (heating and freezing), humidity, or even the impact of the elapsed time on their properties [10], [17], [27]. The change in mechanical properties effected by the above factors is referred to as the so called aging effect, which is sometimes confused with rheological properties. In this context, it is particularly important to protect the material against such types of impact, while from the viewpoint of numerical analyses, there are no realistic models available which would be able to describe such relationships. In engineering practice, such impacts are also taken into account by means of safety factors which do not have any significant reference to the results of experiments.

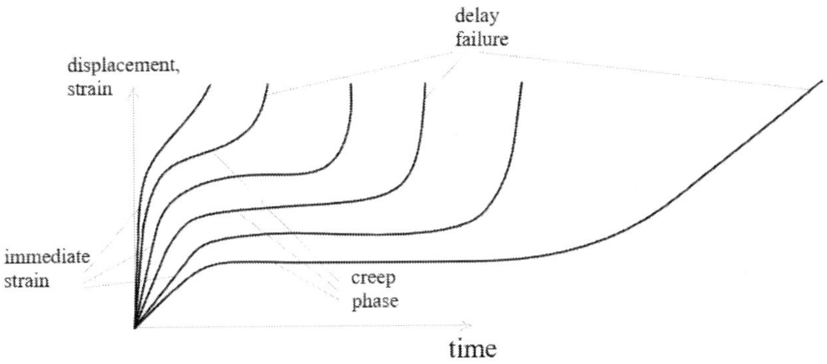

Figure 1. Delayed damage to material in effect of creep.

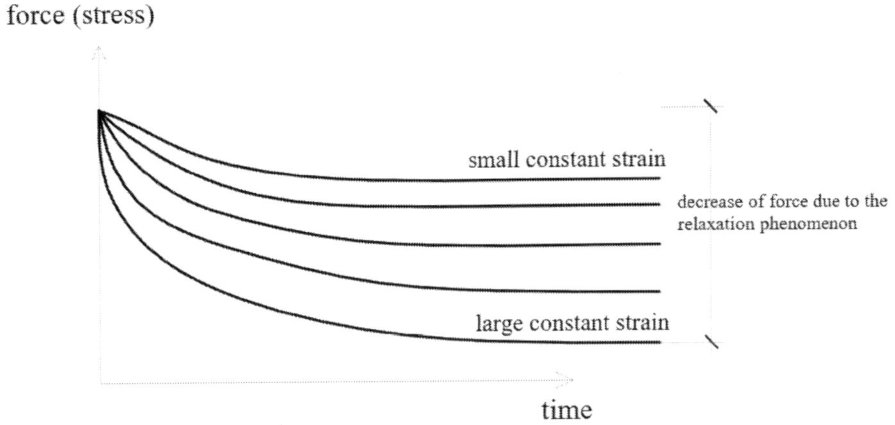

Figure 2. Decrease of stresses over time with constant strain due to relaxation

2. Role of the geomatrix in the reinforcement of subsoil

The discussed in the article structures which make use of geosynthetics are geomatrices. They consist of a layer of compacted aggregate mechanically stabilized by geogrids [12]. Usually the thickness of a single layer is 30 - 40cm, and depending on the needs, we can use a single, double or even triple geomatrix. Since the geosynthetic in a geomatrix transfers tensile stress, it is often believed that it creates a reinforcement, and the entire geomatrix is viewed as a structure from reinforced soil. However, this is not true, because the main task of the geosynthetic is to immobilize (stabilize) the aggregate grains, thanks to which the layer thus formed will provide proper reinforcement of the subsoil and an appropriate support for the structure being built [10], [28]. The most important feature of the geomatrix is its high stiffness, which ensures the correct construction of an object located directly on it (works are carried out on a stable subsoil). By the application of a geosynthetic (in this case a geogrid) we can reduce the thickness of the aggregate layer, which would otherwise have to be even 2-3 times larger [6]. Taking into account the performance characteristics of the object being erected, stiffening the base of the structure built on a geomatrix will result in uniform transmission of loads to a less load-bearing and more deformable subsoil. As a result, we can observe the rise of load-bearing capacity of the subsoil and the minimization of the risk of local stability loss, which increases the safety of the erected structure [29]. In addition, the subsoil for the constructed objects will be characterized by high homogeneity. Although the geomatrix does not provide any significant reduction of the settlement value (due to large dimensions of the object, the weaker layers of the subsoil placed below the geomatrix also have impact on the settlement), the load-bearing capacity of the subsoil together with the applied geomatrix is noticeably increased. With respect to the construction of road embankments or layers of pavement structure on the subsoil reinforced with a geomatrix, the most important is to ensure easy construction of the lower layers having very high parameters. It has positive impact on proper behavior of the structure built on a weaker subsoil. However, when a geomatrix is made under future foundations, it is essential to ensure a stable subsoil on which it is easy to put up formwork and lay reinforcement. The above-mentioned aspects are particularly important in terms of the applied technology of works. They enable proper realization of works in difficult soil conditions.

Due to the character of the geogrid's behavior, the forces transferred by it are generally rather small. This is mainly due to the fact that the occurring strains almost never exceed 0.5%, and hence the force mobilized in geosynthetics usually does not exceed a few to several kN per 1m of band width. Most often, geogrids transfer tensile force, although with proper wedging of aggregate grains by the geogrid, the latter can also participate in taking over the compressive force (analogous to the behavior of reinforcement in reinforced concrete structures). The forces taken over by geogrids are usually of a short-term nature. They occur during the construction stage and shortly after its completion, when settlements are formed effected by dead weight. Subsequently, due to rheological processes taking place in the material, it is excluded from further interaction. It is important that at this stage the structure of the building should rest on a properly compacted aggregate layer, which will constitute a strong and stable subsoil. Therefore, there is no need to analyze the long-term strength of geogrids when designing geomatrices [6], [12]. Only their stiffness matters in terms of the predicted strains as well as rigid structure of the geogrid. When loads are transferred to the geomatrix by a flexible element (e.g. the base of a road embankment), the geomatrix plays a very important role in strengthening the subsoil. Its high stiffness in this case contributes to making the settlements uniform under a high embankment, which also contributes to the decrease in their value. Geomatrices also play an important role in ensuring the stability of high earthworks constructed on weak subsoil. However, when making geogrids to be used under stiff foundations (e.g. pad foundations, strip foundations or foundation slabs), their importance as reinforcement of subsoil is small. Since their stiffness is much lower than that of a concrete element, they can only slightly reduce the values of internal forces in foundations. They are of greater importance only in terms of ensuring homogeneity of the subsoil and compensation of local weakening. Obviously, the importance of geomatrices in the realization of works cannot be overestimated, as they constitute a very good platform without which it would often be impossible to properly perform the works.

In addition to mechanical stabilization, it is also possible to make geogrids in which geosynthetics take over loads (this time only tensile ones) through friction between grains or soil particles and the geosynthetic. In such cases, geomeshes are most frequently applied, and the requirements with respect to the built-in grounds are definitely less restrictive (no need to use sharp-edged aggregates). To make the geomesh carry loads, a lot of pressure is needed from the loading object. However, it must be admitted that such geomatrices will definitely have much worse properties and will not be as effective for reinforcing the subsoil as geomatrices stabilized with geogrids.

3. Modeling the performance of geosynthetics

The analysis of the performance of the geomatrix reinforcing the subsoil was carried out using the finite element method with the application of the program Z_Soil v. 2014 [30]. Two issues were subjected to analysis. The first issue involved the performance of the geomatrix reinforcing weak subsoil under a heavily loaded foundation slab of a building (Fig. 3), while the second one involved the reinforcement for a high road embankment (Fig. 4). In the first case, the load was applied through a stiff element (concrete slab), while in the second example the embankment base was a flexible structure. We investigated one, two, three and four layers of the geomatrix under the objects, and the obtained results were confronted with the behavior of the unreinforced soil. Each layer of the geomatrix consisted of a 35cm thick layer of sharp-edged broken aggregate stabilized in its base with a geogrid having a rigid structure. In the case of several layers of the geomatrix, it was assumed that they were placed directly on top of each other. To simulate the behavior of the soil medium and the aggregate layer, an elastic ideally plastic model with the Coulomb-Mohr plasticity surface was used. The soil medium was modeled as one-component medium, and the parameters describing it define the state of total stress without splitting them into individual components (soil skeleton, pore water and possible air suction pressure). The stiff geogrid was modeled with membrane elements. The said elements transfer only the forces in their plane, and their value is proportional to the produced strains. The membrane elements are able to take over tensile forces that do not reach tensile strength. Due to the way in which the geogrid was installed in the aggregate layer (mutual interlocking of grains and geogrid), it was assumed that the applied membrane elements were capable of transferring certain values of compressive forces not exceeding a few kN (analogous to reinforcing bars taken into account when dimensioning the elements of reinforced concrete for compression). It was assumed in the calculations that the geogrid can take over the compressive force of max. 10 kN /1m. With further increase of compressive stress, the element behaves like an elastic-ideally plastic material (excess load is taken over by other elements through redistribution). When modeling the interaction of the geogrid with the aggregate, it was assumed that the displacements in the nodes of the continual elements being the soil medium were in agreement with the membrane elements. The above assumption is true owing to the way in which the geosynthetic was installed in the aggregate layer (mechanical stabilization). The work of the foundation slab was modeled with coating elements, which apart from the forces in the plane also transfer bending moments.

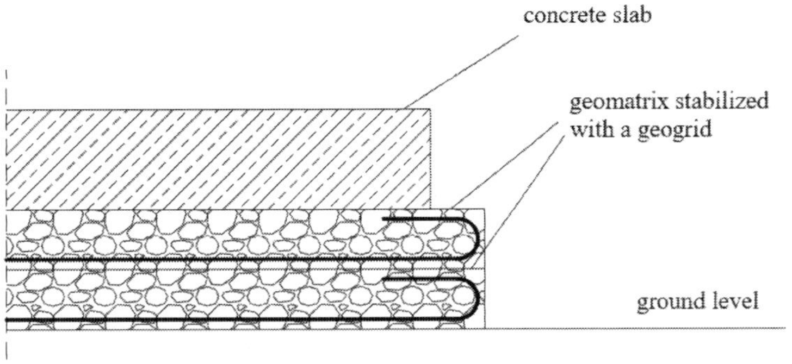

Figure 3. Using of the geomatrixes to subsoil improvement under the concrete slab

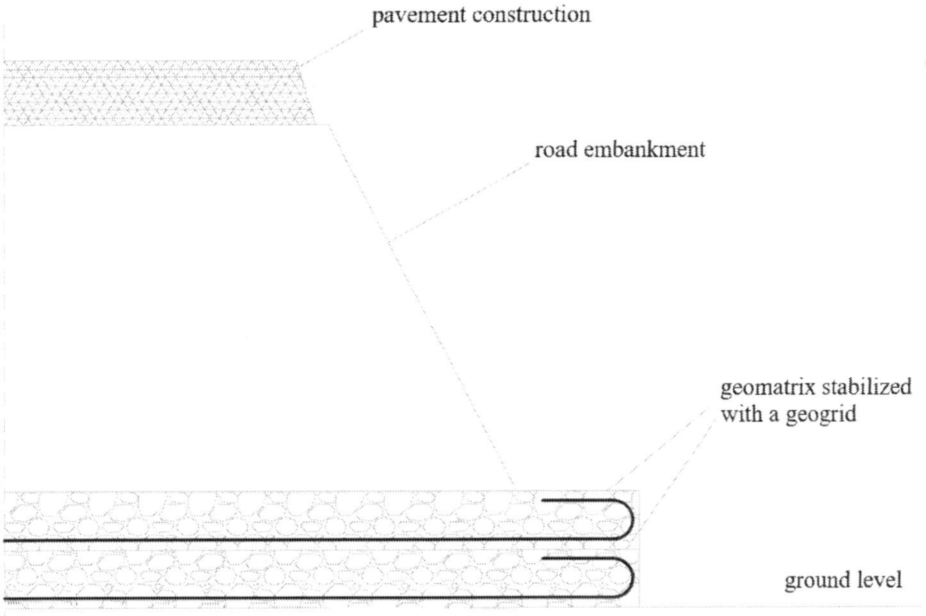

Figure 4. Using of the geomatrixes to subsoil improvement under the road embankment and pavement construction

It is very important to suitably select parameter values used in the calculations. In the article, special attention was paid to the work of geogrid. The characteristics of the membrane simulating its work are defined by the stiffness which expresses the ratio of the mobilized force to the caused strain. The stiffness values should correspond with the range of strains that the geomatrix undergoes, which usually do not exceed 0.5% [31]. They are determined in laboratory tests, but then the geogrid alone is tested, with no built-in aggregate. The obtained stiffness values are usually in such cases within 300 - 600 kN /1m. In order to be realistically able to simulate the work of a geosynthetic, one would have to take into account stiffness together with the interaction of aggregate grains. This means a manyfold higher stiffness, which corresponds to the actual work of the stabilizing element. For this purpose, pull-out tests should be carried out with the local measurement of strains of the geosythetic using strain gauges. In the performed calculations, the theoretical analysis involving the impact of the stiffness value on the obtained results was presented, and the value of the parameter describing this stiffness was analyzed in the range of 1 - 15 MN /1m. Additionally, non-linear geometric relations were taken into account when determining strains characteristic for large displacements.

In the carried out numerical analyses, we adopted a simplification by assuming a flat state of strain. The said assumption is true when analyzing linear objects (e.g. road embankments), or long structures in relation to their width. Despite this simplification, it is possible to analyze the behavior of the geosynthetic stabilizing the geogrid layer, and based on the obtained results, important conclusions useful for designers' analyses can be formulated. For the carried out 2D analysis, the obtained result is the force transmitted by the geogrid in the horizontal direction. Also the coating elements describing the work of foundation slabs, which become flat beams, are simplified.

4. Examples of numerical analyses

The numerical models used in the analyzes are presented in Fig. 5. The analysis involved the behavior of a heavily loaded foundation slab (Fig. 5a) and a 6m high road embankment (Fig. 5b) on the subsoil reinforced with the layers (one, two, three or four) of geomatrix. Also the analysis of the behavior of objects on unreinforced soil was carried out. The symmetry of the defined problem was applied, which allowed half of the model to be analyzed. The subsoil adopted for the analysis consisted of the 5.0m surface layer of weak clay having plastic consistency, which rested on the medium-compacted medium sand. The most important values of the parameters of the subsoil layers model adopted for the analysis are as follows:

Clay in plastic state: $E = 20$ MPa, $\nu = 0{,}30$, $\gamma = 20$ kN/m^3, $\phi = 15°$, $c = 15$ kPa,
Medium sand in medium compacted state : $E = 60$ MPa, $\nu = 0{,}25$, $\gamma = 18$ kN/m^3, $\phi = 32°$, $c = 2$ kPa,
Broken aggregate forming the geomatrix : $E = 120$ MPa, $\nu = 0{,}25$, $\gamma = 19$ kN/m^3, $\phi = 40°$, $c = 2$ kPa.

In the analysis of the road embankment structure, another material zone was used - the soil used for the construction of the embankment, which was characterized by the following parameter values: $E = 60$ MPa, $\nu = 0{,}25$, $\gamma = 19$ kN/m^3, $\phi = 33°$, $c = 20$ kPa. The said parameter values correspond to soils stabilized with binders, and they were selected so as to exclude the mechanism of stability loss which consists in the development of near-surface slope or shallow landslide in which the slip surface does not run through the subsoil (otherwise the analysis of the impact of the geomatrix on the stability of the earth structure would not have any sense). Concrete elements used in the analysis of the geomatrix under the foundation slab (slab and columns) were taken into account as linear-elastic elements ($E = 30$ GPa, $\nu = 0{,}20$, $\gamma = 25$ kN/m^3).

The first stage of calculations involved the generation of primary stresses in the subsoil, which are effected by the dead weight of the subsoil layers. It was assumed that at the time of the generation of those stresses (it is the starting point of the calculations), geomatrices reinforcing the subsoil had already been made (the impact of the construction technology of those structures on the state of stress in the subsoil was not analyzed). As to the analysis of the foundation slab, it was also assumed that at the beginning of the analysis, the slab and the columns transferring loads from the building structure had been also made. The next stage of calculations involved setting loads on the columns in 20 calculation steps, which amounted to 1500 kN /1m for middle columns and 900 kN /1m for the peripheral column. But with respect to the construction of the road embankment, it was assumed to be carried out in 4 stages, in which successive parts of the embankment appeared in the model. After its construction, a load resulting from the operation of the road was set, which was assumed to be evenly distributed on the surface of the embankment crown and equal to 25 kPa. The last stage of calculations involved the analysis of the stability of the embankment slopes on the subsoil reinforced with a geomatrix using the strength reduction method $c\text{–}\phi$ [32], [33].

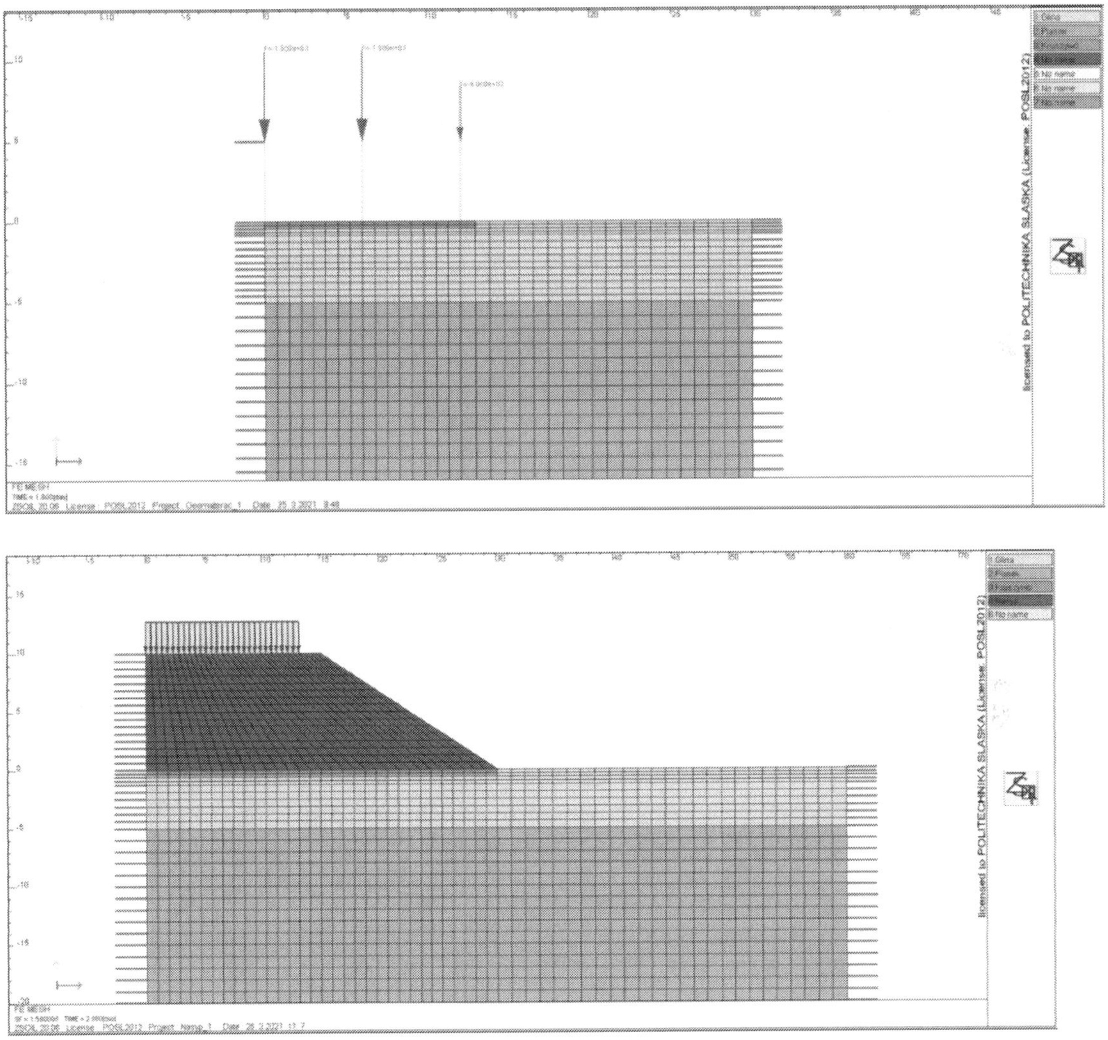

a)

b)

Figure 5. Numerical models of the analyzed issues: a) geomatrix under foundation slab, b) geomatrix under high road embankment

5. Discussion of the obtained results

The most important results subjected to analysis involved the values of internal forces in the geogrid and the impact of the presence of a geomatrix stabilized with a geogrid on the work of the structural elements of the object placed on it. Basing on the above, we can assess the effectiveness of the reinforcement of subsoil and provide guidelines for the use of this type of reinforcement. The results presented below in Figs. 6 and 10 involve the so-called reference issue (a single geomatrix stabilized with a geogrid of the stiffness of 4000 kN /1m). The impact on the obtained results involving the geogrid stiffness and the number of geomatrix layers is presented on the graphs in the further part of the article. As to a foundation slab resting on a geomatrix, bending moments developed in the slab are the most important for structural designers (Fig. 5a). The slab is bent in such a way that lower fibers under the columns and upper fibers in the spans are stretched. In the case of weak subsoil, very high bending moments develop in the slab, which results from the need to distribute the load over a large area. However, when the ground is stiff, the pressure is transferred primarily in the near-column areas, and the bending moments are definitely smaller. This significantly translates into the required thickness of the foundation slab and its reinforcement. The values of the forces in the geosynthetic are presented in Fig. 5b). The largest tensile forces develop under the columns transferring heavy loads, which is analogous to the work of the slab being bent. The values of tensile forces in geosynthetics do not exceed a dozen or so kN/m, which is small in relation to the value of forces analyzed when dimensioning the geosynthetics. In addition, small compressive forces were found on a part of the geogrid, which are transferable if the geogrid is properly installed in the geomatrix and if the aggregate is properly compacted.

a)

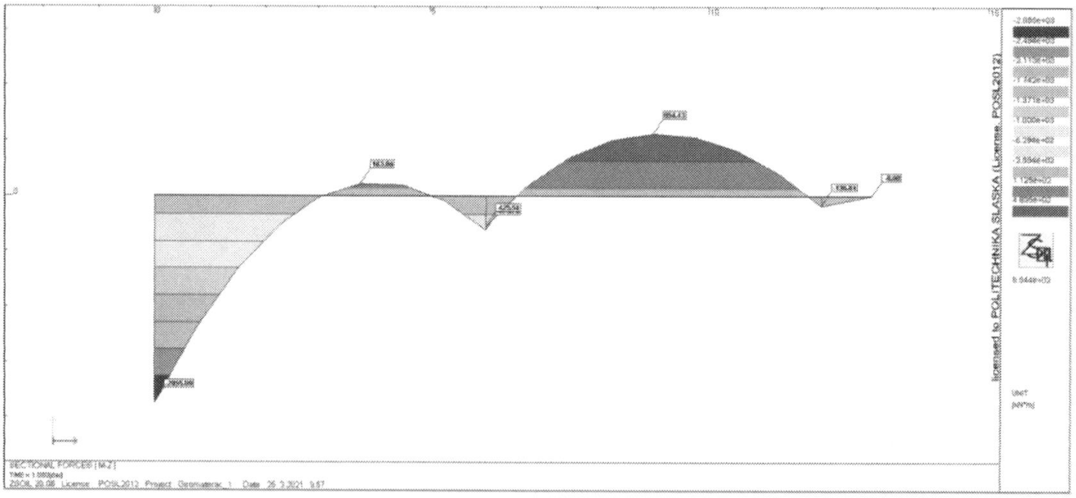

b)

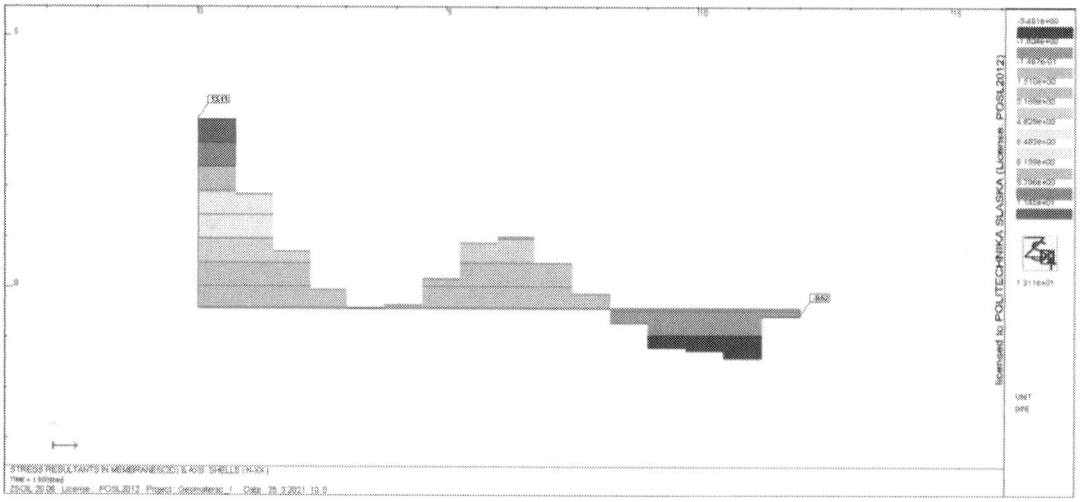

Figure 6. Behavior of the foundation slab on the subsoil reinforced with a geomatrix: a) graph of bending moments in the slab, b) graph of forces in geogrids

The impact of the number of geomatrix layers and that of geogrid stiffness for a single geomatrix is presented in Figs. 7 – 9 and summarized in the Table 1. We can observe a very clear impact of the number of geomatrix layers on the value of bending moments in the foundation slab. Although the positive moment (bottom fibers are stretched) decreases only by a maximum of 6%, in the case of the negative moment this decrease is even 21% (Fig. 7). Equally evident, although definitely smaller (approx. 10%), is also the decrease in settlement values, which is observed with a quadruple geomatrix as compared to the unreinforced soil (Fig. 8). In this case more important is a more uniform distribution of settlements, which results from a more stiff subsoil. However, there are practically no differences in the values of both bending moments and settlements when the stiffness of the geogrid changes. In the latter case, only large values of tensile forces are observed in the geosynthetic, which is a natural consequence of high stiffness of the material, due to which high force is mobilized even at small strains (Fig. 9). The values of forces transferred by geogrids are not too large, which results from their function. The purpose of these elements is to ensure mechanical stabilization of the aggregate forming the geomatrix, and not their work as reinforcement (the analogy to the work of reinforcement in reinforced concrete structures is inappropriate here). When analyzing the results presented in Figs. 7 and 8, we can conclude that the application of geomatrices does not provide a particularly effective reinforcement of weak subsoil for the future foundation. Yet, its most important role is to provide a strong and stiff subsoil on which the works can be carried out properly. And the compliance with the limit state of load-carrying capacity and serviceability is to be ensured by properly designed foundations. Although geomatrices can be classified as structural elements transferring load from the structure to the subsoil, their performance efficiency is definitely lower than that of a reinforced concrete slab. Hence, the use of this type of reinforcement under large building foundations is hardly used in practice.

Table 1. Values of bending moments in the slab, settlement and the forces in the geosynthetics in the mattresses under the foundation slab

Type of geomatrix	Maximal bending moment [kNm/m]	Minimal bending moment [kNm/m]	Maximal tensile forces [kN/m]	Minimal compressible forces [kN/m]	Settlement [mm]
Without geomatrix	2972	954	–	–	228
Single, K=1000 kN/m	2858	855	1,08	0,30	223
Single, K=4000 kN/m	2857	855	4,08	1,14	223
Single, K=8000 kN/m	2856	855	7,67	2,10	223
Single, K=15000 kN/m	2855	855	13,10	3,50	223
Double, K=4000 kN/m	2849	810	9,15	1,76	218
Triple, K=4000 kN/m	2817	774	12,90	2,30	212
Quaternary, K=4000 kN/m	2771	747	15,30	2,29	206

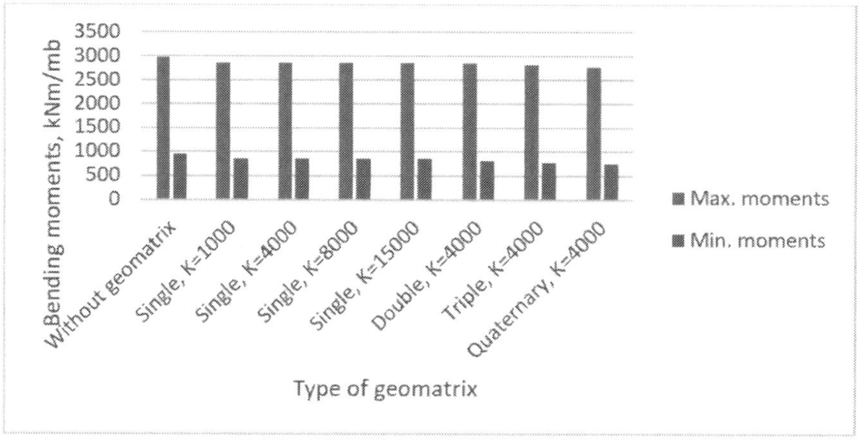

Figure 7. Values of bending moments in the foundation slab

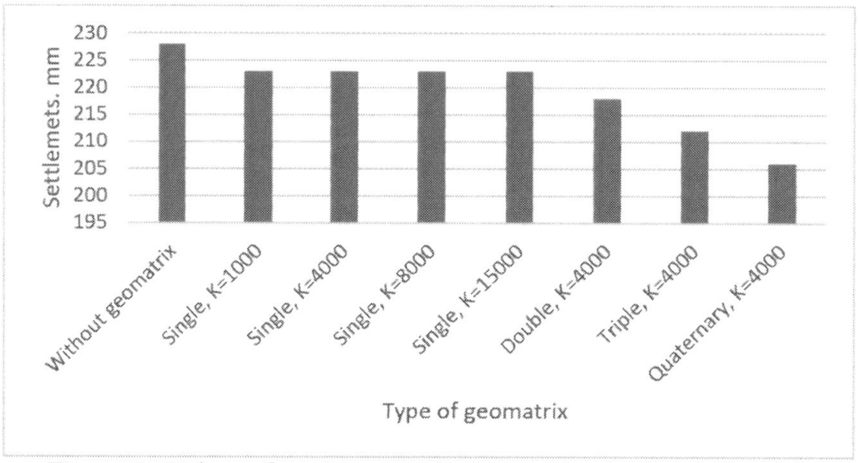

Figure 8. Values of maximum settlements under the foundation slab

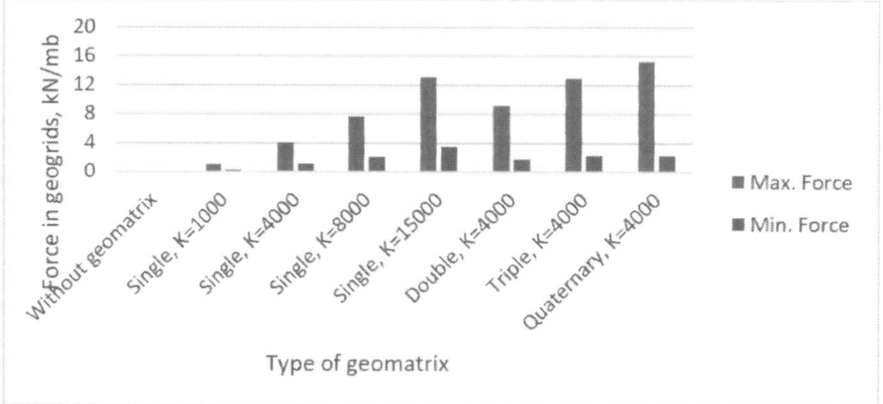

Figure 9. Tensile forces in the geosynthetic

When geomatrices are used under road embankments, their function ensuring the stability of an earth structure built on a weak subsoil is particularly important. In the case of a strong subsoil, the embankment massif is crucial for the stability of the scarps, and the slip surface runs close to the surface of the scarp. If the subsoil is weak, the slip surface runs deeper (see Fig. 10a), and the reinforcement of the subsoil has a large impact on the safety margin expressed by the stability factor. Then, the work of a geogrid stabilizing the geomatrix layer is of key importance for the stability of the object. And thus the force mobilized in the geogrid can assume higher values (Fig. 10b), but they are still not too high as compared to the values obtained when designing structures from the reinforced soil [34].

a) 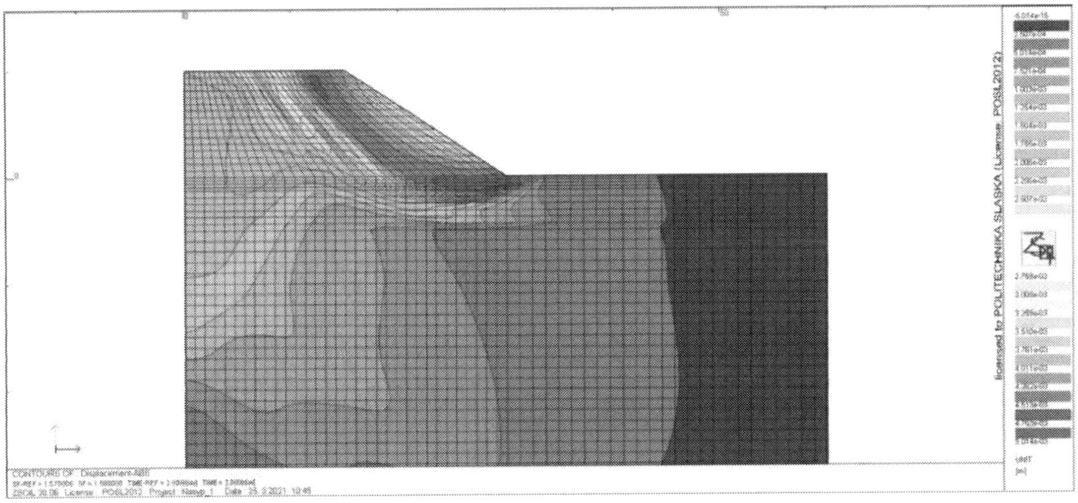

b) 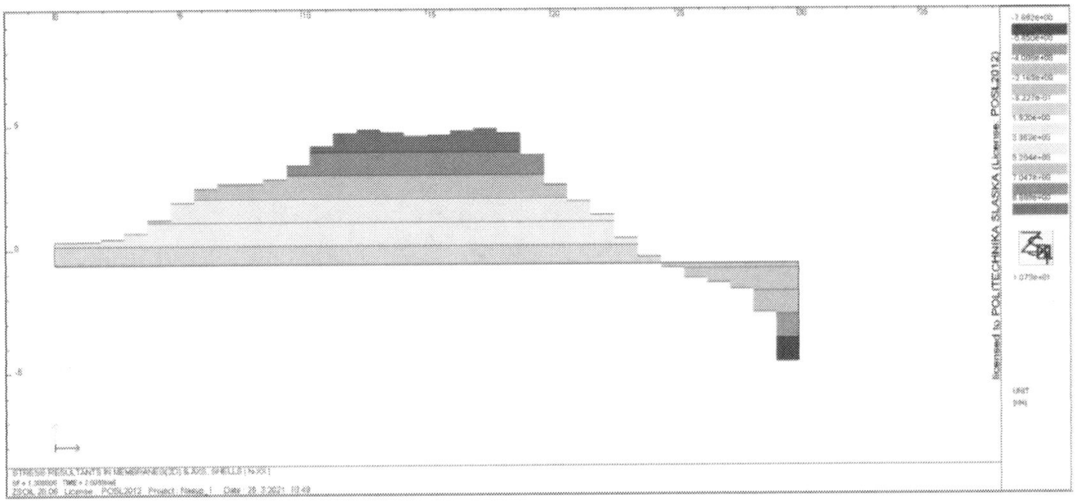

Figure 10. Behavior of road embankment on the subsoil reinforced with a geomatrix: a) slip surface at the moment of stability loss Fs = 1.58, b) graph of forces in geogrids

Fig. 11 presents the dependence of the stability factor on the type of the applied subsoil reinforcement. The value of stability factor is dependent both on the number of geomatrix layers and on the stiffness of the applied geogrid. A larger number of geomatrix layers translates into greater reinforcement thickness, while greater stiffness enables the mobilization of a greater force in the geosytetic. In the case of weak subsoil and when the requirements involving the value of stability factor are very high (e.g. in railroad investment projects currently implemented in Poland), geomatrices under communication embankments are one of the most effective reinforcements of subsoil. Another benefit of using geomatrices is the reduction of settlement values. It may not be as large as the increase in stability, but based on the carried out analyses, a reduction of approx. 13% was reported, with the application of a quadruple geomatrix as compared to the lack of reinforcement. Here, as in the case of the reinforcement of the subsoil for the foundation slab, the settling is also determined by the weak subsoil being under the geomatrix, but due to flexible load, the reduction of settlement is much more visible. Also the reduction of the difference in settlement between individual fragments of the embankment base is not without significance, which may contribute to higher durability of the earth structure and pavement structure made on it.

Fig. 13 presents the values of forces in the geogrids. For each type of reinforcement, the values of maximum tensile forces and maximum compressive forces were compiled. For tensile forces, the values obtained after the construction of the embankment and after setting the operational loads (by F = 1.00) were provided, as well as values obtained with the stability factor of 1.30 (by F = 1.30), which is the value required when designing road embankments. As we can easily observe, ensuring an adequate safety margin requires that specific values of forces in the geosynthetic stabilizing the geomatrix should be enabled to obtain. Such higher values should be adopted for the analyses carried out during design works. Results presented in Figs. 11-13 are summarized in the Table 2.

The application of geomatrices has still another aspect, especially important in road construction. Namely, the acquisition of sufficiently high stiffness of the lower layers of the embankment (measured by test loads with the VSS plate) requires that a suitable subsoil for these layers is ensured. Therefore, without a geomatrices it would not be possible to build an embankment in line with the requirements of the standards and relevant work specifications.

Table 2. Values of the stability coefficient, settlements and the forces in the geosynthetics in the mattresses under the road embankment

Type of geomatrix	Stability factor SF	Settlement [mm]	Maximal tensile forces by SF=1,30 [kN/m]	Minimal compressible forces by SF=1,30 [kN/m]	Maximal tensile forces by SF=1,00 [kN/m]
Without geomatrix	1,48	155	–	–	–
Single, K=1000 kN/m	1,51	150	3,64	1,18	3,95
Single, K=4000 kN/m	1,58	150	11,1	4,50	7,69
Single, K=8000 kN/m	1,62	150	20,1	8,56	7,69
Single, K=15000 kN/m	1,62	150	35,8	15,05	7,69
Double, K=4000 kN/m	1,67	145	9,11	4,45	6,86
Triple, K=4000 kN/m	1,74	140	8,16	4,47	3,88
Quaternary, K=4000 kN/m	1,82	135	8,25	4,39	4,00

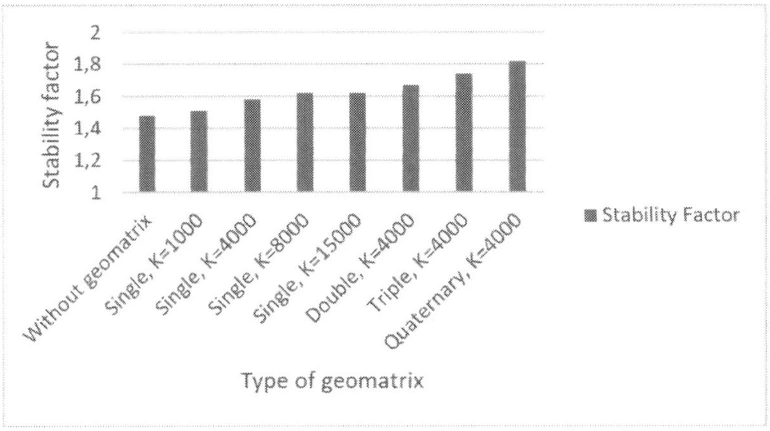

Figure 11. Values of stability factors of the scarp of road embankment

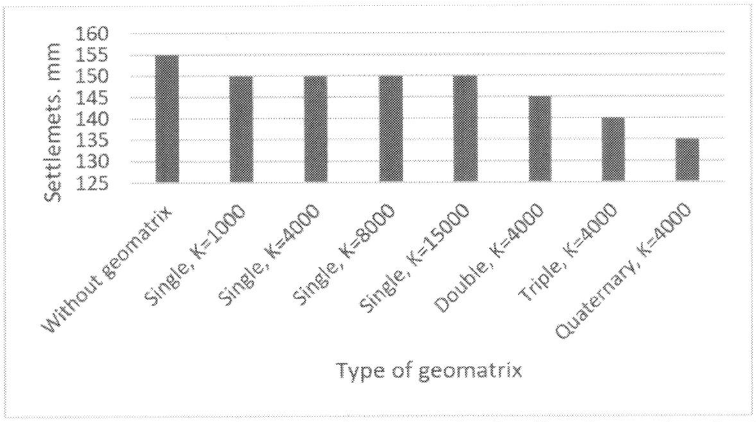

Figure 12. Values of maximum settlements of subsoil under road embankment

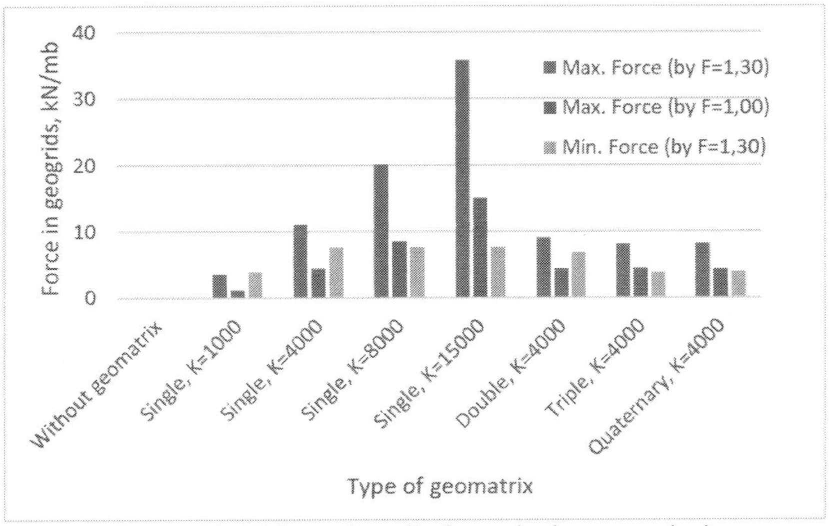

Figure 13. Values of tensile forces in the geosynthetic

6. Summary and conclusions

The article presents a model which allows to carry out the numerical analysis involving the interaction between stiff geogrids and aggregate forming a geomatrix reinforcing weak subsoil. We discussed the method of model preparation, selection of parameters for analysis and the of calculations. It is particularly important to adopt high membrane stiffness that simulates the work of a geosynthetic built into the geomatrix. Otherwise, the impact of this element on load transfer to the subsoil would not be practically taken into account at all.

The results of numerical analyses were also presented, which allowed us to formulate several important conclusions. The first one states that the most important role of the geomatrix as the reinforcement of a weak subsoil is to provide a strong and stable subsoil on which the designed elements can be properly constructed. The second function of the geomatrixa is to ensure the stability of structures having low stiffness (e.g. communication embankments), which are erected on a weak subsoil. When the structure of an object is characterized by much higher stiffness (e.g. a foundation slab), the role of the geomatrix in meeting the limit state conditions of load-carrying capacity is less important. Geomatrices also ensure some reduction of settlements, but this impact is not as significant as with other reinforcements. Such a

situation can be explained by the fact that a geomatrix provides surface reinforcement, and the settlement of subsoil under the objects is also determined by the deformation properties of soils deposited deeper. Finally, a distinctively important role of geomatrices, which cannot be included in the calculations, is that these structures provide a stiff base (platform) that allow proper construction of an object. It is extremely important in communication engineering, where the erected structures are made by means of soil compaction. The presence of a geomatrix in this case allows to obtain an appropriate compaction index, which indicates whether a given element has been made correctly. In the case of reinforced concrete structure, the importance of a geomatrix is lower, yet it is not uncommon to have to lay heavy formwork and reinforcements that require a sufficiently stiff subsoil.
As the last conclusion of the carried out analyses, we can state that the forces transferred by geosynthetics used as mechanical stabilization of geomatrices are usually small as compared to the strength of these materials, and their service life at the maximum loading is short. Their most important function is to provide a suitable base during the construction process, and during the operation, other structural elements are responsible for meeting the limit state conditions. However, with the reinforcement of subsoil for road embankments, the role of geosynthetics is to provide an adequate safety margin, and the element itself does not work for a long time in the state of high effort. Therefore, in this particular case, the rheological properties of geosynthetics can be ignored when designing subsoil reinforcement with geomatrices. There is also no risk of rupture of geosynthetics in geomatrices, because in the long term the phenomenon of relaxation develops, whereby these materials are excluded from the interaction in the transfer of loads. In addition, many designers confuse the effects of rheological phenomena with the aging of material and the loss of its mechanical properties.

7. References

[1] Koerner R M 2005 Designing with Geosynthetics Chapter 5, *NJ Prentice Hall Inc.* 5th ed.
[2] Van Santvoort GP Geosynthetics in civil engineering, *Routledge.*
[3] Choudhary AK, Krishna AM 2016 Experimental investigation of interface behaviour of different types of granular soil/geosynthetics, *International Journal of Geosynthetics and Ground Engineering*, 2(1), 4
[4] Duszynska A 2016 Zbrojenie geosyntetyczne podstawy nasypu (Geosynthetic reinforcement of the embankment base) *Inżynieria Transportowa, Politechnika Gdańska, Wydział Inżynierii Lądowej i Środowiska, Gdańsk* (in Polish)
[5] Yeo SS, Hsuan YG 2007 The short and long term compressive behavior of HDPE geonets and geocomposites under inclined conditions *Journal of Geotextiles and Geomembranes* 14(3),154-164
[6] Kawalec J, Koda E 2016 Polish experiences in using geogrids in earth structures engineered with anthropogenic materials *Conference: 3rd Pan-American Conference on Geosynthetics "GeoAmericas2016"*, at Miami Beach, Vol. 1
[7] Zornberg JG 2017 Functions and applications of geosynthetics in roadways *Procedia engineering* 189, 298-306
[8] Roodi GH, Zornberg JG 2017 Stiffness of soil-geosynthetic composite under small displacements. II: Experimental evaluation, *Journal of Geotechnical and Geoenvironmental Engineering*, 143 (10), 04017076
[9] Rakowski Z., Kawalec J. 2009 Mechanically stabilized layers in road construction *27th International Baltic Road Conference*, Riga, Latvia, *section A2 Road Construction*
[10] Kawalec J, Grygierek M, Koda E, Osiński P 2019 Lessons learned on geosynthetics applications in road structures in Silesia mining region in Poland *Applied Sciences* 9(6), 1122
[11] Hufenus R, Rüegger R, Flum D, Sterba I J 2005 Strength reduction factors due to installation damage of reinforcing geosynthetics, *Geotextiles and Geomembranes*, 23 (5), 401-424
[12] EOTA 2012 Technical Report European Organization for Technical Approvals Non-reinforcing hexagonal geogrid for the stabilization of unbound granular layers by way of interlock with the aggregate, TR41

[13] Kawalec J, Hornicek L, Rakowski Z 2019 Cost effective alternative solution for the renovation of concrete pavements *In IOP Conference Series: Materials Science and Engineering* 471, no. 2, 022041 *IOP Publishing*
[14] Sawicki A, Kazimierowicz-Frankowska K 1998 Creep behaviour of geosynthetics, *Geotextiles and Geomembranes*, 16 (6), 365-382
[15] ASTM D2990-1 2001 Standard Test Methods for Tensile, Compressive, and Flexural Creep and Creep-Rupture of Plastics *West Conshohocken, PA, United States: ASTM International*
[16] Farrag K, Shirazi H. 1997 Development of an accelerated creep testing procedure for geosynthetics – Part I: Testing *Geotechnical Testing Journal* 20(4), 414-422
[17] EBGEO 2011 Recommendations for Design and Analysis of Earth Structures using Geosynthetic Reinforcements *Ernst & Sohn Verlag*, Berlin, Germany
[18] Fannin R J 1995 Factors infuencing the long-term low capacity of geonets *3rd International Conference on Geosynthetics* 267-280
[19] Greenwood J H, Voskamp W. 2000 Predicting the long-term strength of a geogrid using the stepped isothermal method *In: 2nd European Geosynthetics Conference* Italy, 329-331
[20] Lothspeich S E, Thornton J S 2000 Comparison of different long-term reduction factors for geosynthetic reinforcement materials *In: 2nd European Geosynthetics Conference,* Italy 341-346
[21] Allen T M, Bathurst R J 2002 Observed long-term performance of geosynthetic walls and implications for design *Geosynthetics International* 9(5-6).
[22] Jeon H Y 2017 Review of Long-Term Durable Creep Performance of Geosynthetics by Constitutive Equations of Reduction Factors In Creep, *IntechOpen*
[23] Alekseev N, Kovalenko I, Stolyarov O, Melnikov B E 2017 Viscoelastic properties of geosynthetic materials, *Construction of Unique Buildings and Structures*, 56(5), 17-32.
[24] Asha MN, Latha GM 2012 Strength Behaviour of Reinforced Soil-Aggregate Systems under Repeated and Cyclic Loading *GeoCongress 2012: State of the art and practice in Geotechnical Engineering* 1513-1522
[25] Nair AM, Latha GM 2014 Cyclic loading behaviour of reinforced soil–aggregate bases *Proceedings of the Institution of Civil Engineers-Ground Improvement* 167(2), 88-98
[26] Vieira CS, Ferreira FB, Pereira PM, de Lurdes Lopes M 2020 Pullout behaviour of geosynthetics in a recycled construction and demolition material–Effects of cyclic loading, *Transportation Geotechnics*, 23, 100346.
[27] Carneiro JR, Almeida PJ, Lopes MDL 2018 Laboratory evaluation of interactions in the degradation of a polypropylene geotextile in marine environments, *Advances in Materials Science and Engineering*, 2018.
[28] Bathurst RJ, Lin P, Allen T 2019 Reliability-based design of internal limit states for mechanically stabilized earth walls using geosynthetic reinforcement, *Canadian Geotechnical Journal*, 56 (6), 774-788
[29] Łupieżowiec M 2011 Stability analysis of the high motorway embankments contracted on the overflow lands *Czasopismo Techniczne* 108 (3-Ś), 123-136 (in Polish)
[30] Zimmermann Th, Truty A, Urbański A, Podleś K 2014 Z_Soil.PC 2014 3D user manual, Theory, Tutorials and Benchmarks, Data Preparation *Elmepress International & Zace Services Ltd*, Switzerland
[31] Benmebarek S, Berrabah F, Benmebarek N 2015 Effect of geosynthetic reinforced embankment on locally weak zones by numerical approach *Computers and Geotechnics* 65, 115-125
[32] Griffiths DV, Lane PA 1999 Slope stability analysis by finite elements *Geotechnique* 49, 387–403
[33] Łupieżowiec M 2013 The application of c-φ reduction method to estimate the bearing capacity of subsoil *ACEE Architecture Civil Engineering Environment*, 6(4), 35-43
[34] Zheng Y, Patrick JF, McCartney JS 2018 Numerical study on maximum reinforcement tensile forces in geosynthetic reinforced soil bridge abutments *Geotextiles and Geomembranes* 46.5, 634-645

Reinforced Soil Structures: From Standard Design to BIM

F Trovato[1], G Intrevado[1], P Pezzano[1], G Lugli[1]

[1] Officine Maccaferri S.p.A., Bologna, Italy

Email: f.trovato@maccaferri.com; g.intrevado@maccaferri.com; p.pezzano@maccaferri.com; g.lugli@maccaferri.com

Abstract. Building Information Modeling (BIM) emerges as a new process for the architectural, structural, engineering and construction sectors. It is based on three-dimensional (3D) parametric models which centralizes geometrical, physical and project information. The implementation of BIM within the Geotechnical Engineering sector has not reached yet a good level of robustness, frequently appearing as a new concept. BIM methodology in geotechnics and in the design of Reinforced Soil Structures (RSS) represents an opportunity which could enable the AECO world (Architecture, Engineering, Construction, Owner and Operator) to implement the infrastructural BIM, managing complex projects. With BIM, it is possible to improve the construction quality and reduce the wastage of the supplied materials as it allows an overview of the construction models through multidimensional digital design solutions which provide simulation and analysis by using different design software. This paper aims to clarify issues which might arise during the BIM design of RSS. In particular, this paper presents the case of a Mechanically Stabilized Earth (MSE) wall which comprises precast concrete facing panels and discrete high adherence polymeric soil reinforcing strips. The application of BIM technology in this research provides guidelines for design optimization and construction improvement.

1. Introduction

Building Information Modeling (BIM) is a technology which offers integrated platforms able to improve design and construction by giving faster delivery of projects. BIM provides also continuous flow of information which can be exchanged among the stakeholders and which can be used during all the decision-making processes [1].

The use of the BIM software goes beyond the planning and the design phase, being extended to the whole project lifecycle, supporting the construction processes, project management, as well as management of facilities and the maintenance of infrastructures. This allows an optimization of the interventions and a control of the construction processes which bring to cost savings.

While the use of BIM is widely used in the structural engineering and architectural sectors in all the countries, its implementation within the geotechnical engineering sectors has not extensively used, frequently appearing as a new concept. This paper shows how the BIM technology can be used in the design of Reinforced Soil Structures, illustrating the modelling of the S7 Skomielna Biala – Chabowka Mechanically Stabilized Earth (MSE) abutment in Poland.

2. BIM definition

The BIM software defines "parametric objects" as parameters and relationships with other objects therefore if a related object is modified even dependent objects will automatically change. Parameters can be either data or technical specifications. All the operations to develop a BIM object, to implement an interoperable modeling and the data sharing, can lead to a process which takes time in the first phases of the design. The time invested in a good object modelling will be fully rewarded in the later stages thanks to the simplifications given by the "three-dimensional informative model".
Considering the complexity related to the creation and management of public and private works, the BIM methodology is the most suitable tool for the quality control of projects and the planning of maintenance and functionality of the works.
BIM allows a digital construction, an accurate virtual model of a building construction which supports all phases of the construction process, and more efficient analyses and controls with respect to the traditional processes [2].
The BIM technology is a work in progress. BIM will contribute to achieve a higher level of prefabrication, greater flexibility, and variety in construction methods. The creation of fewer documents brings to fewer errors, less inefficiencies and higher productivity. Consequently, construction projects will better meet customer needs, thanks to better analyses and the exploration of more alternatives, lower costs, and lower program overheads. In the future and in the medium term (10 years) the development of the BIM will be guided by technical and economic reasons. The BIM methodology is widely developed within the countries (Figure 1) and it has national and international standard protocols with the goal to improve the infrastructure and building sectors.

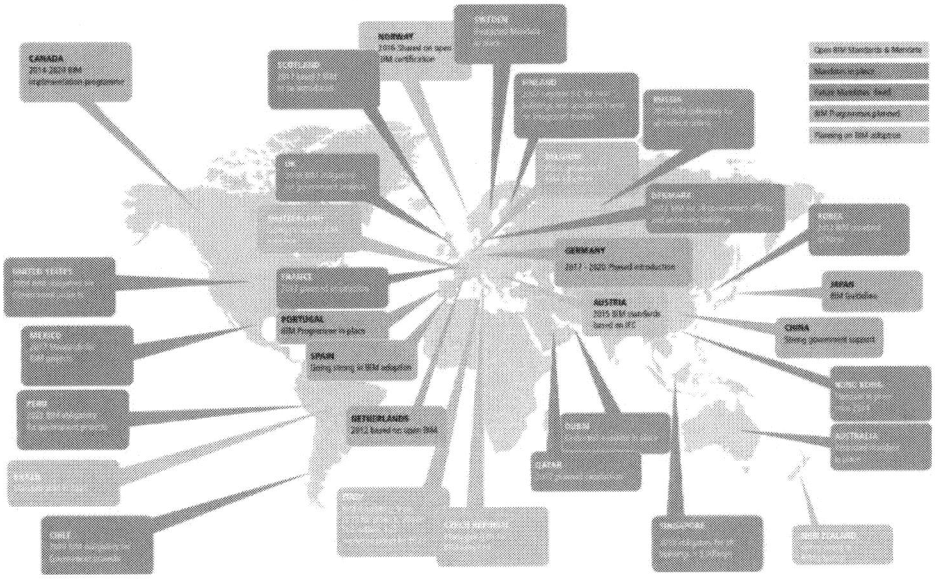

Figure 1. The use of BIM methodology around the world

3. Development of BIM in Italy

Recently, in Italy, public administrations have started to promote initiatives for the use of BIM in both public and private sectors. On the 19th of April 2016 the new Procurement Code (Legislative Decree 50/2016) came into force, with reference to the European Directive, with the purpose of encouraging the progressive use of electronic tools to govern public contracts. Already shortly after the effective entry into force of the mandatory stages envisaged by Decree 560/2017, a detailed analysis on some aspects of the important quantitative evolution coming from the Building Information Modeling offers the starting point for considerations on the potential and growth of this methodology. According to OICE

[3], in these last years of development and implementation of BIM, the number of projects which exploit this methodology has increased dramatically from 4 in 2016 to 86 in 2017, before turning the 2018 with 268 tenders, equivalent to a growth of 219%, that is 104 tenders in the December 2018. In 2019, requests for tenders concerning the use of BIM methodologies reached +58.3% compared to the 2018 (Figure 2). The positive trend coincided with a growth of the entire market of architecture and engineering services in 2019, which recorded a +20.1% in value. In 2019 the requests for architectural and engineering BIM services compared to the total market of services for architecture and engineering showed BIM calls to 8% of the total; a figure in increase compared to the 2018, which recorded a +5,1%. In absolute terms, the calls with requests of BIM are still few compared to the total.
The growth in the number of tenders, where the use of modelling tools is required, is linked to the increasing awareness of the contracting authorities. Indeed, starting from the 1st of January, 2019, the d.m. 560/2017 provided the obligation of BIM for "complex works" concerning works of more than EUR 100 million. This obligation will be gradually applied to all works after 2025.

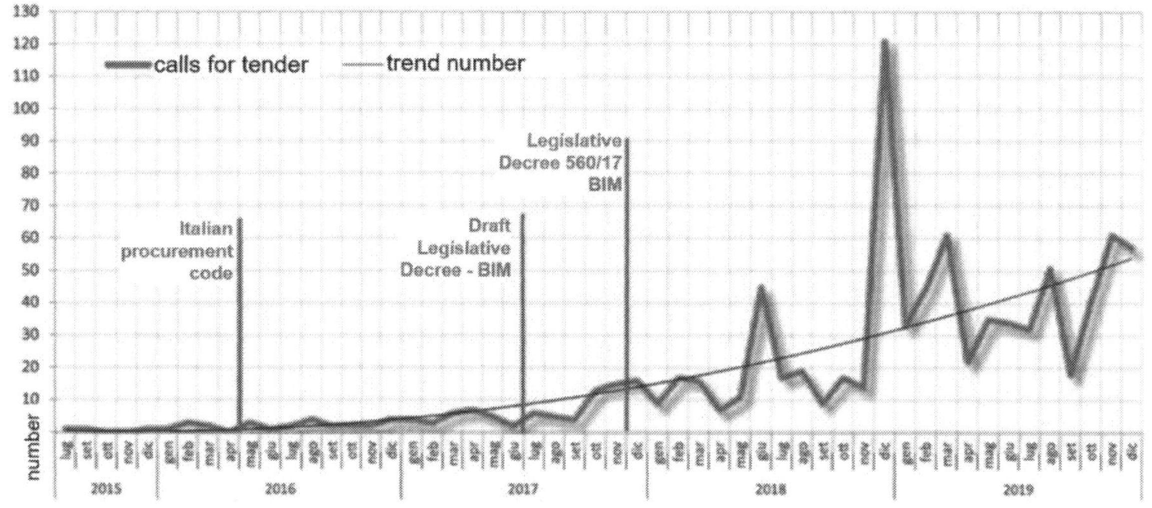

Figure 2. Trend of the issuance of BIM tenders in Italy [3]

BIM tools allow the development of building projects during their life cycle, including the design, construction, maintenance, management, and demolition phases. However, it is still little used in the transport infrastructure sector (roads, bridges, tunnels, railways, airports and ports). It is important to develop BIM software for the transport infrastructure sector and to encourage the application of this methodology in companies involved in this sector. Considering that the global market of the infrastructure industry is showing solid growth, there will be enormous interest in the application of BIM in the case of road infrastructure.
BIM design of infrastructure works is more complicated with respect to the vertical constructions of the building sector. Indeed, in the design of road infrastructures, horizontal works which extend for several kilometers in a geo-referenced manner and which interact with the surrounding environment must be considered. Whereas for a structural artifact or a building or an architectural work, the implementation can be conducted more easily thanks to its verticality and more contained dimensions. Building Information Modeling (or Management) offers great possibilities to revolutionize the approach to design, allowing to: increase the sharing, collaboration and exchange of information among the different design fields; three-dimensionally model the designed object before having built it; reduce risks and unforeseen events on site through the use of software which verifies there are no interferences between architectural, structural and plant engineering objects; make simulations about the energy behavior of the building and optimize the management of the work [4].

One of the most important problems in design lies in the exchange of data and information among the various figures who approach the project, certainly the issuing of standards and guidelines is a clear intervention in the direction of change. Shared procedures and methods, as well as standardized documents, are the basis for the creation of a common language and modus operandi among stakeholders [1]. In this context, interoperability becomes a fundamental requirement. Interoperability is important both among different BIM software and among different disciplines so as to standardize the workflow (Figure 3). For this reason, it was necessary to establish an international interchange format called IFC (Industry Foundation Classes) with the intention of collecting the information contained in all project fields, regardless of the application used.

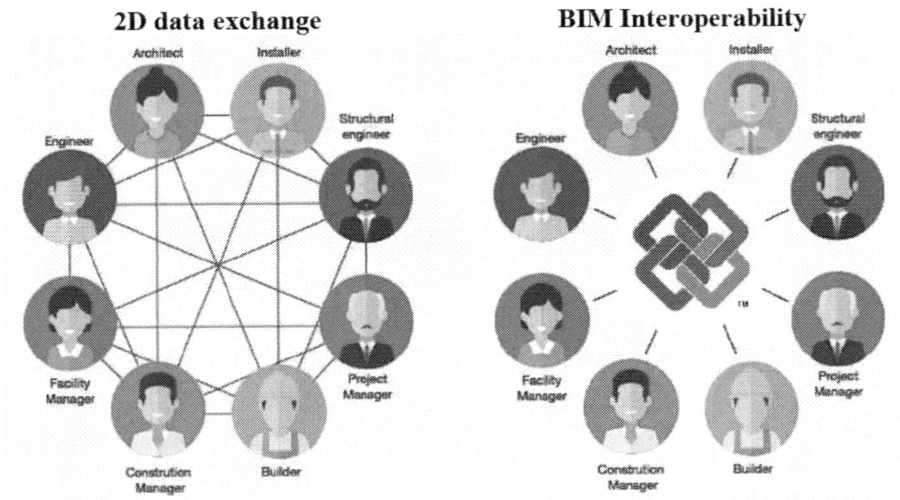

Figure 3. Comparison between traditional information exchange and BIM methodology

4. Case study

The aim of the paper is to show how to model a Mechanically Stabilized Earth (MSE) wall in BIM and how the wall components have been implemented in the 2D standard design of the S7 Skomielna Biala-Chabowka MSE abutment. A comparison between the two designs (2D and BIM design) were carried out, highlighting the advantages in terms of speed, precision and cost-effectiveness.
The project consists of a 164,9 m bridge approaching ramp on the S7 expressway within the Skomielna Biala – Chabowka segment (Figure 4).

Figure 4. MSE bridge approaching ramp

IOP Conf. Series: Materials Science and Engineering 1260 (2022) 012034 doi:10.1088/1757-899X/1260/1/012034

The system, commercially known as MacRes system (Figure 5), is made up of precast concrete facing panels with a standard square shape (1,50m x 1,50 m) and a thickness of 14 cm (Figure 6). The reinforcement system Paraweb is made of high tenacity polyester yarns geostrips extruded with in a linear low-density polyethylene (LLDPE) sheath that protect the reinforcing fibers and enhanced the long term performance of the reinforcement for a durability up to 120 years. The geostrips are connected to the back of the panel through a polymeric loop embedded into the concrete which ensure inclusive connection to the panel itself (Figure 7).

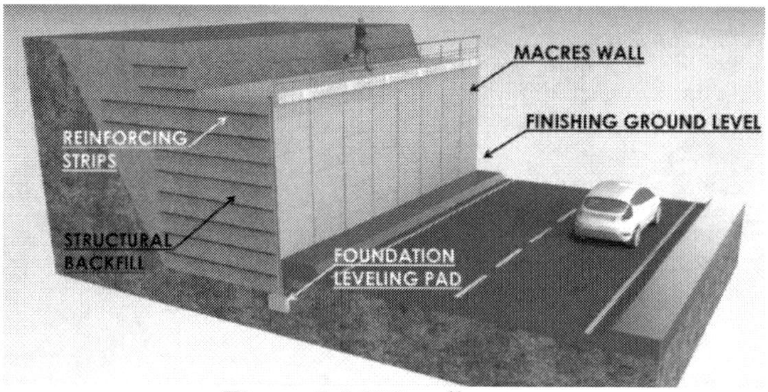

Figure 5. MSE wall system

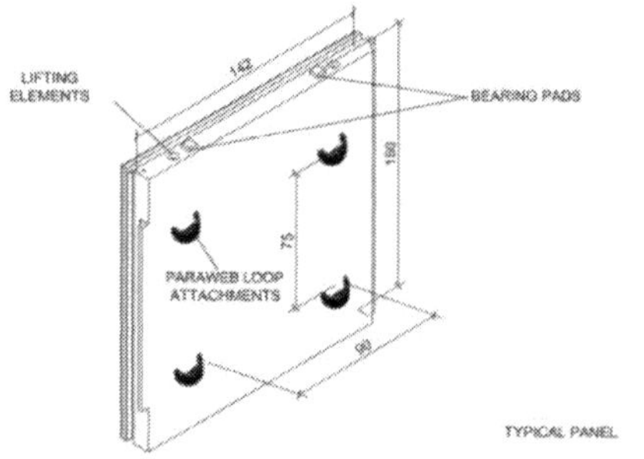

Figure 6. Precast concrete facing panel

Figure 7. Polyester reinforcing geostrips and loop

The other main components of the MSE system are:

- Foundation leveling pad: a smooth horizontal base to receive the first course of panels (Figure 8);
- Nonwoven Geotextile: unrolled behind the panels, used to protect the panels joint preventing soil particle migration (Figure 9);
- EPDM spacers: made by synthetic rubber, placed above every panel avoid a concrete-to-concrete contact (Figure 10);
- Precast coping beam: installed on top of the last row of panels, as a uniform and aesthetical finish, along the top of the wall.

Every MSE wall component has been modelled as a BIM element as explained in the following chapter.

Figure 8. Foundation leveling pad

Figure 10. EPDM Pads

Figure 9. Nonwoven geotextile

5. Digitalization

The creation of the BIM elements requires a deep knowledge of the whole MSE system and how each component is related to each other to set up the wall. Therefore, the first step for the creation of the BIM library was to define which objects had to be considered as an autonomous entity and which objects had to be considered as a combination of several entities. The classification of each entity defined the level of complexity of the BIM objects and the methods of representation as well as the type of three-dimensional tools and templates used. At the same time, the BIM objects were created in order to be easily implemented by the users (designers, builders and manufacturers) into a project avoiding unnecessary information which could unnecessarily burden the models.

The BIM objects were realized by means of AutoCad Revit as 3D parametric model. Since the need was to create families which would fit the concept of "panel" of the MSE wall system, a curtain wall model was used, allowing to create all the different shapes and sizes of the panels. To re-create the MSE structure, the single panel of the curtain wall was extruded and extended along the entire length of the wall. The dimensions and shapes of the panels were defined through reference plans linked to the edges (Figure 11); the reference planes determine the starting and the ending point of the panel geometry (Figure 12).

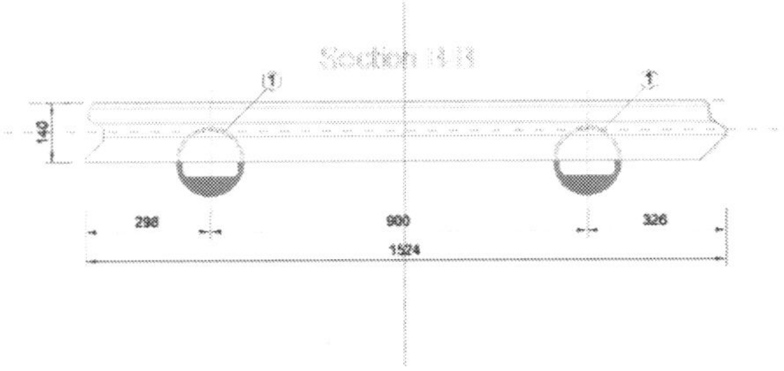

Figure 11. Panels bound to the reference planes

Figure 12. Definition of the panel height through the reference planes

The carvings in the panels contour were generated using the subtraction solid, which is an "empty" extrusion capable of subtracting mass from the previously created element.
To generate the dimensional parameters of the objects, the reference planes were used to create parametric relations starting from a fixed origin. After that, part of the extruded solid were linked to the reference planes, the dimensions of the objects were defined as dimensional parameters (for example "height"), therefore by changing the value of the parameters, the dimensions of the panels were defined. Since the MSE system have a certain number of standard panels (with fixed height, width and components), unique names were associated to these panels so the exact parameters of these are already set into the BIM family. At the same time the parameter which refers to the material of the object were defined as well.
EPDM pads (Figure 13), Nonwoven Geotextile (Figure 14), Loop connection (Figure 15), Coping beam Figure 16. Coping beam) and Polymeric Geostrips (Figure 17) were created with the same technique. Every type of polymeric strip is defined by a material parameter which is referred to the value of the Ultimate Tensile Strength (UTS) of the strip. The polymeric strips 3D elements are color coded according to the UTS grade.

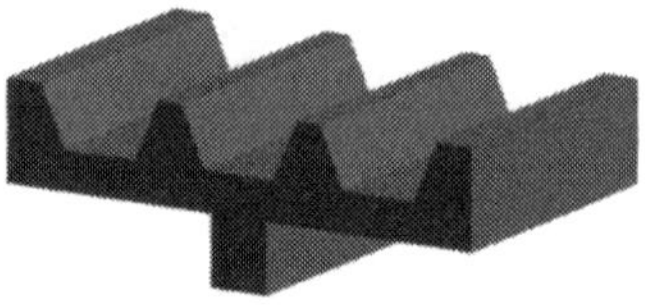

Figure 13. EPDM pad 3D model

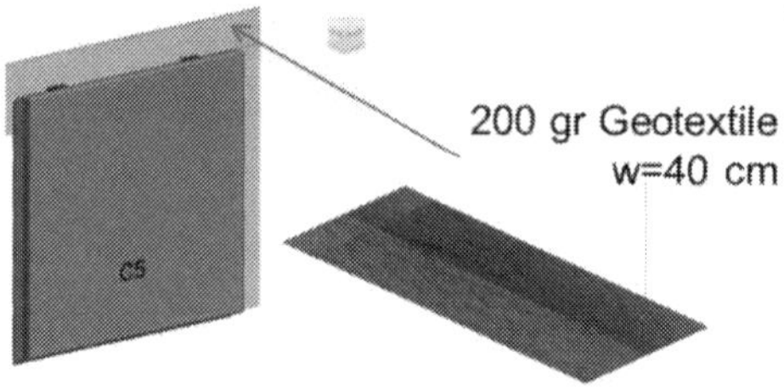

Figure 14. Geotextile 3D model

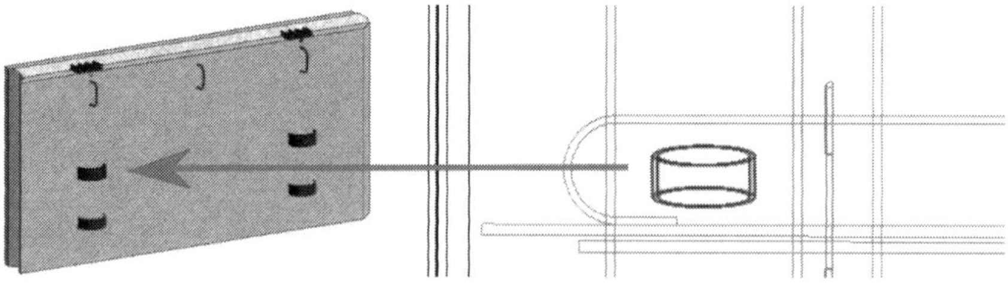

Figure 15. Loop 3D Model

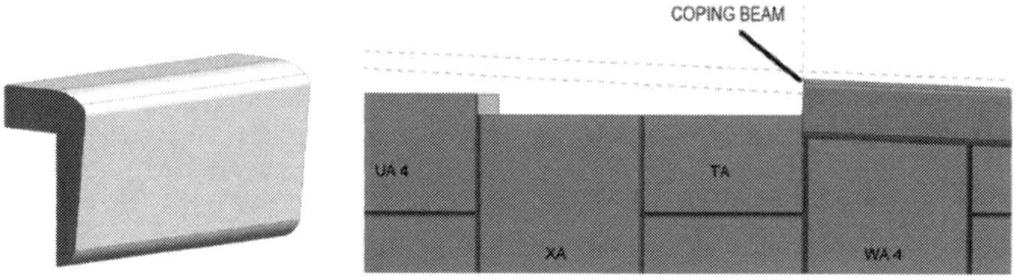

Figure 16. Coping beam

Figure 17. Polimeric Geostrip 3D model

A great effort was made for modeling the polymeric strips; it was necessary to set up a system that, during the design phase, reproduced the real installation of the strips. The extrusion end of the strips was matched in a single point; this position was defined through the reinforcement length parameter. A group of parameters was assigned for each extrusion which defines its starting and ending position, length, height, thickness etc. Different reference planes were created, and the ends of the extrusion were tied to them and fixed with the dimensional parameters. This choice allowed their joining, in the same way, the other end of the extrusion was constrained to the passing reference planes in the center of the loop, on the inner face of the panel (Figure 18).

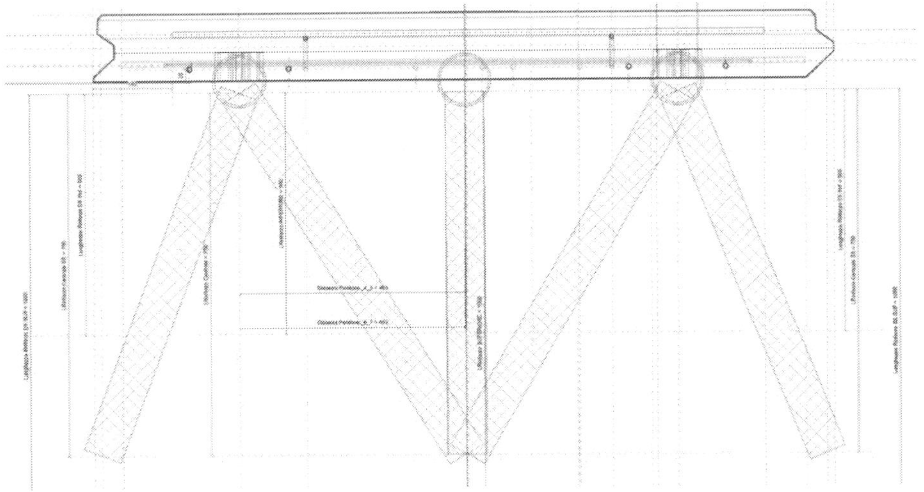

Figure 18. Polymeric strip linked to the reference planes

After creating all these individual families, they were all loaded in a single family called "Standard & cut panels". The main strength of working with parametric modeling is enabling the users to work on pre-established models which can be edited at any step of the process by inserting new parameters with great potential also in the creation of BIM libraries. In this way the user does not model each element from zero but, once the first one is modeled, it can immediately extract the other elements, simply assigning the correct input data in the element properties window, within the project.
The BIM objects were made available to the users through external archives (libraries) which can be downloaded from online BIM platform. The libraries can be uploaded into the project and the designers can select the specific libraries they will intend to use so as not to weigh down the file.

6. Implementation of the BIM project

Once the preliminary modeling of the families included in the BIM library was performed, the operating procedures and management of the project information model were defined.
Using the topographic survey, a digital reconstruction process of the topographic surface was implemented. Once the digital territorial context was defined, the infrastructure project was imported according to the correct georeferenced coordinates. The base point of the project was inserted in the origin. Later the modeling of the excavation of the foundation of the wall started.
From the same base point, the curtain wall began. It consists of the same curtain panel parameterized families previously modeled, in which all the technical characteristics necessary for the computations were defined. The 3D modeling of the geotechnical project allowed the management of the huge amount of data and information which characterize the infrastructural model, inside the same software.
In this way a complete model containing all the necessary information can be delivered to the stakeholders of the project. The implementation of the control activities for the clashes and the parameters gave to the work added value in terms of consistency and feasibility. At the end of this process, quantity schedules were generated to perform a precise and correct Quantity Take Off (QTO).

The main purpose of the BIM implementation is to reproduce an informative 3D model useful for the construction and operational phases, which is equipped with all the useful information for maintenance. The achievement of this aim will allow to build an infrastructure in which the BIM model is the connecting element between the design and construction phase of the project and the certainly more relevant subsequent phases of operations and management.
The quantification phase can reach levels of detail directly proportional to the Level of Development (LOD) and Level of Information (LOI) of the designed artifact [5]. Basically, the more advanced the LOD and the LOI are, the higher the degree of depth of estimated calculation [6]. The chance of being able to anticipate the precise costs of the works is one of the characteristics of the BIM oriented design and it allows to avoid the increase in the costs of the construction due to expenses initially not planned. In this case, it was possible to associate the modeled elements with precise property items by adding specific attributes. The main elements, such as the Standard Panels, can be associated not only with the main material (e.g. precast concrete) but also to additional schedule entries, such as:

- Total Length of the various types of geosynthetic strips (Table 1);
- N. and type of panels;
- N. pf Coping Beam elements;
- N. of loops;
- N. of EPDM Pads;
- Total amount of concrete [m^3];

In additional, with a BIM oriented design it is possible to carry out the bill of the quantities of all the material that is not directly connected to the MSE system itself but (such as panels, loops, polymeric strips etc.) such as the volume of structural soil fill to be used on site [m^3] and other accessories needed by the contractor to set up the wall (such as the n. of virtual connections).

Table 1. Total Length of the various types of geosynthetic strips, schedules from AutoDesk Revit

<Strip Length>		
A	**B**	**C**
Type	Length (m)	Rolls
2D30	3500,15	36
MD27	15578,4	156
MD36	11289,47	113
MD45	25270,55	253

After modeling the main elements three-dimensionally, it is possible to have an updated real-time quantification of all entries present in the computational schedules which refer to the objects present in the model. With specific reports, there is the chance to quickly extract the values and obtain a precise quantification of every object (Table 2). At each modification of the BIM model, the quantitative and economic feedback of what has been modeled can be immediately obtained, together with the graphical identification of the computed objects.

Table 2. Objects quantity, schedules from AutoDesk Revit

<Objects quantity>				
A	**B**	**C**	**D**	**E**
Num Pads	Num Virtual connection	Num loop	Num coping	Geotextile sqm
1444	3384	3384	110	932,90 m2

7. Conclusions

Building information modelling is taking a central position in the digital transformation of the construction industry. By supporting a transparent and seamless flow of information between all stakeholders, thus cutting information loss, BIM facilitates their collaboration throughout the different project phases. In turn, this translates into large efficiency gains, with faster delivery, growing business opportunities and with less costs, miscommunications, inaccuracies, delays, emissions, and waste.
Through the development of the project shown in this paper, it has been possible to understand the potentialities and the criticalities that the BIM methodology offers.
Firstly, it is necessary to change the approach to the project, which must become more and more integrated and with a global vision of the life cycle of the structure. It goes from the conception to the realization, giving importance to the management and maintenance of the building, as well as to the construction, considering the great impact that this has both in economic and environmental terms.
This case study has demonstrated the value and effectiveness of BIM. This powerful tool gives the possibility to manage a fully connected project. As previously explained, there is the chance to modify an element of the model in one view and apply the changes in all the other views and schedules. This means a significant saving in terms of time and increased accuracy.
The BIM brings improvements to the infrastructure sector because it enables a high degree of integration among the various design phases and disciplines. It allows an excellent modeling of all construction phases, starting from the virtual development of the site organization until the end of life. It allows the control of all construction phases, times, and costs, to give to the customer a valid feedback from the early stages of the design and check the costs incurred or to be incurred during the construction phases.
The BIM is an innovative methodology, which allows the user to have total control of the management of the work and for this reason, it could completely transform every field of the construction world, which has been using the traditional design and construction methods for too many years.

References

[1] Eastman C, Teicholz P, Sacks R and Liston K 2008 BIM Handbook: A guide to Building Information Modeling for owners, managers, designer, engineers and contractors *John Wiley & sons*
[2] Garagnani S (2016, 05 23) La misura dell'informazione all'interno del percorso BIM. Retrieved from http://www.ingenio-web.it.
[3] OICE 2019 Rapporto sulle gare BIM 2019 per opere pubbliche. Retrieved from https://www.oice.it/638197/rapporto-sulle-gare-bim-2019
[4] Azhar S 2011 Building Information Modeling (BIM): Benefits, Risks, and Challenges for the AEC Industry *Leadership and Management in Engineering* pp 241 - 52
[5] Ciribini A 2013 Level of Detail e Level of Development: I processi di committenza e l'Information Modeling. *TECHNE-Journal of Technology for Architecture and Environment* pp 90-9
[6] BIMForum 2019 Level of Development Specification

Modelling the impact of landslides on geosyntethics-reinforced barrier using DEM-FDM analyses

S Cuomo [1], M Camusso[2], P Gambardella[1], S Moretti [3], L Frigo [4]

[1] GEG (Geotechnical Engineering Group), University of Salerno, Via Giovanni Paolo II, 132, 84084, Fisciano, Salerno, Italy
[2] ITASCA Consultants Chemin des Mouilles, 64, 69130 Écully, France
[3] Department of Engineering, University of Naples Parthenope", Centro Direzionale, Isola C4, 80133 Naples, Italy
[4] Geosintex s.r.l., Via Leonardo da Vinci, 12, 36066 Sandrigo, Vicenza, Italy

Email: scuomo@unisa.it

Abstract. Geosynthetics-reinforced barriers can be used as protection structures for mitigating the risk posed by fast-moving flow-like landslides such as debris avalanches. In the design of such kind of barriers a crucial role is generally played by the correct analysis of the mutual interaction between the flowing mass and the barrier. This paper is focused on the evaluation of the impact forces and the deformation mechanisms of the barrier. An extensive numerical campaign of dynamic analyses has been performed by means of a coupled 3D Discrete Element Model (DEM) code, namely Particle Flow Code (PFC) and a continuum Finite Difference Method model, named Fast Lagrangian Analysis of Continua (FLAC3D code), both provided by Itasca software. The impacting flow with given initial height and velocity is here simply schematised as a dry granular mass, made of a random distribution of rigid spherical particles. The barrier is conceived as multilayered embankment reinforced by geogrids wrapped around the facing. The geometry of the barrier and the combinations of the materials have been varied to take in account a large variety of factors, also including the size of the impacting mass, the inter-particle friction angle and the initial velocity of the flowing mass. From the numerical results it was learned that the height of the flow may change significantly (or not) during the impact process depending on some of the previously mentioned factors. On the other hand, the global response of the deformable barrier depends on the combined behaviour of the granular soil and the geosynthetics reinforcements installed inside the barrier. Other novelty of the paper is that far from the semi-empirical formulations typically used for a safe design of such barriers, here the time-space dependent mutual interactions are accurately computed along the impact front also providing the chance to adequately take into account the mechanical features of the flowing mass and of the impacted barrier.

1. Introduction

Fast landslides such as debris avalanches could be very dangerous for human safety due to their dynamic features during the propagation, either long run-out distances or high velocity. Indeed, debris avalanches propagate along open slopes with a few meters of thickness in the source areas while increasing much more their volume during the propagation path until reaching the foothill area with large volume and high velocity. In some past cases relevant damage to the structures and even casualties have been recorded in many Countries including Italy [5].

In this work, a geosynthetics-reinforced barrier is proposed as mitigation option against debris avalanches to stop or to constrain the flow in a properly designated area (Figure 1). Recent research works based on SPH landslide propagation simulations have evaluated the kinematical parameters of the impacting flow such as the height, the velocity and the volume [8]. Particularly, for initial soil volumes of few hundreds of cubic meters later increased up to some thousands of cubic meters due to bed entrainment, it was found that the flows may impact the protective structures with flow height of 2-3 m and velocity of 10-15 m/s, which entails impact energy of about some hundreds of MJ. On the other hand, simplified analyses have been carried out to investigate the behavior of Deformable Geosinthetics-Reinforced Barrier (DGRB) under external dynamic horizontal pressure [9]. Such analyses rely on simplified hypotheses: i) the flow height is constant during the impact, while in reality the run-up along the barrier and even the overtopping of the barrier may occur, ii) the impact pressure is evaluated through empirical formulations assuming a constant profile of pressure over the flow height. To overcome such limitations, other recent researches have also used coupled models to simulate the impact of landslides against structures. For instance, [19] coupled DEM and FEM analyses to evaluate the impact of a landslide against a flexible barrier, while [18] coupled the Lattice Bolzman Method (LBM) with the Discrete Element Method (DEM) to simulate the mixture of solid particles and fluid phase in a landslide and also used the Finite Element Method (FEM) to simulate the behavior of a flexible barrier under such impact. Other examples are provided by Ma et al. [20] and Zhou et al. [24].

In this work, an enhancement of the previous contributions is presented considering a coupled 3D model based on Discrete Element Method (DEM) and Finite Difference Method (FDM), where the DEM is used to simulate the moving mass and the FDM to simulate the stress-strain response of a DGRB. The latter is a massive structure composed by granular soil reinforced with polyester (PET) geogrids wrapped around the facing.

The scope of the geogrids is to provide tensile resistance to the soil through the frictional interaction between barrier layers. The flow parameters (height, velocity and volume) are selected from the literature and assigned to the DEM code. The interaction between the flow and the barrier is here analyzed and the variations of impact pressure versus time is evaluated and compared to the values achievable through literature formulations.

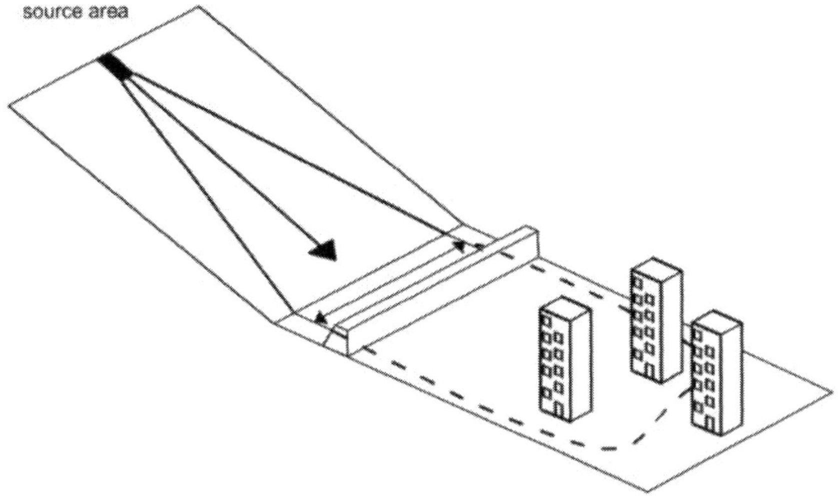

Figure 1. Scheme of an artificial barrier to protect buildings exposed to the threat of debris avalanche

2. Materials and Methods

A 3D discrete-continuum coupling approach, namely FLAC3D-PFC3D, is here used taking advantage of each modelling scheme and minimizing the requirement for computational resources. It is here shortly described while a more comprehensive description is provided in the FLAC3D and PFC3D user's manual.

PFC3D is a code based on the discrete element method particularly suitable to simulate large displacements of granular materials. A PFC3D model consists in a discrete system of spherical particles – treated as rigid bodies – interacting at pair-wise contacts and moving independently. The behaviour at physical contacts uses a soft-contact approach where the rigid particles are allowed to overlap one another at contact point. Contact mechanics is embodied in particle-interaction laws that relate the internal forces and moments to the magnitude of the overlap and/or the relative displacements at contact point. The main governing equations are: i) the motion equation and ii) the contact force equation [17]. Newton's second law is used to determine the motion of each particle arising from the contact and body forces acting upon it, while the force-displacement law is used to update the contact forces arising from the relative motion at each contact. Deformation occurs as a result of the particle packing rearrangement and the deformation in the contacts.

FLAC3D is a program based on the explicit finite difference able to perform small as well as large strain analyses of continuous media. Materials are represented by polyhedral elements within a three-dimensional grid that is adjusted to fit the shape of the object to be modelled. Each element behaves according to a prescribed linear on non-linear stress/strain law in response to applied forces or boundary constrains. As with PFC3D, the calculation first invokes the application of Newton's second law to grid nodes to derive new velocities and displacements from forces acting on them. Then, strain rates for grid zones are derived from velocities and new stresses from strain rates by taking into account the constitutive model of each zone [16]. The material can yield and flow, and the grid can deform (in large-strain mode) and move with the material that is represented.

In this study, PFC3D (DEM) is used to simulate the motion and impact of the debris avalanche against the barrier, while FLAC3D (FDM) is used to evaluate the stress-strain response of the embankment, simulated as a continuum material, reinforced by geogrid structural elements.

The coupling method involves two fundamental assumptions: i) the macro property of the discrete particle assembly is consistent with that of the continuous model [15]; ii) the velocities and forces must be consistent at the contact boundary [3]. For the latter, a conceptual scheme of the coupled FDM-DEM is given in Figure 2. The coupling is activated by the creation, in PFC3D, of wall elements coinciding with FLAC3D zone faces.

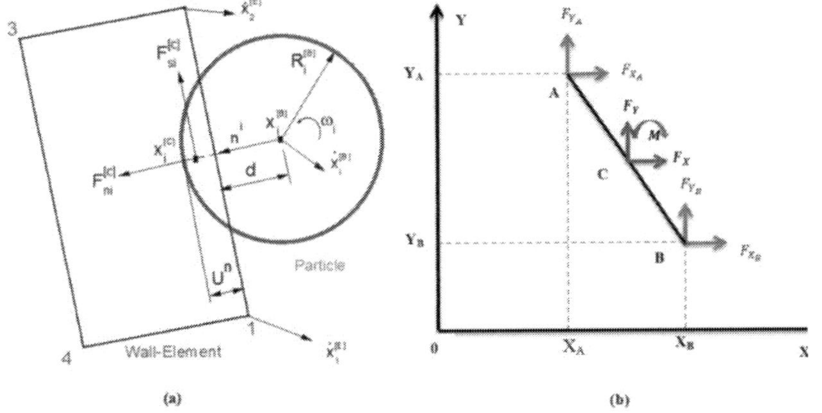

Figure 2. Schematic of DEM-FDM coupling formulation

The motion of the wall vertices is slaved to that of the FLAC3D nodes, and forces exerted on the walls are transferred as externally applied forces (i.e., boundary conditions) to the FLAC3D nodes, resulting in an efficient two-way coupling. More precisely, PFC3D walls consist of edge-connected, triangular faces where vertex velocities and positions can be specified as a function of time. Once the forces and moments of each wall facet are calculated, an equivalent force system at facet vertices is determined and then passed to grid nodes of the FLAC3D model. This coupling scheme is repeated at each calculation cycles, thus resulting in behavior continuity at the interface between the discrete and continuum model regions.

3. Modelling

3.1. Computational schemes and input data

The size and features such as height, velocity and volume of the impact landslides are taken from previous contributions [8]. The landslide volume ranges from 50 m^3 to 500 m^3, the thickness of the landslide body front is equal to 1-3 m, and the average landslide front velocity is equal to 2-10 m/s. The simulation of the landslide propagation from the source to deposition areas being beyond the scope of this paper, only the impact process is analyzed by assigning an initial velocity different than zero to the impacting mass. In addition, the shape of the impacting flow has been changed according to three case scenarios as shown in Figure 3 (rectangular, S1 – triangular, S2 - trapezoidal, S3) to assess the role of spatial distribution of the impacting mass over the time, during and after the impact. In the three cases, the landslide volume was the same.

A single geometry of the barrier is taken from previous contributions, namely the *Type A* barrier analyzed by Cuomo et al. [9] and Cuomo et al. [11]. Such barrier presents a trapezoidal shape (Figure 3) with a 4 m-wide crest, 60° inclined external walls and a 11m-wide base. The reinforcements are polymeric (PET) geogrids, 0.6 m vertically spaced, installed through the wrap-around technique. The analyses are carried out in 3D for a numerical model with a 1 m out of plane thickness. The barrier mesh is refined in the impact zone with elements sizes of about 0.15 m along the horizontal and vertical directions to properly capture the influence of the avalanche impact on the barrier behavior.

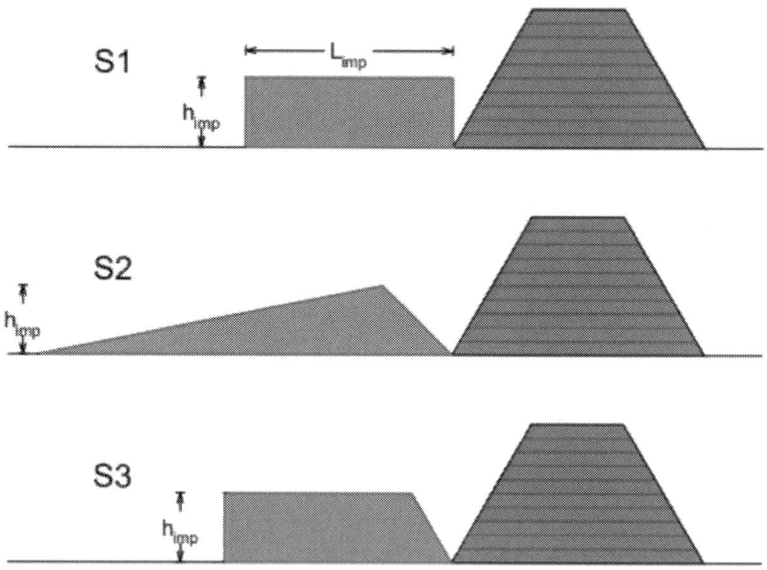

Figure 3. Scheme of landslide impacting the barrier

The barrier foundation is taken sufficiently large to avoid boundary effects on model results, with a size of 50 m x 20 m and more refined elements below near to the barrier (Figure 4).

The geogrids are modeled as shell elements only resistant to tensile axial force (membrane resistance). Metallic anchors, used to keep the design inclination of the slope front during the various construction stages, are also simulated by introducing cable elements. No sliding is considered between soil layers and the geogrids wrapped around the front of the barrier. Thus, the geogrids are interrupted in correspondence of the metallic formwork and rigidly connected to it (Figure5b). The interface between the soil and the geogrids is assumed as elasto-plastic with frictional strength properties reduced to 80% of those of the nearby soil. Rayleigh damping is introduced during the dynamic impact simulation with a center of frequency f_{min} and a critical damping ratio ξ_{min}, representatives of damping coefficients α and β equal to 0.01 and 0.001 [2].

The impacting flow is simulated through an assembly of rigid spheres, each 0.1 m dia. to ensure an accurate spatial description of the flow. A local damping coefficient of 0.7 is used during the generation procedure, as we are only interested in the final equilibrated state of the assembly. During the dynamic analysis, no numerical damping for the balls is used. Tables 1-2 report all the features of materials and how they are combined between them.

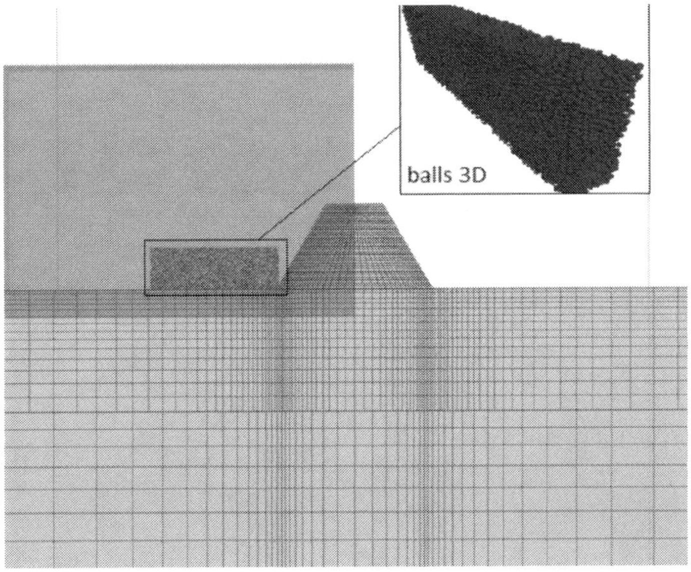

Figure 4. Computational domain

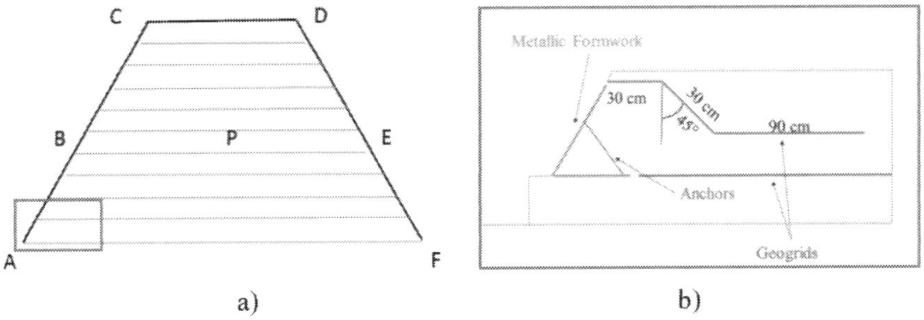

Figure 5. Schematization of the barrier, facing and geosynthetics reinforcement

Table 1. Material properties

FDM parameters		
Height of the mesh barrier	(m)	0.20
Width of the mesh barrier external zone (first ten zones)	(m)	0.15
Width of the mesh barrier internal zone	(m)	0.15 with geometric ratio 1.1
Height of the mesh foundation	(m)	0.20 with geometric ratio 1.1
Width of the mesh foundation	(m)	0.15 with geometric ratio 1.1
DEM parameters		
Height of the grain assembly, h	(m)	3.00
Width of the grain assembly, b	(m)	1.00
Length of the grain assembly, l	(m)	15.0-9.0
Number of grains	(-)	28000-46700
Total mass, m	(t)	38610-81000
Average grain diameter, D	(m)	0.10
Ratio between largest and smallest diameter	(-)	1.00
Particle unit weight, c_s	(kN/m^3)	2600-3272
Initial porosity	(-)	0.45
Initial average velocity, v_0	(m/s)	10.0-5.0-3.0-2.0
Normal stiffness, k_n/D	(MPa)	900
Shear stiffness, k_s/D	(MPa)	225
Inter-particle friction	(-)	0.3-0.4
Interface friction	(-)	0.2-0.267

Table 2. List of variables included in the numerical simulations

Symbol	Description
C1	High resistance core material - Low resistance Geogrid
C2	High resistance core material - High resistance Geogrid
C3	Low resistance core material - Low resistance Geogrid
C4	Low resistance core material - High resistance Geogrid
G1	Unit weight of landslide material equal to 1430 kg/m^3
G2	Unit weight of landslide material equal to 1800 kg/m^3
fi.2	Inter-particle friction equal to 0.3
fi.3	Inter-particle friction equal to 0.4
S1	Rectangular shape of the flow
S2	Triangular shape of the flow
S3	Trapezoidal shape of the flow
I.1	v_{imp}=10 m/s, L_{imp}=9 m, V_{imp}=27 m^3
I.2	v_{imp}=10 m/s, L_{imp}=15 m, V_{imp}=45 m^3
I.3	v_{imp}=5 m/s, L_{imp}=9 m, V_{imp}=27 m^3
I.4	v_{imp}=5 m/s, L_{imp}=15 m, V_{imp}=45 m^3
I.5	v_{imp}=3 m/s, L_{imp}=9 m, V_{imp}=27 m^3
I.6	v_{imp}=3 m/s, L_{imp}=15 m, V_{imp}=45 m^3
I.7	v_{imp}=2 m/s, L_{imp}=9 m, V_{imp}=27 m^3
I.8	v_{imp}=2 m/s, L_{imp}=15 m, V_{imp}=45 m^3

3.2. *Numerical results and discussion*

Amongst the several numerical simulations based on different combinations of variables reported in Table 2, a first case, named "C2-G2-fi2-S1-I3" is discussed below, including high resistance core material combined to high resistance geogrid, landslide unit weight of 1800 kg/m^3, inter-particle friction equal to 0.3, landslide rectangular shape characterized by initial velocity, v_{imp}=5 m/s, length, L_{imp}=9 m and volume, V_{imp}=27 m^3. Snapshots of the computed velocity of the flow at different times are provided in Figure 6.

As expected, despite the initial homogeneous velocity of the landslide, higher displacements are observed in the upper part of the flow (initially 3 m high), where shear interactions between particles are reduced. Consequently, the flow stops first at the bottom corner of the barrier in a short time period (about 0.3 s) and then the flow-structure interaction propagates backwards with some run-up of flow over the barrier (see t=0.68 s), and later final deposition. On the other hand, the barrier is pushed ahead and displace of few centimeters in horizontal direction. Besides this global movement, an internal deformation of the barrier is also observed: while the central bottom part behaves almost rigidly, the front parts deforms under the impact of the flow and the rear part under the effect of the shear resistance at the base which counteracts the displacement of the barrier. This is a quite complex deformation-displacement mechanism which allows the dissipation of the kinetic energy of the flow. Such results are in good agreement with previous FDM analyses performed by Cuomo et al. [10] , in which avalanche impact was modeled as an average pressure time history, characterized by a peak value followed by reduction down to a plateau. It is worth noting that these modelling conditions indirectly accounted for a uniform applied pressure over the whole impact region and a constant front height. As outlined in Figure 6, the flow-structure interaction starts at the front bottom corner of the barrier (see t=0.34 s), and then propagates upwards until the moment when the flow stops completely. The change in impact flow height is almost 1.0 m but indeed a small volume of soil runs-up the barrier, so that the most of the impact forces are applied over the bottom 3 m-high flow front. This also explains why the numerical results are in good agreement with previous literature (e.g. Cuomo et al. [13], which considered the front height constant during the impact.

A second case described here refers to the impact I4 (higher landslide volume). All other features are kept as in the previous case. After the first impact stage, a second pressure peak is recorded, which contributes to the flow-structure interaction. More comments on this issue will be addressed later. The global displacement of the barrier is approximately 25% greater.

The results obtained through the DEM-FDM analyses are compared to those achievable by means of literature formulations and available from previous contributions. As far as the evaluation of the Impact Loading Pressure (ILP) over time is considered, Cuomo et al. [10,13] reported a triangular shape time history with an initial peak pressure reproducing the impulsive action of the impact, followed by a plateau accounting for the passive earth pressure of the material deposited in front of the barrier.

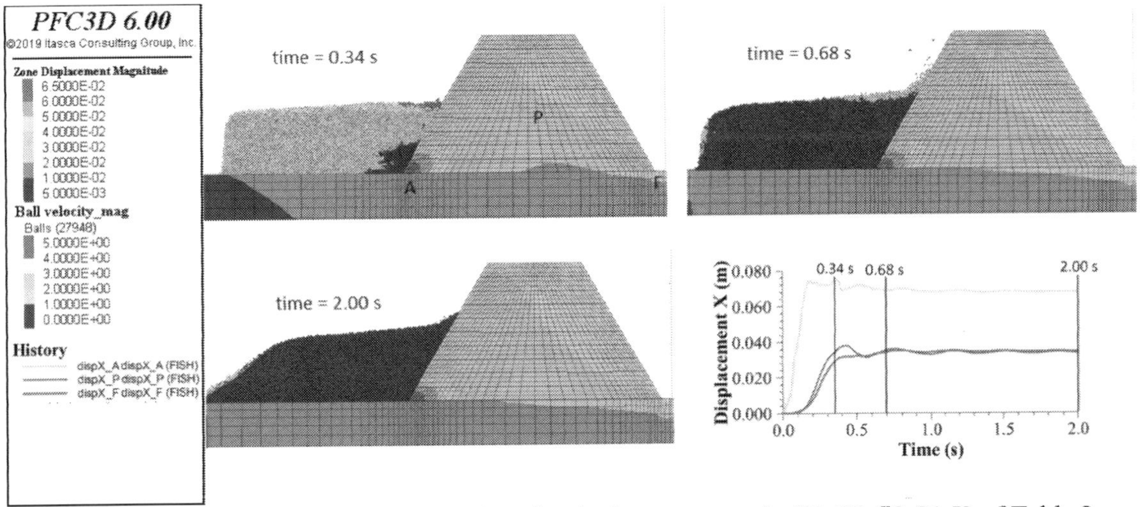

Figure 6. Flow-structure interaction over time for the impact scenario C2-G2-fi2-S1-I3 of Table 2

Figure 8 shows a good agreement of the DEM-FDM numerical results with those reported by [11], particularly for the residual pressure value once the impact terminated. Thus, given the same volume of impact flow mass, the deposits simulated in DEM-FDM analyses is similar to the one considered in ILP method. Numerical analyses also show that the more the initial flow masses is elongated, the lower is the obtained peak pressure because of the avalanche shape influence on the exerted impulsive action. One can thus observe a progressive reduction in peak pressure passing from the case S1 (rectangular shape) to the case S3 (trapezoidal), to the case S3 (triangular).

Another comparison is proposed to show the role of different length of the landslide impacting the barrier, respectively 9 m (*I3*) or 15 m (*I4*), i.e. 3 or 5 times the height of flow (3 m). Such length over height ratios were heuristically proposed by [9]to take into account that the impact process has negligible effect beyond a given distance from the barrier. In the simplified approach used by Cuomo et al [9] and Cuomo et al. [11] the two scenarios present the same peak pressure value, related to same height and velocity of the flow, but the pressure is applied longer for the bigger volume, by scaling-up the pressure history over the time domain.

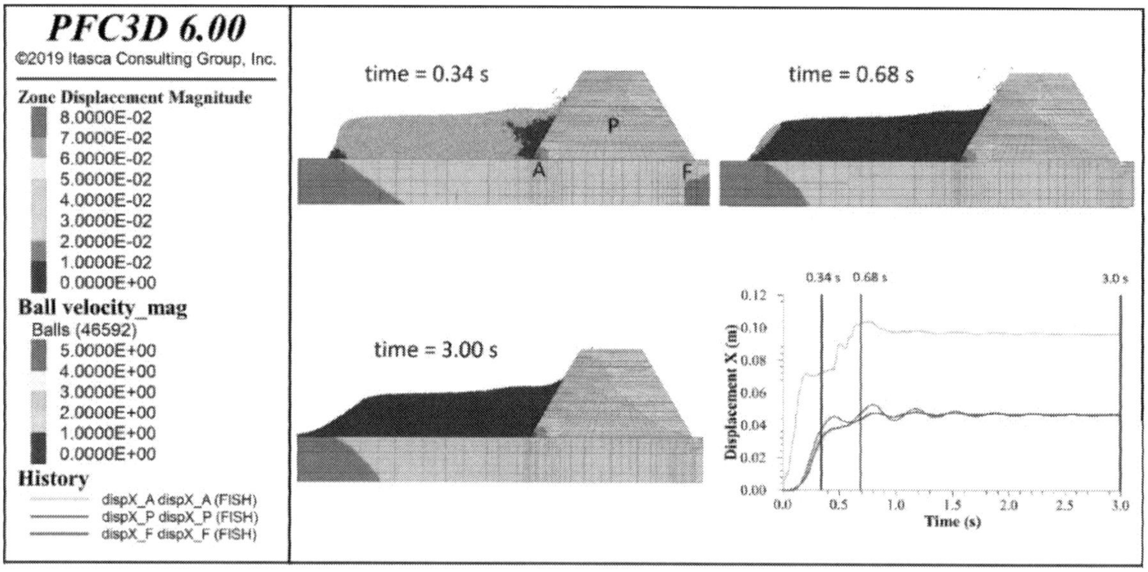

Figure 7. Results for the impact scenario C2-G2-fi2-S1-I4 of Table 2

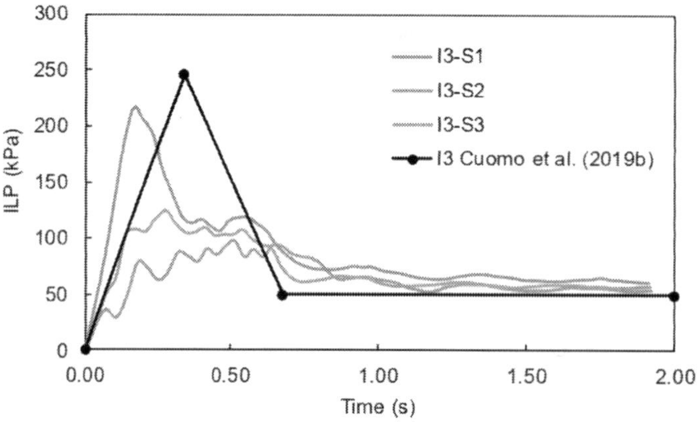

Figure 8. Impact pressure versus time for different shapes (S1-S2-S3) of the impacting volume

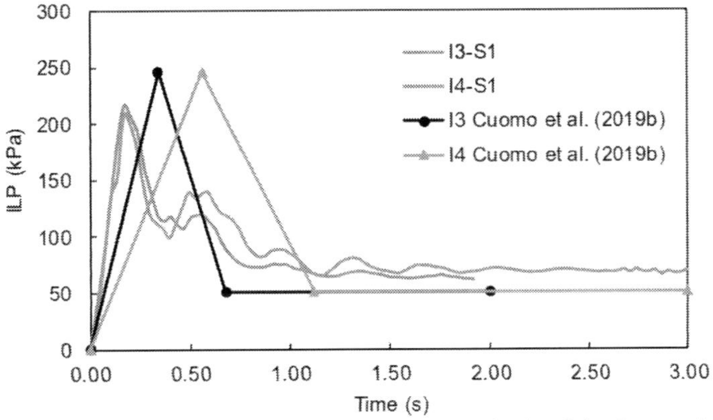

Figure 9. Impact pressure history for different volumes (I3 and I4) of the impacting landslide

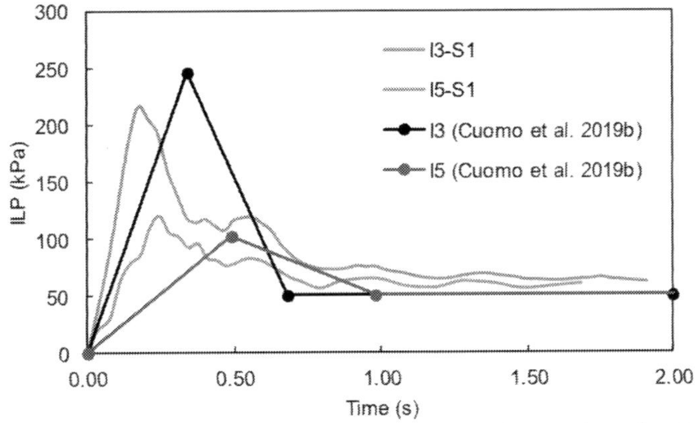

Figure 10. Impact pressure histories for the same landslide impacting the barrier with different velocity

The numerical results achieved here provide similar values of both peak and plateau values, confirming that the simplified procedure is conservative. The main difference observed between the two cases consists in the development of a second pressure peak related to the arrival of the rear part of flow against the front material already stopped by the barrier.

A further validation of the numerical model is provided in Figure 10, by considering the case of the impact I5, with a reduced initial velocity of the flow mass (3 m/s, compared to 10 m/s of case I3). Again, the agreement between the numerical results and the analytical formulations is satisfactory: the impact velocity is confirmed as a key factor in determining the value of the peak pressure.

More generally, when dealing with such complex dynamic interactions between deformable continuous media (the flow and the barrier), it would be recommended to refer to the kinetic energy of the flow by taking into account both the geometry (e.g. volume) and the dynamic features (e.g. velocity). For example, in Figure 11 the case of impact I3 is compared to the case I4, which have same velocity but different volume. As it can be seen, the initial kinetic energy is proportional to the volume, and the dissipation rate of kinetic energy depends on both initial value at the impact and other factors. In fact, for a given set of flow volume and velocity, the energy dissipation is affected by the shape of the impacting mass. This result underline the need for an accurate assessment of the flow propagation features in order to perform a proper flow-structure interaction analysis.

On the same basis, some more considerations can be drawn with reference to Figure 12, where different material properties of the impacting flow are considered. As expected, the greater the unit weight of the material, the higher is the impact pressure. Analogously, the higher is the frictional angle

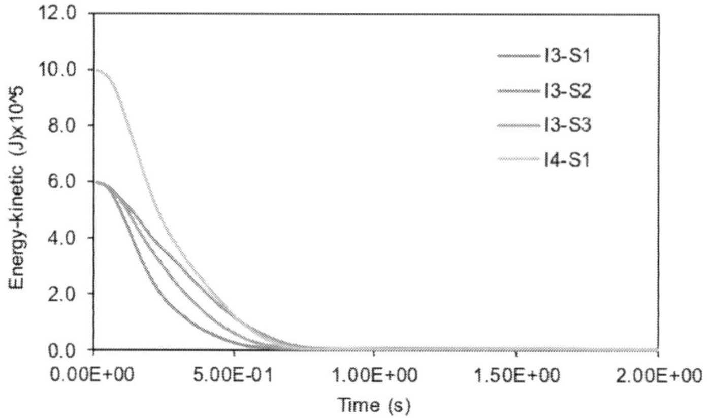

Figure 11. Kinetic energy of the landslide during the impact with the barrier

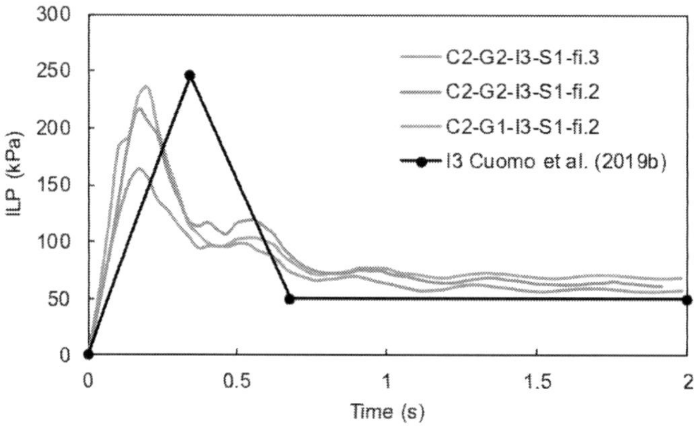

Figure 12. Impact pressure versus time for different unit weight of landslide material and different inter-particle friction

of the impacting material, the higher is the impact pressure. In this case, the higher is the internal shear strength of the material, the shorter will be the deposition length of the material in front of the barrier. Also here is relevant observing that the analytical formulation slightly overestimates the impact pressure over time.

4. Conclusions

This paper analyzed the interaction between a flow-like landslide and a massive deformable structure by taking into account both the deformation of the barrier and the behavior of the flow during the impact. The coupled method DEM-FDM allowed removing a number of simplified hypotheses usually employed while investigating a flow-barrier interaction, such as the constant flow height during the impact, the constant average pressure along the vertical direction or yet the simplified assessment of impact pressure over time.

The simulation campaign has shown that the barrier well reacts to the flow impact with displacements ranging from few centimeters to some ten of decimeters, with both scenarios acceptable in relation to the extremely demanding performance required to the barrier. Interestingly, the numerical results are in agreement with previous FDM analyses in which the flow was simulated as an external pressure, assessed with simplified procedures. Deformation capacity of the barrier has been shown to have a positive effect in dissipating the impact energy and confirmed the overconservative nature of estimations obtained with analytical formulation. Moreover, the analyses have allowed to state on the effect of several factors on the history of the impact pressure developed against the barrier, including the volume and the shape of landslide as well as the mechanical characteristics of the material. In addition, the dissipation of the kinetic energy of the landslide due to the interactions with the deformable barrier was also evaluated in relation to the features of the impacting landslide.

Acknowledgments

The authors would like to thank the two companies that collaborated to this work: i) Itasca Consulting SAS Ecully (France), and ii) Geosintex s.r.l., Sandrigo, Vicenza (Italy).

References

[1] Armanini A, Larcher M, Odorizzi M 2011 Dynamic impact of a debris flow against a vertical wall *In Proceedings of the 5th international conference on debris-flow hazards mitigation: mechanics, prediction and assessment* Padua, Italy pp 1041-9

[2] Aversa S, Maiorano R M S, Tamagnini C 2007 Influence of damping and soil model on the computed seismic response of flexible retaining structures *In 14th European Conference on Soil Mechanics and Geotechnical Engineering* 978-90-5966-055-7

[3] Bertulani C A 2005 Relativistic continuum-continuum coupling in the dissociation of halo nuclei *Phys. Rev. Lett.* **94** 072701

[4] Canelli L, Ferrero A M, Migliazza M, Segalini A 2012 Debris flow risk mitigation by the means of rigid and flexible barriers-experimental tests and impact analysis *Natural Hazards and Earth System Sciences* **12**pp 1693-9

[5] Cascini L, Cuomo S, Della Sala M 2011 Spatial and temporal occurrence of rainfall-induced shallow landslides of flow type: A case of Sarno-Quindici, Italy *Geomorphology* **126**pp 148-58

[6] Ceccato F, Simonini P, di Prisco C, Redaelli I The effect of the front inclination on the impact forces transmitted by granular flows to rigid structures. *In Workshop on world landslide forum* 2017 pp 593-9 Springer, Cham

[7] Cheung A K C, Yiu J, Lam H W K, Sze E H Y 2018 Advanced numerical analysis of landslide debris mobility and barrier interaction *HKIE Trans Hong Kong Inst Eng* **25**pp 76–89

[8] Cuomo S, Moretti S, Aversa S 2019a Effects of artificial barriers on the propagation of debris avalanches *Landslides* **16** pp 1077-87

[9] Cuomo S, Moretti S, Aversa S 2020a Deformation mechanisms of geosynthetics-reinforced barriers impacted by debris avalanches *Bulletin of Engineering Geology and the Environment* **79**pp 659-72

[10] Cuomo S, Moretti S, Aversa S 2020b Modeling of geosynthetics-reinforced barriers under dynamic impact of debris avalanche *XXVII CNG*

[11] Cuomo S, Moretti S, D'Amico A, Frigo L, Aversa S 2020c Modelling of geosynthetic-reinforced barriers under dynamic impact of debris avalanche *Geosynthetics International* **27**pp 65-78

[12] Cuomo S, Moretti S, Frigo L, Aversa S 2020c Performances of geosynthetics-reinforced barriers for protection against debris avalanches *In: Workshop on World Landslide Forum* Springer, Cham 2020 pp 341-7

[13] Cuomo S, Moretti S, Petrosino S, Aversa S 2019c Dynamic impact of debris avalanches on structures. Proceedings overview. *International Conference on Computational Methods in*

Structural Dynamics and Earthquake Engineering 24-26 June 2019, COMPDYN

[14] Cuomo S, Pastor M, Cascini L, Castorino G C 2014 Interplay of rheology and entrainment in debris avalanches: a numerical study. *Canadian Geotechnical Journal* **51**pp 1318-30

[15] Hauer B D, Doolin C, Beach K S D, Davis J P 2013 A general procedure for thermomechanical calibration of nano/micro-mechanical resonators *Ann. Phys.* **339**pp 181–207

[16] Itasca Consulting Group Inc. FLAC3D (Fast Lagrangian Analysis of Continua in 3 Dimensions) User's Manual Version 6.0; Itasca Consulting Group Inc.: Minneapolis, MN, USA, 2017

[17] Itasca Consulting Group Inc. PFC (Particle Flow Code) User's Manual Version 6.0; Itasca Consulting Group Inc.: Minneapolis, MN, USA, 2014

[18] Leonardi A, Wittel F K, Mendoza M, Vetter R, Herrmann H J 2016 Particle–fluid–structure interaction for debris flow impact on flexible barriers *Computer-Aided Civil and Infrastructure Engineering* **31**pp 323-33

[19] Liu C, Yu Z, Zhao S 2020 Quantifying the impact of a debris avalanche against a flexible barrier by coupled DEM-FEM analyses *Landslides* **17**pp 33-47

[20] Ma Y, Sheng Q, Zhang G, Cui Z 2019 A 3D Discrete-Continuum Coupling Approach for Investigating the Deformation and Failure Mechanism of Tunnels across an Active Fault: A Case Study of Xianglushan Tunnel *Applied Sciences* **9** 2318

[21] Moretti S, Cuomo S, Aversa S 2020 Feasibility of Foothill Barriers to Reduce the Propa-gation of Debris Avalanches I*n National Conference of the Researchers of Geotechnical Engineering*, Springer, Cham pp 309-17

[22] Qi X, Yu Z X, Zhao L, Xu H, S C Zhao 2018 A new numerical modelling approach for flexible rockfall protection barriers based on failure modes *Adv Steel Constr* **14**pp 479–95

[23] Vagnon F, Segalini A 2016 Debris flow impact estimation on a rigid barrier *Natural Hazards and Earth System Sciences* **16** pp 1691-7

[24] Zhou J, Jian Q W, Zhang J, Guo J J 2012 Coupled 3D discrete-continuum numerical modeling of pile penetration in sand *Journal of Zhejiang* University SCIENCE , **13** pp 44-55

Durability of EPDM geomembranes - Lifetime estimation by using high pressure autoclave test method

A Armani[1], H Zanzinger[1], A Monami[1], W Hahn[1] and B Amat[2]

[1] SKZ - German Plastics Center, Friedrich-Bergius-Ring 22, 97076 Würzburg, Germany
[2] Firestone Building Products EMEA, Ikaroslaan 75, 1930 Zaventem, Belgium

Email: a.armani@skz.de

Abstract. For determination of end of service life different accelerated ageing tests are available. However, ageing of modern polymers in conventional oven ageing tests takes many years. In high-pressure autoclave test (HPAT), thermo-oxidative ageing is caused by elevated temperature and by elevated oxygen availability. This results in significantly shorter test times of around 8 months. An EPDM geomembrane was submitted to thermo-oxidative ageing in high pressure autoclaves (HPA) at 3 temperatures (75 °C, 80 °C and 85 °C) and 3 elevated oxygen pressures (5 bar, 10 bar and 51 bar). During exposure in HPAT the oxidation progress was monitored via tensile tests and oxygen pressure measurements. Tensile strength decreases steadily with increasing exposure time, temperature and oxygen pressure. Based on 75% residual tensile strength as theoretical failure criteria and Arrhenius based 3D extrapolation model, time till failure for service temperature at atmospheric condition was calculated. As a result, service life of about 360 years at 40 °C and atmospheric pressure was estimated.

1. Introduction

Oxidation is a major ageing factor for all polymeric geomembranes including those made of ethylene propylene diene (EPDM). Especially in the field of long-term applications, it is essential to know the moment when the geomembrane cannot fulfil its function any longer.

SKZ (Süddeutsches Kunststoff Zentrum – German Plastics Center) has experience in accelerated ageing tests with HPA on products made of Polyethylene (PE), Polypropylene (PP) and Polyvinyl alcohol (PVA) for more than one decade. Until now, no EPDM membrane had been exposed to HPAT.

The knowledge of resistance against thermo-oxidation of EPDM waterproofing sheets for roofing is mainly based on studies with conventional oven ageing tests like for example a study carried out by SKZ, and commissioned by WDK (Wirtschaftsverband Deutsche Kautschukindustrie – German Rubber Industry) and the VESP (Vereniging EPDM Systeem Producenten Benelux – United EPDM System Manufacturers) from 1999 to 2004. The conventional oven tests were performed on several new and already installed roofing membranes from different roofs. They took a test time of many years with the result of expected service life of over 50 years when exposed to Western Europe normal weathering conditions based on the defined failure criterion of 150% elongation at break. There is also an important feedback on the ageing process of EPDM geomembranes in irrigation reservoirs installed in Spain, coming from several studies published by Blanco et al. [1,2,3].

As a comparison to the oven ageing tests, in this study the longest HPAT time until failure with mildest test condition (75 °C / 5 bar) lasted approx. 8 months and shortest test time was less than one month.

EPDM rubber is a synthetic elastomer characterized by a saturated backbone structure providing excellent ozone, heat and weather resistance. During the vulcanization process, the polymer chains cross-link together to create a 3-dimensional network that will give the elasticity and flexibility of the EPDM geomembrane.

2. Experimental part

2.1. High-pressure autoclave tests (HPAT)
One smooth EPDM membrane, black coloured, with 1.1 mm thickness was submitted to a test program with five HPAT test conditions. Test method was EN ISO 13438, method C. In HPAT test specimens were immersed in a pressure vessel in aqueous $NaHCO_3$-solution (c = 0.01 mol/l) (Figure 1). The solution was prepared for pH 10 with NaOH (c = 1 mol/l) and was stirred. The HPAT were performed at 3 different temperatures (75 °C, 80 °C and 85 °C) at a constant oxygen pressure of 5 bar. Additionally, the HPAT were performed at 3 different oxygen pressures (5 bar, 10 bar and 51 bar) at a constant temperature of 80 °C.

2.2. Tensile tests
After defined time intervals, specimens for tensile testing were removed from HPA and tested according to EN 12311-2. Five test specimens in cross machine direction were tested with 100 mm/min test speed, 80 mm distance, optical extensometer as strain measurement and the mean value was calculated.

In accordance with EN ISO 13438, five specimens were immersed in the aqueous medium under atmospheric conditions for 24 h at (85 ± 1) °C, (80 ± 1) °C and (75 ± 1) °C and used as reference. Before tensile testing, the specimens of all withdrawals (and reference samples) were dried according to EN 14415 (max. 96 h at 50 °C). Additionally, oxygen pressure development in the HPA was recorded continuously.

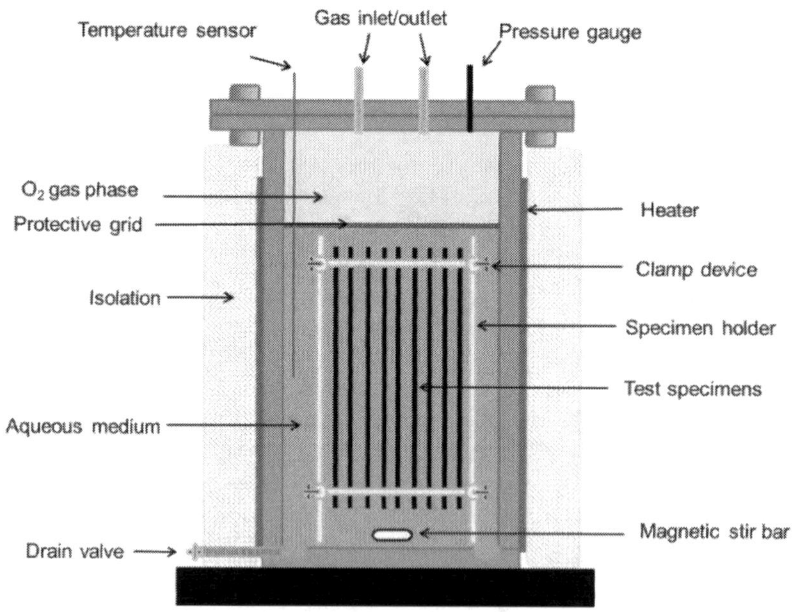

Figure 1. Schema of high-pressure autoclave (HPA)

3. Test results and evaluation

3.1. HPAT at various temperatures and 5 bar oxygen pressure

In Figure 2 the time-dependent change of residual stress at break is presented graphically. The samples were exposed at 3 different temperatures (75 °C, 80 °C and 85 °C) while oxygen pressure was kept constant at 5 bar.

During the test, contrary to what it is expected in oven and natural ageing, it was observed that the tensile strength at break decreased faster than the elongation at break. Therefore, it was decided to use the tensile strength as an indicator of the evolution of the test. The test was stopped when the samples lost over 50% of their original tensile strength at break.

For lifetime estimation, 75% residual tensile stress at break with regard to the control specimens was defined as the theoretical service life criterion. The actual residual tensile stress after different times and for each test temperature is plotted in Figure 2. Those values are fitted exponentially and the test time when the stress decreases by 25% in comparison to the reference sample is determined as the time to reach the membrane service life. Figure 2 shows that the higher the temperature, the faster the 75% residual stress at break is reached.

The elongation at break also decreases with time and temperature (Figure 3). However, the reduction of the elongation at break is delayed compared to the residual stress at break.

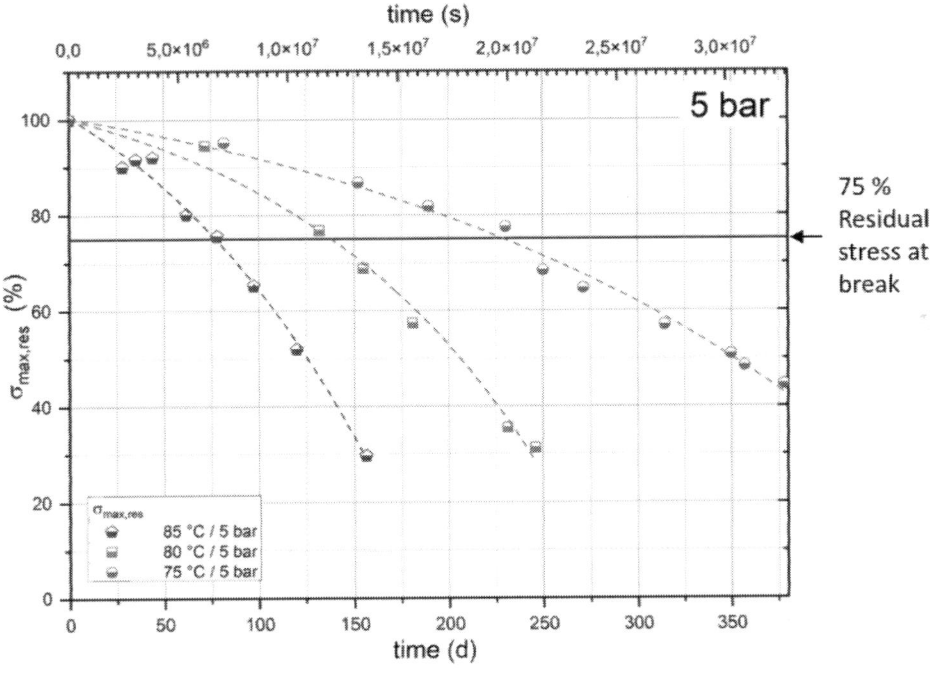

Figure 2. HPAT - Time dependent change in residual tensile stress at break $\sigma_{max,res}$ at 5 bar oxygen pressure and 85 °C, 80 °C and 75 °C test temperature

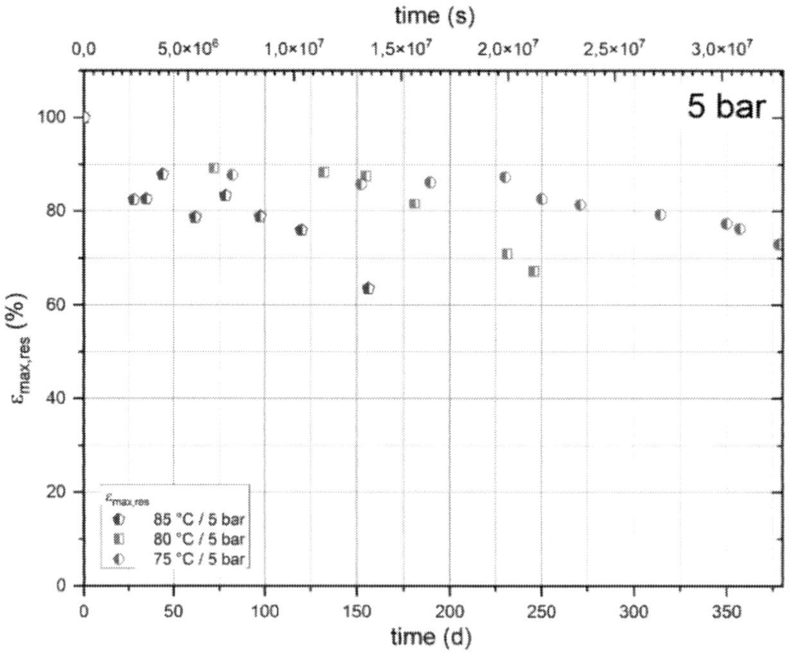

Figure 3. HPAT - Time dependent change in residual elongation at break $\varepsilon_{max,res}$ at 5 bar oxygen pressure and 85 °C, 80 °C and 75 °C test temperature

Table 1 shows the test time in HPA till 75% residual stress at break with regard to the control specimens is reached at test conditions and based on the fit. This test time is the time to reach the membrane service life.

Table 1. Measured time to reach the service life at 5 bar oxygen pressure and different temperatures

Test condition	Service life time [d] based on 75% $\sigma_{max,res}$
at 85 °C / 5 bar $p(O_2)$	77
at 80 °C / 5 bar $p(O_2)$	138
at 75 °C / 5 bar $p(O_2)$	229

Based on these results, the Arrhenius extrapolation was applied. The result is shown in Figure 4. It can be seen that the points are on a linear line, which is a precondition for an estimation of service life at 5 bar oxygen pressure.

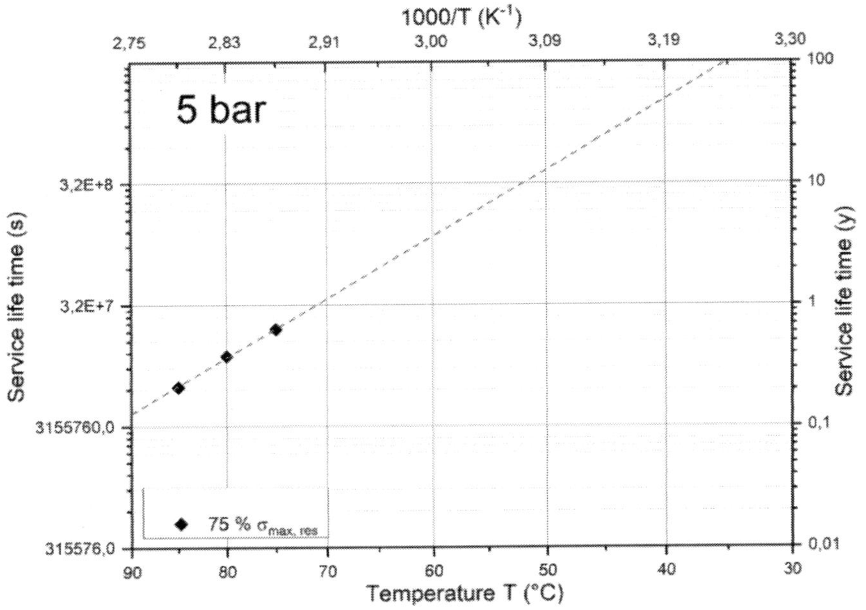

Figure 4. Arrhenius-typical extrapolation of times-till-failure based on 75% $\sigma_{max,res}$ at 5 bar oxygen pressure

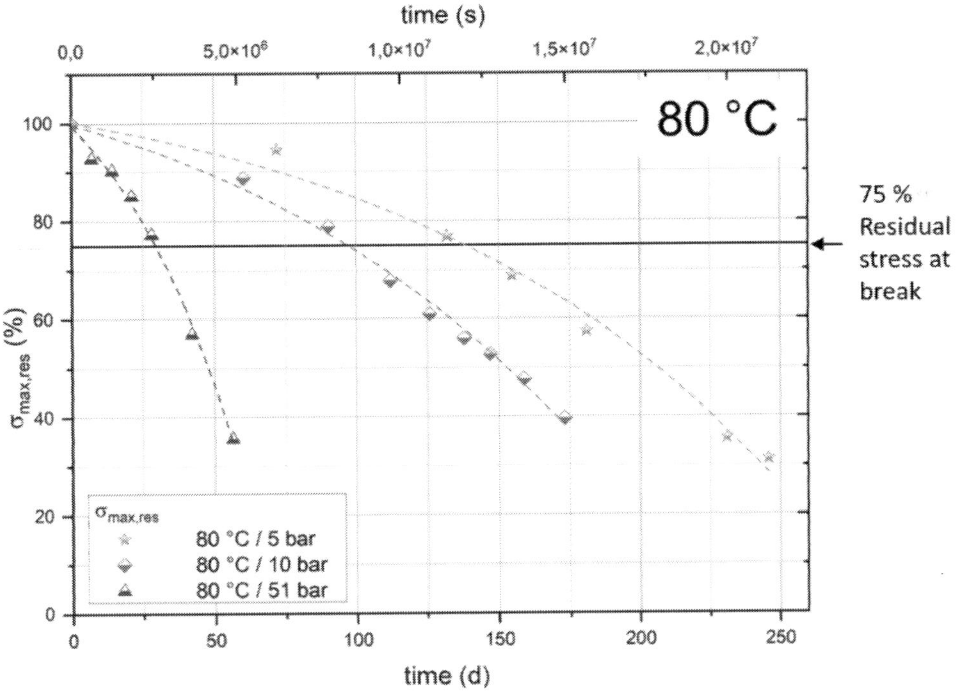

Figure 5. HPAT - Time dependent change in tensile stress $\sigma_{max,res}$ at 80 °C test temperature and 51 bar, 10 bar and 5 bar oxygen pressure

3.2. HPAT at various oxygen pressures and 80 °C

In Figure 5 time-dependent change in residual tensile stress at break is shown graphically for HPAT. The exposure oxygen pressures were 5 bar, 10 bar and 51 bar while the temperature was kept constant at 80 °C. The higher the oxygen pressures the shorter the test times until 75% residual stress is reached. Again, the residual stress at break decreases earlier than the residual elongation (Figure 6).

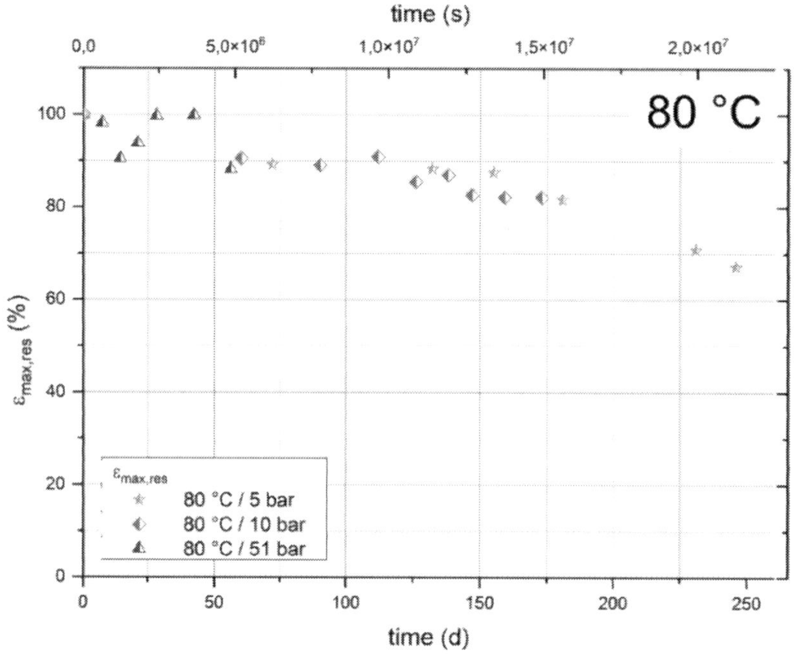

Figure 6. HPAT - Time dependent change in elongation at break $\varepsilon_{max,res}$ at 80 °C test temperature and 51 bar, 10 bar and 5 bar oxygen pressure

The time to reach the service life defined criterion based on HPAT at 80 °C and 51 bar, 10 bar and 5 bar oxygen pressure and based on 75% $\sigma_{max,res}$ were evaluated and are summarized in table 2.

In Figure 7, time to reach the service life regarding the mentioned tests are plotted double logarithmically versus the amount of dissolved oxygen.

During HPAT, sample material is not directly in contact with gaseous oxygen but surrounded by an aqueous solution. The amount of dissolved oxygen depends on the oxygen pressure above the solution and on the temperature. An overview of dissolved oxygen in water at different temperatures and pressures is given by Geng & Duan [4]. Those values were reduced to the relevant parameter range (i.e. $p \leq 50$ bar and $0\ °C \leq T \leq 90\ °C$) and fitted. Based on this fit, the dissolved oxygen can be calculated for any pressure and temperature in this range. For pressures below 1 bar equations given by Battino [5] are used.

Table 2. Measured time to reach the service life at 80 °C and different oxygen pressures

Test condition	Service life time [d] based on 75% $\sigma_{max,res}$
at 80 °C / 51 bar $p(O_2)$	29
at 80 °C / 10 bar $p(O_2)$	97
at 80 °C / 5 bar $p(O_2)$	138

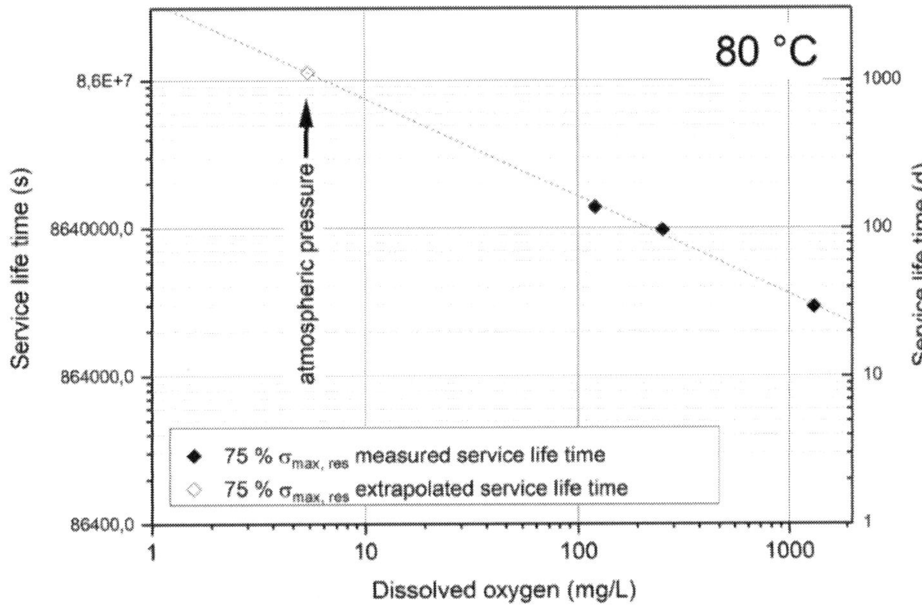

Figure 7. Double logarithmic extrapolation of service life time based on 75% $\sigma_{max,res}$ at 80 °C

3.3. Extrapolation to service conditions

Based on the extrapolations already shown in Figures 4 and 7, a three-dimensional plot was generated. Therefore, both extrapolations – regarding temperature and oxygen pressure/ dissolved oxygen – were carried out simultaneously (Figure 8). Time is plotted logarithmically versus dissolved oxygen (also logarithmically) and inverse temperature (Arrhenius-typically). This results in a plane-shaped fit providing the service life at service conditions.

The 3D-extrapolation allows to estimate the service life at different conditions of temperature and oxygen pressure. In most applications, the oxygen pressure will be the atmospheric pressure. The average temperatures in service conditions can be quite variable. In exposed conditions, it is assumed that the average service temperature of the liner in Western Europe weather conditions will be around 40 °C (conclusions of the SKZ study mentioned in chapter 1). In covered conditions (e.g. 1.0 m of ground), the average service temperature will be most probably much lower and the expected service life therefore higher.

In general, extrapolations are carried out over a range of maximum 40 °C below test temperature. Therefore, the extrapolation to service conditions have been made at 40 °C and 50 °C and 6.9 mg/L which corresponds to the atmospheric pressure of 0.21 bar $p(O_2)$. The results at these conditions are presented in Table 3. Depending on the temperatures the assumed service life lies between 100 years and 360 years.

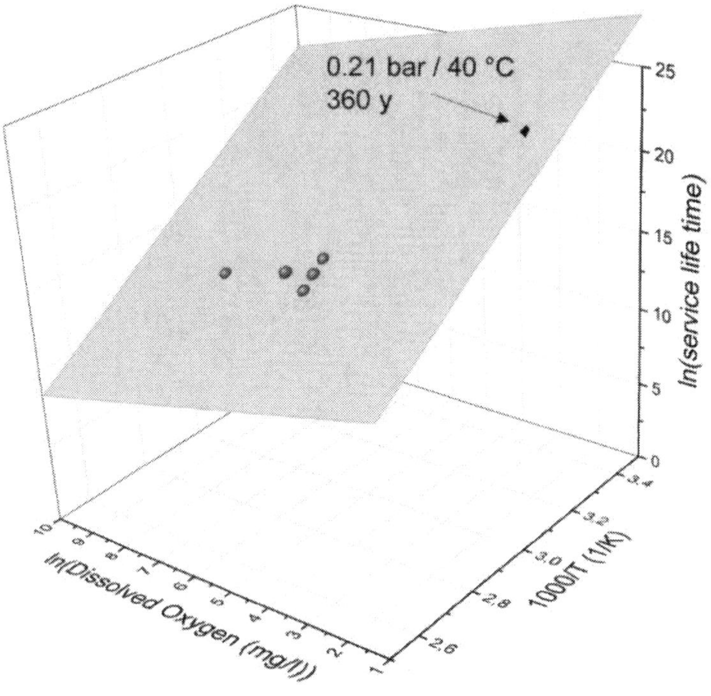

Figure 8. 3D extrapolation of service life based on 75% $\sigma_{max,res}$. The five (red) balls are measured failures at test conditions. Black triangular pyramid is the extrapolated service lifetime at service condition

In table 3 the extrapolated service lifetimes at 40 °C and 50 °C service temperature regarding 75% residual stress at break are presented.

Table 3. Extrapolated service lives

Service condition	Service live (y) based on 75% $\sigma_{max,res}$
40 °C / atmospheric pressure	360
50 °C / atmospheric pressure	100

3.4. *Oxygen consumption*

In this study with EPDM, it was found that the oxygen pressure and the mechanical properties decreases slightly but continuously from the beginning of the HPAT. The typical oxygen pressure decrease is shown in the right schema of Figure 9. The actual pressure curve looks different, because the pressure was readjusted many times after small decreases. Readjustment was necessary to maintain a nearly constant test pressure. That means that oxygen was consumed continuously during ageing.

The interpretation is that degradation of EPDM under thermo-oxidative conditions is a continuous process and no sharp changes of mechanical properties must be expected. This is certainly related to the fact that there are no antioxidants in the composition of EPDM membranes.

The experience from other HPAT on phenolic-stabilized samples for example PE-HD plastic pipes [6,7] or PP geotextiles [8] is that oxygen pressure and mechanical properties are nearly constant for a long test time. At the very end of the test period, the pressure drops, and the mechanical properties decrease. The oxygen pressure decrease is shown in Figure 9, left schema.

Hsuan and Koerner [9] and Hsuan and Li [10] describe the oxidation of phenolic-stabilized PE and PP samples in a three stage-oxidation-mode. In the oxidation process of phenolic-stabilized PE and PP samples, first the stabilizer is consumed (first stage) than there is an induction stage (second stage) and a third stage with the decrease of mechanical properties.

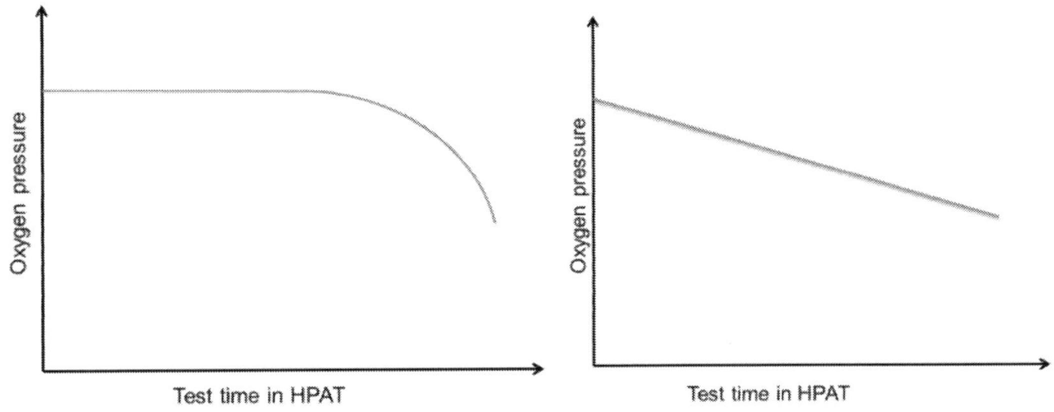

Figure 9. Schema of oxygen pressure decrease during HPAT: Left: Phenolic-stabilized PE or PP sample. Right: EPDM sample

3.5. Polymer degradation process
Samples in conventional oven ageing and taken from old infrastructures (irrigation reservoirs) show that the elongation at break tends to decrease over time while the tensile strength at break is quite stable [3]. This evolution of the mechanical properties is explained by the fact that, in the case of EPDM, during the oxidation phenomenon (induced by the temperature and UV exposure), the combination reactions of the carbon chains (cross-linking) prevail upon the cleavage reaction of the principal carbon chain [11]. Consequently, a decrease in molecular mobility and an increase in molecular mass can be seen [11]. These reactions are similar to the process of curing used during EPDM membrane production. This is one of the reasons why EPDM geomembranes don't need antioxidants in their formulation.

On the contrary, in the case of the HPAT, the drop of the elongation at break happens in a much slower pace than the tensile strength at break. This difference is more pronounced in the case of increasing the oxygen pressure than by increasing the temperature. This evolution of the mechanical properties could be explained by the fact that at high oxygen pressure, the cross-linking effect doesn't prevail anymore upon the cleavage of the principal carbon chains. Therefore, the tensile strength of the liner is directly affected and decreases much faster than in normal oxygen pressure. The HPAT accelerates chemical reactions that would take place in a very slow pace in real site conditions and can therefore give an estimation of the service life of the liner in a testing time that is quite fast compared to other ageing tests.

4. Conclusion
In this study an EPDM membrane was submitted to thermo-oxidative ageing in high pressure autoclaves at temperatures of 75 °C, 80 °C and 85 °C and elevated oxygen pressure of 5 bar, 10 bar and 51 bar. During exposure in HPAT the progress of oxidation was monitored by tensile tests and oxygen pressure record. Tensile strength and oxygen pressure decrease steadily with increasing exposure. Based on 75% residual tensile strength as theoretical service life criteria and Arrhenius based 3D extrapolation model, service life for service temperature at atmospheric condition was calculated.

This study shows that using HPAT as acceleration ageing method results in reliable lifetime estimation in short test times. The result is an oxidative service life estimation of 100 years and 360 years at respectively 50 °C and 40 °C average service temperature for the investigated EPDM membrane. The

EPDM membrane seemed to consume the oxygen molecules from the beginning of exposure. Also, the mechanical properties decrease from the beginning due to chain degradation and no rapid failure of EPDM at the end of service life must be expected.

References

[1] Blanco M, Aguiar E, Castillo F, Soriano J, Noval A, Touze-Foltz N, Pargada L, Rico G, Aguiar E 2011 Comparative study of three different kinds of géomembranes (PVC-P, HDPE, EPDM). *5th European Geosynthetics Congress* **2** 25 030 Besancon

[2] Blanco M, Aguiar E, García F, Vara T, Soriano J, Castillo F 2011 État de la géomembrane en EPDM de la retenue d' "El Golfo" dans l'île d'El Hierro (Îles Canaries). *Proc. 8eme Rencontres Géosynthétiques* pp 283-93. Tours

[3] Blanco M, Aguiar E, Castillo F, Touze-Foltz N, Aguiar E, Amat B 2014 Behaviour of EPDM geomembrane 18 years after its installation in a water reservoir. *10th International Conference on Geosynthetics*, Berlin

[4] Geng M, Duan Z 2010 *Geochimica et Cosmochimica Acta* Prediction of oxygen solubility in pure water and brines up to high temperatures and pressures **74** pp 5631-40

[5] Battino R, Rettich T, Tominaga T 1983 *Journal of Physical and Chemical Reference Data* The solubaility of Oxygen and Ozone in Liquids **12** pp 163-78

[6] Martin A, Zanzinger H 2018 Comparative study: Oven tests and high-pressure autoclave tests (HPAT) on one commercially available PE-HD material. *11th International conference on Geosynthetics, Seoul, Korea*

[7] Zanzinger H, Engelsing K and Hausmann S 2014 Durability of polyethylene geopipes for landfill applications after several years in service. *10th International conference on Geosynthetics*, Berlin, Germany, 21-25 September 2014 p 8

[8] Hausmann S, Zanzinger H, MacKenzie C, and Els C 2016 Accelerated ageing of polypropylene geotextiles in autoclaves at elevated temperature and oxygen pressure. *EuroGeo6, 6th European Geosynthetics Congress*, 25-28 September 2016, Ljubljana, Slovenia

[9] Hsuan Y G and Koerner R M 1998 Antioxidant Depletion Lifetime in High Density Polyethylene Geomembranes, *Journal of Geotechnical and Geoenvironmental Engineering* **124** pp 532-41

[10] Hsuan Y G and Li M 2005 Temperature and pressure effects on the oxidation of high-density polyethylene geogrids, *Geotextiles and Geomembranes* **23** pp 55-75

[11] Kumar A, Commereuc S,Verney V 2004 Ageing of elastomers: a molecular approach based on rheological characterization *Polymer Degradation and Stability* pp 751-7

Standards

- EN ISO 13438:2004 "Geotextiles and geotextile-related products - Screening test method for determining the resistance to oxidation"
- EN 12311-2:2013 "Flexible sheets for waterproofing - Determination of tensile properties - Part 2: Plastic and rubber sheets for roof waterproofing"
- EN 14415:2004 "Geosynthetic barriers - Test method for determining the resistance to leaching"

IOP Conf. Series: Materials Science and Engineering 1260 (2022) 012037 doi:10.1088/1757-899X/1260/1/012037

Numerical analysis of stress concentration around bridge bank seat in Geosynthetic Reinforced Soil – Integrated Bridge Systems (GRS-IBS)

Bahia S El Refai[1] and Dr. Patrick J Naughton[2]

[1] PhD Research Student, Department of Civil Engineering, Institute of Technology Sligo, Ireland.
[2] PhD, Senior Lecturer, School of Engineering & Design, Institute of Technology Sligo, Ireland.
naughton.patrick@itsligo.ie

Abstract. Geosynthetic Reinforced Soil – Integrated Bridge Systems involve directly supporting the bridge abutment bank seat on the top of a reinforced soil structure without piles or bridge bearings. Analysis, using Plaxis 2D, found that the arrangement of geogrid in the upper part of the reinforced soil abutment, the distance of the abutment bank seat from the wall face and the direction of horizontal load application were important considerations in controlling the deformational response of the abutment. Placing the bank seat further away from the wall face reduced wall displacement, peak vertical stress and settlement under the bank seat and peak strain in the geogrid. Adding secondary reinforcement in the top of the wall reduced wall deformations, peak vertical stress and settlement under the bank seat. The secondary reinforcement redistributed the tension in the geogrid immediately under the bank seat. Longer secondary reinforcement was more efficient at controlling displacement, while shorter secondary reinforcement reduced the peak vertical stress under the bank seat. The line of maximum tension, which was largely independent of horizontal load direction, originated close to the bottom-rear corner of the bank seat and was log spiral in shape, terminating at the toe of the wall.

1. Introduction

Geosynthetic Reinforced Soil – Integrated Bridge Systems (GRS-IBS) is an amalgamation of reinforced soil and integral bridge technology with the explicit aim of overcoming several inherent problems with conventional abutment arrangements, such as: high construction costs, long construction periods, durability and maintenance issues due to water ingress to bridge bearings and joints including differential settlement with the adjoining road construction and low stability in high seismic areas (Abu-Farsakh *et al.* 2018). Unlike conventional integral abutment systems, GRS-IBS involves the placement of the bridge deck directly onto a reinforced soil structure, which is typically a segmental block wall, without the use of vertical piles or bridge bearings.

The Federal Highways Administration (FHWA) (Adams *et al.*, 2011) have developed a dedicated design method for GRS-IBS. However, Adams *et al.* (2011) showed that this method is heavily empirically based and provided a 'recipe' for design.

The main objective of this study was to develop a finite element model using Plaxis 2D, that was capable of simulating the deformational and stress response of GRS-IBS under typical bridge loadings. The validated model was then used to investigate the impact of geogrid arrangement at the top of the wall and the location of the abutment bank seat relative to the wall face. The data is discussed in terms of displacement of the wall face, settlement under the bank seat, strain in the reinforcement and location of the line of maximum tension in the geosynthetic reinforced structure (GRS).

2. Methodology

Plaxis 2D was used to investigate the impact of geogrid arrangement and location of the bank seat on the deformational response of GRS-IBS.

The Plaxis 2D model was initially validated against data from a full-scale instrumented GRS wall and finite difference analysis reported by Hatami and Bathurst (2005). This was to ensure that Plaxis 2D would adequately capture the deformational response of a GRS. The validation model consisted of a 3.6m high segmental block wall, with 0.15m high by 0.3m thick concrete facing blocks and a face batter of 8° from the vertical. The geogrid was installed at a vertical spacing of 0.6m (every fourth layer of facing blocks) starting at 0.3m above the base of the wall. In the validation model the geogrid was 2.3m long. The horizontal distance from the rear of the geogrids to the model boundary was nominally 5.8m.

In the parametric model the wall height was 6.0m, with a vertical facing constructed from 0.15m high by 0.3m thick concrete facing blocks. Similar to the validation model, the geogrid, which was 7m long, was installed at a vertical spacing of 0.6m (every fourth layer of facing blocks) starting at 0.3 m above the base of the wall. The horizontal distance from the rear of the geogrids to the model boundary was nominally 15m. The front face of the abutment bank seat, which was 2m wide by 1.4m high, was located at 0.3m and 1.3m away from the back of the wall facing blocks. A schematic of the Plaxis 2D model is shown in Figure 1.

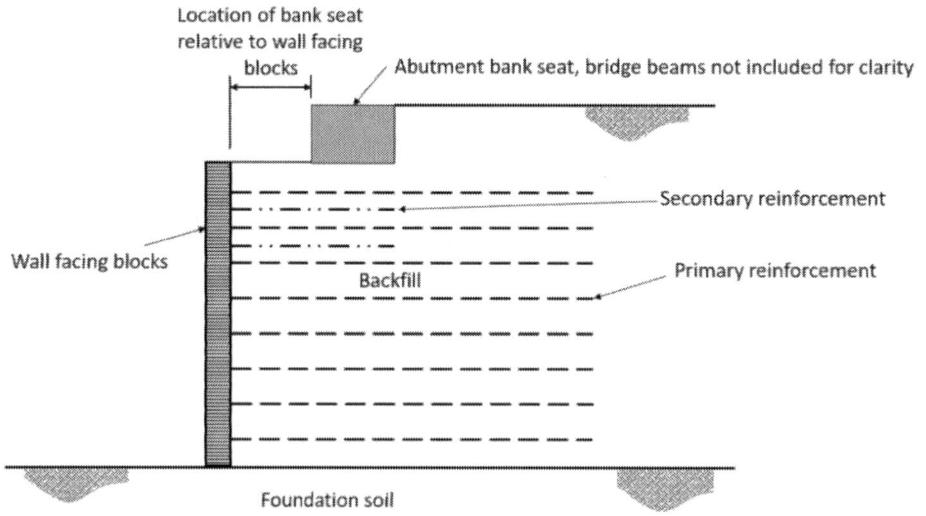

Figure 1. Schematic of GRS-IBS investigated in this study.

2.1. *Material properties*

The structural backfill was modelled in Plaxis 2D using the hardening soil model. The concrete facing blocks were modelled using a linear elastic model. The model parameters for the backfill and facing blocks used in the Plaxis 2D analysis are summarized in Table 1. The structural backfill had a cohesion of 1kPa, to prevent premature soil yielding in locally confining pressure zones, and to take account of any possible apparent cohesion due to moisture in the soil (Cazzuffi et al., 1993 & Rowe and Skinner, 2001).

Table 1. Model parameters for the backfill and facing blocks used in the Plaxis 2D analysis.

Material	γ (kN/m^3)	E_{50}^{ref} (MN/m^2)	E_{oed}^{ref} (MN/m^2)	E_{ur}^{ref} (MN/m^2)	ν	c (kN/m^2)	φ_{peak} (°)	ψ (°)
Backfill	22	50	28	100	0.15	1	44	11
Facing block	16	100	-	-	0.15	-	-	-

The geogrid was modelled using the geogrid element in Plaxis 2D, which is a linear elastic-perfect plastic planar element. In the validation model the reinforcement had a stiffness, EA, of 500kN/m, while in the parametric analysis the stiffness, EA, was 2000kN/m in both the primary and secondary reinforcement layers.

2.2. *Mesh, interfaces and boundary conditions*

Plaxis 2D automatically created the mesh in all models, using 15-node triangular elements. These elements were chosen as they provide a 4th order interpolation of displacement. The model geometry was automatically divided into finite elements by Plaxis 2D. In this study, a medium mesh was selected with local refinement around the geogrid layers at the front face of the wall. This analysis also produced a refined mesh along the face of the wall which provided a high-quality displacement field to assess horizontal movement of the wall face whilst retaining model computational time at a manageable level. A tolerated error of 0.03 was used in all analysis.

Interfaces were created between the facing blocks and the backfill soil, between the facing blocks themselves and between the backfill and the geogrid elements. The interface properties between the backfill and the facing blocks and between the facing blocks themselves were modelled using a Mohr-Coulomb model, similar to Hatami and Bathurst (2005). The model parameters used are summarized in Table 2.

The interaction between the geogrid and the surrounding soil depends on the friction between the geogrid portion and also the soil–soil contact in the apertures of the geogrid. In this study, a reduction in interface strength of 5% was used in all models, corresponding to Rinter=0.95, (Gouw, 2018).

Table 2. Backfill – facing block and facing block – facing block interface properties used in the Plaxis 2D model.

Interface	E (MPa)	c (kPa)	φ (°)	ψ (°)	k_n (MN/m^3)	k_s (MN/m^3)
Backfill – facing block	50	0	44	11	100	1
Facing block – facing block	100	46	57	0	1000	40

The boundary conditions at the extremities of the model were similar to those used by Hatami and Bathurst (2005); a fixed boundary condition was applied along the bottom of the backfill to ensure full fixity. Along the vertical boundary at the rear of the backfill the boundary conditions were taken as hinged, were vertical displacement was permitted but horizontal displacement was prevented. This allowed the backfill to deform vertically.

The boundary conditions under the facing blocks consisted of a stiff steel plate with a free boundary condition in the horizontal and a fixed boundary in the vertical direction to allow the facing blocks to slide in the horizontal direction. A steel plate was place at the toe to allow the wall to slide forward slightly. The stiffness of the steel plate was 189MPa. Hatami and Bathurst (2005) used a similar plate in their numerical analysis in the finite difference programme, FLAC.

Fixed anchor plates, in the horizontal direction only, were used at the toe of both the validation and parametric models and also at the top of the bank seat in the parametric model. The axial stiffness of the fixed anchor plate at the wall toe was taken as 4MN/m, based on measured loads and displacements reported by Hatami and Bathurst (2005) and Huang *et al.*, (2009). The fixed anchor plate at the top of the bank seat represented the bridge beams and was required to restrain the abutment from unconditional outward movement when the bridge loading was activated in the model. The axial stiffness of this fixed anchor plate was 16MN/m, which was selected to limit the horizontal displacement of the bridge beams to 1cm. Without this anchor the GRS wall would have no restraint to horizontal displacement.

2.3. *The GRS-IBS parametric models*

Four parametric models were used to investigate the impact of geogrid arrangement and location of the bank seat on the deformational response of the GRS-IBS in this study, Table 3. In each model the backfill and facing blocks had the parameters presented in Tables 1 & 2. However, each model had a different configuration of geogrid, consisting of primary reinforcement only or both primary and secondary reinforcement. Primary reinforcement was 7m long and installed at 0.6m vertical spacing over the full height of the wall in all models. Secondary reinforcement was placed mid height between the upper two layers of primary reinforcement, immediately below the bank seat, Figure 1. Two lengths of secondary reinforcement were investigated, the first had a similar length to the primary reinforcement (7m long), while the second extended just to the rear of the bank seat (3.3m long), Table 3. Using secondary reinforcement reduced the vertical spacing between reinforcement layers to 0.3m in the upper part of the GRS. The location of the abutment banks seat was investigated at 0.3m and 1.3m behind the facing blocks to determine the impact of load location on the performance of the structure.

Table 3. Geometric details of the models investigated in this study.

	Abutment location, measured from behind the facing blocks (m)	Number of primary geogrid layers	Length of primary geogrid (m)	Number of secondary geogrid layer	Length of secondary geogrid (m)
Model 1	1.3	10	7	-	-
Model 2	1.3	10	7	2	7
Model 3	1.3	10	7	2	3.3
Model 4	0.3	10	7	-	-

2.4. Bridge load cases

In each model the reinforced segmental block wall bridge abutment was constructed in stages within Plaxis 2D, with an analysis undertaken at each stage. The stages used in this study consisted of:

- Stage 1: The first construction phase consisted of modelling the backfill, geogrid and facing blocks without the abutment bank seat or bridge loadings.
- Stage 2: The second construction phase involved adding the abutment bank seat and backfill immediately behind the bank seat. The analysis of this stage modelled the impact of the self-weight of the bank seat and bridge beams only.
- Stage 3: The third phase involved applying the vertical load and braking force to the bank seat, and traffic load on the backfill to the rear of the bank seat.
- Stage 4: The fourth and final phase involved the application of the horizontal load, which was applied in a cyclic manner, both towards (inward) and away (outward) from the bridge abutment.

Load Model 1 (LM1) - tandem axle and general uniformly distributed case in accordance with the design standard for traffic loads on bridges (EN 1991-2:2003+NA, 2009) was used to calculate the vertical traffic loads on the bridge. The shrinkage, creep, temperature and braking force loads were calculated using EN 1992-2:2005 (2005) for Irish conditions consisting of a temperature of ± 25°C, an ambient temperature of 15°C and a relative humidity of the ambient environment of 70%.

The vertical and horizontal loads used in this analysis are shown in Table 4 and were determined from the analysis of a single lane bridge deck, 15m long, supported on a 2m wide and 1.4m high bank seat.

Table 4. Magnitude of vertical and horizontal load used in the analysis.

	Self-weight	Variable traffic load	Traffic load LM1	Shrinkage	Creep	Temperature	Braking force
Vertical load(kN)	185.3	31	72	0	0	0	4.3
Horizontal load(kN)	123.4	36.6	10.5	-100	-44.2	±73.5	±23

3. Model validation

The predicted horizontal displacement of the wall in the validation model was in good agreement with the values reported by Hatami and Bathurst (2005), Figure 2. The displacement started at the base of the wall and increased to a peak value at close to mid-height in the wall, before decreasing again towards the top of the wall.

The predicted axial strain in the reinforcement from this study was also in good agreement with the values reported by Hatami and Bathurst (2005), Figure 3. The axial strain was found to reach a maximum value behind the block facing units and to decrease to zero at the end of the reinforcement. Based on the agreements shown between the trend and magnitude of predicted horizontal wall face displacement and axial reinforcement strain and that reported by Hatami & Bathurst (2005), it was considered that the Plaxis 2D model was successfully validated.

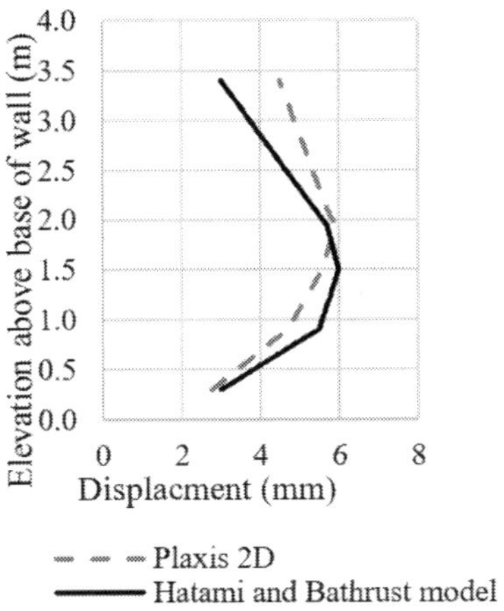

Figure 2. Comparison of predicted facing displacement from Plaxis 2D analysis with values reported by Hatami and Bathurst (2005).

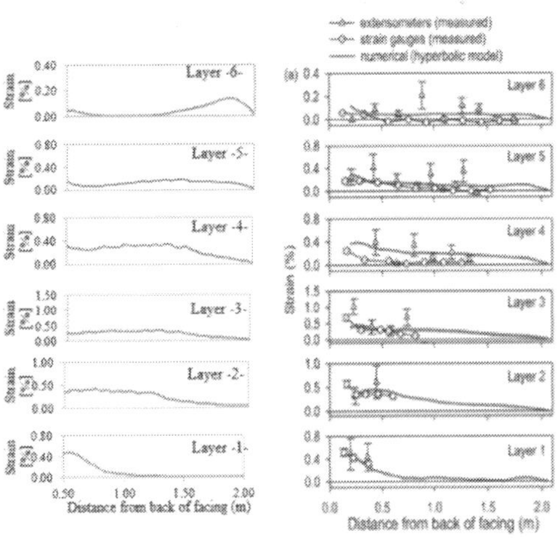

Figure 3. Comparison of predicted axial strain from Plaxis 2D analysis with values reported by Hatami and Bathurst (2005).

4. Results of numerical analysis

The results from the four models investigating the impact of the geogrid configuration and location of the abutment bank seat are discussed in the following sections in terms of the total wall displacement, settlement under the bank seat, strain in the geogrid and the vertical stress distribution under the abutment.

4.1. Total wall displacement

The total wall displacement under the four stages of loading, for Models 1 – 4 are presented in Figure 4. In all cases the wall displacement reached a maximum value at approximately 4m above the base of the wall, corresponding to 2/3 of the wall height, before wall displacement reduced near the top of the wall. The wall displacements increased progressively from Stage 1, construction of the wall, to Stage 3, application of vertical and braking loads. Then, depending on the direction of the horizontal load the wall displacement either reduced, for inward horizontal load, or increased further, for outward horizontal load. The pattern of predicted total wall displacement under different geometrical configurations was similar.

The impact of adding two additional layers of secondary reinforcement near the top of the GRS-IBS resulted in a reduction in the predicted total wall displacement by about 0.5cm, both at the top and 2/3 up the wall face, Figure 4(a), (b) and (c). Shortening the length of the secondary reinforcement from 7m, Model 2 Figure 4(b), to 3.3m, Model 3 Figure 4(c), had no significant impact on the wall displacement.

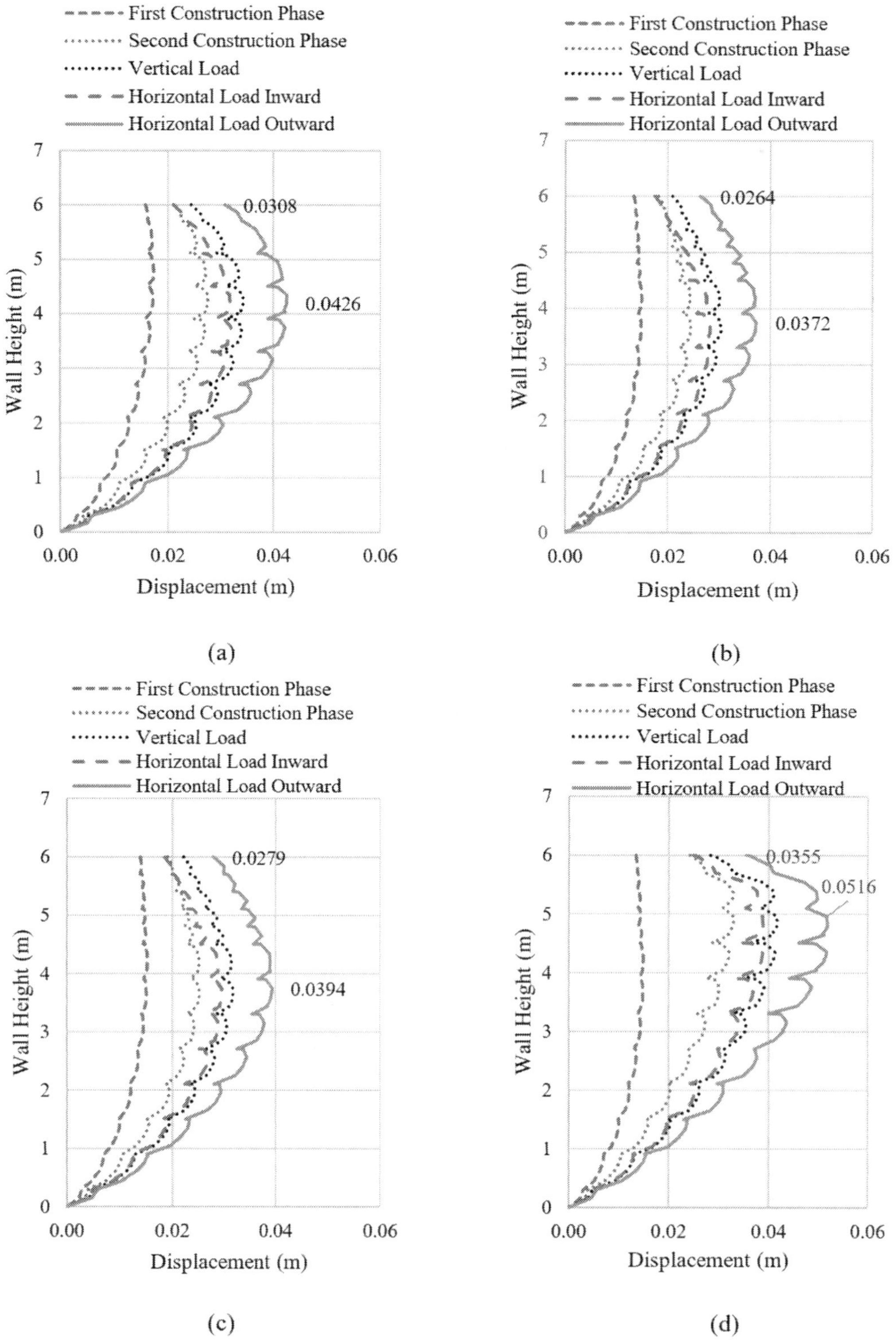

Figure 4. Wall displacement (a) Model 1, (b) Model 2, (c) Model 3, (d) Model 4.

The magnitude of the predicted total wall displacement and the location of the maximum displacement both changed when the bank seat was moved closer to the back of the facing blocks, Figure 4(a) and (c). When the bank seat was closer to the back of the facing block the wall displacement increased by 0.9cm and the maximum displacement occurred at 5m above the base of the wall, rather than 4m when the bank seat was located 1.3m from the back of the facing blocks.

4.2. Settlement of the abutment bank seat

The predicted settlement of the bank seat is presented in Figure 5 for the four models investigated in this study. Similar to the wall displacement, the predicted settlement of the bank seat increased for both construction phases, Stages 1 & 2, and for the application of the vertical bridge loads, Stage 3. Then, depending on the direction of the horizontal load the settlement either reduced, for inward horizontal load, or increased further, for outward horizontal load. The pattern of predicted settlement of the bank seat was similar in all cases examined.

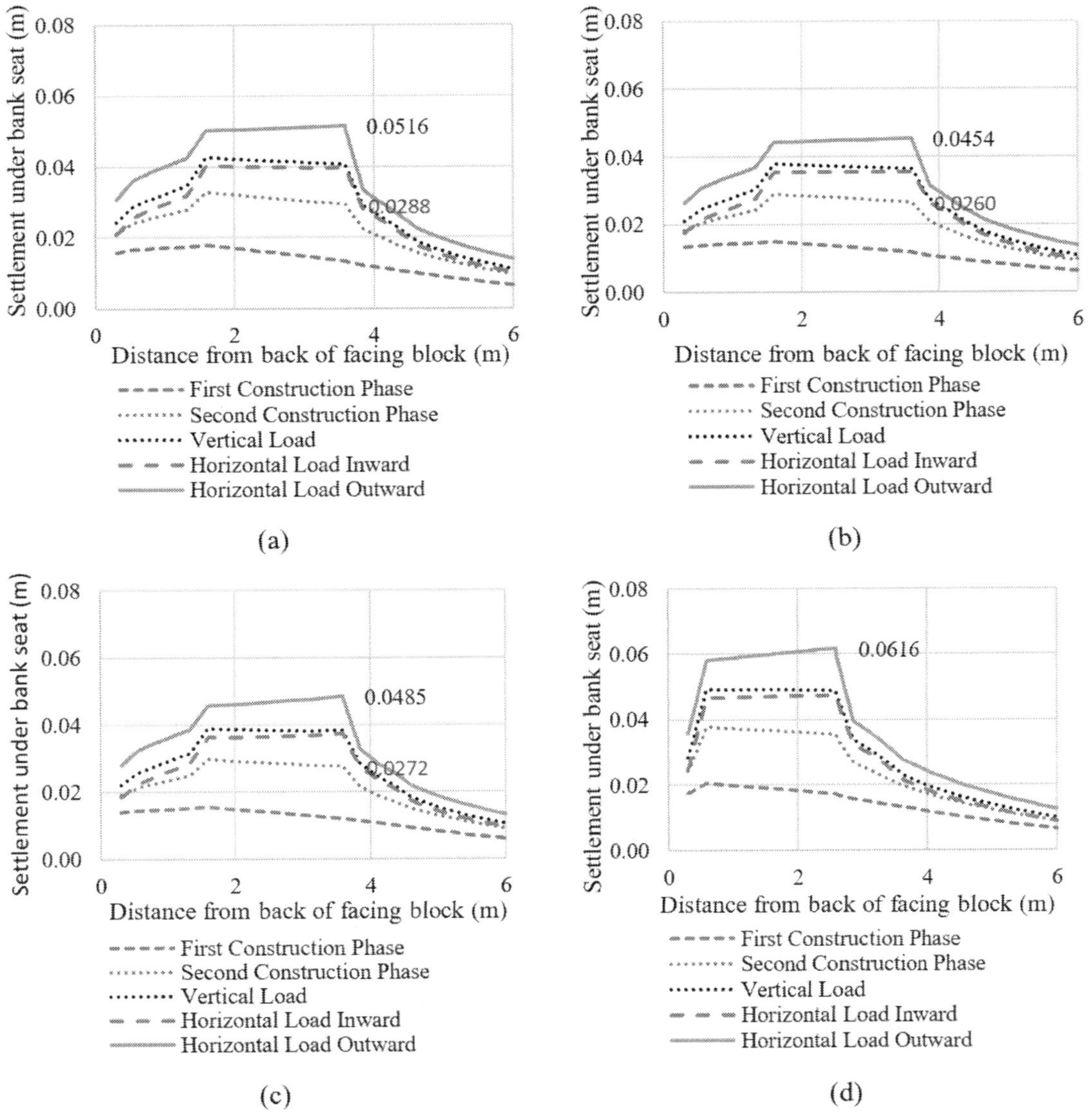

Figure 5. Total displacement under abutment (a) Model 1, (b) Model 2, (c) Model 3, (d) Model 4.

The addition of the two layers of secondary reinforcement reduced the predicted settlement by about 0.6cm at the edge of the bank seat, Figure 5(a), (b) and (c). As with the wall displacement, only a marginal change in the settlement of the bank seat was observed when the length of the secondary reinforcement was reduced from 7m to 3.3m, Figure 5(b) and (c).

Moving the bank seat closer to the back of the facing block, Model 4, resulted in a significant increase (about 1cm) in the predicted settlement of the bank seat, Figure 5(d). This was consistent with the observed increase in the predicted wall displacements, where larger wall displacements were observed when the bank seat was closer to rear of the the facing blocks.

4.3. Vertical stress under the abutment bank seat

The predicted vertical stress under the bank seat for the four cases investigated are presented in Figure 6. The vertical stress increased during the construction phase, Stages 1 and 2, and the application of the vertical bridge loads, Stage 3. Again, depending on the direction of the horizontal load, the stress either decreased (inward horizontal load) or increased further (outward horizontal load). This response was consistent with both the predicted wall displacement and the settlement of the bank seat. In all cases investigated, the maximum stress was observed to occur directly under the bottom-rear corner of the bank seat, closest to the retained fill behind the bank seat.

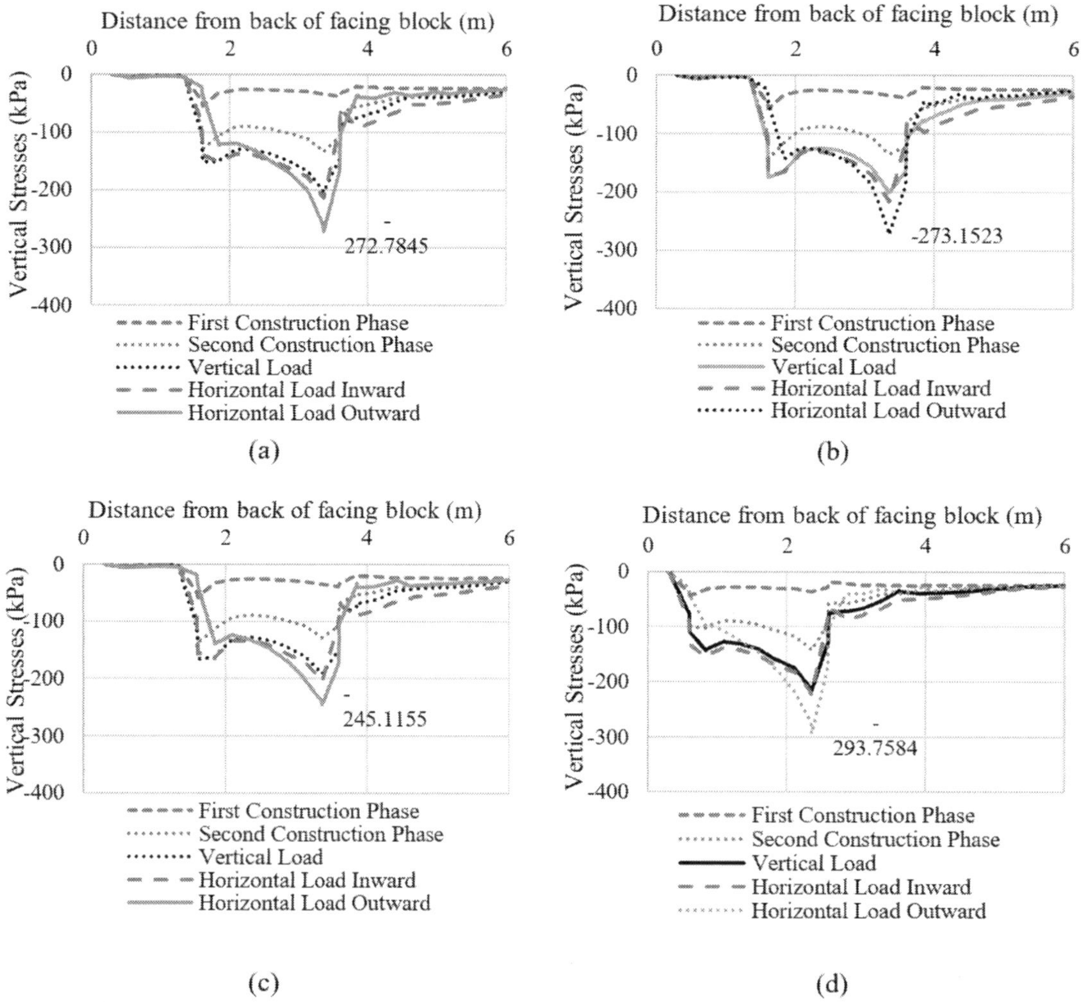

Figure 6. Vertical stresses under the abutment (a) Model 1, (b) Model 2, (c) Model 3, (d) Model 4.

The addition of 7m long secondary reinforcement had no significant effect on the stress distribution beneath the banks seat, Figure 6(a) and (b). Reducing the length of the secondary reinforcement from 7m to 3.3m, resulted in an approximately 8% reduction in the peak predicted stress under the banks seat.

Moving the bank seat closer to the facing blocks increased the peak vertical stress by approximately 7%, Figure 6(a) and (d).

Both the length of the secondary reinforcement and the location of the bank seat, relative to the back of the facing blocks, were observed to be important contributors to the vertical stress distribution under the bank seat. Secondary reinforcement extending just past the rear of the bank seat and having the bank seat located further away from the face of the wall both contributed to reducing the vertical stress.

4.4. Strain in the geogrid

The predicted strain in the geogrid layers 1 (bottom layer), 2, 4, 6, 8 and 10 (top layer), in Models 1, 2 and 3 in the final loading phase, Stage 4, where the horizontal load acts in the outward direction is presented in Figure 7. Model 1 had no secondary reinforcement, while Models 2 and 3 had secondary reinforcement of length 7m and 3.3m respectively. No appreciable difference in geogrid strain, for the different geogrid configurations, was observed in the lower layers of reinforcement, layers 1, 2, 4 and 6, where the strains were broadly the same under the different configurations investigated. In the upper reinforcement layers, layers 8 and 10, the addition of secondary reinforcement resulted in a reduction in the predicted geogrid strain, with the addition of the 7m long secondary reinforcement resulting in the largest reduction.

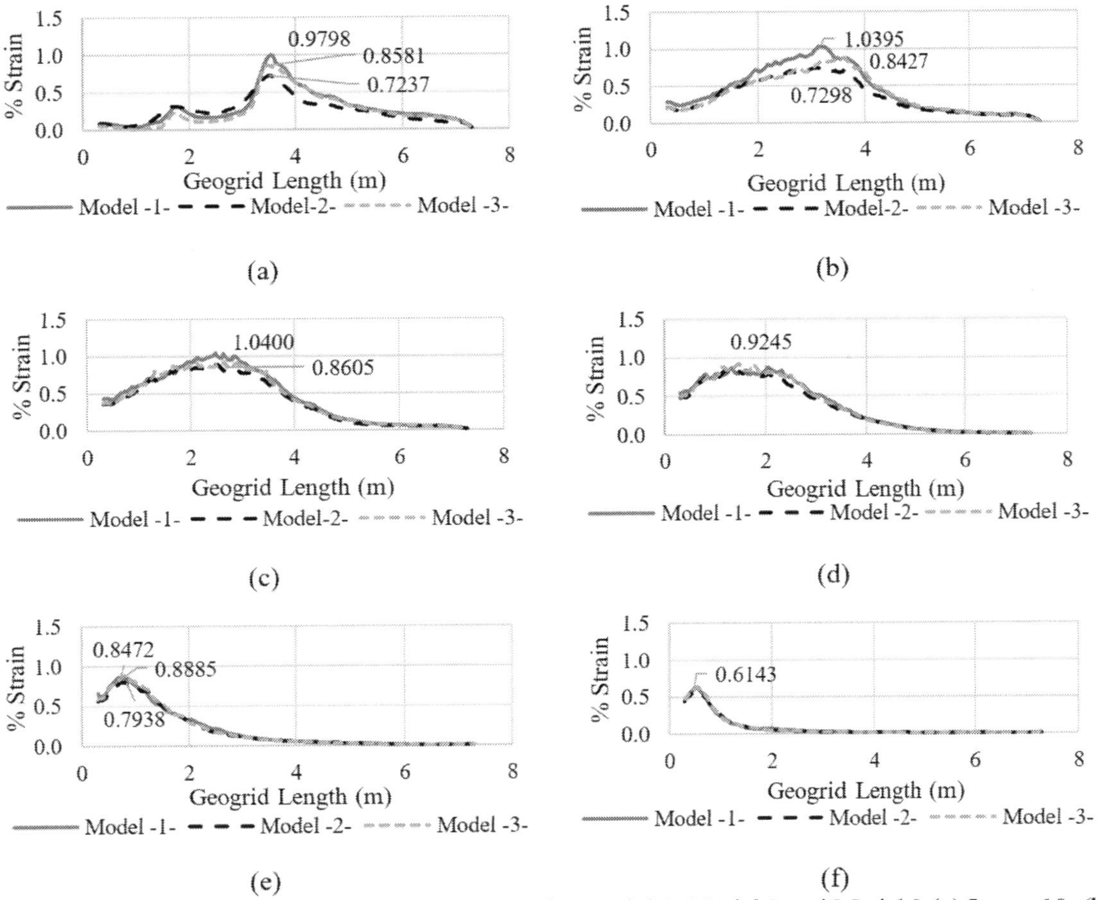

Figure 7. % Strain in geosynthetic reinforcement for Model 1, Model 2 and Model 3 (a) Layer 10, (b) Layer 8, (c) Layer 6, (d) Layer 4, (e) Layer 2 and (f) Layer 1.

The tension in the reinforcement is directly related to the strain, through the stiffness of the reinforcement. Therefore, the addition of secondary reinforcement appeared to redistribute the tension in the geogrid at the top of the wall directly under the bank seat. The greatest redistribution occurred in the longer secondary reinforcement (geogrid length of 7m), Model 2.

Figures 8 and 9 present the distribution of strain along all reinforcement layers for Stage 4 in Model 1, where the bank seat was 1.3m from the back of the facing blocks and Model 4, where the bank seat was 0.3m from the back of the facing blocks. No secondary reinforcement was included in either model. The following was observed:

- The predicted strain in the upper layers of reinforcement, layers 5 -10 only, were significantly impacted by the location of the bank seat. The predicted strains in the lower half of the wall, Layers 1-4, did not change significantly under different geometric conditions.
- The location of peak strain in layers 5 -10 was directly related to the location of the bank seat. In both cases the peak strain in the upper most layer, layer 10, occurred directly under the bottom-rear corner of the bank seat,
- The magnitude of the peak strain was higher in Model 4, where the bank seat was at its closest to the back of the facing blocks.

The lines of maximum tension (corresponding to maximum strain in the geogrid) predicted by Models 1 and 4 are shown in Figures 8 and 9 for inward and outward application of the horizontal load. In all cases the line of maximum tension was approximately log spiral in shape and originated, at or close to the bottom-rear corner of the bank seat. This was consistent with the predicted lines of maximum tension assumed in the coherent gravity method for a vertically loaded area on top of the wall, where the maximum tension line 2 originates from the bottom-rear corner of the loaded area and is log spiral in shape (BS 8006-1:2010+A1, 2016). In Figures 8 & 9, the Rankine failure plane, drawn at (45^0 + ϕ/2) from the bottom corner of the wall is included for reference.

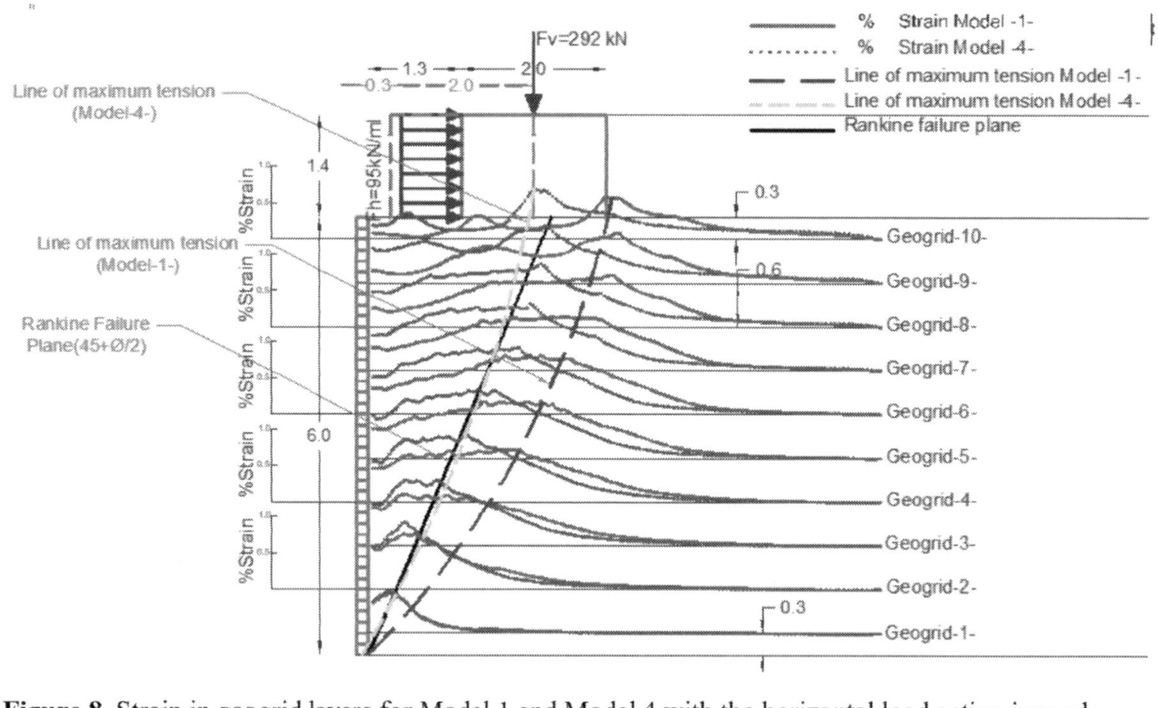

Figure 8. Strain in geogrid layers for Model 1 and Model 4 with the horizontal load acting inward.

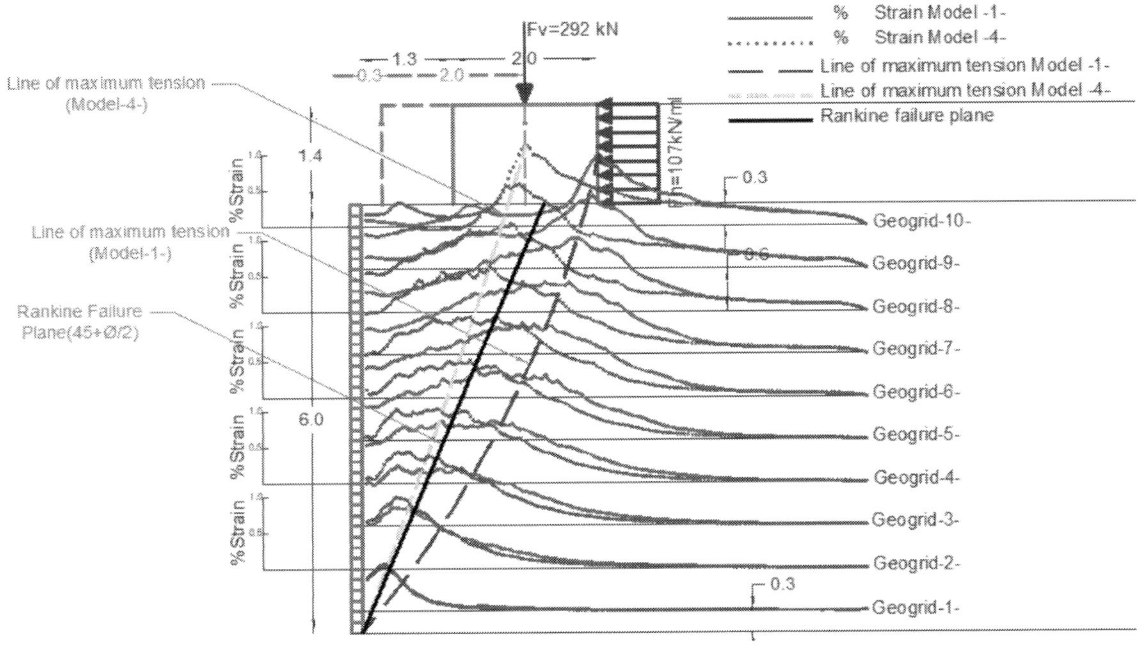

Figure 9. Strain in geogrid layers for Model 1 and Model 4, with the horizontal load acting outward.

5. Discussion

Placing the abutment bank seat 1.3m compared to 0.3m from the back of the facing blocks reduced wall displacement, the peak vertical stress and settlement of the bank seat and peak strain in the reinforcement.

The magnitude of wall deflection, vertical stress and settlement of the bank seat and strain in the reinforcement were all influenced by the direction of application of the horizontal load, reducing when the horizontal load was acting inward and increasing when the load was acting outward.

Adding secondary reinforcement in the upper layers of the GRS, reduced wall displacement, reduced settlement under the banks seat and redistributed the tension between the layers of reinforcement in this zone. However, the required length of the secondary reinforcement is not yet understood. Longer secondary reinforcement appeared to be ideal for reducing wall displacements and settlement under the bank seat and redistributing the tension between reinforcement layers. Shorter secondary reinforcement, extending to the rear of the bank seat only, appeared to be optimal for reducing the peak vertical stress under the bank seat.

The line of maximum tension in the GRS wall originated close to the bottom-rear corner of the bank seat and was log spiral in shape, terminating at the toe of the wall. The line of maximum tension was found to be independent of the direction of application of the horizontal load.

6. Conclusions

Plaxis 2D was used to predict the deformational response of a GRS-IBS abutment under different geometric considerations, involving the addition of secondary reinforcement and different locations of the abutment bank seat from the face of the wall.

The magnitude of wall deflection, vertical stress and settlement of the bank seat and strain in the reinforcement were all influenced by the direction of application of the horizontal load, reducing when the horizontal load was acting inward and increasing when the load was acting outward.

The analysis indicated that both the arrangement of geogrid in the upper part of the GRS abutment and the distance of the abutment bank seat from the face of the wall were very important elements in controlling wall displacement, vertical stress and settlement under the bank seat and strain in the geogrid. Placing the abutment bank seat further away from the face of the wall reduced wall displacement, peak vertical stress and settlement under the bank seat and peak strain in the geogrid. The addition of secondary reinforcement in the top of the wall also reduced wall deformation, peak vertical stress and settlement under the bank seat. The secondary reinforcement also redistributed the tension in the geogrid immediately under the bank seat. Longer secondary reinforcement resulted in greater reduction in wall displacement, settlement of the bank seat and strain in the geogrid. However, shorter secondary reinforcement was more efficient at reducing the peak vertical stress under the bank seat. Further analysis is required to identify the optimum length of the secondary reinforcement. The line of maximum tension in the GRS wall originated close to the bottom-rear corner of the bank seat and was log spiral in shape, terminating at the toe of the wall. The line of maximum tension was found to be independent of the direction of application of the horizontal load.

7. Acknowledgments

The financial support of this study was provided by the Institute of Technology of Sligo, Ireland (President Bursary). The authors also acknowledge the helpful comments by Dr. Moustafa Soubra whose comments helped the authors clarify a number of points.

References

[1] Adams, M., Nicks, J., Stabile, T., Wu, J., Schlatter, W. and Hartmann, J., 2011. Geosynthetic Reinforced Soil Integrated Bridge System Interim Implementation Guide, Publication No. FHWA-HRT-11-026, Federal Highways Administration, Washington.

[2] Abu-Farsakh, N., Wang, T. and Zornberg, J. G. (2000) 'Performance of geosynthetic-reinforced walls supporting bridge and approaching roadway structures', in *Geotechnical Special Publication*.

[3] BS 8006-1:2010+A1, 2016, Code of practice for strengthened/reinforced soils and other fills. British Standard Institution, London.

[4] D. Cazzuffi, L. Picarelli, A. Ricciuti, and P. Rimoldi (1993) 'Laboratory Investigations on the Shear Strength of Geogrid Reinforced Soils', in Cheng, S.-C. (ed.) *Geosynthetic Soil Reinforcement Testing Procedures*. West Conshohocken, PA: ASTM International, pp. 119–137.

[5] EN 1991-2:2003+NA:2009. Eurocode 1: Actions on structures - Part 2: Traffic loads on bridges (Including Irish National Annex), National Standards Association of Ireland.

[6] EN 1992-2 (2005). Eurocode 2: Design of concrete structures - Part 2: Concrete bridges - Design and detailing rules, National Standards Association of Ireland.

[7] Gouw, T. (2018) 'International conference', *Journal of the Society of Archivists*, 7(5), pp. 341–342.

[8] Hatami, K. and Bathurst, R. J. (2005) 'Development and verification of a numerical model for the analysis of geosynthethic-reinforced soil segmental walls under working stress conditions', *Canadian Geotechnical Journal*. doi: 10.1139/t05-040.

[9] Huang, B., Bathurst, R. J. and Hatami, K. (2009) 'Numerical study of reinforced soil segmental walls using three different constitutive soil models', *Journal of Geotechnical and Geoenvironmental Engineering*, 135(10), pp. 1486–1498.

[10] Rowe, K. R. and Skinner, G. D. (2001) 'Numerical analysis of geosynthetic reinforced retaining wall constructed on a layered soil foundation', *Geotextiles and Geomembranes*, 19(7), pp. 387–412.

Reclamation of fertilizer sludge pond: the Azomures case study in Romania

L Carbone[1], OSyllwasschy[1], T Hasslacher[1] and U Hein[2]

[1]HUESKER Synthetic GmbH, Fabrikstraße 13-15, 48712, Gescher, Germany
[2]Geiger Transilvania SRL

E-mail: carbone@huesker.de

Abstract. Residues coming from the production of fertilizer of the Azomures company were released into 300.000 m² pond close to Tirgu Mures in Romania. To limit the exposition of the pollutants with the close environment and population and to reduce the leakage of contaminated leachate into the river, the environmental remediation of the pond was conducted. An impermeable cover system consisting of the succession of soil and geosynthetic layers was constructed on top of the contaminated residues. Key point for the success of the project was the use of geosynthetic reinforcement to increase the bearing capacity of the system as well as the development of an innovative method for the construction of intermediate embankments. In this paper, the remediation work for Azomures pond in Romania from its initial concept to the design and execution is fully described and can be used as a reference for alternative solutions in similar projects.

1. Introduction

Azomures company is one of the most important producers of mineral fertilizers for agricultural use in Romania, supplying farmers with mineral Nitrogen (N) and complex NPK fertilizers over the past 50 years. The Azomures pond is located near the town of Tirgu Mures in Romania. The construction of the Tirgu Mures plant began in 1962, it was completed in several stages and it covers now an area of approximately 100 hectares in the south-western part of the city.

During the operating years of the plant, sludge was produced as a by-product of the fertilizer production and released into the pond which covers an area of approximately 300.000 m². The chemical composition of the sludge presented high contamination of heavy metals, nitrogen, phosphorus, chlorine and sulphur compounds with a pH value varying from 4 to 7.

Near the pond lies the second largest river in Romania. To limit the exposition of the pollutants with the close environment and population and to reduce the leakage of contaminated leachate into the river, the sealing of the entire pond surface was foreseen.

The Geiger Transilvania SRL was awarded to carry out the design and execution of the remediation of Azomures pond. The remediation concept foresaw a cover system composed of the sealing system, drainage layer and top soil for vegetation. Key design issues for such systems were the management of waters, the stability of the sludge, its consolidation behaviour and the trafficability for construction equipment. To enable the mechanical placement of the capping system on top of an extremely soft and weak subgrade geosynthetics for reinforcement were used in the project. In particular, due to the large surface area to be covered the construction of intermediate access roads was required. An innovative construction method involving the combined use of geosynthetics and soils to induce a controlled ground failure was specially developed for this project. This paper presents the conception and the design of the cover system of the Azomures pond in Romania. Particular attention is given to the innovative construction method and its related challenges.

2. Capping system of sludge ponds

2.1. *Generalities*

The design of the cover system of a sludge/tailings pond normally depends on the requirements of the customer or the appropriate authority and the intended use after capping. In the design of such systems, the use of geosynthetics stabilizes the soft subsoil and therefore enables workers and construction machinery to work carefully on the tailings pond without treating the tailings further [1].

Generally speaking, the cover system may consist of a simple geosynthetic layer for reinforcement and fill material or a multi-layer system with geosynthetics, soils and qualified liner system with gas and water drainage [2].

According to the characteristics of the tailings, the regulations in force in the specific lands and the final use of the site after remediation, the thickness of the different layers may vary but in general, it is not thinner than 1,5 m – 2 m. The main challenge by the design of such structure is the construction of such a system on top of a very soft and extremely heterogeneous soil. In fact, according to the deposition history, dewatering and weather conditions, the hydraulic and mechanical characteristics of the tailings may differ in-depth and across the area. In this case, geosynthetics for reinforcement as geogrids or woven geotextiles with adapted design strength might be used.

For the technical design of the capping system, an intensive investigation of in situ soils is required. The following data are requested:

- Geotechnical parameters of sludge/fill material
- Stratification of the subsoil
- Tailings pond size
- Free water level
- Live load of construction machinery

2.2. *Settlement estimation of the sludge*

Settlement due to consolidation may be significant in such subgrade. Therefore, they should be assessed and they should be consistent with the allowable deformation of the structure especially of the sealing system of the capping.

Soil investigation should contain consolidation tests to specify settlement during and after the construction period. In case no data are available, a settlement assumption has to be done based on experience in comparison with similar soils and the prediction should be verified during the construction phase and afterwards, for example, employing measuring points on the surface.

2.3. *Geosynthetic reinforcement design*

Typically, in the capping systems design, woven fabric or geogrids or combinations thereof like woven/geogrid or geogrid/non-woven are selected. The main function of woven fabrics and geogrids is to transfer tensile stresses resulting from the construction of the cover system i.e., soil and traffic load, into the anchor trench. Besides, non-woven can also work as separator and filter to keep sludge in place below the geotextile. The chemical characteristics of the sludge should also be taken into account and different raw materials should be selected. Normally polypropylene (PP), polyester (PET, PES) and polyvinylalcohol (PVA) can be used in a normal pH-range from 4 to 9.5. In areas with pH 2-4 or 10-13 only PP and PVA may be used, if long term stability has to be considered in the design.

The required tensile strength of the geosynthetic reinforcement should verify the stability of the system. Currently, there is no established method available to design the geosynthetic reinforcement according to these membrane-like loading effects. Edil and Aydilek [3] described a design procedure or Espinoza et al. [4] presented a case history and a more sophisticated design method. These designs are based on bearing capacity (rutting) analyses, the latter in combination with membrane contribution of the reinforcement. Additionally Bishop's method can be used for the stability analysis where the mud waving at the edge of the geosynthetic during the filling process can be considered. It is worth to point out that the stability of the system, the stratification of the cover layers and their soil parameters have to be checked carefully in each stage of filling. The first soil layers up to 0.3 to 1.0 m may be the most critical since the shear resistance of the weak sludge as well as the counter-pressure activated by surcharge is negligible small to prevent ground failure.

Another key aspect of the design is the installation of the geosynthetic layers. This depends on the size of the pond and the characteristics of the tailings. It can be done by sewing a large panel pulled by external edges into the pond (Figure 1a) or by unrolling and overlapping them (Figure 1b).

Figure 1. Examples of installation methods of geosynthetics in tailings ponds: a) assembling of large panels; b) unroll and overlap of geosynthetic reinforcement elements.

3. Azomures sludge pond cover system design

3.1. Sludge deposit characterization

The sludge deposit of the Azomures company is located near the town of Tirgu Mures in Romania. During the operation activities, the sludge coming from the production of fertilizers was released in a settling basin which covers an area of approximately 300.000 m^2 (Figure 2). The chemical composition of the sludge presented high contamination of heavy metals, nitrogen, phosphorus, chlorine and sulfur compounds with a pH value varying from 4-7.

An intense geotechnical investigation was carried out on-site and in the laboratory to characterize the mechanical properties of the sludge during different phases of the project.

In situ tests comprised static cone penetration tests (CPT), Marchetti dilatometer test (DMT), vane shear test and dynamic probing-light (DP-L) tests. To enable workers to carry out safely the in situ investigation a customized platform pulled from the embankment with the aid of a crane was specially developed and assembled.

Results obtained by the in situ tests showed that the sludge mechanical characteristics in terms of the undrained shear strength varied substantially in the longitudinal and transversal direction as well as with the depth. A characteristic value of about c_u = 5 kPa was recommended to be used in the design.

Figure 2. Sludge deposit Azomures: a) aerial view (from Google Earth - December 2014); b) sludge deposit topographical map – top view.

Geotechnical laboratory tests were carried out to further characterize the sludge. It is worth noting that due to the nature of the sludge which is different from soil composition, the classical geotechnical approach for the interpretation of the results was in some cases unsuitable.

The unit weight of the sludge was determined out of the oedometer and direct shear tests. Values obtained were in the range between γ = 12.2 - 14.4 kN/m^3 with a characteristic value of γ= 13.5 kN/m^3 to be used in the design.

Furthermore, oedometer tests were performed giving oedomeric modulus in the range of E_{oed} = 0.29 – 1.17 MPa, values consistent with the weak nature of the sludge. It is worth noting that during the execution of the oedometer test, over 50% of the settlement took place within the first 10 minutes of loading out of the total loading time of 24 days.

During the site supervision visual inspection, it was observed that the sludge surface in a large part of the pond was composed of a crust (ca. 0.2 - 0.6 m) which showed sufficient bearing capacity to be safely walkable by the operators. Further, the company Geiger Transilvania SRL carried out several in situ trials to test the behaviour of the subgrade under different loadings. The trials consisted of building small dams of different heights which corresponded to different load levels and to observe the sludge behaviour.

In cases of high loads acting on the surface, ground failure occurred to the side of the dam, in which the backfilled dam lost almost its entire original height above the sludge surface. In this case, the sludge laying underneath the crust entered in contact with the soil of the dam causing the loss of the stabilizing effect provided by the crust and its bearing capacity, especially in areas with greater sludge depths. (Figure 3).

Figure 3. Azomures sludge deposit: a) crust on the surface of the pond; b) ground failure after the construction of small dams during the field trials.

The use of geosynthetics for reinforcement has been made therefore essential to guarantee the safe execution of the works and at the same time to optimize work performance. In fact, through the reinforcement, the acting loads could be transferred to the product as tensile forces and distributed over a larger area. In this way, the safety against ground failure was increased and at the same time, more load could be brought to the system.

3.2. Azomures cover system generalities

The cover system of "Azomures sludge pond" consisted of the succession of mineral soils and geosynthetics designed according to the storage and evaporation principle. The aim is to reduce the exposure of residues and therefore reducing the exposure of the contaminated materials and to prevent new leakage formation by limiting the infiltration of water into the pond. Therefore, the surface needs to be properly sealed and the rainfall water has to be properly collected through a drainage layer. In this regard, geosynthetics were also used in the system for sealing and drainage purposes.

The final cover was therefore composed of the following layers (from the bottom to the top) (figure 4):

- sludge
- biaxial high tensile strength woven geotextile big panel
- mineral drainage and bearing layer (d_{90} < 63mm)
- geosynthetic clay liner
- geosynthetic drainage layer
- top soil
- geosynthetic erosion protection mattress

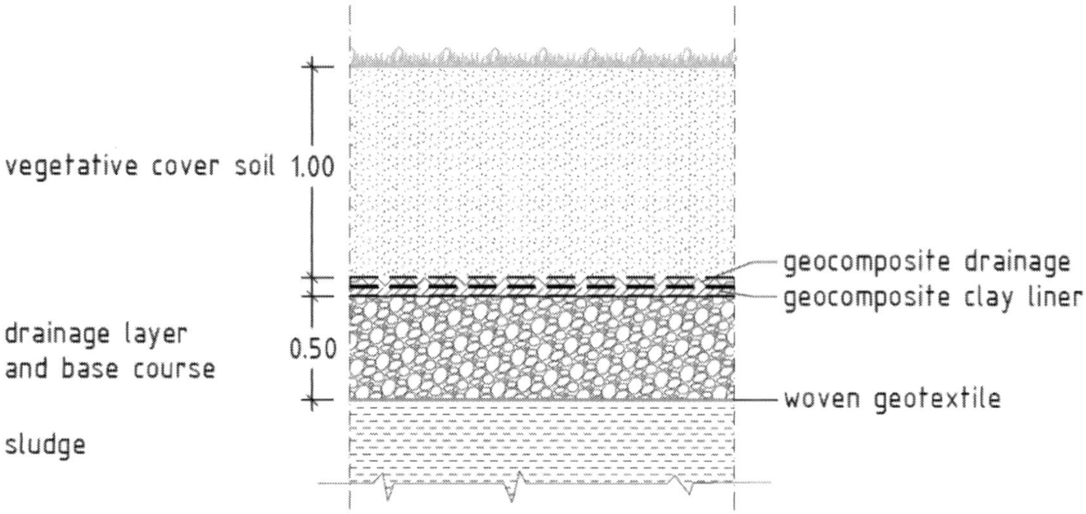

Figure 4. Typical cross section of the Azomures cover system

3.3. Design of the geosynthetic reinforcement: required tensile strength, anchoring and installation method

For the Azomures sludge pond, the installation of a large panel made by woven geotextiles was selected as the best solution to cover the large area under the specific site conditions.

The reinforcement geosynthetic selected for this project was a high biaxial tensile strength woven geotextile (i.e., Sefitec® PP 80) made from polypropylene (PP) which presented the required tensile strength in both machine and transversal direction as well as the chemical stability adapted to the specific site conditions. The use of this product not only enhanced the load-bearing capacity and the load distribution of the system but also served as a separation and filtration layer.

The tensile strength of the geosynthetic reinforcement was determined by carrying out the stability analysis of the system. The design takes into account the verification of the bearing capacity of the combined system (soft subgrade and geosynthetics). This analysis enables the determination of the design tensile strength of the bearing layers while the actual tensile strength of the reinforcement (i.e., geogrids or woven geotextile) will be determined by applying reduction factors that take into account creep, installation damage, joints and connections, environmental chemical impacts and dynamic effects if present. The worst-case scenario taken into account during the calculation is taking place during the construction of the first lift of soil with the light machinery driving on top.

The traffic loads were set to $p = 8$ kN/m² and the soil thickness varied from 0.5 m in the first lift to 1,5 m in the final design phase. The design was performed according to the Eurocode 7 for the temporary load case.

Since the sludge mechanical properties varied significantly according to the region of the pond, a very low undrained shear strength equal to $c_u = 5$ kN/m² was chosen to characterize the sludge behaviour in the whole pond.

Once the stability check was carried out, the dimensioning of the anchor trench was performed by verifying the stability analysis against the pull-out and/or sliding of the geosynthetic reinforcement.

3.4. Drainage capacity design

In the Azomures project, the drainage of the rainwater of the cover system was carried out by the use of geocomposite drainage.

The planned drainage system is composed of a geocomposite draining into a central drainage ditch which runs through the length of the pond with a gradient of 1%.

The design and selection of the appropriate geocomposite drain was carried out by considering the long term drainage capacity of the material drain under the project boundary conditions. The drainage capacity verification was carried out considering the German "GDA E 2-20" [5] recommendation guideline for the design of drainage systems in surface sealing systems.

According to the available data, the amount of precipitation in Tirgu Mures was between 550 - 800 mm/year with a peak value of 856 mm/year in 2005.

Based on field measurements and corresponding water balance calculations (i.e., the influence of evapotranspiration, surface runoff and storage) the calculation was run considering a discharge of 10 mm/d which corresponded to 10 times the average value of the maximum amount of water expected in Azomures.

For the long term design of the geocomposite drain drainage capacity, reduction factors are applied to take into account the influence of creep (thickness reduction over time), scattering of measured data, installation stress, local cross-sectional changes as well as other effects such as precipitation, soil input or roots.

The calculation showed that the selected product had a sufficient drainage capacity of approximately 0.083 l/s·m at a gradient of 1% with a maximum drainage length of 205 m.

Therefore, the required drainage capacity according to the project cross-section was exceeded by a factor of 11.5.

4. Installation of the cover system

4.1. Innovative method for the construction of separation embankments

The "Azomures" sludge pond covers an area of approximately 300.000 m² with a maximum width of about 400 m.

Due to the large surface area to be covered, an innovative construction method involving the combined use of geosynthetics and soils to induce a controlled ground failure was carried out to be able to install safely the capping system on top of the extremely soft subgrade. Therefore, the first step of the installation consisted of dividing the very large basin into individual sub-basins with a width of approximately 100 m and a length between 300 m and 400 m using separation embankments. The embankments are created by filling gravel starting from the bedrock up to 1 m on top of the sludge level.

To prevent excessive spreading and the mixing of the gravel in the slurry, panels made of high tensile strength woven geotextiles serving as a separating and reinforcing layer were used. The embankments served not only as stable intermediate access roads but also were used as anchorage for the successive reinforcement layers. The first step was to remove the surficial crust with a floating long-arm excavator and successively to lay down the woven geotextile, which was preassembled in panels of 1000 m². Afterwards, the gravel was brought in the system which started to sink and provoked a controlled ground failure (Figure 5).

a)

b)

Figure 5. Innovative construction method involving the combined use of geosynthetics and soils to build intermediate embankments over the entire width of the pond: a) installation of the woven geotextile panels; b) backfilling the dams with soil.

4.2. Installation of the cover layers

Once the construction of the intermediate embankments was finalised, the installation of the cover layers could take place. The separation embankments were used as a track for construction machinery as well as anchor area for big panel of the woven geotextile installed on top of the pond.

The high strength woven geotextile used in the project could be stitched to form big panels to cover large areas in a short time. The individual fabric panels were sewn together in the factory to form large panels of approximately 2500 m². The panels were folded and rolled into an accordion shape and delivered to the construction site and afterwards spread out along the edges of the area to be covered.

The panels could be easily handled on-site to be sewn together to form larger panels of approximately 10.000 m² and pulled into the final position by operators and excavators.

The absence of overlaps offered the great advantage of avoiding possible weak points with the risk of sludge leakage.

After laying, the panels were secured against the wind with sandbags until they were covered with soil and anchored in anchor trenches in the edge area (Figure 6).

Figure 6. Woven geotextile 10.000 m² large panel installed at Azomures pond

The installation of the woven geotextile big panel enables the further safe construction of the 0.5 m soil bearing layer as well as successive sealing layer (i.e., geocomposite clay liner), geocomposite drainage as well as the top soil.

Figure 7. Sludge deposit Azomures aerial view after completion of final cover system construction (from Google Earth - September 2019)

5. Conclusion

Azomures is one of the most important producers of mineral fertilizers for agricultural use in Romania. By-products coming from the production of the fertilizers were released in the form of sludge into a settling basin covering an area of approximately 300.000 m². In 2014, to limit the exposition of the pollutants with the close environment and population and to reduce the leakage of contaminated leachate into the river, the sealing of the entire pond surface was foreseen.

The Geiger Transilvania SRL was awarded to carry out the design and execution of the remediation of Azomures pond. Due to the large surface area to be covered the construction of intermediate access roads was required. An innovative construction method involving the combined use of geosynthetics and soils to induce a controlled ground failure was specially developed for this project.

The conceptual design, the selection of the materials and the installation phases are fully described in this paper and can be used as a reference for alternative solutions in similar projects.

References

[1] Langseth J, Luger D, Jacobs W, Sisson R, Hedblom E, Greenleaf T 2015 Placement of caps on soft and fluid tailings. *Proceedings Tailings and Mine Waste*October 26-28 2015 Vancouver, BC

[2] Carbone L, Syllwasschy O, Schimmel L 2019 Remediation case study of uranium ore residues settling basin *Geosynthetics* **37** n 5

[3] Edil T B, Aydilek A H 2001 Geotechnics of Capping Very Soft Wastes. *Proceedings of the Thirteenth Int. Conf. on Soil Mechanics and Geotechnical Engineering,* Istanbul, Turkey August 2001 **3** pp 1903-06

[4] Espinoza R D, Steier, W M, Steiner M 2012 Construction of cover soil over soft sludge lagoons *Proceeding of Geoamericas 2nd Panamerican Geosynthetics Conf.* May 1-4 2012 Lima, Peru

[5] GDA-Empfehlungen E2-20 2015 Entwässerungsschichten in Oberflächenabdichtungssystemen GDA-Empfehlungen **3** Auflage 1997 p 185 *Bautechnik 9/2000, Bautechnik 9/2003 Überarbeitung 5/2015* (https://www.gdaonline.de/empfehlungen)

IOP Conf. Series: Materials Science and Engineering 1260 (2022) 012039 doi:10.1088/1757-899X/1260/1/012039

CE marking for geosynthetics

Ph. Delmas [1], **H. Ehrenberg** [2]

[1] Convenor of the CEN TC189 WG1, Chairman of French Standardisation Committee on Geosynthetics, Paris, France
E-mail: philippe.delmas589@orange.fr

[2] Member of CEN TC189 WG1, Member of EAGM, Germany

Abstract. 1st October 2002, CE Marking has been implemented for Geosynthetics and from 1st July 2013, it has been reinforced by the Construction Products Regulation (CPR 305/2011) making mandatory in all EEA Countries for manufacturers to apply CE Marking to any products covered by a harmonised European Standard (hEN) or a European Technical Assessment (ETA). Main principles of the CPR and of the harmonised European Standards are presented, including a rapid overview of the general rules imposed by the CPR, like Standardisation Request, position versus references to national requirements and assessment method(s). Secondly, the paper shows how to deal with CE Marking in practical use of Geosynthetics for the placement on the market and for specific requirements in construction jobs, for control of incoming products and for fit for purpose.

1. Introduction

After the implementation of the Construction Product Directive (CPD) in 1989, the CE Marking of Geosynthetics became obligatory since 2002. The harmonised product standards (hENs) describe the corresponding rules to be followed for CE Marking for each of the different applications of Geosynthetics. At that time, the corresponding level Attestation of Conformity (AoC) had been fixed to 2+ by the Mandate M/107, amended by M/386, of the European Commission

The CE Marking has been reinforced from 1st July 2013 by the Construction Products Regulation (CPR 305/2011) making mandatory in all EEA Countries for manufacturers to apply CE Marking to any products covered by a harmonised European Standard (hEN) or a European Technical Assessment (ETA). The CPR is a regulation. This gives to the CE marking an explicit meaning, for which the Declaration of Performance (DoP), which replaces the Declaration of Conformity, is a key concept. This means that the CE marking and the (DoP), based on the list of essential characteristics, are now mandatory in all countries, for Construction products, including Geosynthetics.

From the 1st July 2013, the manufacturers shall comply with the new European CPR. The declared objective of the CPR is to simplify and improve the competitiveness of the Construction sector.

2. Principles of the CPR and of the harmonised European Standards (hENs)

2.1. Principles of the CPR

A clear explanation of the principles of the CPR and of the harmonised European Standards has been presented by Katsarakis (2018) during the "CPR" Workshop in Brussels. Presented from the legal perspective, the document shows how the Harmonised European Standards provide a solid technical basis for manufacturers for testing the performance of their products.

Already in 1985, the White Paper on the creation of the Internal Market recognised the importance of the construction products sector. To create the Internal Market for construction products it was necessary (1) to abolish existing technical barriers to trade of products (Directive 89/106/EEC replaced by Regulation (EU) 305/2011), (2) to establish an effective market surveillance by the Member States, (3) to avoid apparition of new unjustified barriers and (4) to open the public procurement to competition.

The basic mechanism to abolish trade barriers for construction products is defined in the Regulation (EU) 305/2011). This Regulation foresees the elaboration of harmonised European standards (hENs), the implementation of the hENs in all EU Member States, the manufacturers to affix the CE marking and to issue a Declaration of Performance in line with these hENs and the use of the hENs by all actors (authorities, design engineers, producers, contractors) in a "common European technical language".

In this mechanism, the following basic principles apply:

- each Member States remains competent to define for construction works the level of the Basic Works Requirements to be satisfied in his territory
- the product performance declared by the manufacturer allows engineers / authorities to verify that construction works built with these products will satisfy the Basic Requirements for Construction Works as applied in the Member State.

2.2. The harmonised European Standards (hENs)

The elaboration of harmonised European standards (hENs) is realised by European Standardisation Organisations through mandates for specific families of construction products.

For Geosynthetics, CEN TC189 works since 1994 under the Mandate M/107 (1994) amended by M/386 (2006). For the future revisions of the harmonised European standards (hENs), CEN TC189 starts to work in the new frame of a standardization request (Implementing Act) to be issued under Regulation (EU) 1025/2012.

The standardisation request contains the essential characteristics i.e. those characteristics of the construction product which relate to the Basic Requirements for Construction Works. For Geosynthetics, the Basic Requirements for Construction Works are: (1) Mechanical resistance and stability, (3) Hygiene, health and environment (through the topic Dangerous substances studies through the CEN TC351) and (7) Sustainable use of natural resources, recently under study.

The hENs for Geosynthetics provide methods and criteria for assessing the product performance in relation to the essential characteristics and the intended use as identified in the Mandate. These intended uses are defined by the functions of the Geosynthetics will perform in the different applications. The Table 1 presents the present hENs according the different applications of Geosynthetics. The functions considered for Geosynthetics by the Mandate are fluid & gas barrier, separation, protection, reinforcement, drainage and filtration. For each function/application, the hENs define the essential characteristics and the corresponding test procedure.

Table 1. hENs according the different applications of Geosynthetics

EN 13249:2016 - Geotextiles and geotextile-related products - Characteristics required for use in the construction of roads and other trafficked areas (excluding railways and asphalt inclusion)
EN 13250:2016 - Geotextiles and geotextile-related products - Characteristics required for use in the construction of railways
EN 13251:2016 - Geotextiles and geotextile-related products - Characteristics required for use in earthworks, foundations and retaining structures
EN 13252:2016 - Geotextiles and geotextile-related products - Characteristics required for use in drainage systems
EN 13253:2016 - Geotextiles and geotextile-related products - Characteristics required for use in erosion control works (coastal protection, bank revetments)

EN 13254:2016 -Geotextiles and geotextile-related products - Characteristics required for the use in the construction of reservoirs and dams
EN 13255:2016 -Geotextiles and geotextile-related products - Characteristics required for use in the construction of canals
EN 13256:2016 - Geotextiles and geotextile-related products - Characteristics required for use in the construction of tunnels and underground structures
EN 13257:2016 - Geotextiles and geotextile-related products - Characteristics required for use in solid waste disposals
EN 13265:2016 - Geotextiles and geotextile-related products - Characteristics required for use in liquid waste containment projects
EN 13361: 2004 - Geosynthetic barriers — Characteristics required for use in the construction of reservoirs and dams - EN 13361: 2004/A1:2006
EN 13362: 2005 - Geosynthetic Barriers - Characteristics required for use in the construction of canals
EN 13491: 2004 - Geosynthetic barriers - Characteristics required for use in the construction of tunnels and associated underground structures / EN 13491:2004/A1:2006
EN 13492:2004 – Geosynthetic barriers - Characteristics required for use in the construction of liquid waste disposal sites, transfer stations or secondary containment / EN 13492:2004/A1:2006
EN 13493:2005 - Geosynthetic barriers - Characteristics required for use in the construction of solid waste storage and disposal sites
EN 15381:2008 - Geotextiles and geotextile-related products - Characteristics required for use in pavements and asphalt overlays
EN 15382:2013 - Geosynthetic barriers - Characteristics required for use in transportation infrastructure

The hENs for Geosynthetics don't specify classes or thresholds of performance as this is not requested in the present Mandate.

During the elaboration of hENs, the CEN TC189 has determined the applicable Factory Production Control (FPC), which takes into account the specific conditions of the manufacturing process of the construction product concerned.

The harmonised standard includes also technical details necessary for the implementation of the system of Assessment and Verification of Constancy of Performance (AVCP).

3. Assessment and Verification of Constancy of Performance (AVCP)

To overcome national differences which fragment the EU market, the Commission defines by a Delegated Act the Assessment and Verification of Constancy of Performance (AVCP) system for the products.

The CEN policy on conformity aspects in standards is based on the neutrality principle, which requires that all documents shall be written in a way such that conformity with the specified requirements can be assessed by a manufacturer or supplier (first party), a user or purchaser (second party), or an independent body (third party).

3.1. Basic principle

This is the product which has to be evaluated, and not the production facility. Provisions should be applicable for any involved party on any individual product. Provisions should apply to large, medium, small and micro-enterprises; manufacturers manufacturing the products completely and manufacturers (partly or even completely) subcontracting manufacturing, series and non-series, continuous and discontinuous manufacturing processes, prototypes, etc.

Assessment and Verification of Constancy of Performance comprises 2 parts:

- Assessment of Performance corresponds in initial testing, calculation, use of tabulated values and descriptive documentation by which the performance of the construction product is

determined; based on the Assessment of Performance, the manufacturer declares the performance of the construction product
- Verification of Constancy of Performance corresponds in the set of on-going activities to ensure that the manufactured construction products have the performance declared.

3.2. *Assessment of Performance*
The Assessment of Performance shall be repeated only in case of changes which could affect the conformity of the product with the declared performance, e.g.
- changes to the construction product, its constituents, the manufacturing equipment, the Factory Production Control (FPC) system and/or the manufacturing process; and/or
- changes to the harmonised specification with regard to methods and criteria for the assessment of performance, including changes to supporting standards called up by the harmonised specification.

3.3. *Verification of constancy of Performance*
The Verification of constancy of Performance is based on the Factory Production Control (FPC). The FPC corresponds to the documented, permanent and internal control of production in a factory, in accordance with the relevant harmonised technical specifications.

The objective is to ensure the constancy of performance of the manufactured construction products. Manufacturers should not neglect the practical implementation including personnel, equipment and other resources used for the controlling of the production (not only documentation) and the effectiveness of the FPC with regard to the objective.

3.4. *Responsibilities*
The Commission determines the AVCP systems in accordance with CPR Article 28(2); the manufacturer remains solely responsible for the conformity of the construction product with the declared performance and for the compliance with any other requirement defined by CPR; the tasks for manufacturers and the notified bodies are specified in CPR, Annex V; the tasks for notified bodies are specified in the GNB's document 17/722 (regularly updated).

The different systems of AVCP are presented in Table 2.

Table 2. Different systems of Assessment and Verification of Constancy of Performance (AVCP)

TASKS	System				
	1+	1	2+	3	4
PRODUCER					
Declaration of performance	yes	yes	**yes**	yes	yes
Assessment of product performance			x		x
Factory production control (FPC)	x	x	x	x	x
Further testing samples	x	x	x		
NOTIFIED BODY					
Assessment of product performance	x	x		x	
Initial inspection of factory + FPC	x	x	x		
Continuous surveillance + assessment + evaluation of FPC	x	x	x		
Audit testing samples	x				

For Geosynthetics Mandate M/107 (1994) amended by M/386 (2006) define the AVCP system 2+. This means that there is:
- NO Assessment of performance carried out by the Notified Body

- NO Audit-testing of samples taken by the Notified Body at the manufacturing plant or at the manufacturer's storage facilities.

3.5. The citation of hENs for construction products in the Official Journal of the European Union

After the elaboration of the standards on Geosynthetics, the Commission assesses the conformity of harmonised standards established by the CEN TC189 with the relevant Mandates.

The importance of the decision to cite hENs in the OJEU is due to the legal effects of the citation:
- from the date of the beginning of the coexistence period it shall be possible to use a harmonised standard to make a declaration of performance for a construction product covered by it
- from the date of the end of the coexistence period, the harmonised standard shall be the only means used for drawing up a declaration of performance for a construction product
- at the end of the coexistence period, conflicting national standards must be withdrawn, and Member States must terminate the validity of all conflicting national provisions.

3.6. The important obligations for Member States

A Member State shall not prohibit or impede, within its territory or under its responsibility, the making available on the market or the use of construction products bearing the CE marking, when the declared performances correspond to the requirements for such use in that Member State.

A Member State shall ensure that the use of construction products bearing the CE marking shall not be impeded by rules or conditions imposed by public bodies or private bodies acting as a public undertaking, or acting as a public body on the basis of a monopoly position or under a public mandate, when the declared performances correspond to the requirements for such use in that Member State.

4. Short synthesis on the CPR and CE marking

4.1. Some simple statements on the Legal Framework (CPR and CE marking):
- Essential requirements (ERs) apply on the Construction products (like Geosynthetics)
- CE Marking shows that the product satisfies the Essential requirements (ERs), which are expressed via the Essential characteristics
- Basic Requirements apply on the works using the products:
  in the case of Geosynthetics the products shall provide the following Basic requirements:
  (1) Mechanical resistance and stability,
  (3) Hygiene, health and environment (through the topic Dangerous substances studies through the CEN TC351) and
  (7) Sustainable use of natural resources
- Declaration of Performance (DoP) is defined in the hENs
- hENs must be used by manufacturer to declare performance of the products in the DoP (at least one essential characteristic shall be declared)
- CE marking shows conformity with declared performance according to the Assessment, Verification of Constancy of Performance system defined in the Mandate.

For Geosynthetics this means that the DoP is declared by the producer using the hENs in accordance with the CPR. Nevertheless, it shall be kept in mind that NO Assessment of performance is carried out by the Notified Body and that NO Audit-testing of samples are taken by the Notified Body at the manufacturing plant or at the manufacturer's storage facilities.

4.2. Some complementary statements to understand CE Marking:
- The CE marking shows that the product can be used if the declared performance corresponds to the requirements for such use in the Member State
- It is the manufacturer who undertakes the responsibility for the DoP and affixes the CE Marking on the product

- This is not the Notified Body who gives the CE Marking to the manufacturer
- This is the design engineer who must specify the product with the required performance and the contractor who must use the product which has the performance specified by the design engineer.

5. CE Marking (DoP) in practical use of Geosynthetics

5.1. CE Marking for the placement of Geosynthetics on the market

A shown previously (paragraphs 1., 2.1, 4.2), CE Marking is required for the placement of Geosynthetics on the market.

The CE marking product shows conformity with the declared performance and compliance with all applicable requirements laid down in the Regulation 305/2011 and in other relevant EU legislation.

5.2. Need of specific requirements defined by the users for specific construction jobs

As expressed by Winnepenninckx (2019), historically in 80's the experts have written European standards without influence of the European Commission. When CPD has been implemented, the annex ZA (rereferring to the clauses and requirements linked to CE Marking) has been added at the end of the standards. The European Commission rules express that the standards are expected not to exclude products on the market in EU Member States: this means that some thresholds (and certain voluntary characteristics) have to disappear from the standards.

With time, due to the requirements linked, first to the CPD, and then to the CPR requirements, the primary European product standards have changed to standards dealing only with the legal requirements linked to the CPR (terminology, relevant characteristics, thresholds and classifications), rendering European product standards less useful for other uses than for the CPR.

This statement shall not be misunderstood: CE Marking is extremely important for the internal market of the European construction products which corresponds approximatively to 500 billion € and CE marking permits free circulation of construction products in more than 30 countries, but CE Marking don't give information on their fit for purpose for the specific works.

Nevertheless, as shown by E. Winnepenninckx (2019), the European Construction sector's is not only selling construction products, but it is doing more than placing them on the market. The Construction sector does something more with these construction products (like building structures, etc.) and the European construction sector represents approximatively 1.500 billion € and the international activities of European construction sector represents approximatively 100 billion €. In this respect, for its competitiveness, the European construction sector needs European product standards which satisfy its needs which are larger that the ones linked to the CPR.

It can be noted that presently discussions are ongoing between the European Commission and the CEN to evaluate the best way to take care of all the needs expressed: the ones linked to the CPR and the ones coming from all stakeholders in the Construction process. In this respect, the CENCENLEC Conference "Harmonised standards supporting the Construction Products Regulation (CPR)" (Brussels, 4 December 2019) was an important opportunity of exchanges on this topic.

5.3. Specific requirements defined by the users for specific construction jobs using Geosynthetics

Concerning specifically Geosynthetics, and based on the experience of more than 40 years of application of these products, in Europe and in the world, it appears that the users have, with the time, precised and expressed their requirements in term of quality control of the performance of the products (Delmas et al., 2014).

These needs have led to the implementation of several product quality certification systems in different European countries. Nevertheless, at this stage it is important to evaluate these "users" requirements in the context of the CE marking under the CPR. Some of these requirements are linked to the performance of the products and its verification.

As example, the following points may be raised as they are not covered by the present harmonised product standards (hEN) on Geosynthetics:

(a) there is no verification of the performance of the product by the Notified Body, his task is the verification of the Factory Production Control and its continuous surveillance, assessment and evaluation, but not of the values announced (see paragraph 5.1)

(b) there are no regular and random controls planned (see paragraph 5.1) and

(c) the harmonised characteristics don't cover all used characteristics.

Some other "users" requirements are linked to the organization and the qualification of the main actors of the CE certification system. As example, it can be mentioned that there are not specific requirements for the laboratories making the performance evaluation (e.g. accreditation); as an example, when testing geotextiles and related products, the high dispersion of the results obtained between several European laboratories have shown the need of establishing round-robin testing procedure between the laboratories.

These differences between the present definition of the CE marking for Geosynthetics and the needs of the users can certainly be explained firstly by the type of system of Assessment and Verification of Constancy of Performance (AVCP) chosen in the Mandate (2+), where the users and the market seems to need a system closer to (1+). Secondly, the fact that, in the present harmonised product standards (hENs), the harmonised characteristics, as defined in the Mandate, don't cover the so called "voluntary" characteristics explains the needs expressed by the users for some complementary quality systems.

The Figure 1 shows the different areas (characteristics / systems of (AVCP) assessment and verification of constancy of performance) covered by:

- the CE Marking according to the Mandates M/107 amended by M/386 (Figure 1a).
- the so called "voluntary" part covered by the CEN TC189 (Figure 1b); this part, not covered by the CE Marking, corresponds to the characteristics relevant to all conditions of use (A) and the ones relevant to specific conditions of use (S), not included in the harmonised characteristics;
- the areas corresponding to a system of assessment and verification of constancy of performance (AVCP) higher than 2+ (e.g. 1+), for all characteristics harmonised (Figure 1c)
- or voluntary (Figure 1d).

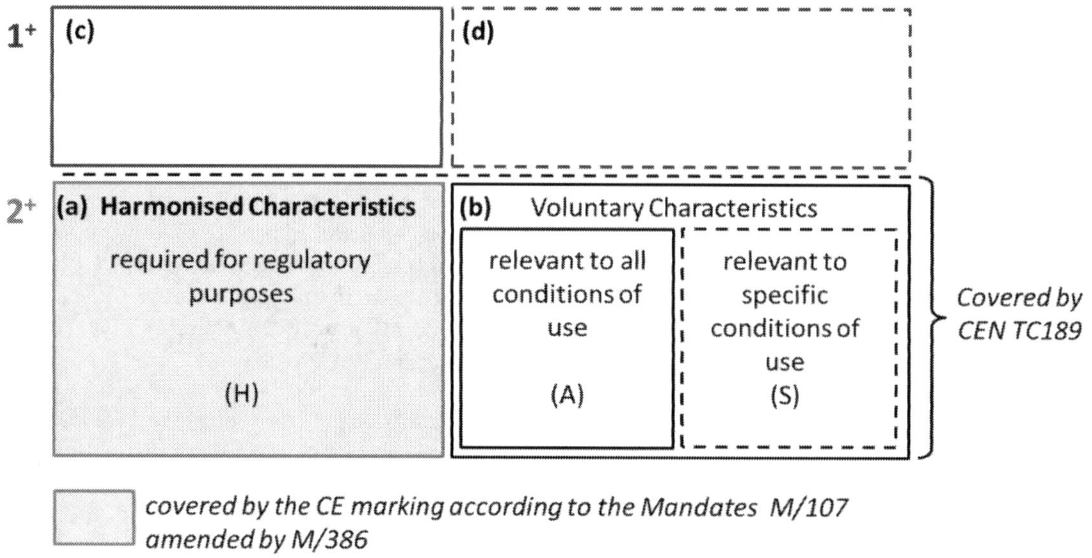

Figure 1. Different areas (characteristics / systems of (AVCP) assessment and verification of constancy of performance for geotextiles and related products (e.g. 2+ and 1+).

It is clear that the area not covered by the CE Marking, according to the present Mandates, depends on both the level coverage of the characteristics by the harmonization and the type of system of Assessment and Verification of Constancy of Performance defined by the Mandates.

This explains the existence in some areas in Europe of Specific Quality Marks for the control of incoming products and for the control of the fit for purpose. Nevertheless, these complementary Specific Quality systems have the obligation to be in accordance with the CPR.

6. Practical experience: responsibilities of manufacturers and of the users (owners, contractors, etc.)

It shall be remembered that it is up to the manufacturer to undertake the responsibility for the DoP (CE Marking) and that the manufacturer remains solely responsible for the conformity of the construction product with the declared performance and for the compliance with any other requirement defined by CPR. In the case of Geosynthetics, this means that there is:

- NO Assessment of performance carried out by the Notified Body
- NO Audit-testing of samples taken by the Notified Body at the manufacturing plant or at the manufacturer's storage facilities.

This confirms that the CE marking is not and can't be considered as a Product Quality Mark and that CE marking don't cover characteristics not mentioned as Essential characteristics, also if they are required by the user.

On another hand, this is the user responsibility to define the requirements on the products for the application/function in his specific project.

This is in line with the basic principles of the CPR (paragraph 2.1):

- each Member States remains competent to define for construction works the level of the Basic Works Requirements to be satisfied in his territory; this can easily be understood considering the large area of Europe with total different climate conditions (from permafrost areas up to 50°C), with different natural constructions materials (clayey soils, gravel, etc.) and the different traffic and loading conditions for the different types of structures.
- the product performance declared by the manufacturer allows engineers / authorities to verify that construction works built with these products will satisfy the Basic Requirements for Construction Works as applied in the Member State.

7. Principle of the market surveillance (policing of the CE system on national markets) and the complementarity use of quality systems in addition to CE Marking

According to the CPR, for the purposes of ensuring an equivalent and consistent enforcement of Union harmonisation legislation, effective market surveillance should be operated by the Member States, as defined in the chapter VIII of the CPR.

Until now for Geosynthetics, the effectiveness of the control of the CE Marking on the Market by the Member states seems not really proven for much Member states, also if some case of misuse may be reported. Parallel, as expressed before (paragraph 6), for Geosynthetics the Declaration of the DoP is solely the responsibility of the manufacturer, without any control by an external independent body (no Assessment of performance carried out by the Notified Body and no Audit-testing of samples taken by the Notified Body at the manufacturing plant or at the manufacturer's storage facilities).

This explain the reason why the Construction partners (authorities, manufacturers, distributors, public and private procurers, designers, specifiers, engineers, contractors, installers, conformity assessment bodies, construction works owners, clients, research institutes, universities) in many countries in Europe support the complementary existence of "Specific Quality systems", like for instance, but not exhaustively, ASQUAL in France, NorGeoSpec in Norway, Sweden, Finland & Estonia, IVG in Germany. These systems include

Nevertheless, "Specific Quality system" and CE marking shall not be mixed and the "Specific Quality system" shall, in any way, not be in contradiction with the CPR which is mandatory (Table 3).

The consequences are:
- the results of product testing by "Specific Quality system" certification and random testing will be compared only with the "Specific Quality system" control limits.
- the manufacturer is flexible in changing the DoP, as long as the "Specific Quality system" control limits are fulfilled; otherwise, the "Specific Quality" certificate will not be valid anymore.

The manufacturer has in any case the full control of his DoP and of the corresponding expressed values.

Table 3. Complementary of CE Marking and "Specific Quality system"

CE marking (DoP)	"Specific Quality system"
Responsibilities	Responsibilities
manufacturer undertakes responsibility for DoP	*user* undertakes responsibility for requirements on products for specific application, function and project
DoP	"Specific Quality system" certificate
manufacturer expresses the values mean values and tolerance values (95% confidence level)	*users* express the values to fulfil the "Specific Quality system" requirements for *specific projects* *These values may express* · *the quality requirements for the specific project* · *the technical specification for the specific project*
Values mentioned on the DoP is the *responsibility of the manufacturer*	Verification by tests of values mentioned on the DoP is *not the task of* "Specific Quality system"
Verification of the FPC is the *responsibility of Notified Body*	Certification body *checks only* that the *"Specific Quality system" requirements* are fulfilled

As a conclusion, in the "Specific Quality system" Certificate, mean values shall correspond to DoP mean values. In the "Specific Quality system" Certificate, tolerances correspond to specific requirements of "Specific Quality system" and as "Specific Quality system" characteristics can't be in contradiction with DoP, the "Specific Quality system" tolerances of the product shall be equal or smaller than the ones in the DoP.

8. Conclusion

The paper firstly highlights the main principles of the CPR and of the harmonised European Standards: solid technical basis for manufacturers for testing the performance of their products, describing the expression of performance versus the essential characteristics.

It also presents a rapid overview of the general rules imposed by the CPR, like (1) Mandate/Answer to the Mandate/Standardisation Request, (2) Position versus references to national requirements or (3) Assessment method(s).

In the second part, the communication has discussed how to deal with CE Marking (DoP) in practical use of Geosynthetics: (1) need of CE Marking for the placement of Geosynthetics on the market and (2) need of specific requirements defined by the users for the construction jobs, for control of incoming products and for fit for purpose.

It has been remembered that it is up to the manufacturer to undertake the responsibility for the DoP (CE marking) and it is the user responsibility to define the requirements on products for specific application, function and project.

The principle of the market surveillance (policing of the CE system on national markets) and the complementarity use of quality systems in addition to CE has been presented, while it has to be accepted, that the policing of the CE system on national markets is more in the beginning and not an implemented standard today.

References

[1] CPR 2011 *Regulation (EU) N° 305/2011 of the European Parliament and of the Council of 9 March 2011*, published on 4th April 2011 in the Official Journal of the European Union p 39.

[2] Katsarakis G 2018 *CPR: the legal perspective. Presentation at the CPR Workshop Harmonized European Standards under the construction Products Regulation (CPR) – Essentials for drafting standards"* 22 October 2018 Brussels, Belgium p 18

[3] M/107 1994 *Mandate to CEN/CENELEC concerning the execution of standardization work for harmonized standards on geotextiles – European Commission – 1996* p 16 modified by mandate M/386 2006 p 3

[4] Winnepenninckx E 2019 *Harmonised standards supporting the CPR; European standards supporting the construction sector. Presentation at CENCENLEC Conference the Harmonised standards supporting the Construction Products Regulation (CPR)* 4 December 2019Brussels, Belgium p 18

[5] Delmas P, Mailler P, Recker C 2014 *Geosynthetics quality systems: evolutions with the implementation of the construction product regulation* - 10th ICG, DGGT & IGS Berlin, Germany p 7

Design guide for geosynthetic barrier systems as groundwater protection in road construction

K P von Maubeuge[1], A Müller-Kirchenbauer[2] and C Schlötzer[2]

[1]Naue GmbH & Co. KG, Gewerbestraße 2, 32339 Espelkamp, Germany
[2]MKP Müller-Kirchenbauer Ingenieurgesellschaft mbH, Eilveser Hauptstraße 56, 31535 Neustadt, Germany
Email: kvmaubeuge@naue.com; amk@mkp-ingenieurgesellschaft.de

Abstract. Geosynthetic clay liners (GCL) are used for groundwater protection in road constructions in water protection areas, RiStWag [1]. Geosynthetic clay liners (GCLs) are listed as an approved barrier and a long-term permittivity of $\leq 1 \cdot 10^{-7}$ [1/s] is required as an essential characteristic value. In order to ensure the effectiveness of the sealing system during the entire service life, however, further characteristic values should be described. In the following, the protection concept of RiStWag is explained and the effectiveness, reliability, cost-effectiveness and control as well as the maintenance of the road systems are described in more detail.

1. Introduction

Solid, liquid and gaseous emissions from motor vehicles contribute to risks to the soils, stretch of water (rivers, lakes, ponds, etc.) and groundwater. The protection of these waters is anchored in the German Water Resources Act. Declared zones, especially public water catchment facilities and state-approved mineral springs require special protection. Water extraction areas and therapeutic springs and their declared zones are therefore designated as water and therapeutic spring protection areas.

Water protection areas (WPA) for groundwater and drinking water reservoirs are generally divided into three protection zones. Their size is derived from the local hydrogeological conditions. Protection zone I (extraction area) encompasses the immediate vicinity of the well or spring. Protection zone II (narrower protection zone) corresponds to an area around the well or spring within which groundwater takes up to 50 days to reach the well. Protection zone III (wider protection zone) covers the entire groundwater zone and can be subdivided into sub-zones III A and III B. Required regulations applicable to the respective protection zones are also stated in the German Ordinance on Protected Areas.

If it is not possible to avoid a road route through a water protection area, despite all the considerations in the line identification process, a number of measures are necessary in order to exclude a risk to the water resources as far as possible. They are described in the German Guidelines for Structural Measures on Roads in Water Protection Areas, RiStWag [1].

On the one hand, traffic engineering measures are planned, such as simplified road layout and the use of crash barriers at the edge of the road. In addition, geotechnical measures are included to ensure temporary protection against the effects of accidents involving water-polluting substances in order to achieve a sufficient time buffer until the occurred damage can be removed. If the retention effect of the existing subsoil is not high enough, the use of geosynthetic barrier systems is necessary. In the case of near-surface barrier systems (typically approx. 0.80m deep), structure penetrations such as man holes, collectors and foundations must be water-tight connected. In particular, the connection to concrete structures can be easier achieved with geosynthetic clay liners than with geomembranes. This design guide provides the designer with valuable information on the selection and dimensioning of geosynthetic clay liners (GCLs).

2. The protection concept of RiStWag

2.1. *The risk-based prevention concept*

The prevention concept of the German RiStWag is designed to ensure permanent protection against permanent and temporary material penetration from road construction and road traffic and temporary protection against the effects of accidents involving water-polluting substances. The factors "protection zones of the WPA", "retention effect of the groundwater cover soil", "barrier efficiency of the sealing system as a structural measure", "risk of release of pollutants" and "severity of possible consequences for the drinking water extraction" are jointly incorporated into a risk-based precautionary concept. Complete protection would be neither economically nor ecologically justifiable [2].

2.2. *Retention effect of the groundwater cover soil*

The retention period of the environmentally unfriendly substances penetrating into the unsaturated zone (groundwater cover) and their three-dimensional spreading in the saturated zone essentially depend on the permeability and structure of soil material as well as on the distance to the groundwater (Table 1). The retention effect of the groundwater cover soil is related to the combination of soil permeability and soil thickness.

Table 1. Retention effect of the cover soil over the groundwater [1]

Soil permeability	Layer thickness	Retention effect
$k_s < 1{\cdot}10^{-7}$ m/s	> 2m	Large
	1m – 2m	Medium
	< 1m	Low
$k_s < 1{\cdot}10^{-6}$ m/s to $1{\cdot}10^{-7}$ m/s	> 4m	Large
	2m – 4m	Medium
	< 2m	Low
$k_s < 1{\cdot}10^{-4}$ m/s to $1{\cdot}10^{-6}$ m/s	> 8m	Large
	4m – 8m	Medium
	< 4m	Low
$k_s < 1{\cdot}10^{-3}$ m/s to $1{\cdot}10^{-4}$ m/s	> 15m	Large
	5m – 15m	Medium
	< 5m	Low
$k_s > 1{\cdot}10^{-3}$ m/s	-	Low

2.3. *Structural measures – Wider protection zone III*

In zone III (wider protection zone), the retention effect of the groundwater cover soil and the amount of daily traffic are taken into account in order to determine the required protective measures, which lead to different solution options for the rain water and road run-off collection measures (Table 2). A distinction is made between levels 1 (lowest) to 3 (highest).

In level 1, the protective effect of the soil as groundwater cover is sufficient enough, so that no other constructive measures are required. The road surface water should be drained or spread over a wide area over the banquets and embankments. In level 2, the retention effect of the groundwater cover is less effective than in level 1. Therefore, additional constructional measures are required (e.g., Figure 1). The rainwater that runs off from the traffic areas at the lower edge of the road with the transverse slope is to be collected and led out of the water protection area. In level 3, the protective effect of the groundwater cover soil is not efficient enough. The structural measures for the protection of the groundwater largely correspond to those of level 2. However, the artificial barrier must be continued beyond the water collection ditch and an inclination towards the ditch must be constructed. In the case of terrain incisions, the entire area under the traffic area must always be sealed.

Table 2. Retention effect (large/medium/low) of the cover soil over the groundwater and the necessary design levels for constructions in zones III [1]

Average daily traffic [vehicles/24h]	Zone III/ IIIa Protection level of soil over groundwater level			Zone IIIb Protection level of soil over groundwater level		
	Large	Medium	Low	Large	Medium	Low
< 2,000	1	1	1	1	1	1
2,000 – 15,000	1	1	**2**	1	1	**2**
More than 15,000	1	**2**	**3**	1	1	**2**

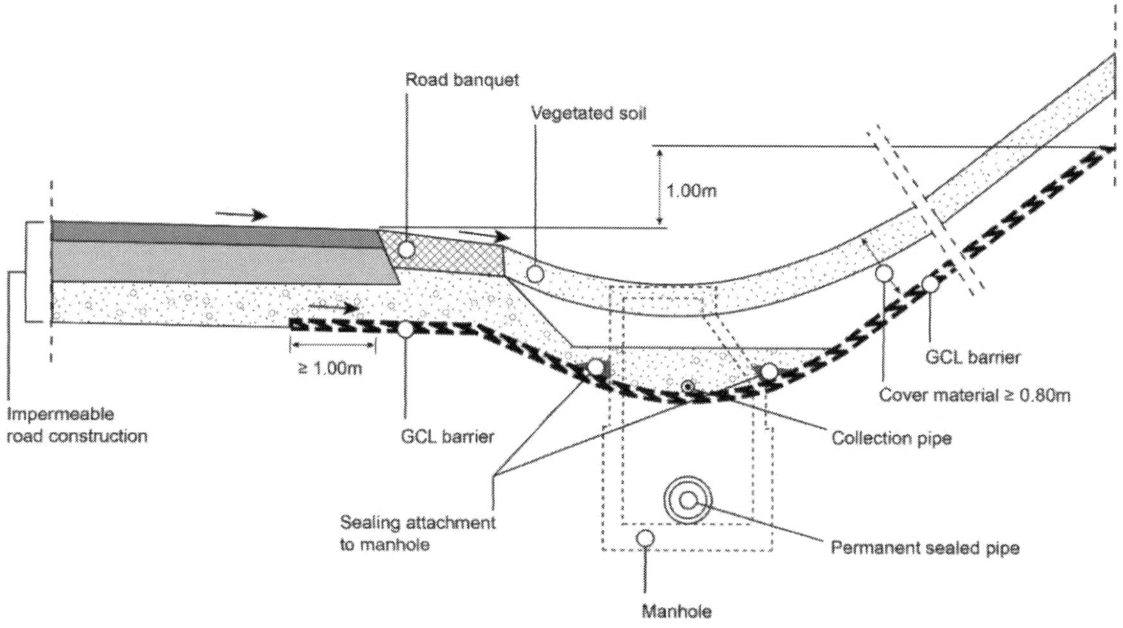

Figure 1. Zone III – Groundwater protection, terrain incision, level 2/3 (cross-section 5b) [1]

2.4. *Structural measure – Narrower protection zone II*

If a road layout through zone II (narrower protection zone) cannot be avoided, all necessary design and constructional measures must already be considered in the planning phase (e.g., Figure 3). In general, passive protective devices (crash barriers, concrete walls, car run-off barriers as in Figure 2) are necessary to prevent vehicles from running off the roadway. The banquets have to be designed rigid and stable. Rainwater must be collected via side drains, gutters or pipes and moved out of the protection zone.

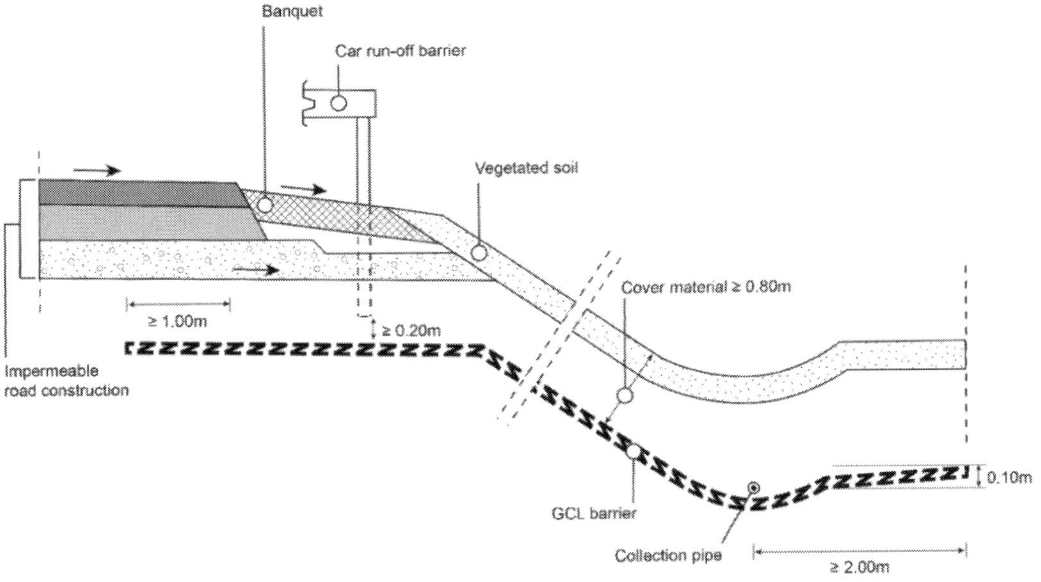

Figure 2. Zone III – Groundwater protection, dam construction, level 2/3 (cross-section 4d) [1]

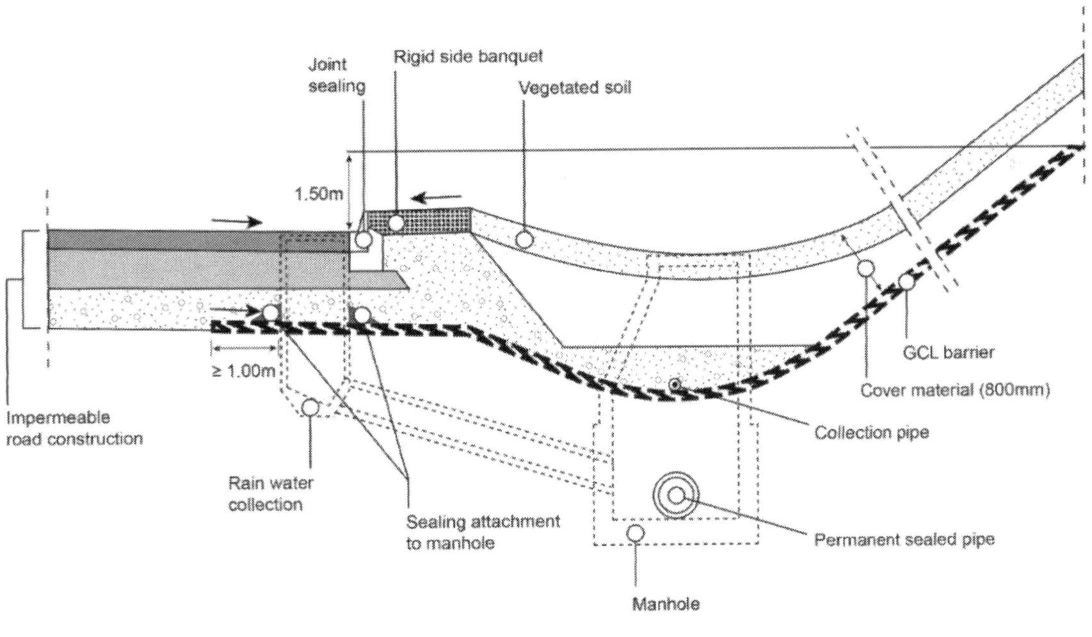

Figure 3. Zone II – Groundwater protection, terrain incision (cross-section 8b) [1]

The artificial barrier, e.g., geosynthetic clay liner (GCL), should be lined at least 1m under the paved road. Should the cover soil over the groundwater only fulfill level 1 itself, then the barrier should be lined entirely under the paved road and run under the side drain ditch and continue for another 4m into the side terrain, with a slope inclination towards the ditch of at least 10%. If a terrain incision of larger than 2m is necessary for the road construction, the artificial barrier has to be installed sideward at least 1.5m higher than the top side of the road pavement.

2.5. *Structural measure – Water extraction area*

In general, road and associated structures are not allowed to be built in protection zones I (water extraction area). If this is not possible, groundwater extraction facilities shall be closed and built elsewhere. In the case of drinking water reservoirs, maximum water protection measures are necessary.

2.6. *Structure of a geosynthetic clay liner*

Geosynthetic clay liners (GCL; or GBR-C for Geosynthetic Barrier - Clay) - also known as bentonite mats – are needle-punched, reinforced composites that combine two durable geotextile outer layers and an erosion-stable, uniform core of high-swelling powder sodium bentonite clay (Figure). This construction forms a shear-resistant hydraulic barrier with self-sealing and re-healing characteristics. When hydrated with fresh water, the bentonite swells and forms a low permeability gel layer, with a hydraulic performance equal to or better than traditional, thick compacted clay liners.

The needle-punching (also known as fibre reinforcement) firmly bonds the three layers of the GCL creating a high internal shear strength between all components which allows steeper or safer designs on slopes.

Sodium bentonite is characterized by the lowest known permeability of naturally occurring geological materials and a high absorption capacity. For GCLs the swelling capacity and "self-sealing" characteristic of the sodium bentonite core truly separate them from other barrier materials. Swelling occurs during contact with moisture and fresh water. If the GCL is punctured during construction, this swelling reaction is known as self-sealing and is a characteristic not found in any other single or composite geosynthetic material and unlikely to be found in traditional clay barriers. This geosynthetic/clay composite material creates a single, engineered, high-performance GCL barrier that utilises the advantages of both synthetic and natural materials.

A comparable sealing efficiency of a compacted clay liner (CCL) and a geosynthetic clay liner (GCL) can be assumed if they have the same permeation rate or permittivity. The permittivity ψ is the amount of water V_W passing through the GCL per time unit Δt, height difference Δh and area unit A. Considering a permeability value $k \leq 5 \cdot 10^{-9}$ m/s and a thickness of $d \geq 0.5$m for a CCL, the continuity equation gives at a constant water of 0.3m a permeation rate of $q \leq 8 \cdot 10^{-9}$ m³/(m²·s). This then allows to calculate the permissible permittivity ψ_{per}:

$$\psi_{per} = V_W / (A \cdot \Delta t \cdot \Delta h) = q / \Delta h = 2.7 \cdot 10^{-8} \text{ [1/s]} \quad (1)$$

Typically, commercially available GCLs have a permittivity of approx. $5 \cdot 10^{-9}$ [1/s] when delivered and are therefore more effective (impermeable) than a mineral clay seal (as calculated above - ψ_{per}).

The success of GCLs has been proven by hundred million square metres of installed material. However, there might be situations where an improved GCL is the right choice, e.g., if gas can flow through in the first phase of an application when the bentonite is not yet hydrated; or desiccation can influence the performance of the GCL's bentonite layer. This is not an issue for every installation, but is for some, e.g., waterproofing in areas where radon gas occurs, applications with low confining stresses, or with ponds, lagoon, and canal applications. The most recent GCL development is a multi-component GCL where a polyethylene barrier is additionally attached to the GCL. These multi-component GCLs are an advance for GCLs in some specific applications.

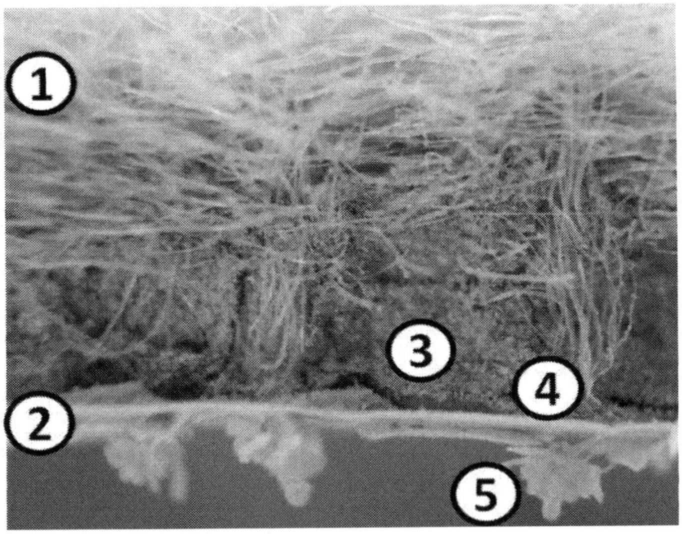

(1) Cover layer: nonwoven geotextile
(2) Carrier layer: woven geotextile
(3) Sealing bentonite layer
(4) Shear-strength transmitting fibre reinforcement
(5) Additional thermal locking of needle-punched fibres

Figure 4. Cross-section of needle-punched geosynthetic clay liner

2.7. *General information on GCLs*
In areas of roads and railways, barrier systems for groundwater and environmental protection have the task of protecting the subsoil and groundwater from water-polluting substances.

As a basis for the evaluation of geosynthetic clay liners (GCLs) with regard to the recommended requirements, the target issues of performance efficiency, system reliability, economic efficiency, local characteristics and operational aspects should be considered.

2.8. *GCL performance efficiency*
According to the German RiStWag guideline, the GCL must have a long-term permittivity of $\psi_{GCL} \leq 2.7 \cdot 10^{-8}$ [1/s] during the expected service life in the structure. Converted to a typically 1cm thick bentonite layer (based on a sodium bentonite mass per unit area of approx. 4,000g/m^2), this corresponds to a permeability coefficient of the GCL of $k \leq 1 \cdot 10^{-9}$ m/s (tested according to [3]).

However, it is also necessary to investigate the performance efficiency of the barrier system with regard to possible effects against dry-wet changes, frost-thaw cycles, exposure to salt solutions and hydrocarbons or other relevant liquids. Therefore, when using any sealing system, e.g., a GCL, the required long-term permittivity must also be achieved against other media such as petrol, heating oil, diesel fuels and salt solutions.

At present, efforts are being made at the European level to standardise the long-term permittivity of a GCL against a liquid salt solution. This standard will be based on Annex 3 of BQS 5-5 [4], so that the testing of such solutions can be done in accordance with that Annex 3 method.

2.9. *GCL Reliability*
In order to achieve a reliable, permanently efficient barrier system, further requirements are necessary in addition to prove performance efficiency.

According to the German guideline RiStWag [1], the minimum soil cover thickness over the GCL is stated with 0.80m. This thickness is considered as sufficient to protect the GCL from desiccation and frost effects.

A multicomponent GCL is a GCL with an attached film, coating, or membrane decreasing the hydraulic conductivity or protecting the clay core or both [5]. It can be added, e.g., as a coating (a layer of a synthetic substance which is allowed to solidify onto the GCL). This additional barrier can significantly increase the performance and reliability of the sealing system. This GCL development opens more applications where GCLs can be used in and significantly improves its performance. Advantages of the polymer coated barriers are [6]:

- Prevention of root penetration
- Increasing resistance against desiccation
- Bentonite piping resistance under high water gradients
- Lower permeability
- Barrier against ion exchange
- Durable and immediate gas barrier

To ensure the long-term performance of a polymer coated GCL, the following design issues might be of concern and should be considered prior to the installation of the multi-component GCL [6]:

- Durability of the coating
- Resistance against installation stress
- Overlapping of polymer-coated GCLs
- Transmissivity between coating and GCL
- Interface and internal shear stress
- Durable peel performance of the coating barrier
- Resistance against damage during installation (optional with a protection nonwoven)

The geosynthetic clay liner (GCL) must be installed with the polyethylene (PE) coating side facing up as only then it can act as a protection of the bentonite layer against desiccation and as a root barrier. Further, the PE coating layer increases the efficiency of the bentonite against possible, non-naturally occurring critical media, e.g., leakage of pollutants in the event of a traffic accident or accident.

The application of a GCL is considered as uncritical, if the soil pH value is within a range of 5 - 11. In any other cases an investigation of the suitability of the GCL is recommended.

Barrier systems are not designed to handle downhill tensile forces like a geogrid. For that reason, tensile strength properties for a GCL are more a quality control measure and are only needed to handle installation stresses. The recommended minimum tensile strength (mainly for installation purposes) should be $\geq$ 10kN/m.

When installed, the GCL can adequately follow sub-soil deformations. Without reducing the sealing performance, needle-punched GCLs can deform to 10% in machine and cross-machine direction.

If the GCL is used on slopes, it is necessary to investigate all friction properties (internal and interface). This is also valid for multicomponent GCLs. In the case of, e.g., PE extrusion coated GCLs, it may be necessary that the PE coated side has to have a surface structure for shear strength purposes.

In applications with sloped embankments, all shear forces parallel to the embankment must be investigated both on the interface as well as in the GCL (internal shear strength) in the hydrated state of the bentonite. Alternatively, if the necessary shear strengths are not achieved, geogrid reinforcement can be used to ensure slope stability.

Unfortunately, the current version of the German RiStWag guidelines [1] does not list any further GCL requirements. The guideline rather refers to the concept that the designer should follow the approach on looking over the entire situation and consider the following evaluation criteria (further details listed in Figure 5):

- Effectiveness of the system
- Performance reliability
- Cost effectiveness
- Road system - control and maintenance

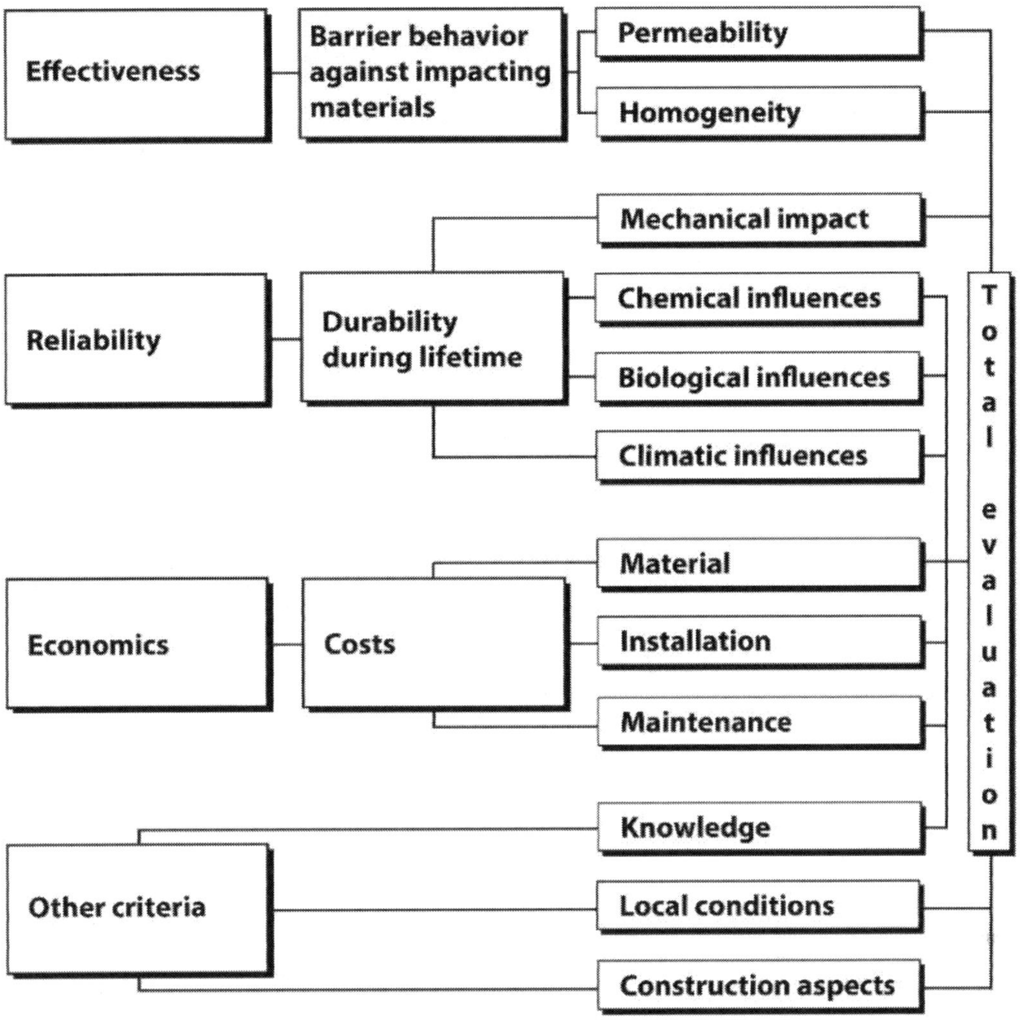

Figure 5. Evaluation criteria for a construction in groundwater protected areas (translated from [1])

Therefore, it is recommended to set minimum requirements for a GCL in order to ensure long-term performance. The following listed specification values are therefore recommended:

Geosynthetic clay liner (GCL):

- GCL panel consisting of two bentonite-encapsulating geosynthetics: a polypropylene (PP) carrier woven (≥ 100g/m²) and cover nonwoven (≥ 200g/m²)
- layer of sodium bentonite powder encapsulated between two geosynthetic layers; bentonite mass per unit area ≥ 3,700g/m² at max. 15% water and a free swell value ≥ 24ml following ASTM D5890 [11]
- uniform shear-stress transferring needle-punching through the two bentonite-encapsulating geosynthetics creating a shear strength transmitting GCL with a peel strength [7] of ≥ 360N/m
- GCL tensile strength ([8] or [9]) longitudinal/transverse 10kN/m
- self-sealed overlap width 30cm
- required service life over 25 years
- permittivity max. $5 \cdot 10^{-9}$ [1/s] ([10] or [3])

GCL with additional polymer-coated barrier:

- PE extrusion-coated geosynthetic clay liner (GCL) consisting of two bentonite-encapsulating geosynthetics: a polypropylene (PP) carrier woven (≥ 100g/m²), cover nonwoven (≥ 200g/m²) and a smooth PE barrier (≥ 200g/m²) firmly attached to the woven geosynthetic
- in inclined embankment areas the PE barrier firmly attached to the woven geosynthetic should be structured and have a mass per unit area of ≥ 500g/m²
- layer of sodium bentonite powder encapsulated between two geosynthetic layers; bentonite mass per unit area ≥ 3,700g/m² at max. 15% water and a free swell value ≥ 24ml following [11]
- uniform shear-stress transferring needle-punching through the two bentonite-encapsulating geosynthetics creating a shear strength transmitting GCL with a peel strength [7] of ≥ 360N/m
- GCL tensile strength ([8] or [9]) longitudinal/transverse 10kN/m
- self-sealed overlap width 30cm
- required service life over 25 years
- permittivity max. $5 \cdot 10^{-9}$ [1/s] ([10] or [3])

2.10. *GCL – Economic efficiency and ecological aspects*

The comparative energy balance of different barrier systems shows that compacted clay liner systems require high energy consumption for extraction, transport and installation due to their large masses. In contrast, geosynthetic clay liners generally require comparatively little energy for production, transport and installation [12].

From the point of view of a comparative energy balance, the decisive criterion for the decision between the creation of a seal with GCLs (bentonite mats) or compacted clay liners is the availability of suitable clay material in the vicinity of the construction site. The causative factors of the higher energy consumption of the compacted clay liner clearly result from the necessary use of trucks to transport the mineral sealing material and downstream of the installation vehicles, whereby the energy consumption of the installation vehicles is to be regarded as approximately fixed. The transport-related energy consumption decreases nearly linearly with a shorter distance between the paving site and the installation site.

In the case of the GCL, the transport distance from the manufacturer to the construction site has only a comparatively small influence on the energy balance due to the low specific mass of the GCL. The paver's energy consumption is also of little significance. The main balance factors for GCLs are the high energy content (feedstock) of the high-quality plastics polyethylene and polypropylene and the manufacturing process. For bentonite, the energy input for drying and transport is the largest factor.

Egloffstein et al [12] compared a geosynthetic clay liner (GCL) with a 625mm thick compacted clay liner for an area of 36,000m² and calculated a lower cumulated energy input (CEI) for the GCL (71.8MJ/m²). The CEI for the compacted clay liner was 122.3MJ/m² (70% higher than with a GCL). The average clay transportation distance of 35km was assumed for the compacted clay liner, totaling to 45,000t of clay for the project. The transportation distance resulted to be the largest share for the total calculated CEI. In contrast, the transportation distance for the geosynthetic clay liner (GCL) from the manufacturing plant to the construction site was 580km and did not have the largest impact for the GCL CEI calculation.

The CO_2 output shows a similar percentage when comparing the two barrier systems. The produced CO_2 emission for the GCL was calculated with a bentonite mass per unit area of 4.0kg/m². The compacted clay liner had a 2.58 higher CO_2 output (9.9kg/m²) than the GCL solution. Both barrier systems had the same cover soil thickness of 1m, which had a CO_2 output of 7.9kg/m². Therefore, a design with a CO_2 output could be a solution with only 500mm of cover soil, but to reduce any possible risks to the sealing system a feasible solution could only be realised with a multi-component GCL.

2.11. *Control and maintenance during service life*

The operation of roads in water protection areas also includes continuous monitoring and maintenance after construction. In terms of the barrier, this includes regular checks during road maintenance (at least four times a year). Attention should be paid to possible signs of damage to the barrier element, e.g., damage to the cover, ramming or boreholes, growth of larger shrubs or deeper small animal enclosures.

In the event of an accident, an inspection must always be carried out on site to ensure that the sealing system is also permanently effective after the accident.

2.12. *Notes on installation*

GCL products may only be used with valid CE marking. Overlapping areas must be sealed with the same bentonite as contained in the bentonite mat. Some manufacturers offer factory-prefabricated edge infill or full-surface infill GCL types. This makes the laying process on the construction site considerably easier, as overlapping areas on the construction site do not have to be additionally sealed with bentonite.

Basically, the installation instructions of the manufacturer or the project-specific installation instructions must be observed, as these describe suitable measures to ensure that the GCL can be installed properly.

A laying plan should also be drawn up prior to installation, which not only shows the GCL positioning, but also specifies the time schedule for the laying.

Mechanical stresses that occur during the installation phase and after completion can influence the effectiveness of any sealing system.

The highest static and dynamic loads for the GCL usually occur during floor application. When applying the bulk material, care must be taken to ensure that the floor material is not poured directly onto the GCL, but in front of the head and onto previously distributed material, and then distributed with a suitable device.

A widely graded material with the smallest possible maximum grain size is recommended and round grain is preferable to broken material.

In principle, a GCL must not be built over in a hydrated state. As a guideline, the water content of bentonite must not exceed 50% by mass.

In the overlap area, care must be taken to ensure that the soil is distributed in the overlap direction, preventing that soil material penetrates into the overlap area.

In order to ensure protection against damage by vehicles leaving the roadway, a minimum overlap of 30cm is required, if the covered area is only crossed once. In the case of multiple traffic, it is advisable to increase the thickness of the cover to at least 60cm, depending on the load.

The minimum overlap of the first layer depends on the particle size distribution and the GCL components. An upper fabric component, for example, is more susceptible to damage than a nonwoven component. However, even with nonwoven components of the GCL top layer, a higher area-related mass is more robust and thus more resistant, if coarse-grained bulk material is used.

A covering at least 30cm thick must be carried out every working day. Overlaps required for the following day must be wrapped overnight with foil to prevent free swelling.

2.13. *Conclusions*

Groundwater protection is generally required where a road enters a groundwater sensitive area, to avoid damage from winter maintenance with de-icing salt, everyday pollution arising from motor vehicles, and to protect the area from accidents with the possible release of polluting substances (chemical/petroleum tankers/transporters). The German RiStWag (Guidelines for Construction Projects in Waterways of Protected Areas) [1] from the Research Society for Road and Traffic was one of the earlier guidelines on this topic. This German guideline describes among other topics geosynthetic sealing systems such as geomembranes and geosynthetic clay liners (bentonite mats). Covered with at least 80cm of soil, geosynthetic clay liners (GCL) are equated with the previous standard solution, a clay liner [13].

However, GCLs exhibit various advantages compared to other RiStWag sealing systems: The GCL is a homogenous sealing layer that can be tested for its quality prior to installation. The use of sodium bentonite provides a self-healing effect of the GCL, especially during installation. A needle-punched GCL exhibits a direction-independent shear resistance over the entire surface and, correctly designed, it can be installed on steep slopes. Multi-component GCLs allow safer and economic designs, act as a double-lined system with very low permeability values and are less affected by desiccation or critical liquids.

GCL products can easily be delivered to the site (approx. 4,500m² per load) and easily be installed with light equipment. Therefore, GCL products are economic and ecological solutions. Needle-punched geosynthetic clay liners following the described specification values are recommended to comply with the new RiStWag requirements. Not only are GCLs cost-saving barrier products, they also preserve our natural environment.

3. References

[1] FGSV 2016 Forschungsgesellschaft für Straßen- und Verkehrswesen (Hrsg): Richtlinien für bautechnische Maßnahmen an Straßen in Wasserschutzgebieten (RiStWag), Ausgabe 2016, Köln

[2] Kocher B 2016 Neuerungen in den RiStWag 2016 – Richtlinien für bautechnische Maßnahmen in Wasserschutzgebieten *Erd- und Grundbautagung 2016* (FGSV C 13) pp 76-87

[3] DIN EN 16416:2013-12: Geosynthetische Tondichtungsbahnen – Bestimmung der Durchflussrate – Triaxialzellen-Methode mit konstanter Druckhöhe. Deutsches Institut für Normung e. V. (Hrsg.), Beuth Verlag GmbH, Berlin

[4] LAGA 2018: LAGA Ad-hoc-AG „Deponietechnik“. Bundeseinheitlicher Qualitätsstandard 5-5 - Oberflächenabdichtungskomponenten aus geosynthetischen Tondichtungsbahnen vom 12.06.2018; BQS 5-5, Anhang 3: Bestimmung der Permittivität nach Erstquellung unter Salzeinwirkung und Ionenaustausch

[5] ASTM D4439: Standard Terminology for Geosynthetics. ASTM International, West Conshohocken, PA, USA

[6] von Maubeuge K, Sreenivas K and Pohlmann H 2011 The New Generation of GEOSYNTHETIC CLAY LINERS *Geosynthetics India 2011 and an Introductory course on Geosynthetics* (Chennai, Tamil Nadu, India)

[7] ASTM D6496: Test Method for Determining Average Bonding Peel Strength Between the Top and Bottom Layers of Needle-Punched Geosynthetic Clay Liners. ASTM International, West Conshohocken, PA, USA

[8] EN ISO 10319: Geosynthetics - Wide-width tensile test. Deutsches Institut für Normung e. V. (Hrsg.), Beuth Verlag GmbH, Berlin

[9] ASTM D6768: Test Method for Tensile Strength of Geosynthetic Clay Liners. ASTM International, West Conshohocken, PA, USA

[10] ASTM D5887: Test Method for Measurement of Index Flux Through Saturated Geosynthetic Clay Liner Specimens Using Flexible Wall Permeameter. ASTM International, West Conshohocken, PA, USA

[11] ASTM D5890: Test Method for Swell Index of Clay Mineral Component of Geosynthetic Clay Liners. ASTM International, West Conshohocken, PA, USA

[12] Egloffstein T, Heerten G and von Maubeuge K 2010 Comparative life cycle assessment (LCA) for clay geosynthetic barriers versus clay liners and other sealing systems used in river dykes, canals, storm water retention ponds and landfills *Proc. 3rd International Symposium on Geosynthetic Clay Liners, GBR-C 2k10* (Würzburg, Germany)

[13] von Maubeuge K 2018 Geosynthetic barriers in regulations and recommendations in line with the ISO design guide? *Innovative Infrastructure Solutions* **3** n 75 https://doi.org/10.1007/s41062-018-0177-9

IOP Conf. Series: Materials Science and Engineering 1260 (2022) 012041 doi:10.1088/1757-899X/1260/1/012041

Basal reinforcement of embankment over voids and soil subsidence area – new standard XP G38065

A Nancey [1], **P Delmas** [2]

[1] TenCate Geosynthetics, Bezons, France
[2] Chairman of French Standardisation Committee on Geosynthetics, Paris, France
Email: a.nancey@tencategeo.com

Abstract. The use of geosynthetic reinforcement to limit the consequences of a localized collapse of the ground under road and railway structure started in the 90s. The new French standard XP G 38065 for basal reinforcement establishes the rules for justifying works on potential cavity risk areas and will provide users with a tool adapted to this use. This paper shows the different steps of a design, considering the Eurocode approach with deformation and stress criteria, and the lateral displacement design model.

1. Introduction

Of natural origin (internal erosion of the ground) or resulting from human activities (pumping of water, unstable underground mines and quarries, buried trenches), soil subsidence presents a risk toward the safety and the use of the structures. Geosynthetic as basal reinforcement of the embankment to prevent collapse is a solution that has proven effective for more than thirty years. The experience acquired mainly in full-scale experiments, but also on real structures, allowed a better understanding of the mechanisms in reinforced embankments and lead to develop relevant calculation methods.

The design of reinforcement spanning void is not limited to a simple calculation of tensile strength but must incorporate all the constraints related to the safety or the usability of the structure, such as the allowable deformation for the reinforcement duration, when the geosynthetic is under tension. This publication shows the new French standard XP G 38065 « Basal reinforcement of embankment prone to soil subsidence », describes the methodology for the design and the justification of the structure. It follows the Eurocodes approach with the service limit states (SLS) and ultimate limit states (ULS) to define the geosynthetic fulfilling the criteria of failure and allowable deformation.

2. The structure and the safety and serviceability criteria

A cavity rising to the ground surface is characterized by its shape and size, essential information for the design. For the area that needs to be secured, the design value will be the width of the trench in the case of a linear collapse or the diameter for a circular cavity. Delmas et al. [1] showed the very high sensitivity of surface settlement with respect to cavity size. The selected dimension value must therefore consider the variation of the values in place and be carefully defined as the maximum value.

The first objective of the reinforcement is to prevent the full collapse of the structure. When a cavity goes up to the level of the reinforcement, the geosynthetic must support the loads coming from the traffic as well as the permanent surcharges. Under the effect of these loads, the geosynthetic will deform into a membrane above the cavity inducing stresses in the reinforcement. The tensile forces in the geosynthetic are balanced thanks to a progressive mobilization of anchors on both sides of the cavity.

During this phenomenon of contained subsidence, the subjacent soil will deform resulting in a settlement of the surface structure depending on the thickness of the embankment. While avoiding the failure of the geosynthetic, a proper design of the reinforcement aims to define the mechanical characteristics needed to limit the deformations and particularly the surface settlement in an affordable value for the users. Figure 1 shows how a geosynthetic reinforced structure works during time, from the installation until the void come at the level of the ground, and over time.

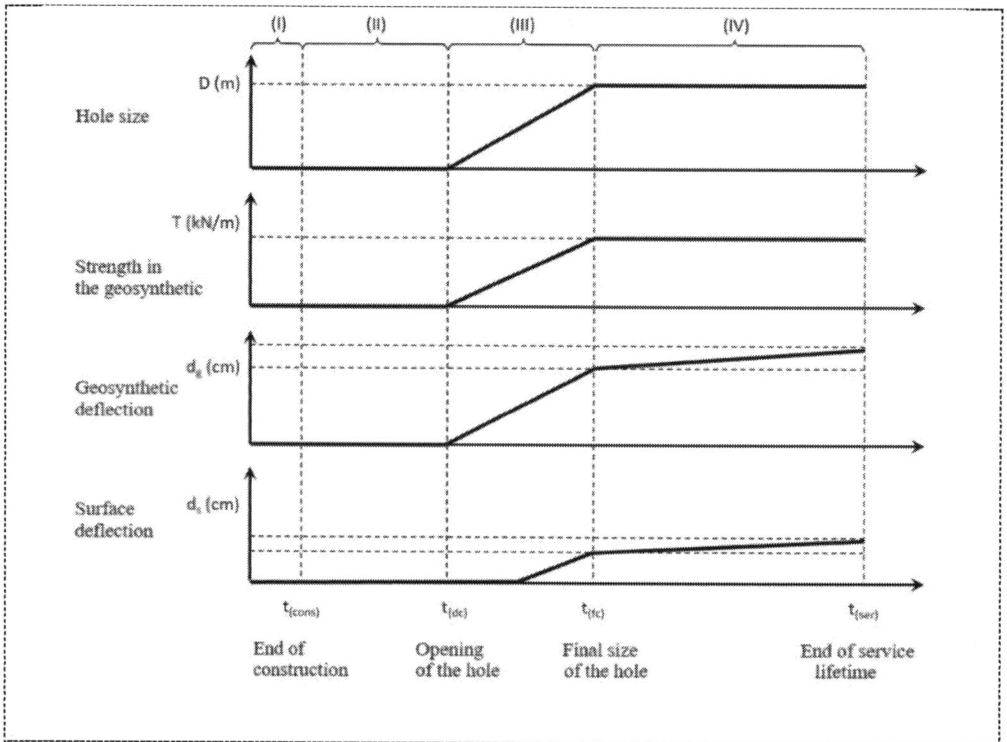

Figure 1. principle of a geosynthetic reinforced structure before and after a cavity goes up to the ground surface

Several periods of time can be distinguished:

- period (I): installation of the geosynthetic and construction of the structure;
- period (II): the cavity is not raised to the surface; the structure is fully support by the subground;
- period (III): opening of the cavity up to the nominal dimension; this phase can be brutal or extend over a longer or shorter period;
- Period (IV): the cavity has reached its final dimension assumed until the end of the service life of the structure; it may correspond either to the initial expected life of the structure (for example 100 years) or to the length of time required to repair the structure (usually few months maximum).

2.1. Geometrical model and loads

The geometry of a reinforced embankment spanning void, can be described in a simple way by using two characteristics represented in Figure 2: the length to span D and the height of the embankment H.

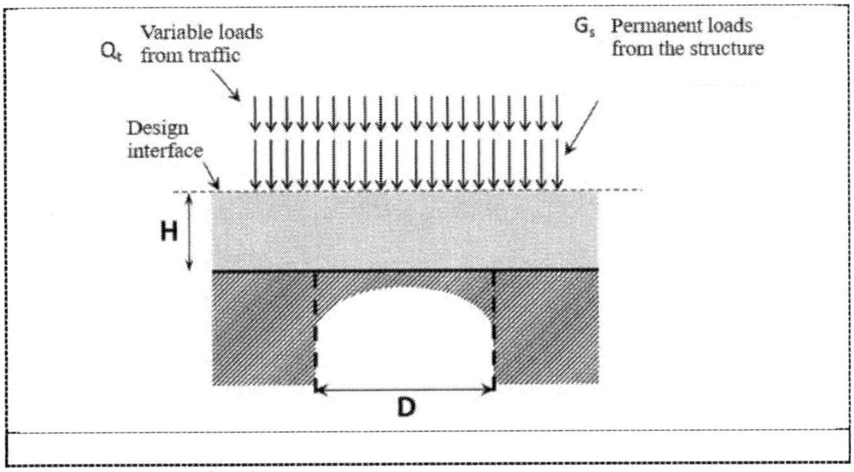

Figure 2. Design interface and loads

The length D is related to the size of the cavity and its shape. In the case of a circular cavity, the span is equal to the diameter of the collapse. Cavities that have a greater dimension in one direction compared to the other can be described as "linear cavities". The design value must consider the orientation of the reinforcement with respect to the axis of the collapse.

A reinforced embankment on cavity is represented by a layer of height H resting on the geosynthetic reinforcement, generally installed at the level of the natural ground. The height H corresponds to the thickness of granular materials constituting the body of the embankment. Pavement layers for roads and ballast, rails, sleepers or for railway tracks are represented in the form of a permanent surcharge G_s. The case of the treated materials (cement or lime) or cohesive fill is not addressed in this standard. Such layers may be considered as a permanent load if appropriate. Traffic is represented by a variable load Q_t. The design interface is defined as the surface of the embankment to which the loads apply (Figure 2). When the conditions make it possible, it may be advantageous to increase the height H to allow a better load transfer to the edges of the cavity and then to limit deformation.

2.2. Allowable settlement at the top of the embankment

Settlement can cause the failure of the structure by excessive deformation, even if the geosynthetic does not break. This settlement corresponding to an ultimate limit state that would need the application of a partial safety factor which is difficult to define. In practice, the settlement is set at the serviceability limit state, making it possible to maintain the operation of the structure, even if reduced. This maximum allowable settlement applies at the design interface.

The allowable settlement value must be chosen for the reinforcement duration determined according to the maintenance strategy defined for the structure. The duration of effective stress is the duration during which the geosynthetic is in tension. It corresponds to the period between the appearance of the cavity at the level of the reinforcement and the repair of the structure (case of temporary reinforcement), or by default to the duration of use of the project (permanent reinforcement). Consideration of creep is relative to this duration of stress. Chemical aging of the geosynthetic begins during installation and operates over the whole life of the structure.

This requirement is often expressed in the form of relative settlement (d_s/D_s), defined as the ratio between the maximum allowable settlement d_s and the size of the disorder D_s at the surface. Table 1 presents examples of commonly used surface settlement values.

Table 1. Examples of allowable differential settlements d_s/D_s

Structures	d_s/D_s
Railways	0 %
Highways (high speed)	1,0 % - 1,7 %
Secondary roads (low speed)	1,7 % - 2,5 %
Other urban roads, parking area, etc.	2,5 % - 7 %
Park and pedestrian ways	up to 10 %

2.3. *Ultimate Limit States and Service Limit State*

The XP G 38065 standard defines two ultimate limit states which are related to the failure of the geosynthetic and the anchorage. When a geosynthetic span void, a failure no longer allows to support the loads from the embankment and causes the collapse of the structure. It corresponds to a STR type ultimate limit state in which the resistance of the materials contributes significantly to the resistance of the structure as defined in Eurocode 7, reference EN 1997-1 [2]. Insufficient anchoring (anchor or trench length) causes the geosynthetic to slip, which can lead to excessive deformation or complete collapse of the structure. The failure of the anchorage corresponds to a STR / GEO type ultimate limit state. Verification of the ultimate limit states is carried out using calculation approach 2 according to the Eurocodes and his national annex:

- the tensile strength of the reinforcing geosynthetic for the service life $R_{t;d}$ is greater than the value of the design strength $T_{max\ ;ELU}$;
- the resistance of the anchor $R_{A;d}$ is greater than $T_{max\ ;ELU}$, the design strength of the geosynthetic in anchoring zone at the ULS.

The main service limit state is defined by the settlement at the design interface, cf. § 2.2. At the service limit state, it must be verified that:

- the deformation of the geosynthetic ε_{calc} remains less than or equal to the maximum allowable deformation ε_{adm}, related to the maximum allowable settlement criterion.
- the tensile strength of the reinforcing geosynthetic at the allowable deformation for the service life $R_{t\ ;d}\,(\varepsilon)$ is greater than the value of the design strength $T_{max}(\varepsilon)$
- the resistance of the anchorage $R_{A;d}$ is greater than $T_{max}(\varepsilon)$, the geosynthetic strength in anchorage zone at SLS.

3. Geosynthetic design

The geosynthetic results from a model considering the vertical load applied on the reinforcement layer and from a deformation criterion. The vertical load depends on the surcharge, permanent and variable, and from the granular layer on top of the reinforcement. A general model, as shown on Figure 3 consist to define the collapsing zone with θ the angle of draw. The friction at the interface between the collapsing zone and the stable part of the embankment may be also considered.

3.1. *Embankment behaviour – failure mode above geosynthetic reinforcement*

To interpret the surface settlement (at the design interface) into a geosynthetic allowable deformation criterion of reinforcement, the knowledge of the fill layer behaviour is necessary. In the XP G 38065 design standard, the approach chosen was proposed by Giroud et al. [3] and which assumes a vertical collapse (θ =90°) with lateral confinement reducing the vertical forces on the geosynthetic. The RAFAEL experiments [4], showed a vertical collapse of the embankment for the configurations studied above the reinforcement geosynthetic. The synthesis of recent research confirms this model for a granular material [1].

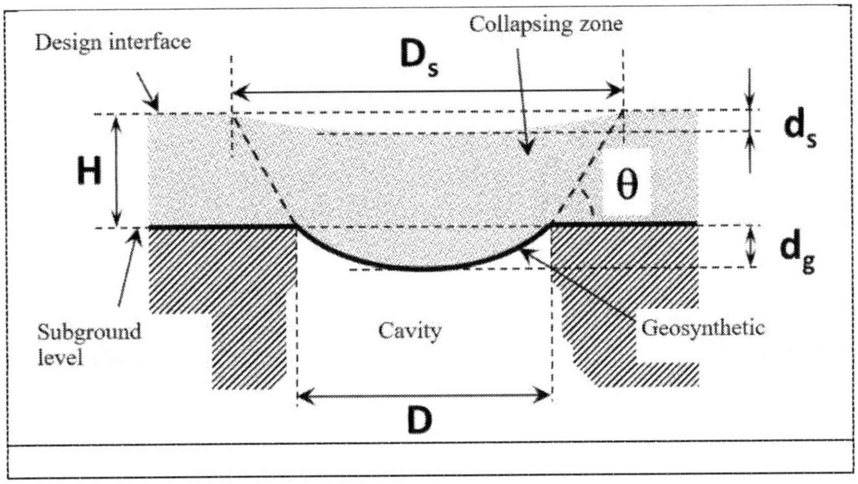

Figure 3. Design model

3.2. Embankment behaviour - expansion coefficient

In the design method proposed by Blivet et al.[5] , this approach consider the expansion of the soil above the geosynthetic. Expansion is the increase in soil volume above the reinforcement that can occur during collapse.

The expansion coefficient C_e depends mainly on the type of material and the initial level of compaction of the embankment. It can reach a value of 1.15 for highly granular backfill soils, but "values of 1.03 to 1.1 are more appropriate for common backfill materials [6] . In the XP G 38065 standard [7], to be on the safe side, indicative values of the expansion coefficient are given: 1.03 for round materials, 1.05 for angular materials.

The maximum deflection of the geosynthetic $\boldsymbol{d_{g,max}}$ is related to the admissible surface deformation (design interface) $\boldsymbol{d_{s,max}}$ by equation 1:

$$d_{g,max} = d_{s,max} + 2.H\,(C_e - 1) \tag{1}$$

3.3. Calculation of the average vertical stress on reinforcement

The value of the average stress q acting on the geosynthetic can be obtained by considering the cylindrical collapse mechanism proposed by Terzaghi [8] and the overload p from equation (2) for a circular collapse or equation (3) for a linear collapse. with the parameters K the coefficient of thrust, φ', the angle of friction and γ the density of the backfill material.

$$q = \gamma_G \frac{D.\gamma}{4\,K.\tan\varphi'}\left(1 - exp\left(\frac{-4\,K.\,(H.\tan\varphi')}{D}\right)\right) + p\,.\,exp\left(\frac{-4\,K.\,(H.\tan\varphi')}{D}\right) \tag{2}$$

$$q = \gamma_G \frac{D.\gamma}{2\,K.\tan\varphi'}\left(1 - exp\left(\frac{-2\,K.\,(H.\tan\varphi')}{D}\right)\right) + p\,.\,exp\left(\frac{-2\,K.\,(H.\tan\varphi')}{D}\right) \tag{3}$$

where p is the load combining the permanent and temporary loads:

$$p = \gamma_G.\,G_{s;k} + \gamma_Q.\,Q_{t;k} \tag{4}$$

Table 2 gives the partial coefficients that apply to the Ultimate Limit State (ULS) and Service Limit State (SLS).

Table 2. partial coefficients applied on actions or resistance

Action or resistance	Symbol	ULS	SLS
permanent and unfavourable surcharge	γ_G	1.35	1.0
variable and unfavourable surcharge	γ_Q	1.50	1.0
interface friction	$\gamma_{R;f}$	1.35	1.0

3.4. Calculation of the tensile strength and deformation at the allowable settlement

Briançon and Villard [9] developed a calculation method to determine the deformation and tension of the geosynthetic, considering the lateral displacement of the geosynthetic on both sides of the cavity, during its opening and under the effect of the loading. This displacement is necessary to tensile strength and mobilize forces at the soil / geosynthetic interface. An elastoplastic interaction law with a Mohr-Coulomb type failure criterion is used.

This method, whose mechanisms are represented in Figure 4, considers that the stress distribution on the geosynthetic is vertical and uniform. The application of a uniform charge leads to a parabolic deformation of the reinforcement.

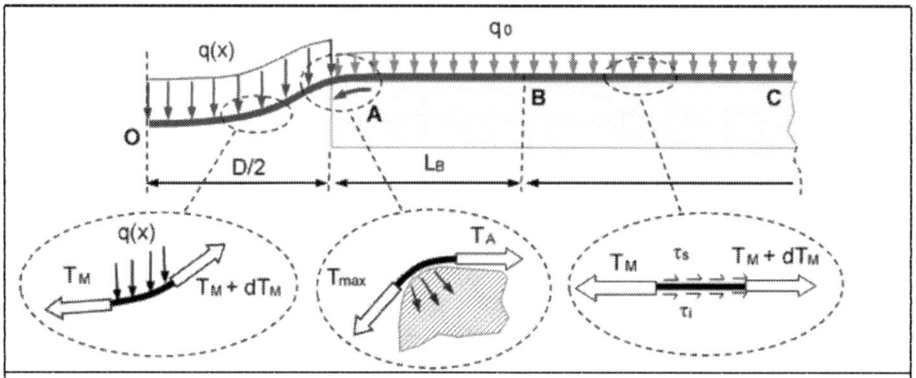

Figure 4. Mechanisms considered in the design

Solving these the equilibrium equations makes possible to evaluate the geosynthetic characteristics considering the settlement criterion at the service limit state:

- The deflection of the geosynthetic $d_{g,max}$ given by equation (1) leads to calculate the parameter β, the angle between the reinforcement and the horizontal at the point A (Figure 4):

$$\beta = \frac{4.d_{g;max}}{D} \quad (5)$$

- The maximal tensile strength in the geosynthetic T_{max} is calculated from the load q, the span length D and β:

$$T_{max} = \frac{q.D}{2.\beta}(1+\beta^2)^{1/2} \quad (6)$$

- Geosynthetic deformation is obtained from T_{max} and the stiffness J:

$$\varepsilon_g = \frac{T_{max}}{J} \quad (7)$$

- T_A , the tensile strength in the anchorage part in both side of the void is calculated from the load q, the span length D and β:

$$T_A = \frac{q.D}{2.\beta}\left(1+\beta^2\right)^{1/2}\left(exp\,(\alpha.Atan\beta.tan\delta_i)\right)^{-1} \quad (8)$$

Where :

- δi is the friction angle at the lower interface of the geosynthetic
- $\alpha=U_A/U_0$ if $U_A\geq U_0$ or $\alpha=1$ if $U_A>U_0$ with U_A the displacement at point A and U_0 the displacement to needed to mobilize the anchoring.

At the ultimate limit state, the maximum tensile strength of the geosynthetic is calculated at the ultimate deformation of the geosynthetic:

$$T_{max} = \frac{q.D}{2}\left(1+\frac{1}{6\,\varepsilon_{ult}}\right)^{1/2} \tag{9}$$

3.5. *Justification of the tensile strength at the ultimate limit state*

The design value of the tensile strength of the geosynthetic $R_{t;d}$ must be greater than the value of the design tensile $T_{max;ULS}$. The design value also known as long-term tensile strength $R_{t;d}$ is calculated from the ultimate resistance $R_{t;k}$ of the geosynthetic, resulting from the tensile test EN ISO 10319 [10]:

$$R_{t;d} = \frac{1}{\Gamma_{geo}} R_{t;k} \tag{10}$$

The coefficient Γ_{geo} integrates the partial safety factor on the tensile strength ($\gamma_{M;t}$) and the reducing coefficients related to the long term behaviour of the geosynthetic:

$$\Gamma_{géo} = \gamma_{M;t}\,.\,\Gamma_{flu}\,.\,\Gamma_{vieil}\,.\,\Gamma_{instal} \tag{11}$$

with

- $\gamma_{M;t}$ partial factor on the tensile strength.
- Γ_{flu} reduction factor on creep, determined for the service duration.
- Γ_{vieil} reduction factor related to aging of the geosynthetics and depending on the environmental conditions. This coefficient is determined for the lifetime of the structure.
- Γ_{instal} factor corresponds to the reduction of the tensile strength due to the damage during installation and the compaction of the embankment.

3.6. *Justification of the tensile strength at the service limit state*

The design value of the geosynthetic strength at the allowable strain ε and the service life (t_{ser}), $R_{t;d(\varepsilon)(t_{serv})}$ must be greater than the value of the design tensile $T_{max(\varepsilon);SLS}$. The design value of the tensile strength $R_{t;d(\varepsilon)(t_{serv})}$ is calculated from the isochronous curves and the characteristic short-term resistance $R_{t;k}$ measured according to the EN ISO 10319 standard [10]. The ratio $T(\varepsilon)/T_{ult}$ at strain ε is determined on the isochronous curve corresponding to the service life (t_{ser}), as shown on Figure 5.
The tensile strength $R_{t;d\,(\varepsilon)}$ at the service life (t_{ser}) for the deformation (ε) is obtained from the geosynthetic resistance $R_{t;k}$ by applying C_{geo} coefficient which integrates the partial factor and the reduction coefficients at strain (ε).

$$R_{t;d}\,(\varepsilon)(t_{ser}) = \left(\frac{1}{C_{géo}(\varepsilon)}\right).\frac{T(\varepsilon)}{T_{ult}}(t_{ser}).\,R_{t;k} \tag{12}$$

With

$$C_{geo}\,(\varepsilon) = \gamma_{M;t}\,.\,C_{vieil}\,(\varepsilon).\,C_{instal}\,(\varepsilon) \tag{13}$$

$\gamma_{M;t}$	partial factor on geosynthetic strength $\gamma_{M;t}$ is set to 1 to at SLS
$C_{vieil}(\varepsilon)$	reduction factor related to aging of the geosynthetics and depending on the environmental conditions, determined at strain (ε) for the lifetime duration.
$C_{instal}(\varepsilon)$	reduction factor related to the damage during installation and the compaction of the embankment determined at strain (ε)

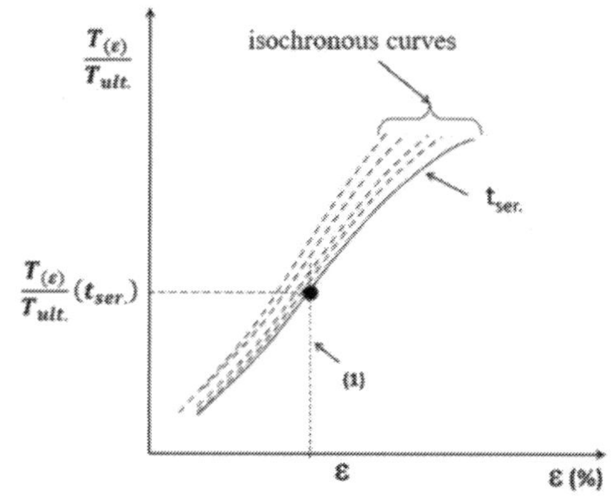

Figure 5. Ratio $T(\varepsilon)/T_{ult}$ obtained from isochronous curve at (t_{ser})

3.7. Reinforcement bond strength

The design value of the bond strength shall be greater than $T_{max}(\varepsilon)$ and T_{max} the design tensile values of the geosynthetic at the ULS and SLS. It is necessary to check the minimum anchorage length to be respected on each side of the cavity for a flat anchorage and the minimum overlap length between two consecutive layers.

4. Conclusion

The new French standard XP G 38065 offers users, contractors and design offices, a designing assistance for reinforced embankments where there is a risk of subsidence. The proposed method is based on the latest results of research and experiments carried out over the past twenty years and incorporates the approach developed in Eurocodes for taking safety into account.

References

[1] Delmas P, Villard P, Huckert A 2015 Dimensionnement à court terme et à long terme de structures renforcées par géosynthétique sur cavités potentielles : prise en compte de la sécurité *CR 10ème Rencontres géosynthétiques* pp 13-34

[2] EN 1997-1 2004 (English) Eurocode 7: Geotechnical design - Part 1: General rules

[3] Giroud J P, Bonaparte R, Beech J F, Gross B A 1990 Design layer- geosynthetics system overlying voids *Geotextiles and Geomembranes* **9** pp 11-50

[4] Gourc J P, Villard P, Giraud H, Blivet J C, Khay M, Imbert B, Morbois A, Delmas P 1999 Sinkholes beneath a reinforced earthfill – A large scale motorway and railway experiment Geosynthetics, Boston, Massachusetts, USA, 28-30 April 1999 **2** pp 833-46

[5] Blivet J C, Khay M, Gourc J P, Giraud H 2001 Design considerations of geosynthetic for reinforced embankments subjected to localized subsidence *Proceeding of the. Geosynthetics Conference*, February 12-14, 2001, Portland, Oregon, USA pp 741- 54

[6] Briançon L, Villard P 2006 Dimensionnement des renforcements géosynthétiques de plateformes sur cavités *Revue Française de Géotechnique* n 117 4ème trimestre 2006 pp 51-61

[7] XP G 380635 Geosynthetics and geotextile related products — Basal reinforcement of embankment prone to soil subsidence — Justification of dimensioning and design elements.

[8] Terzaghi K 1943 Theoretical Soil Mechanics. J. Wiley and Sons, New York

[9] Briançon L, Villard P 2008 Design of geosynthetic reinforcements of platforms subjected to localised sinkholes *Geotextiles and Geomembranes* **26** pp 416-28

[10] EN ISO 10319 2015 Geosynthetics - Wide-width tensile test

Investigation progress of geogrid-soil interaction using DEM

Z Wang[1, 2, 3*], B Ma[3], G Yang[1, 2, 3], M Ziegler[4]

[1] State Key Laboratory of Mechanical Behavior and System Safety of Traffic Engineering Structures (Shijiazhuang Tiedao University), Shijiazhuang 050043, China

[2] Key Laboratory of Roads and Railway Engineering Safety Control (Shijiazhuang Tiedao University) , Ministry of Education, Shijiazhuang 050043, China

[3] School of Civil Engineering, Shijiazhuang Tiedao University, Shijiazhuang 050043, China

[4] ZAI Ziegler und Aulbach Ingenieurgesellschaft mbH, Aschaffenburg 63741, Germany

E-mail: zwang@stdu.edu.cn

Abstract: The discrete element modeling, which has particular advantages of capturing detailed insights into the geogrid–soil interaction, provides a new perspective to study the geogrid reinforcement mechanism especially at a microscopic scale. To summarize the current investigation achievements, discrete element modeling of geogrid and soil as well as their interaction under different testing conditions were reviewed in this paper. The DEM investigations provide not only qualitative visualization of the geogrid–soil interaction but also quantitative response of the geogrid and its surrounding soil. Such investigation achievements provide researchers helpful hints for the design of geogrid reinforced soil structures.

1. Introduction

As an important geosynthetic material, geogrids have been widely used in dealing with various geotechnical problems, e.g. stabilizing earth structures, increasing the bearing capacity of base courses and reducing rut depths in flexible pavements, etc. [1-7]. Geogrid reinforcement effects are performed via the interaction of geogrids together with the surrounding soil. Hence, the geogrid–soil interface behavior has been regarded as a significant factor to investigate the geogrid reinforcement mechanisms [8].

In order to investigate the reinforcement mechanisms of geogrids, many researchers have conducted various laboratory tests on the interface behavior between geogrid and soil as well as the general compound behavior of the composite material [9-11]. They investigated the influencing factors and their effects on the composite material, which provides useful hints to the design and construction of real geogrid reinforced structures.

Numerical approaches, such as the finite element method (FEM) and the finite difference method (FDM), have also been used to evaluate the geogrid–soil interface behavior. The FEM and the FDM can be regarded as important supplements to laboratory tests for the investigation of reinforcement mechanisms at macroscopic scales. However, with the continuous mechanical approach, the FEM and the FDM are not able to model discrete soil particles and hence they do not take the particle rolling and sliding within geogrid reinforced soil structures into consideration. As the geogrid reinforcement mechanisms highly depends on discrete soil particle shape and distribution, they cannot be satisfactorily investigated only by FEM/FDM, especially at a microscopic scale. The discrete element method (DEM), which was introduced by Cundall [12] for analyzing rock mechanics problems and then applied to soils by Cundall and Strack [13], has been regarded as an ideal tool for investigating the geogrid–soil interaction [14].

To summarize the current investigation achievements of geogrid-soil interaction using DEM, this study first summarizes the discrete element modeling of geogrid and soil, and then introduces the research status of interaction between geogrid and soil under different loading conditions.

2. Discrete Element Modeling of Soil

In discrete element modeling, the shape of soil particles will influence the response of geogrid-reinforced systems [15, 16]. For sand particles, they were always modeled as circular or spherical particles [17-21]. Since the computational time in the DEM simulations is highly dependent on the particle numbers, the technique of "up-scaling" for soil particles has been used in DEM studies to balance the computational cost against the scaling effect on the sample response. The up-scaling factor in those studies ranged from 1.25 to 15. In order to compensate the lack of angularity for such circular particles, large friction coefficients have been used to reach an equivalent internal friction angle of the soil [19, 21].

Another possible way to reach the equivalent internal friction angle of the soil is to simulate irregular shapes of the real particles by bonding circular or spherical particles together to form clusters or clumps, as shown in Figure 1. Overlaps may exist in each cluster and each clump. The difference between a cluster and a clump is that the cluster can be used to simulate grain crushing since the intragranular bonds can be broken, while the intragranular bonds in the clump are always considered to be infinitely stiff and strong [22]. Without using large friction coefficients for the clusters and clumps, high values of the macroscopic internal friction angles of the soil can be achieved. However, due to the high requirement of computing power, this technique is currently only applied for simulating medium/coarse gravel or ballast [23-29].

In order to get the equivalent internal friction angle of the soil with less computational time, a clump consisting of limited numbers of circular discs or spheres provides a good solution. This technique has been successfully used in the numerical modeling of geogrid–ballast interaction under cyclic loads and shear band development in granular materials as well as 2D DEM calibrations of large size plane strain compression tests [30-32]. Figure 2 shows examples of 2-ball, 3-ball and 4-ball clumps reported in the literature. The shape of soil particles was formed by 3-ball clumps, which can effectively balance the calculation accuracy of numerical simulation according to the studies of Rui et al. [33].

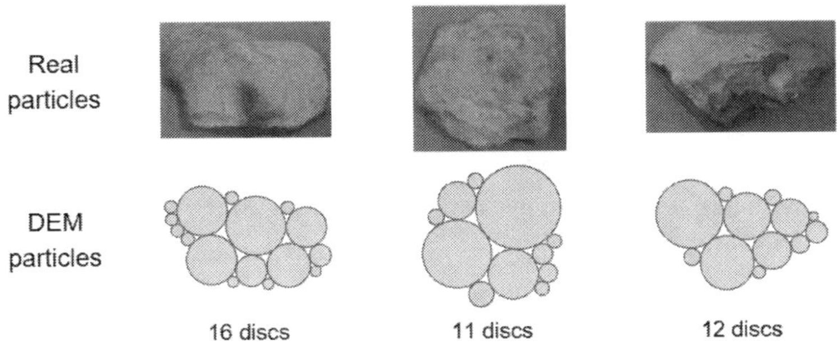

(a) Discrete element modeling of real particles with 2D clusters [34]

(b) Discrete element modeling of real particles with 3D clumps [26]

Figure 1. Discrete element modeling of real particles in 2D and 3D

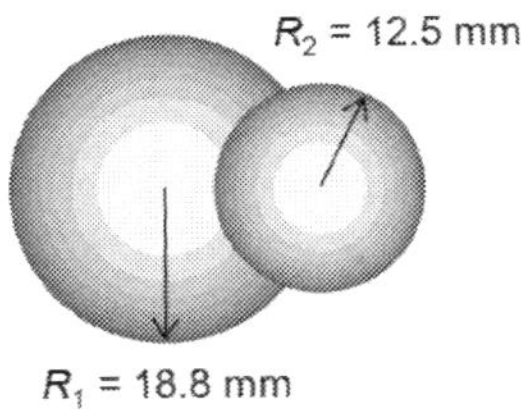

(a) a 2-ball clump in 3D [30]

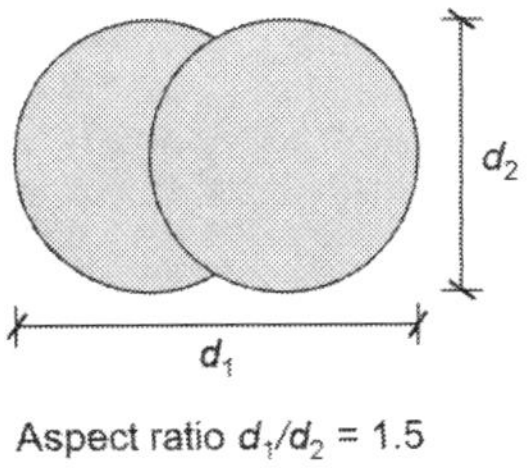

(b) a 2-ball clump in 2D [31]

(c) 3-ball and 4-ball clumps in 2D [32]

Figure 2. Examples of clumps with limited numbers of circular discs or spheres in 2D and 3D

3. Discrete Element Modeling of Geogrids

In discrete element modeling, geogrids were modeled by bonding and/or overlapping circular or spherical particles to form similar geometries of real geogrids as those used in laboratory tests [26, 28-

30, 35-38]. Quite similar geometries of real geogrids, including transverse members and knots at the junctions, can be easily simulated in 3D modeling, as shown in Figure 3. In 2D modeling, however, the geometries of real geogrids has to be simplified. The longitudinal member was simulated with one row of bonded particles without any overlaps. Extra bonded particles on and beneath the longitudinal member or particles with larger diameters at the positions of junctions were simulated to investigate the contributions of geogrid transverse members, as shown in Figure 4. Therefore, the bearing resistances caused by the transverse members in these two-dimensional investigations can be transferred to the longitudinal member via the knot particles [39].

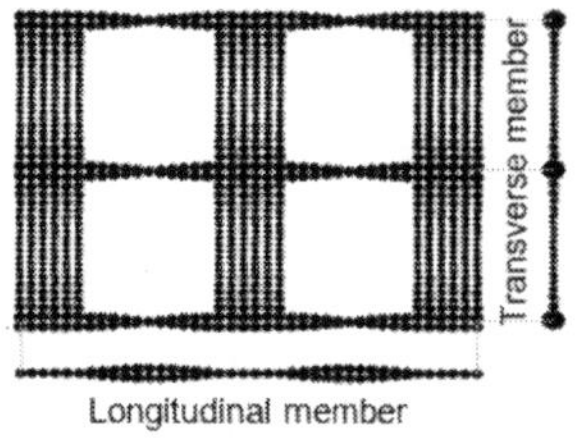

(a) Woven geogrid [26]

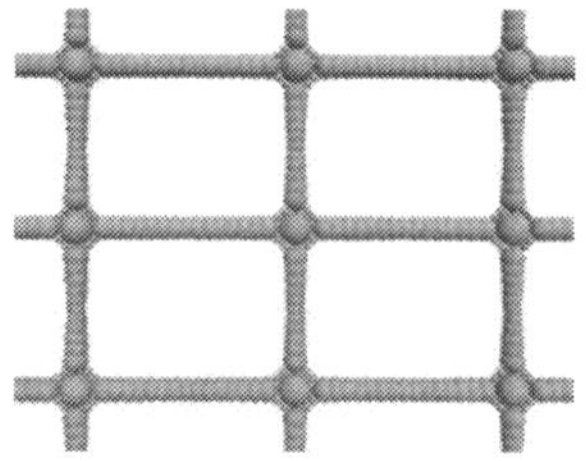

(b) Extruded biaxial geogrid [29]

(c) Extruded biaxial and triaxial geogrids [30]

Figure 3. Examples of DEM models of geogrids in 3D

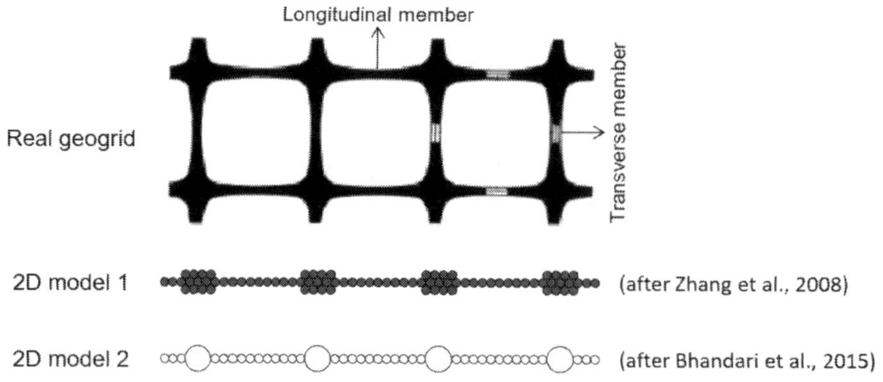

Figure 4. Examples of DEM models of geogrids in 2D

Compared to the geometric modeling of geogrids, the mechanical modeling of geogrids is equally important. The tensile strength of a geogrid is determined by the bond forces between geogrid particles in discrete element modeling [30, 35]. With the parallel bond model provided by Itasca [40], it is easily possible to represent the tensile strength properties of a geogrid with a linear force–strain relationship [28, 36, 37]. However, most laboratory tensile test results of geogrids show nonlinear tensile strength behavior, as shown in Figure 5. In addition, it should be noted that the tensile rates have great effects on the tensile behaviour of PP geogrids [41, 42].

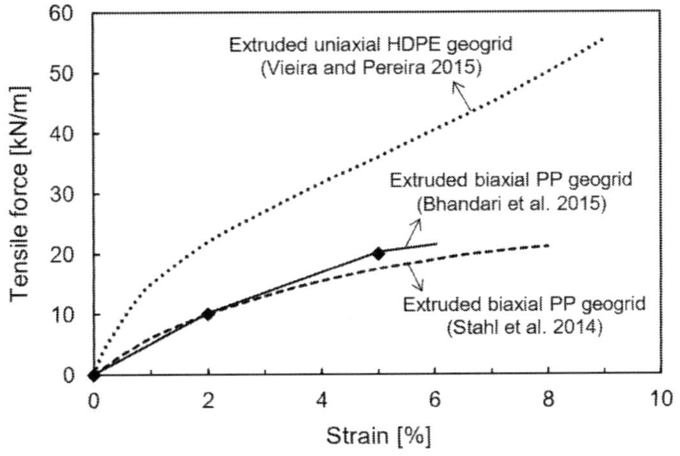

Figure 5. Examples of nonlinear tensile force–strain behavior of geogrids in laboratory tests

In order to describe the nonlinear tensile behavior of geogrids under tensile loads, a parallel bond radius decreasing law was implemented into the discrete element modeling [29]. With increasing parallel bond radius, the parallel bond stiffness was reduced. The parallel bond radius decreasing law mimicked the effects of reducing the amount of cement-like material joining the two bonded particles. The calibrated numerical tensile test results showed good agreement with the corresponding experimental data [29].

4. Discrete Element Modeling of Geogrid–Soil Interaction

4.1. Pullout Tests

Konietzky et al. [14] carried out 3D discrete element modeling of pullout tests using spheres to investigate the interlocking effect of geogrids. Qualitative contact force distributions within the specimen at the beginning and the end of the test visualized the influencing zones of the geogrid under pullout loads, as shown in Figure 6. Moreover, the recorded normal and shear forces inside the model at the end of the test showed that the interlocking effect was restricted to a relatively small height at both sides of the geogrid (i.e. approximately 20 cm).

The subsequent study conducted by McDowell et al. [35] focused on the effect of the ratio of geogrid aperture size to ballast particle diameter under pullout loads, using clumps by overlapping spheres to represent the actual shape of ballast. For the ballast particles simulated in their study (40 mm in diameter), a ratio of aperture size to particle diameter of around 1.4 showed optimum results in terms of peak resistance mobilized at the smallest displacement.

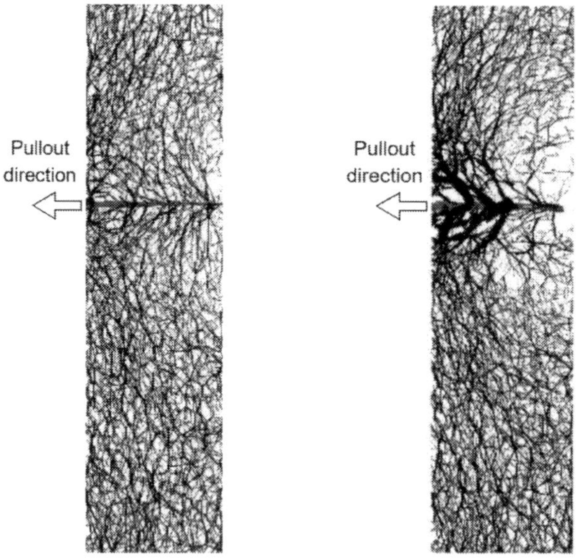

(a) At the beginning of the test (b) After pulling of 12 mm

Figure 6. Contact force distributions in pullout tests [14]

Zhang et al. [44] conducted 2D numerical pullout tests on loose and dense soil samples to study the effect of compaction on geogrid pullout response. The development of contact forces and displacement vectors of soil particles in both samples showed that the density had great influence on the 2D dilatancy behavior of geogrid reinforced soil under pullout loads.

Ferellec and McDowell [45] investigated the influence of ballast shape (i.e. spheres and clumps with angular shapes) on ballast–geogrid interlocking behavior under pullout loads. The clumps used in their study were simulated by overlapping spherical particles. The pullout force was larger for the clumps than that for the spheres, which can be explained by the better interlocking of the geogrid with the clumps than that with the spheres. However, the maximum pullout forces in both simulations were much smaller than the values based on laboratory tests. A higher value of the pullout resistance, which was closer to the experimental results, could be achieved by modeling clumps with sharper edges. However, it required many more spheres for each clump and therefore much more computing power.

Based on 3D discrete element modeling, Chen et al. [46] reproduced experimental pullout tests using 2-ball clumps. The DEM simulation results showed good agreement with the corresponding experimental data and the distribution of contact forces in the geogrid reinforced ballast system was provided. Their further DEM investigations focused on the influence of clump shapes on both the pullout resistance and the distribution of contact forces [27]. The numerical modeling with 8-ball tetrahedral clumps showed the best fitting results with the experimental data.

A coupled finite-discrete framework has been developed by Tran et al. [20] to investigate the behavior of a biaxial geogrid embedded in granular soil subjected to pullout loading. The geogrid was modeled with finite element and interface elements, while the soil was modeled using spherical particles. Response of both the geogrid and the surrounding backfill soil has been obtained numerically to visualize the geogrid–soil interaction under pullout loads.

Based on 3D discrete element modeling, Zheng et al. [47, 48] carried out numerical pullout tests using biaxial geogrids and 3-ball clumps. They investigated the influence of soil particle sizes on the pullout resistance as well as the contribution of frictional resistance and bearing resistance to the total pullout force. The shear strength of the geogrid–soil interface with big particles was larger than that with small particles. At a large clamp displacement, the contribution of bearing resistance to the total pullout force was larger than that of frictional resistance.

Miao et al. [49, 50] conducted 3D discrete element modeling of pullout tests using triaxial geogrids. The development of fabric anisotropy (e.g. coordinate number, orientation of contact forces) in the pullout model has been illustrated with increasing clamp displacements. Moreover, they also investigated the influence of joint protuberances on the geogrid pullout behavior. The shear strength of the geogrid–soil interface was increased by using joint protuberances.

Biaxial geogrids and granular soil with similar geometries as those used in the laboratory tests have been modeled by overlapping and/or bonding spherical particles in 3D DEM investigations [29]. Using the above developed geogrid and soil, numerical pullout tests were carried out. DEM analyses of soil mobilization and geogrid deformation have been presented in their study to illustrate the significance of geogrid interlocking effect as a key mechanism for soil stabilization.

Wang et al. [39] carried out numerical pullout tests on geogrid with different number of transverse members. The geogrid-soil interaction has been visualized by the force distributions along the geogrid and in the specimen with different numbers of transverse members, as shown in Figure 7. The quantitative force, displacement and strain distributions along the geogrid with different numbers of transverse members have been obtained, which also illustrated the load transfer behaviour between geogrid and granular soil. Moreover, the contributions of geogrid transverse members have been obtained numerically, which could be used to explain the different failure modes in the laboratory pullout tests.

Wang et al. [51] used PFC^{2D} to simulate the pullout tests of geogrid under rigid and flexible top boundary conditions, and analyzed the influence of top boundary conditions on the interface behaviour between geogrid and soil from a mesoscopic level. The results showed that the rigid and flexible top boundaries had almost no effects on the pullout test results when the pullout displacement was small. With the increasing pullout displacement, the influences of the top boundaries on the pullout test results became obvious. The maximum pullout resistance under the rigid top boundary conditions was slightly larger than that under flexible ones. Moreover, the influences of the top boundaries on the pullout test results could be visualized by the tensile force distribution of geogrids, the contact force distribution of soil particles in the specimens, the rotations of soil particles and the vertical displacement distribution of top loading particles.

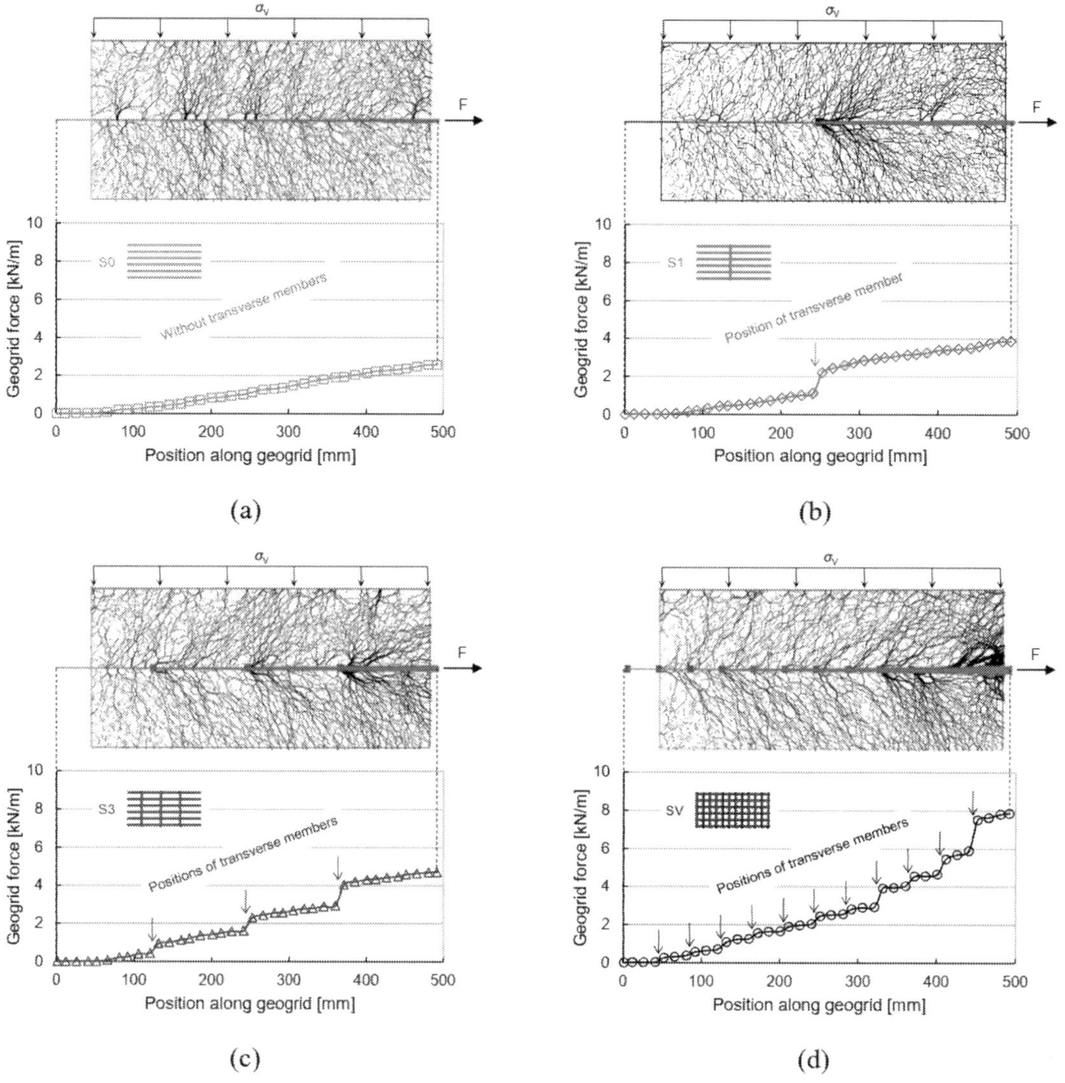

Figure 7. Force distributions along the geogrid and in the specimen

4.2. Tensile Tests in Soil

Wang et al. [52] studied the load transfer behaviour between geogrid and sand with a numerical compound tensile test, in which the models and the input parameters of the geogrid and sand were calibrated by tensile tests and direct shear tests, respectively. The load transfer behaviour between the geogrid and sand has been visualized in their study and an improved understanding of the geogrid reinforcement mechanism has been provided at a microscopic scale. A piecewise linear model was developed to represent the nonlinear force-strain relation of geogrid tensile members and the model has been successfully used in the numerical compound tensile test.

Based on the unconfined tensile test in the laboratory, Wang et al. [53] established a 3D discrete element model of geogrid tensile test under lateral restraint conditions. The sensitivity of the tensile speed and overlying load to the tensile properties of the geogrid were analyzed. The interaction between the soil displacement and tensile deformation of the geogrid under the lateral restraint was discussed. The development of the porosity at the interface between the reinforcement and the soil was revealed. The results showed that the ultimate tensile strength of the geogrid can be increased by increasing the tensile speed and the overlying load.

4.3. Large-scale Triaxial Tests

Based on 3D discrete element modeling, Konietzky et al. [14] modeled large-scale triaxial tests using spheres to study the confinement zones of geogrids. For aggregates with particle radii between 2 and 11 mm, a single extruded biaxial geogrid layer had a confinement influence zone of approximately 100 mm on both sides of the geogrid. They also conducted further DEM modeling with three geogrid layers (see Figure 8a to investigate the effect of multiple geogrid layers on the displacement response of the sample. The numerical results showed that the vertical and radial displacements of the sample with three geogrid layers were reduced by approximately 50 % compared with those with only one geogrid layer. The geogrid interlocking effect and the soil arching effect could be clearly observed by the contact force distribution in the specimen, as shown in Figure 8b.

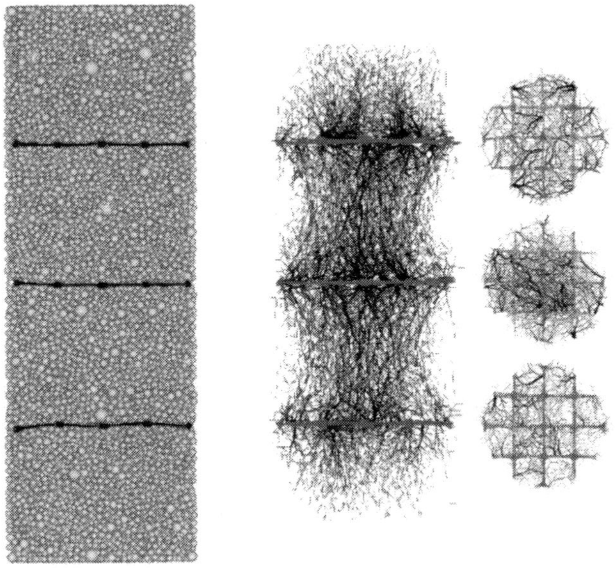

(a) Deformed sample with 3 geogrid layers (b) Contact force distributions in specimen and in geogrid layers

Figure 8. Deformed sample and contact force distributions in large-scale triaxial tests [14]

Mishra et al. [54] carried out numerical large-scale triaxial tests using an imaging-based DEM modeling approach to identify the optimal position for geogrid reinforcement in terms of the maximum shear strength. Specimens with five different reinforcement configurations using one geogrid or two geogrids were tested (see Figure 9). Placing two geogrids at 254 mm from top and bottom of the specimen (Configuration F in Figure 9) resulted in the maximum shear strength for both biaxial and triaxial geogrids.

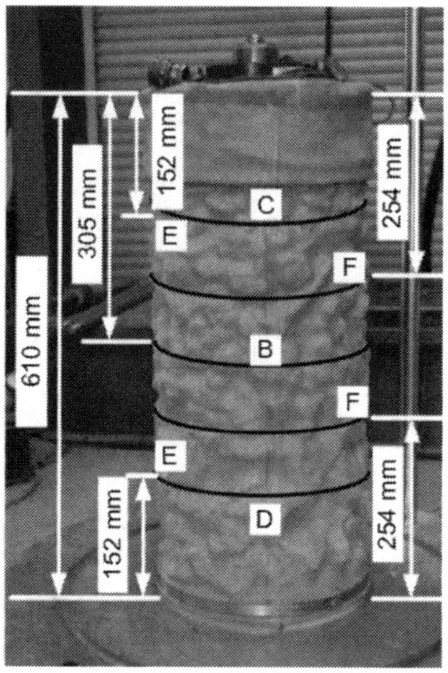

Configuration A: unreinforced,

Configuration B: one geogrid placed at specimen mid-height,

Configuration C: one geogrid placed at 152 mm from top of the specimen,

Configuration D: one geogrid placed at 152 mm from bottom of the specimen,

Configuration E: two geogrids placed at 152 mm from top and bottom of the specimen,

Configuration F: two geogrids placed at 254 mm from top and bottom of the specimen.

Figure 9. Reinforcement configuration in large-scale triaxial tests [54].

Wang et al. [55] established a numerical model of triaxial test using PFC^{3D}, and compared its modeling results with the experimental data. Their study analyzed the influence of reinforcement layers on the shear strength. The results showed that the peak strength, the cohesion and the internal friction angle increased with increasing reinforcement layers. The reinforcement effect of triaxial specimens under low confining pressures was higher than that of the high confining pressure. In the study of Yu et al. [56], monotonic triaxial shear strength tests were conducted on both new and degraded ballast materials with and without geogrids. The DEM modelling results showed the shear strength behavior of geogrid-reinforced ballast specimens prepared with both new and degraded ballast samples.

4.4. Direct Shear Tests

Direct shear tests were modeled by Bussert [57] to study the load transfer mechanisms in both unreinforced and reinforced systems. In both systems, the vertical stress was maintained constant and the lower shear box was moved horizontally. In the unreinforced model, the contact forces developed between the lower "shifted" wall and the opposite upper wall in an "S" shape with the main principal stress being horizontal in the shear plane. However, a different load transfer mechanism has been observed in the reinforced system, in which the geogrid was placed in the upper shear box near the shear plane. The horizontal forces acting on the lower "shifted" wall were restrained by the geogrid and thus almost no horizontal stresses developed in the composite material above the geogrid layer even at a large horizontal shear displacement (e.g. 10 mm), as shown in Figure 10. Similar observations have also been reported by Ngo et al. [28] in the DEM investigations of large-scale direct shear tests with unreinforced and geogrid reinforced ballast (see Figure 11).

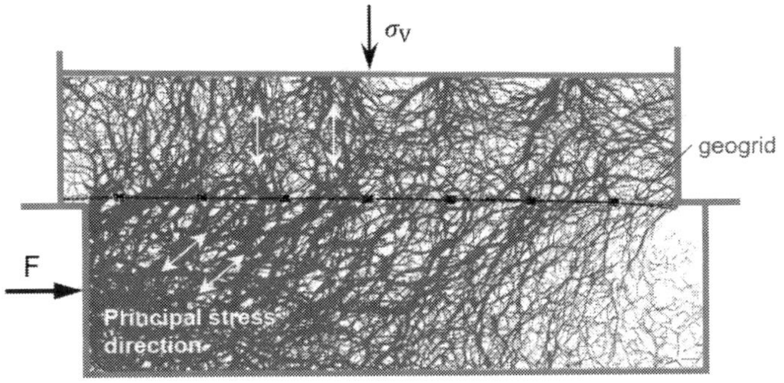

Figure 10. Contact force distribution in a direct shear test at a horizontal displacement of 10 mm [57]

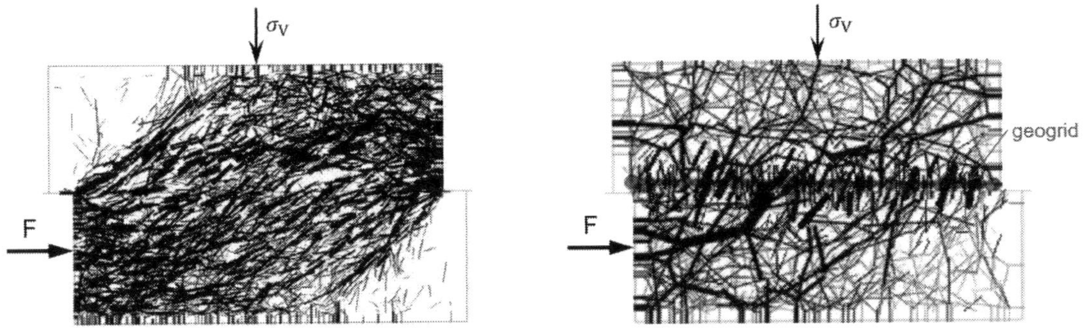

(a) Unreinforced ballast (b) Geogrid reinforced ballast

Figure 11. Contact force distributions in direct shear tests at a horizontal displacement of 18 mm [28]

In order to analyze the reinforcing performance of geogrid reinforced ballast, Shao et al. [58] established numerical models of direct shear tests. Their results indicated that lateral and vertical displacements were greatly reduced by the geogrid reinforcement. The geogrids could provide ballast particles with horizontal restraint and help to maintain the track geometry.

Ngo et al. [59] used large scale direct shear tests to study the shear stress-strain response of fresh and polluted ballast reinforced with geogrids. The experimental results showed that the geogrid provided extra internal confinement and interlocked the aggregates in its apertures, hence reduced ballast deformation. In addition, Ngo et al. [60] presented a study of the interface of geogrid reinforced subballast through a series of large-scale direct shear tests and discrete element modelling. Their investigation results shows that the shear strength of the interface was governed by the geogrids.

4.5. Plate Load Tests

A plate load test is commonly used to evaluate the benefits of placing geosynthetics in foundations [61]. Chen et al. [30] carried out 3D discrete element modeling of cyclic loads on ballast reinforced with biaxial and triaxial geogrids under confined and unconfined conditions. The ballast particles were simulated with 2-ball clumps. Compared with unreinforced samples, the settlement of geogrid reinforced specimens were greatly reduced. For confined box tests, an optimum location of a single biaxial geogrid layer was found to be 100 mm from the base and the triaxial geogrid showed better performance than the biaxial geogrid. Under unconfined conditions, the optimum location of a single geogrid layer was 50 mm from the sub-ballast layer. The use of two geogrid layers located both near the base and at mid-depth led to a slight improvement compared with the sample reinforced at 50 mm above the sub-ballast.

4.6. Wheel Load Tests

In order to develop the concept of prestressed reinforced soil during compaction, 3D discrete element modeling of geogrid–soil interaction under a compaction roller was carried out by Lackner and Semprich [62], as shown in Figure 12. Their DEM simulation results showed that the interaction between soil and geogrid could be defined as a composition of frictional and interlocking effects. The frictional effect was influenced by the surface roughness of soil particles and geogrid, while the interlocking effect depended on the ratio of geogrid aperture size to soil particle size.

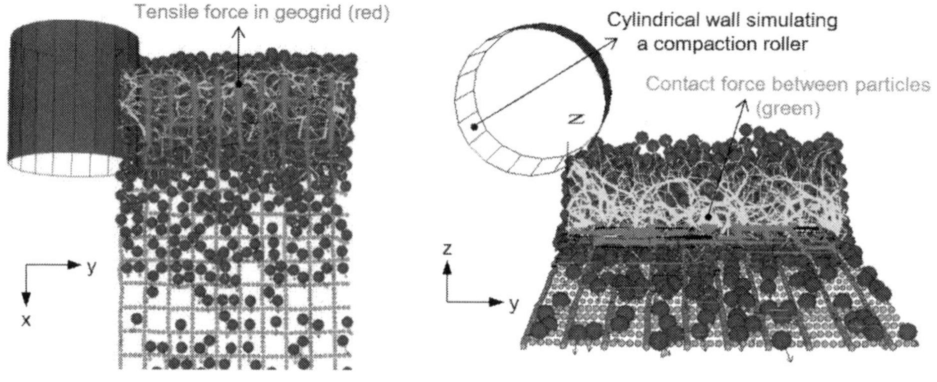

Figure 12. Discrete element modeling of geogrid–soil interaction under a compaction roller [62]

Bhandari et al. [38] carried out 2D discrete element modeling of unpaved geogrid reinforced granular bases under a vertical cyclic load. The geogrid reinforcement was placed at the mid-depth or the bottom of the base course. A vertical wheel load was applied cyclically on top of the base course. The reinforced bases showed better performance than the unreinforced base. Moreover, the geogrid placed at the mid-depth of the base performed better than that placed at the bottom of the base. Their numerical results showed that the geogrid tensile force was mobilized under the wheel load, which helped to widen the distribution of contact force chains in the geogrid reinforced system.

Based on 3D discrete element modeling, Jas et al. [63] investigated the performance of a geogrid stabilized sub-base under a running wheel load. Their simulation results showed that the rutting behavior of the sub-base was greatly improved by using a triaxial geogrid. The development of geogrid tensile forces and displacements were visualized at different loading cycles.

5. Summary

The DEM, which has particular advantages of capturing detailed insights into the geogrid–soil interaction, provides a new perspective to study the geogrid reinforcement mechanism, especially at a microscopic scale. Current achievements of discrete element modeling of geogrid and soil as well as their interaction under different testing conditions has been summarized in this study. The DEM investigations provide not only qualitative visualization of the geogrid–soil interaction but also quantitative response of the geogrid and its surrounding soil. Such results provides helpful hints to optimize the design of geogrid reinforced soil systems.

However, due to the varying elastic-plastic properties of geogrids and soils as well as the high sensitivity of their interaction to many influencing factors, the complicated geogrid–soil interaction as well as the compound stress–strain behavior of geogrid reinforced soil have not been described conclusively up to now. Therefore, it is still necessary to conduct further experimental and DEM investigations to describe the reinforcement mechanisms of geogrids embedded in soil.

Acknowledgement

The authors appreciate the financial support from the National Natural Science Foundation of China (No. 51709175), the Science and Technology Research Projects of Universities in Hebei Province (BJ2020045).

References

[1] Yoo C, Kim S-B 2008 Performance of a two-tier geosynthetic reinforced segmental retaining wall under a surcharge load: full-scale load test and 3D finite element analysis *Geotextiles and Geomembranes* **26** pp 460-472

[2] Leshchinsky D, Imamoglu B, Meehan CL 2010 Exhumed geogrid-reinforced retaining wall. *Journal of Geotechnical and Geoenvironmental Engineering* **136** pp 1311-23

[3] Al-Qadi I L, Dessouky S, Tutumluer E, Kwon J 2011 Geogrid mechanism in low-volume flexible pavements: accelerated testing of full-scale heavily instrumented pavement sections. *International Journal of Pavement Engineering* **12** pp 121-135

[4] Yang G, Liu H, Lv P, Zhang B 2012 Geogrid-reinforced lime-treated cohesive soil retaining wall: Case study and implications *Geotextiles and Geomembranes* **35** pp 112-8

[5] Koerner R M, Koerner GR 2013 A data base, statistics and recommendations regarding 171 failed geosynthetic reinforced mechanically stabilized earth (MSE) walls *Geotextiles and Geomembranes* **40** pp 20-7

[6] Allen T M, Bathurst R J 2014 Design and performance of 6.3-m-high, block-faced geogrid wall designed using k-stiffness method *Journal of Geotechnical and Geoenvironmental Engineering* **140** 04013016

[7] Han B, Ling J, Shu X, Song W, Boudreau R L, Hu W, Huang B 2019 Quantifying the effects of geogrid reinforcement in unbound granular base. *Geotextiles and Geomembranes* **47** pp 369-76

[8] Palmeira E M 2009 Soil–geosynthetic interaction: Modelling and analysis. *Geotextiles and Geomembranes* **27** pp 368-90

[9] Moraci N, Recalcati P 2006 Factors affecting the pullout behaviour of extruded geogrids embedded in a compacted granular soil. *Geotextiles and Geomembranes* **24** pp 220-42

[10] Ruiken A 2013 Zum Spannungs-Dehnungsverhalten des Verbundbaustoffs „geogitterbewehrter Boden". RWTH Aachen University (in German)

[11] Peng X, Zornberg J G 2019 Evaluation of soil-geogrid interaction using transparent soil with laser illumination *Geosynthetics International* **26** pp 206-221

[12] Cundall PA 1971 A computer model for simulating progressive, large scale movements in blocky rock systems *Symposium of Rock Fracture*, Nancy, France

[13] Cundall P A, Strack O D L 1979 A discrete numerical model for granular assemblies. *Géotechnique* **29** pp 47-65

[14] Konietzky H, te Kamp L, Groeger T, Jenner C 2004 Use of DEM to model the interlocking effect of geogrids under static and cyclic loading. 2nd International PFC Symposium, Kyoto, Japan

[15] Miao C-x, Zheng J-j, Zhang R.-j, Cui L 2017 DEM modeling of pullout behavior of geogrid reinforced ballast: the effect of particle shape *Computer and Geotechnics* **81**pp 249-61

[16] Gao G, Meguid M A 2018 Effect of particle shape on the response of geogrid-reinforced systems: insights from 3D discrete element analysis *Geotextiles and Geomembranes* **46** pp 685-98

[17] Wang Y-H, Leung S-C 2008 A particulate-scale investigation of cemented sand behavior. *Canadian Geotechnical Journal* **45** pp 29-44

[18] Bhandari A, Han J 2010 Investigation of geotextile–soil interaction under a cyclic vertical load using the discrete element method *Geotextiles and Geomembranes* **28** pp 33-43

[19] Lin Y L, Zhang M X, Javadi A A, Lu Y, Zhang S L 2013 Experimental and DEM simulation of sandy soil reinforced with H–V inclusions in plane strain tests *Geosynthetics International* **20** pp 162-173

[20] Tran V D H, Meguid M A, Chouinard L E 2013 A finite-discrete element framework for the 3D modeling of geogrid–soil interaction under pullout loading conditions *Geotextiles and Geomembranes* **37** pp 1-9

[21] Zhang M X, Qiu C C, Javadi A A, Lu Y, Zhang S L 2013 Discrete-element method simulation of a model test of an embankment reinforced with horizontal–vertical inclusions *Geosynthetics International* **20** pp 238-51

[22] Li L, Holt R M 2002 Particle scale reservoir mechanics *Oil & Gas Science and Technology* **57** pp 525-58

[23] Lu M, McDowell G R 2007 The importance of modelling ballast particle shape in the discrete element method *Granular Matter* **9** pp 69-80

[24] Ferellec J, McDowell G 2010a Modelling realistic shape and particle inertia in DEM *Géotechnique* **60** pp 227-32

[25] Ferellec J-F, McDowell G R 2010b A method to model realistic particle shape and inertia in DEM *Granular Matter* **12** pp 459-67

[26] Lackner C 2012 Prestressed reinforced soil – concept, investigations and recommendations. Graz University of Technology

[27] Chen C, McDowell G R, Thom N H 2014 Investigating geogrid-reinforced ballast: experimental pull-out tests and discrete element modelling. *Soils and Foundations* **54** pp 1-11

[28] Ngo N T, Indraratna B, Rujikiatkamjorn C 2014 DEM simulation of the behaviour of geogrid stabilised ballast fouled with coal *Computers and Geotechnics* **55** pp 224-31

[29] Stahl M, Konietzky H, te Kamp L, Jas H 2014 Discrete element simulation of geogrid-stabilised soil *Acta Geotechnica* **9** pp 1073-84.

[30] Chen C, McDowell G R, Thom N H 2012 Discrete element modelling of cyclic loads of geogrid-reinforced ballast under confined and unconfined conditions. *Geotextiles and Geomembranes* **35** pp 76-86

[31] Gu X, Huang M, Qian, J 2014 Discrete element modeling of shear band in granular materials. *Theoretical and Applied Fracture Mechanics* **72** pp 37-49

[32] Rui R., van Tol A F, Hu G, Xia X L 2015 Large size plane strain compression tests on sands and 2D DEM calibrations. XVI European Conference on Soil Mechanics and Geotechnical Engineering, Edinburgh, UK

[33] Rui R, van Tol F, Xia X-L, van Eekelen S, Hu G, Xia Y-Y 2016 Evolution of soil arching: 2D DEM simulations *Computers and Geotechnics* **73** pp 199-209

[34] Indraratna B, Thakur P K, Vinod J S 2010 Experimental and numerical study of railway ballast behavior under cyclic loading *International Journal of Geomechanics* **10** pp 136-44

[35] McDowell G R, Harireche O, Konietzky H, Brown S F, Thom N H 2006 Discrete element modelling of geogrid-reinforced aggregates. *Proceedings of the ICE - Geotechnical Engineering* **159** pp 35-48

[36] Zhang J, Yasufuku N, Ochiai H 2008 Discrete element modelling of geogrid pullout test. 4th Asian Regional Conference on Geosynthetics, Shanghai, China

[37] Han J, Bhandari A, Wang F 2012 DEM analysis of stresses and deformations of geogrid-reinforced embankments over piles *International Journal of Geomechanics* **12** pp 340-50

[38] Bhandari A, Han J, Parsons R L 2015 Two-dimensional DEM analysis of behavior of geogrid-reinforced uniform granular bases under a vertical cyclic load *Acta Geotechnica* **10** pp 469-480

[39] Wang Z, Jacobs F, Ziegler M 2016 Experimental and DEM investigation of geogrid-soil interaction under pullout loads. *Geotextiles and Geomembranes* **44** pp 230-46

[40] Itasca 2008 Particle Flow Code in 2 Dimensions. Itasca Consulting Group Inc. *Minneapolis*, USA

[41] Shinoda M, Bathurst R J 2004 Lateral and axial deformation of PP, HDPE and PET geogrids under tensile load *Geotextiles and Geomembranes* **22** pp 205-222

[42] Ezzein FM, Bathurst R J, Kongkitkul W 2015 Nonlinear load–strain modeling of polypropylene geogrids during constant rate-of-strain loading. *Polymer Engineering & Science* **55** pp 1617-27

[43] Vieira C S, Pereira P M 2015 Damage induced by recycled construction and demolition wastes on the short-term tensile behaviour of two geosynthetics *Transportation Geotechnics* **4** pp 64-75

[44] Zhang J, Yasufuku N, Ochiai H 2007 A few considerations of pullout test characteristics of geogrid reinforced sand using DEM analysis *Geosynthetics Engineering Journal* 22pp 103-10

[45] Ferellec J-F, McDowell G R 2012 Modelling of ballast-geogrid interaction using the discrete-element method *Geosynthetics International* **19** pp 470-9

[46] Chen C, McDowell G R, Thom N H 2013 A study of geogrid-reinforced ballast using laboratory pull-out tests and discrete element modelling *Geomechanics and Geoengineering* **8** pp 244-53

[47] Zheng J, Miao C, Zhang J, Xie M 2013a Study on mesoscopical properties of interface between geogrid and sand by DEM *Journal of Huazhong University of Science and Technology (Natural Science Edition)* **41** pp 5-9 (in Chinese)

[48] Zheng J.-j, Miao C-x, Xie M.-x, Zhang J 2013b Interface properties and influence of particle size on geogrid reinforcement performance by DEM *Chinese Journal of Geotechnical Engineering* **35** pp 1423-8 (in Chinese)

[49] Miao C-x, Zheng J-j, Cui M-j, Xie M-x, Zhao J-b 2014a Discrete element simulation of interface characteristic and frictional properties of triaxial geogrid *Rock and Soil Mechanics* **35** pp 423-30 (in Chinese)

[50] Miao C-x, Zheng J-j, Cui M-j, Xie M-x, Zhao J-b 2014b Study of influence of joint protuberance on geogrid reinforcement performance by discrete element method *Rock and Soil Mechanics* **35** pp 1181-6 (in Chinese)

[51] Wang Z J, Yang GQ, Wang H, Liu W C 2019 Mesoscopic numerical studies on geogrid-soil interface behavior under rigid and flexible top boundary conditions *Chinese Journal of Geotechnical Engineering* **41** pp 967-73 (in Chinese)

[52] Wang Z J, Jacobs F, Ziegler M 2014 Visualization of load transfer behaviour between geogrid and sand using PFC2D *Geotextiles and Geomembranes* **42** pp 83-90

[53] Wang J Q, Huang W Q, Zhang L L, Liu Z Q 2018 Discrete element simulation of tensile test of geosynthetics subjected to lateral confinement *Journal of Guangxi University (Natural Science Edition)* **43** pp 1482-9 (in Chinese)

[54] MishraD, Qian Y, Kazmee H, Tutumluer E 2014 Investigation of geogrid-reinforced railroad ballast behavior using large-scale triaxial testing and discrete element modeling *Transportation Research Record* (2462) pp 98-108

[55] Wang J Q, Zhang L L, Chen Y J, Shi C H 2017 Mesoscopic analysis of reinforced sand triaxial test using PFC3D *Journal of Hydraulic Engineering* **48** pp 426-34. (in Chinese)

[56] Yu Q, Mishra D, Tutumluer E, Kazmee H A 2015 Characterization of geogrid reinforced ballast behavior at different levels of degradation through triaxial shear strength test and discrete element modeling *Geotextiles and Geomembranes* **43** pp 393-402

[57] Bussert F 2009 Recent research into the actual behaviour of geogrids in reinforced soil *Jubilee Symposium on Polymer Geogrid Reinforcement*, London, UK

[58] Shao S, Yan Y, Ji S Y 2016 Discrete-finite element analysis of dynamic behaviors of ballasted railway with geogrid reinforcement *Chinese Journal of Solid Mechanics* **37** pp 444-55 (in Chinese)

[59] Ngo N T, Indraratna B, Rujikiatkamjorn C 2016 Modelling geogrid-reinforced railway ballast using the discrete element method *Transportation Geotechnics* **8** pp 86-102

[60] Ngo N T, Indraratna B, Rujikiatkamjorn C 2017 A study of the geogrid-subballast interface via experimental evaluation and discrete element modelling *Granular Matter* **19** pp 1-16

[61] Tingle J S, Jersey S R.2005 Cyclic plate load testing of geosynthetic-reinforced unbound aggregate roads *Transportation Research Record* (1936) pp 60-9

[62] Lackner C, Semprich S 2010 Prestressed geosynthetic reinforced soil by compaction. 9th International Conference on Geosynthetics, Brazil

[63] Jas H, Stahl M, te Kamp L, Konietzky H, Oliver T 2015 Discrete element simulation: modelling and analysis of a geogrid stabilized sub-base while loaded with a moving wheel. XVI European Conference on Soil Mechanics and Geotechnical Engineering, Edinburgh, UK

Finite Difference Modelling of Deformable Geosynthetics-Reinforced Barriers (DGRB) under the dynamic impact of debris avalanches

S Moretti[1], S Cuomo[2] and L Frigo[3]

[1] Department of Engineering, University of Naples Parthenope", Centro Direzionale, Isola C4, 80133 Naples, Italy
[2] GEG (Geotechnical Engineering Group), University of Salerno, Via Giovanni Paolo II, 132, 84084, Fisciano, Salerno, Italy
[3] Geosintex s.r.l., Via Leonardo da Vinci, 12, 36066 Sandrigo, Vicenza, Italy

Email: sabrina.moretti@uniparthenope.it

Abstract. The behaviour of Deformable Geosynthetics-Reinforced Barrier (DGRB) is investigated with reference to the dynamic impact of debris avalanches. The aim of these barriers is to protect life of inhabitants and reduce the threat for the exposed structures. A Finite Difference Method code (FLAC, Itasca) is used and the dynamic impact is simulated as an external pressure applied over time to the barrier. The peak lateral pressure exerted by a debris avalanche is evaluated by a literature formulation including a flow-height dependent static component and a flow-velocity dependent dynamic component. Real values of height and velocity of debris avalanches are taken from previous research works. The trend of the pressure during the impact is obtained from the landslide volume and the impulse theorem. The DGRBs are composed of: i) coarse-grained material, ii) bidirectional PET geogrids; iii) external formwork. The soil-geosynthetics contact is simulated through a frictional interface. The several components used in the wrap-around technique are simulated following two different schematizations and the differences between them are highlighted. The used mathematical formulation allows reproducing the impacted barrier both in small and large displacement cases. The local and global displacements of the barrier, its failure mechanisms and the soil-geogrid interaction inside the barrier are investigated. The results outline the differences between the two ways of modelling the wrap-around technique. In particular, the computed global displacements are comparable in the two computational schemes while the simulated stress-strain responses of the structural elements are somehow different.

1. Introduction

Debris avalanches are a type of fast landslides that largely increase in volume during the propagation stage and hence can be highly destructive for structures and very hazardous for people. In fact, some past debris avalanches caused significant loss of life all over the world [1,2]. Mitigation works are important to reduce the risk posed by debris avalanches, which Hungr et al. [3[defined as "very rapid to extremely rapid shallow flow partially or fully saturated debris on a steep slope without confinement in an established channel".

The main features of debris avalanches are: (i) limited height of unstable soil inside the source areas (generally a few meters); (ii) significant lateral spreading, favored by the open slope – i.e. absence of channel, drainage line or ravine, where debris avalanches propagate; (iii) relevant entrainment of further material from the ground surface, which increases more the lateral spreading [4,5]..

It is worth noting that the specific location of each potential landslide source areas is very difficult to individuate at territorial scale, being the susceptible slopes not easily recognizable without detailed field investigations and accurate slope analyses. However, we can also notice that, compared to other types of flow-like landslides, debris avalanches have lower mobility and generally stop at the toe of the slope (Figure 1).

From those specific features, recent researches have shown that a linear-type protection barrier installed at toe of the slope could be a valid option for reduce the landslide propagation inside designated areas and thus the consequence [6,7]. Such barrier should: (i) be located downslope the slope angle break, which often exists in many geo-environmental contexts, and (ii) run along the piedmont area at the base of those slopes individuated as the most susceptible to fail. In those hypotheses, what can happen is that the barrier: (i) stops the flow, or (ii) reduces the landslide propagation but then it is overpassed by the flow. In both cases, the barrier must resist to the impulsive impact of flow arriving at high velocity.

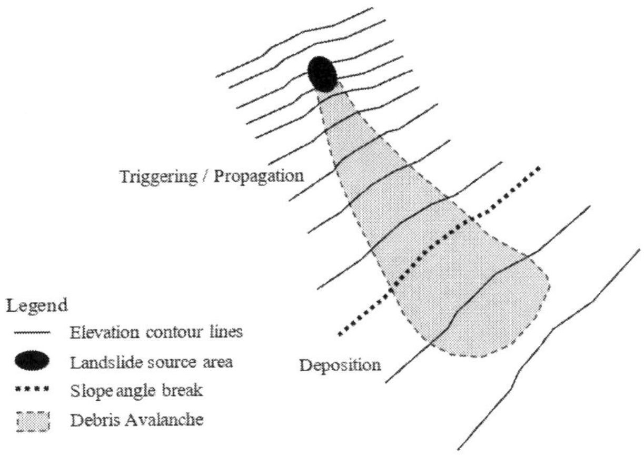

Figure 1. Scheme of debris avalanche [8]

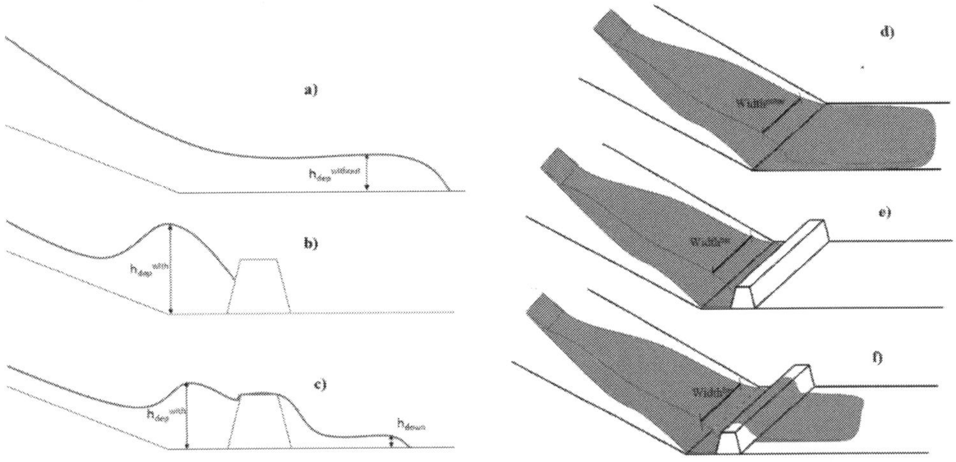

Figure 2. Different scenarios of flow propagation: a, d) natural slope; b, e) slope engineered with a barrier capable to stop the flow; c, f) flow overpassing the barrier [6].

In this paper, we will refer to the scenario that the barrier is not overtopped and contain the flow behind (Figure 2b,e). For the sake of the simplicity, we will also assume that the flow height is constant during the impact process. Indeed, it depends on the impact mechanism of the flow against what we can define as an "obstacle". On that topic there is much literature based on analytical studies, experimental works or field evidence. For instance, Faug [9] indicate that the flow-obstacle interaction for flow-like landslides is driven by the combination of: (i) the Froude number and (ii) the obstacle height relative to the flow height. The Froude number is equal to $v/\sqrt{gh}$ (with v and h the flow velocity and propagation height, respectively and g the gravity acceleration), and represents the ratio between inertial forces and gravity forces inside the moving mass. Thus, fundamental features of the landslide to be considered are v and h. But another factor is also important, i.e. the volume of the impacting mass, or better said, the volume of the aliquot of the landslide effectively contributing to the impact mechanism. This assessment is not easy because a landslide of the flow-type propagates over a 3D topographic surface, and so during the impact some parts of the mass are diverted elsewhere or stopped or overloaded by further material arriving from behind.

Taking into account those features, the barrier should have some requisites and ensure some performances. (i) The barrier should be sufficiently high to not be overpassed but not excessively wide at the base in order to reduce the cost of the work, it means to have a steep front. (ii) The barrier should be deformable to dissipate the impact energy of the flow. (iii) A global displacement of the barrier could be also allowed within certain values, as this helps dissipating an aliquot of impact energy. (iv) The barrier could be done of coarse grained material able to eventually dissipate some of excess pore water pressure developed in the flowing mass during the impact. (v) Due to the linear extension of such barrier system, three fundamental issues to be tackled are: saving of construction material, using of well known construction method and limitation of the costs.

Geosynthetics, and specifically geogrids, may serve to accomplish all these requirements if the barrier is conceived as a GRS (Geosynthetic Reinforced Soil) wall. In fact, geogrids installed in coarse-grained soil and wrapped around the front can easily allow having high structures with external steep front. As we will see a barrier of geogrid-reinforced granular material is capable to resist, keep its integrity but at the same time deform as a whole under an impulsive impact. Such barrier could be also shifted along the geogrid at the base, so dissipating some part of the impact energy.

Aimed to indicate such a novel solution for mitigation of debris avalanche propagation Cuomo et al. (2019b, 2019c) proposed the acronym of DGRB (Deformable Geosynthetics-Reinforced Barriers). In this paper we will focus on the stress-strain response of the geogrids under the impact.

2. Methods and input data

An example of barrier is here analysed in a 2D cross-section and the geometry is taken from Cuomo et al. [10]. The barrier is trapezoidal, 6 m high and equipped with 10 geogrids 0.6 m vertically spaced. The base width is 11 m and the top equal to 4 m (Figure 3a). The Finite Difference Method (FDM) commercial FLAC [11] code is used for running the dynamic analyses of the impact process, here schematized as the application of an external horizontal pressure over the impact height, being the latter assumed not to change during the impact. The average mesh size is equal to 0.5 m for the barrier, and the base computational domain is equal to 50 m x 20 m with the maximum element size equal to 0.8 m. The time step is selected equal to 0.1 s taking into account both the properties of the materials and the mesh size considered (Figure 3b). A control point P is taken into account in the core of the barrier order to track the displacements of the barrier (Figure 3a). The wrap-around technique is here considered due to the higher tensile stress obtainable compared to other reinforcement techniques [12].

Several combinations of materials are taken into account. In particular, two different granular soils are considered for the core of the barrier and a non-associative Mohr-Coulomb model is used for modelling. Two types of bidirectional geogrids are also implemented in the numerical model, with the maximum tensile strength equal to 50 kN/m and 100 kN/m and respectively named *Geogrid 50* and *Geogrid 100*.

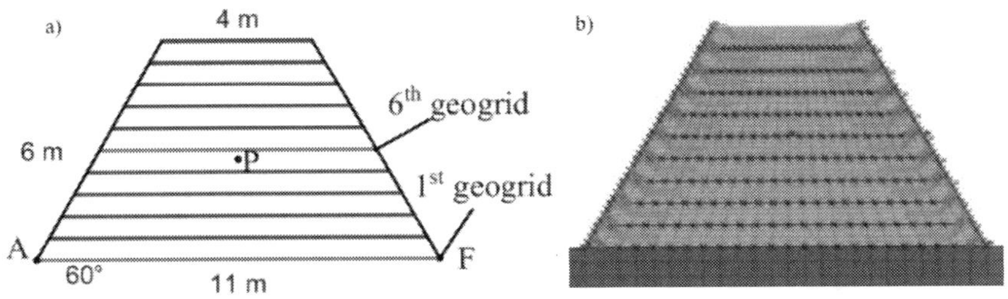

Figure 3. a) Geometry considered in the analyses; b) FDM mesh used for calculation

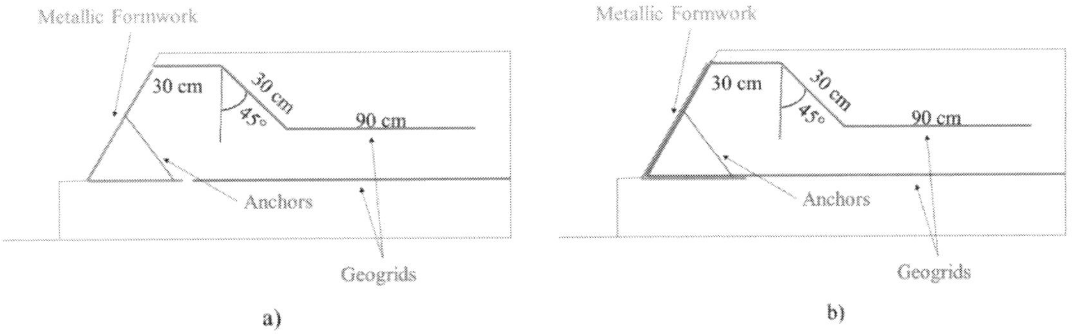

Figure 4. Details of the models implemented: a) continuous geogrid wrapping around the soil (*Type A*); b) geogrid interrupted in correspondence of the metallic formwork (*Type A**)

The geogrids are schematized as linear elements only resistant to tensile axial force. The soil-geogrid interface is assumed elasto-plastic with frictional strength properties reduced to 80% of those of the nearby soil. The metallic formwork is schematized as "beam" element both traction and bending resistant. The metallic anchors, generally one 8 mm dia. bar each 3 m, used in the practice to keep the design slope angle of the front during the construction stages, are also simulated in the model as "anchor" elements resistant to traction.

The wrap-around technique is simulated following two different schemes. A sketch of the two different models here implemented is shown in Fig. 4. In the first model (Figure 3a, *Type A*), it is assumed that the geogrid does not slide along the metallic framework and so the geogrid is interrupted at both the lower and the upper extremes of the metallic framework. In the second model (Figure 3b, *Type A**), it is allowed that geogrid can slide respect to the metallic framework, and hence the horizontal geogrid is continuous and wraps around the soil. The first model is easier to implement and less time consuming while the second model should be in principle more accurate. So, it make sense to compare the results of both models also to acquire an estimate of the possible effect of relative shifting of the geogrids along the front of the barrier during the impact action.

All the mechanical properties of the materials and their combinations are reported in Table 1. The Rayleigh approach is used to simulate the damping and the representative coefficient are the center of frequency f_{min} and the critical damping ratio ξ_{min}, in turn related to the coefficient α and β assumed respectively equal to 0.01 and 0.001 [13].

The impact pressures have been computed as the sum of a height-dependent component (Eq. 1) and a velocity-dependent component (Eq. 2) [14]. Two simplified hypotheses are taken into account for the impact: i) the barrier is not overtopped by flow; ii) the height of the impacting flow is constant during the impact process and flow-barrier interaction.

$$ILP_{peak}{}^{stat} = 0.5\ K_p\ \gamma\ h_{impact} \quad (1)$$

$$ILP_{peak}{}^{dyn} = k\ v^2\ \gamma/g \quad (2)$$

where Kp is the passive earth pressure coefficient of landslide; h_{impact} is the landslide height considered constant during the impact action, b is the landslide width, γ is the unit weight of debris flow here assumed equal to 18 kN/m^3, k is an impact coefficient assumed equal to 5, v is landslide velocity. ILP_{peak} is the peak pressure give by the sum of ILP_{peak}^{stat} and ILP_{peak}^{dyn}.

The variation of ILP over time (Figure 5) is obtained through the impulse theorem as shown in Cuomo et al. [15] and the several scenarios here analyzed are listed in Table 2.

Table 1. Material properties

	γ_{dry}	φ	k_0	ν	E	Tu	EA	EI	Combo			
	(kN/m^3)	(°)	(-)	(-)	(kPa)	(kN/m)	(kN/m)	(kN/m^2)	1	2	3	4
Soil #1	18	33	0.45	0.25	2.0e4				X	X		
Soil #2	20	38	0.38	0.25	1.5e4						X	X
Geogrid 50						50	500		X		X	
Geogrid 100						100	1000			X		X
Subsoil	18	30	0.5	0.29	1.0e4			35	X	X	X	X
Formwork							6.63e4		X	X	X	X
Anchor							2.11e4		X	X	X	X

γ_{dry} is the unit weight of granular soil; c' is the cohesion equal to 0.1; φ is the friction angle; k_0 is the dilatation angle; ψ is the earth pressure; ν is the Poisson coefficient; E is the soil Young modulus; Tu is the ultimate tensile stress of geogrids; EA is the extensional stiffness; EI is the bending stiffness. Combo: Combination of materials; * interface friction assumed equal to 0.8 tan(φ) of the confining material

Table 2. List of scenarios

	v (m/s)	h_{impact} (m)	b_{impact} (m)	L_{impact} (m)	V_{impact} (m^3)	ILP_{peak} (kPa)	T_{peak} (s)	Impact Energy (MJ)
I1	10	3	1	9	27	921	0.18	2.47
I2	10	3	1	15	45	921	0.30	4.13
I3	5	3	1	9	27	246	0.34	0.62
I4	5	3	1	15	45	246	0.56	1.03
I5	3	3	1	9	27	102	0.49	0.23
I6	3	3	1	15	45	102	0.81	0.37
I7	2	3	1	9	27	57	0.58	0.01
I8	2	3	1	15	45	57	0.97	0.16

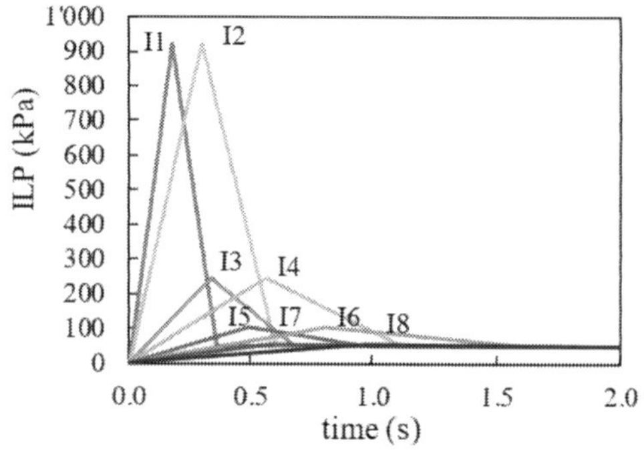

Figure 5. Impact Lateral Pressure (ILP) over time for the 8 scenarios of Table 2

3. Dynamic analyses

3.1. Computed displacements

The results of the dynamic analyses in terms of the displacements are showed in Figure 6. The displacements change in a wide range for the list of scenarios analysed. Some comments can be first done in relation to the impact considered: for the high energy impact scenarios, the two models, *Type A* and *Type A**, do not differ in terms of displacements (namely impact *I1* and *I2*, Figure 6a) while the combinations of materials used change the final computed displacement. The maximum displacement is equal to about 12 m and as expected it is simulated for the impact *I2*, irrespective of if *Combo 1* and *Combo 2* are considered. The same impact causes a displacement equal to about 9 m for the *Combo 4* (the combination of the best materials). In the other cases, the displacements of *I1* are comprised between 3 m and 5 m that presents the same peak value of *ILP* of the *I2*, but different peak time. For *I4* displacement ranges between 0.8 m and 1.4 m, and for *I3* displacements are between 0.4 m and 0.6 m (Fig. 6b).

On the other hand, the maximum displacement of the barrier is influenced by the modelling scheme used once we focus on the lowest energetic impact scenarios (*I5* to *I8*). The displacements of *I5* and *I6* are comparable between them such as the displacements of *I7* and *I8* that reach the same ILP_{peak} for the different combinations of material. The *Type A* barrier has higher displacement than *Type A**, in particular, the displacements of *Type A* are equal to about 0.12 m and 0.08 m respectively, for impacts *I5* (or *I6)* and *I7* (or *I8)*; the displacements of *Type A** are equal to 0.02 for all impacts (*I5* to *I8*).

The displacements are more influenced by the filling material used than the geogrid used. Indeed, the displacements are the same for combinations of material that present the same filling soil such as *Combo 1* and *Combo 2;* the same applies also for *Combo 3* compared to *Combo 4*..

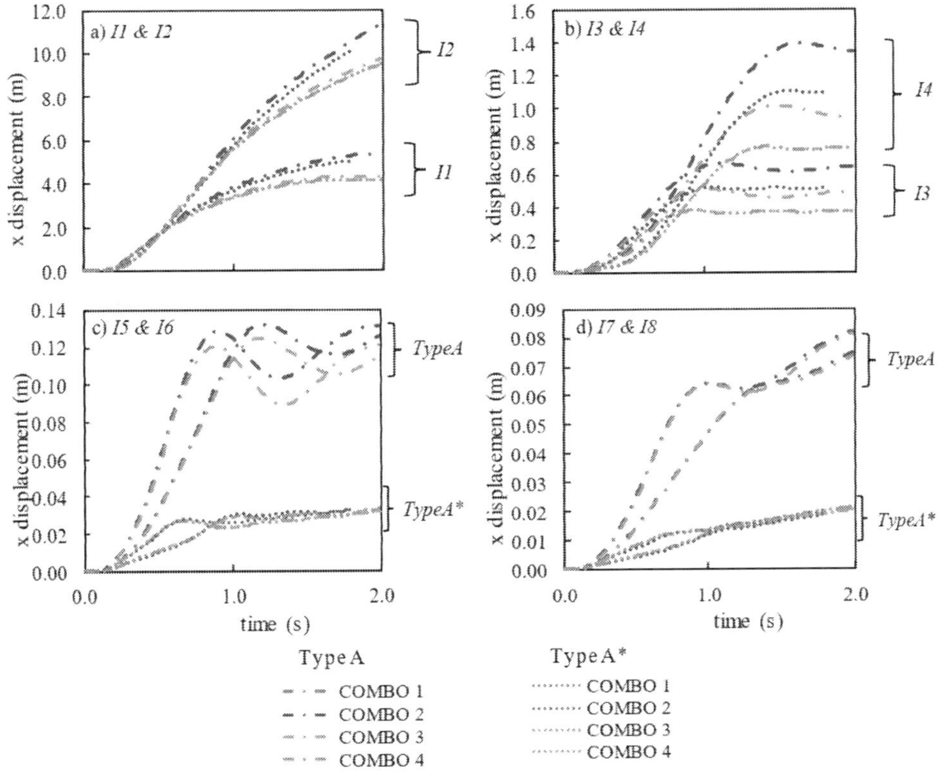

Figure 6. Horizontal displacements computed versus time for the two geometries (*Type A* and *Type A**), for different combinations of materials (Combo 1-4) and impact scenarios (*I1-I4*).

3.2. Soil-geosyntetics interaction

The selection between the two modelling schemes does not affect the computed displacements for the impacts *I1* to *I4*, while the stress distribution computed inside the geogrids are different. The impact *I4* is chosen to evaluate the axial force of geogrids and to evaluate the differences in the interaction soil-geogrid. Figure 7 shows the axial forces mobilized in the geogrids in both the barriers *Type A* and *Type A** considering the *Combo 1* of materials. It can be noted that at the impact side, the computed axial forces are higher in *Type A** barrier than in *Type A* barrier, while the axial forces are comparable in the remaining part of the barrier. This result can be explained observing that the geogrid in *Type A* negligibly contributes to the response of the barrier under the impact. However, the maximum value of the axial force is about 30 kN/m in both the barriers (about 60% of the maximum value available of *Geogrid 50*). Similar results are recorded for *Combo 2*, which differs from *Combo 1* only for the geogrid used (*Geogrid 100*). Also in terms of axial stresses, there is no difference between the combination of material with the same filling soil and different geogrids. The maximum axial force is still about 30 kN/m in both the combinations. The explanation is given by the fact that the limit state of the whole barrier would not be reached for the ultimate axial force into the reinforcements, but for the frictional failure at the contact between the soil and the geogrids. During the impact the geogrid allows the relative shifting of the reinforced soil layers. This internal deformation of the barrier surely dissipates a part of the energy of the debris avalanche.

Similar results are recorded also for low impact energy scenarios where the impact energy is lower than 0.37 MJ and the displacements are lower than 0.15 m. The *Combo 1* of materials is taken into account to show the results. The distribution of the stresses inside the geogrids is opposite to the high impact energy scenarios; in particular, the most stressed geogrids are recorded in the upper part of the geometry; instead, the lowest values are in correspondence of the first layers.

The most stressed geogrid in *Type A** is in correspondence of the impact side at the ground level; while in the *Type A*, the same geogrid is stressed less, this because the geogrid is interrupted in correspondence of the formwork. The values of tensile stress reach 15 kN/m in *I5* and *I6* where the maximum recorded tensile stress is equal to 20 kN/m (Figure 8). For impacts *I7* and *I8* the tensile stress is not reached due to the low value of impact energy (less than 0.16 MJ); therefore, also for impact *I4* (with impact energy equal to 1.03 MJ) the tensile stress is lower than 15 kN/m; this is due to the deformation of the geogrid and the subsequently shifting that reduces the stress in that segment and move the stress along the geogrid in the follow segment. This mechanism allows dissipating the impact energy of the flow for the several impacts.

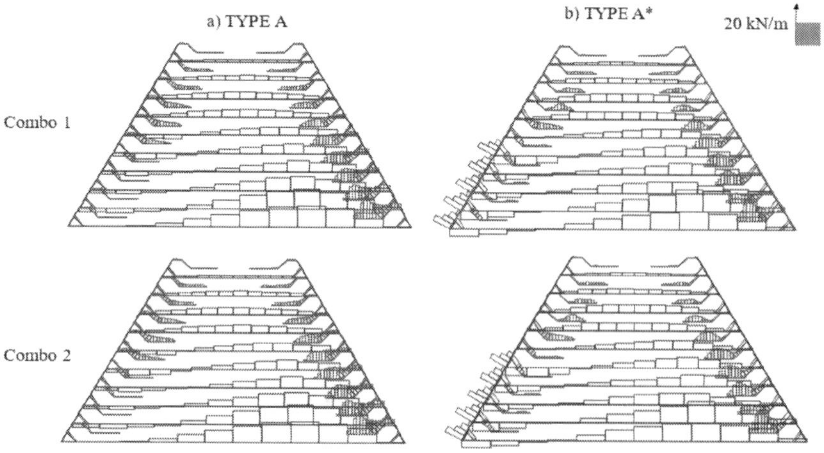

Figure 7. Axial forces of geogrids for barrier a) *Type A* and b) *Type A** related to *Combo 1* and *Combo 2* and impact *I4* at the end of the third phase

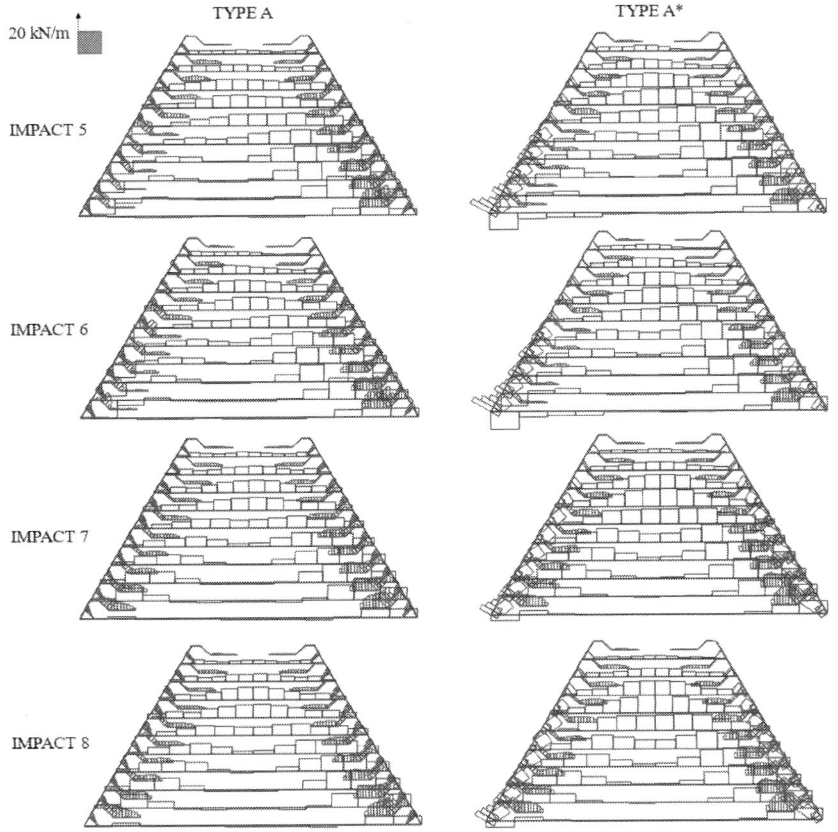

Figure 8. Axial forces of geogrids for barrier for low impact energy for *Combo 1*

4. Conclusions

This paper analyzed the results obtained from two different modelling schemes of the wrap around technique considering the same geometry of barrier. Firstly, the displacements of an inner point were evaluated, and it was ascertained that the different modelling scheme does not influence the maximum displacement of the barrier for the highest energy impact scenarios (*I1* to *I4*). On the contrary, the displacement is influenced by the modelling option for the lowest impact scenarios (*I5* to *I8*).

In all cases, the impact energy of the flow is totally dissipate by the deformation of the barrier; for high values of impact energy, the displacements varies between 0.4 m and 10 m; while the low values of impact energy give displacements comprised between 0.01 m and 0.14 m.

However, the modelling influences the results obtained for soil-geosynthetics interaction as well, being the maximum axial forces of geogrids higher in correspondence of the impact side in *Type A** barrier than in *Type A* if *Combo 1* or *Combo 2* of materials are taken into account. The use of different geogrids did not influence the final displacement and the axial forces of geogrids; the choice of the geogrid is linked to the design under static condition of the barrier; while, the geogrids contribute to dissipate the energy during the impact.

Acknowledgments

This research was developed within the framework of different projects: i) Industrial Partnership PhD Course (Dottorato Industriale Regione Campania, Italia); ii) project FARB 2017 "Modellazione numerica e analisi inversa per frane tipo flusso" funded by the Italian Education and Research Ministry. The authors would like to thank Geosintex s.r.l. for support to the research.

References

[1] Revellino P, Guerriero L, Grelle G, Hungr O, Fiorillo F, Esposito L, Guadagno F M 2013 Initiation and propagation of the 2005 debris avalanche at Nocera Inferiore (Southern Italy) *Italian Journal of Geosciences* **132** pp 366-79

[2] Cascini L., Cuomo S, Pastor M, Rendina I 2016 SPH-FDM propagation and pore water pressure modelling for debris flows in flume tests *Engineering Geology* **213** pp 74-83

[3] Hungr O, Evans S G, Hutchinson I N 2001 A Review of the Classification of Landslides of the Flow Type *Environmental & Engineering Geoscience* **7** pp 221-38

[4] Cuomo S, Pastor M, Cascini L, Castorino G C 2014 Interplay of rheology and entrainment in debris avalanches: a numerical study *Canadian Geotechnical Journal* **51** pp 1318-30

[5] Cuomo S, Pastor M, Capobianco V, Cascini L 2016 Modelling the space–time evolution of bed entrainment for flow-like landslides *Engineering geology* **212** pp 10-20

[6] Cuomo S, Moretti S, Aversa S 2019 Effects of artificial barriers on the propagation of debris avalanches *Landslides* **16**, pp 1077-87

[7] Moretti S, Cuomo S, Aversa S 2020 Feasibility of Foothill Barriers to Reduce the Propa-gation of Debris Avalanches. I*n National Conference of the Researchers of Geotechnical Engineering*, Springer, Cham pp 309-17

[8] Cuomo S 2020 Modelling of flowslides and debris avalanches in natural and engineered slopes: a review. *Geoenvironmental Disasters* **7**

[9] Faug T 2015 Macroscopic force experienced by extended objects in granular flows over a very broad Froude-number range. *The European Physical Journal* **38** p 34

[10] Cuomo S, Moretti S, Aversa S 2020 Modeling of geosynthetics-reinforced barriers under dynamic impact of debris avalanche *XXVII CNG*. In print

[11] Itasca Consulting Group 2016 FLAC Fast Lagrangian Analysis of Continua **8**

[12] Lajevardi S H, Silvani C, Dias D, Briançon , Villard P 2015 Geosynthetics anchorage with wrap around: experimental and numerical studies *Geosynthetics International* **22** pp 273-87

[13] Aversa S, Maiorano R M S, Tamagnini C 2007 Influence of damping and soil model on the computed seismic response of flexible retaining structures. *In 14th European Conference on Soil Mechanics and Geotechnical Engineering* ISBN 978-90-5966-055-7

[14] Gioffrè D, Mandaglio M C, Di Prisco C, Moraci N 201. Evaluation of rapid landslide impact forces against sheltering structures. *Rivista Italiana di Geotecnica* **51** pp 64-76

[15] Cuomo S, Moretti S, Aversa S 2019 Deformation mechanisms of geosynthetics-reinforced barriers impacted by debris avalanches. B*ulletin of Engineering Geology and the Environment* pp 1-14

IOP Conf. Series: Materials Science and Engineering 1260 (2022) 012044 doi:10.1088/1757-899X/1260/1/012044

Current condition of recovery method of disaster embankments using geo-cell and soil reinforcement bar

K Kojima[1], G Noda[2], S Yazaki[3], M Harata[4], K Yokota[5]

[1]Railway Technical Research Institute, 2-8-38 Hikari-cho, Kokubunji-shi, Tokyo, Japan
[2] Sanriku Railway Co., LTD., 4 Sakae, Miyako-shi, Iwate Prefecture, Japan
[3]Integrated Geotechnology Institute, LTD., 1-23-5 Yotsuya, Shinjyuku-ku, Tokyo, Japan
[4]Tokyo Printing Ink Mfg. Co., LTD., 1-12-4, Oji, Kita-ku, Tokyo, Japan
[5]Raito Kogyo Co., LTD., 4-2-35, Kudan-Kita, Chiyoda-ku, Tokyo, Japan

Email: kojima.kenichi.03@rtri.or.jp

Abstract. Recently, many embankment disasters have occurred. The main causes of these disasters are natural hazards, such as earthquakes, rainfalls and the combination of these. We are proposing a sustainable seismic and rainfall countermeasures for railway and road embankments by means of geo-cells (made of HDPE) and soil reinforcement bars (Gc-Rb embankment). The soil reinforcement bar is inserted into an embankment, and the geo-cell is laid on the slope of the embankment. The characteristics of this method are the convenience of construction and the economy, in addition to the reinforcement against the natural hazards. The embankment of the Sanriku Railway Riasu Line was seriously damaged by the Tohoku Region Pacific Offshore Earthquake occurred on March 11, 2011 in Japan. Many civil structures were damaged not only in the Sanriku railway but also in Pacific ocean side of Tohoku area. In the restoration works of these structures, the shortage of the construction machine, the machine operator and related materials was the problem. Therefore easiness of construction and usual materials were demanded (such materials other than concrete and so on). The proposed method satisfies these required conditions. The operation of this railway line was restarted on April 3, 2013, two years after the earthquake. In this paper we report the current state of the embankment using this recovery method, including the maintenance, after 6 years from the restart of the service.

1. Introduction

This method consist of geo-cells and reinforcement bars (See Figure 1). The construction of it is as follows. The first step of the construction is to insert reinforcement bars into the embankment, and the second is to lay geo-cells on the slope.

The point is the work connecting soil reinforcement bars with geo-cells using connecting steel bars, by which the whole structure becomes as a unified body. This unified structure produces a synergistic effect of internal reinforcement and external reinforcement. The synergistic effects mean that prevention of sliding by the soil reinforcement bar, prevention of 3-dimensional sliding between bars by bars and geo-cells, and prevention of localized erosion by geo-cells.

There are various kinds of soil reinforcement bars, from a rock bolt to a pile of soil cement used for ground improvement, where we can choose any raw materials, construction methods and diameters. This method was selected by next four reasons.

1) If the slope protection is not damaged by the earthquake, this method can be used without modify the existing slopes. 2) If the part of slope protection is damaged, this method can be applied, leaving the existing slope protection untouched. 3) This method uses the geo-cell manufactured by HDPE (See Figure 2). Therefore, the amount of concrete can be reduced. 4) The easiness of construction of the geo-cells is evaluated. Especially, the removal of existing slope protection that is only partially collapsed requires a lot of time, works and costs (see Figure 3). On the bad case, the stability of embankment is lost by these removals. Therefore, the no-removal method is very effective.

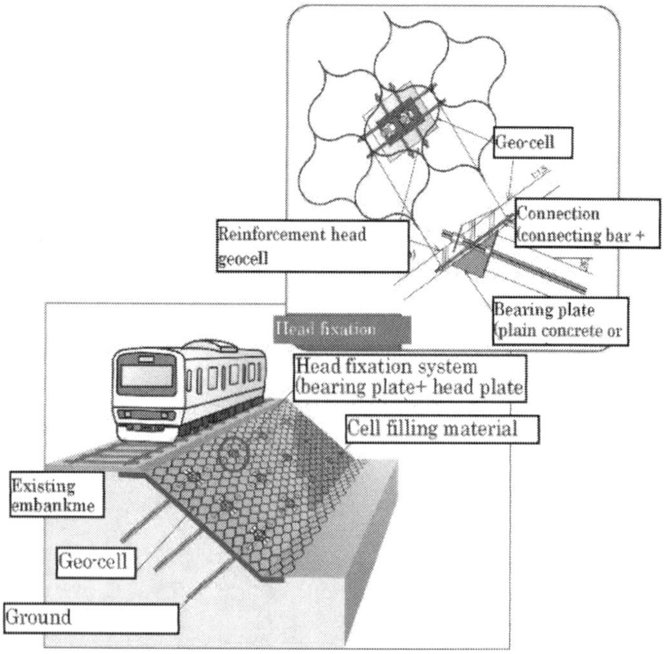

Figure 1. Outline

Figure.2 Geo-cells

Figure 3. Removal of existing slope protection materials

2. Slope protection by geo-cell and reinforcement bar

In this method, the geo-cells are the role of external protection and the reinforcement bars are the role of internal reinforcement. As described above, there is no restriction on the type of reinforcement bars. Selection of the diameter, length and materials are determined, based on the require performance of embankment, workability and economy of the reinforcement method.

There are many kinds of materials which can be filled into geo-cells. As the filling material, for example, soil, crushed stone, shotcrete can be applied according to required performance. If we need to consider the environment and want to green, we can use the vegetative gabions.

Since this method can be used in various combinations, it is a highly flexible method that functions can be arbitrarily adjusted so as perform the maximum protection against earthquake, rainfall and environmental protection.

3. Disaster occurred after construction of this method

3.1. Earthquake

This method for railway embankment is served for about six years since April, 2013 (See Figure 4, 5). Table 1 shows major earthquakes (over seismic intensity 5) occurred between April, 2013 and January 2020. The number of earthquakes over seismic intensity 5 is five. The maximum seismic intensity is upper 5. The maximum earthquake in this term is the earthquake occurred in May 2015. The level of this earthquake is seismic intensity upper: 5 and magnitude 6.8. In this area, middle level earthquakes occur frequently. But, structures (including embankments) of Sanriku railway are not damaged. These structures proved to have the high resistance of earthquake.

Figure 4. Re-operation on April 2013

Table 1. Major earthquakes

Date	Level (Max. Seismic intensity)	Condition (Sanriku railway)
5th July 2014	Lower 5, M5.9	No accident
17th Feb. 2015	Upper 5, M5.7	No accident
13th May 2015	Upper 5, M6.8	No accident
10th July 2015	Upper 5, M5.7	No accident
19th Dec. 2019	Lower 5, M5.5	No accident

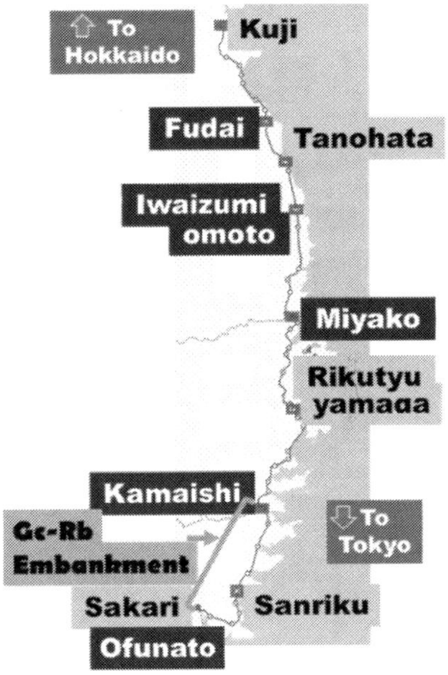

Figure 5. Sanriku railway

3.2. Rainfall

A disaster that is much more frequent than an earthquake is rainfall. Figure 6 and 7 show the annual rainfall and maximum 24 hour rainfall since 2013 at two representative observatories in the area (Ofunato and Miyako).

Annual rainfalls at these observatories are similar. Amounts of rainfall are between 1300mm and 1800mm. These rainfalls are relatively high from the average of worldwide. However, they are almost the same as the average annual rainfall in Japan. Except for 2019, there is no significant difference between the two areas in 24 hour maximum rainfall. This value is not unique in Japan.

However the amount of 24 hour maximum rainfall at Miyako in October 2019 was approximately 400mm (82.5mm of maximum 1 hour rainfall). This value was an extremely large rainfall. The cause of this rainfall was Typhoon No.19. This typhoon caused severe damages to many areas of eastern Japan, because of heavy rainfall. For example, Hakone area recorded 942.5mm of 24 hour maximum rainfall. This rainfall flooded many rivers and broke 140 embankments. Railways suffered with a lot of damages on structures on mountainous areas. Major damages were the sediment inflow and the bridge outflow.

At Fudai and Omoto which situated far about 30km north from Miyako on Sanriku railway operation areas, the amount of rainfalls were recorded over 90mm of rainfall per one hour (Fudai: 91.5mm, Omoto: 91.0mm). At these areas, there were the damages of over 70 point. (i.e. the embankment destruction, the embankment slip, the scouring of base structure and the sediment inflow) As these rainfalls were occurred on short time, the water accompanied with soil flooded into the railway structures. Therefore, the operation of Sanriku railway was stopped.

Figure 8 shows the view of Sanriku railway after this rainfall. This site is situated at the middle from Miyako to Kamaishi (55.4km). The road embankment is located at mountainside against the railway embankment. Many fluidized soil have entered the area between the road embankment and the railway embankment. Structures based on soil were destroyed and finally it washed away railway embankments, resulting in the picture shown as Figure 8, where only remains are rails and sleepers in a shape of ladder.

The recovery embankments with the proposed method in 2013 are situated on the south area against the heavy rainfall area. These sites are nearby Ofunato area shown in Figure 5. The rainfall in these areas was 50-100mm/h more than the average rainfall per hour in the past. But it was about 200mm/h less than the rainfall in disaster areas of typhoon 19. The number of these damages at the areas from Sakari to Kamaishi were seven. Typical damages were slip of embankment and scouring of base structure. Figure 9a and b show the condition of Gc-Rb embankment after two weeks from the passing to Typhoon 19. In spite of the occurring to damages (erosions or collapses and so on) of other embankments, Gc-Rb embankments had no damages. They were healthier at all parts.

These embankments had suffered many middle level earthquakes and the huge typhoon with heavy rainfall. Because of non-damages of them, we were able to confirm that they had high performance.

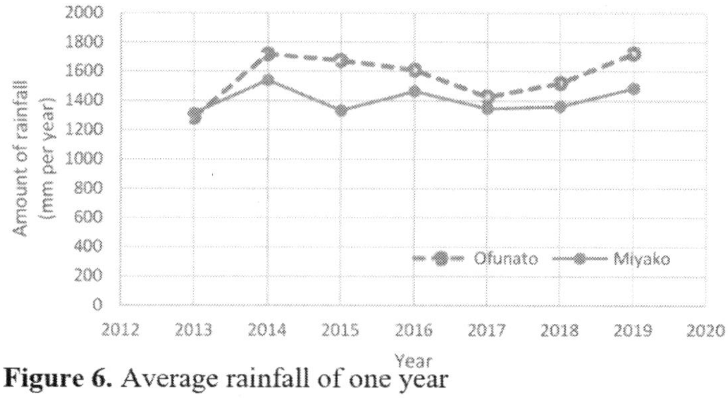

Figure 6. Average rainfall of one year

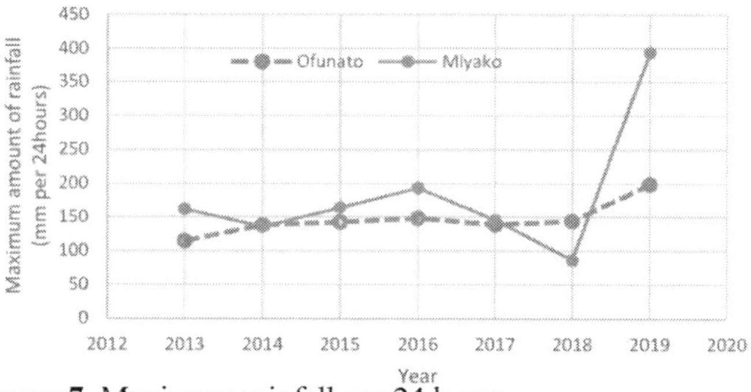

Figure 7. Maximum rainfall per 24 hours

Figure 8. Embankment of Sanriku railway after this rainfall

Figure 9a Condition of Gc-Rb embankment

Figure 9b Condition of Gc-Rb embankment

4. Maintenance

The geo-cells covered the embankment are protected against ultraviolet rays and human troubles, because the cells are filled with the rocks or sands. Basically, we think that this method requires no maintenance. In addition, these sites have used the shotcrete for filling materials. The grass by seeds that have scattered does not root on the surface of these embankment slopes. Generally, the surface of embankment slopes needs mow the grass, however the surface of Gc-Rb embankment slopes does not need mow.

The slope surfaces existing before Tohoku region Pacific ocean earthquake have grown in many trees and plants. These slope surfaces require to mow regularly. From this point of view, the Gc-Rb embankment is convenient and economical. Figure 10a, b and c show the Gc-Ri embankment and the conventional embankment (with rock facing). The conventional embankment has overgrown many plants and we think to need mowing. As using the

Figure 10a. Gc-Rb embankment and conventional embankment

Figure 10b. Gc-Rb embankment and conventional embankment

Figure 10 c Gc-Rb embankment and conventional embankment

shotcrete facing, the deterioration such as cracks might occur over time, the cracks did not occur at this location. In this method, the durability of rainfall was given to the reinforcement by reinforced bars. Therefore, the rainfall seepage from the surface of slope was not influenced by the strength of this embankment. In addition, the shotcrete facing was able to decrease the rainfall seepage. This effect was one reason that the Gc-Rb embankment was not destroyed.

5. Conclusion

This paper describes the recent condition of the Gc-Rb embankments. The conclusion of paper is shown below.

(1) In this six years, the middle earthquakes and heavy rainfall by typhoon (ex. Typhoon 19,2019) are occurred, but there are no damaged.
(2) These embankments are healthy despite six years have passed since its construction.
(3) In addition, the maintenance of these embankments are not required for six years.

As a result, we think that they have high seismic and rainfall resistance performance. We accumulate the data and estimate the influence of aging.

References

[1] Mochizuki M, Noda G 2012 Damage and restoration plan of Sanriku railway, Japan Society of Civil Engineering, Japan Society of Civil Engineering magazine (in Japanese)
[2] Ueno K, Nakamura H, Suzuki T, Amemiya S, Aratame Y Frost heaving countaermeasure of soft slope protection using non-frost heaving materials *The Journal of 25th Region Technology Conference* pp 410-3 (in Japanese)
[3] Rosen A *Proceedings of the International Symposium on Design ane Practice of Geosynthethic-Reinforced Soil Structures*, Bologna
[4] Ohki M, Shoji T, Seki M 2011 Development of new reinforced soil method by nailing *The Journal of Japan Society of Civil Engineers Annual Meeting* **66** pp 927-8 (in Japanese)
[5] Kojima K, Sato T 2017 Seismic and rainfall resistance method with geo-cells and reinforcement bars *The Journal of Japan Railway Civil Engineering Association* **55** pp752-75 (in Japanese)
[6] Kojima K, Yazaki S, Harata M, Yokota K 2017 Sustainable seismic and rainfall countermeasures for railway and road embankments using geo-cells and soil reinforcement bars -Proposal of new sustainable method *International conference on sustainable infrastructure*, ASCE
[7] Kojima K, Yazaki S, Harata M, Itou M, 2018 Construction and design of seismic and rainfall reinforced countermeasure using geo-cell and soil reinforcement bar for embankment –Sustainable and economical method *11th International conference on geosynthetics*, IGS
[8] Japan Meteorological Agency, Homepage, www.jma.go.jp/jma/index.html

Use of ASTM D8058 Flexural Strength Test Method to assess overall GCCM performance

M Kujawski, L Church and W Crawford

Concrete Canvas Ltd, Cowbridge Road, Talbot Green, Pontyclun CF72 8HL

Email: Marcin.kujawski@concretecanvas.com

Abstract. ASTM D8058, a flexural strength test, is currently used as the main index test for Geosynthetic Cementitious Composite Mats (GCCMs) within Factory Production Control and is also used to assess comparative performance of GCCMs for durability and chemical resistance testing. This paper provides an overview of the suitability of ASTM D8058, three-point flexural strength test as an indicative method for assessing overall GCCM performance. Specifically, this paper will look at the relationship between GCCM Initial Flexural Strength (IFS) values obtained using ASTM D8058 and characteristics associated with the cementitious fill material. It also recommends further research to assess the relationship between Final Flexural Strength (FFS) values and characteristics associated with the geosynthetic components of the GCCM.
IFS is obtained at the end of the first stage of a three-point bending test in the form of the first peak that is observed when the cementitious material first cracks in the tested specimen. The value of that first peak is mainly governed by the strength of the cementitious component of the GCCM. The relationship between IFS and the following GCCM properties is examined: Dry (uncured) mix density, Hardened (cured) density, Comressive strength development, Resistance to abrasion, Freeze and thaw durability and Resistance to chemicals.

1. Introduction

Geosynthetic Cementitious Composite Mats (GCCMs) are flexible concrete filled geosynthetics that harden on hydration to form thin, durable, fibre reinforced concrete layers. GCCMs enable geosynthetic technology to replace conventional poured, sprayed and precast concretes for erosion control and weed suppression applications such as channel lining, slope protection, hydraulic concrete structure remediation and containment bund lining. The material technology has been defined by ASTM as *'a factory-assembled geosynthetic composite consisting of a cementitious material contained within a layer or layers of geosynthetic materials that becomes hardened when hydrated'*.

GCCMs consist of four essential components:

i) A cementitious material (dry powder mix) that sets on hydration to form a hard wearing, chemically resistant and durable core that provides the material with its structural rigidity in service.
ii) A top polymeric geotextile layer that acts to contain the cementitious material during transportation and aids hydration by wicking water into the cementitious material.
iii) A 3-dimensional fibre matrix to control the material thickness and prevent the cementitious material from dislocating during transport/installation, therefore maintaining the cementitious material density and helping to control the water:cementitious material ratio of the dry powder mix on

hydration. Once set the fibres act to prevent crack propagation and provide a 'semi-ductile failure' mode.

iv) A polymeric backing layer that also contains the cementitious material during transportation and aids hydration by preventing the hydration water from seeping through the material and to the underlying substrate. Once set, the polymeric backing provides a low permeability liner, which in addition to the set concrete acts to provide a waterproof layer.

Unlike most geosynthetics, GCCMs are left exposed for their service life and are rarely buried for protection. Therefore, in order for the backing layer to provide waterproofing properties for the life of the structure the overlying fibre matrix and cementitious material must act to protect the backing from damage, from actions such as impact/puncture, abrasion, UV degradation and deformation due to excessive strain. It is therefore important to understand the physical properties of the cementitious material and its behaviour under such loading/damage conditions.

GCCMs are relatively unique to the geosynthetics industry as their physical properties transition from flexible to rigid once hydrated. Being composite materials, this also presents some unique challenges when characterising GCCM properties.

Tensile strength testing has long been established within the geosynthetics industry as the standard index test for evaluating geomembranes and geotextiles for the purposes of quality control, quality assurance and as a performance indicator (for durability testing for example). Whilst these test methods are applicable and appropriate for a flexible sheet product, often designed to be loaded in tension during use and whose material properties such as density, puncture resistance and durability might be linked to tensile performance; this relationship does not hold for GCCMs.

When testing the tensile strength of GCCMs, the performance of the polymeric geotextile component of the material typically governs the ultimate strength, whereas the cementitious material component of the GCCM makes relatively little contribution. Tensile testing is therefore unsuitable as an index test for GCCMs as it does not capture the performance of all components within the composite material.

2. GCCM Flexural Behaviour

Flexural strength testing provides a means to evaluate the performance of both the polymeric and cementitious components of a GCCM. A suitable flexural strength test method was formalised in the ASTM standard D8058, first published in 2017, which defines the 3-point flexural bending test method for assessing the performance of GCCM materials. This provides a suitable index test for GCCMs when assessing comparative performance for properties such as chemical resistance and long-term environmental durability, since it provides a better representation of the material's overall performance.

This performance can be characterized by measuring and recording 4 distinct points from the stress-displacement graph of the ASTM test, as shown in Figure 1 and listed below:

- Initial flexural strength
- Initial deflection
- Final flexural strength
- Final deflection

In addition, analysis of the graph provides a large amount of information about how the composite layers within GCCMs behave and interact.

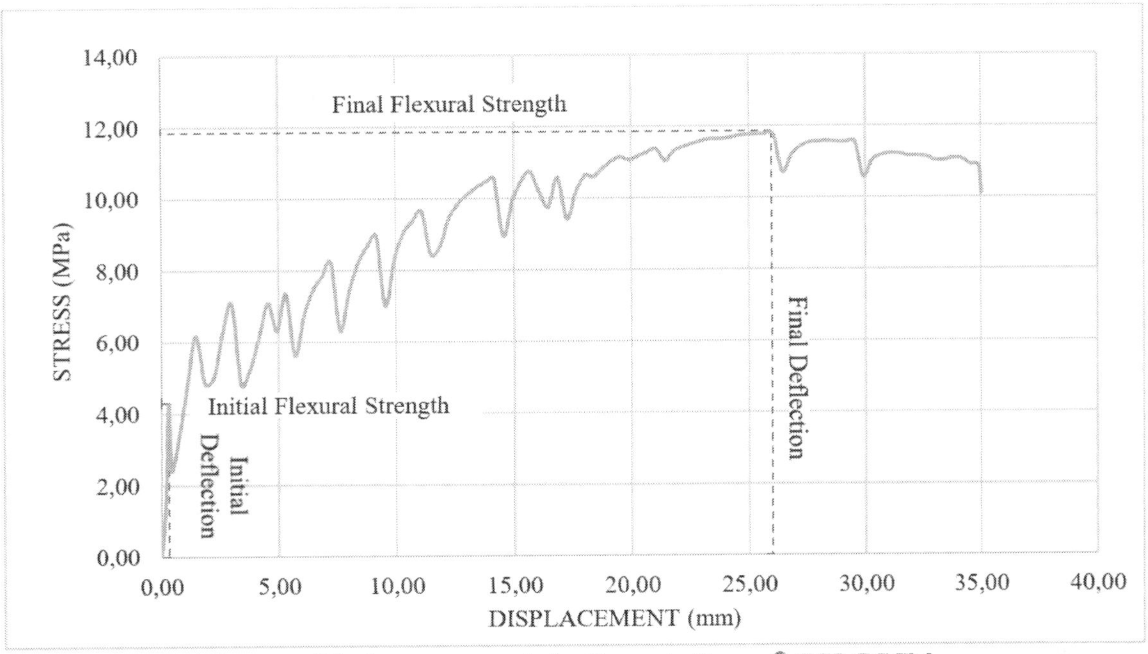

Figure 1. Typical Stress vs Displacement graph for Concrete Canvas® CC5 GCCM

Flexural testing of GCCM material to ASTM D8058 demonstrates a semi ductile failure in 3 phases. First, the initial elastic phase (up to a stress of approximately 4MPa for the material tested in Figure 1), before the initial break of the cementitious material occurs, known as the Initial Flexural Strength (IFS). Second, the GCCM ruptures in a saw tooth pattern of progressive cementitious material failure and fibre loading, and finally ultimate failure as the geosynthetic fibres themselves reach their ultimate loading limit, known as Final Flexural Strength (FFS). This semi ductile failure has practical benefits, particularly in areas of differential settlements or ground heave, as the GCCM can crack and deform locally. The polymeric backing layer is typically more elastic and is therefore still protected by the protective cover layers of the composite.

ASTM D8058 3-point bending tests are used by GCCM manufacturers to check material quality as part of factory production control procedures. This is defined in the European Assessment Document EAD 080009-00-0301, requiring GCCM manufacturers to report IFS according to ASTM D8058 at 1 day from initial hydration. It is therefore necessary to understand the factors that can affect the IFS of a GCCM.

3. Uncured GCCM Properties that can affect IFS – Density

One of the key features of GCCMs is that unlike conventional concrete mixes which are very sensitive to water:cement ratio, GCCMs must be manufactured so that they cannot be over-hydrated. This is to remove a key element of skill required on site when preparing conventional concrete for application, so there is no need to carefully monitor and measure the water addition in the GCCM. When hydrating GCCMs the hydration water is likely to pool on horizontal or concave GCCM surfaces, meaning that the material must be manufactured to cure under fully immersed conditions. It is therefore necessary for GCCMs to be manufactured to control hydration water absorption into the cementitious material. ASTM D8030 'Standard Practise for Sample Preparation for GCCM', specifies the procedure for hydration and curing of GCCM material for flexural strength testing. Hydration is by full immersion for a period of 24 hours to represent in-field immersed conditions.

By fixing the internal volume and controlling, through the manufacturing process, the fill density and particle size distribution, the void space within the cementitious material of a GCCM can be set within carefully prescribed bounds to subsequently limit the maximum water:cementitious material ratio achievable in the field.

In 2019, research was carried out on trial samples of the Concrete Canvas® CC5 GCCM material to compare the relationship between uncured dry density and 1-day IFS, which is presented in Figure 2.

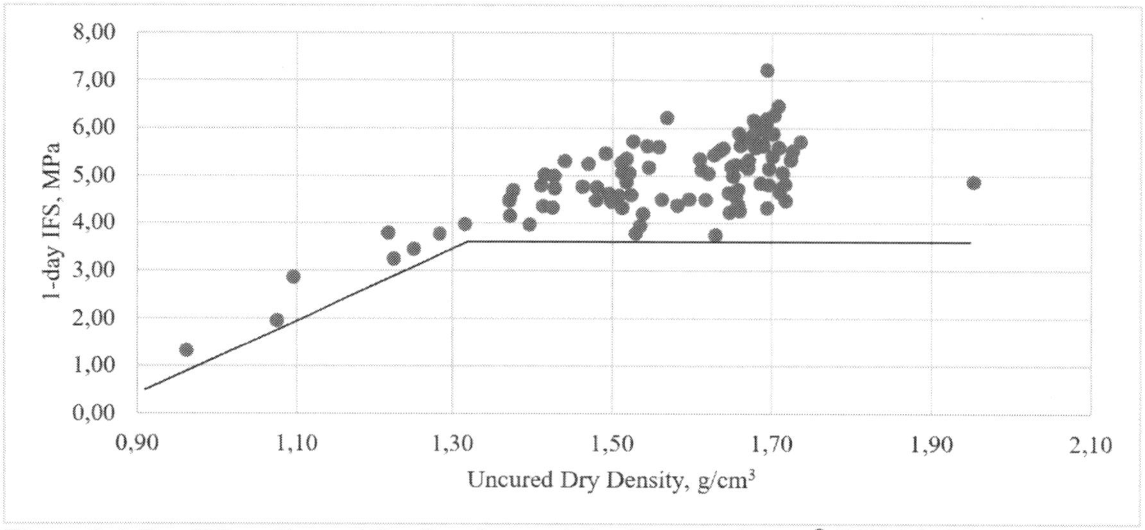

Figure 2. 1-Day IFS vs Uncured Dry Density graph for Concrete Canvas® CC5 GCCM

The results demonstrate a relationship between uncured dry density and 1-day IFS. Increasing uncured dry density increases the 1-day IFS in a linear fashion, up until a critical density of approximately 1.3g/cm^3 when an increase in density has a less clear relationship with 1-day IFS, but a minimum IFS of approximately 4MPa is consistently achieved. This can be explained as the uncured dry density of a GCCM is related to the void space in the cementitious material. Lower GCCM uncured dry densities contain greater void space, which on hydration allows more water to enter the cementitious material, therefore increasing the water:cementitious material ratio beyond optimum and weakening the strength of the set cementitious layer. At an uncured dry density above 1.3g/cm^3 the void space in the cementitious material provides an optimum condition for hydration, by limiting the available space for hydration water to enter the cementitious material, so that the water:cementitious material ratio remains close to optimum and 1-day IFS's above 4MPa are achieved.

GCCM Manufacturers can therefore define a minimum uncured dry GCCM density as part of their quality control measures, to increase the likelihood that minimum 1-day IFS's can be achieved when hydrated by full immersion. This will not guarantee that 1-day IFS will be achieved, since other factors such as the quality and composition of the cementitious material will be influential. In addition, the density of the cementitious material prior to hydration could be reduced by manipulation/folding of the GCCM during installation, so IFS should still be used to provide the best indicator of in service GCCM performance instead of uncured dry density.

4. Cured GCCM Properties and their Relationship with IFS

4.1. Relationship with 1-Day IFS

Test standards for cementitious materials typically require 28 days curing for evaluation of their characteristics. This can provide limitations when assessing the performance of the cementitious material in GCCMs and determining acceptance criteria, considering GCCMs are typically manufactured to reach performance strengths in 1 day from hydration. Understanding the relationship of a GCCMs 1-day IFS with other cementitious material properties can enable designers to understand the likely long-term performance of a GCCM by considering 1-day data alone.

4.1.1 Hardened Density

The GCCM hardened (dry) density will be greater than that of the pre-hydrated material uncured dry density since the cementitious material in the GCCM combines with water molecules during the hydration process.

The relationship between hardened (cured dry) density and 1-day IFS for the Concrete Canvas® CC5 GCCM material is presented in Figure 3.

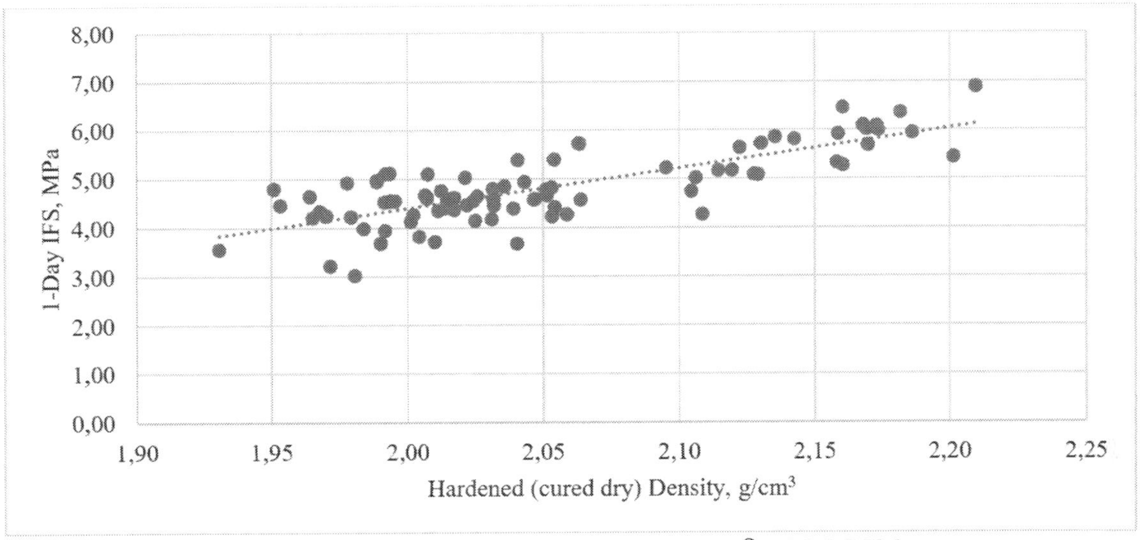

Figure 3. 1-Day IFS-Hardened Density graph for Concrete Canvas® CC5 GCCM

A relationship is seen as 1-day IFS increases as hardened density increases. A higher hardened density means that a higher proportion of cementitious material to water is contained in the cementitious core. Increasing the cementitious material content in the GCCM therefore increases the 1-day IFS of the material.

4.1.2 Compressive Strength Development

In most applications, cementitious materials are subjected to compressive loading. There are several standards for determining the compressive strength of a cementitious material, such as EN 12390-3 'Testing hardened concrete. Compressive strength of test specimens' that requires 100mm cubes of the material to be formed for testing. As GCCMs are typically 5-13mm thick, the material itself cannot be tested using such standards. A process for assessing the compressive strength of the cementitious material within a GCCM is to remove the dry powder mix within the material, mixing it with water (using a water:cementitious material ratio representative of the GCCM when hydrated by full immersion to ASTM D8030) and preparing cubes for testing to EN12390-3.

As discussed in section 3, the 1-day IFS of a GCCM is related to the uncured dry density of the material. A link between uncured dry density and the compressive strength of the cementitious material can also be made, since lower dry density materials tend to have increased porosity which in turn increases the water:cementitious material ratio of the GCCM on hydration. Considering the purpose of this paper is to understand the relationship between 1-day IFS and other GCCM properties, the 28-day compressive strength of the cementitious material is compared to 1-day IFS and not uncured dry density. 2019 research by Concrete Canvas Ltd compared the relationship between 1-day IFS and EN12390-3 28-day compressive strength of the cementitious core material (using a water:cementitious material ratio representing hydration to ASTM D8030) for a number of commercially available GCCM materials produced by different manufacturers. The GCCM materials tested had a range of uncured dry densities, therefore adjusting the water:cementitious material ratio on hydration and corresponding 1-Day IFS values. The results are presented in Figure 4.

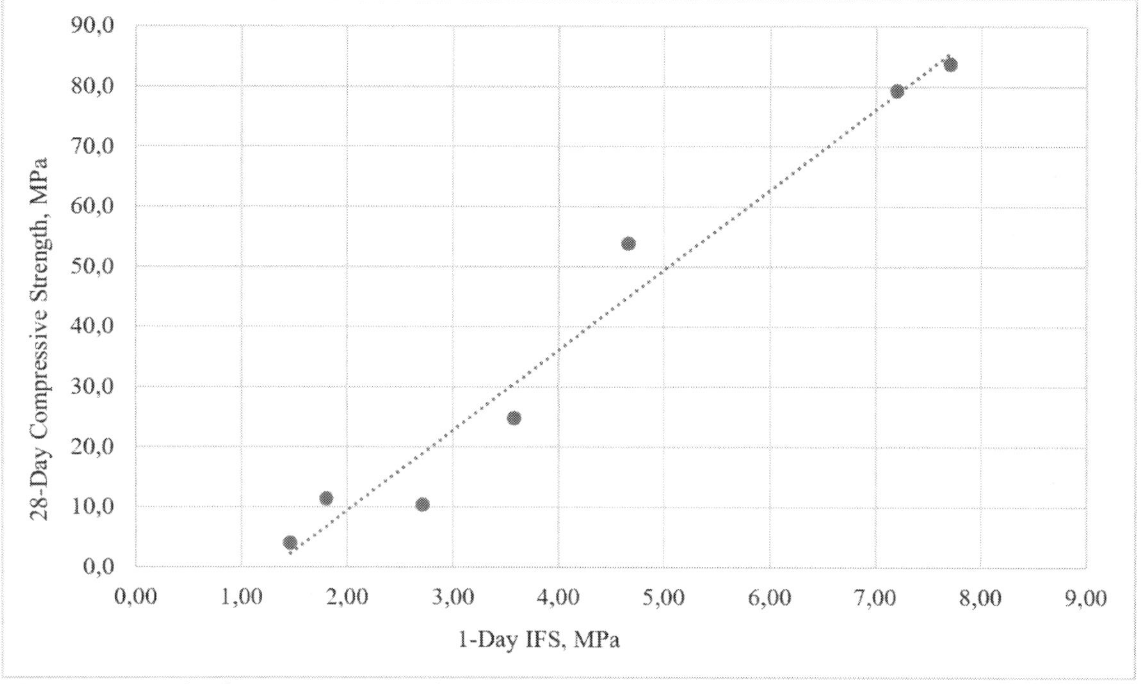

Figure 4. Graph of 28-Day Compressive Strength vs 1-Day IFS for commercially available GCCMs

A clear relationship between 1-day IFS and the 28-day compressive strength of the cementitious material is observed. GCCMs with higher 1-day IFS have a higher compressive strength development. GCCMs with a lower 1-day IFS have a higher ratio of voids to cementitious material and therefore higher water:cementitious material ratio on hydration. When casting cubes of the lower 1-Day IFS material at field deployed water:cementitious material ratios, the increased water weakens the strength of the set cementitious core material resulting in a lower 28-day compressive strength of the cementitious core.

4.1.3 Resistance to Abrasion

Materials used for erosion control purposes will be subject to the effects of scour, both from wind-blown particles but also more significantly sedimentation carried in surface water flow. GCCM's most common application is to line water channels both for drainage and irrigation. Their resistance to abrasion is therefore an essential characteristic when assessing their performance. A product with low abrasion resistance will have a significantly shorter operational life than one with a high abrasion resistance and will also deposit contaminants into the environment and water course at a higher rate as it degrades. Failure of these products in service may lead to significant consequential damage, for example if GCCM sheets fail they can become dislodged and block culverts potentially causing a high level of costly flood damage.

Since the major contributor to abrasion resistance in GCCMs is the cementitious material component, this can be characterized using a rotary platform abrader to ASTM C1353-15 'Test Method for Using Taber Abraser for Abrasion Resistance' with modifications in accordance with EAD 080009-00-0301.

Results are presented as thickness lost (mm) per 1000 cycles, therefore a high value represents a poor level of abrasion resistance and a low value a high level of abrasion resistance. Results are taken after 1000 cycles to ensure comparison of the cementitious material component of the samples rather than the fibrous surface.

Included in the research conducted by Concrete Canvas Ltd in 2019 and as discussed in 4.1.2, the abrasion resistance of a number of commercially available GCCM materials produced by different manufacturers has been tested to ASTM C1353. The relationship between 1-day IFS and the 28-day abrasion resistance is presented in Figure 5.

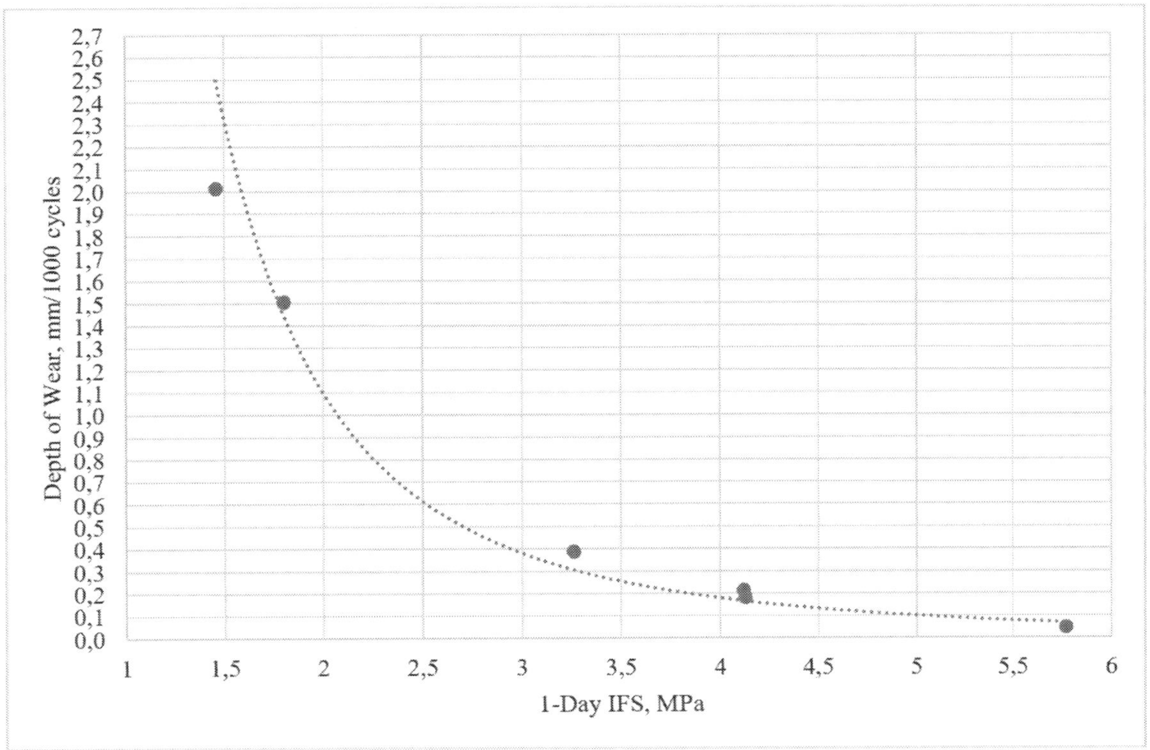

Figure 5. Graph of 28-Day Abrasion Resistance vs 1-Day IFS for commercially available GCCMs

The results show a correlation between 1-day IFS and 28-day abrasion resistance of the cementitious core material. Since a lower 1-day IFS is related to an increased water:cement ratio within the cementitious core material, the resulting hardened GCCM has a weaker and lower density cementitious core which is easier to abrade. This results in a greater depth wear of the cementitious material under the same number of abrasion cycles. A GCCM with a higher 1-Day IFS is therefore likely to have a higher resistance to abrasion.

4.2. 28-Day IFS

The durability of GCCMs in degrading conditions can be understood by determining the retained IFS following exposure. As GCCMs continue to cure and IFS increases from 1 day to 28 days following hydration, specimens are typically aged for 28 days before testing. To test the durability of a GCCM, the IFS of specimens exposed to degrading conditions can be compared to unexposed control samples to calculate the retained IFS following exposure. A low retained IFS would demonstrate that the cementitious material within a GCCM is significantly affected by the degrading condition. The relationship between 28-day IFS of control specimens and the retained IFS of exposed specimens can be assessed.

4.2.1 Freeze and Thaw Durability

Freeze-thaw weathering can significantly affect the working life of concrete structures, particularly when used in water erosion control applications. Water can enter a crack or void in the concrete and expand as it freezes, forcing the crack to widen. When the ice melts, water can make its way deeper into the crack before freezing again, forcing the crack to widen further and causing spalling of concrete surfaces and water loss through the structure.

When installed in susceptible climates, the cementitious material core of a GCCM is also subject to freeze-thaw weathering. Given the thin sheet nature of GCCMs and the fact that they contain polymeric fibre reinforcement, the most relevant test method for assessing the durability of the cementitious material component can be obtained from within EN 12467:2012 'Fibre-cement flat sheets. Product specification and test methods' which covers environmental cyclic testing including freeze-thaw, soak-dry and heat-rain.

When testing GCCM freeze-thaw resistance, hardened specimens are subject to freezing for over 2 hours at -20C, then thawing (immersed in water) at +20C for over 2 hours. This is repeated for 100 cycles and then a flexural strength test to ASTM D8058 is performed. The IFS is then compared to the 28-day IFS of control samples and a retained flexural strength is calculated. The relationship between 28-day IFS of control specimens and the retained IFS of specimens subjected to freeze-thaw testing for Concrete Canvas® CC5 GCCM material is presented in Figure 6.

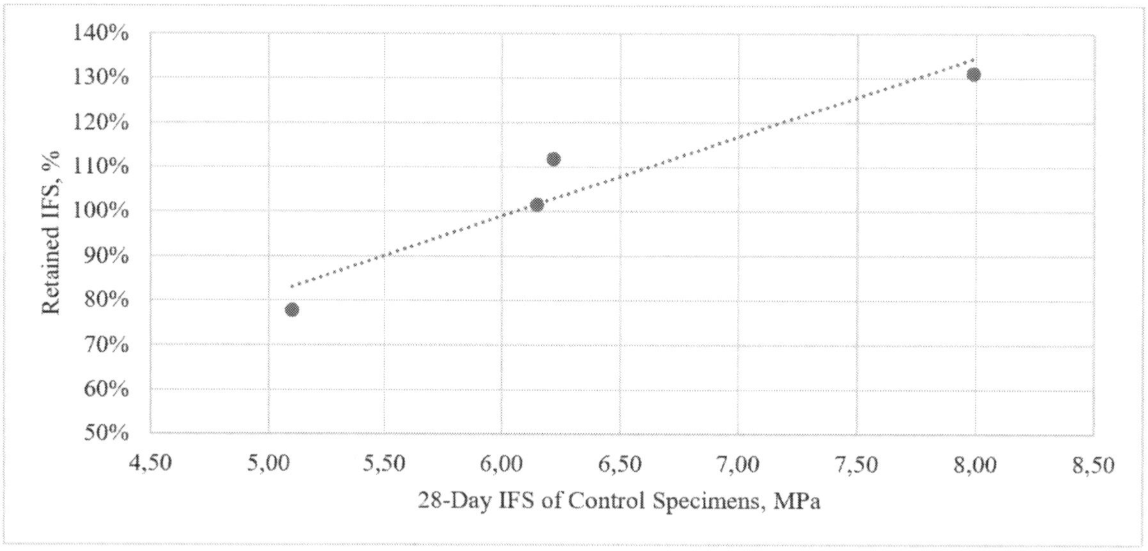

Figure 6. Graph of Retained IFS after Freeze-Thaw Exposure vs 28-Day IFS of Control Specimens for Concrete Canvas® CC5 GCCM

The results show a correlation between 28-day IFS and retained IFS after freeze-thaw durability testing. This can be explained as lower 28-day IFS is linked to a lower cured density which will create a cementitious material core with higher porosity, which in turn will enable more water to enter the GCCM and initiate freeze-thaw weathering.

4.2.2 Resistance to Chemicals

GCCMs are composite materials, so it is important to understand the chemical resistance of both the polymeric and cementitious material components of GCCMs. Each component may react differently when exposed to specific chemicals, such as acids which often have a detrimental effect on the performance of cementitious materials.

The chemical resistance of GCCMs can be assessed based on testing to BS EN 14414:2004 'Geosynthetics. Screening test method for determining chemical resistance in landfill applications'. The test method involves full immersion of 28-day hardened GCCM sample bars in the test chemical over a period of 56 days at an elevated temperature of 50°C to accelerate any signs of deterioration. Following the 56-day immersion period, the samples are subjected to visual inspection and an index strength test is conducted to determine any loss of performance as a result of the immersion. Instead of performing tensile testing (which as discussed in section 1 is dominated by the performance of the polymeric component of the material), flexural strength testing is performed to ASTM D8058 and compared to the IFS of 28-day aged control samples to determine the retained IFS (%).

The acidic resistance of the Concrete Canvas® GCCM material has been tested to BS EN 14414:2004 Method A: hydrolysis under acidic conditions (modified), requiring immersion in sulfuric acid with a starting pH of 1. Further research by Concrete Canvas Ltd assessed the effect of reducing the starting acidity of the sulfuric acid to pH 0 and adjusting the duration of immersion between 14 days and 70 days. The relationship between the IFS of the exposed samples and the retained IFS after immersion is presented in Figure 7.

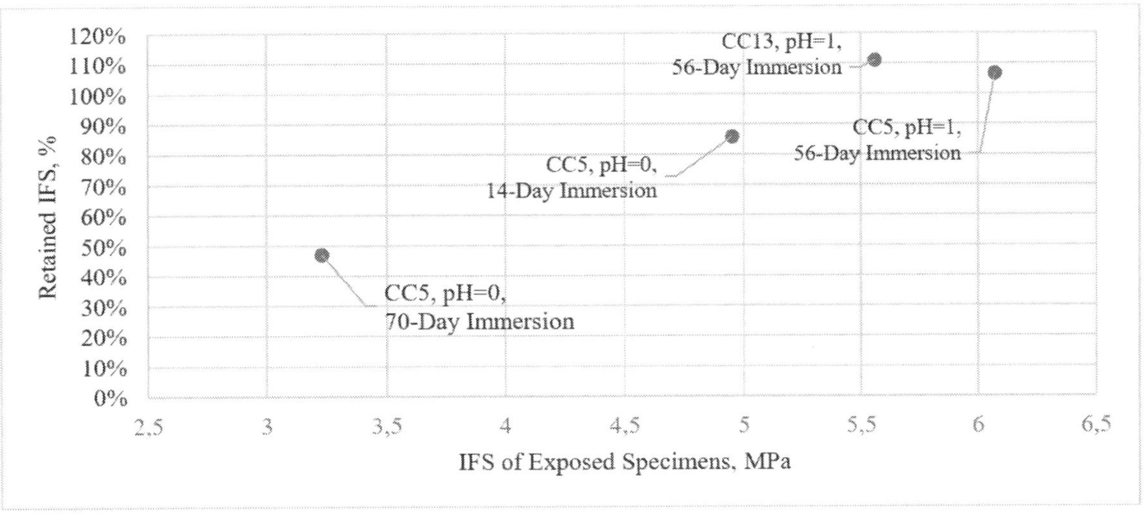

Figure 7. Graph of Retained IFS vs IFS of Exposed Concrete Canvas® Specimens after Immersion in Sulfuric Acid

The results show that the retained IFS of GCCMs can be affected by both the acidity and duration of immersion in an acid. The results demonstrate that performing ASTM D8058 flexural strength testing on exposed test specimens can provide a clear indication on the effect of chemical immersion on the cementitious material contained within a GCCM. However, comparison of 28-day IFS does not necessarily provide an indication of the chemical resistance of the polymeric geosynthetic components of the GCCM.

5. Proposed Further Research

Further research is needed to compare Final Flexural Strength (FFS) of GCCMs to their performance characteristics, particularly when assessing chemical resistance where immersion may have a detrimental effect on the polymeric components of the GCCM, which may not be indicated when assessing IFS alone. Understanding FFS may help to understand the performance of the polymeric components and the performance of the GCCM as a whole, but IFS still provides the best indication on performance of the GCCM cementitious material component.

6. Conclusion

GCCMs are a unique geosynthetic material as they change from being flexible during installation to being rigid in service. Understanding the flexural strength behaviour of the material (in particular the initial break properties of the cementitious material core) provides a more detailed understanding of the overall performance of a GCCM.

Using ASTM D8058 to report the IFS of a GCCM can provide important information on the likely performance of the material in service, as there is a clear relationship between IFS and key performance criteria such as hardened density, compressive strength, abrasion resistance and freeze-thaw durability. IFS can also be used to determine resistance of the cementitious core of the GCCM to chemical immersion. It is therefore important for GCCM manufacturers to report the IFS of their material, hydrated according to ASTM D8030 and tested to ASTM D8058. 1-day IFS can be used for GCCM acceptance criteria, to provide designers with a clearer understanding on the likely long-term performance of the material in field conditions.

References

[1] Crawford W and Kujawski M 2018 Geosynthetic Cementitious Composite Mats – Essential Characteristics and Properties. *11th International Conference on Geosynthetics*, Seoul

[2] ASTM D 8030/D8030M. Standard Practice for Sample Preparation for GCCM, American Society for Testing and Materials.

[3] ASTM D 8058. Standard Test Method for Determining the Flexural Strength of a Geosynthetic Cementitious Composite Mat (GCCM) Using the Three-Point Bending Test, American Society for Testing and Materials.

[4] ASTM D 8173. Standard Guide for Site Preparation, Layout, Installation and Hydration of Geosynthetic Cementitious Composite Mats, American Society for Testing and Materials.

[5] ASTM C1353-15. Test Method for Using Taber Abraser for Abrasion Resistance American Society for Testing and Materials.

[6] EAD 080009-00-0301. Geosynthetic Cementitious Composite Mats and Barriers, European Organisation for Technical Assessment

[7] BS EN12390-3. Testing hardened concrete. Compressive strength of test specimens.

[8] BS EN 12467:2012. Fibre-cement flat sheets - Product specification and test methods.

[9] BS EN 14414:2004. Geosynthetics. Screening test method for determining chemical resistance in landfill applications.

[10] Hewlett P C, Liska M Eds 2019 Lea's Chemistry of Cement and Concrete (fifth ed.), Butterworth-Heinemann

Field measurements in a partly submerged woven geotextile-reinforced pile-supported embankment

S J M van Eekelen[1], R A Zwaan[1], A Nancey[2], M Hazenkamp[3] and Y H Jung[4]

[1] Deltares, Boussinesqweg 1, 2629 HV Delft, Netherlands,
suzanne.vaneekelen@deltares.nl
[2] TenCate Geosynthetics, 9 rue Marcel Paul - B.P. 40080, 95873 Bezons Cedex, France
[3] TenCate Geosynthetics, Europalaan 206, 7559SC Hengelo, Netherlands
[4] Kyung Hee University, 1732 Deogyeong-daero, Giheung-gu, Yongin-si, Gyeonggi-do, Republic of Korea

Abstract. The geosynthetic reinforcement in a pile-supported (GRPS) embankment can be designed using the CUR226 (2016) design guideline. This design guideline adopted the Concentric Arches model (Van Eekelen et al, 2013, 2015). The validated application of this model is limited to GRPS embankments whose geometry and materials meet the conditions of the field cases and experiments used for the validation of the Concentric Arches model. This means for example that: the whole embankment should be located above the groundwater table and that at least one geogrid layer should be used as reinforcement. If these requirements are not met, additional measurements are requested by CUR226 to demonstrate that the system comes within the framework of the guideline. This paper presents and discusses field measurements that were therefore conducted in a partially submerged GRPS embankment. The embankment was reinforced with two layers woven geotextile and no geogrid. It was concluded that the Concentric Arches model matches the measured geotextile strains very well, so that we may conclude that the CUR226 is also applicable for geotextile-reinforced piled embankment without geogrids. Furthermore, an increasing groundwater level seems to result in a reduction of the geotextile strain. It is recommended to conduct more research to further analyse the influence of water in GRPS embankments.

1. Partly submerged geosynthetic-reinforced piled embankment without geogrids

The CUR226 (2016) design guideline for geosynthetic-reinforced pile-supported (GRPS) embankments uses the Concentric Arches model of Van Eekelen (2015) for the design of the geosynthetic reinforcement. This model has been validated using more than 100 measurements from many experiments and field measurements conducted in several countries (Van Eekelen et al 2015). All these measurements were conducted in GRPS embankments with at least one geogrid layer as reinforcement.

In addition, all measurements used by Van Eekelen (2015) were taken in unsaturated embankments, with the whole embankment being located above the groundwater table. To the authors' best knowledge, the same is true for most other measurement projects presented in the literature. Only Van Eekelen et al. (2012), Briançon and Simon (2012), Sloan (2011), and Van Eekelen et al. (2020) showed that heavy rainfall affects arching behaviour. Song et al. (2018) presented 2D trapdoor tests with sand under different water tables. They concluded that the arching mechanism can degrade if the water table is located inside a zone of two times the trapdoor width above the trapdoor.

The validated application of the design rules described in the CUR226 (2016) guideline is limited to embankments whose geometry and materials used meet these conditions (unsaturated embankment, at least one geogrid layer). If these requirements are not met, additional measurements are requested by CUR226 (2016) to demonstrate that the system comes within the framework of the guideline. These measurements should be conducted in representative practical cases.

For this purpose, field measurements were conducted in a partially submerged GRPS embankment, that was reinforced with two layers woven geotextile and no geogrid. This paper reports on these field measurements and compares the measured geotextile strains with the calculation rules in CUR226. The purpose of these measurement is to make available measurement data to validate the calculation rules in CUR226 for GRPS embankments with only woven geotextile reinforcement, and for partially submerged GRPS embankments.

2. GRPS embankments in the national road N210, Netherlands

In 2011, a 14 km section of the national road N210 in the Netherlands was reconstructed (Haring et al 2008). As the entire national road was constructed on a thick extremely compressible layer of organic peat, it was decided to construct it on a GRPS embankment. Eight years later, in 2019, a new industrial area close to the village Bergambacht was connected to the N210, which was, again, built on a GRPS embankment.

2.1. Description of the geotextile-reinforced pile-supported embankment

Table 1 lists the key dates of the building work and the monitoring of the GRPS embankment. Figure 2 shows a plan view and a cross-section of the GRPS embankment at the monitoring location. The system consists of square precast concrete piles with sides measuring 0.29 m and an average of transverse x longitudinal = 2.28 m x 2.27 m centre-to-centre-(ctc) spacing. Square precast concrete pile caps (0.75 m x 0.75 m) with smoothly rounded edges were installed on top of the piles. The top of the pile caps was installed at NAP -2.42 m (average) ± 0.05 m, where NAP is a reference level in the Netherlands. A protective nonwoven geotextile on top of the pile caps ensures that no damage happens to the geosynthetic reinforcement. Two layers of woven geotextile (TenCate Geolon® PET 400/50) were applied directly on top of the nonwoven. The bottom geotextile layer was applied across the road axis, the second one along the road axis.

Table 1. Key dates for construction and monitoring at the location of the N210 Roundabout de Wetering.

Before 27 February 2019	Installation piles and pile caps; termination drainage
27 February 2019	Excavation of 0.35 m sand between the pile caps and installation of soft peat instead (only the measurement area, Figure 1)
28 February 2019	Installation settlement tube 4 underneath the geotextile, geotextile layers and strain transducers
1 March 2019	Installation settlement tubes 1 to 3 on top of the geotextile and pore pressure transducers on top of the geotextile 11:00-12:00 Installation first 0.30 m layer of granular fill
3 March 2019	First measurement settlement tubes.
4 March 2019	7:00-10:00 Installation second 0.30 m layer of granular fill, wrapping back and anchorage geotextile.
4 March 2019	Strain transducers on wrap-back geotextile
4 March 2019	Pore pressure transducers on wrap-back geotextile
12 March 2019	14:00-16:00 Installation 0.40 m sand
Until 25 March 2019	Finishing berm North side and installation trees
1 April 2019	7:00-13:00 0.30 m hydraulic granular fill
3 to 5 April 2019	Asphalt pavement.
6 April 2019	Road is open
3 May 2019	Measurement settlement tubes
8 January 2020	Water level measurement in ditch
4 March 2020	Measurement settlement tubes

The embankment consisted of three fill layers (Figure 2) and an asphalt pavement. These layers were top – down:

- 0.21 m pavement of asphalt
- 0.3 m base course of well-compacted good-quality hydraulic broken demolition waste (mainly broken concrete)
- 0.4 m subgrade 1, consisting of well-compacted sand
- 0.6 m subgrade 2, consisting of partly well-compacted broken demolition waste (mainly masonry).

For the sand of subgrade 1, an average friction angle of 35 deg is a good assumption. The demolition waste of subgrade 2 mainly consisted of broken masonry. The grain shape was subangular and the grading was average to good, so that BSI (2015) and CEN (2008) recommends a characteristic residual value for the friction angle of 34 to 36 deg, giving an average value of 36.5 to 38.5 deg. The weighted average value of the fill friction angle with the arch height is 35.8 to 36.9 deg. In this paper, the fill friction angle is varied in the calculations: the average value friction angle (subgrade 1 + subgrade 2) is chosen to be 34 and 38 deg.

Figure 1. Monitoring area of the GRPS embankment under construction. At the monitoring location, the sandy subsoil was replaced by soft peat. The pile caps were covered by a protective non-woven geotextile.

The national road and the new connection were constructed on a thick extremely compressible layer of organic peat. However, the subsoil conditions at the monitoring area happened to be relatively good, as the monitoring area was located exactly on a crossing old road body. To ensure that the geosynthetic reinforcement was going to deflect sufficiently within the time planned to monitor, the top 0.35 m of the sand body between the pile caps was excavated and replaced by uncompacted, very soft and compressible local peat, as shown in Figure 1. Therefore, virtually no subsoil support underneath the reinforcement was to be expected. Two samples of the peat were tested, showing an average unit weight of 11.2 kN/m^3 and a natural water content of 320%.

3. Monitoring programme

Figure 2 shows the top view and a cross section of the monitoring area. The following sections describe the monitoring programme.

3.1. Settlements

Table 1 and Figure 2 show the timing of the installation and the position of settlement tubes 1 to 4. Tubes 1, 2 and 3 were installed on and attached to the geotextile reinforcement. Tube 4 was installed below the reinforcement and not attached to the geotextile. The tubes were Polvalene PE100, SDR11, PN16)) with an external diameter of 50 mm and a skin of 4.2 mm. The tube settlement is measured with a liquid levelling system: a sensor in a thin tube with water that is pulled through the tubes. The accuracy of the height measurements in the settlement tubes is 0.01 to 0.02 m, depending on the weather: more wind means more scatter in the measurements, and therefore it is always tried to find a windless day.

3.2. Pore pressures

Six pore pressure transducers (ppt) were installed, ppt1-4 are located directly on a pile cap (on average at NAP -2.38 m) and ppt5-6 are located 0.66 m higher (on average at NAP -1.72 m).

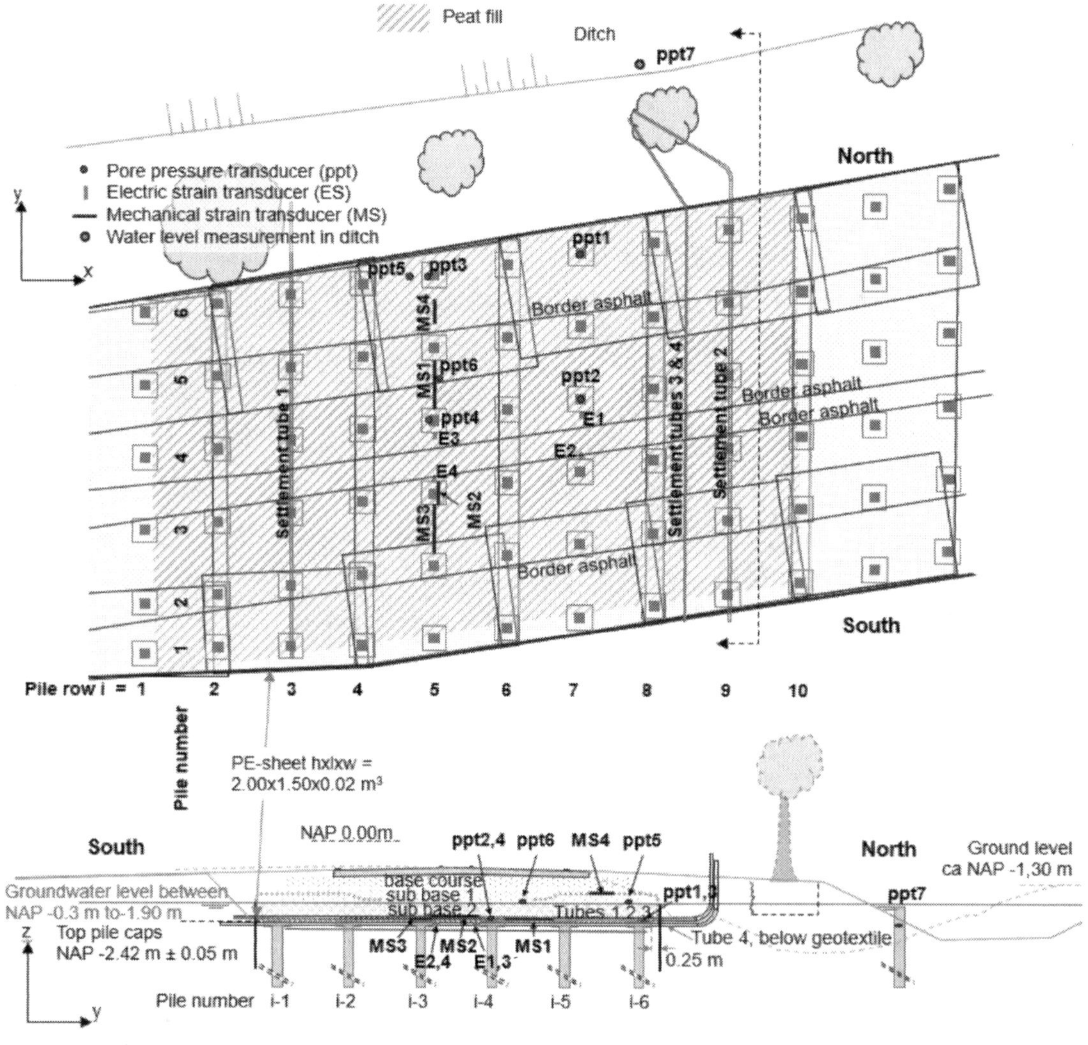

Figure 2. Top view and cross section monitoring area. The reinforcement along the road axis is left out for clarity reasons.

3.3. *Strains in the geotextiles*

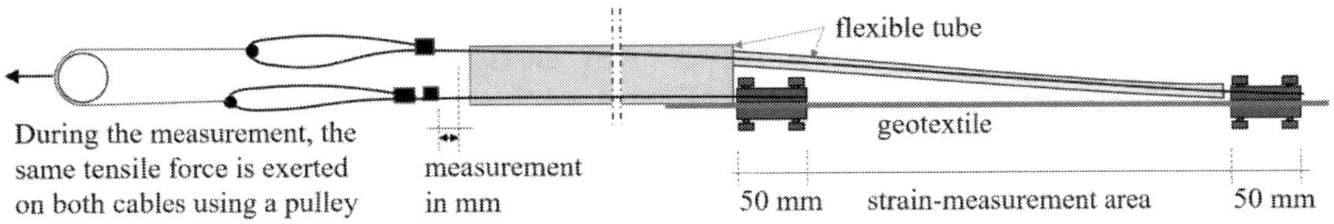

Figure 3. Mechanical strain (MS) measurement system.

Van Eekelen et al. (2012, 2020) reported the use of bicycle gear-type cables to measure strains in a geosynthetic reinforcement in experiments or field tests. This mechanical strain (MS) measurement system was also applied in this field test. Each transducer consists of a pair of two similar steel cables with a diameter of 1.5 mm, installed in a PE-tube and attached alongside to each other to the geotextile, as shown in Figure 3. The MS-transducers were applied alongside the piles in row 5, as indicated in Figure 2. The length of this strain measurement system varied between 55 cm and 139 cm.

In addition, four electronic (E) strain transducers were installed that are comparable to the MS transducers. They also measure the change of the distance between two points on the geotextile, but they work continuously. The electronic strain transducers consist of two PVC geotextile clamps with a deformation transducer of AE Sensors in between, type PCST24-M18-0100-I1-P1A-L02-KAB30M.

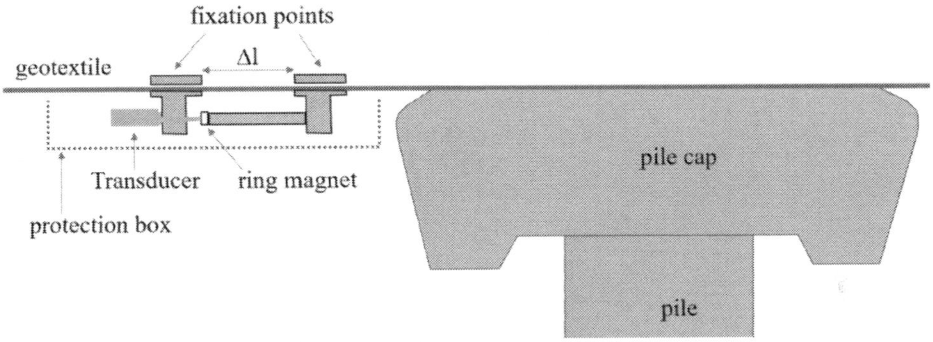

Figure 4. Electronic (E) strain measurement system.

3.4. *Weather conditions and groundwater*

The groundwater table in the pile-supported embankment was measured with the pore pressure transducers described in section 3.4. In addition, the daily amount of rain was measured in the weather station Groot Ammers (KNMI, 2020a), which is around 2 km from the monitoring location. The average of the evaporation values measured at the weather stations of Rotterdam, de Bilt and Herwijnen was adopted for the evaporation at our monitoring location (KNMI, 2020b). The precipitation shortage (m or mm) on a given day was determined as follows: the precipitation shortage of the previous day + evaporation – precipitation. In this paper, the precipitation shortage was not set to zero during the winter season (this is usually done for use for farming but not useful for our project).

4. Measurement results

4.1. Settlement tubes

The position of the settlement tubes was measured three times: in March and May 2019 and in March 2020. The location of the piles is easy to see in the measurements in tube 1 and 2 that were installed across the pile caps (Figure 5). The geotextile between the piles settles more than the geotextile on top of the piles. In March 2020, the deflection between the piles was on average 0.10 m, and max 0.13 m. The geotextile deflection between the piles is necessary for soil arching to develop. In practice, it is best if most geotextile deflection takes place before the road pavement is completed. Figure 5 indeed shows that the geotextile already has a large deflection in the first measurement, due to the compressibility of the uncompacted peat fill, and the geotextile not being installed taut or prestressed. The major part of the settlement and the geotextile strain after the first measurement occurred in the first two months, between March and May 2019. The measurements show that there was very little residual settlement after the measurement in May 2019. In April 2019, the road was opened, just in between the two first measurements. It is therefore expected that no or nearly no settlement of the road surface has occurred since May 2019.

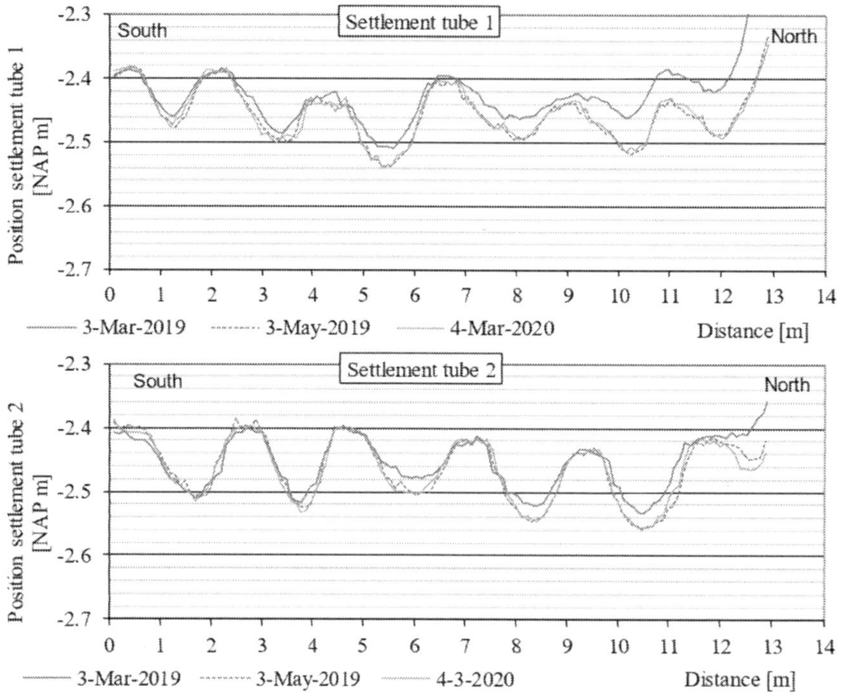

Figure 5. Settlement tubes 1 and 2.

Figure 6 shows the measurements from tubes 3 and 4, that are located on top of and below the geotextile respectively. Both tubes are located in a line between pile rows 8 and 9 (Figure 2). Figure 6 also shows that most settlement occurred before May 2019. During the first two months, the maximum settlement of tube 3 is 7 cm and this increases to 8 cm in the rest of the year (less than the measurement accuracy of 1 to 2 cm). The maximum settlement of tube 4 in the first two months is 8 cm and this increases to 10 cm in the rest of the year. This shows that the subsoil settled more than the geotextile, leaving the geotextile without subsoil support, as expected.

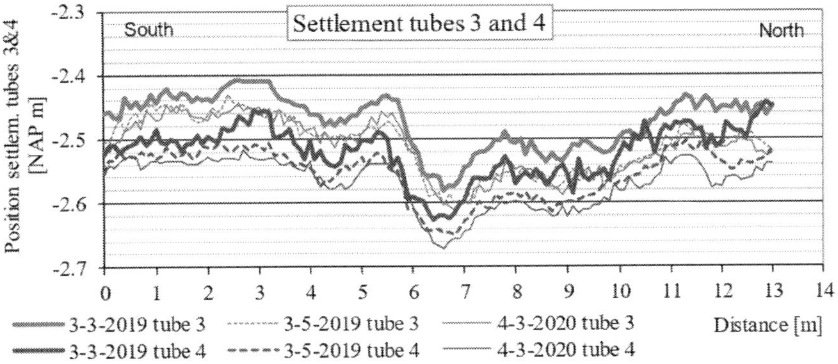

Figure 6. Settlement tubes between piles, tube 3: above geotextile, tube 4: below geotextile.

4.2. *Strain of the geotextile*

The strain is measured or estimated in three different ways:

- the mechanical strain (MS) measurement system, from which only one gave results;
- the four electronic (E) strain transducers;
- estimating from the measurements in the settlement tube, which is done for each geotextile span between each pair of adjacent piles in tubes 1 and 2.

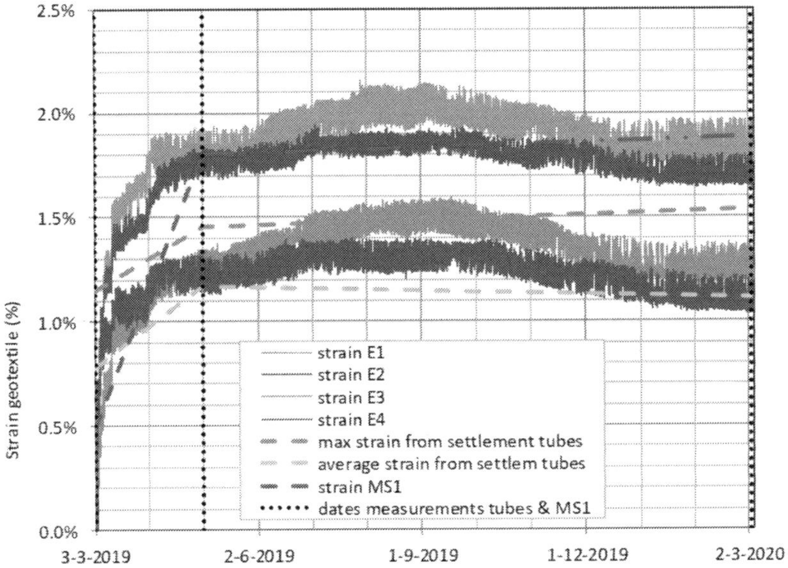

Figure 7. Measured geotextile strains; see Figure 2 for the locations of the transducers.

Figure 7 gives the measured strains of the geotextiles. As E1 and E2 give the larger values, and these values correspond with the measurement of MS1, these values are considered as governing. The strains show a daily cycle that was seen in earlier projects, like in Van Eekelen et al (2007, 2020). In contrary to Van Eekelen et al. (2007), the authors of this paper cannot explain the daily cycle presented in this paper. An explanation may exist with the daily traffic, electrical influences or daily temperature changes.

4.3. Pore pressure, ground water level and precipitation shortage

The pore pressure transducers measure a pore pressure in kPa, and this is converted into the level of the groundwater table in Figure 8. The figure shows that most pore pressure transducers stopped working during the year. Two pore transducers, ppt1 and ppt6 continued working properly and gave the same results. Pore pressure transducer ppt 7 was installed in the ditch in January 2020. The figure shows that the embankment is partially submerged; half the embankment lies below groundwater table.

The figure also shows the precipitation curve (see section3.4). The monitoring location is located in a polder, which means that the level of the open water (the ditches) is maintained artificially between certain pre-set levels. When necessary, water is pumped from the ditches. This is sometimes done when heavy rain or melting snow is expected. However, it is not possible to control the groundwater level completely. During heavy rain, the ground water level can increase considerably, which is illustrated clearly by the pore pressure transducers in Figure 8. This shows that the pore pressure transducers give reliable results.

The figure shows that 2019 was a dry year, resulting in an increasing precipitation shortage. By the end of September 2019, it started raining. Until the early days of October, 90 mm rain fell in 8 days. This resulted in a sharp increase of the water table, and a decrease of the precipitation shortage. After a while, the pumping affected the water table: it started to decrease although it was still raining frequently, shown by a still decreasing precipitation shortage.

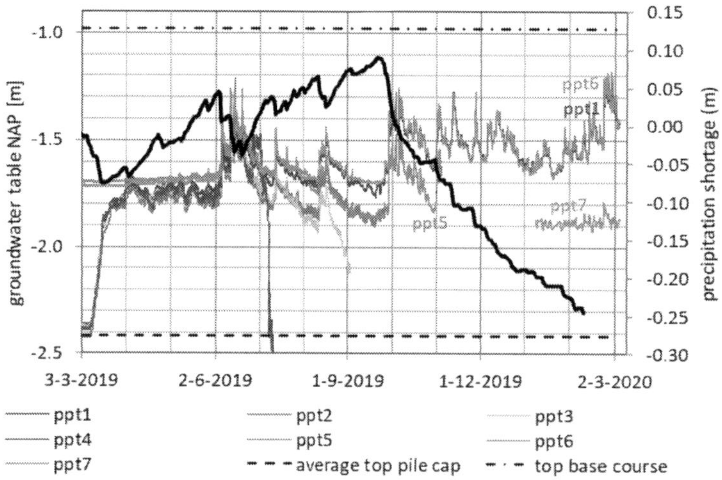

Figure 8. Measured groundwater table and precipitation shortage. The groundwater level was registered once every hour.

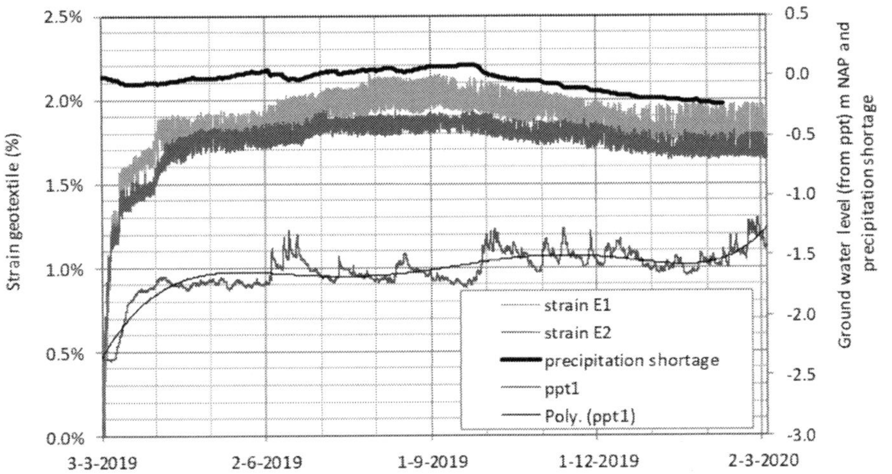

Figure 9. Comparison measured geotextile strains, precipitation shortage and groundwater table. The strains were measured once every hour.

Figure 9 shows the relation between the groundwater table, precipitation shortage and the strain in the geotextile. By the end of September 2019, a wet period started. The rain resulted in a decreasing precipitation shortage and an increase of the water table. This was accompanied by a decreasing strain in the geotextile. Figure 10 confirms the conclusion that the geotextile strain decreases in an embankment with a higher water table. The reduction in soil weight due to the higher water level may have resulted in a decrease of the strain in the geosynthetic. However, in the last month of the measurements, the water table had an extra increase, but this time the geotextile strain did not respond. Both responses on the water table changes are not in agreement with the findings of Song et al 2018. It is therefore recommended to further study the influence of water in the embankment.

Figure 10, which is a detail of Figure 9 also shows a daily scatter of the measured strains. As discussed before, the authors cannot explain the daily cycle in this case.

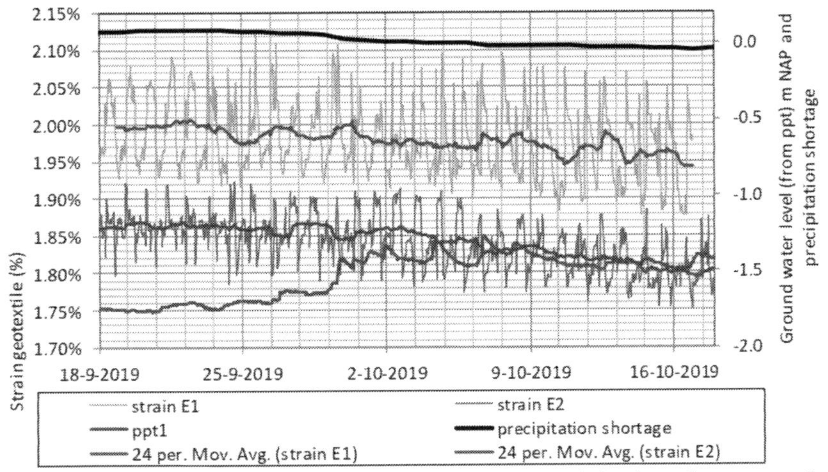

Figure 10. Comparison measured geotextile strains, precipitation shortage and groundwater table, detail of Figure 9, with a moving average through the measured strains.

5. Calculations

The strains in the geotextiles were calculated using the Concentric Arches model (Van Eekelen, 2015, adopted in the CUR226 (2016) design guideline). No partial factors were used. Table 2 lists the parameters used in the calculations with the Concentric Arches model. Some values were varied:

- The friction angle, as discussed before;
- The traffic load; without traffic load because traffic is not present at most times, and with 25% of the design value of the traffic load, because traffic has some permanent influence;
- With and without reduction of arching. CUR 226 requires reducing the arching (following Heitz, 2006) if the traffic load $p > \gamma H$. This reduction will probably not heal fully after truck passages. The arching coefficient in this case is $\kappa = 1.58$ (CUR2226).

Table 2. Parameters used in the calculations with the Concentric Arches model.

date	ctc-distance piles s_x / s_y (m)	width square pile caps a (m)	height fill H(m)	unit weight fill γ(kN/m^3)	friction angle fill φ (deg)	subgrade reaction k (kN/m^3)	stiffness geotextile $J_x = J_y$ (kN/m)	surcharge load p (kPa)	Arching reduction κ (-)
28-2-2019			0.00				3200		
1-3-2019			0.30				3200		
5-3-2019			0.60				3200		
12-3-2019	2.27 / 2.28	0.75	1.00	19.0	34 / 38	0	2961	0 / 11.5	1.00 (no reduction of arching) / 1.58 (reduction of arching)
24-4-2019			1.51				2722		
29-2-2020			1.51				2544		
25-8-2020			1.51				2426		

6. Comparison measured and calculated strain in the geosynthetic

Figure 11 and Table 3 compare the calculated and measured strains in the geosynthetic. The graphs and the table show the calculated maximum strains, that occur at the edge of the pile caps. This maximum value corresponds with the measured strains, that were taken close to the pile cap edges. The calculated values of the strains should be larger or equal to the measured values to show that CUR226 gives a safe prediction of the strains. The figure shows that lowest calculated values ($\varphi = 38$ deg, $p = 0$, $\kappa = 1.0$) for the strain agree reasonably well with the average measured values of E1 to E4. All other calculations give higher values than measured. This is a safe result.

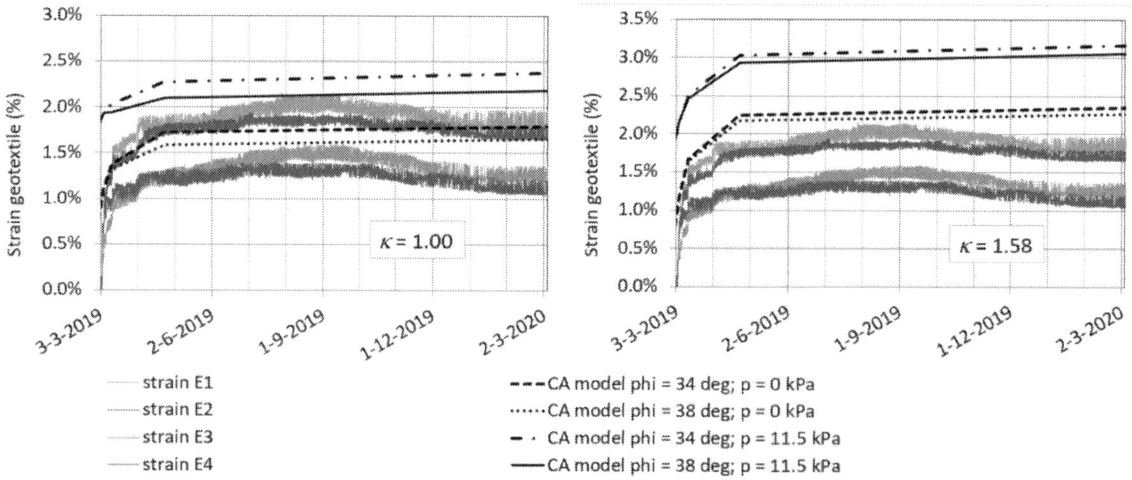

Figure 11. Comparison measured and calculated geotextile strains. Calculations were conducted using the Concentric Arches (CA) model of Van Eekelen et al. (2013, 2015), adopted in CUR226 (2016). Left: no reduction of soil arching ($\kappa = 1.0$); right: reduction of arching ($\kappa = 1.58$).

Table 3. Comparison measured geotextile strains and calculations with the Concentric Arches model.

date	CA model $\varphi = 34$ deg $p = 0$ kPa	CA model $\varphi = 38$ deg $p = 0$ kPa	CA model $\varphi = 34$ deg $p = 11.5$ kPa	CA model $\varphi = 38$ deg $p = 11.4$ kPa	E1 [a]	E2 [a]	E3 [a]	E4 [a]
No reduction of soil arching ($\kappa = 1.0$)								
12-3-2019	1.38%	1.32%	2.03%	1.94%	1.35%	1.25%	0.82%	1.00%
24-4-2019	1.72%	1.58%	2.28%	2.09%	1.78%	1.77%	1.24%	1.22%
29-2-2020	1.79%	1.65%	2.37%	2.18%	1.81%	1.69%	1.26%	1.09%
Reduction of soil arching ($\kappa = 1.58$)								
12-3-2019	1.67%	1.63%	2.51%	2.46%	1.35%	1.25%	0.82%	1.00%
24-4-2019	2.25%	2.18%	3.03%	2.93%	1.78%	1.77%	1.24%	1.22%
29-2-2020	2.35%	2.27%	3.16%	3.06%	1.81%	1.69%	1.26%	1.09%

[a] The measurements were taken each hour. The given values are 24 hours - average values of the measurements.

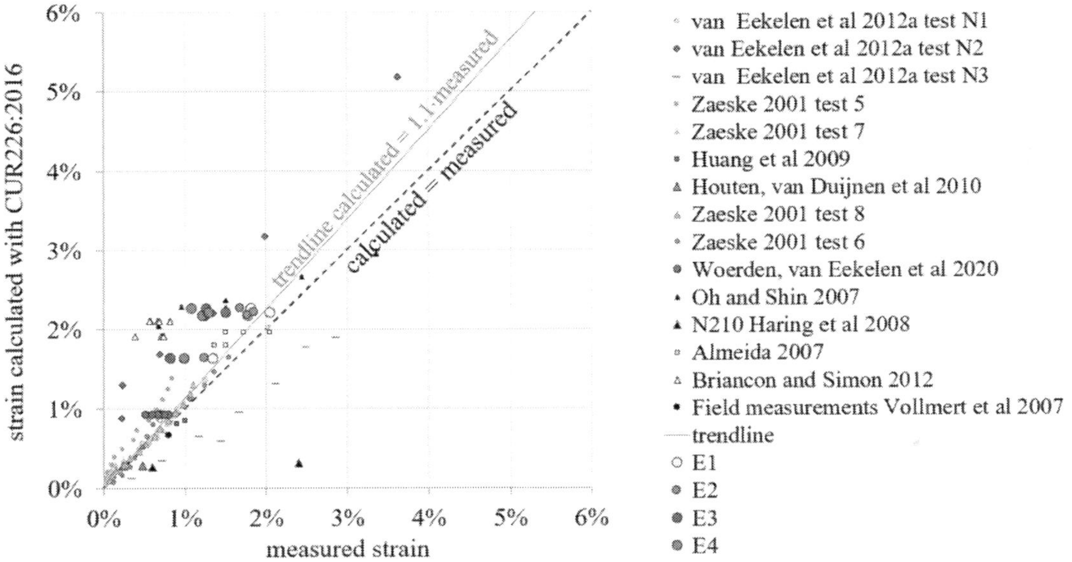

Figure 12. Validation of the Concentric Arches model: comparison measured and calculated geosynthetic strains using data from a large number of projects. Sources are given in Van Eekelen et al., 2015. Calculations were conducted using the Concentric Arches (CA) model of van Eekelen et al. (2013, 2015), adopted in CUR226 (2016).

Figure 12 extends the validation of the Concentric Arches model as given by Van Eekelen et al. (2015) with the values given in Table 3 (for $\varphi = 38$ deg; $p = 0$ kPa and $\kappa = 1.58$). The figure shows that the measurements of E1 and E2 match the calculations very well, and that the measured strains of E3 and E4 are lower than the calculated values, which is on the safe side. From this, it may be concluded that for this case, which is a woven geotextile-reinforced pile-supported embankment, CUR226 would give a safe design.

7. Discussion and Conclusions

The Concentric Arches model was developed to design the geosynthetic reinforcement for pile-supported embankments. The validated application of the Concentric Arches model, adopted in CUR226 (2016), however, was limited to embankments whose geometry and materials meet the same conditions as the embankments used for the validation of the Concentric Arches model. This meant that the embankment should not be located below the groundwater table and at least one geogrid layer should be used. If these requirements were not met, additional measurements were requested by CUR226 (2016) to demonstrate that the system comes within the framework of this guideline. These measurements had to be conducted in representative practical cases.

For this purpose, field measurements were conducted in a partially submerged GRPS embankment, that was reinforced with two layers woven geotextile and no geogrid. This paper presents these measurements and compares the measured strains with the values calculated with CUR226. It is concluded that:

- CUR226 matches the measurements well. This means that it is feasible to apply the Concentric Arches model and therefore the CUR226 (2016) guideline for GRPS embankments where the reinforcement consists of two layers of woven geotextile without geogrid. This conclusion is valid for woven geotextile like the geotextile applied in this case.
- A high groundwater level resulted in less strain in the geotextile, or in no response of the strain in the geotextile. The reduction in soil weight due to the higher water level may result in a decrease of the strain in the geotextile. A reduction in the arching because of a lower effective stress in the embankment fill does not seem important in this case. Further research on the influence of water in a geosynthetic-reinforced pile-supported embankments is recommended.

Acknowledgments

The authors are grateful for the support of the TKI-PPS funding of the Dutch Ministry of Economic Affairs, TenCate Geosynthetics, the Kyung Hee University in Seoul, Republic of Korea, BAM and Deltares, Netherlands.

References

[1] Briançon L and Simon B 2012 Performance of Pile-Supported Embankment over Soft Soil: Full-Scale Experiment. *J. of Geotechnical and Geoenvironmental Engineering* **138**, 551–561. https://doi.org/10.1061/(ASCE)GT.1943-5606.0000561

[2] BSI (British Standards Institution) 2015. *Code of practice for earth retaining structures.*

[3] CUR226 2016. see Van Eekelen and Brugman 2016.

[4] CEN 2008 European Committee for Standardization, Techn. Com. CEN/TC 1 54 "Aggregates.", British Standards Institution. *Aggregates for unbound and hydraulically bound materials for use in civil engineering work and road construction.* British Standards Institution, London.

[5] Haring W, Profittlich M and Hangen H 2008 Reconstruction of the National Road N210 Bergambacht to Krimpen a.d. IJssel, NL: Design Approach, Construction Experiences and Measurement Results., *Proc. of EuroGeo 4,* Edinburgh, UK, 8 p. paper number 259.

[6] Heitz C 2006 Bodengewölbe unter ruhender und nichtruhender Belastung bei Berücksichtigung von Bewehrungseinlagen aus Geogittern. Schriftenreihe Geotechnik, Uni Kassel, Heft 19.

[7] KNMI (Royal Netherlands Meteorological Institute) 2020a Dagwaarden neerslagstations https://www.knmi.nl/nederland-nu/klimatologie/monv/reeksen#G (accessed 3.9.20).

[8] KNMI (Royal Netherlands Meteorological Institute) 2020b Daggegevens van het weer in Nederland, https://www.knmi.nl/nederland-nu/klimatologie/daggegevens (accessed 3.9.20).

[9] Sloan J A 2011 *Column-Supported Embankments: Full-Scale Tests and Design Recommendations* PhD Thesis. Virginia Polytechnic Institute and State University, Blacksburg, Virginia, USA.

[10] Song J, Chen K, Li P, Zhang Y and Sun C 2018 Soil arching in unsaturated soil with different water table. *Granular Matter* 20, 78. https://doi.org/10.1007/s10035-018-0849-3.

[11] Van Eekelen, S.J.M., Van, M.A., Bezuijen, A., 2007. The Kyoto Road, a full-scale test. Measurementes and calculations. In: the proceedings of ECSMGE14, Madrid, Spain 1533-1538.

[12] Van Eekelen, S J M, Bezuijen A, Lodder H J, van Tol A F, 2012 Model experiments on piled embankments. Part I. *Geotextiles and Geomembranes* **32**, 69–81. https://doi.org/10.1016/j.geotexmem.2011.11.002

[13] Van Eekelen S J M 2015 *Basal Reinforced Piled Embankments* PhD Thesis Delft University of Technology, Delft, Netherlands. ISBN 978 94 6203 825 7.

[14] Van Eekelen S J M Bezuijen A van Tol A F 2015 Validation of analytical models for the design of basal reinforced piled embankments. *Geotextiles and Geomembranes* **43**, 56–81. https://doi.org/10.1016/j.geotexmem.2014.10.002

[15] Van Eekelen S J M Bezuijen A van Tol AF 2013 An analytical model for arching in piled embankments. *Geotextiles and Geomembranes* **39**, 78–102. https://doi.org/10.1016/j.geotexmem.2013.07.005

[16] Van Eekelen S J M and Brugman M H A (Eds.) 2016. *Design Guideline Basal Reinforced Piled Embankments*. CRC Press, Delft, Netherlands.

[17] Van Eekelen S J M and Han J 2020. Geosynthetic-reinforced pile-supported embankments: state of the art. *Geosynthetics International Special Issue Geosynthetic-reinforced pile-supported embankments*. https://doi.org/10.1680/jgein.20.00005

[18] Van Eekelen S J M, Venmans A A M, Bezuijen A, Van Tol A F 2020. Long term measurements in the Woerden geosynthetic-reinforced pile-supported embankment. *Geosynthetics International*. https://doi.org/10.1680/jgein.17.00022

Application of Eurocode 7 for the analysis and design of veneer cover soils in static and seismic load conditions

P Rimoldi[1], P Pezzano[2], A Gebely[2]

[1]Consultant, Milano, Italy
[2]Officine Maccaferri S.p.A, Bologna, Italy

Email: pietro.rimoldi@gmail.com

Abstract. Veneer is intended as a relatively thin cover soil layer placed on a slope. Veneer cover soil systems in containment facilities such as landfills, liquid disposal and reservoirs involve geosynthetic-soil interfaces. The main design criterion for protective veneer covers is to ensure that no sliding occurs above the geosynthetic or the existing soil slope. The design of the protective veneer covers should include the stability of the lining system under the effect of the gravity forces, equipment loadings, seismic actions and pore water pressures. The limit equilibrium method used in the present paper for the evaluation of veneer stability has been derived from the method of Koerner and Soong (1988) and adapted to the limit state design offered by Eurocode 7 with reference to the Resistance Factor Approach (RFA). Serviceability limit state considerations are included in the design method for completeness.

1. Introduction

Veneer is intended as a relatively thin cover soil layer placed on a slope.

The stability analysis for sliding of a veneer above a geosynthetic or a natural soil layer, including sliding above liners, like geomembranes (GBR), geosynthetic clay liners (GCL) and compacted clay liners (CCL), is the object of the present paper.

There are two specific applications in which cover soil stability needs to be checked:

1) Leachate collection soil placed above a GBR, GCL and/or CCL along the sides of a landfill or a heap leach pad before waste or ore is placed and stability achieved accordingly.

2) Final cover soil placed above a GBR, GCL and/or CCL in the cap or closure of a landfill or a heap leach pad.

Veneer stability above a liner represents a major challenge due to the following reasons [1]:

(a) The underlying barrier materials generally represent a low interface shear strength boundary with respect to the soil placed above them.

(b) The liner system is oriented precisely in the direction of potential sliding.

(c) The potential shear planes are usually linear and are essentially uninterrupted along the slope.

(d) Liquid (water or leachate) cannot continue to percolate downward through the cross section due to the presence of the barrier material.

Since both situations (leachate collection and final covers) present the same technical issues, the method presented in the present paper is applicable to both. Anyway, the paper is particularly addressed to the final cover situation, which presents much higher challenges.

1.1. Interface Shear Testing

The interface shear strength of a cover soil with respect to the underlying material (geomembrane or natural soil) is critical so as to properly analyze the stability of the cover soil. This value of interface shear strength is obtained by laboratory testing of the project specific materials, that is by sampling of the candidate geosynthetics to be used at the site, as well as of the cover soil at its targeted density and moisture conditions, at the site specific conditions, that is under project specific normal stresses, strain rates, peak or residual shear strengths and temperature extremes (high and/or low). Hence values of interface shear strengths from the literature should be used only for preliminary design.

The test that has been developed to evaluate geosynthetics shear strength parameters are ASTM D5321 [2] in the USA, while internationally reference has to be made to ISO 12957-1 [3], and to ISO 12957-2 [4]. Usually in the direct shear test relatively high compressive loads are applied, in the range 50 – 200 kPa. But a veneer soil layer on a capping applies a very small load: a 1 m thick veneer applies a uniform pressure of maximum 20 kPa. Therefore, the inclined plane test appears to be the most suitable for obtaining the friction characteristics at the interface between the veneer soil and the layer below it [5], at least for capping design.

With either direct shear or inclined plane tests, two fundamental shear strength parameters are obtained:
δ = the angle of shearing resistance, peak and/or residual, of the two opposing surfaces (usually called the interface friction angle)
C_a = the adhesion of the two opposing surfaces, peak and/or residual.

The upper limit of δ, when soil is involved as one of the interfaces, is the friction angle of the soil component, φ. The upper limit of the C_a value is the cohesion of the soil component, c. In the following slope stability analyses, the C_a term will be included for the sake of completeness, but to utilize an adhesion value there must be a clear physical justification for the use of such values when geosynthetics are involved, like textured geomembranes with physical interlocking of soils having cohesion.

1.2. Types of Loadings

There are a large variety of slope stability problems that may be encountered in analyzing and/or designing final covers of landfills or heap leach pads as well as leachate collection soils covering geomembranes beneath the waste or ore. The most common situation is a uniformly thick cover soil on a geomembrane placed over the soil subgrade at a given and constant slope angle. Analyses will include loads produced by equipment moving up the slope and then moving down the slope during placement of cover soil on the geomembrane.

Since cover soil slides have occurred due to seepage forces, often drainage above a geomembrane (or other barrier material) in the cover soil cross section must be accommodated to avoid the possibility of seepage forces. This situation requires specific analyses of slope stability problems.

If an earthquake occurs in the vicinity of a landfill or heap leach pad, the seismic wave travels through the solid waste or ore mass reaching the upper surface of the cover. It then decouples from the cover soil materials, producing horizontal and vertical forces which must be appropriately analyzed.

All of the above actions are destabilizing forces tending to cause veneer instability. To increase the stability of the veneer soil, the designer can use geogrids or high strength geotextiles within the cover soil acting as reinforcement materials. This technique, usually referred to as veneer reinforcement, will be specifically analyzed.

2. Analytical approach to veneer stability

The analytical approach used in the present paper for the evaluation of veneer stability has been derived by the method of Koerner and Soong [1], adapted to limit state design approach according to EuroCodes. The present method for veneer stability analyses includes evaluations of toe wedge forces, seismic forces, and tensile forces of geosynthetic reinforcements,

The model assumes that the veneer soil cover can be divided in the active wedge of trapezoidal shape, limited by a tension crack at the crest, whose sliding down the slope is contrasted by the passive triangular wedge at toe and by the geosynthetic reinforcement anchored at top of the slope, as shown in the scheme of Figure 1.

Final cover systems for a landfill or a heap leach pad can be characterized as a veneer soil layer of thickness h on a surface sloped at angle β to the horizontal, as shown in Figure 1.

The critical slip surface is considered to occur on the liner below the veneer cover. The tangential weight of the soil mass, $W_A \cdot \sin \beta$, is the driving force that might cause the soil veneer to slip.

The resisting forces against slippage are:

(a) the shear strength between the soil mass and the slip surface
(b) the toe buttressing force at the toe of the slope
(c) the tensile force, T_D, provided by the geosynthetics.

The goal of veneer stability design is to provide all of the slope resistance with shear forces and not to put the geosynthetic barriers in tension. Only the tensile strength contribution provided by a high tensile resistant geosynthetic, such as a geogrid or woven / knitted geotextile, is considered for the support of the cover system, while any tensile strength contribution from other geosynthetics (geomembranes, nonwoven geotextile filters, etc.) shall be ignored.

The shear strength along the critical interface is often a function of the effective normal force acting on this interface. As well known, the relationship between shear strength and normal force is called the shear strength envelope, and it is usually expressed in terms of the Mohr-Coulomb parameters of friction angle φ and cohesion c. Since the actual shear strength envelope is often not linear, it is important that the Mohr-Coulomb parameters are defined within a specific normal load range, as it is always unconservative to extrapolate the envelope, either to lower normal loads or to higher ones. The shear strength on the interface can be reduced by pore pressures, u, because pore pressures reduce the effective normal force acting on the interface. Thus, flowing water pressures would produce a destabilizing force.

In Ultimate Limit State (ULS) conditions, according to EuroCodes, loads are amplified by Amplification Factors, while resistances are reduced by Reduction Factors.

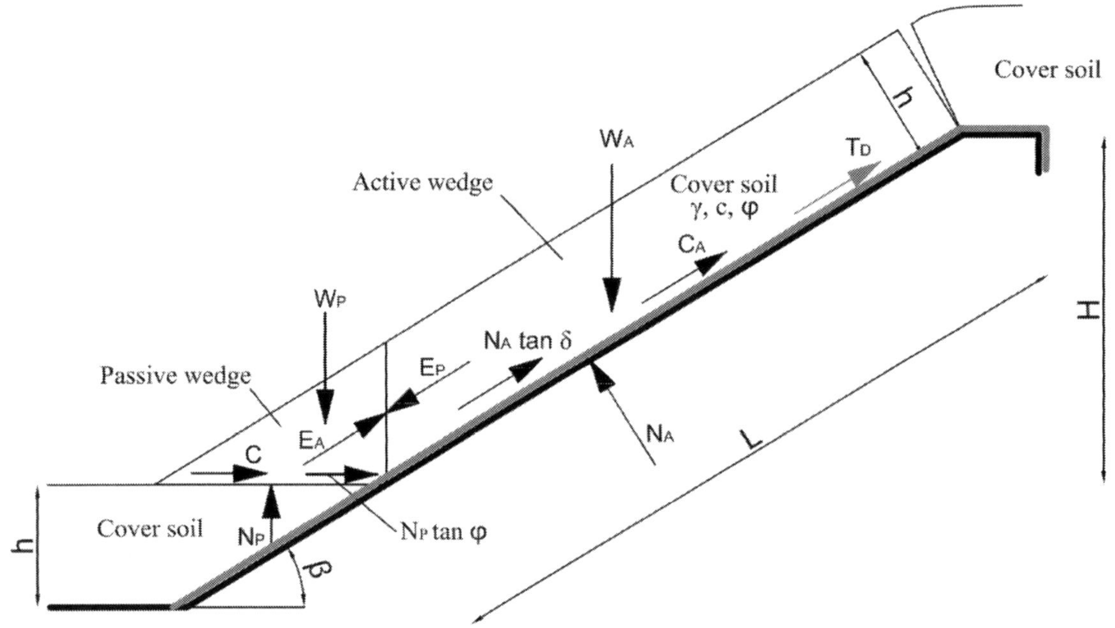

Figure 1. Scheme of veneer stability analysis with only gravitational static loads applied, including the use of veneer reinforcement (modified from [1])

The analysis of forces in the free body diagrams of the active and passive wedges (Figure 1) allows to calculate the active and passive interwedge forces, E_a and E_p. The stability check is satisfied if:

$$FS = (E_p / E_a) \geq \gamma_R \quad (1)$$

where FS is the Factor of Safety and γ_R is the partial factor R3 required by the EuroCode norms for sliding analysis (see later).

In practice the EC7 approach can be easily implemented by performing the "traditional" analyses, like sliding along the slope, pull-out, etc., but using as input values the factorized values of loads and geotechnical parameters. Analyses are satisfied when the minimum Factor of Safety, now called R3, or γ_R, is achieved.

3. Design approach according to Eurocodes

3.1. Eurocodes framework

In Eurocode 7 [6] (usually abbreviated as EC7), design by calculation involves the definition of:
- actions, which may be either imposed loads or imposed displacements, e.g. from ground movements;
- properties of soils, rocks and other materials;
- geometrical data;
- limiting values of deformations, crack widths, vibrations etc.;
- calculation models.

According to EC7, the Ultimate Limit State (ULS) analyses of any geotechnical structure (including veneer soil layers) shall address both failure of the ground and of the structure.

In EC7, currently there are three optional design approaches in combination with the Material Factoring Approach (MFA), the Load Factoring Approach (LFA) and the Resistance Factoring Approach (RFA). They differ in the way in which partial factors are taken into account.

For the ULS analysis of reinforced soil veneers, the Resistance Factoring Approach (RFA) appears to be the best option, and it will be used thereinafter.

In RFA: loads are amplified through the set of factors M1; geotechnical parameters are not reduced, according to the set of factors M1; resistances are reduced through the factors γ_R according to the factor R3.

The condition of Eq. (1) says that, according to EC7, the ratio between the design resistance (R_d) and the design effect of the actions (E_d) shall be larger than the coefficient γ_R (that is, the Factor R1, R2, or R3 of the specific approach that has been chosen); hence the coefficient γ_R can be considered equivalent to the Factor of Safety FS of traditional methods.

In this paper, for the Factors A1, M1, and R3 reference has been made to the Italian Technical Norm for Construction issued in 2018 [7], which implement both Eurocode 7 [6] and Eurocode 8 [8]. Anyway, all National Annexes are applicable, it is enough to use the specific national factors.

3.2. Actions

In EC7:
- An action is given the general symbol, F.
- Actions can be permanent (persistent), permanent non-structural, or variable (transient), accidental, or seismic.
- Persistent actions are denoted by F_G. Transient actions are denoted by F_Q.
- Actions can be either "favourable" or "unfavourable".
- Load Coefficients (A) shall be applied to actions in each Ultimate Limit State (ULS) analysis.
- Partial coefficients (A) are denoted by γ_{G1}, γ_{G2}, γ_{Qi}, for permanent, permanent non-structural, and variable loads, respectively.

The combinations of factors to be applied to the actions (A1) in the ultimate limit state analyses, according to Italian Norm [7], are reported in Table 1.

3.3. Ground properties

In EC7:

• Geotechnical parameters should be established with consideration given to published data and local and general experience
• Material properties are given the general symbol, X.
• Characteristic values of material properties are given the general symbol, X_k.
• The selection of characteristic values for geotechnical parameters shall be based on results and derived values from laboratory and field tests.
• Reduction coefficients (M) shall be applied to characteristic values of material properties in each Ultimate Limit State (ULS) analysis.
• Partial coefficients (M) are denoted by $\gamma_{\phi'}$, $\gamma_{c'}$, γ_{cu}, γ_{γ}, for tangent of friction angle, drained cohesion, undrained cohesion, and unit weight, respectively.

The combinations of factors to be applied to the soil parameters (M1) in the ultimate limit state analyses, according to Italian Norm (NTC 2018), are reported in Table 2.

Table 1. Combinations of factors to be applied to the actions (A1) in the ultimate limit state analyses, according to Italian Norm [7]

TYPE OF LOAD	EFFECT	PARTIAL COEFFICIENT	A1
PERMANENT	FAVOURABLE	γ_{G1}	1.0
	UNFAVOURABLE		1.3
PERMANENT NON-STRUCTURAL	FAVOURABLE	γ_{G2}	0.8
	UNFAVOURABLE		1.5
VARIABLE	FAVOURABLE	γ_{Qi}	0.0
	UNFAVOURABLE		1.5

Table 2. Combinations of factors to be applied to the soil parameters (M1) in the ultimate limit state analyses, according to Italian Norm [7]

PARAMETER	APPLIED TO	PARTIAL COEFFICIENT	M1
Tangent of friction angle	$\tan \varphi'_k$	$\gamma_{\varphi'}$	1.00
Drained cohesion	c'_k	$\gamma_{c'}$	1.00
Undrained cohesion	c_{uk}	γ_{cu}	1.00
Unit weight	γ	γ_{γ}	1.00

3.4. Stability Verification

According to EC7, the ratio between the design resistance (R_d) and the design effect of the actions (E_d) shall be larger than the coefficient γ_R (that is, the Factor R3 in case of RFA), according to Eq. (1).

The γ_R coefficient for sliding, to be used for ultimate limit states verification purposes (R3) according to Italian Norm [7], is reported in Table 3.

Therefore, according to EC7, for veneer cover stability the following analyses shall be carried out:
- Direct sliding, tensile strength of reinforcement, pull out of reinforcement and/or other geosynthetics (for anchorage in top trenches)
- Verification shall be carried out according to RFA Approach with Combination (A1+M1+R3)
- For direct sliding, verification is satisfied if: $FS = R_d / E_d = (E_p / E_a) \geq R3$

Table 3. Partial coefficients γ_R (R3) for verification of the ultimate limit states, applicable to veneer soil, according to Italian Norm [7]

ANALYSIS	PARTIAL COEFFICIENT R3
Sliding	$\gamma_R = 1.1$

4. Sliding stability analyses for veneer covers

4.1. *Basic case of only gravitational forces in static conditions*

Figure 1 illustrates the common situation of a finite length, uniformly thick cover soil placed over a liner material at a slope angle β. It includes a passive wedge at the toe and has a tension crack at the crest. The analysis that follows is derived from [1] and [9], but comparable analyses are available from [10], [11], [12], and others.

Note that all analyses refer to 1.0 m running slope length, hence all forces are expressed in kN/m.

The symbols used in Figure 1 are defined below:

W_A = total weight of the active wedge (kN/m)
W_p = total weight of the passive wedge (kN/m)
N_A = effective force normal to the failure plane of the active wedge (kN/m)
N_p =effective force normal to the failure plane of the passive wedge (kN/m)
γ = unit weight of the cover soil (kN/m^3)
h = thickness of the cover soil (m)
H = vertical height of the slope measured from the toe of the active wedge (m)
L = length of slope measured along the soil slope (m)
β = angle of the slope (°)
φ = friction angle of the cover soil (°)
δ = friction angle between cover soil and liner (deg)
C_a = adhesive force between cover soil of the active wedge and the liner (kN/m)
c_a = adhesion between cover soil of the active wedge and the liner (kPa)
C = cohesive force along the failure plane of the passive wedge (kN/m)
c = cohesion of the cover soil (kPa)
T_D = design value of reinforcement strength (kN/m)
E_A = interwedge force acting on the active wedge from the passive wedge (kN/m)
E_P = interwedge force acting on the passive wedge from the active wedge (kN/m)
FS = factor of safety against cover soil sliding along the slope

Note that, to avoid long writing, the above geotechnical parameters represent the characteristic values of soil and interface properties already reduced with the partial coefficients of the set M1 (see Table 2), that is:

$$\tan\varphi = \tan\varphi_{soil} / \gamma_{\phi'},\quad c = c_{soil} / \gamma_{c'},\quad \tan\delta = \tan\delta_{interface} / \gamma_{\phi'},\quad c_a = c_{interface} / \gamma_{c'},\quad \gamma = \gamma_{soil} / \gamma_{\gamma} \tag{2}$$

The expression for determining the factor of safety can be derived as follows.

From Figure 1, without the geosynthetic reinforcement, considering the active wedge:

$$W_A = \frac{1}{2}\gamma \cdot h \cdot \left(\frac{2H}{\sin\beta} - h\tan\beta\right) \tag{3}$$

$$N_A = W_A \cdot \cos\beta \tag{4}$$

$$C_A = c_A \cdot L = c_A \cdot \frac{H}{\sin\beta} \tag{5}$$

From Figure 1, considering the passive wedge:

$$W_P = \frac{1}{2}\gamma \cdot h^2 \cdot \left(\frac{1}{\sin\beta \cdot \cos\beta}\right) \tag{6}$$

$$N_P = W_P + E_P \cdot \sin\beta \tag{7}$$

$$C = c \cdot \frac{h}{\sin \beta} \tag{8}$$

From the equilibrium of vertical forces on the active wedge $(\Sigma F_v = 0)_{active\ wedge}$:

$$E_A = \frac{1}{\sin \beta} \cdot \left[\gamma_{G1u} \cdot W_A \cdot (1-\cos^2 \beta - \cos \beta \cdot \sin \beta \cdot \tan \delta) - \gamma_{G1f} \cdot C_A \cdot \sin \beta\right] \tag{9}$$

where:
γ_{G1u} = Amplification Factor for permanent unfavourable loads (see Tab. 1)
γ_{G1f} = Amplification Factor for permanent favourable loads (see Tab. 1).
From the equilibrium of horizontal forces on the passive wedge $(\Sigma F_h = 0)_{passive\ wedge}$:

$$E_P = \frac{\gamma_{G1f} \cdot C + \gamma_{G1f} \cdot W_P \cdot \tan \varphi}{\cos \beta - \sin \beta \cdot \tan \varphi} \tag{10}$$

Stability is then verified if, for the combinations (A1 + M1 + R3), where from Tab. 3 for sliding analyses it is R3 = 1.10, the Factor of Safety results:

$$FS = \frac{E_P}{E_A} \geq 1.10 \tag{11}$$

If the above condition is not verified, it means that the veneer cover is unstable for sliding down the slope: in this case a geosynthetic reinforcement shall be introduced and the stability analysis including reinforcement shall be performed until the condition is verified.

4.2. Veneer Reinforcement

The best way of increasing the Factor of Safety for a slope is to reinforce it with a geosynthetic. Such reinforcement can be either intentional or non-intentional [1].

By intentional reinforcement it is meant to include a geogrid or high strength geotextile within the cover soil to purposely reinforce the system against instability. Depending on the type and amount of reinforcement, the majority, or even all, of the driving, or mobilizing, stresses can be supported, resulting in major increase in FS value.

By non-intentional reinforcement reference is made to multicomponent liner systems where a low shear strength interface is located beneath an overlying geosynthetics. In this case, the overlying geosynthetics are inadvertently acting as veneer reinforcement to the composite system. The difference is that the geogrids and/or high strength geotextiles give a major increase in the FS value, while a protection geotextile (or other low strength geosynthetics) only nominally increases the FS value at the cost of high strains in the geosynthetics. Therefore, it is evident that non-intentional reinforcement shall be avoided.

Hence we will refer only to intentional reinforcement, that is a geogrid or high strength geotextile placed below the veneer cover and above the multicomponent liner system; while the tensile strength of any component of the lining system (filter geotextiles, draining geocomposites, geomembranes, etc.) will be considered equal to zero.

The scheme of the stability analysis of the veneer cover subject only to gravitational forces in static conditions, with a reinforcing geosynthetic below the veneer cover, is shown in Figure 1.

All forces have the same expressions as in the previous analysis, but a force T_D from the reinforcement, acting parallel to the slope, provides additional stability. This force acts only within the active wedge.

The design tensile force T_D is obviously less than the ultimate strength of the geosynthetic reinforcement. Considering the ultimate strength T_{ult}, the value should be reduced by such factors as installation damage, creep and long-term degradation. Note that if seams are involved in the reinforcement, a reduction factor should be added accordingly. Koerner [13], among others, provides recommended numeric values.

Hence the expression of T_D is:

$$T_D = \frac{T_{ult}}{RF_{ID} \cdot RF_{cr} \cdot RF_{ch} \cdot RF_b} \tag{12}$$

where:
T_D = design value of reinforcement strength (kN/m)
T_{ult} = ultimate value of reinforcement strength (kN/m)
RF_{ID} = reduction factor for installation damage
RF_{cr} = reduction factor for creep
RF_{ch} = reduction factor for long term chemical degradation
RF_b = reduction factor for long term biological degradation

Note that for analyses at short term (e.g. at the end of construction) the Reduction Factors for creep and for chemical and biological damage shall be set as: $RF_{cr} = RF_{ch} = RF_b = 1.0$; while, for seismic analyses, only the Reduction Factor for creep shall be set as: $RF_{cr} = 1.0$.

With reference to Figure 1, the interwedge force E_A is obtained from the equilibrium of vertical forces on the active wedge $(\Sigma F_v = 0)_{active\ wedge}$:

$$E_A = \frac{1}{\sin\beta} \cdot \left[\gamma_{G1u} \cdot W_A \cdot (1-\cos^2\beta - \cos\beta \cdot \sin\beta \cdot \tan\delta) - \gamma_{G1f} \cdot C_A \cdot \sin\beta - \gamma_{G1f} \cdot T_D \cdot \sin\beta\right] \tag{13}$$

The interwedge force E_P has the same expression as in Eq. (10).

Stability is verified if, for combination (A1 + M1 + R3), the Factor of Safety results as per Eq. (11).

For this analysis and for all the following ones, the tensile strength T_D (and therefore the ultimate tensile strength T_{ult}) shall be varied by trials and errors until the condition of Eq. (11) gets verified.

For high values of T_D the interwedge force E_A may become negative, and as a consequence also FS becomes negative: this is just a sign that T_D has been overdesigned; then the designer may try with a lower value of T_D, or the designer may keep this high T_D value, in favour of safety.

4.3. Tracked Construction Equipment Forces

The placement of cover soil on a slope with a relatively low shear strength inclusion (like a liner) should always be from the toe upward to the crest. In so doing, the gravitational forces of the cover soil and live load of the construction equipment are compacting the previously placed soil and working with an ever-present passive wedge and stable lower portion beneath the active wedge. For soil placement down the slope, however, a stability analysis cannot rely on toe buttressing and also a dynamic stress should be included in the calculation. These conditions decrease the FS value and in some cases to a great extent. Unless absolutely necessary, it is not recommended to place cover soil on a slope in this manner. If it is necessary, the design must consider the unsupported soil mass and the dynamic force of the specific type of construction equipment and its manner of operation.

For the second case of a bulldozer pushing cover soil down from the crest of the slope to the toe, or in any case for the bulldozer moving down the slope, the analysis uses the scheme of Figure 2.

The analysis uses a specific piece of tracked construction equipment (like a bulldozer characterized by its ground contact pressure) and dissipates this force or stress through the cover soil thickness to the surface of the liner. A Boussinesq analysis is used to transfer the bulldozer weight at the veneer surface as an equivalent vertical force on the liner interface [14].

Moreover, we have to consider an additional force due to acceleration (or deceleration) of the equipment along the slope. This analysis again uses a specific piece of construction equipment operated in a specific manner. The acceleration or deceleration produces a force parallel to the slope, equivalent to the weight of the bulldozer multiplied by the acceleration of the bulldozer and divided by the acceleration due to gravity. Its magnitude is equipment operator dependent and related to both the equipment speed and time to reach such a speed.

It is clear that the case of the equipment moving down the slope is the critical one; it is also obvious that, even if the soil is placed upward, when the equipment has reached the top of the slope it has to go down moving on the soil just placed.

Therefore, only the case of the equipment moving down the slope will be considered for stability analyses.

The equipment force per unit width (see Figure 2) can be calculated as follows:

$$W_e = q \cdot w \cdot I \tag{14}$$

$$q = \frac{W_b}{2 \cdot w \cdot b} \tag{15}$$

where:
W_e = equivalent equipment force per unit width at the liner interface (kN/m)
W_b = weight of the equipment force (kN)
q = uniform pressure applied by equipment tracks on the veneer surface (kPa)
w = length of equipment tracks (m)
b = width of equipment tracks (m)
I = influence factor at the liner interface

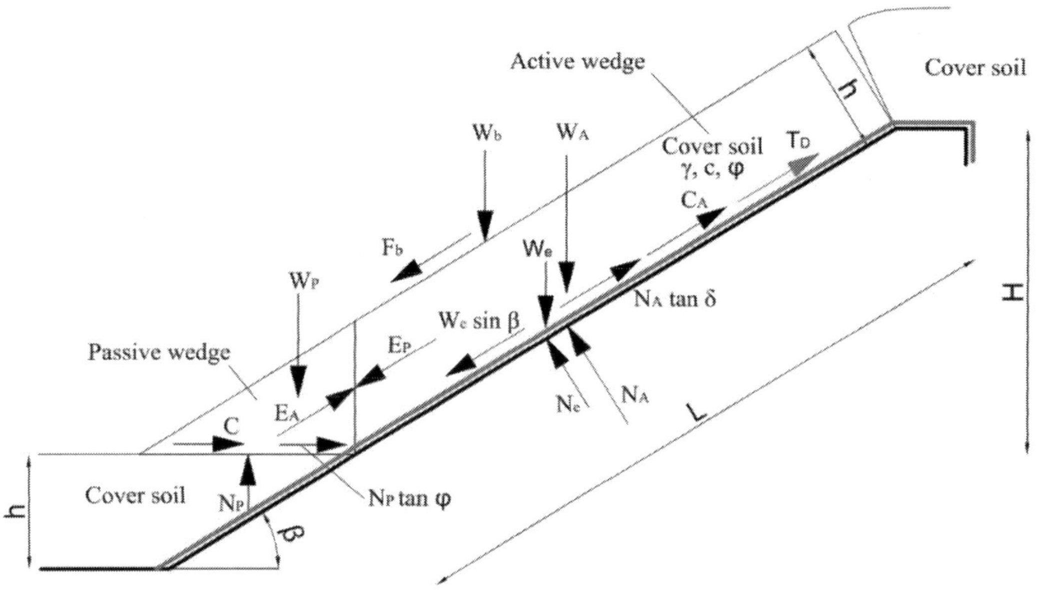

Figure 2. Limit equilibrium forces involved in a finite length slope analysis for a uniformly thick cover soil with equipment moving down the slope, including a geosynthetic reinforcement below the veneer cover (modified from [1])

The influence factor I can be obtained from Figure 3 (modified from [14]).

The curve in Figure 3 can be approximated with the polynomial expression:

$$I = 0.0127\ (b/h)^5 - 0.1661\ (b/h)^4 + 0.8380\ (b/h)^3 - 2.0307\ (b/h)^2 + 2.3709\ (b/h) - 0.1059 \tag{16}$$

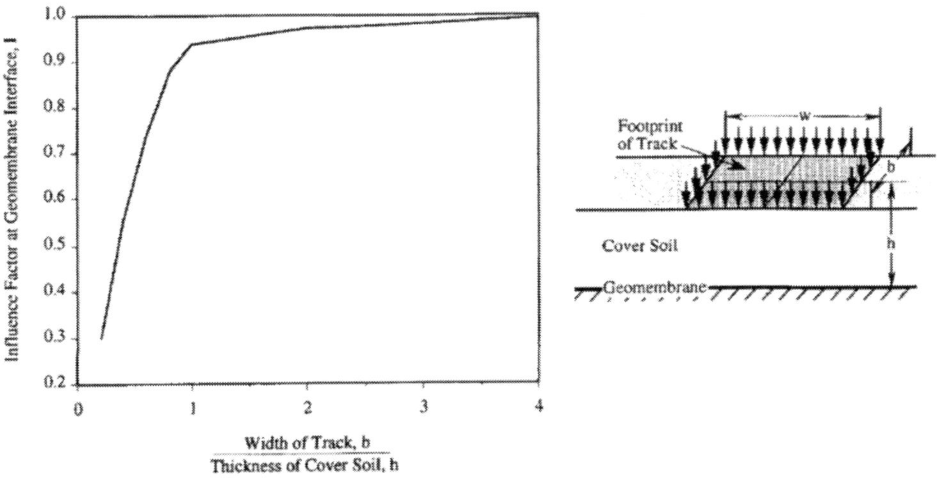

Figure 3. Values of the influence factor, I, for use to dissipate surface force of tracked equipment through the cover soil to the liner interface (modified from [14])

The acceleration of the bulldozer, coupled with the influence factor I from Figure 3 or Eq. 16, results in the dynamic force per unit width at the cover soil to liner interface, F_b (Figure 3):

$$F_b = W_b \cdot \frac{a_b}{g} \tag{17}$$

where
F_b == dynamic force per unit width parallel to the slope at the liner interface (kN/m)
a_b = acceleration of the bulldozer (m^2/s)
g = acceleration due to gravity = 9.81 m^2/s
The acceleration of the bulldozer is equipment operator dependent and related to both the equipment speed and the time to reach such a speed; hence the value of acceleration, a_b, can be evaluated from Figure 4 (from [1]).

When there is equipment moving down the slope, the critical situation occurs when the veneer cover has been laid up to the top of the slope: then the scheme of the stability analysis, including a geosynthetic reinforcement below the veneer cover, becomes as shown in Figure 2.

The force N_e, normal to the sliding plane and produced by the equipment on the liner interface, is:

$$N_e = W_e \cdot \sin\beta \tag{18}$$

Considering that the equipment produces an unfavourable variable load, the expression of the interwedge force E_A can be obtained from the equilibrium of vertical forces on the active wedge $(\Sigma F_v = 0)_{active\ wedge}$:

$$E_A = \frac{1}{\sin\beta} \cdot \left\{ \begin{array}{l} \left[\gamma_{G1u} \cdot W_A \cdot (1 - \cos^2\beta - \cos\beta \cdot \sin\beta \cdot \tan\delta) - \gamma_{G1f} \cdot C_A \cdot \sin\beta\right] \\ -\gamma_{G1f} \cdot T_D \cdot \sin\beta + \left[\gamma_{Qiu} \cdot W_e \cdot (1 + \frac{a}{g} - \cos\beta \cdot \sin\beta \cdot \tan\varphi\right] \end{array} \right\} \tag{19}$$

where:
γ_{Qi} = amplification factor for variable unfavorable loads, from Table 1.

The expression of E_P remains the same of Eq. (10).

Stability is verified if, for combination (A1+M1+R3), the Factor of Safety results as per Eq. (11).

Also in this case, for high values of T_D the interwedge force E_A may become negative, and as a consequence also FS becomes negative: this is just a sign that T_D has been overdesigned; then the designer may try with a lower value of T_D, or the designer may keep this high T_D value, in favour of safety.

Note that the critical situation occurs at the end of construction of the veneer cover, hence analyses are considered at short term; since the probability of an earthquake occurring just at the end of construction is negligible, analyses with equipment loads are only carried out in static conditions, without considering any seismic condition.

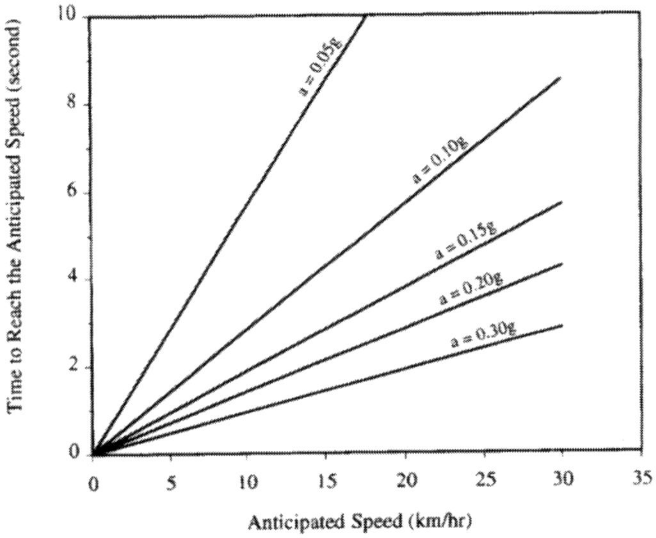

Figure 4. Graphic relationship of construction equipment speed and rise time to obtain equipment acceleration (from [1])

Being a short term analysis, if reinforcement is used, the Reduction Factors for creep, chemical and biological damage shall be set equal to 1.0, while only the Reduction Factor for installation damage shall be applied.

The inherent danger of a bulldozer moving down the slope is readily apparent. Note that the same result comes about by the bulldozer decelerating instead of accelerating. The sharp breaking action of the bulldozer is arguable the more severe condition due to the extremely short times involved when stopping forward motion. Hence acceleration should be estimated from Figure 4 with reference to the condition of sudden breaking.

Since acceleration and breaking of the equipment is unavoidable, an analysis should be made of the specific stability situation and the construction specifications should reflect the exact conditions made in the design. The maximum allowable weight and ground contact pressure of the equipment should be stated along with suggested operator movement of the cover soil placement operations.

From above analysis it is clear that the best construction solution is to avoid the forces produced by equipment moving on the slope: hence, whenever possible, placement of veneer soil cover using long reach boom excavators operating at the toe and the top of the slope should be preferred.

Truck traffic on the slopes can also give as high, or even higher, stresses than tracked bulldozers, and should be avoided unless adequately designed. Additional details are given in [15].

The issue of access ramps is a unique subset of this analysis and one which deserves focused attention due to the high loads and decelerations that often occur.

4.4. Seismic Forces

In areas of anticipated earthquake activity, the slope stability analysis of a final cover soil over a landfill or a heap leach pad site must consider seismic forces.

According to EuroCodes, seismic forces shall always be considered wherever the seismic design acceleration is more than zero. EuroCode 8 [8] specifies how to determine the seismic parameter for a specific project.

The EuroCode 8 approach is now presented with reference to Italian Norm [7], being Italy one of the most seismic Countries in Europe.

The motion generated by an earthquake at a site depends on the particular local conditions, that is, it depends on the topographic characteristics and stratigraphic deposits of soil and rock mass and the physical and mechanical properties of the materials that constitute them. The seismic motion at the surface of a site, associated with each category of the subsoil, is defined by the maximum acceleration (a_{max}) at the surface. Once the acceleration ag expected at the bedrock has been defined, it is possible to calculate the maximum seismic acceleration at the surface by the following formula:

$$a_{max} = S \cdot a_g = S_S \cdot S_T \cdot a_g \qquad (20)$$

where:
a_g = acceleration expected at the bedrock (m/s^2)
a_{max} = maximum acceleration at the surface (m/s^2)
S_S = stratigraphic amplification coefficient
S_T = topographic amplification coefficient.

The acceleration at bedrock ag is provided by the National Agency in charge of seismic security, as a function of the seismic history and records for the site, and of the reference period for the seismic action. Seismic maps, charts, tables, and software are usually available for determining the value of ag on a grid of few km.

The topographic amplification coefficient S_T shall be evaluated based on the slope: for veneer covers it should be set equal to 1.0 for slope angle $\beta < 15°$, and equal to 1.20 for slope angle $\beta \geq 15°$.

The definition of S_S depends on the category of the soil below the structure under consideration: it varies from 1.00 for rocky outcrops or very rigid soils characterized by values of $V_{s,30}$ (average velocity of shear waves in the first 30 m of subsoil) greater than 800 m/s, to 1.80 for deposits of low density coarse soils or poorly consistent fine grained soils characterized by values of $V_{s,30}$ lower than 180 m/s.

In the case of landfill capping, between the bedrock and the top surface there is the waste, which has similar geotechnical properties to low density coarse soils or poorly consistent fine-grained soils characterized by values of $V_{s,30}$ lower than 180 m/s: therefore, when the height of waste below the capping system is of the order of 20 – 30 m, it is safe to assume S_S = 1.80.

Stability analyses in seismic conditions can be carried out using pseudo-static limit equilibrium methods, where the seismic action is represented by an equivalent static force equal to the product of gravity forces by proper seismic coefficients for the horizontal and vertical components, given by:

$$K_h = \beta_m \cdot \frac{a_{max}}{g} = \beta_m \cdot S_S \cdot S_T \cdot \frac{a_g}{g} \qquad (21)$$

$$K_v = \pm 0.5 \cdot K_h \qquad (22)$$

where:
K_h = horizontal coefficient for pseudo-static analyses
K_v = vertical coefficient for pseudo-static analyses
g = acceleration due to gravity = 9.81 m/s^2
β_m = dumping coefficient for reduction of the maximum acceleration at the site (see Table 4)

When the height of waste below the capping system is of the order of 20 – 30 m, it is safe to assume the soil category E.

Table 4. Values of the dumping coefficient β_m according to Italian Norm [7]

Expected acceleration at bedrock	β_m for soil category A	β_m for soil category B, C, D, E
$0.2 < a_g / g \leq 0.4$	0.30	0.28
$0.1 < a_g / g \leq 0.2$	0.27	0.24
$a_g / g \leq 0.1$	0.20	0.20

In seismic conditions it is important to perform analyses not only for ULS but also for Serviceability Limit States (SLS), that is to evaluate the maximum displacements and deformations of the veneer soil.

The SLS analysis for maximum displacement in seismic condition could be performed using the well-known Newmark method.

Anyway, Anderson and Kawazanjian [16] argue that results of Newmark deformation predictions just indicate whether movements are large or small. Accuracies less than 15 to 30 mm are meaningless in many cases, given the simplifying assumptions and uncertainties associated with the prediction. As an example of these assumptions, the yield acceleration as well as the deformation prediction in a Newmark analysis assume that the failure mass moves as a rigid block. Shearing stresses and resistances at the base of the block are assumed to be uniformly distributed and the base is assumed to be horizontal and smooth. Such assumptions may be numerically gratifying, but they have little resemblance to conditions likely existing at a landfill base. Typical construction practice is to have cross slopes to collect leachate above the liner, resulting in a series of small, relatively flat ridges and valleys. Given the differences in soil confinement from the head to the toe of the landfill, it is also unreasonable to expect soil resistance to develop uniformly during seismic loading. Furthermore, the acceleration time history used in the deformation analysis is typically calculated assuming no relative displacement and is then used to calculate a relative displacement. While the net effect of all of the inaccuracies may well be conservative, caution is warranted in interpreting the results of Newmark seismic deformation analyses, whether from design charts or computer analysis.

Therefore, based on the arguments by Anderson and Kawazanjian [16], it is not recommended to use the Newmark method for the SLS analysis for maximum displacement of the veneer soil layer of a landfill capping in seismic condition.

The EPA document [17] reports the results of Newmark seismic deformation analyses performed by Hynes and Franklin [18] using 387 strong motion records and 6 artificial accelerograms, shown in Figure 5. Based upon this data and their experience with seismic response analyses of slopes and embankments, Hynes and Franklin [18] concluded that slopes and embankments with a yield acceleration equal to half the peak ground acceleration would experience permanent seismic deformations of less than one meter in any earthquake, even for embankments where amplification of acceleration by a factor of three occurs. Hence EPA [17] states that, in the absence of amplification, or if amplification is taken into account in determining the peak acceleration, the Hynes and Franklin data (Figure 5) suggest that deformations will remain less than 0.3 m for yield accelerations less than or equal to one-half the peak acceleration.

Therefore, based upon the work of Hynes and Franklin [18] and the EPA guidance [17], it appears that the maximum value of the horizontal coefficient for pseudo-static analyses in seismic conditions, K_s, may be determined as $K_s = (0.5\ a_{max}/g)$ to limit permanent seismic deformations to less than 0.3 m, where a_{max} is the peak horizontal acceleration at the ground surface for analyses of the liner system and at the top of the landfill for analyses of the capping system.

This means that assuming $\beta_m = 0.5$ in Eq. (21) would allow to satisfy both the ULS and SLS analysis of the veneer soil layer of a landfill capping in seismic condition.

It has to be noted that the vertical seismic acceleration shall be considered both as downward ($+K_v$) and upward ($-K_v$), while the horizontal seismic acceleration shall always be considered as outward; hence each soil veneer shall be analysed with both values, positive and negative, of the vertical coefficient K_v; that is, two stability analyses are required for each value of K_h.

Once the seismic coefficients K_h and K_v have been determined, the stability analyses of the veneer cover in seismic conditions, with only gravitational forces applied, follow the scheme shown in Figure 6.

Note that for seismic analyses all load amplification factors A1 shall be set equal to A1-seismic = 1.00.

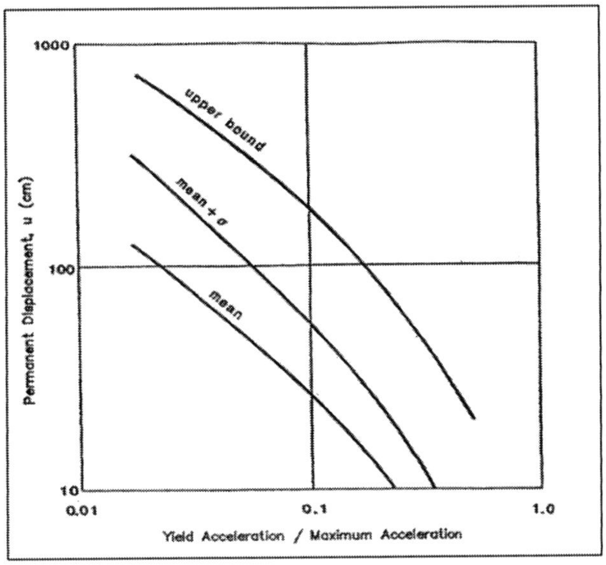

Figure 5. Permanent Seismic Displacement Chart (from [18])

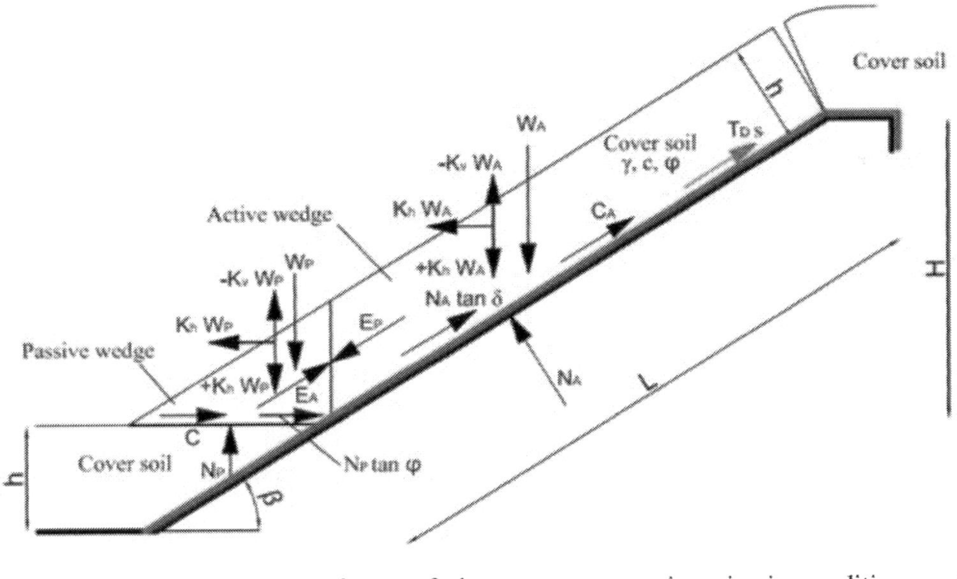

Figure 6. Scheme for stability analyses of the veneer cover in seismic conditions, with only gravitational forces applied, including a geosynthetic reinforcement below the veneer cover

It is assumed that in seismic conditions the geotechnical parameters φ, δ, c, c_a have the same values as in static conditions.

Therefore, with reference to Fig 6, the forces assume the same expressions as for the static case.

Considering that in seismic conditions all Amplification Factors A1-seismic are equal to 1.0, the expression of the interwedge force E_{AS} in seismic conditions, from the equilibrium of vertical forces on the active wedge $(\Sigma F_v = 0)_{active\ wedge}$, becomes:

$$E_{AS} = \frac{1}{\sin\beta} \cdot \left[W_A \cdot (1 \pm K_v) \cdot (1 - \cos^2\beta - \cos\beta \cdot \sin\beta \cdot \tan\delta) - C_A \cdot \sin\beta - T_{DS} \cdot \sin\beta \right] \quad (23)$$

where:
T_{DS} = design tensile strength of reinforcement in seismic conditions, with no Reduction factor for creep applied (kN/m).

The expression of the interwedge force E_{PS} in seismic conditions, from the equilibrium of horizontal forces on the passive wedge $(\Sigma F_h = 0)_{passive\ wedge}$, becomes:

$$E_P = \frac{C + W_P \cdot (1 \pm K_v) \cdot \tan\varphi - K_h \cdot W_P}{\cos\beta - \sin\beta \cdot \tan\varphi} \quad (24)$$

Stability is then verified if, for the seismic combination (A1-seismic+M1+R3), the Factor of Safety results as per Eq. (11).

4.5. *Seepage Forces*

Slope stability analyses of veneers can be carried out in drained conditions if either permeable soil or a drainage layer was placed above the barrier layer with adequate flow capacity to efficiently remove permeating water safely away from the cross section. The amount of water to be removed is obviously a site-specific situation. Anyway, adequate drainage of final covers may not be available.

If seepage forces occur, a variation in slope stability design methodology is required. Additional discussion on this issue is given in [19,20,21,22].

Since seepage forces may impair the stability of the veneer cover, requiring a stronger reinforcement to reach the prescribed Factor of Safety, it is recommended to design a geocomposite drainage layer below the veneer cover. Design of such drainage layer can be accomplished with the methods presented in ISO/TR 18228-4 [23].

If, for any reason, a drainage layer below the veneer cover is not available, the formation of a phreatic surface due to water ponding at the toe or flow along the slope on the liner interface shall always be considered.

For the analysis of the veneer cover in presence of seepage forces, let's consider a cover soil of uniform thickness placed directly above a liner at a slope angle β. Different from previous analyses is that within the cover soil exists a saturated soil zone for part or all of the veneer thickness.

The saturated boundary can develop with two possibly different phreatic surface orientations:

- a horizontal buildup from the toe upward;
- a parallel-to-slope buildup outward.

When analyzing the stability of slopes using the limit equilibrium method, free body diagrams of the passive and active wedges are taken with the appropriate forces (now including pore water pressures) being applied.

4.5.1. *The Case of the horizontal seepage buildup.* Horizontal seepage buildup can occur when toe blockage occurs due to inadequate outlet capacity, contamination or physical blocking of outlets, or freezing conditions at the outlets.

In case of horizontal pressure buildup there are two possible situations:
a) phreatic surface above the passive wedge (that is, above the interwedge vertical line),
b) phreatic surface partially in the passive wedge (that is, intercepting the interwedge vertical line).

a) Phreatic surface above the passive wedge

When the phreatic surface is above the passive wedge (that is, above the interwedge vertical line), the free body diagram of both the active and passive wedge is shown in Figure 7.

Hence this situation occurs when $H_w \geq (h / \cos \beta)$.

All symbols used in Figure 7 were previously defined except the following:

γ_{sat} = saturated unit weight of the cover soil (kN/m3)
γ_m = total (moist) unit weight of the cover soil (kN/m3)
γ_w = unit weight of water (kN/m3)
H_w = vertical height of the free water surface measured from the toe of the active wedge (m)
u = pore pressure at toe (kPa)
U_h = resultant of the pore pressures acting on the interwedge surfaces (kN/m)
U_n = resultant of the pore pressures acting perpendicular to the slope (kN/m)
U_v = resultant of the vertical pore pressures acting on the base of the passive wedge (kN/m)

Above the phreatic surface the veneer soil will be dried only if the soil is free draining, but in general it will be in moist condition; hence, on the safety side, for the active wedge above the phreatic surface it is better to consider the total unit weight of soil in moist condition, γ_m.

When the water pressure acts on an inclined surface, the water pressure along the slope assumes the distribution shown in Figure 7. The water pressure shall then be calculated as shown at bottom of Figure 7.

Therefore, the maximum value of the water pressure, at the toe of the veneer cover, is:

$$u = \gamma_w \cdot h \cdot \cos \beta \tag{25}$$

With reference to Figure 7, the following forces assume different expressions than without the horizontal seepage:

$$W_A = \frac{1}{2} \cdot \gamma_{sat} \cdot h \cdot \left[\left(\frac{2 \cdot H_w}{\sin \beta} \right) - h \cdot \left(\frac{1}{\tan \beta} + \tan \beta \right) \right] + \frac{1}{2} \cdot \gamma_m \cdot h \cdot \left[\left(\frac{2 \cdot (H - H_w)}{\sin \beta} \right) + \frac{h}{\tan \beta} \right] \tag{26}$$

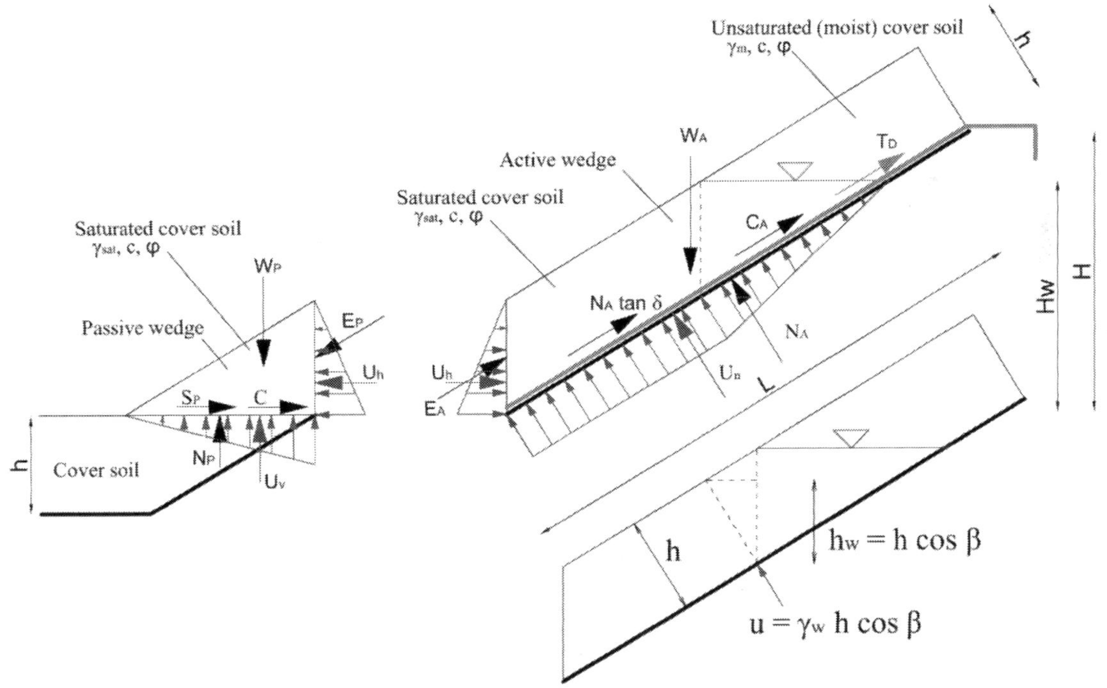

Figure 7. Limit equilibrium forces involved in a finite length slope of uniform cover soil with horizontal seepage buildup when the phreatic surface is above the passive wedge, including a geosynthetic reinforcement, and scheme for calculation of water pressure on a slope

$$U_n = \gamma_w \cdot h \cdot \cos\beta \cdot \left[\left(\frac{H_w}{\sin\beta} \right) - \frac{h}{2} \cdot \left(\frac{1}{\tan\beta} + \tan\beta \right) \right] \tag{27}$$

$$U_v = \frac{1}{2} \cdot \gamma_w \cdot h^2 \cdot \frac{1}{\tan\beta} \tag{28}$$

$$U_h = \frac{1}{2} \cdot \gamma_w \cdot h^2 \cdot \cos\beta \tag{29}$$

$$N_A = W_A \cdot \cos\beta + U_h \cdot \sin\beta - U_n \tag{30}$$

$$W_P = \frac{1}{2} \gamma_{sat} \cdot h^2 \cdot \left(\frac{1}{\sin\beta \cdot \cos\beta} \right) \tag{31}$$

$$N_P = W_P + E_P \cdot \sin\beta - U_v \tag{32}$$

The expression of the interwedge force E_A can be obtained from the equilibrium of vertical forces on the active wedge $(\Sigma F_v = 0)_{\text{active wedge}}$:

$$E_A = \frac{1}{\sin\beta} \cdot \begin{Bmatrix} \gamma_{G1u} \cdot W_A \cdot (1 - \cos^2\beta - \cos\beta \cdot \sin\beta \cdot \tan\delta) - \gamma_{G1f} \cdot C_A \cdot \sin\beta - \gamma_{G1f} \cdot T_D \cdot \sin\beta \\ + \gamma_{G1u} \cdot \left[U_h \cdot \left(\sin\beta \cdot \cos\beta + \sin^2\beta \tan\delta \right) + U_n \cdot \left(\cos\beta + \sin\beta \cdot \tan\delta \right) \right] \end{Bmatrix} \tag{33}$$

From the equilibrium of horizontal forces on the passive wedge $(\Sigma F_h = 0)_{\text{passive wedge}}$:

$$E_P = \frac{\gamma_{G1f} \cdot C - \gamma_{G1u} \cdot U_h + (\gamma_{G1f} \cdot W_P - \gamma_{G1u} \cdot U_v) \cdot \tan\varphi}{\cos\beta - \sin\beta \cdot \tan\varphi} \tag{34}$$

Stability in static conditions is verified if, for the combination (A1+M1+R3), the Factor of Safety results as per Eq. (11).

In seismic conditions with horizontal seepage the expressions of E_A and E_P become:

$$E_A = \frac{1}{\sin\beta} \cdot \begin{Bmatrix} \left[W_A \cdot (1 \pm K_v) \cdot (1 - \cos^2\beta - \cos\beta \cdot \sin\beta \cdot \tan\delta) - C_A \cdot \sin\beta \right] - T_{DS} \cdot \sin\beta \\ + \left[U_h \cdot \left(\sin\beta \cdot \cos\beta + \sin^2\beta \tan\delta \right) + U_n \cdot \left(\cos\beta + \sin\beta \cdot \tan\delta \right) \right] \end{Bmatrix} \tag{35}$$

$$E_P = \frac{C - U_h + W_P \cdot (1 \pm K_v) \cdot \tan\varphi - U_v \cdot \tan\varphi - K_h \cdot W_P}{\cos\beta - \sin\beta \cdot \tan\varphi} \tag{36}$$

Note that the seismic acceleration is not applied to pore water forces.

Stability in seismic conditions is verified if, for the combination (A1-seismic+M1+R3), the Factor of Safety results as per Eq. (11).

b) Phreatic surface partially in the passive wedge

When the phreatic surface is partially in the passive wedge (that is, crossing the interwedge vertical line), the free body diagram of both the active and passive wedge is shown in Figure 8.

Hence this situation occurs when $H_w < (h / \cos\beta)$.

In this case the maximum value of the water pressure, at the toe of the veneer cover, is:

$$u = \gamma_w \cdot H_w \tag{37}$$

Referring to Figure 8, the following forces assume different expressions than in the previous case:

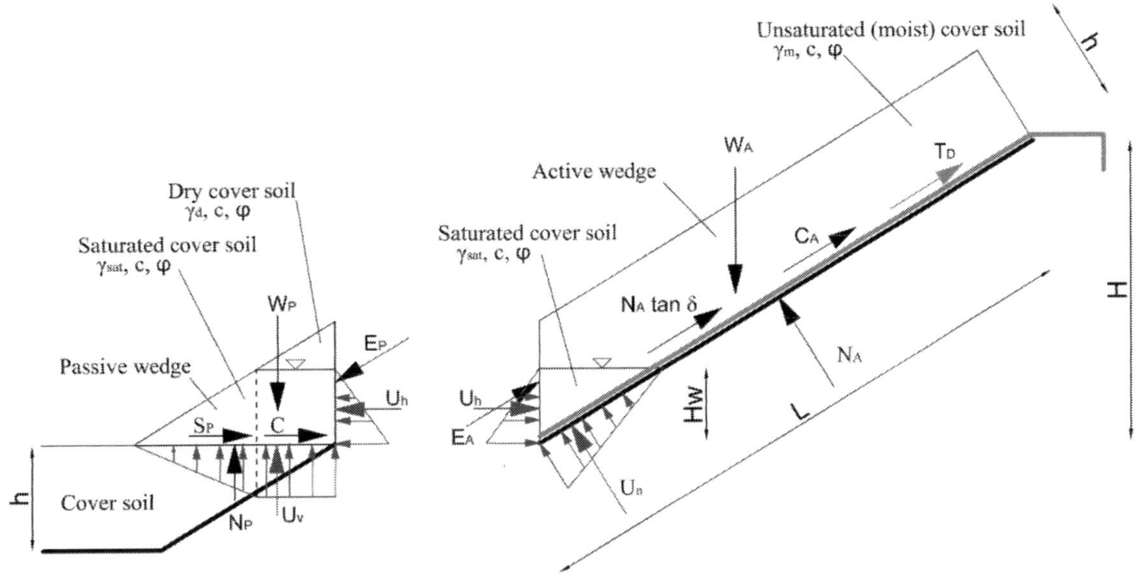

Figure 8. Limit equilibrium forces involved in a finite length slope of uniform cover soil with horizontal seepage buildup when the phreatic surface is partially in the passive wedge, including a geosynthetic reinforcement below the veneer cover

$$W_A = \frac{1}{2}\cdot\gamma_{sat}\cdot\frac{H_w{}^2}{\tan\beta}+\frac{1}{2}\cdot\gamma_m\cdot\left[\left(\frac{2h\cdot(H-H_w)}{\sin\beta}\right)-h^2\cdot\tan\beta+\frac{H_w}{\tan\beta}\cdot\left(\frac{2h}{\cos\beta}-H_w\right)\right] \quad (38)$$

$$U_n = \frac{1}{2}\gamma_w\cdot\frac{H_w{}^2}{\sin\beta} \quad (39)$$

$$U_v = \gamma_w\cdot H_w\cdot\left(\frac{h}{\cos\beta}-H_w\right)\cdot\frac{1}{\tan\beta}+\frac{1}{2}\cdot\gamma_w\cdot H_w{}^2\cdot\tan\beta \quad (40)$$

$$U_h = \frac{1}{2}\cdot\gamma_w\cdot H_w{}^2 \quad (41)$$

$$W_P = \frac{1}{2}\gamma_{sat}\cdot H_w\cdot\left(\frac{2h}{\sin\beta}-\frac{H_w}{\tan\beta}\right)+\frac{1}{2}\gamma_d\cdot\left(\frac{h}{\cos\beta}-H_w\right)^2\cdot\frac{1}{\tan\beta} \quad (42)$$

The expressions of E_A and E_p in static and seismic conditions can be obtained from Eqs. (33) – (36) by introducing the expressions of forces in Eqs. (38) – (42).

Stability in static and seismic conditions is then verified if, for the static combination (A1+M1+R3) and for the seismic combination (A1-seismic+M1+R3), the Factor of Safety results as per Eq. (11).

4.5.2. *The Case of Parallel-to-Slope Seepage Buildup.* Figure 9 shows the free body diagrams of both the active and passive wedges with seepage buildup in the direction parallel to the slope. Parallel seepage buildup can occur when soils placed above a liner are initially too low in their hydraulic conductivity or become too low due to long term clogging from overlying soils which do not have a filter.

Also in this case, above the phreatic surface the veneer soil will be dried only if the soil is free draining, but in general it will be in moist condition; hence, on the safety side, above the phreatic surface it is better to consider the total unit weight of soil in moist condition, γ_m.

Identical symbols as defined in the previous cases are used here, with an additional definition:
h_w = height of free water surface measured in the direction perpendicular to the slope (m).

In this case the maximum value of the water pressure, at the toe of the veneer cover, is:

$$u = \gamma_w \cdot h_w \cdot \cos\beta \tag{43}$$

Referring to Fig. 9, the following forces assume different expressions than in the previous case:

$$W_A = \frac{1}{2}\cdot\gamma_{sat}\cdot h_w\cdot\left(\frac{2H}{\sin\beta} - h_w\cdot\tan\beta\right) + \frac{1}{2}\cdot\gamma_m\cdot(h-h_w)\cdot\left(\frac{2H}{\sin\beta} - h_w\cdot\tan\beta - h\cdot\tan\beta\right) \tag{44}$$

$$U_n = \frac{\gamma_w\cdot h_w\cdot H}{\tan\beta} \tag{45}$$

$$U_h = \frac{1}{2}\cdot\gamma_w\cdot h_w^{\,2} \tag{46}$$

$$U_v = \frac{1}{2}\gamma_w\cdot\frac{h_w^{\,2}}{\tan\beta} \tag{47}$$

$$W_P = \frac{1}{2}\gamma_{sat}\cdot\frac{h_w^{\,2}}{\sin\beta\cdot\cos\beta} + \frac{1}{2}\gamma_d\cdot\frac{\left(h^2 - h_w^{\,2}\right)}{\sin\beta\cdot\cos\beta} \tag{48}$$

The expressions of E_A and E_p in static and seismic conditions can be obtained from Eqs. (33) – (36) by introducing the expressions of forces in Eqs. (44) – (48).

Stability in static and seismic conditions is then verified if, for the static combination (A1+M1+R3) and for the seismic combination (A1-seismic+M1+R3), the Factor of Safety results as per Eq. (11).

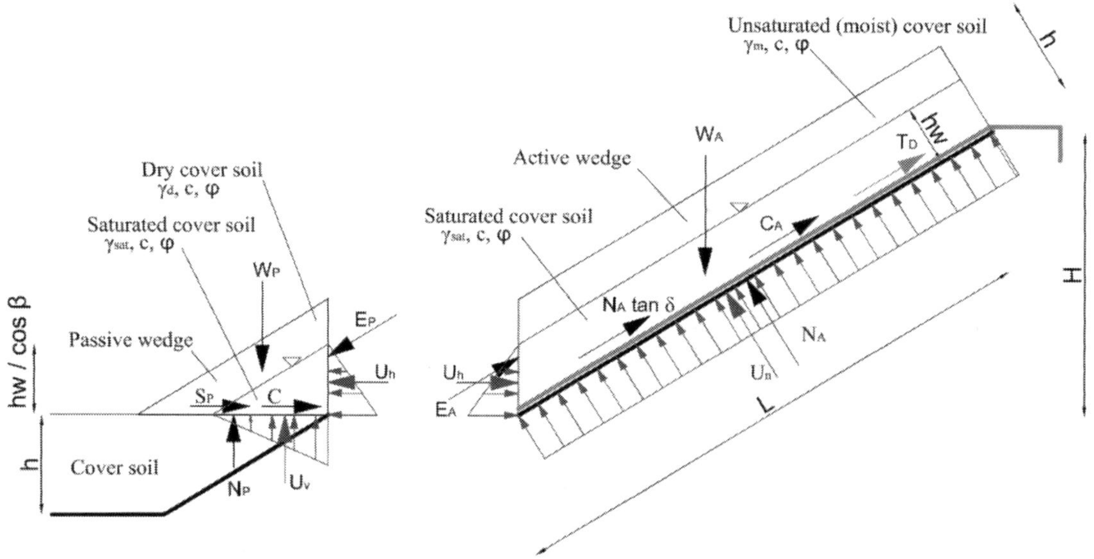

Figure 9. Limit equilibrium forces involved in a finite length slope of uniform cover soil with parallel-to-slope seepage buildup, including a geosynthetic reinforcement below the veneer cover

5. Required stability analyses

The stability of the veneer cover shall be analysed in the following Ultimate Limit State (ULS) conditions:

1) At the end of veneer construction, with equipment moving down the slope, in static conditions, without seismic actions nor seepage forces; being a short-term analysis, if reinforcement is used, the Reduction Factors for creep, chemical and biological damage shall be set equal to 1.0.

2) At the end of the design life, both in static and in seismic conditions, with seepage forces if horizontal seepage buildup or parallel to slope seepage buildup is anticipated; for the long term analysis in static conditions, if reinforcement is used, all Reduction Factors for creep, installation damage, chemical and biological damage shall be applied; for the long term analysis in seismic conditions, the vertical seismic acceleration shall be considered both as downward ($+K_v$) and upward ($-K_v$), while the horizontal seismic acceleration shall always be considered as outward; being a long term analysis in seismic conditions, if reinforcement is used, the Reduction Factors for installation damage, chemical and biological damage shall be applied, while the Reduction Factors for creep shall be set equal to 1.0.

Note that assuming $\beta_m = 0.5$ in Eq. (21) would allow to satisfy both the ULS and SLS analysis of the veneer soil layer in seismic condition.

Note also that, if reinforcement is used, the tensile strength T_D (and therefore the ultimate tensile strength T_{ult}) shall be finally set as the highest value from all analyses.

Finally, note that, for considering the tensile strength T_D, the reinforcing geosynthetic shall be properly anchored at the top of the slope; methods for designing the anchorage at top are not included in the present paper but can be found in [13].

References

[1] Koerner R M. and Soong T Y 1998 Analysis and Design of Veneer Cover Soils. *Proc. Sixth International Conference on Geosynthetics*, IFAI, Atlanta, Ga., USA pp 1–23

[2] ASTM D5321-2017 Standard Test Method for Determining the Shear Strength of Soil-Geosynthetic and Geosynthetic-Geosynthetic Interfaces by Direct Shear. *ASTM International*, West Conshohocken, PA, USA

[3] EN ISO 12957-1 - Geosynthetics – Determination of the friction characteristics – Part 1: Direct shear test. *ISO*, Geneve, Switzerland

[4] EN ISO 12957-2 - Geosynthetics – Determination of friction characteristics – Part 2: Inclined plane test. *ISO*, Geneve, Switzerland

[5] Pavanello P, Carrubba P, Moraci N, Pezzano P, Miuzzi M 2016 Parameters and conditions affecting friction angles in geosynthetic interfaces. *Proc. EUROGEO 6, 6th European Geosynthetics Congress*. Ljubljana, Slovenia

[6] EN 1997-1 Eurocode 7: Geotechnical Design. *CEN*, Bruxelles, Belgium

[7] NTC 2018 Norme tecniche per le costruzioni (Italian Technical Norms for Construction). DM 17 Gennaio 2018. *Gazzetta Ufficiale Nr 29 del 20 Febbraio 2018, Supplemento Ordinario Nr 42* (in Italian)

[8] EN 1998-1 Eurocode 8: Design of structures for earthquake resistance – Part 1: General rules, seismic actions and rules for buildings. *CEN*, Bruxelles, Belgium

[9] Koerner R M and Hwu B L 1991 Stability and Tension Considerations Regarding Cover Soils on Geomembrane Lined Slopes. *Geotextiles and Geomembranes* **10** pp 335-55

[10] Giroud J P and Beech J F 1989 Stability of Soil Layers on Geosynthetic Lining System *Proc. Geosynthetics* IFAI, St. Paul, MN, USA pp 35-46

[11] McKelvey J A and Deutsch W L 1991 The Effect of Equipment Loading and Tapered Cover Soil Layers on Geosynthetic Lined Landfill Slopes *Proceedings of the 14th Annual Madison Avenue Conference*, Madison, WI, USA, University of Wisconsin pp 395-411

[12] Ling H I and Leshchinsky D 1997 Seismic Stability and Permanent Displacement of Landfill Cover Systems *Jour. of Geotechnical and Geoenvironmental Engrg. ASCE* **123** No. 2

[13] Koerner R M 2012 Designing with Geosynthetics. *6th Ed., Xlibris LLC,* Bloomington, IN, USA
[14] Poulos L G and Davis E H 1974 Elastic Solutions for Soil and Rock Mechanics. J. Wiley & Sons, Inc., NewYork, NY, USA, 411 pgs.
[15] McKelvey J A 1994 Consideration of Equipment Loadings in Geosynthetic Lined Slope Design. *Proc. 8th Intl. Conf. of the Intl. Assoc. for Computer Methods and Advancements in Geomechanics*, Morgantown, WV:Balkema, pp 1371-7
[16] Anderson D G. and Kavazanjian E J 1995 Performance of Landfills Under Seismic Loading *International Conferences on Recent Advances in Geotechnical Earthquake Engineering and Soil Dynamics*. St. Louis, Missouri, USA
[17] EPA 1995 RCRA Subtitle D (258) Seismic design guidance for municipal solid waste landfill facilities. EPA/600/R-95j051. *United States Environmental Protection Agency (EPA).* Washington, DC, USA
[18] Hynes M E and Franklin A G 1984 Rationalizing the Seismic Coefficient Method. Miscellaneous Paper GL-84-/3, *U.S. Army Engineer Waterways Experiment Station*, Vicksburg, Mississippi, USA
[19] Cancelli A and Rimoldi P 1989 Design Criteria for Geosynthetic Drainage Systems in Waste Disposal *Proc. Sardinia, 2nd Intl. Landfill Symposium*, Porto Conte, Sassari, Italy
[20] Thiel R S and Stewart M G 1993 Geosynthetic Landfill Cover Design Methodology and Construction Experience in the Pacific Northwest *Proc. Geosynthetics* IFAI, St. Paul, MN, USA pp 1131-44
[21] Giroud J P, Bachus R C. and Bonaparte R 1995 Influence of Water Flow on the Stability of Geosynthetic-Soil Layered Systems on Slopes *Geosynthetic International* **2**pp 1149-80
[22] Soong T-Y and Koerner RM 1996 Seepage Induced Slope Instability *Geotextiles and Geomembranes* **14** No. 7/8 pp 425-45
[23] ISO/TR 18228-4 Design using geosynthetics - Part 4: Drainage. *ISO*, Geneve, Switzerland

IOP Conf. Series: Materials Science and Engineering 1260 (2022) 012048 doi:10.1088/1757-899X/1260/1/012048

Geogrid stabilisation application – review of strain measurement methods, research and implications on design approaches

K Zamara[1], G Fowmes[2], J Kawalec[1,3]

[1] Tensar International Ltd UK,
[2] Warwick University UK,
[3] Silesian University of Technology, Poland

E-mail: kzamara@tensar.co.uk

Abstract. Geogrids have been used in the construction industry for stabilisation applications for decades, however, there is still a lot of conflicting information relating to the mechanical properties required to achieve this stabilisation function. Many designers adopt traditional tensile strength against tensile strain approaches that reflect membrane effect and ignore the actual mechanism of interlock and confinement. Recent ISO 10318-1:2015 defines geogrid stabilisation function by means of aggregate interlock and confinement, whereas tensile strength is seen as a function of reinforcement. Researchers across the globe have attempted to quantify the strain developed in geogrid used in a stabilising function exposed to heavy loads. Instruments used for monitoring typically comprise of standard wire strains gauges, fibre optics or foil strain gauges. Considering the harsh construction environment where the geogrids, and hence the strain gauges are exposed to damage or penetration by sharp aggregate edges, installation of sophisticated and sensitive monitoring systems can be challenging. There are several issues that can affect monitoring results such as the size of the monitoring instrument or the ability to achieve an adequate bond to the grid. Monitoring results have a major influence on geotechnical design and how practitioners perceive stabilisation mechanism in geogrids. Whilst the mechanisms of interlock and increased confinement have been studied, there still exists few design tools to provide designers with a straightforward analysis. This paper reviews research on the methods for in-situ strain measurement of geogrid installed in a stabilisation function, summarises results obtained by researchers and provides a summary of how these are reflected in geotechnical design approaches for geogrid stabilised soil.

1. Introduction

Geogrids have been used in the construction industry for stabilisation applications for decades. However, within the industry there is still a lot of conflicting information and terminology relating to the mechanical properties that are required to perform in a stabilisation function, and design approaches that accurately model their effect on stabilised soil.

This paper presents a brief literature review, summarising recent research publications on geogrid applications in a stabilisation function. The first part of the paper looks at the typical methods of strain measurement in laboratories or field conditions. The second part of the paper provides a summary of the strain force or tensile strength recorded either through full scale experiments, scaled down experiments or discrete element modelling (DEM).

This paper attempts to evaluate general trends and degree of strain development in geogrid under operating conditions. Data is taken from international research and is not limited to any specific function hence records from pavement, railway and shallow footing projects are used. For each of the applications geogrid was installed in a stabilisation function forming a mechanically stabilised layer (MSL) with one or more layers of geogrid.

The mechanisms involved in developing full stabilisation function has been a topic of discussion by researchers, specialists and practitioners for many years. Recently ISO 10318-1:2015 [1] defined geogrid stabilisation function by means of aggregate interlock and confinement, as opposed to reinforcing function being corelated to geogrid tensile strength (e.g. in mechanically stabilised earth walls applications). This is reinforced by the fact that in finite element method (FEM) and limit equilibrium software, geogrid can be introduced to the system only as a tensile element, representing product properties that were tested 'in air' but not as an actual in-situ element where combined influence of geogrid and soil takes effect.

Research shows that in order to fully represent geogrid in FEM, FEM models should include the effect of the geogrid on soil that varies with distance from the geogrid [2]. It is considered by the Author's that stabilisation application is still an under-researched area of geogrid development, where technology has not fully followed the typical steps of new design models development (Figure 1). It has also been observed by the Author's that many practitioners still use inadequate tools and incorrect mechanisms to describe geogrid effect on the soil.

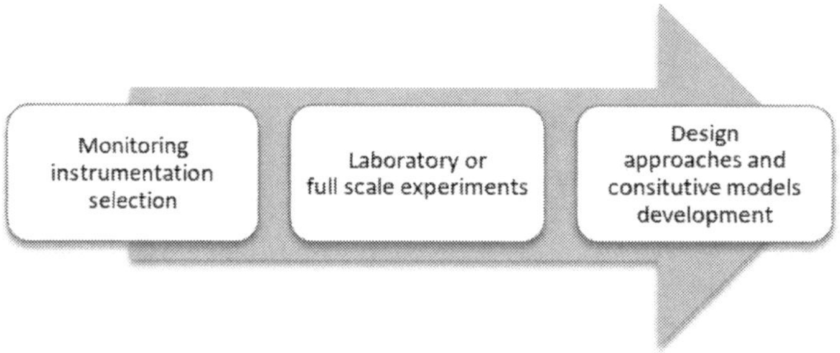

Figure 1. Generic research process for design approaches development

2. Geotechnical monitoring programme

The geotechnical monitoring programme first and foremost should establish what degree of accuracy and precision is required from the monitoring system, and secondly what the expected magnitude of the observed parameter is. Some other factors that should be considered are the test duration options for recording data or endurance of the instrumentation. This is required to correctly select the required monitoring system.

One of the most important aspect of strain gauge application is their fixture to the grid. If the sensors are too large, this may affect the reading. The method of bonding sensor to the grid should also be thoroughly evaluated and calibrated to ensure that bonding does not introduce local stiffening points or that it is too soft allowing the bond to strain without much strain on the gauge. Whilst use of epoxy raisin is a common method of gluing fibre optics to polymers, others develop clamp systems for strain meters (Figure 2) where polymers surface energy and small available area prohibits typical adhesives from using.

2.1. Typical strain gauges used in combination with geogrids

Accurate measurement of strains and axial forces developing within geogrids is a challenging task. Much research is available on the topic that focuses on mechanically stabilised earth walls, where high tensile forces develop, and creep characteristics of geogrids are investigated (e.g. [3]). However, many researchers recognise the importance of discrete strains measurements in the geogrids applications to investigate the level of tensile strain. Table 1 presents a summary of methods used to measure strains developing within geosynthetics.

The selection of the appropriate instrument should reflect the location, i.e. section length and magnitude of strain that is to be measured. Foil resistance strain gauges (Figure 1) are reliable and used in various industries for over 70 years, however they measure changes in strains only in the location where they are bonded to. Strand meters (Figure 3) measure an average strain between two points of attachment. This was reported to be approx. 200mm by [4]. In terms of fibre optics (Figure 4), there are two types readily available on the market: Fibre Bragg gratings (FBG) which allow strain measurement as an average value between main point of attachment to geogrid. Section length can vary, e.g. [5] reported 500mm. Then distributed optical fibre sensors allow a continuous strain measurement along the monitored material; however, the precision and accuracy are known to be reduced when compared with FBG (e.g. [6]).

Table 1. Method used commonly to measure strains in geogrids (modified after [7])

Gauge	Transducer Type	Principle
Foil or wire grid	Resistance Strain Gauges	The straining of wire changes its cross-sectional area and consequently its electrical resistance (here we have bonded gauges, encapsulated gauges or unbonded, encapsulated gauges. These were invented in late 1930s.
Strand meter	Vibrating wire devices	A vibrating steel wire with a stress relieved spring are attached to the object on which a measurement has to be made. The tension in the wire is directly proportional to the extension. Deformation is determined based on vibrating wire frequencies calibrated to given strains.
LVDT sensor	Linear Variable Differential Transformer (LVDT)	This type of device is typically used within extensometers, where an actual displacement of a discrete point is measured and recorded.
Fibre	Fibre Optic	Change of wavelength of the light travelling in the fibre optic core is measured. Fibre Bragg Gratings (FBG) are frequently used in conjunction with geosynthetics (e.g. [6], [20]). Bragg gratings positioned within a fibres central core enables light to reflect. Shift in wavelength described by Bragg law is calibrated with changes in strain of the monitored structure. Chen et al 2015 also reports use of distributed optical fibre sensors that allow a continuous strain measurement along the monitored material.
Some novel technologies under development:		
Tensoresistive PVC Coating [8]	Resistance Strain Gauges	Relatively novel technology developed that possess built-in strain-sensing capability in addition to the primary function of reinforcement or stabilisation.
3D printing with piezoresistive (Warwick and Monach University)	Resistance Strain Gauges	3D printing offers the opportunity to embed sensors within a typical geosynthetic during the manufacturing stage, making the sensors much more effective.

In addition to conventional strain measurement methods, recently [8] researched the topic of so-called sensor-enabled geosynthetics. His study investigates how geogrid yarns coated with tensoresistive composites can provide continuous information on strain state of geogrids. In addition, researchers from Warwick and Monash University has investigated 3D printed geogrids with embedded strain and temperature sensors. The printing system allows incorporation of a range of materials within polymeric mass. Prototype with data retrieving connections is shown in Figure 5.

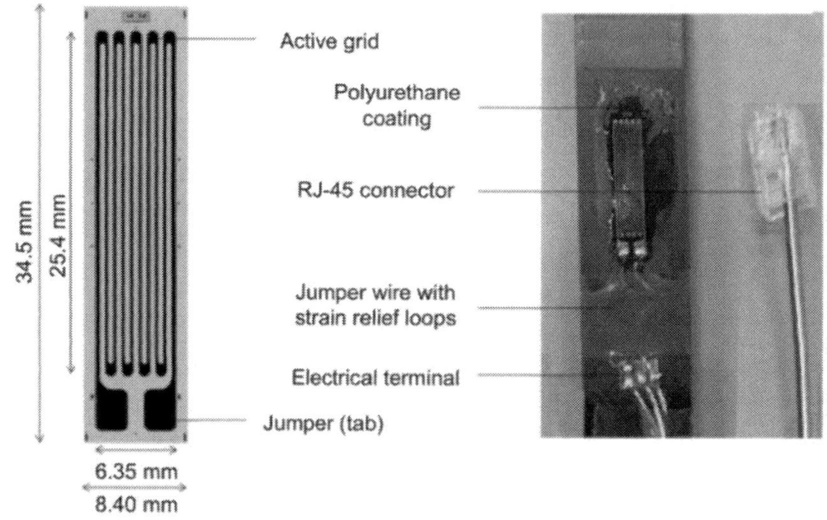

Figure 2. Foil Strain Gauges [9]

Figure 3. Strand meters (Left: courtesy of Silesian University of Technology, Right: [4])

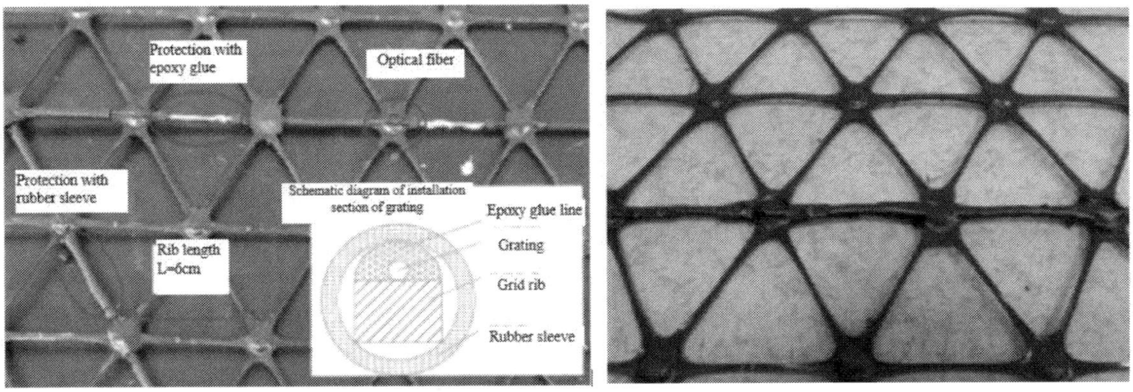

Figure 4. FBGs attached to geogrid before (left) and after (right) installation of protective coat [10]

Figure 5. 3D printed polymeric grid with embedded sensing elements (courtesy of Warwick and Monash Univ Alliance)

3. Monitored geogrids

It was observed that typically strain measurement appears to be a by-product of studies focusing on global performance of a system, where focus would be rather on another aspect such total settlements and system deformation, rather than solely geogrid axial performance study in soil. However, in the last decades many researchers attempted to measure strains developing in geogrids in both laboratory and field experiments.

Table 2 summarises the initial review undertaken for the purpose of this paper. Reported studies in majority include railway applications, i.e.:[12], [6], [4] and [10], others include pavement application, i.e. [12] and [13] and research into shallow footing behaviour [14] and [15].

Reported railway studies took a variety of forms, e.g. a laboratory box test researching plate load settlement and lateral pressures imposed on the box under dynamic load and 2 000 000 load cycles, where strand meters were installed on the geogrid samples [4]. This study recorded max strain developing in the geogrid to be 0.028%. Another study [11] reported a field trial where under a life railway line different types of instrumented geogrids were installed (type of strain gauge was not reported). This study recorded the highest strains developing in the geogrids reviewed in this study. Cumulative strains ranged between 0.6% and 1.8%, in addition it was reported that under an actual load of a traveling train strains were between 0.14 to 0.17%. An interesting study on high speed railway was reported by [6], where load transfer platform with induced consolidation was monitored.

Other studies researching pavement's performance, reported up to 1.3% strain under a plate load test after 700 000 cycles [12], whilst more recent study [13] using model mobile load simulator (MMLS3), reported strains of 0.2% under the load of 2.7kN and 100 000 cycles. Interestingly recorded strains were only in transverse direction to the direction of the wheel movement. Maximum reported tensile strength was at 0.65kN.

In addition, a couple of studies reported strains developing under a load of a footing. These include a laboratory study [14] and discrete element method analysis [15], with maximum recorded strains of up to 2.5% for a footing located close to a slope. However, this example due to close slope location may include elements of slope reinforcement generating additional straining within the geogrid.

4. Design implications

Many commercially available geotechnical engineering design software packages allow for geosynthetic element inclusion within the model. However, these geosynthetic inclusions are simplistically modelled by interface/bond friction and maximum tensile capacity in limit equilibrium software that varies available tensile strength based on required bond length or in numerical modelling software' by elastic or elastoplastic elements. These methods to Author's knowledge do not represent geogrid stabilised structures performance accurately.

Whilst, numerical modelling software has an ability to calculate strains developing in the geosynthetic element, it should be noted that geogrids can exhibit very little strain development within geosynthetic in stabilisation application as indicated in Table 2. Moreover, even for applications such MSE walls, these strains still may stay within lower values 0.2% (e.g. [16]). There has been a considerable research attempt that aimed to verify principal mechanisms involved in performance of geogrids in stabilisation function. Adequate representation of this effect within stabilised layer is of significant importance for practitioners/designers, as this allows accurately predict stabilised structure behaviour.

Some experienced practitioners choose not to use readily available geosynthetic elements within the software but instead to adopt a composite block with improved strength and stiffness parameters that is assigned to the stabilised block of soil. However, research shows that the effect of the increased confinement that geogrid offers to stabilised soil diminishes with distance from the geogrid, as shown by [17]. The effect of geogrid is not constant across the stabilised layers. Hence, others [18] choose to vary stabilised soil block strength parameters with distance from the geogrid (see Figure 6).

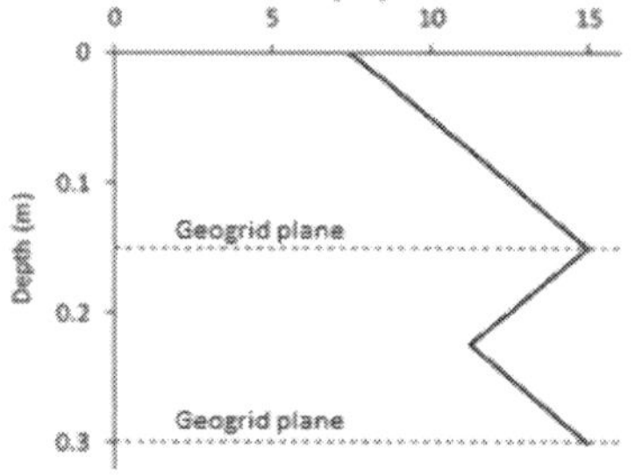

Figure 6. c '-profile adopted for stabilised sandy gravel layer in FEA of plate load test [19]

Table 2. Summary of some research incorporating strain measurement on geogrids

References	Gauge Type	Application	Results Maximum	Comment
[11] Anantanasakul et al (2012)	n/a	Railway	Max cumulative strains between 0.6% to 1.8% Transient strains induced by 30 tons axial load traveling at 40km/hr 0.14%-0.17%	Field trial. 90 days of monitoring $2.3x10^5$ load cycles. 6 sections various grids monitored
[6] Chen et al (2016)	Fibre Bragg Gratings & distributed optical fibre sensors	High speed rail on load transfer platform	0.33kN/m @ static load of 40kPa 5kN/m @ subsoil soil settlement of 20mm at pile cap location	Load transfer platform test. Large scale laboratory test with 8 actuators. Static test with load varying from 4kPa to 40 kPa, then induced consolidation in soft soils, lastly dynamic load with speed of 324km/h
[14] Gnanendran & Selvaduarai (2001)	Foil gauge	Footing	Max strains of 2.5% for footing load 320kPa	Laboratory. Small scale. Footing settlement and geogrid strains monitored close to a slope.
[12] Haas et al 1988	Resistance strain gauges	Flexible pavement	Max 1.3% @ 700 000 cycles <0.5% @ 250 000 cycles	Laboratory. Cyclic plate load test in a box; 550kPa load (80psi), 700 000 loading cycles
[4] Hornicek & Rakowski (2019)	Strand meter	Railway	Max strain 0.028%	Laboratory. Cyclic plate load test in a box; Up to 100kN applied to 0.79m DIA plate; Up to 2 000 000 loading cycles, tests run with 25, 50, 75 and 100kN load;
[10] (IGS 2019)	Fibre optic	Railway	Max strain 0.15% @ 21 kN load Max strain 0.2% @ 31 kN load	Laboratory. Cyclic load of a single sleeper. Load between 21kN (passenger train) to 31kN (freight train) 100 000 loading cycles
[13] Tang et al (2013)	An electrical-resistance-type strain gauge (KFG-5-120-C1-11L3M3R), was selected.	Flexible pavement	Max 0.2% Max 0.65kN	Reduced scaled accelerated pavement testing. 100,000 passes MMLS3 applying load of 2.7kN The strain gauge had a backing material constructed from polyimide, and the measurement grid is made of a constantan alloy that can sustain strains up to 5%.
[15] Wang et al (2020)	DEM	Footing	Below 0.6kN of tensile force reported	Wang et al 2020 performed DEM analysis to examine footing loaded to 80kPa modelled in three options: control, 1 layer of biaxial geogrid and two layers of biaxial geogrids

5. Conclusions

Since a relatively new and still controversial term has appeared in geosynthetic terminology, namely stabilisation, much discussions have been happening about its adequacy. Mechanism involved in stabilisation application of geosynthetics differ from typical reinforcement mechanism where tensile strength and strains developed within the product are the design parameters. Stabilisation manifests itself by increased confinement and geogrid interlocking aggregate [1], where developed strains and tensile strength do not perform a critical role.

There is still much contradicting information in the literature and within the industry, as the development in stabilisation has evolved from being reinforcement to improved understanding of interlock and confinement.

This paper summarises typical strains monitoring methods, used in the last two decades, to evaluate strains developing in geogrids, provides a summary of some research that reported strains measurement in applications that can be categorised as stabilisation, however in many cases is called reinforcement due to historical reasons. Typical values recorded by researches globally, focusing specifically on strains and axial tensile forces development were reported. The aim of this exercise was to examine the common argument around required magnitude of tensile strength within stabilising geogrid. The values recorded by various researchers worldwide range from 0.01% to 1.8%. The higher values typically would be associated with membrane effect developing in the geomembrane where reinforcement mechanisms are triggered, and higher deformations occur. Relatively low axial strain values correspond with relatively low axial tension values ranging in the reported data between.

In addition, this paper gives a short overview on the design approaches that are typically employed by practitioners/ designers used in commercially available geotechnical design software packages. It is evident that up to date there has been very limited research allowing validation or developing fully compatible methods that represent the actual geogrid effect on the stabilised soil mass. Hence, more independent studies are needed to establish effect of different geogrids on stabilised soil mass.

References

[1] ISO 2015 Geosynthetics – Part 1: Terms and definitions 10318-1:2015. ISO/TC 221 Geosynthetics

[2] Lees A S, Clausen J 2019 A new approach to characterise the benefit of multiaxial geogrid on soil. Proceedings of the XVII ECSMGE-2019 *Geotech. Eng. Found. of the Future ISBN 978-9935-9436-1-3*

[3] Baral P, Bergado D T and Duangkhae S 2016 The use of polumeric and metallic geogrid on a full scale MSE wall/embankment on hard foundation: comparison of field data with simulation *Inter. J. of Geo-Engineering* 7n 20

[4] Horníček L and Rakowski Z 2019 Mechanically Stabilized Granular Layers - An Effective Solution for Tunnel Projects. In: Badr M., Lotfy A. (eds) *Sustainable Tunneling and Underground Use*. GeoMEast 2018. Sustainable Civil Infrastructures. Springer, Cham

[5] Zamara K A, Dixon N, Jones D R V and Fowmes G 2012 Monitoring of a landfill side slope lining system: Instrumentation selection, installation and performance *Geotextiles and Geomembranes* **35** pp 1-13

[6] Chen R P, Wang Y W, Ye X W, Bian X C and Dong X P 2016 Tensile force of geogrids embedded in pile-supported reinforced embankment: A full-scale experimental study *Geotextiles and Geomembranes* **44** pp 157-169

[7] Hunt R E 1984 Geotechnical engineering investigation handbook. 2nd Edition ISBN 0-8493-2182-4

[8] Hatami K, Hassanikhah A, Yazdani H and Grady B P 2014 Tensoresistive PVC Coating for Sensor-Enabled Geogrids; *Journal of Nanomechanics and Micromechanics* **4**

[9] Tano B F G, Stoltz G, Coulibaly S S, Bruhier J, Dias D, Olivier F and Touze-Foltz N 2018 Large-scale tests to assess the efficiency of a geosynthetic reinforcement over a cavity; Geosynthetics

International **25**n 2

[10] IGS 2019 10 – 13 November 2019, the IGS Technical Committee on Stabilization (TC-S), prof Bian of Zhejiang University presentation, Prague, Czech Republic

[11] Anantanasakul P, Indraratna B, Nimbalkar S S and Neville T 2012 Field monitoring of performance of ballasted rail track with geosynthetic reinforcement. In G. A. Narsilio A

[12] Haas R, Walls J and Carroll R G 1988 Geogrid Reinforcement of Granular Bases in Flexible Pavements *Transportation Research Record* 1188

[13] Tang X, Stoffels S and Palomino A M 2013 Reinforcement Tensile Behavior Under Cyclic Moving Wheel Loads; *Transportation Research Record Journal of the Transportation Research Board* **2363**pp113-21

[14] Gnanendran C T and Selvaduarai A P S 2001. Strain measurement and interpretation of stabilising force in geogrid reinforcement. Geotextiles and Geomembranes **19** pp 177-94

[15] Wang Z, Jacobs F, Ziegler M and Yang G 2020 Visualisation and qualification of geogrid reinforcing effects under strip footing loads using discrete element method *Geotextiles and Geomembranes* **48** pp 62-70

[16] Bussert F and Cavanaugh J 2010 Recent research and future implications of the actual behaviour of geogrids in reinforced soil. *Earth Retention Conference* Bellevue, Washington 1-4 August 2010

[17] Cook J and Horvat F 2014 Assessment of particle confinement within a mechanically stabilised layer. *10th International Conference on Geosynthetics*, Berlin, Germany 21-25 September 2014

[18] Lees A S and Clausen J 2020 Strength envelope of granular soil stabilised by multiaxial geogrid in large triaxial tests *Canadian Geotechnical Journal* **57** pp 448-52

[19] Lees A S 2017 Simulation of geogrid stabilisation by finite element analysis Proceedings of the 19th Intern. Conf. on Soil Mechanics and Geotechnical Engineering, Seoul 2017

[20] Wang Z-F, Wang J, Sui Q-M, Liang X-M, Jia L, Li S-C and Lu S-D 2015 Development and Application of Smart Geogrid Embedded with Fiber Bragg Grating Sensors *Journal of sensors* **2015**

Two mechanisms in reinforced soil structures subjected to seismic forces

S Sharma[1], A Prashant[2]

[1] PhD Scholar, Indian Institute of Technology, Gandhinagar, Gujarat, India
[2] Professor, Indian Institute of Technology, Gandhinagar, Gujarat, India

E-mail: sukrit.sharma@iitgn.ac.in, ap@iitgn.ac.in

Abstract. Internal stability analysis of reinforced soil structures assumes different failure surface generally applicable to geosynthetic reinforcement and steel reinforcement, considering those as extensible and inextensible, respectively. The same assumption is applicable to seismic design too. However, a wall reinforced with geosynthetic reinforcement of high stiffness or small spacing and comparatively rigid facia can depict behaviour similar to walls reinforced with inextensible reinforcement. It is imperative to understand this behaviour and see applicability of different mechanisms depending on relative extensibility of reinforcements during seismic loading. A 2D Finite Element model of a wrap-around reinforced soil structure is developed in OpenSees and subjected to seismic forces. Analysis is focused on the development of failure plane, defined on the basis of the locus of maximum reinforcement forces. The reinforcements with low stiffness in general showed a failure plane that follows Rankine's active condition and with relatively high stiffness it showed roughly vertical failure mechanism. The locus of maximum reinforcement forces was also observed to be coincident with the zone of localized shear strains. These observations are useful in developing guidelines for safe and economical seismic design of reinforced soil structures.

1. Introduction

Huang [1] and Ling et al. [2] reported several incidents where GRS walls had experienced significant displacements and connection failures with the facia during seismic events. These cases indicate that there is a need to study more about the behaviour of these structures under seismic forces. Currently, the seismic design of GRS walls is performed using Mononobe-Okabe (M-O) method, based on Coulomb's theory of earth pressure, by considering the earthquake force as a pseudo-static force. Geosynthetics are considered extensible in nature and expected to strain enough so as to reach a plastic equilibrium or active state. Therefore, for internal stability analysis a failure plane is defined using Rankine's active condition for analysis. In the case of internal stability analysis for seismic forces, FHWA [3] recommends to use the same failure plane as adopted for static design. NCMA [4] on the other hand recommends to use M-O method to determine the seismic earth pressure at the facing for internal stability analysis. It is calculated by considering a pseudo-static force acting on a sliding wedge in the reinforced zone and finding the critical failure plane where the seismic earth pressure is maximum. Lack of sufficient literature related to development of failure planes during earthquakes motivated many researchers to perform numerical and experimental study to explore more. Bhattacharjee and Krishna [5] observed three failure modes for a wrap faced reinforced wall using a numerical study. First mode of failure was due to the horizontal movement of soil in the outward direction which led to shear deformations in reinforced zone. The increase in vertical settlement of soil at the end of reinforcements led to the formation of second failure mode. The third mode extended into the retained backfill leading to a compound failure. Liu and Ling [6] also reported similar observations for a numerical study on seismic analysis of 5.2 m high wrap faced wall. Huang [7] performed reduced scale shake table tests for wrap faced reinforced walls. The failure plane was vertical in the top half and inclined along Rankine active wedge in the bottom half of the wall. This was defined using the locus of maximum reinforcement forces, which was found coinciding with the location of maximum shear strains. Liu et al. [8] suggested to use Rankine active failure surface using similar criteria, after analyzing response of reinforced wall with panel facia subjected to seismic forces. El-Emam and Bathurst [9,10,11] observed that displacements in wall reduced when stiffer reinforcements were used or spacing between the reinforcements was decreased. El-Emam and Bathurst [10] and Sakaguchi [12] studied the influence of facia on the response of reinforced soil walls and found that light weight facia lead to lower wall displacements compared to heavier facia.

Walls reinforced with geosynthetic reinforcement of high stiffness may not deform sufficiently to develop Rankine's active failure wedge. It may behave similar to a wall reinforced with an inextensible reinforcement. It is an unconservative proposition for pullout safety if the designer adopts Rankine active failure plane for a wall reinforced with higher overall stiffness. The pullout safety depends on the forces developed in reinforcement and the length of reinforcement in the resisting zone. Hence, it is necessary to understand the failure mechanism in GRS walls with due consideration to reinforcement stiffness. Therefore, finite element (FE) model of a wrap faced facia wall is developed in OpenSees and validated using full scale model results. The FE model with varying reinforcement stiffness is subjected to earthquake forces. Subsequently, analysis is performed to understand the influence of reinforcement stiffness on the development of the failure planes in the reinforced zone of wall.

2. Finite Element Model

A 2D plane-strain finite element model of wrap faced geosynthetic reinforced wall was developed using an open source FE software OpenSees v2.5. The pre- and post-processing was done using GiD v11. The models wall was adopted from Yu et al. [13] which had welded wire mesh and PP geogrid as reinforcements. Schematic representation of the wall is illustrated in Figure 1. The backfill soil domain was discretized using four noded quad elements which followed non-linear elasto-plastic material model - Pressure Dependent Multi Yield (PDMY) model. The material parameters were calibrated using the plane strain test data provided by Hatami and Bathurst [14], as listed in Table 1. Yu et al. [13] in their simulation considered lower friction angle and shear modulus for the soil in the proximity of the facing, so as to account for reduced confinement near the facing. The PDMY being pressure dependent model gives due consideration to confinement effect at each step of analysis, and hence required no adjustments. The unit weight of soil in the model wall with WWM reinforcement was adopted as 1.72 t/m^3.

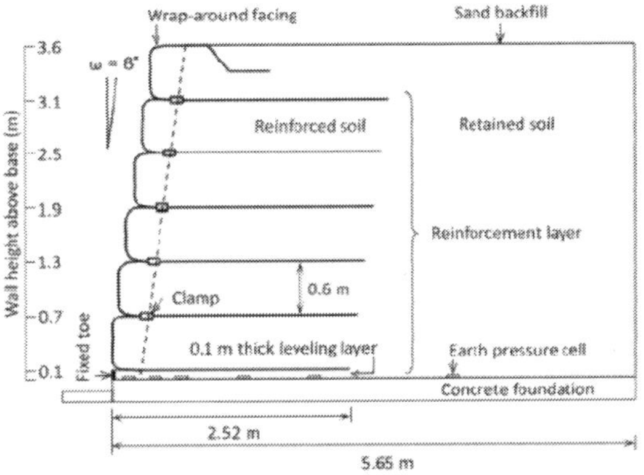

Figure 1. Schematic of wrap faced reinforced model wall (adopted from Yu et al. [13])

Table 1. Material Properties used for backfill soil

Parameter	Value
Unit Weight	1.68 t/m^3
Reference Shear Modulus	22.1 MPa
Reference Bulk Modulus	48 MPa
Friction Angle	44°
Peak Shear Strain	0.03
Reference Pressure	80
Pressure Dependent Coefficient	0.24
Phase Transformation Angle	27°
Contrac	0.03
Dilat1	0.8
Dilat2	5
e	0.45

A levelling soil layer was provided at the base to prevent direct contact between concrete foundation and reinforcement. It was modelled as isotropic elastic material with modulus of elasticity (E) of 20 MPa and poisson's ratio (ν) of 0.3. The unit weight was taken same as of the backfill soil. Concrete foundation was also assumed to behave as an isotropic elastic material with $E = 32$ GPa and $\nu = 0.15$. The reinforcement was discretized into uniaxial truss elements with elastic perfectly-plastic behaviour. The geometric and material properties of reinforcement elements are listed in Table 2.

Table 2 Geometric and Material Properties of PP Geogrid and WWM reinforcement

	Cross-sectional area (m^2)	Exposed Perimeter (m)	Initial Stiffness (kN/m)	Ultimate Strength (kN/m)	Yield Strain in compression, ε_c (%)	Initial Strain, ε_0 (%)
PP Geogrid	2x10^{-3}	2	122	14	0	0
WWM	2x10^{-3}	2	3100	7	0	0

The interface between soil and reinforcement was modelled using zero length node to node contact element. Its behaviour is governed by Coulomb's friction law. The contact algorithm is based on penalty method. The normal penalty stiffness is provided so as to limit the penetration between the contacting nodes to 0.2 mm. The tangential penalty stiffness was adopted after Yu et al. [13]. The interface angle of friction was taken 44° as suggested by Huang et al. [15] and Yu et al. [16]. This interface friction angle is large as compared to the general recommendation of ${}^{2}/_{3}\,\varphi$. The interface behaviour would significantly influence the global response of the GRS wall. However, for the present study the interface parameters have been directly adopted from Yu et al. [13]. Similar considerations were made for the soil-concrete foundation interface. A mesh size of 0.084x0.15 m was adopted for validation and subsequent analysis under gravity loading, after appropriate mesh convergence study.

The FE model was validated by comparing the maximum reinforcement forces and facia deformation with the response observed by Yu et al. [13], as shown in Figure 2. In the experimental study of wall with WWM reinforcement, the maximum reinforcement force was observed in the third reinforcement from base and its magnitude was 4.5 kN/m. The FE results indicated its value as 5.1 kN/m, which shows a reasonable match with the experimental results. In the case of wall reinforced with PP geogrid, the response obtained was close to the simulation performed by Yu et al. [13], but there was some deviation with the experimental results due to roll out and sag of flexible facing walls on removal of formwork, which is complicated to model [13].

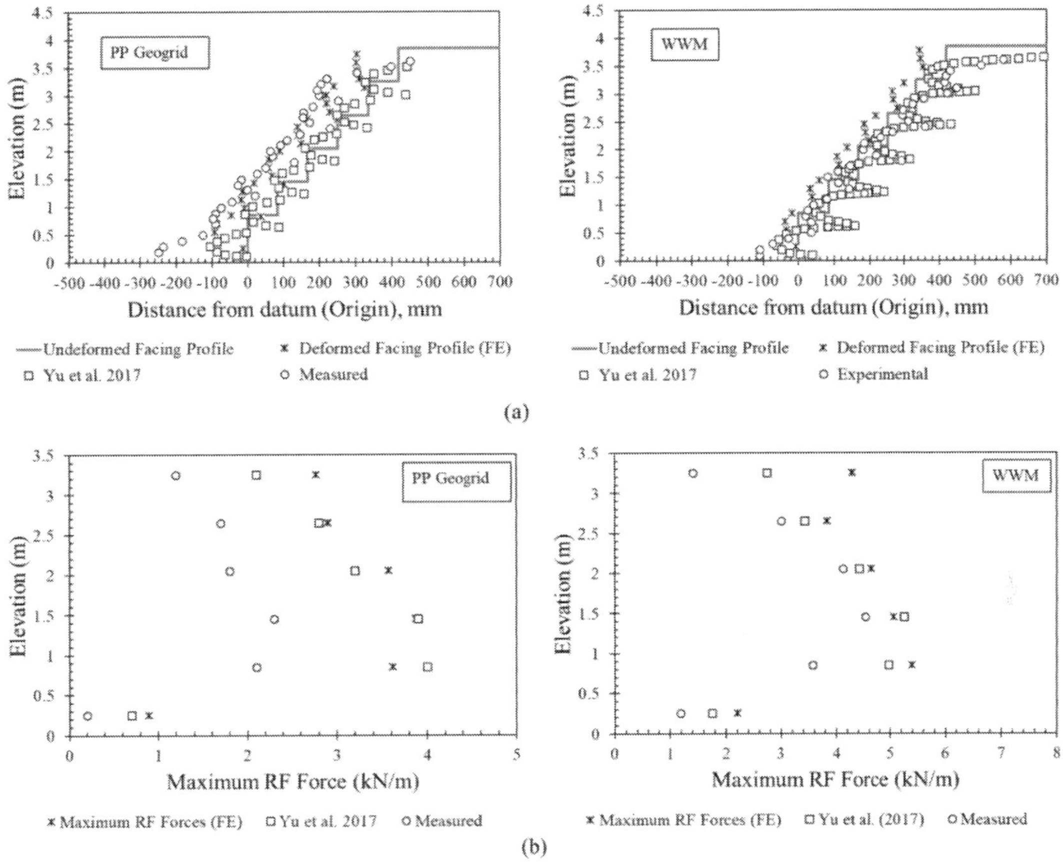

Figure 2. Validation of FE model by comparing (a) Facia Deformation profile (b) Maximum Reinforcement (RF) Force obtained from Finite Element with Yu et al. [13] simulation and experimentally measured results

Seismic analysis in FE models required modifications to the computational domain. Rigid boundaries reflect the waves back into the domain which should have been transmitted. Therefore, transmitting or absorbing boundary conditions recommended by Lysmer-Kuhlemeyer [16] were used at the horizontal extents of the domain. Dashpots were defined using zero length elements at the boundaries in both normal and shear direction to absorb compressional and shear waves respectively, to absorb the incident wave and act as infinite medium. The earthquake forces were applied using equivalent nodal shear forces at the base of the computational domain as suggested by Zhang et al. [17]. Free field analysis of a soil domain was performed, in order to verify the efficacy of modified boundaries and application of equivalent shear forces. The height of soil domain was kept constant at 3.6 m in all the cases. But, the horizontal extent of soil domain was increased to 46.85 m on the backfill side and 40 m on the other side (no fill side) to avoid effect of seismic reflections. The El-Centro earthquake acceleration-time history scaled to 0.3g was used for free field analysis. The acceleration response at the centre of the FE model was compared with 1D response analysis using the software DeepSoil v7.0 [18], as shown in Figure 3 for points at two different heights and duration of the first five seconds. The comparison shows a satisfactory match for the provided domain size.

3. Influence of Reinforcement stiffness on failure mechanism

The FE model was subjected to earthquake forces to study the development of failure plane in the backfill. Time history recorded during Northridge earthquake was used as shown in Figure 4, which had dominant frequency as 1.22 Hz. The time history was scaled to peak horizontal acceleration (PHA) of 0.12g, 0.24g, 0.36g and 0.48g. The acceleration time history was converted to equivalent force-time history and applied at the base of the model. The time step of analysis was kept equal to the time step of earthquake time history. To study the influence of extensibility of reinforcement, three different models were considered with the reinforcement stiffness of 150 kN/m, 500 kN/m and 1000 kN/m, respectively. The yield strain was kept constant at 11% in all the cases.

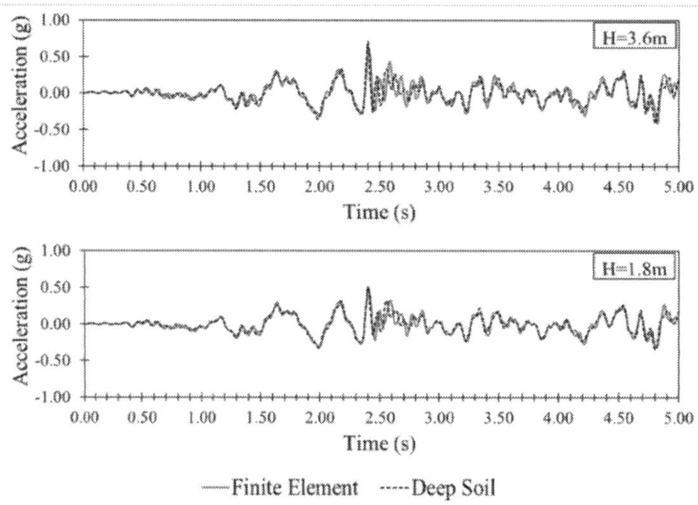

Figure 3 Comparison of response acceleration-time history at 1.8m and 3.6m obtained from Free Field analysis in FE with 1D response analysis in DeepSoil

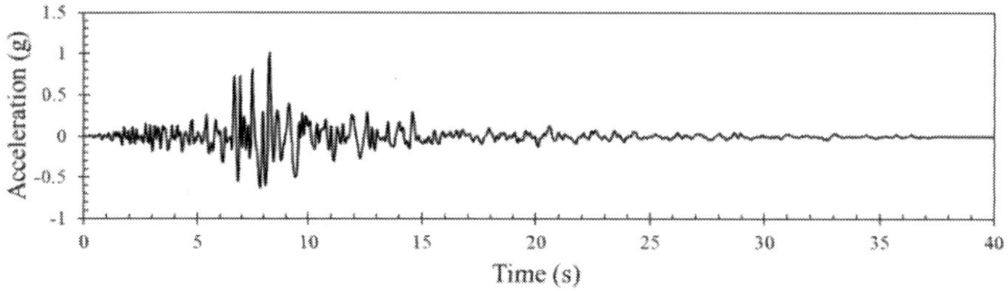

Figure 4 Acceleration-time history of Northridge Earthquake (Scaled to 1.0g)

It is observed that the failure plane for extensible reinforcements is coincident with locus of maximum reinforcement forces. The locus of the maximum reinforcement force separates the active and the passive zone, which is taken at the instant when the force in each reinforcement is maximum, i.e., they are not coincident. Figure 5 shows this locus from results of the FE model subjected to seismic forces due to Northridge earthquake. The offset due to batter is not included in the loci presented in Figure 5. The earth pressure distribution in the reinforced zone at the end of gravity stage for three reinforcement stiffness is depicted in Figure 6. This was observed to be closer to at rest condition in case of reinforcement stiffness 1000 kN/m while it was closer to active state condition in case of wall with reinforcement stiffness 150 kN/m. In essence, the locus of maximum reinforcement force shown in Figure 5a is approximately coincident with a Rankine failure plane inclined at $45+\phi/2$ even at low shaking levels of 0.12g. With use of low stiffness reinforcement significant plastic deformations occur at low shaking levels itself. In contrast, greater seismic force would be required for the wall with higher reinforcement stiffness of 1000 kN/m to develop sufficient deformations and reach active state. The behaviour of wall with reinforcement stiffness 500 kN/m depicted in Figure 5b was along similar lines as that for 1000 kN/m. The deformations in wall were not that significant at low shaking levels in case of relatively stiff reinforcement. Figure 5c depicts that the locus is vertical throughout the height of wall at lower shaking levels of 0.12g and 0.24g at a distance of 0.3-0.4H from the facing (0.5L or 0.5*0.7H=0.35H). The increase in seismic forces at higher shaking levels of 0.36g and 0.48g caused higher deformations and the locus was observed to be a bit inclined from vertical. But it is pertinent to note that the failure plane is not yet inclined to Rankine's failure plane at 0.48g. It can be reasonably argued that with seismic forces greater than 0.48g, the deformation would be sufficient for the wall to be closer to active condition and the failure plane might align with the Rankine's active failure surface. The locus varies significantly at lower shaking levels (0.12g and 0.24g) than at higher shaking levels (0.36g and 0.48g).

The contours of localized octahedral shear strains for these cases are depicted in Figure 7. There are two distinct zones of localization observed in the contour plots. One of the zones is observed close to the end of reinforced zone, which can be seen even at lower PHAs. While the other localized one is observed inside the reinforced zone which evolves as the shaking levels increase. The present study focuses on the localized shear strains inside the reinforced zone. In the case of wall reinforced with reinforcement stiffness 150 kN/m, the shear strains are localized roughly parallel to the Rankine's active failure wedge at an inclination of $45+\phi/2$. On the other hand. for reinforcement stiffness of 500 kN/m and 1000 kN/m, the localized shear strains were observed to evolve from low PHA of 0.12-0.24g to higher PHAs of 0.36-0.48g. The soil elements in the centre of the reinforced zone depicted more strains at 0.12-0.24g. Subsequently, at higher PHAs the strains began to localize along the Rankine's active failure plane. This is along similar lines to the observations reported using locus of maximum reinforcement forces.

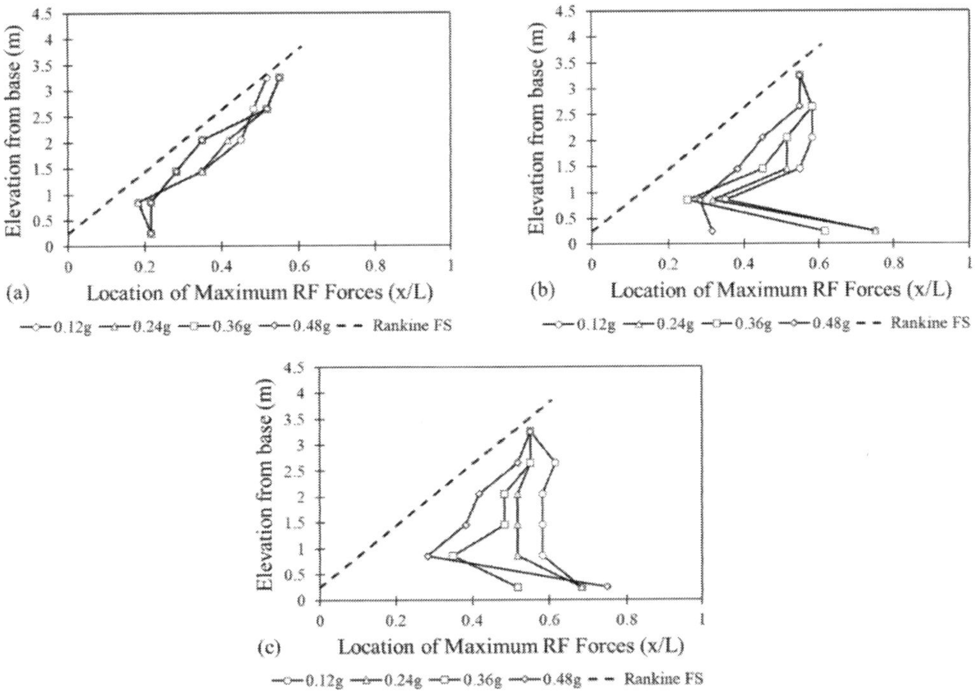

Figure 5. Locus of maximum reinforcement forces when FE model with reinforcement stiffness **a** 150 kN/m, **b** 500 kN/m and **c** 1000 kN/m was subjected to Northridge earthquake

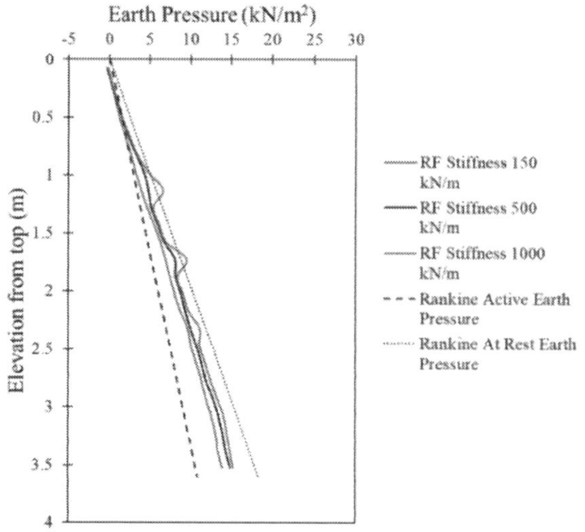

Figure 6 Earth Pressure at a plane inside the reinforced zone after application of gravity forces (Static)

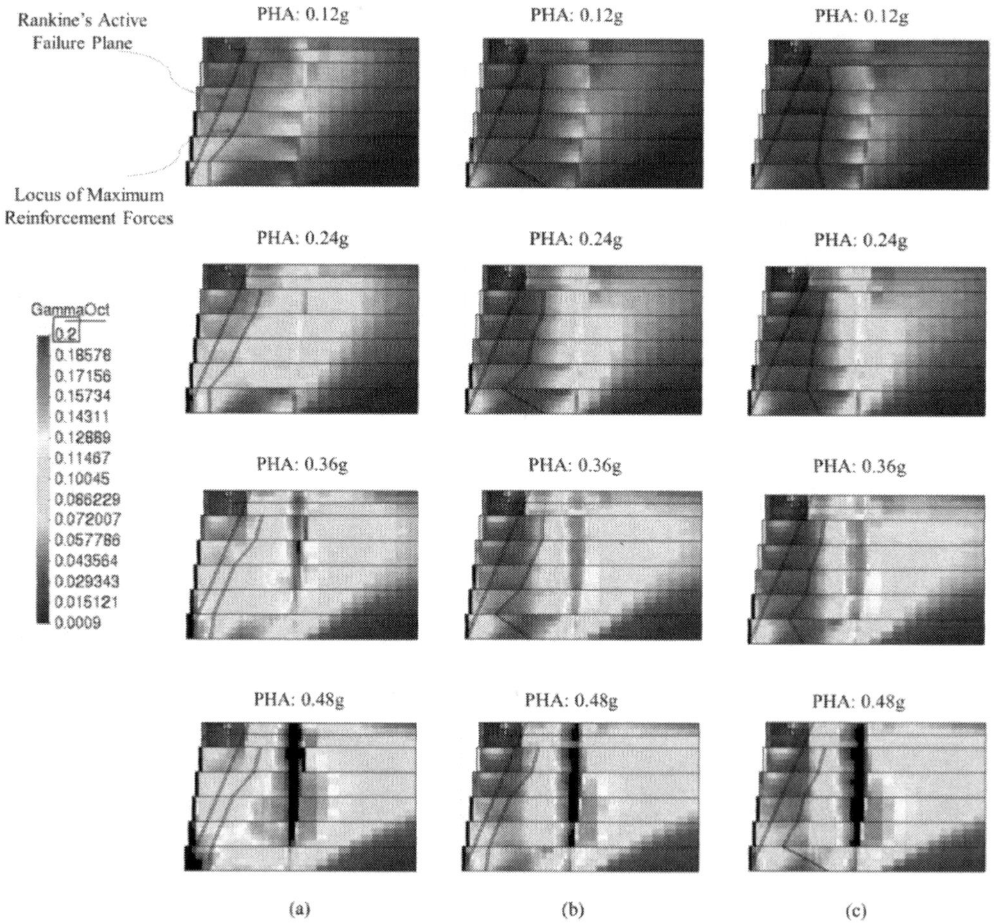

Figure 7. Octahedral Shear Strain Contours for wall with reinforcement stiffness of **a** 150 kN/m **b** 500 kN/m **c** 1000 kN/m, at different shaking levels for Northridge earthquake

4. Influence of earthquake characteristics on failure mechanism

The earthquake characteristics, i.e., dominant frequency and significant peaks in acceleration-time history can significantly influence the response of a structure. Therefore, response analysis was also conducted for Chi-Chi earthquake with dominant frequency of 1.12 Hz and only a few intermediate peaks in the acceleration time history (Figure 8), which were different from Northridge earthquake. The locus of maximum reinforcement forces observed in the FE model subjected to Chi-Chi earthquake is shown in Figure 9. In the case of wall reinforced with low reinforcement stiffness of 150 kN/m, the locus was vertical in the top half at low shaking levels of 0.12g and inclined in the bottom half, as shown in Figure 9a. With increase in seismic forces, the locus was inclined closer to Rankine active failure

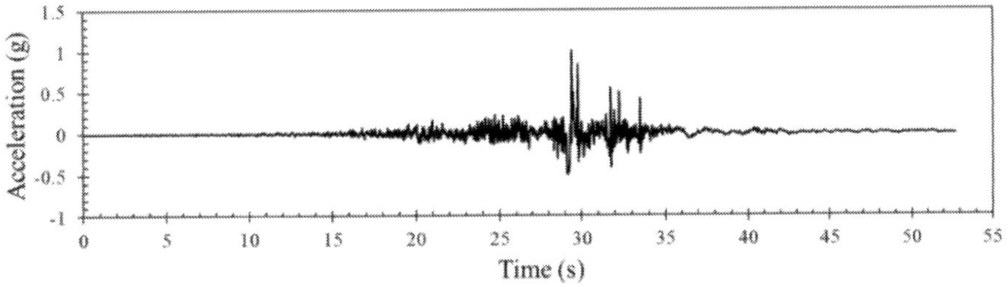

Figure 8. Acceleration-time history of Chi-Chi Earthquake (Scaled to 1.0g)

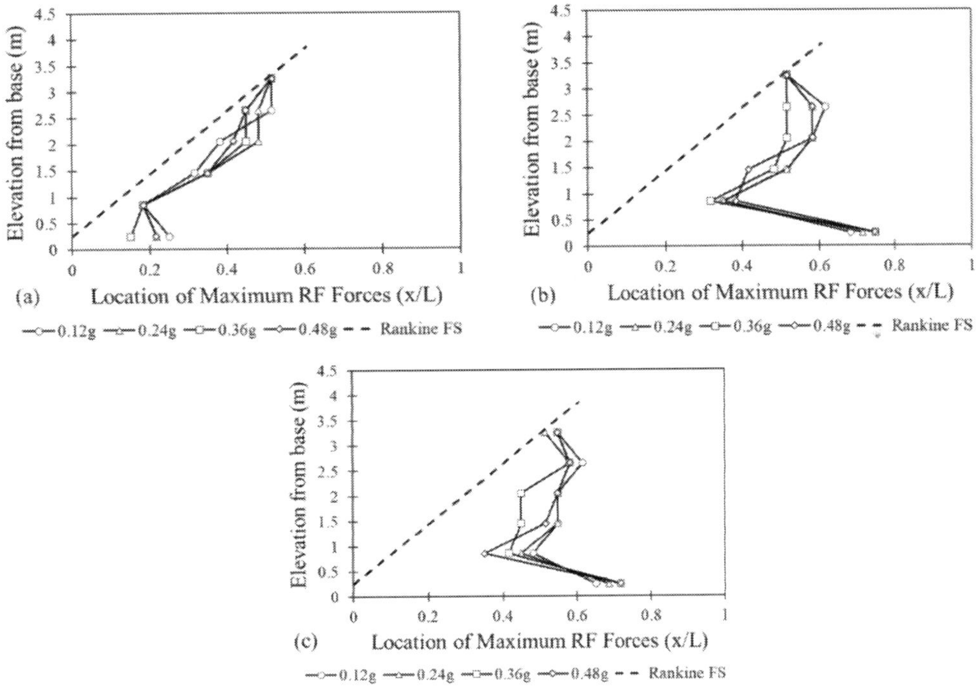

Figure 9: Locus of maximum reinforcement forces when FE model with reinforcement stiffness **a** 150 kN/m **b** 500 kN/m and **c** 1000 kN/m was subjected to Chi-Chi earthquake

plane. This signifies that a wall with less stiff reinforcement might experience significant deformations, even for a low intensity earthquake. On the contrary, in case of wall with stiffer reinforcement (500-1000 kN/m) the locus is vertical through the height at all the shaking levels and that lies at a distance of 0.3-0.4H from the facing, as shown in Figures 9b and 9c. This can be attributed to the fact that the intermediate peaks in acceleration time history were not significant enough to cause sufficient deformations in the system for it to develop active condition at even high PHAs of 0.48g.

The discussion presented in the previous two sections signifies that the failure plane defined using locus of maximum reinforcement forces is dependent upon the stiffness of reinforcement. It influences the overall stiffness of the wall, and hence the deformation response. A wall with low stiffness reinforcement was observed to develop active state at low shaking levels itself. While a wall reinforced with stiffer reinforcement did not develop an active condition even at high shaking levels (0.48g). But a transition in failure plane from almost vertical at low shaking levels to an inclined one at higher shaking levels was observed. The failure plane for walls with higher reinforcement stiffness was observed to be vertical at a distance of 0.3-0.4H from facing. Therefore, the failure plane is influenced by the overall stiffness of the wall which is further dependent on the reinforcement properties, spacing and facia rigidity. The earthquake characteristics also play an important role in the development of the failure plane. The effect of these factors on the failure plane should be studied extensively to understand the development of critical failure surface. The study would help make required modifications to the seismic design procedure of reinforced walls.

5. Conclusion

The selection of an appropriate failure plane is of utmost importance in seismic design of GRS walls. An FE analysis of wrap faced GRS wall subjected to seismic forces was performed to study the failure mechanism. In case of walls reinforced with low stiffness reinforcements, the failure plane was approximately aligned to Rankine active state even when subjected to seismic forces at low shaking levels. It was observed that failure plane can deviate from the generally assumed Rankine active state in the case of stiffer reinforcements. It was observed to be vertical at a distance of 0.3-0.4H from the facing at low shaking levels and it was a bit inclined (towards active failure surface) at high shaking levels. It was therefore concluded that it might be inappropriate to exclusively adopt a Rankine failure plane for GRS walls reinforced with higher reinforcement stiffness.

References

[1] Huang C C 2000 Investigations of soil retaining structures damaged during the chi-chi (Taiwan) earthquake *Journal of the Chinese Institute of Engineers* **23** pp 417-28

[2] Ling H I, Leshchinsky D, Chou N N 2001 Post-earthquake investigation on several geosynthetic-reinforced soil retaining walls and slopes during the Ji-Ji earthquake of Taiwan *Soil Dynamics and Earthquake Engineering* **21** pp 297-313

[3] Berg R R, Samtani N C, Christopher B R 2009 Design of mechanically stabilized earth walls and reinforced soil slopes–Volume I (No. FHWA-NHI-10-025) United States Department of Transportation Federal Highway Administration

[4] National Concrete Masonry Association (NCMA) 2009 Design manual for segmental retaining walls, 3rd Ed., Herndon, VA

[5] Bhattacharjee A, Murali Krishna A 2012 Development of numerical model of wrap-faced walls subjected to seismic excitation *Geosynthetics International* **19** pp 354-69

[6] Liu H, Ling H I 2012 Seismic responses of reinforced soil retaining walls and the strain softening of backfill soils *International Journal of Geomechanics* **12** pp 351-6

[7] Huang C C 2019 Seismic responses of vertical-faced wrap-around reinforced soil walls *Geosynthetics International* **26** pp 146-63

[8] Liu H, Wang X, Song E 2011 Reinforcement load and deformation mode of geosynthetic-reinforced soil walls subject to seismic loading during service life *Geotextiles and Geomembranes* 29 pp 1-6

[9] El-Emam M M, Bathurst R J 2004 Experimental design, instrumentation and interpretation of reinforced soil wall response using a shaking table *International Journal of Physical Modelling in Geotechnics*. **4** pp 13-32

[10] El-Emam M M, Bathurst R J 2005 Facing contribution to seismic response of reduced-scale

reinforced soil walls *Geosynthetics International*. **12** pp 215-38
[11] El-Emam M M, Bathurst R J 2007 Influence of reinforcement parameters on the seismic response of reduced-scale reinforced soil retaining walls *Geotextiles and Geomembranes* **25** pp 33-49
[12] Sakaguchi M A 1996 Study of the seismic behavior of geosynthetic reinforced walls in Japan *Geosynthetics International* **3** pp 13-30
[13] Yu Y, Bathurst R J, Allen T M 2017 Numerical modelling of two full-scale reinforced soil wrapped-face walls *Geotextiles and Geomembranes* **45** pp 237-49
[14] Hatami K, Bathurst R J 2005 Development and verification of a numerical model for the analysis of geosynthetic-reinforced soil segmental walls under working stress conditions *Canadian Geotechnical Journal* **42** pp 1066-85
[15] Huang B, Bathurst R J, Hatami K 2009 Numerical study of reinforced soil segmental walls using three different constitutive soil models *Journal of Geotechnical and Geoenvironmental engineering* **135** pp 1486-98
[16] Lysmer J, Kuhlemeyer R L 1969 Finite dynamic model for infinite media *Journal of the Engineering Mechanics Division* **95** pp 859-78
[17] Zhang Y, Conte J P, Yang Z, Elgamal A, Bielak J, Acero G 2008 Two-dimensional nonlinear earthquake response analysis of a bridge-foundation-ground system *Earthquake Spectra* **24** pp 343-86
[18] Hashash Y M A, Musgrove M I, Harmon J A, Ilhan O, Xing G, Numanoglu O, Groholski D R, Phillips C A, Park D 2020 DEEPSOIL 7.0, User Manual Urbana, IL, Board of Trustees of University of Illinois at Urbana-Champaign

AUTHOR INDEX